3rd Asia Conference on Power and Electrical Engineering (ACPEE 2018)

IOP Conference Series: Materials Science and Engineering
Volume 366

Kitakyushu, Japan
22 – 24 March 2018

ISBN: 978-1-5108-6481-8
ISSN: 1757-8981

Printed from e-media with permission by:

Curran Associates, Inc.
57 Morehouse Lane
Red Hook, NY 12571

Printed by Curran Associates, Inc. (2018)

Institute of Physics
Dirac House, Temple Back
Bristol BS1 6BE UK

Phone: 44 1 17 929 7481
Fax: 44 1 17 920 0979

techtracking@iop.org

Additional copies of this publication are available from:

Curran Associates, Inc.
57 Morehouse Lane
Red Hook, NY 12571 USA
Phone: 845-758-0400
Fax: 845-758-2633
Email: curran@proceedings.com
Web: www.proceedings.com

Table of contents

Volume 366

2018 3rd Asia Conference on Power and Electrical Engineering (ACPEE 2018)

22–24 March 2018, Kitakyushu, Japan

Accepted papers received: 10 May 2018
Published online: 13 June 2018

Preface

Papers

Renewable Energy & Smart Grids

Power and Electrical Engineering

PREFACE

ACPEE 2018 was held in Kitakyushu, Japan during Mar.22-24, 2018. ACPEE 2018 was organized by Hong Kong Society of Mechanical Engineers, supported by Kitakyushu Convention and Visitors Association, Auckland University of Technology, American University of Madaba (AUM). The conference provides a useful and wide platform both for display the latest research and for exchange of research results and thoughts in broad Power and Electrical Engineering, Renewable Energy and Smart Grid. The participants of the conference were from almost every part of the world, with background of either academia or industry, even well-known enterprise. The success and prosperity of the conference is reflected high level of the papers received.

The proceedings are a compilation of the accepted papers and represent an interesting outcome of the conference. This book covers 3 chapters: Renewable Energy & Smart Grids; Power and Electrical Engineering; Power electronics.

We would like to acknowledge all of those who supported ACPEE 2018. Each individual and institutional help were very important for the success of this conference. Especially we would like to thank the organizing committee for their valuable advices in the organization and helpful peer review of the papers.

We sincerely hope that ACPEE 2018 will be a forum for excellent discussions that will put forward new ideas and promote collaborative researches. We are sure that the proceedings will serve as an important research source of references and the knowledge, which will lead to not only scientific and engineering progress but also other new products and processes.

Prof. Tek-Tjing Lie, Auckland University of Technology, New Zealand
Prof. Moustafa Eissa, Helwan university, Egypt

IOP Conf. Series: Materials Science and Engineering **366** (2018) 011002 doi:10.1088/1757-899X/366/1/011002

COMMITTEES

General Chairs

Prof. Tek-Tjing Lie, Auckland University of Technology, New Zealand

Conference Committee Co-Chairs

Dr. Nopbhorn Leeprechanon, Thammasat University, Thailand

Prof. Emanuele Calabrò, Institute of Industrial Technology, Italy

Local Chair

Prof. Yaser Qudaih, American University of Madaba (AUM), Jordan

Advisory Committee Chair

Prof. Man Chung WONG, University of Macau, Macau

Program Chairs

Prof. Moustafa Eissa, Helwan university, Egypt

International Technical Committee

Dr. Khoa Dang Hoang,University of Sheffield, UK

Prof. Zulfiqar Khan,Bournemouth University, UK

Dr. Shuheng Chen,University of Electronic Science and Technology of China, China

Dr. Tosak Thasananuyariya, Metropolitan Electricity Authority, Thailand

Dr. Prakornchai Polratanasak, North Eastern University, Khonkaen, Thailand

Dr. Michael Bernard, University of Alberta, Canada

Dr. Mohamed Yahia Edries, Space Division National Authority for Remote Sensing and Space Science, Egypt

Dr. Thongchart Kerdphol, Kyushu Institute of Technology, Japan

IOP Conf. Series: Materials Science and Engineering **366** (2018) 011002 doi:10.1088/1757-899X/366/1/011002

IOP Conf. Series: Materials Science and Engineering **366** (2018) 011003 doi:10.1088/1757-899X/366/1/011003

Peer review statement

All papers published in this volume of *IOP Conference Series: Materials Science and Engineering* have been peer reviewed through processes administered by the proceedings Editors. Reviews were conducted by expert referees to the professional and scientific standards expected of a proceedings journal published by IOP Publishing.

IOP Conf. Series: Materials Science and Engineering **366** (2018) 012001 doi:10.1088/1757-899X/366/1/012001

Research on Information Organization Method of Power Intelligent Test System Based on CPS

Hongyan WU,Jiang GUO,Ke MENG and Wei CHAI

Wuhan University, Wuhan Hubei

1401453335@qq.com

Abstract. In this paper, we propose the idea of the power intelligent test system by combining the concept of Cyber Physical System. And research the ways to organize information to make the intelligent test system of power is more heterogeneous, high degree of autonomy, adaptability and scalability. At the same time, to achieve high utilization and the Intelligence of the test data.

1. Introduction

In response to the call of the smart grid, the power test information should run through the entire test process, including the collection of test data, preservation and diagnostic analysis. In order to realize the intelligent diagnosis and analysis of intelligent grid test data, it is urgent to study the information organization method in the test process. In order to solve the shortcomings of traditional experimental data recording and processing technology, the concept of power intelligent test system is put forward. The method of information organization in power intelligent test system is studied, and the data is the center. Data classification and statistics can ensure the integrity, accuracy and validity of the test data. It also provides a powerful and complete data base for the realization of intelligent diagnosis and analysis of equipment test data.

In this paper, by explaining the characteristics of the test data in the electric power test and comparing the drawbacks of the traditional experimental data processing technology, this paper introduces the method of hierarchical information organization, improves the utilization rate of the test data by standardizing the storage format and organization method of the test information, Based on the research of physical system, this paper puts forward the research framework of power intelligent test system, and verifies the realization of its function. Finally, it summarizes the research significance of information organization method of power intelligent test system based on CPS.

2. Power Intelligent Test System

2.1.Test Data Characteristics

There are a wide variety of equipment that need to be tested for electricity in the power system. The equipment structure is different and the test items are different. The data size and type of the test are not the same, and the data are not completely independent, but are interrelated to affect the operation of the equipment [1]. Transformer and GIS (including SF6 circuit breaker) as an example [2], transformer and GIS detection items, characteristic parameters, detection of defects such as Table 1 (table for some data).

Table 1 Test item and content

Device	Test item	Parameters	Defect
Transformer	Analysis of dissolved gases in oil	Characteristic gas, micro water	Internal overheating, internal discharge, internal moisture
	High frequency partial discharge detection	Discharge map	Partial Discharge
	Core ground current detection	Core current	Core multi-point grounding
	Infrared thermal image detection	Temperature	Temperature anomaly
GIS（With SF6 circuit breaker）	Ultrasonic partial discharge detection	Local mass	Partial discharge, suspended particles, joints loose
	UHF partial discharge detection	Local mass	Partial discharge, solid insulation within the bubble discharge
	SF6 gas humidity detection	Micro water content	Water exceeded
	SF6 gas decomposition products	Gas composition	Partial Discharge
	SF6 gas leak imaging detection	Laser / infrared imaging	Gas leaks
	SF6 gas purity detection	Gas composition	Detection of purity
	Infrared thermal image detection	temperature	Temperature anomaly

From the above table we can see that different equipment has different test items, different detection items corresponding to different characteristic parameters (ie test data) and detection of defects (ie test results), the same equipment detection defects can also be different test items Out, the same feature parameters can determine the different detection defects. Therefore, the integrated power system equipment, the contents of the test can be seen, the power system equipment test data have the following characteristics [3]:

- data scattered widely: scattered in different equipment;
- large amount of data: the need to collect more data;
- the relationship between the data complex: the data between each other;
- data types: including structured, semi-structured and unstructured.

2.2.Traditional data processing means

The traditional test data is based on the test object, a test data is only used for this test once, the same data in different test items have been collected several times, the use of historical test data again when the query trouble, classification is not clear, And other shortcomings, so follow the traditional data acquisition and analysis methods to deal with data can not meet the intelligent diagnostic requirements. How to make the above information reasonable organization, to facilitate the system more clearly understand the diagnostic test data is the power of intelligent test and test system research focus and key [4].

2.3.Power intelligent test system

Power intelligent test system is an intelligent system, through the construction of the test site and the central side of the exchange between the "transit agent", with the cloud system call center side resources, on-site work to provide support and guidance, and has a certain data analysis , Can achieve "one-click" data upload.

The system diagram of the power intelligent test system is shown in figure1.

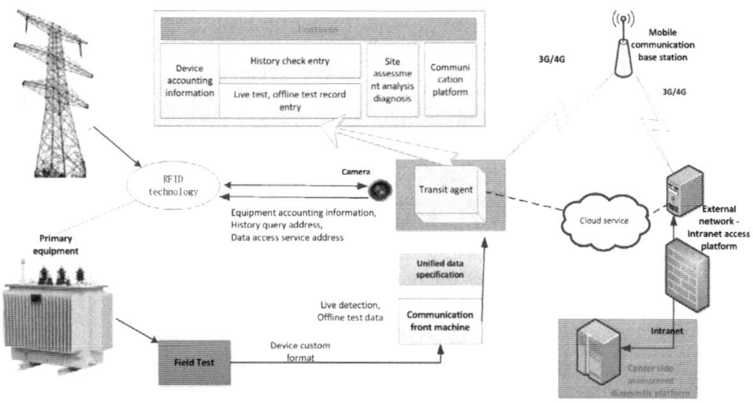

Figure 1 System architecture diagram

Through the establishment of power intelligent test and test system, theoretically, the role of the sensor forward, the information system of physical computing, communication, control the integration of the three functions of the depth of integration into the power system, the computing unit and the physical unit closely The system has the characteristics of controllable, credible and scalable, and can realize the dynamic control and real-time perception of complex system.

3. Research On Information Origanization Method

3.1.Research on hierarchical structure

In order to ensure the orderliness, simplicity and expansibility of the system information, the information organization of the information physical system is adopted, and the information is organized with the hierarchical structure, such as the company - power - station - system - equipment - characteristic parameter, the staff can be based on the actual situation of data to add, modify or delete, to ensure that all information system adaptability and scalability, hierarchical model is shown in figure 2.

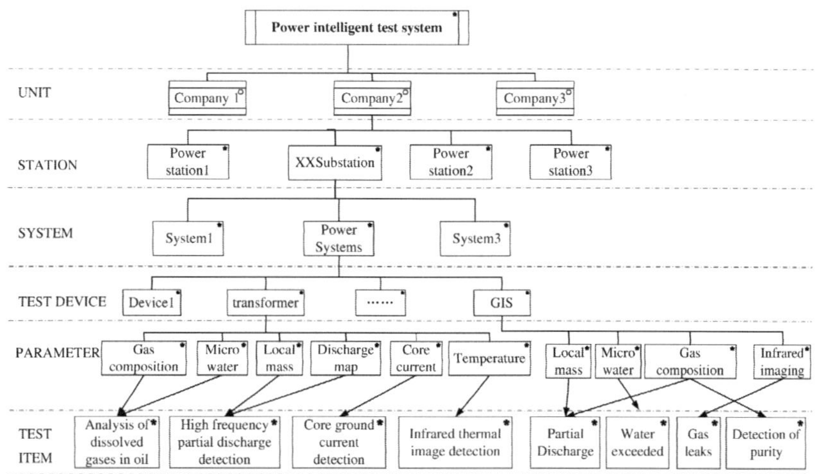

Figure 2 Information structure diagram

On the basis of the hierarchical structure, all the information needed by the comprehensive test of the power intelligent test system is unified and collected by the related technology. When the test results are analyzed, the experimental data can be automatically called according to the existing algorithms, norms or formulas, and the relevant results can be calculated intelligently, and the results

IOP Conf. Series: Materials Science and Engineering **366** (2018) 012001 doi:10.1088/1757-899X/366/1/012001

of the test analysis are also stored hierarchically as the empirical data A similar case reference is provided for subsequent trials. After the data has been used, it can be used as historical data for other test analysis to continue to use, you can also use the same data on the equipment for longitudinal, horizontal statistical analysis and comparison.

In short, the experimental characteristics of the use of hierarchical structure of the unified management and division and then save, can avoid duplication of data collection, to achieve "a collection, multiple use, centralized management, efficient query" target, but also save the test data collection Time, improve work efficiency; can also make full use of the various test data for the operation of the equipment to conduct a more comprehensive, accurate and effective diagnostic analysis.

3.2.Logical model design

Based on the research of the level and content of the test object of the electric power intelligent test system, the information organization logic model of the power intelligent test and test system is carried out according to the characteristics such as the heterogeneity of the information physical system, the high degree of autonomy and the data center. Design, the conceptual structure into a relational model, so that each of the physical relations to meet the requirements of various types of data integrity, but also to meet the basic requirements of each data item minimization. Finally, in combination with the needs of the system, the logic model is further optimized to improve the performance of the database system [5].

In this paper, ER Studio database modeling software for logical model design, using ER Studio design of the power intelligent test test system logic model shown in Figure3.

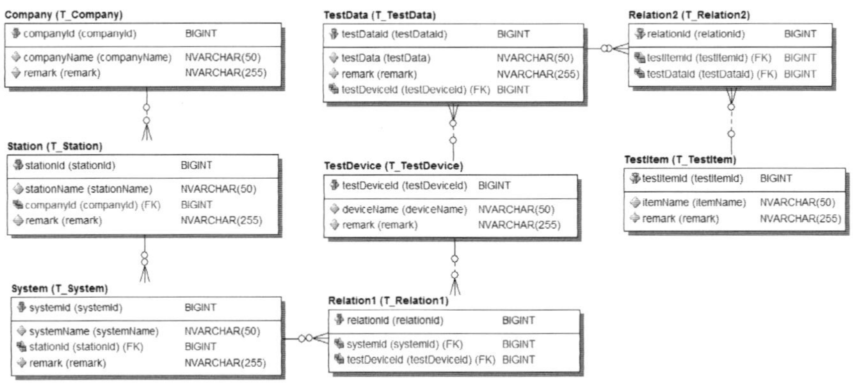

Figure 3 Logical model diagram

Through the design of the information logic model of the power intelligent test and test system, all the data systems involved in the pilot project of the power system are managed in a comprehensive and orderly manner, and the staff can inquire statistics, trend analysis and so on. Provide bottom-end data support.

4. Case Study

4.1.Data analysis

To verify the hierarchical information organization method using transformer oil chromatographic test as an example. The oil chromatographic test mainly analyzes the content of the dissolved gas in the oil, including the total hydrocarbon(TH), C_2H_2, CH_4, C_2H_2, C_2H_6, H_2, CO, CO_2 and other gas components.

This paper selects the data of No.6 transformer of Wusheng Substation in Guang'an City, Sichuan Province as an example to verify the feasibility of hierarchical information organization method. the

Wusheng 6 transformer 2014 7/8 two months of the gas content of some of the data (In μL / L) is shown in Table 2.

Table 2 Oil chromatographic data

TH	C_2H_2	H_2	CH_4	C_2H_6	C_2H_4	CO	CO_2
132.50	16.40	135.00	52.60	10.80	52.70	228.00	590.00
144.80	16.70	137.00	57.10	11.90	59.10	224.00	628.00
136.00	16.90	144.00	53.20	11.30	54.60	239.00	627.00
132.30	17.10	137.00	53.80	10.20	51.20	218.00	604.00
137.50	17.50	139.00	58.10	11.00	50.90	248.00	605.00
129.50	17.50	120.00	52.00	10.00	50.00	210.00	536.00
118.80	17.80	112.00	46.00	9.00	46.00	190.00	491.00
125.00	18.00	124.00	48.00	10.00	49.00	210.00	520.00
117.60	18.00	123.00	45.00	9.60	45.00	198.00	500.00
126.30	18.30	113.00	47.00	11.00	50.00	195.00	510.00

Table 3 Three ratio method to calculate the results

NO.	C_2H_2/C_2H_4	CH_4/H_2	C_2H_4/C_2H_6
1	0.3426	0.3994	4.8521

Table 4 Three-ratio method Encoding Rules

Ratio	C_2H_2/C_2H_4	CH_4/H_2	C_2H_4/C_2H_6
<0.1	0	1	0
0.1-1	1	0	0
1-3	1	2	1
>3	2	2	2

Table 5 List of fault properties

NO.	Fault nature	C_2H_2 $/C_2H_4$	CH_4 $/H_2$	C_2H_4 $/C_2H_6$
1	No fault	0	0	0
2	Partial discharge of low energy density	0	1	0
3	Partial discharge of high energy density	1	1	0
4	Low-energy discharge	1-2	0	1-2
5	High energy discharge	1	0	2
6	Low temperature overheating (<150°C)	0	0	1
7	Low temperature overheating (150-300°C)	0	2	0
8	Medium temperature overheating (300-700°C)	0	2	1
9	High temperature overheating (>700°C)	0	2	2

The August and July data were averaged in accordance with the three ratio method to calculate the calculation results, which was shown in Table 3.

Combining the coding rules as shown in Table 4, we can see C_2H_2/C_2H_4, CH_4/H_2, C_2H_4/C_2H_6 coding analysis of 1,0,2.

From Table 5, we can see that,analysis by the nature of the failure list shows that the failure of the transformer is a high energy discharge

Based on the above calculation process, the oil chromatographic improved three-ratio coding analysis of 1,0,2, to determine the fault belongs to the arc discharge.

Because CO growth is stable (-3% / month), CO content is less than 300μL / L (257μL / L), thus judging the discharge failure does not involve solid insulation, there may be oil flow, suspension discharge, magnetic shielding contact, Bubble discharge, unstable core multi-point grounding, etc. Fault:

- Because C_2H_2 content is less than 40% (14%) of total hydrocarbon content, does not belong to single acetylene growth type failure, so basically can rule out oil flow failure;
- According to the oil chromatographic analysis of high energy discharge, and hydrogen hydrocarbon content of less than 80% (51%), CH_4 total hydrocarbon content of less than 80% (39%), the basic can rule out the oil bubble discharge.

The final cause of the possible failure is: suspension discharge, fuel tank magnetic contact bad, unstable core multi-point grounding.

Using the data measured in Table 2, the power intelligent test system in addition to the use of three ratio method, but also can be used to analyze the David triangle method, compared with the three ratio method, combined with the known norms of total hydrocarbons, C_2H_2, H_2 content of the impact of the judge, the results obtained in the following table.

Table 6 Oil chromatographic data analysis results

Three ratio method	David Triangle Law	TH	C_2H_2	H_2
Arc discharge	Overheating and discharge	Normal	Level 4 deterioration	Normal
Arc discharge	Overheating and discharge	Level 3 deterioration	Level 4 deterioration	Normal
Arc discharge	Overheating and discharge	Normal	Level 4 deterioration	Level 2 deterioration
Arc discharge	Overheating and discharge	Level 3 deterioration	Level 4 deterioration	Level 2 deterioration
Arc discharge	Overheating and discharge	Level 3 deterioration	Level 4 deterioration	Level 2 deterioration
Arc discharge	Overheating and discharge	Level 3 deterioration	Level 4 deterioration	Normal
Arc discharge	Overheating and discharge	Normal	Level 4 deterioration	Normal
Arc discharge	Overheating and discharge	Normal	Level 4 deterioration	Normal
Arc discharge	Overheating and discharge	Normal	Level 4 deterioration	Normal
Arc discharge	Overheating and discharge	Level 3 deterioration	Level 4 deterioration	Level 2 deterioration

Combining the above analysis results and the results in Table 6, we can see, through the hierarchical storage structure, for the transformer in the gas data is provided by the basis of multiple calls, the initial implementation of the test data "a collection, multiple use, centralized management, efficient query" concept.

4.2.System implementation

The idea of CPS and power intelligent test system integration, proposed hierarchical information organization method to achieve the system's automatic diagnostic analysis function, in order to verify the feasibility of the method can be a simple system development, the system to achieve The basic functions are shown in Figure 4.

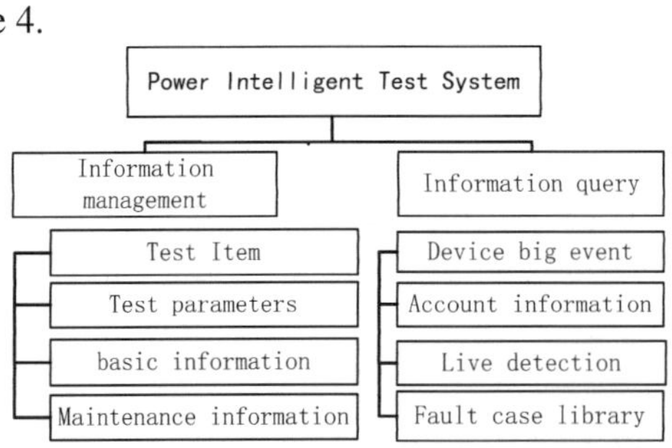

Figure 4 System function design

The intelligent intelligent test system adopts the modularization thought to carry on the development, the basic function of which is divided into two kinds of modules of information management and information inquiry. Information management is mainly used for the management and maintenance of test information, all data fields are from the design of the data table. The information query module is used to analyze and diversify the data stored in the database.

Information management is mainly based on the experimental data design, which is mainly in accordance with the hierarchical structure of the project information management. The left side of the management page is a tree list, the data information by different levels of division, both for the convenience of staff management, but also to facilitate the statistics of information. Specific examples shown in Figure 5, in each information management page can be added to the data, modify, delete and other functions to ensure that the system has good adaptability and scalability.

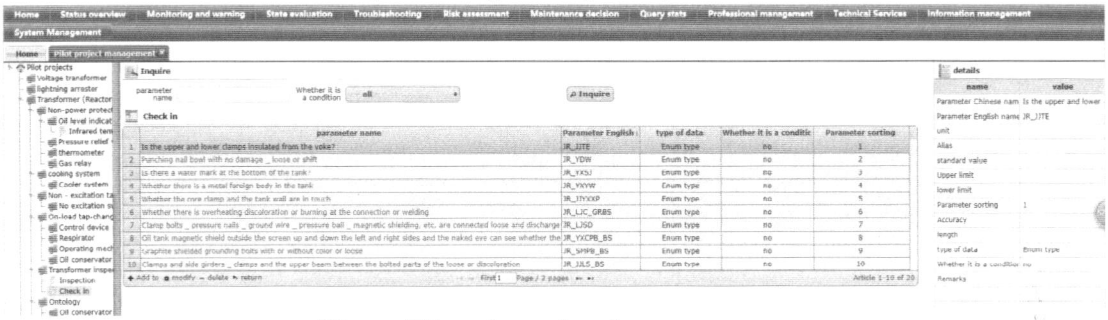

Figure 5 Experimental project management

The information query page mainly displays the information of the equipment of the power station, the memorandum, the test data of the live test, and the standard case database. In the hierarchical structure based on the collection of equipment data, developed a different information query display, as shown in Figure 6 to large equipment, for example, the equipment test results for statistical analysis of the equipment can query all the test results. This example is shown in the form of a timeline chart, or it can be displayed with a fishbone chart, a Gantt chart, or the like.

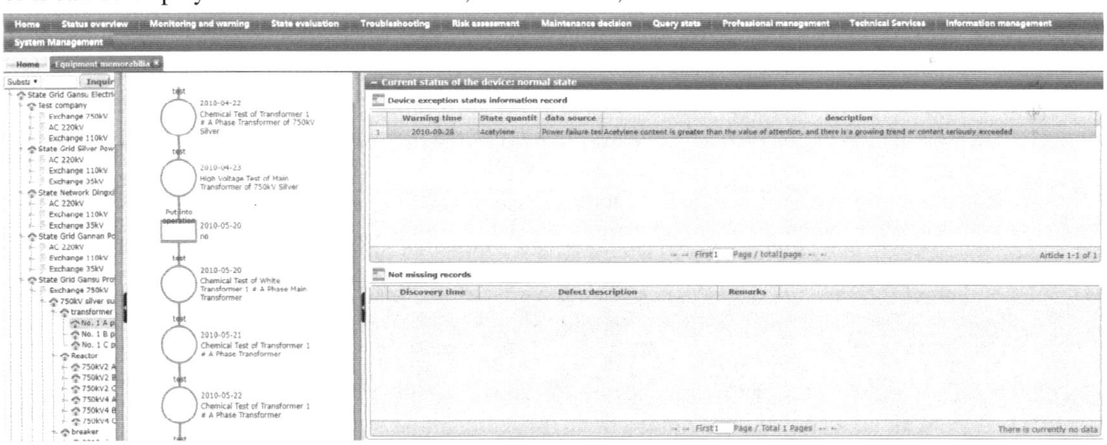

Figure 6 Device big event

As shown in Figures 7, Fault diagnosis combined with the hierarchical information organization method to complete the equipment data collection, storage, and the data for the initial analysis and diagnosis, to provide system and system users and experts of the human-computer interaction interface, the system automatically diagnosed on the basis of improved diagnostic analysis Flow, rich fault tree information. The expert's diagnostic results and test recommendations take effect on the operation and maintenance of the equipment.

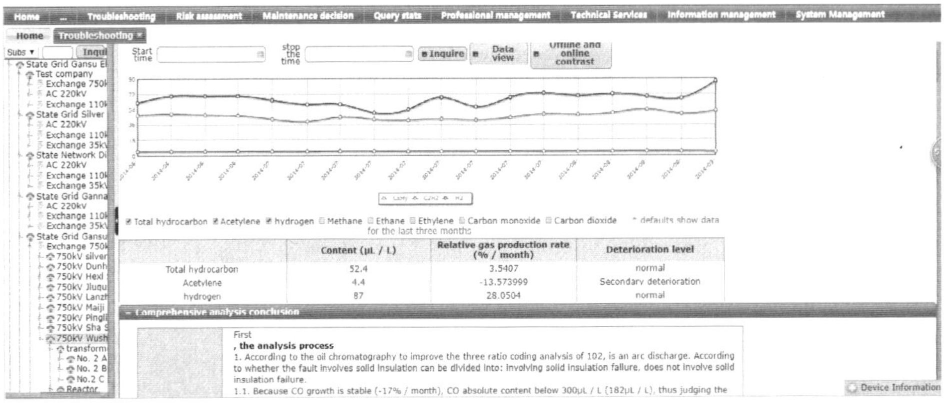

Figure 7 Diagnostic analysis

The system realizes some of the main functions of the system, and through the application of the concept of CPS, it can verify that the data-centric hierarchical information organization can realize the intelligent diagnosis and other functions of the system.

5. Conclusion

In this paper, according to the requirement of developing intelligent power test system, this paper deeply studies the information organization method of power test with the characteristics of data center, heterogeneity and high degree of autonomy of information physics system, and puts forward the hierarchical information organization And the design of the corresponding logical structure model, the model has a good self-adaptability and scalability; and developed a power intelligent test and test system to verify the system information organization of the rationality and feasibility.

By studying the information organization of CPS-based power intelligent test and test system, it provides a good theoretical and practical basis for the acquisition, expansion, preservation, utilization, automatic analysis and intelligent diagnosis of power system test data, which can be realized Intelligent test of power test system.

Reference

[1]. Wang Shaohua, Ye Ziqiang, Mei Bing laugh. Application of on - line monitoring and electrification technology for power transmission and transformation equipment in power grid [J]. High voltage electrical appliances, 2011,47 (4): 84-90.

[2]. Liu Yunpeng, Law Fang Cheng, Li Yanqing, Chen Zhiyang. Design and implementation of transformer partial discharge measurement system [J]. High Voltage Technology, 2002, 28 (1): 32-34.

[3]. Zhang Dongxia, Miao Xin, Liu Liping, Zhang Yan, Liu Kelian. Research on the development of large data technology of smart grid [J]. Journal of China Electromechanical Engineering, 2015,35 (1): 2-11.

[4]. Zhang Yongyue. Several Wiring Methods for Transformer Field PD Test [J]. Transformers, 2001,38 (9): 26-31.

[5]. Wen Jingrong, Wu Muqing, Su Jingfang. Information physical fusion system [J]. Journal of Automation, 2012,38 (4): 507-517.

IOP Conf. Series: Materials Science and Engineering **366** (2018) 012002 doi:10.1088/1757-899X/366/1/012002

Feasibility Study of Renewable Energy System for an Island in Terengganu

W M W Muda[1, 2] and N A S Salleh[2]

[1]Special Interest Group of Eastern Corridor Renewable Energy.
[2]School of Ocean Engineering, Universiti Malaysia Terengganu, 20130 Kuala Terengganu, Malaysia

Corresponding email: w.mariam@umt.edu.my

Abstract. Malaysia has started encouraging renewable energy technology in 7th Malaysia Plan. However, inadequate power distribution in the island makes diesel generator important power sources which cause pollution. Some initiatives have been made by installing wind turbine and solar panel to reduce the dependency on generator. The hybrid PV/wind system is a good combination as solar radiation is high during the Southeast Monsoon, and strong wind during the Northeast Monsoon. Since the wind speed and wave height has a strong tie, during windy season, the wave height can go up to two meters. However, the feasibility of wave energy as power supply hasn't been explored. This paper focuses on the feasibility of using wave energy system to supply electricity to an island using HOMER. The obtained results show that a hybrid wave energy system has lower potential with cost of electricity (COE) is RM 1.303/kWh compared to the most optimal system, hybrid PV-battery generator, with COE is RM 1.262/kWh.

1. Introduction

The consumption of fossil fuels lead to environmental problems especially CO_2 emissions. According to International Energy Agency (IEA), nearly 68% of the worldwide electricity was generated from fossil fuels including coal (40.6%), gas (21.4%) and oil (5.1%) in 2009 [1]. The electricity production from fossil fuels is higher in the developing countries. Thus, cleaner sources of energy are required to secure the future electricity supply. The advantages of renewable energy are no global warming emission, improves public health and environmental quality. In 2009, UCS analysis found that a 25% by 2025 national renewable electricity standard would lower power plant CO_2 emission 277 million metric tons annually by 2025, equivalent of the annual output from 70 typical (600MW) new coal plants [2].

As renewable energy is greatly exploited for various technologies, Malaysia also not left behind in using these gifts by implementing Sustainable Energy Authority Development (SEDA). Malaysia is ideal for large scale solar power installations due to its location in the equatorial region. The continuous supply of sunlight, low maintenance cost, independent of fuels source, and contributions to lower carbon emissions, made solar is the best choice for future energy power generation [3]. Another potential source is wind energy. The wind speed in Malaysia is light and varies from season to season in the range of 2 m/s to 13 m/s [4]. The Northeast Monsoon which is from September to March recorded the strongest wind comes from the South China Sea to the East Coast. Despite from that, ocean wave energy also has high power density and generating significant wealth [5]. But, the energy has not yet been explored to any significant extent in Malaysia.

Terengganu is popular with beautiful islands. A well-known island that can be found at Peninsular Malaysia on the east side is Pulau Perhentian. Due to the popularity of this place, tourists attract to go there. In Pulau Perhentian Kecil, tourist can experience more environment-friendly excursions as the island has been commissioned by Tenaga Nasional Berhad (TNB) to develop wind and solar technologies [6]. But, the power demand keeps increases in future. Mentioned in [7], 4.8 MW of power needed in future to fill the energy demand in Pulau Perhentian Kecil especially for the resorts and hotels in tourism purpose. An issue arise as the owner of the resorts and hotels are requesting continuous electricity supply from the government for their business. So far, the resorts and hotels in the islands still are depending on the diesel generator to supply electricity because TNB only supplying the power to the residential area [7]. Therefore, it became concerned to the owners. The uses of diesel generator also causing pollution especially to the underwater ecosystem and the price of diesel is unstable.

With the presence of renewable technologies in Pulau Perhentian Kecil and abovementioned potential of renewable energy, this paper is prepared to add wave energy in the power system. The disadvantages of diesel generator also presented and an off-grid simulation-based hybrid renewable energy system to supply electricity in Terengganu is developed in HOMER.

2. Pulau Perhentian System Configuration

The proposed system configuration illustrates in Figure 1. It consists of four sources of electivity which are photovoltaic (PV), wind turbine (WT), hydro power plant to represent wave generator (WG) as used in [8] and diesel generator (DG). Diesel generator is added as backup unit of the system. Meanwhile, battery (B) is needed to store the excess electricity from sources to be used later especially at night as this system is an off-grid system.

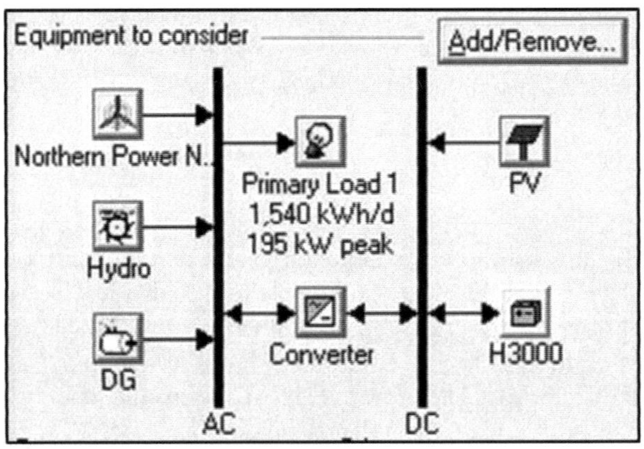

Figure 1. System configuration of proposed system.

2.1. Load Profile

Load in the system is the electrical appliance used in Pulau Perhentian Kecil. The hourly average load profiles in the island for two seasons are shown in Figure 2 and Figure 3, respectively. Only one residential area located in Kampung Pasir Hantu, Pulau Perhentian Kecil with population approximately 2023 people [7]. Most of the residents involved in tourism activities and build their own resorts/homestays/chalets. The rest are doing fisheries activities. In the fishing village, some facilities are provided such as police stations, health centre, jetty and shops. The load profile of Pulau Perhentian Kecil increases approximately twice in peak seasons which from Feb to October compared to low season happens from November until January.

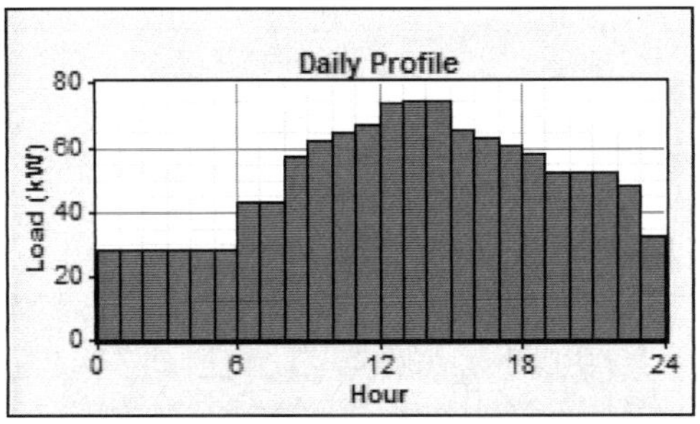

Figure 2. Load profile on low season.
(Nov – Jan)

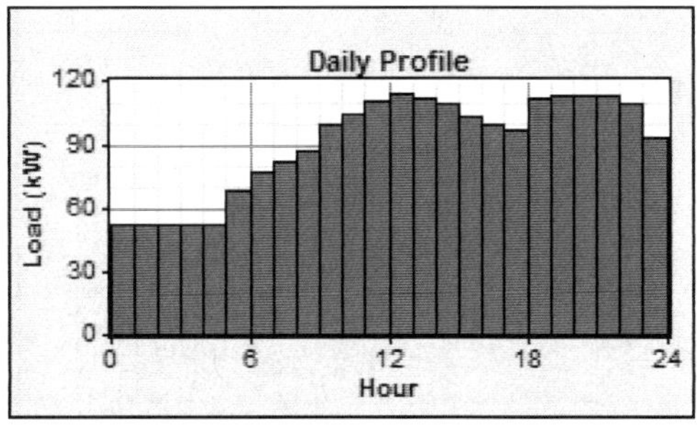

Figure 3. Load profile on peak season.
(Feb – Oct)

2.2 Renewable energy resources and specifications

Solar radiation and wind speed data were collected at Universiti Sultan Zainal Abidin (UniSZA) weather station and the monthly average of solar radiation shown in Figure 4. According to the figure, the highest radiation is in March with radiation of 5.532 kWh/m^2/day and lowest on November with 3.069 kWh/m^2/day. The decreasing of solar radiation from October to January is due to monsoon season as Terengganu is located at East of Peninsular Malaysia facing to South China Ocean.

Otherwise, the wind speed data can be seen in Figure 5, shows the highest reading during January at 3.213 m/s. The significant wave height and wave period were obtained from Institute of Oceanography and Environment (INOS), Universiti Malaysia Terengganu (UMT). From the data, flow rate can be calculated using equation (6) of [8] and the annual wave flow rate is depicted in Figure 6. For the cost of components in this research were referred to the previous research. Specification of PV panel, wind turbine, converter, diesel generator, battery are obtained from [9] while wave generator from [8] as shown in Table 1.

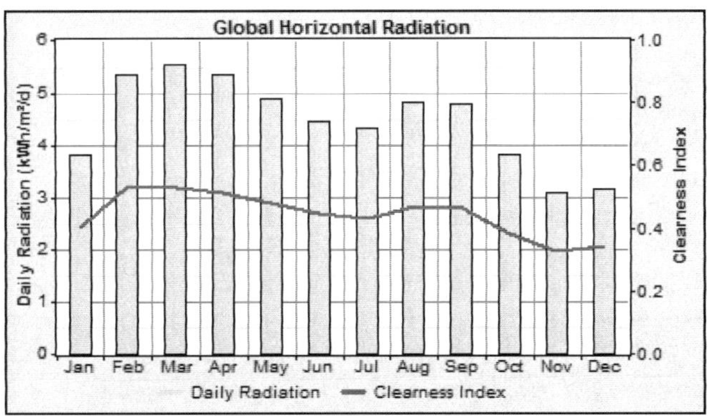

Figure 4. The annual solar radiation data.

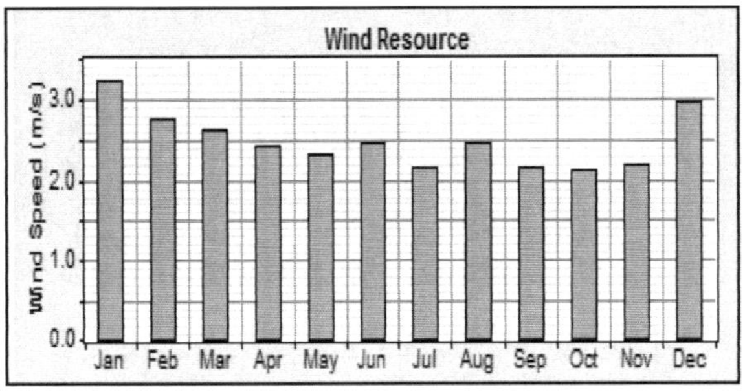

Figure 5. The annual wind speed data.

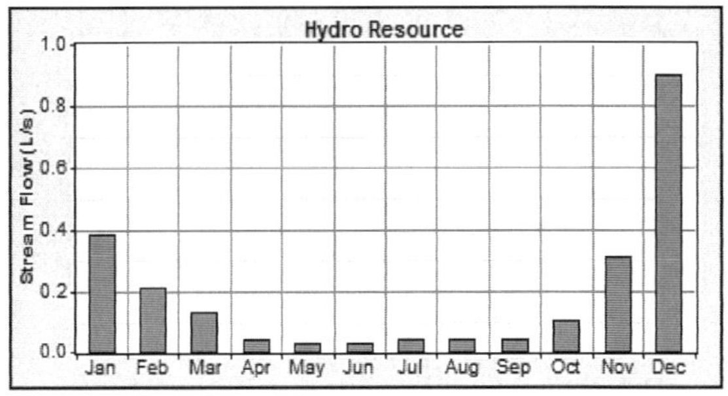

Figure 6. The annual wave flow rate data.

Table 1. List of components specification.

Component	Sizing (kW)	Lifetime (years)	Capital cost (MYR)	Replacement cost (MYR)	Size to consider
PV panel [9]	0.055	20	2030	1544	0 – 198 kW
Wind turbine [9]	100	20	1,763,840	1,664,000	0 – 5 unit
Wave generator [8]	9.81	12.5	210,300	168,240	-
Diesel generator [9]	80.0	~1.14	85, 118	78 694	0 – 80 kW
Battery [9]	120V[a]	10	6071	5460	0 – 5 strings
Converter [9]	1.0	15	2573	2, 412	0 – 130 kW

[a] batteries per string

3. Results and Discussions

The simulation results generated by HOMER which rank all possible configurations for the proposed system from the least to highest net present cost (NPC) shown in Figure 7. PV/DG/B system is the most optimal system with configuration of 143 kW PV panel, 80 kW diesel generator, two strings (240 V) of battery and 120 kW converter. The total NPC is RM 8,855,308.00 with cost of electricity (COE) RM 1.262/kWh which is the lowest than other configurations.

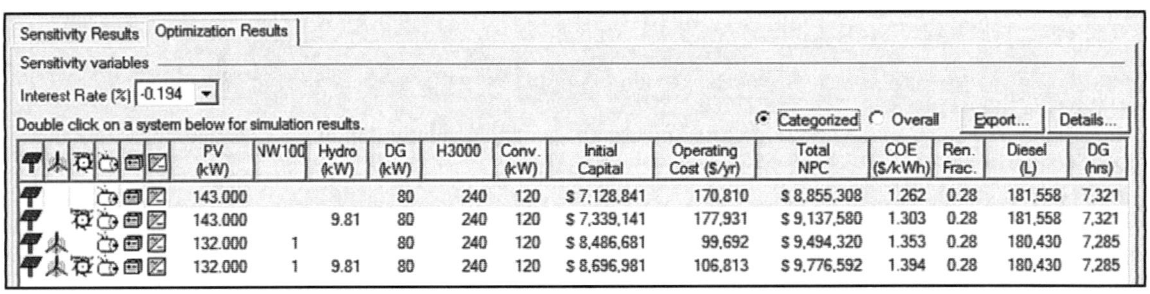

PV (kW)	NW100	Hydro (kW)	DG (kW)	H3000	Conv. (kW)	Initial Capital	Operating Cost ($/yr)	Total NPC	COE ($/kWh)	Ren. Frac.	Diesel (L)	DG (hrs)
143.000			80	240	120	$ 7,128,841	170,810	$ 8,855,308	1.262	0.28	181,558	7,321
143.000		9.81	80	240	120	$ 7,339,141	177,931	$ 9,137,580	1.303	0.28	181,558	7,321
132.000	1		80	240	120	$ 8,486,681	99,692	$ 9,494,320	1.353	0.28	180,430	7,285
132.000	1	9.81	80	240	120	$ 8,696,981	106,813	$ 9,776,592	1.394	0.28	180,430	7,285

Figure 7. Optimization results of hybrid-RE system.

From the result, PV is the most potential renewable energy component to be developed as the initial cost to install the PV is cheaper compared to wind turbine [10] and wave generator. In addition, the solar radiation in Terengganu is high while the wind speed is still considered low. Still, diesel generator is needed to supply electricity for the power system. For the use of wave generator in generating electricity, Figure 7 shows that the second best combination is PV/WG/DG/B with the total

NPC is RM 9,137,580.00 and COE RM 1.303/kWh followed by PV/WT/DG/B and PV/WT/WG/DG/B.

The production of electricity for different configurations is shown in Figure 8. In PV/DG/B system, DG produces most electricity for the system which is 538,804 kWh/yr, while the consumption of electricity is 694,025 kWh/yr. Similar trend in the rest of configuration which shows high electricity production by DG at 416, 441 kWh/yr, 420,800 kWh/yr and 424,652 kWh/yr, respectively. WT only contributes 17, 401 kWh/yr of electricity. WG produces smallest amount of energy compared to the other renewable energy resources which at 6, 148 kWh/yr.

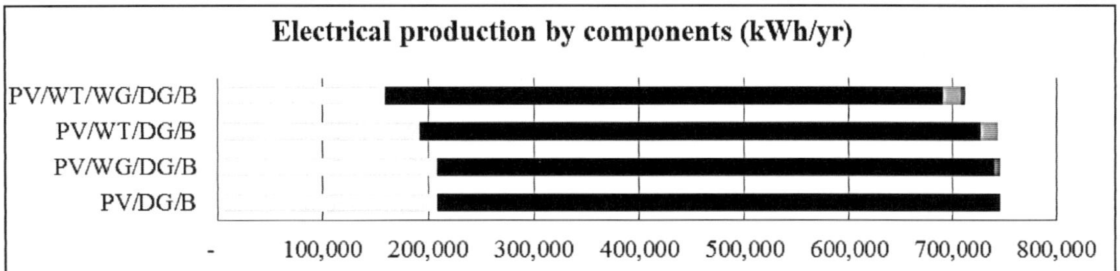

	PV/DG/B	PV/WG/DG/B	PV/WT/DG/B	PV/WT/WG/DG/B
PV	208,256	208,256	192,236	160,197
■DG	538,804	533,416	535,215	529,827
■WT			17,401	17,401
■WG		5,388		5,388

Figure 8. The electrical production by component of each system configurations.

When the system is hybrid with diesel generator, surely the system produced emission. Yet, with the presence of renewable energy components, the emission factor can be reduced. Figure 9 shows the carbon emission released by each system configurations compared to DG system. As expected, the DG system produces the highest carbon dioxide (CO_2) and RE resources helps in reducing carbon emission of the system.

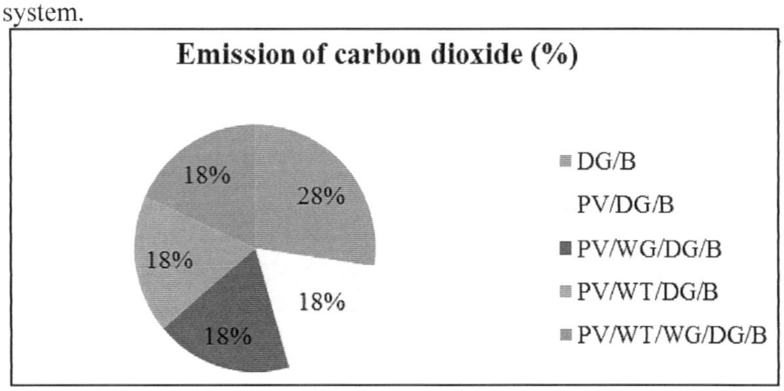

Figure 9. Percentage of carbon emission of hybrid-RE system and DG system.

4. Conclusion

HOMER optimizes hybrid-RE configuration for power system in Pulau Perhentian Kecil. This paper marked that PV is the best RE resources to be implemented in Pulau Perhentian Kecil. The other resources also possible to be developed by more advance renewable technologies because the uses of electricity is important especially in tourism spot as demand is high during peak season. In addition, the presence of RE in power system may reduce the emission of pollutants and dependency on diesel generator which is widely used on remote area. As there is no wave power plants developed on the island, this idea is presented. Even though the results for the wave potential is not convincing but, wave generator still producing power. Improvement in terms of suitable component uses can be studied. Therefore, it is the best to find the alternative to increase the feasibility of RE power system.

References

[1] H Y Chong and W H Lam 2013 Ocean renewable energy in Malaysia: The potential of the Straits of Malacca. *Renewable and Sustainable Energy Reviews* **23** 169-178
[2] 2009 Clean Power Green Jobs.
[3] S Mekhilef, et al. 2012 Solar energy in Malaysia: Current state and prospects. *Renewable and Sustainable Energy Reviews* **16** 386-396
[4] Z M Darus, et al. 2009 The Development of Hybrid Integrated Renewable Energy System (Wind and Solar) for Sustainable Living at Perhentian Island, Malaysia *European Journal of Social Sciences* **9** 557 - 563
[5] L Duckers 2004 *Wave Energy*, in *Renewable Energy*, G Boyle, Editor, Oxford University Press
[6] *Pulau Perhentian Kecil now powered by the sun and wind*, in *The Star*. 2007, Star Media Group Berhad.
[7] 2015 *Study on the repositioning of Pulau Perhentian as a premier ecotourism destination* Universiti Teknologi Malaysia (UTM)
[8] Jones S. Silva, Alexandre Beluco, and L E B d Almeida 2014 Simulating an ocean wave power plant with HOMER. *International Journal of Energy and Environment* **5** 619-630
[9] N A A Razak, M M Othman, and I Musirin 2010 Optimal sizing and operational strategy of hybrid renewable energy system using homer. *Power Engineering and Optimization Conference (PEOCO), 2010 4th International* 495-501
[10] A Singh and P Baredar 2016 Techno-economic assessment of a solar PV, fuel cell, and biomass gasifier hybrid energy system. *Energy Reports* **2** 254 - 260

IOP Conf. Series: Materials Science and Engineering **366** (2018) 012003 doi:10.1088/1757-899X/366/1/012003

Non-Intrusive Appliance Load Monitoring using Genetic Algorithms

D Hock[1,a], M Kappes[1,b] and B Ghita[2,c]

[1] Frankfurt University of Applied Sciences, Nibelungenplatz 1, D-60318 Frankfurt am Main, Germany

[2]Plymouth University, Drake Circus, Plymouth, Devon PL48AA, UK

[a]dehock@fb2.fra-uas.de; [b]kappes@fb2.fra-uas.de; [c]bogdan.ghita@plymouth.ac.uk

Abstract. Smart Meters provide detailed energy consumption data and rich contextual information which can be utilized to assist energy providers and consumers in understanding and managing energy use. Here, we present a novel approach using genetic algorithms to infer appliance level data from aggregate load curves without a-priori information. We introduce a theoretical framework to encode load data in a chromosomal representation, to reconstruct individual appliance loads and propose several fitness functions for the evaluation. Our results, using artificial and real world data, confirm the practical relevance and feasibility of our approach.

1. Introduction

Energy providers need to react fast and efficient on variable energy demand while maintaining grid stability – a problem which arises partially due to the large-scale use of decentralized volatile and renewable energy, such as wind and photovoltaic energy. Many energy providers aim to face this challenge by utilizing the fine-grained demand measurements of smart meters to control and monitor devices on residential level or foresee and respond to otherwise unexpected energy demand. The wide availability of load data and rapid advancements in statistical methods, akin to machine learning techniques, resulted in many concepts to extract information from aggregated load data as provided by a smart meter. A particular approach to generate sophisticated load profiles from smart meter readings, is Non-Intrusive Appliance Load Monitoring (NIALM) [1], which aims to estimate appliance level data without the intrusive deployment of sensors on individual end-uses. The idea to disaggregate the data from a single point of measurement, is cost efficient and convenient while the estimated load profiles can still provide rich information on residential households. We propose a genetic algorithm to infer appliance level load curves without a-priori information or relying on historic information. Inferring information about the energy consumption and presence of individual devices from energy demand is referred to as energy disaggregation. The usage of genetic algorithms in combination with machine learning algorithms is not new. However, while the machine learning methods are well described, many aspects of the genetic algorithm have not been studied yet. Our contribution is the chromosomal representation to solve NIALM as combinatorial optimization problem. Moreover, we present and evaluate different fitness functions and introduce a feature extraction method to reduce the number of genes of our problem set. We introduce a straightforward

IOP Conf. Series: Materials Science and Engineering **366** (2018) 012003 doi:10.1088/1757-899X/366/1/012003

approach to in-cooperate the likelihood of appliances state changes (on/off). The remainder of the paper is organised as follows. In the next section, we introduce our NIALM algorithm and identify a set of features, extracted from the aggregate load curve, to model the problem in a chromosome representation. Subsequently, we present our genetic algorithm as well as a fitness function to assess the similarity of the approximated individual loads to the original load. Furthermore, we estimate the most probable state for each individual appliance. We conclude this paper by evaluating the results using real world data.

2. Related Work

G.W. Hart [1] initiated the research on NIALM to distinguish appliances from aggregated load. The idea of NIALM as an optimization problem, which was taken up by Liang et al. [2] and Suzuki et al. [3], with soft computing methods, such as integer programming, to identify appliances. Other machine learning approaches were recently summarized by Zoha et al. [4]. Several articles introduce genetic algorithms to compute smart grid related problems: Baranski et al. [5] used genetic algorithms to identify certain appliance types by modelling the power steps of the appliance. A comprehensive overview of genetic approaches for demand side management was presented by Barbato et al. [6].

3. Energy Disaggregation

Here, we provide a brief overview of energy disaggregation and the reasoning behind our feature extraction algorithm as well as our fitness function to evaluate the feasibility of each generated solution. Colloquially speaking, a load curve consists of the sum of the loads of all active individual appliances and measurement errors. Hence, if a set of appliances and their load curves are given, the assignment to find the correct state (active/inactive) at each time can be formulated as a combinatorial optimization problem – choose the set of active appliances for all times to minimize the error – which is an NP-complete weighted set problem, similar to the knapsack problem [1]. Hart et al. [1] was the first to describe Load Disaggregation as combinatorial optimization problem as follows:

Algorithm 1. Load Disaggregation as Combinatorial Optimization Problem

for each t **in** time **do**
  activeAppliances ← loads[active$_t$ == true]
  error ← error + (totalLoad[t] – *sum(*activeAppliances*)*)^2
end for
fitness ← error

Algorithm (1), adapted from Hart's [1] publication, evaluates the fitness of estimated loads using the deviation between the sum of individual loads and the total load. Under the assumption that we find the correct amplitude of each appliance, the load curve of each individual appliance can be reconstructed. We can further minimize this combinatorial problem by considering only turn on and off events instead of all measurements. *Note, that we use the term 'event' synonymous for a transition between the on and off state of an appliance.* Our reasoning here is, that all measurements without a state change do not contain any additional information since everything besides the sum of active appliance amplitudes is filtered by the algorithm. Hence, modelling the change-points is sufficient. However, when extracting the amplitude by heuristic means, we need to address another problem: Our set of amplitudes can contain measurement errors or detect several individual devices as one. Hence, there may be several cases of invalid amplitudes which should be terminated. Hart [1] introduced a problem related to this case as follows: Suppose a load curve contains the unique amplitudes 100W, 200W, 300W and 401W. If the measured total load at time t_i is 500W, the best estimate is t_i = *(200W,300W)*. If a moment later, at time t_{i+1} the measured load increases slightly to 501W, the best estimate would be t_{i+1} = *(100W,401W)*, which implies that every appliance changed state between t_i and t_{i+1}. While the problem was originally due to measurement errors, we think that addressing a transition likelihood may also help to minimize the number of invalid appliances.

3.1. Chromosome Representation

Genetic algorithms are based on the evolutionary ideas of natural selection and genetics. As such they represent an intelligent implementation of a random search to solve a given problem. The problem represented by a genetic algorithm is encoded as binary 'genes'. Whereas a string of n genes represents a so called chromosome, which is one solution candidate. The genetic algorithm maintains a population of m chromosomes, with associated fitness values. Parents are selected for reproduction on the basis of their fitness, which represents the abilities of an individual to solve the given problem. By inheriting the properties of a chromosome from parents to children, the reproduction improves the solution candidates in a population from iteration to iteration. The operations of genetic algorithms are *(i)* selection, *(ii)* crossover and *(iii)* mutation. Where *(i)* selects the parents based on the fitness, *(ii)* represents the mating of two individuals and *(iii)* introduces random modifications. In our case, the search space are all possible combinations of active appliance for each event. Hence, the chromosome is constructed as a bit string, which encodes the state of each appliance for each event. While the initial population of chromosomes is usually randomly generated, the genetic algorithms applies mutation and crossover on these chromosomes to create a population of different solutions. Since the length of the chromosome is directly related to the complexity of the genetic algorithm, we aim to limit the chromosome length by providing a set of amplitudes, which appeared in the original load curve. In order to extract these amplitudes we utilize the edge detection algorithm as introduced by Hart [1], which aims to finds step like changes. The algorithm categorizes the load curve in phases of steady and changing demand. A steady phase is defined as a specified number of measurements without change (in a given tolerance). The remaining periods, between steady phases, are defined transition phases. Each steady phase is averaged to minimize noise. The differences between two averaged steady phases result in the amplitude of one or several appliances switched at this moment.

3.2. Fitness Function

In each generation, the fitness of all individual chromosomes is evaluated and represents the ability to solve the problem, whereas a small values represents good fitness. The best candidates get selected for the reproduction process and, due to the crossover operation, inherit their genes to the next generation. Our fitness function models the likelihood for a state change, because we argue that it is more likely for an appliance to stay in the same state than to change its state. For illustration, assume a load curve with the unique amplitudes of 20W, 30W and 50W – hence, a transition from 20W to 50W, can have two possible solution sets:

a) $t_i = (20W),\ t_{i+1} = (50W)$

b) $t_i = (20W),\ t_{i+1} = (20W, 30W)$

While the amplitude at t_{i+1} is equal, the transition from t_i to t_{i+1} in *a)*, can be seen as two events (a turn-off event + a turn-on event), whereas *b)* is only one event. Following this reasoning, a load curve with a total load 20W, 50W and 30W, can could be disaggregated as follows:

a) $t_i = (20W),\ t_{i+1} = (50W),\ t_{i+2} = (30W)$

b) $t_i = (20W),\ t_{i+1} = (20W, 30W),\ t_{i+2} = (30W)$

Where *b)* is consists of 4 events and two appliances and *a)* of 6 events and three appliances. We assume that the sequence featuring the smallest effort is most likely to appear and utilize this fact to create a fitness function which considers this smallest effort. We model the above, using Algorithm (1) with the extension of an additional weight for the cost of an event. This weight is given by the number of state changes required for a given event and is defined by the difference of active appliances between the current and previous event. Assuming, the state t_i and t_{i+1} is given as binary string (which is the case in our genetic algorithm), the weighting can be computed using the sum of the XOR operation. The actual weight should be in range of [1, n+1], where n is the number of appliances – otherwise a constant demand would result in a fitness value of zero. Following this reasoning, the fitness functions derived from the previously mentioned combinatorial optimization problem can be defined as:

Algorithm 2. Fitness Function

for each t **in** events **do**
  activeAppliances ← loads[$gene_t$ == true]
  error ← error + (1+ω[$gene_{t-1}$, $gene_t$])*(totalLoad[t] – *sum(*activeAppliances*)*)^2
end for
fitness ← **error**

Note, that there are still limitations in this approach: We use the unique amplitudes to build the chromosome. Hence, we cannot distinguish appliances that have the same amplitude.

4. Results

In this section, we evaluate the feasibility of our approach to reconstruct individual appliance load curves. We first use an artificially generated load curve to evaluate our proposed fitness function in optimal conditions without measurement errors. In the latter section, we repeat this process on a real world smart meter dataset and present the accuracy of our results.

Next, we evaluated a load curve with five artificial appliances, each with a unique amplitude – two appliances with permanent rectangular pattern and three single On/Off signals. We compare our fitness function, as described in the previous section to Hart's initial algorithm and the Euclidean function.

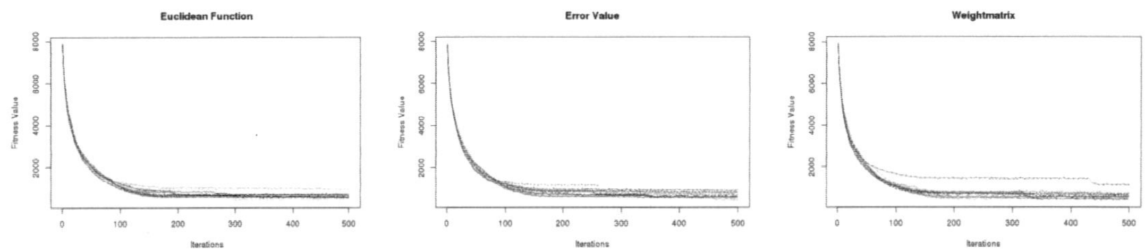

Figure 1. Comparison of Fitness Functions

Figure 1 shows the improvement of the fitness over time for each function (left to right, Euclidean distance, Algorithm (1), Algorithm (2)). The genetic algorithm was executed for 500 iterations and the best 10 chromosomes were used to evaluate the similarity to the ground truth chromosome, which we computed from the individual signals. The quality of each function has been evaluated using the Area Under Curve (AUC), which is a metric based on the true and false positive rate showing the statistical accuracy. Repeating each experiment 10 times, we received for each function a mean AUC greater 0.90. The equally good performance, even without the weighting function, is due to the fact that there is no event included were several appliances are switched at the exact same time. Hence, we were able to extract the exact values of each individual appliance with amplitudes 50W, 120W, 500W, 300W and 20W. However, this is unlikely to happen in real world data.

For our next experiment, we used the Electricity Consumption & Occupancy (ECO) data set, published by Beckel et al. [7], which provides - apart from declared exceptions - measurements in Watt with 4 decimal places and offers individual appliance readings every second (Hz). The measurements used here are real power (W) values of smart meters, showing the mean-cycle-power of a period. Figure 2 depicts one day of the original energy demand (dotted) and the expected result (line) of the genetic algorithm.

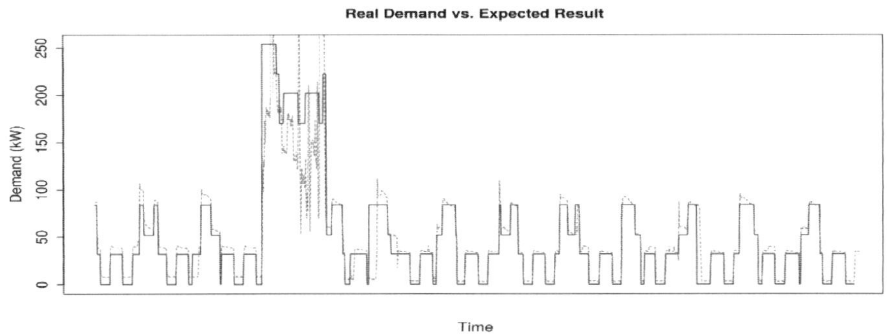

Figure 2. Original Energy Demand and Expected Result of the Genetic Algorithm

In contrast to the artificial energy demand, the real energy demand includes significant noise due to appliances in stand-by mode and characteristic saw-tooth pattern. In our experiment, with a total of four appliances (fridge, kettle, dyer and coffee machine) the noise resulted in a significant decrease of the performance for two reasons: First, the genetic algorithm considered the aggregated amplitude as possible appliance due to the simultaneous switch events and, furthermore, the saw-tooth pattern resulted resulted in several similar amplitudes; for example, while the real amplitudes of the appliances were 30W and 50W, Hart's edge detection algorithm extracted the values 80W, 50W, 30W, 40W, 20W and 10W. Using our weight as a penalty for the number of switch operations, some combinations should be excluded in favour of values the occur more often. However, since the extracted values define the gene length and larger error scope, the real energy demand still requires more computation time and, hence, it is reasonable to exclude as many of them in the pre-processing phase as possible. Our results, using measurements over 1 month (September) as provided by one of the six houses (House 1), indicate that our weighted fitness function performs significantly better with realistic demand then the previous error function. Using the same number of iterations (500), we obtained an AUC of 0.7 while the unweighed function resulted in little more than random results (mean AUC ~0.5). A closer analysis revealed that the performance of our algorithm shows huge diversity throughout the day. In the following experiment, we divided the energy demand in equal sized time windows of 180 minutes and tried to reveal the individual appliances using only 500 iterations – the experiment was carried out for 10 days and 5 repetitions for each day using the fitness function as described in Algorithms (2).

Table 1. Resulting AUC for 500 Iterations

	0:00	**3:00**	**6:00**	**9:00**	**12:00**	**15:00**	**18:00**	**21:00**
Day 1	1.00	0.70	0.55	0.60	0.63	0.56	0.54	0.50
Day 2	0.88	0.81	0.70	0.53	0.73	0.61	0.65	0.65
Day 3	1.00	0.71	0.57	0.67	1.00	0.60	0.57	0.66
Day 4	0.90	0.64	0.59	0.70	0.62	0.79	0.77	0.58
Day 5	0.99	0.50	0.53	0.59	0.92	0.54	0.55	0.68
Day 6	0.70	0.57	0.58	0.53	0.60	0.58	0.5	0.59
Day 7	0.88	0.54	0.57	0.65	0.52	0.68	0.61	0.53
Day 8	0.68	0.65	0.56	0.58	0.53	0.55	0.57	0.88
Day 9	0.68	0.54	0.54	0.69	0.81	0.60	0.61	0.57
Day 10	0.94	0.66	0.64	0.62	0.70	0.58	0.66	0.63

Table 1 shows the resulting AUC for each time window. The results indicate that the genetic algorithm has problems to disaggregate sections with many appliances stacked or with appliances that are rarely used, which need more iterations. This problem could be encountered with a dynamic number of iterations per time window, but is out of the scope of this paper.

5. Conclusion

In this paper, we provided a comprehensive overview of non-intrusive appliance disaggregation as combinatorial optimization method. We first presented a method to encode appliances in a chromosome encoding and described algorithms to evaluate the fitness of individual appliance loads generated by a genetic algorithm. The resulting best fitting chromosome can be used to read the amplitude as well as turn on and off timings of each appliance. The proposed approach does have a number of inherent limitations, such as the inability to model simultaneously running appliances with equal amplitude or to implement so called state-machines with several amplitudes, but represents also a robust alternative, given its simplicity and the fact it does not require historical information. The presented experiments aimed to evaluate the feasibility of genetic algorithms for NIALM in general and point at several challenges – in future work, we aim to test several parameters, such as customized mutation and crossover operations, which can heavily influence the performance of genetic algorithms, but were out of scope this work.

References

[1] Hart G W (1992). Nonintrusive appliance load monitoring *Proc. of the IEEE* **80** 1870-91
[2] Liang J, Ng K and Cheng J 2007 Identifying appliances using load signatures and genetic algorithms *ICEE*
[3] Suzuki K, Inagaki S, Suzuki T, Nakamura H and Ito K 2008 Nonintrusive appliance load monitoring based on integer programming *IEEE SICE* **2008** 2742-47
[4] Zoha A, Gluhak A, Imran M and Rajasegarar S 2012 Non-intrusive load monitoring approaches for disaggregated energy sensing: a survey *Sensors* **12** 16838-66
[5] Baranski M and Voss J 2004 Genetic algorithm for pattern detection in nialm systems *IEEE SMC* **4** 3462-68
[6] Barbato A and Capone A 2014 Optimization models and methods for demand-side management of residential users: A survey *Energies* **7** 5787-824
[7] Beckel C, Kleiminger W, Cicchetti, Staake T and Santini S 2014 The ECO data set and the performance of non-intrusive load monitoring algorithms *ACM BuildSys* **2014**

Short-term hydro generation operation of cascade hydropower stations in Han River, China

Rui Zhang , Xuemin Wang, Shan Yu and Qiang Zou

Changjiang Survey, Planning, design and Research CO., LTD., Wuhan, 430010, P R China

E-mail address: ruiz6551@foxmail.com

Abstract. Hydropower is a natural renewable energy with outstanding technological superiority in short-term operation. Short-term hydro generation operation (SHGO) requires not only lord dispatch in single station, but also the coordination of hydraulic and electric connection for different stations. In this paper, a coordinated peak shaving operation method of cascade hydropower stations is proposed. Firstly, regular pattern of peak load regulation is analyzed using the load characteristics of power grid. The daily power output of cascade hydropower stations should meet the needs of typical load pattern. In order to satisfy comprehensive utilizations of water resources, the outflow of each reservoir should be no less than the demand of water supply or irrigation. To verify the feasibility and effectiveness of the proposed method, an application of Pankou and Xiaoxuan cascade hydropower stations, located in Hanjiang River in China, is presented in this paper. Simulation results indicate that the proposed method can obtain the maximum peak shaving amplitudes and suggested peak shaving amplitudes at the same time, which satisfies different operation demands of cascade hydropower stations.

1. Introduction

The operation of cascade hydropower stations is an operation problem for more than one hydropower stations. The target of this problem is to improve waterpower utilization rate and power generation, while requirements of flood control, water supply and output of each hydropower station are satisfied[1]. The operation of cascade hydropower stations can be divided into three main categories: mid-long-term operation, short-term operation and real-time operation. As a short-term optimal operation problem of electric system, short-term hydro generation operation (SHGO) requires the coordination of different departments and multiple stations. Hydropower stations have to allocation water resources on the basis of the runoff and storage capacity of reservoir. On the other hand, power output generated by water resources should meet the requirement of power demand and grid management. Therefore, technical and managerial challenges remain to be solved for the coordination of these departments and water resources allocation in different stations.

The SHGO problem is a typical nonlinear mixed optimization problem is to figure out the daily power output, as well as the maximum amplitude of load shaving, which is determined by limited water resources and power demand. In the past few years, many studies on short-term scheduling operation mainly focused on the new energy construction, for example, developing pumped-hydro energy storage stations[2–4], gas-fired power stations[5–7] and energy storage[8,9], or optimization models and methods for peak shaving on the power grid side[10,11]. Other studies concentrate on algorithm and application of short-term hydro generation scheduling, such as particle swarm optimization[12,13], genetic algorithm[14], differential evolution[15], and the application of world largest hydropower plants, Three Gorges-Gezhouba cascade[16] and Xiluodu-Xiangjiaba cascade[17]. However, traditional short-term hydro generation scheduling aims to minimize the total water consumption while simultaneously meeting some hydraulic and electrical constraints. The requirement of power demand is ignored especially when the hydropower stations play a significant role in power grid. On the other hand, peak shaving capability of other new energy is limited by their distinguishing features and energy policy in China. Therefore, the research of SHGO problem is meaningful and crucial to enhance the waterpower utilization rate of the station and dependability of the power system.

In order to solve the SHGO problem, a method of peak shaving operation for cascade hydropower stations is proposed in this paper. Firstly, regular pattern of peak load regulation is analyzed using the system and load characteristics. The daily power output of cascade hydropower stations should meet the needs of typical load pattern. Then the comprehensive utilizations of each reservoir transfer into constraints which should be considerde. To verify the feasibility and effectiveness of the proposed method, an application of Pankou and Xiaoxuan cascade hydropower stations is presented for case study. Pankou and Xiaoxuan cascade hydropower stations are located along the Duhe River, which flows through Hubei Province in China and joins the Hanjiang River at about 127.2 km upstream of Danjiangkou Reservoir. Hanjiang River is the water source of Mid-Route of South-To-North Water Transfer Project, which is diverting water from Hanjiang River in middle of China to the north[18,19]. The project characteristics of Pankou and Xiaoxuan cascade hydropower stations are shown in Table 1.

Table 1 Project characteristics of Pankou and Xiaoxuan hydropower stations

Project	Pankou hydropower station	Xiaoxuan hydropower station
Basin area (km^2)	8950	9040
Normal pool level (m)	355	264
Flood control level (m)	347.6	264
Dead water level (m)	330	261.3
Total storage capacity (billion m^3)	2.34	0.04
Installed capacity (MW)	500	50
Number of units	2	3
Yearly average power output (billion kW•h)	1.05	0.15

Since Pankou and Xiaoxuan cascade hydropower stations are put into use, some concerns have been raised about the increasing conflict between short-term scheduling and high efficient utilization of water resources. In this study, we analyzed the load characteristics of Hubei provincial power grid and Siyan regional power grid that connected power from Pankou and Xiaoxuan hydropower stations in Hubei Province. So the typical load pattern of peak load regulation that cascade hydropower stations followed was proposed. Then, the coordinated peak load regulation of cascade hydropower stations is analyzed, which could provide some important support to the solution of SHGO problem.

2. Typical load pattern of peak shaving for cascade hydropower stations

In general, the power energy generated by Pankou and Xiaoxuan cascade hydropower stations is consumed primarily by local users in the northwest of Hubei Province. During the flood season from May to September, the power energy is consumed mainly by Shiyan city, and the surplus power is fed to Hubei power grid by a double-circuit 500kV line. Thus the typical load pattern of peak shaving that cascade hydropower station followed should consider the load characteristics of both Hubei provincial power grid and Siyan regional power grid.

2.1. Load characteristics of Hubei provincial power grid and Shiyan regional power grid

Hubei power grid is the passage of the West-to-East Electricity Transmission Project and the North-to-South Electricity Transmission Project[20,21] in China. Hubei power grid shows a double-peak curve of daily load. One peak is in the morning and a second peak in the evening. According to the daily load of Hubei power grid measured in recent years and the 12th "Five-year" Development Plan of

Hubei Power Grid[22], the typical maximum daily load of Hubei power grid is shown in Fig.1.Shiyan power grid is responsible for delivering the electrical energy to local users in Shiyan City. Similar to the Hubei power grid, a double-peak curve of daily load is observed for Shiyan power grid, with one peak in the morning and a second peak in the evening. A typical daily load curve of Shiyan regional power grid is shown in Fig.2.

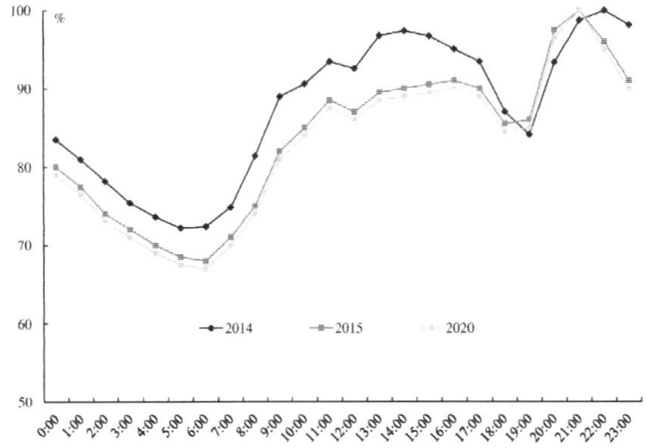

Figure.1 Typical dailwy load curve of Hubei power grid **Figure.2** A typical load curve of Shiyan power grid

2.2. Load pattern of peak shaving for cascade hydropower stations
As it is estimated the daily load curve of both Hubei power grid and Shiyan power grid will remain largely unchanged for years to come. Considering that the evening peak of Hubei power grid occurs later than that of Shiyan power grid, two distinct regulation models should be adopted. We call them "Peak shaving model A"(Shiyan peak load regulation model) and "Peak shaving model B" (Hubei peak load regulation model). Therefore, the typical load pattern of peak shaving that cascade hydropower station followed can be figure out by the load characteristics of both Hubei provincial power grid and Siyan regional power grid.

2.2.1. Peak shaving model A Shiyan peak load regulation model. According to the peak-load characteristics of Shiyan power grid, two peak loads are observed at 8:00~12:00 and 14: 00~18: 00 during which peak-load regulation is required, and valley load lasts for at least 8 hours, typically from 23:00 to 7:00 of the next day. The rest of the day is usually governed by intermediate load.

2.2.2. Peak shaving model B: Hubei peak load regulation model. The peak load occurs at 8:00~12:00 in the morning and 18:00~22:00 in the evening, while the valley load occurs from 23:00 to 7:00 of the next day.

3. Coordinated peak shaving operation

The peak shaving operation of cascade hydropower stations, as the main model of short-term operation, depends mainly on the load characteristics of the power grid and the regulation capacity of the reservoir. The peak shaving operation of hydropower stations should be in accordance with the typical load pattern of power grids which stations supply. Therefore, the generalized procedures of coordinated peak shaving operation for the cascade hydropower stations can be expressed as follows:

Step 1: Read the started water level $\underline{Z}$ and average water discharge of one day $\underline{Q}$ by the result of by mid-long term operation of hydropower station.

Step 2: Calculate the average power generation of hydropower stations by the formula as follows. Set the intermediate load L_{inter}=N_{avg}, iteration coefficient i=1.

$$N_{avg} = \sum_{i=1}^{N} p_i = \sum_{i=1}^{N} 9.81 \bullet \eta \bullet q \bullet h \tag{1}$$

Where the p_i is the power generation of ith unit. η is the turbine-generator efficiency, which is indexed by the efficiency curve of the hydropower unit. q and h are the water discharge and effective water head, respectively.

Step 3: Use the typical load pattern of peak shaving extracted by the analysis of power grid to generate the daily output of hydropower stations. Calculate the peak-load and valley-load as follows:

$$L_{peak} = L_{inter} + \frac{N_{max} - N_{min}}{IC} \times i \quad , \quad L_{valley} = L_{inter} - \frac{N_{max} - N_{min}}{IC} \times i \tag{2}$$

Step 4: If $L_{peak} \geq N_{max}$ or $L_{valley} \leq N_{min}$, turn to **Step 7**; otherwise turn to **Step 5**.

Step 5: If total discharging water $W_d = \sum_{k=1}^{K} w_k > 0$, turn to **Step 7**; otherwise turn to **Step 6**.

Step 6: If $i = IC$ (the maximum of the iteration number) turn to **Step 7**; otherwise go back to **Step 2**.

Step 7: The search is terminated and output the solutions as the final results. Otherwise, set $i = i + 1$ and go back to **Step 2**.

4. Implement of coordinated peak shaving operation for SHGO problem

4.1. Coordinated peak shaving operation scheme

The purpose of Pankou and Xiaoxuan cascade hydropower stations are flood control, water supply and power generation. The tasks of flood control and water supply take precedence over power generation, as well as peak shaving operation. Therefore, considered the requirements of cascade hydropower stations for peak shaving, the scheme of coordinated peak shaving operation is proposed:

Case 1: If the power generating flow is small, cascade hydropower stations can take full responsibility for peak shaving operation on the premise that the water demand in the downstream area can be well satisfied;

Case 2: If the power generating flow is large, it is necessary to avoid abandoned water caused by peak load regulation. So peak shaving amplitude of upstream station should be limited. The maximum amplitude occurs when the water discharge of upstream station is just the full-load flow of downstream station;

Case 3: If the power generating flow is moderate, peak load regulation scheme should be selected prudentially to avoid radical scheme and to take into account the transition between different schemes.

4.2. Simulation results

In order to ensure a uniform discharge and avoid abandoned water as much as possible during the anti-regulation of Xiaoxuan hydropower station, it is necessary to restrict the peak shaving amplitude of Pankou hydropower station. In this way, the daily power generating flow can be determined. The full-load flow of Pankou hydropower station is 680 m^3/s, which is set as the upper limit of the power generating flow. As mentioned above, the water supply of downstream is the requirement of water supply in peak shaving. Thus an important constraint for the peak shaving operation of Pankou hydropower station is that the minimum discharge of Xiaoxuan hydropower station should not be less than 16.7 m^3/s. Considering the future trend and safety margin, the minimum power generating flow of Pankou hydropower station is determined to be 20 m^3/s. The suggested peak shaving amplitudes is the allowance range of amplitudes that Pankou station satisfies all the requirements of water resources, as shown in Table 2.

Coordinated peak shaving results of cascade hydropower stations appears different as the daily average power flow of Pankou station Q changes. Firstly, if Q is more than 500 m^3/s, the discharge exceeds the regulation capacity of Xiaoxuan hydropower station, resulting in abandoned water on Xiaoxuan. In this case, Pankou hydropower station is suggested not taking part in the peak shaving operation. Secondly, if Q is between 300 and 500 m^3/s, the lord shaving amplitude should be within the range of suggested amplitudes. Otherwise, there may be abandoned water as the discharge of Pankou station exceeds. Furthermore, if Q is between 70 and 300m^3/s, Xiaoxuan hydropower station can totally anti-regulate Pankou hydropower station. In this case, Pankou hydropower station can take full part in the peak shaving operation and lord shaving amplitude can reach the maximum one. Finally, if Q is lower than 70 m^3/s, there may exist an unstable operation region for hydraulic generator units, which may probably result in the aggravation of unit vibration. In this case, Pankou hydropower station should not take part in peak shaving operation and it is important to go over the unstable region as quickly as possible to ensure the safe operation of units.

Table 2 Suggested peak shaving amplitude of Pankou hydropower station

Pankou reservoir water level (m)	Amplitude for different power generating flow rates (m^3/s) (ten MW)											
	Peak shaving model A						Peak shaving model B					
	100	200	300	400	500	600	100	200	300	400	500	600
330	7.60	22.80	36.00	35.85	0.00	0.00	7.60	22.80	36.00	35.85	0.00	0.00
335	8.60	24.60	39.42	39.34	0.00	0.00	8.60	24.60	39.42	39.34	0.00	0.00
340	9.60	26.60	42.80	32.20	0.00	0.00	9.60	26.60	42.80	32.20	0.00	0.00
345	10.40	28.60	45.80	34.20	0.00	0.00	10.40	28.60	45.80	34.00	0.00	0.00
350	11.40	30.20	48.60	36.80	0.00	0.00	11.40	30.20	48.60	36.60	0.00	0.00
355	12.20	32.00	50.91	37.77	0.00	0.00	12.20	32.00	50.91	37.77	0.00	0.00

4.3. Peak shaving operation of Xiaoxuan hydropower station

There are several choices for Xiaoxuan hydropower station when Pankou and Xiaoxuan cascade hydropower stations take part in the peak shaving operation, such as in synchronization with Pankou in peak shaving operation, no peak shaving operation and appropriate peak shaving operation are options.

Compared the advantage and disadvantage among three strategies above, Xiaoxuan hydropower station operates in synchronization with Pankou in short-term operation is preferred. This strategy is security and stable, which has been adopted by administration section in engineering application. So the power generating flow of Xiaoxuan hydropower station is preferably regulated in synchronization with Pankou hydropower station in this paper.

5. Conclusions

Short-term hydro generation scheduling is a good choice to carry out peak shaving in the power grid because of various outstanding properties. In order to deal with the SHGO problem, we developed a coordinated peak shaving operation method of cascade hydropower stations in this paper. This method is aim to maximize the amplitude of peak shaving while lord demand of power grid and water requirement of downstream are satisfied. The proposed method is applied to solve Pankou and Xiaoxuan cascade hydropower stations in Hanjiang River in China. Simulation results indicate that the proposed method can obtain suggested peak shaving amplitudes, which satisfies different operation demands of cascade hydropower stations in practical decision-making.

Acknowledge

This work is supported by The National Key Research and Development Program of China (No. 2016YFC0400907).

References

[1] Zhang R, Zhou J, Zhang H, Liao X, Wang X. Optimal operation of large-scale cascaded hydropower systems in the upper reaches of the Yangtze River, China. Journal of Water Resources Planning and Management 2014; 140:480-495.
[2] Zeng M, Zhang K, Liu D. Overall review of pumped-hydro energy storage in China: status quo, operation mechanism and policy barriers. Renew Sustain Energy Rev 2012;17:35–43.
[3] Cheng C, Cheng X, Shen J, Xi. Short-term peak shaving operation for multiple power grids with

pumped storage power plants. Electr Power Energy Syst 2015;67:570–87.
[4] Cao F, Zhang L. Determination of pumped-storage plant capacity with peak-regulation proportion. Electr Power Autom Equip 2007;27(6):47–50.
[5] Ding N, Duan J, Xue S, Zeng M, Shen J. Overall review of peaking power in China: status quo, barriers and solution. Renew Sustain Energy Rev 2015;42:503–16.
[6] Sigrist L, Lobato E, Rouco L. Energy storage systems providing primary reserve and peak shaving in small isolated power systems: an economic assessment. Electr Power Energy Syst 2013;53:675–83.
[7] Yang T, Wang W, Zeng D, Liu J, Cui C. Closed-loop optimization control on fan speed of air-cooled steam condenser units for energy saving and rapid load regulation. Energy 2017; 135:394-404.
[8] Lv S, He W, Zhang A, Li G, Luo B, Liu X. Modelling and analysis of a novel compressed air energy storage system for trigeneration based on electrical energy peak load shifting. Energy Conversion and Management 2017; 135:394-401.
[9] Chua K H, Lim Y S, Morris S. A novel fuzzy control algorithm for reducing the peak demands using energy storage system. Energy 2017; 122:265-273.
[10] Sadineni S B, Boehm R F. Measurements and simulations for peak electrical load reduction in cooling dominated climate. Energy 2012;37:689–697.
[11] Shen J, Cheng C, Wu X, Cheng X, Li W, Lu J. Optimization of peak load among multiple provincial power grids under a certain dispatching authority. Energy 2014;74:494–505.
[12] Zhang J, Tang Q, Chen Y, Lin S. A hybrid particle swarm optimization with small population size to solve the optimal short-term hydro-thermal unit commitment problem. Energy 2016;109:765–780.
[13] Zhang R, Zhou J, Ouyang S, Wang X, Zhang H. Optimal operation of multi-reservoir system by multi-elite guide particle swarm optimization. Electr Power Energy Syst 2013;48:58~68.
[14] Nazari-Heris M, Mohammadi-Ivatloo B, Haghrah A. Optimal short-term generation scheduling of hydrothermal systems by implementation of real-coded genetic algorithm based on improved Mühlenbein mutation. Energy 2017;128:77–85.
[15] Arnel G, Adnan G, Peter K, Joze P, Igor T. Optimal Optimization of hydro energy storage plants by using differential evolution algorithm. Energy 2014;77:97-105.
[16] Mo L, Lu P, Wang C, Zhou J. Short-term hydro generation scheduling of Three Gorges-Gezhouba cascaded hydropower plants using hybrid MACS-ADE approach. Energy Conversion and Management 2013; 76:260-273.
[17] Lu P, Zhou J, Wang C, Qiao Q, Mo L. Short-term hydro generation scheduling of Xiluodu and Xiangjiaba cascade hydropower stations using improved binary-real coded bee colony optimization algorithm. Energy Conversion and Management 2015; 91:19-31.
[18] Yan B, Chen L. Coincidence probability of precipitation for the middle route of South-to-North water transfer project in China. Journal of Hydrology 2013; 499:19-26.
[19] Office of the South-to-North Water Diversion Project Construction Committee, State Council, PRC. The South-to-North Water Diversion Project. Engineering 2016; 2:265-267.
[20] Wang J, Dong Y, Jiang H. A study on the characteristics, predictions and policies of China's eight main power grids. Energy Conversion and Management 2014; 86:818-830.
[21] Li Y, Lukszo Z, Weijnen M. The impact of inter-regional transmission grid expansion on China's power sector decarbonization. Applied Energy 2016; 183:853-873.
[22] Hubei electric power company, State Grid. The 12th "Five-year Development Plan of Hubei Power Grid. 2010.

IOP Conf. Series: Materials Science and Engineering **366** (2018) 012005 doi:10.1088/1757-899X/366/1/012005

Optimal Operation of Cascade Hydropower Stations Based on Chaos Cultural Sine Cosine Algorithm

Zou Qiang, Li Anqiang, He Xiaocong and Wang Xuemin

Changjiang Institute of Survey, Planning, design and Research CO., LTD., Wuhan 430010, China

E-mail (Zou Qiang): zouqianghust@163.com

Abstract. Due to the fact that Sine Cosine Algorithm(SCA) may easily premature convergence, by taking cultural algorithm as the framework of population space and using chaotic search to lead the evolvement of belief space, a new Chaos Cultural Sine Cosine Algorithm (CCSCA) was present in this paper. And the results of typical benchmark functions show that CCSCA has better global searching ability than SCA to avoid the premature phenomenon. Finally, CCSCA was employed for optimal operation of cascade hydropower stations (OOCHS), and case study showed that CCSCA could get satisfying results with high accuracy and fast efficient, thus providing a new efficient method to solve OOCHS.

1. Introduction

In order to respond to the green development requirements for energy conservation and emission reduction, it is very important to optimize the optimal dispatching of cascade hydropower stations and the research of OOCHS has attracted wide attention [1]. The main purpose of OOCHS is to obtain the maximum power generation benefit during the scheduling period, but the cascade hydropower stations have complex hydraulic and electric connections, and the water level and output has a complex coupling at any time with multi dimensions and multi constraints [2]. The traditional optimization methods include dynamic programming (DP), progressive optimality algorithm (POA) are difficult to solve OOCHS. In recent years, plenty of relative algorithms have been present to obtain achievements.

SCA is a population-based stochastic optimization algorithm, and its principle is simple with few controlled parameter [3], but SCA has not yet been employed to OOCHS. This research attempts to apply SCA in OOCHS after clarifying its operation mechanism, but found SCA is similar with other algorithms to fall into local optimum, so it is necessary to improve the global convergence ability. Therefore, SCA combined with culture algorithm [4] and chaotic search [5] is proposed to form the CCSCA, which used cultural algorithm as the main space in SCA and using chaotic search with local traversal search performance to lead the evolvement of belief space. The test results of several typical test functions show that compared with SCA, differential evolution algorithm (DE) as well as Particle swarm optimization (PSO), the CCSCA proposed in this paper can effectively avoid precocious convergence, and the ability of searching the global optimal solution is significantly improved. Finally, the optimal scheduling of CCSCA is applied to cascade reservoirs, and the feasibility and effectiveness of CCSCA are verified through the comparison and analysis of the scheduling results.

2. Optimal operation model of cascade hydropower stations

According to the main principle of maximizing the total power generation benefit, the mathematical objective function of OOCHS is described as bellow [1][2]:

$$F = \max \sum_{i=1}^{N} \sum_{t=1}^{T} A_i \cdot Q_{i,t} \cdot H_{i,t} \cdot \Delta t \tag{1}$$

where N is the number of power station; T is the total periods; A_i is output power coefficient of i -th power station; $Q_{i,t}$, $H_{i,t}$ are respectively the flood discharge volume and water head of i -th power station at t -th period; Δt is the length of period's time. And the constrains of OOCHS are as follows:
(1) Water balance equation:

$$V_t = V_{t-1} + (I_t - Q_t)\Delta t \tag{2}$$

(2) Reservoir upstream water level limit:

$$Z_{i,t}^{\min} \le Z_{i,t} \le Z_{i,t}^{\max} \tag{3}$$

(3) Discharge volume limit:

$$Q_{i,t}^{\min} \le Q_{i,t} \le Q_{i,t}^{\max} \tag{4}$$

(4) Output power limit:

$$N_{i,t}^{\min} \le N_{i,t} \le N_{i,t}^{\max} \tag{5}$$

in Eqs.(2)-(5), $Z_{i,t}^{\min}$ and $Z_{i,t}^{\max}$ are the minimum and maximum upstream water level limit at t -th period; $Q_{t,\min}$ and $Q_{t,\max}$ are the minimum and maximum discharge volume limit at t -th period; $Q_{t,\min}$ and $Q_{t,\max}$ are the minimum and maximum output power limit at t -th period.

3. Chaos Cultural Sine Cosine Algorithm (CCSCA)

3.1. Sine Cosine Algorithm (SCA)

The population of SCA contains N D-dimensions real-valued parameter vectors in the population [3]. In t generation, the position of i individual is the vector $\mathbf{X}_i(t) = (x_{i1}(t), x_{i2}(t), \cdots, x_{iD}(t))$, and the global optimal location in population so far is expressed as $\mathbf{P}_g(t) = (p_{g1}(t), p_{g2}(t), \cdots, p_{gD}(t))$. And the vector updating is determined by the following equations:

$$\begin{aligned} \mathbf{X}_i(t+1) &= \mathbf{X}_i(t) + r_1 \times \sin(r_2) \times \left| r_3 \cdot \mathbf{P}_g(t) - \mathbf{X}_i(t) \right|, \ r_4 < 0.5 \\ \mathbf{X}_i(t+1) &= \mathbf{X}_i(t) + r_1 \times \cos(r_2) \times \left| r_3 \cdot \mathbf{P}_g(t) - \mathbf{X}_i(t) \right|, \ r_4 \ge 0.5 \end{aligned} \tag{6}$$

where r_1, r_2, r_3 and r_4 are random numbers, and $r_1 = 2\times(1 - t/G)$, $r_2 \in [0, 2\pi]$, $r_3 \in [0, 2]$, $r_4 \in [0, 1]$, and their specific utilities are mentioned as follows: r_1 defines the next position's region, which could be either in or outside the space between the solution and destination; r_2 indicates the movement towards or outwards the destination; r_3 indicates a random weight for the destination in order to stochastically emphasize ($r_3 > 1$) or deemphasize ($r_3 < 1$) the effect of destination in defining the distance; r_4 equally switches between the sine and cosine components.

3.2. Chaos Cultural Sine Cosine Algorithm (CCSCA)

The framework of Cultural algorithm is shown in Fig. 1, which includes three components, i.e. a population space for producing knowledge, a belief space for using knowledge as well as a communication protocol for exchanging information of the two spaces [4]. Any population based algorithm can be adopted as the evolution process of population space. Then, belief space selectively accepts the knowledge, and uses the knowledge to adjust its knowledge structure so as to guide population space evolution in next generation. Here chaotic search [5] is adopted in the belief space. Finally, the communication protocol is implemented by the accept operation and influence operation.

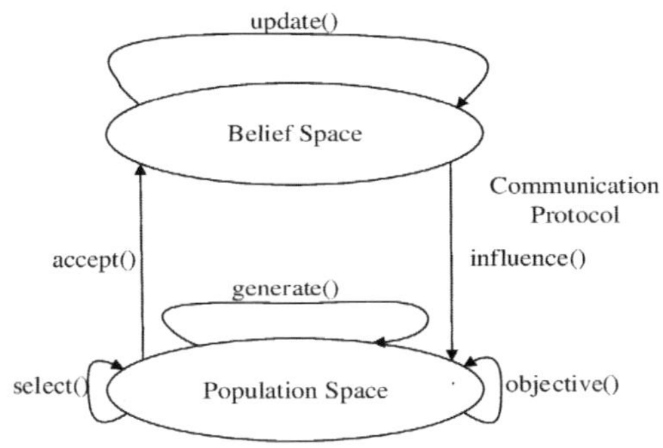

Figure.1 Framework of cultural algorithm

The specific procedures for accept operation and influence operation are as follows: (1) accept operation. Set the size of belief space is $M = 0.2N$, population space provides the better M individuals for belief space, then the belief space takes the better M individuals to replace itself by comparing the fitness value. (2) Influence operations. The belief space chooses the better M individuals to replace the individuals in population space. In this paper, CCSCA employs the two operations every 20 generations.

For any individual $\mathbf{X}_i(t) = (x_{i1}(t), x_{i2}(t), \cdots, x_{iD}(t))$ in belief space, when the maximum number $k_{\max}$ of local chaotic searches is given, the specific flow of chaotic local search is as follows [5]:

Step 1: Set $k = 1$, a chaotic sequence $\mathbf{cx}^k = (cx_1^k, cx_2^k, ..., cx_D^k)$ is produced with improved logistic chaotic map written as $cx_j^{k+1} = 1 - 2 \cdot (cx_j^k)^2$, here $j = 1, 2, ..., D, k = 1, 2, ..., k_{\max}$ and $cx_j^k \in [-1, 1]$.

Step 2: Two new individuals $\mathbf{c}_i(t) = (c_{i1}(t), c_{i2}(t), \ldots, c_{iD}(t))$ and $\mathbf{r}_i(t) = (c_{i1}(t), c_{i2}(t), \ldots, c_{iD}(t))$ are achieved based on $\mathbf{X}_i(t)$ and the optimal location $\mathbf{P}_g(t)$ for chaotic search as follows:

$$c_{ij}(t) = cx_j^k \cdot p_{gj}(t) + (1 - cx_j^k) \cdot x_{ij}(t), \quad j = 1, 2, \ldots, D \tag{7}$$

$$r_{ij}(t) = cx_j^k \cdot x_{ij}(t) + (1 - cx_j^k) \cdot p_{gj}(t), \quad j = 1, 2, \ldots, D \tag{8}$$

Step 3: Calculate the fitness values of the two new individual and find the better individual from them called $\mathbf{cr}_i(t)$. Moreover, compare the fitness values between $\mathbf{X}_i(t)$ and $\mathbf{cr}_i(t)$, and replace $\mathbf{X}_i(t)$ by $\mathbf{cr}_i(t)$ if the fitness value of $\mathbf{cr}_i(t)$ is better than $\mathbf{X}_i(t)$.

Step 4: Set $k = k + 1$, if $k > k_{\max}$, then the chaotic local search is stopped, otherwise turn to **Step 1**.

3.3. Flowchart of CCSCA

With the basic SCA and the improvements, the flowchart of CCSCA proposed is as follows:

Step 1: Initialization. Set generation $g = 0$, and initialize the population space and belief space.

Step 2: Implement the SCA operation for each individual with Eq.(6) in population space.

Step 3: If the generation reaches the specified item, accept operation and impact operation would be proceeded between population space and belief space.

Step 4: If g equal the given condition, export the optimal solution, otherwise g=g+1, turn **Step 2**.

4. Benchmark Function Tests

In order to demonstrate the effectiveness of CCSCA, four well-known Benchmark functions with their global optimum value are mentioned in Table 1. Additionally, SCA, DE and PSO and are used to show the improvement of the proposed algorithm.

Table1 benchmark functions

Function	Expression	Dim	Shift position	optimal value
Sphere	$f_1(x)=\sum_{i=1}^{D}x_i^2$	30	[-100,100]	0
Schwefel	$f_2(x)=\sum_{i=1}^{D}(\sum_{j=1}^{i}x_j)^2$	30	[-100,100]	0
Rastrigin	$f_3(x)=\sum_{i=1}^{D}[x_i^2-10\cos(2\pi x_i)+10]$	30	[-5.12,5.12]	0
Griewank	$f_4(x)=\frac{1}{4000}\sum_{i=1}^{D}x_i^2-\prod_{i=1}^{D}\cos(\frac{x_i}{\sqrt{i}})+1$	30	[-600,600]	0

The parameters of SCA and CCSCA are *NP*=100, *D*=100, *Gmax*=1000. The parameters of DE is $F=CR=0.4$. The parameters of PSO is $w_{\max}=0.9$, $w_{\min}=0.2$, $c_1=c_2=2$. The test results are shown in table 2 and each function is independently processed 50 times. And the convergence curves for each function are respectively shown in Fig.2-Fig.7.

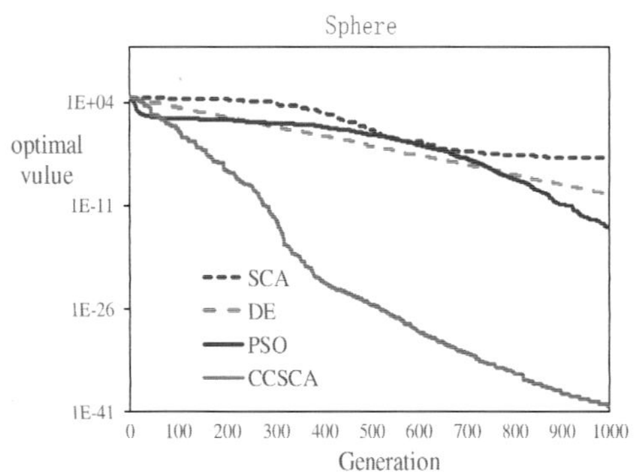

Figure.2 Convergence curve of function Sphere

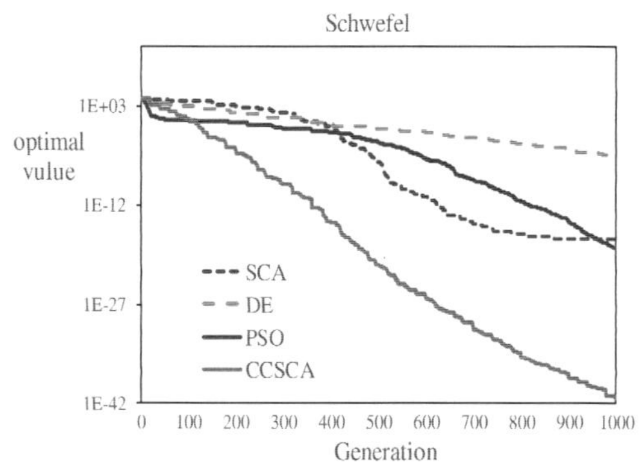

Figure.3 Convergence curve of function Schwefel

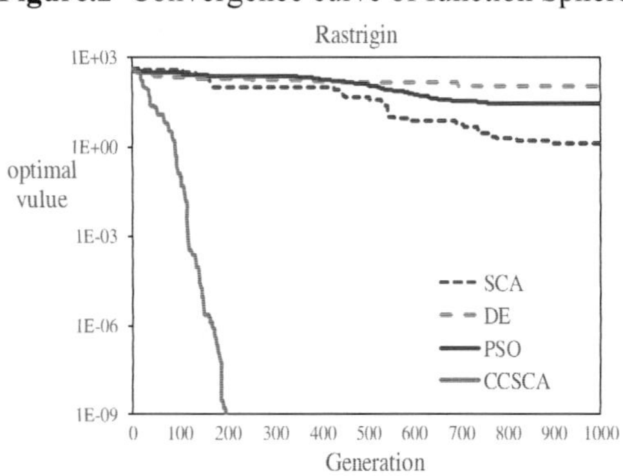

Figure.4 Convergence curve of function Rastrigin

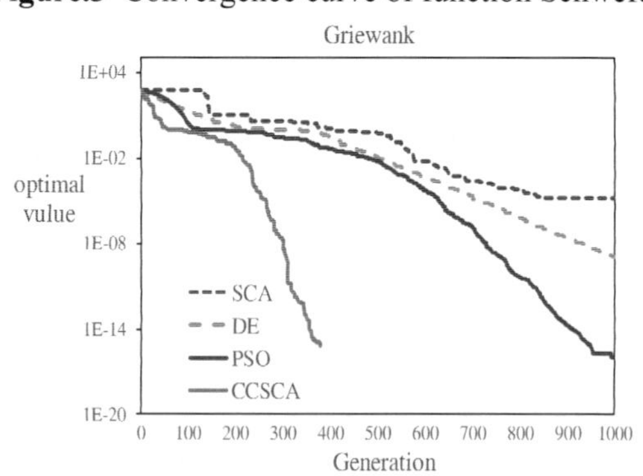

Figure.5 Convergence curve of function Griewank

Seen from **table 2** and **Fig.2-Fig.5**, CCSCA outperforms SCA, DE and PSO in solving all functions, and CCSCA can avoid premature effectively and get better convergence precision.

Table2 Test function results for SCA, DE, PSO and CCSCA

function	Algorithm	best	worst	average	standard deviation
Sphere	SCA	5.1467e-08	1.0642e-04	3.8457e-05	4.3636e-05
	DE	2.3862e-10	5.2235e-10	3.4644e-10	1.4131e-10
	PSO	2.1439e-17	6.4129e-15	2.5669e-15	3.3807e-15
	CCSCA	**7.0495e-46**	**1.4926e-39**	**4.8565e-40**	**6.6848e-40**
Schwefel	SCA	3.4311e-24	9.4549e-18	3.2041e-18	5.4139e-18
	DE	2.9124e-05	3.9115e-05	3.3415e-05	5.1423e-06
	PSO	1.3736e-21	6.5816e-19	3.2738e-19	3.2842e-19
	CCSCA	**6.8903e-42**	**1.5377e-34**	**5.1270e-35**	**8.8767e-35**
Rastrigin	SCA	9.7101e-08	29.2436	11.7733	15.5131
	DE	117.8569	146.8859	129.0434	13.9066
	PSO	29.8487	74.8625	53.9890	17.0591
	CCSCA	**0**	**0**	**0**	**0**
Griewank	SCA	6.5608e-06	0.0027	8.9970e-04	0.0015
	DE	7.0845e-10	5.1065e-09	2.3265e-09	2.4183e-09
	PSO	0	0.0172	0.0057	0.0099
	CCSCA	**0**	**0**	**0**	**0**

5. Case study of OOCHS based on CCSCA

Yalong river is the largest tributary of Jinsha river, which is extremely rich in hydraulic resources. It has planned 22 hydropower stations with annual capacity of about 150 billion kW·h, and has outstanding features, such as abundant water resources, good regulation performance and superior economic indicators. Moreover, the lower reaches of Yalong river is the key section of hydropower development of Yalong river, including five hydropower stations with installed capacity 14.7 million kW from upstream to downstream, i.e. Jinping1, Jinping2, Guandi, Ertan and Tongzilin, which have been put into operation since 2015, and their characteristic parameters are put forward in Table 3, thus the management mode of cascade hydropower stations in modern watershed is basically formed.

Table 3. Characteristic parameters of cascade hydropower stations

Number	Hydropower station	Normal level (m)	Regulation volume ($10^8 \cdot m^3$)	Regulation performance	Guaranteed output(million kW)
1	Jinping1	1880	49.1	Year	3.6
2	Jinping2	1646	0.0496	Day	4.8
3	Guandi	1330	1.232	Day	2.4
4	Ertan	1200	33.7	Season	3.3
5	Tongzilin	1015	0.146	Day	0.6

In order to verify the efficiency of CCSCA, we have a case study on the OOCHS of 5 hydropower stations located in the lower reaches of Yalong river, and water level of each hydropower station at each period is adapted as the optimal variable in SCA and CCSCA. The scheduling period is one year, the schedule time is one month, so the total time period is 12, and the beginning and end water level are normal level. For the daily regulation power station, the storage capacity of Jinping2, Guandi, and Tongzilin are not considered, only their water heads are considered for generation. And the parameters of SCA and CCSCA are NP=100, D=24, $Gmax$=1000.

Based on the basic data and runoff data, the typical inflow process in the wet year was selected for maximum power generation calculation, and the results of the two algorithms are shown in table 4. The calculation result of total power generation of CCSCA is better than SCA, which improves the solution precision and indicates that CCSCA can be better applied to practical engineering practice.

Furthermore, Fig. 6 and Fig.7 show the water level and output power of Jinping1 and Ertan hydropower station, it is obvious that the optimization results of CCSCA meet the various constraints, and the scale efficiency and compensation efficiency of cascade hydropower stations are fully utilized.

Table 4. The comparing power generation benefit of SCA and CCSCA

Algorithm	Jinping1	Jinping2	Guandi	Ertan	Tongzilin	Total(million kW·h)
SCA	195.6	284.3	129.6	199.1	34.5	843.1
CCSAC	195.9	285.1	129.8	200.2	35.8	846.8

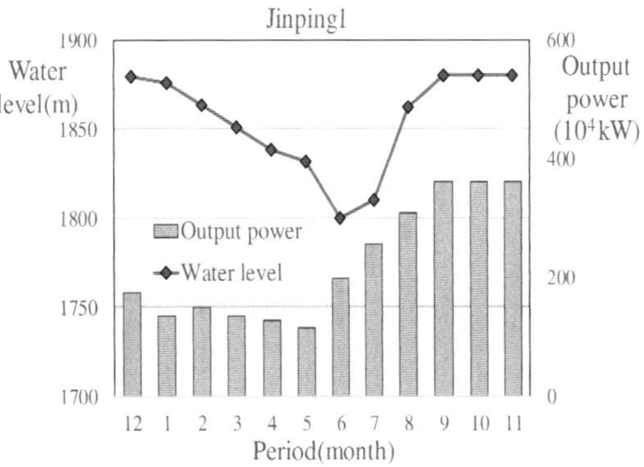

Figure.6 Optimal results of CCSCA for Jinping1

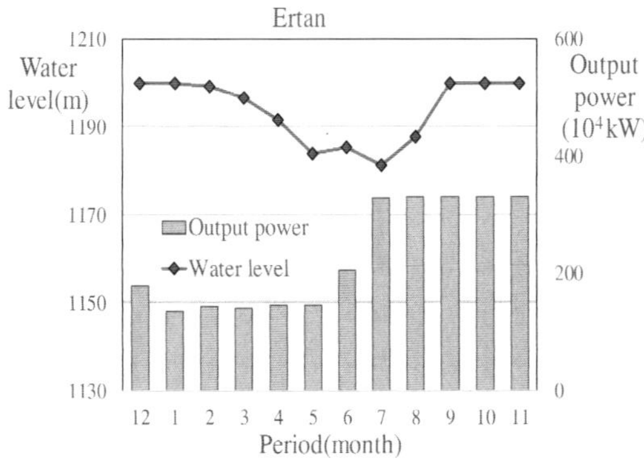

Figure.7 Optimal results of CCSCA for Ertan

6. Conclusions

In this paper, Chaos cultural sine cosine algorithm (CCSCA) is proposed by taking cultural algorithm as the framework of population space and using chaotic search to lead the evolvement of belief space. benchmark function tests prove that CCSCA has excellent ability and strong robustness. And case study on the lower reaches of Yalong river indicates that CCSCA provides a new way for optimal operation of cascade hydropower stations (OOCHS).

Acknowledgement

This work is supported by the National Key R&D Program of China (Item Nos. 2016YFC0402202 and 2016YFC0402209).

References

[1] Ming B, Chang J, Huang Q, et al. Optimal operation of multi- reservoir system based-on cuckoo search algorithm [J]. Water Resour Manag, 2015, 29(15):5671–5687.

[2] Sun Ping, Jiang Zhiqiang, Wang Tingting, et al. Research and Application of Parallel Normal Cloud Mutation Shuffled Frog Leaping Algorithm in Cascade Reservoirs Optimal Operation[J]. Water Resour Manage, 2016, (30):1019–1035.

[3] Seyedali Mirjalili. SCA: A Sine Cosine Algorithm for solving optimization problems [J]. Knowledge-Based Systems, 2016, (96):120–133.

[4] Reynolds, R.G., Mostafa, Z., Ali, T.J.: Mining the Social Fabric of Archaic Urban Centers with Cultural Algorithms [J]. IEEE Computer, 2008, 41(1), 64–72

[5] He Yaoyao, Zhou Jianzhong, Kou Pangao, et al. A fuzzy clustering iterative model using chaotic differential evolution algorithm for evaluating flood disaster[J]. Expert Systems with Applications, 2011, (38): 10060-10065.

IOP Conf. Series: Materials Science and Engineering **366** (2018) 012006 doi:10.1088/1757-899X/366/1/012006

Joint Operation of Reservoirs on Upper Yangtze River Facing new challenges

Yan Lingzhi[1], Chen Fang[2], Li Anqiang[1] and Lu Jun[1]

[1]Changjiang Survey, Planning, Design and Research CO., LTD., Wuhan 430010, China;

[2] Bureau of Hydrology, Changjiang Water Resources commission, Wuhan 430010, China

E-mail (YAN Lingzhi): yanlz@mail.dlut.edu.cn

Abstract. Suffering strong El Nino, the middle and lower reaches of the Yangtze River experienced extreme rainfall anomalies during the Bai-U rainy period (19th Jun. to 20th Jul.) in 2016. The 21 reservoirs including Three Gorges reservoir in upper Yangtze River were operated together to control flood. Most of these reservoirs are distributed on Jinsha River, Ya-lung River, Min River, Chia-ling River, and Wu River. They totally offered about 37.3 MAF available storage with 29.2 MAF for flood storage. This paper analysed the operation performance of the 21 reservoirs during the Bai-U rainy period. It evaluated the joint operation performance of 21 reservoirs, and proposed suggestions for multi-reservoir management of Yangtze River upstream in the future. The experience of operating multiple large reservoirs would inspire others to utilize reservoirs crossing multiple rivers solving flood problems in a better way, facing the new challenges from climate change.

1. Introduction

Last summer, a strong EL Niño-fueled Bai-U rainy period (19th Jun. to 20th Jul.) generated unusually huge precipitation among Yangtze River Basin (YRB). The precipitation presented characteristics with long rainy period, huge rainfall, high intensity, concentrated raining locations. During the Bai-U rainy period, mostly rainfall struck the lower Yangtze River where accumulative precipitation (584.3 mm) is twice as much as the average precipitation of past 30 years (281.0 mm). The precipitation ranks 3rd among other historical precipitation during Bai-U rainy period, after CY1954 (811.5 mm) and CY1996 (695.7 mm) [1]. On 1st July 2016, the river inflow of Three Gorges Reservoir reached 50,000 m^3/s, the flood was labelled as "the No.1 Flood 2016" of Yangtze River.

There were 21 reservoirs on upper Yangtze River participated in the joint operation for flood protection. The article presents the results of a comprehensive assessment of their operational performance during the Bai-U rainy period 2016. The assessment included an evaluation of unregulated flows without flood control by these 21 reservoirs. It also identified opportunities to better utilize the flood resource.

2. The Yangtze River Basin

The river above Yichang Station is called the upper Yangtze River and its watershed are 247 M acres. Generally, all running reservoirs in YRB whose maximum allowable storage is larger than 0.08 MAF should be included in the joint operation to control flood, i.e., 102 reservoirs with more than 64.9 MAF storage capacity and 32.1 MAF flood control pool. Based on comprehensive consideration of flood control ability and structural situations, 21 reservoirs were selected to be included in the joint operation. Their storage capacity is 37.2 MAF and flood control pool is 29.4 MAF. These reservoirs are Liyuan, Ahai, Jinanqiao, Longkaikou, Ludila, and Guanyinyan on the midstream of Jinsha River, Jinping I and Ertan on Ya-lung River, Pubugou and Zipingpu on Min River, Xiluodu and Xiangjiaba on the lower Jinsha River, Bikou, Baozhusi, Tingzikou and Caojie on Chia-ling River, Goupitan, Silin, Shatuo and Pengshui on Wu River, and Three Gorges, with details in Figure 1.

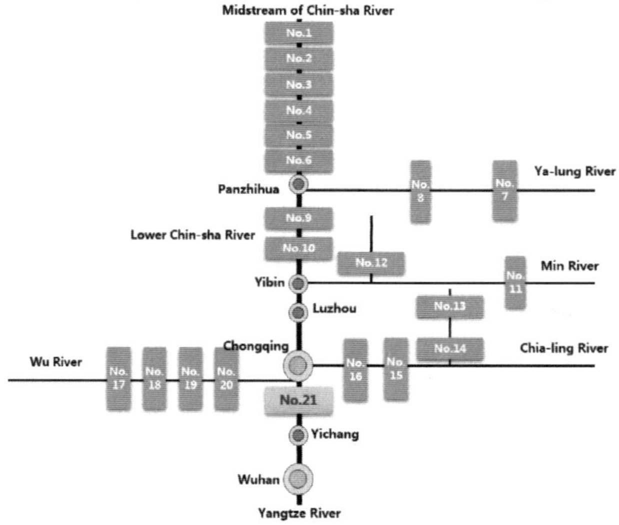

Figure 1. The 21 reservoirs in YRB for joint operation to control flood in 2016.

3. Unregulated flows

3.1. The model

The data covers inflows, outflows, and storage of 21 reservoirs and three hydrometrical stations (Cutan, Wulong, and Yichang) from 1st June to 19th July in 2016 with 6h as a time interval.

There are two kinds of inflows to Three Gorges because of its huge channel-like reservoir zone. One is the total reservoir inflows at the ending points of its backwater zone. And another (the dam-site inflow) is the river runoff at the dam site of Three Gorges.

The stored water during a time interval by each reservoir was calculated respectively by water balance method and was turned into the stored flow. The travel time of outflows from one reservoir to another inside a cascaded-reservoir system were ignored, which means that one cascaded-reservoir system was considered as one virtual reservoir. There are two virtual reservoirs on Jinsha River, and one virtual reservoir on Ya-lung River, Min River, Chia-ling River, and Wu River respectively.

3.2. The methods

Hydrometrical stations Cutan and Wulong station were selected as the ending points of backwater zone of Three Gorges [2]. Cutan and Wulong represent the reservoir inflows of Three Gorges.

6 hydrometrical stations named Panzhiuha, Xiaodeshi, Pingshan, Gaochang, Beibei, and Wulong were selected to represent the discharges of 6 virtual reservoirs on Jinsha River, Ya-lung River, Min River, Chia-ling River, and Wu River.

The stored flow routing from Pingshan station and Gaochang station to Cuntan station utilized the Muskingum Method. And the stored flow routing in other parts of rivers utilized a linear time-delay method.

The unregulated reservoir inflow of Three Gorges without storage by other 20 reservoirs is calculated by:

$$\Delta Q_c(t) = f_{p,c}\left(\sum_{i=1}^{8} \Delta Q_i(t-8\Delta t) + \sum_{i=9}^{10} \Delta Q_i(t)\right) + f_{g,c}\left(\Delta Q_{11}(t-5\Delta t) + \Delta Q_{12}(t-4\Delta t)\right) + \sum_{i=13}^{14} \Delta Q_i(t-10\Delta t) + \Delta Q_{15}(t-9\Delta t) + \Delta Q_{16}(t-\Delta t) \quad (1)$$

$$\Delta Q_w(t) = \sum_{i=17}^{20} \Delta Q_i(t) \quad (2)$$

$$\Delta Q_i(t) = (V_i(t) - V_i(t-\Delta t))/\Delta t \quad (3)$$

where $Q_{21}^*(t)$ is the unregulated reservoir inflow of Three Gorges in cms, $Q_c(t)$ and $Q_w(t)$ are actual flows at Cutan station and Wulong station respectively, $Q_{c,w,21}(t)$ is actual runoff of Three Gorges' reservoir zone, $\Delta Q_c(t)$ is the total stored flows of five virtual reservoirs on the upper river of Cutan station, $Q_w(t)$ is the total stored flows of five virtual reservoirs on the upper river of Wulong station, $f_{p,c}(*)$、 $f_{g,c}(*)$ is the Muskingum function of flow routing from Pingshan to Cutan and from Gaochang to Cutan respectively, $\Delta Q_i(t)$ is the average flow of reservoir i stored at time t, and Δt is the time interval (6 hours).

The unregulated flow at Yichang station without storage by 21 reservoirs is calculated by:

$$Q_y^*(t) = q_{21}(t) + \Delta Q_{21}(t) + \Delta Q_c(t-10\Delta t) + \Delta Q_w(t-9\Delta t) \quad (4)$$

where $Q_y^*(t)$ is the unregulated flow at Yichang station in cms, and $q_{21}(t)$ is the actual discharge of Three Gorges.

3.3. Result analysis

Bureau of Hydrology(BOH) of Changjiang Water Resources Commission (CWRC) utilized a similar way to calculate the reservoir inflow of Three Gorges [2]. As for the unregulated flow at Yichang station, BOH utilized hydrological models of MIKE 11 to calculate the flow routing of reservoir inflow from Cutan and Wulong to Three Gorges' dam site and rainfall run-off model NAM to calculate the runoff of Three Gorges' reservoir zone [3, 4]. The proposed method was testified by floods event in August 2009 and July 2010, and results were compared to reference [3] and [4], which is shown in Table 1.

Table 1. The unregulated flow at Yichang station by different methods.

Flow at Yichang station（m^3/s）	Flood event in August 2009		Flood event in July 2010	
	Actual max. flow	Unregulated max. flow without storage by Three Gorges	Actual max. flow	Unregulated max. flow without storage by Three Gorges
Reference [3]	40200	56100	-	-
Reference [4]	-	-	42000	65000
Proposed method	39800	56100	41500	67200
Deviation	0.7%		4.2%	

The BOH method is complicated and requires lots of input data. The proposed method is relatively easy and needs less input data. The comparisons illustrate that proposed method is acceptable to calculate the unregulated flow at Yichang station for academic study. But the reservoir type and time interval length will affect the result reliability of the proposed method. If the reservoir inflow is low, the result will follow a sawtooth pattern. To the Three Gorges as a channel reservoir, its generated unregulated flow would be affected obviously by its backwater and show a sawtooth pattern (Figure 2, 3).

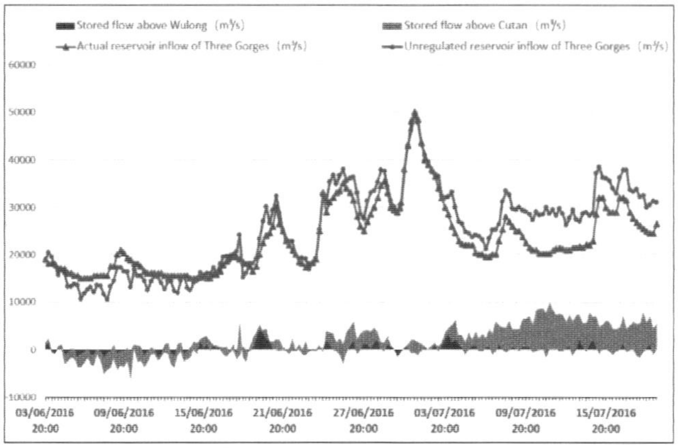

Figure 2. The unregulated reservoir inflow of Three Gorges without storage by 20 reservoirs.

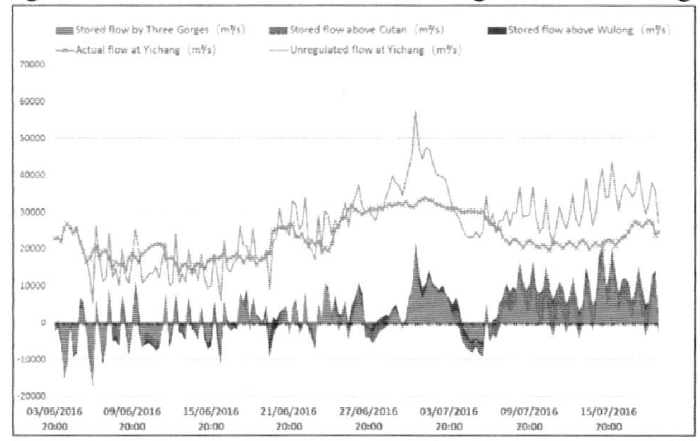

Figure 3. The unregulated flow at Yichang station without storage by 21 reservoirs.

4. Joint operation of 21 reservoirs for flood control

4.1. Operation manual

CWRC has been in charge of operating reservoirs in YRB for flood control. At 30th June 2016, CWRC issued an annual operation manual (OM) about joint operation of 21 reservoirs on upper Yangtze River for flood control. The OM was drafted by the Changjiang Institute of Survey, Planning, Design and Research (CISPDR), which was made to offer suggestions to CWRC and other stakeholders for water management. It divided 21 reservoirs into 3 categories (one heart, two backbones, and five hands) according to their storage capacity and locations. According to the OM, Three Gorges as the heart should undertake most work for flood control to protect the middle and downstream of Yangtze River. Two backbones are Xiluodu reservoir and Xiangjiaba reservoir, which should be the backbone to help Three Gorges with reducing flood flowing to the downstream. Five hands are referring to five cascaded reservoir systems on Jinsha River, Ya-lung River, Min River, Chia-ling River, and Wu River respectively. These systems were supposed to help reduce reservoir inflow of Three Gorges, so the Three Gorges could store more flood for its downstream.

During the Bai-U rainy period 2016, the Three Gorges was operated by CWRC during four subperiods:1) from 5th June to 25th June, the reservoir stage had been kept around 144.9 m to 146.5 m, 2) from 25th June to 3rd July, the first and strongest flood event in 2016 occurred and the Three Gorges had been kept a discharge at 31000 m^3/s, 3) from 3rd July to 16th July, Three Gorges released 1.12 MAF over three days and then kept its discharge around 20000 m^3/s during last days, and 4) from 16th July to 19th July, Three Gorges released water with rate around 25000 m^3/s but its outflows were made smaller than its inflows. Other reservoirs on Jinsha River, Min River and Ya-lung River had been operated to reduce inflows of Three Gorges from 25th June to 19th July.

4.2. Operation performance

From 5th June to 19th July 2016, 21 reservoirs on the upper Yangtze River accumulatively stored 11.30 MAF flood (as shown in Figure 4), of which the Three Gorges reduced 4.11 MAF flowing to its downstream. As the heart of flood control system in YRB, the Three Gorges stored 36% of totally reduced flood to lower Yangtze River from 5th June to 19th July. During the No.1 Flood 2016 event, Three Gorges decreased the peak value from 50000 m^3/s to 31100 m^3/s, once stored 2.4 MAF flood for 6 hours. Without Three Gorges, the river stage of Chenglingji station at the lower Yangtze River will be 34.5m exceeding its insuring stage with 0.1 m and the high stage may last for two days [5].

Other 20 reservoirs made the reservoir inflow of the Three Gorges kept as about 30000 m^3/s from 3rd July to 19th July, with a maximum cut of inflows at 9860 m^3/s, which let the Three Gorges could safely keep a low discharge to its downstream where the river stages had been higher than their warning stages for a long time.

- From 25th June to 10th July, reservoirs on the midstream of Jinsha River and Ya-lung River were the main force of storing runoff from the upper YRB.
- From 10th July to 19th July, reservoirs on the lower Jinsha River, Ya-lung River and Min River were the main force of storing runoff from the upper YRB. Reservoirs on Chia-ling River and Wu River also helped a little with reducing reservoir inflows of Three Gorges.

These 21 reservoirs together protected 9 provinces and about 100 million people in YRB. More than 82,000-acre farm lands were saved from flood.

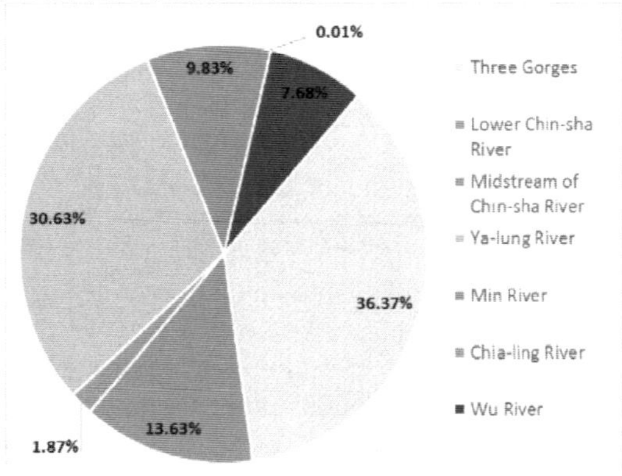

Figure 4. Accumulatively stored flood water by 21 reservoirs on different rivers.

4.3. Challenges and principles

- Dynamical allocation of cascaded flood control spaces

From 10th July to 19th July, the Three Gorges had been stored huge flood since 25th June and its reservoir stage was increasing rapidly. If the reservoir stage of Three Gorges is higher than 155m, it should stop store flood for the lower Yangtze River to leave enough flood control space for the midstream. However, the lower Yangtze River were still in need of help from Three Gorges. Since the river runoff in upper Yangtze River was low-risk, other 20 reservoirs on the upstream were supposed to store water that would reduce reservoir inflows of Three Gorges.

The OM suggested reservoirs on the midstream of Jinsha River and on the Ya-lung River should help Three Gorges store flood at first, then other 13 reservoirs would do a favor if necessary. So, during the beginning 10 days of July 2016 when heavy rainfall struck the upper Yangtze River, reservoirs on the midstream of Jinsha River and Ya-lung River had been storing flood water together with Three Gorges. The operation of cascaded reservoir systems should be dynamic and flexible, so flood control pools of reservoirs on different rivers could exchange with each other.

- Dynamical reservoir stage during flood season

Normally Xiluodu and Xiangjiaba should keep their stages not higher than their flood control pool level from 1st July to 10th September. So, reservoirs could leave enough space to fight against possible huge flood events. During last Bai-U rainy period, the peak values of Xiluodu and Xiangjiaba's inflows were 10,500 m^3/s and 12,500 m^3/s respectively and both were smaller than the peak values of their designed flood events with 5-year return period (28,200 m^3/s). So, it is acceptable for these two reservoirs to keep their stages higher than their flood control pool levels.

The limiting stage of one reservoir during flood season normally was designed without considering help from other reservoirs' flood control spaces. The truth is that keeping all reservoirs' stages under the limiting stages may be not necessary, especially when the flood is not big. To a cascaded reservoir system on one river, let the upstream large reservoirs keep all flood control spaces of the system that other reservoirs could running with higher water stages. Optimize the order of storing and releasing water among these reservoirs could make it possible to reutilize the water from upstream to downstream for hydropower generation with higher average head water.

- Forecast informed reservoir operations under climate change

To insure enough flood control space, Tingzikou is supposed to keep its reservoir stage below 447m from 21th June to 31th August every year. Last April and May, the runoff in Chia-ling River was about 40% larger than its contemporaneous average runoff. So, it is understandable that Tingzikou released water earlier in May and its stage was about 5 meters lower than its flood control pool level (447 m) on 1st June. At the beginning of June, Tingzikou stored some water and released them immediately since 5th June because that its inflows were increasing obviously by then. It was supposed to release more space for possible big flood in June. But Chia-ling River's runoff in last June and July was actually about 40% smaller than its contemporaneous runoff. If it had stored water in the beginning of last June, Tingzikou could have been operating safely with a higher stage which means more available water and better electricity generation efficiency.

The rainy patterns are changing under climate change. It is necessary to improve integration of weather forecast, hydrological forecast, and reservoir operation. Based on the study of dynamical reservoir stage during flood season, reservoir should learn to store, and release water smartly, even break the rule curve sometimes.

5. Conclusion

Anthropogenic influences are more apparent than ever. Climate change affects the rainfall patterns during the flood season. Recent flood events illustrated that storage abilities of rivers and lakes are weakening [6]. The river stage is higher in the same river during a same flood event with poorer storage capacity. Furthermore, there are numerous large reservoirs on the upstream and midstream of Yangtze River. They all need to release water to restore their full flood control spaces since April to July each year. Thus, huge water will flow to the downstream and rise downstream river stage. The long-term high river stage at the lower Yangtze River could be common in near future.

In comparison with single operation of Three Gorges, joint operation of cascaded reservoirs could reduce 5.3 MAF unsolvable flood [7]. So fewer residents around the downstream would be transferred in case of huge flood. The operation performance of 21 reservoirs in 2016 proved that joint operation could certainly reduce flood control pressure for the middle and lower reaches of the Yangtze River.

Dynamical allocation of cascaded flood control spaces makes it possible for Three Gorges to store water for its downstream without reducing its storage space for other cities around the midstream. Meanwhile, with numerous large reservoirs on the same river, the reservoir stage during flood season could be higher than its limiting level if the flood is not big. But threshold value of stage during flood season with acceptable risk should be studied further. To increase resiliency for droughts and floods, it is necessary to study forecast-informed reservoir operations. Just like what has been done for Lake Mendocino in California [8]. And the better utilization of water is, the more hydropower will be obtained, which also help keep the world be sustainable.

Acknowledgments

This work is supported by the National Key R&D Program of China (Item Nos. 2016YFC0402202).

References

[1] F. Zhang, L. Zi, H. Qiu. Climate characteristics analysis of rainstorm and flood in Changjiang River in 2016[J]. Yangtze River, 2017(04): 62-65.

[2] Y. Min, J. Wang, L. Chen. Discussion on calculation method of flood routing and inflow of Three Gorges Reservoir [J]. Yangtze River, 2011(06): 49-52.
[3] B. Feng, Y. Li, Y. Chen, et al. Influence of flood control regulation of Three Gorges Reservoir on flood of upper Yangtze River in August, 2009[J]. Yangtze River, 2011(06): 72-74,79.
[4] B. Zou, Y. Li, B. Feng. Analysis on dispatching influence of Three Gorges Reservoir on water level of main stream in mid-lower reaches of Yangtze River: a case study of flood in July, 2010 [J]. Yangtze River, 2011(06): 80-82,100.
[5] J. Zhang, L. Chen. Regulation effects of Three Gorges Reservoir and flood forecast for No.1 flood in 2016 of Changjiang River [J]. Yangtze River, 2017(04): 13-15,36.
[6] J. Wang. Characteristics and enlightenment of 2016 Changjiang River flood [J]. Yangtze River, 2017(04): 54-57,65.
[7] X. Jin. Study on co-regulation of reservoirs in upper reaches of Changjiang River in 2016 for flood control [J]. Yangtze River, 2017(04): 22-27.
[8] C. Talbot. Forecast informed reservoir operations: research on increasing resiliency for droughts and floods [J]. Flood Risk Management Newsletter, 2017.10(1):9-11.

IOP Conf. Series: Materials Science and Engineering **366** (2018) 012007 doi:10.1088/1757-899X/366/1/012007

Brief Discussion on the Design of Decision Support System for Flood Control Operation in Yangtze River

YU Shan* and LI Xiaonan

Changjiang Survey, Planning, Design and Research CO., LTD , Wuhan 430010, China

*yushan@cjwsjy.com.cn

Abstract. Flood control operation is the top priority for the management of the Yangtze River. To improve the timeliness and reliability of operation decision, it is of great significance to establish a set of intelligent decision system for flood control management, which can serve multi-level users. In order to achieve the goal, the key is to modularize the core business functions through the construction of topology relations of river network and control sites for the Yangtze River. In this paper, it would integrate the functions of hydrological forecast, flood propagation, flood control operation and flood risk analysis for the core business. Based on the core business model and information expansion interface, we develop flood control decision support system of the Yangtze River by virtue of GIS visualization and interactive WebService technology. The system can realize the real-time query of water-rainfall information and the fast-time simulation of flood control scheduling scheme, so as to provide the professional, scientific and accurate technical support for the group decision in the flood control.

1. Introduction

Flood control is of extreme significance in maintaining the stability of the Yangtze River. However, remarkable changes in hydrological features, as well as immediate pressure from socio-economic development in the catchment of the Yangtze River, pose great challenges for the river training. Especially, the situation of river training is becoming more and more complex, and training task is becoming more difficult as well as training requirements are getting higher and higher. Thus it calls for the hydraulic informatization to reinforce and become an crucial tool in the river training, which is also an essential condition for water conservancy modernization[1]. With the improvement of water conservancy modernization and development of hydraulic informatization, the rapid, real-time, accurate, automatic and intelligent processing in the flood control, to improve the flood control efficiency and accuracy, is the key to enhance flood control and management ability[2]. Moreover, it is also an important issue in the construction of flood control informatization.

In the flood control work, establishing a complete flood control decision-making support technology is necessary for flood prediction, which collects the relevant information quickly and automatically. With the real-time analysis and prediction of flood process, it can provide fast and effective technical support for management decision according to the requirements in the reservoir scheduling schemes[3,4]. This paper discusses the design of decision support system for flood control operation of the Yangtze River Basin from the aspects of the principles, ideas, system framework and system integration.

2. System design principles

The overall objective of decision support system for flood control dispatching is to provide decision support for correctly analysing and judging flood control situation and scientifically formulating flood control dispatching scheme[5]. At the same time, with the modern information technology, in order to strengthen the scientific nature of flood control command, improve the quickness, real-time and accuracy for information acquisition, transmission, processing and flood control scheduling decision-making, non-engineering measures are increasingly becoming an important means of flood control. Therefore, the following principles should be referred to in the design of flood control dispatching system.

2.1. Rapidity

In the decision-making process of flood control and dispatching, it is necessary to provide a scientific and reasonable scheduling scheme for the decision maker through the repeated interaction between the decision maker or the analyst and the computer system. Therefore, flood control software system should be able to improvise together with the computer rapid and accurate calculation, ordered logic judgment, high speed and high capacity data storage capacity and a series of algorithms, tools and people's creativity, providing convenient and fast operating environment for scheduling decision makers or business personnel.

2.2. Real-time performance

The occurrence of flood disaster usually has obvious diachronic process, and it is necessary to monitor the relevant information in real time from the initial state to the dangerous situation and the disaster situation. In order to ensure the scientific and effective command and decision-making, all kinds of flood control information need to be monitored in real time, such as warning, emergency response, command, dispatch, emergency rescue and disaster relief. Therefore, the flood control dispatching system should have strong actual combat and high real-time requirement.

2.3. Accuracy

According to the analysis of the business process in flood control work, combined with the actual conditions of the application requirements, the flood control dispatching system should combine with the principle of accuracy, to ensure the correctness and safety of the dispatching, and reduce the risk caused by the inaccurate scheduling scheme. On the premise of satisfying the accuracy, the performance of the scheduling model is optimized and the operation speed of the model can be improved. Correspondingly, the accuracy of flood dispatching is ensured.

2.4. Simulation

The occurrence, development and influence of flood disaster are generally distributed in space. Meanwhile the rapid response, such as command, dispatch, emergency rescue and disaster relief, is usually carried out in a certain space. Therefore, the spatial characteristics are important aspects of flood control work. To evidently demonstrate the spatial distribution, the emergency command and dispatch level will be greatly improved when three dimensional simulation is involved in flood control decision support. Using geographic information and 3D visualization technology to simulate reservoir scheduling, downstream channel flood evolution and inundation effect after the dispatching decision could be simulated vividly for decision-makers, thus help to make a scientific decision, and minimize flood damage.

2.5. Technical

Professional functions such as information query, forecast, scheduling calculation and scheme comparison, flood risk assessment, etc., can provide technical guarantee for the fast and accurate simulation and calculation. Additionally, it can pave the way for the easy man-machine interaction and multi-objective, multi-project convenient analysis and comparison.

2.6. Persistence
In the design and implementation of flood control system, to follow the existing flood control business, it is envisioned to give full consideration to the next period of flood control requirements. The practical and mature technology is adopted to make the developed flood control dispatching system, which has favourable practicability and long life cycle. The system should be more open and expansible, so that the system can be updated, function upgraded, and the scheduling scheme can be extended in the future.

2.7. Friendliness
Flood control dispatching system is a man-machine interactive system, and the process of supporting flood control decision must be realized through the interaction between the decision-maker and the computer system. The system should provide a friendly man-machine interface with clear layers, flexible input and convenient operation. The decision makers can input operation instructions at any time to control the model calculation. System interface design as far as possible to provide function "navigation" or "one button" service, to improve the user experience, simplify the interface, and benefit the users.

3. System design ideas
The design and implementation of the flood control decision support system for the Yangtze River Basin should be based on the requirements of the flood control business application in the Yangtze River Basin[6]. According to the on-going trend of the information technology and the software engineering method, the design and implementation should be carried out under the norms and guidance of the unified technical framework.

The system design is based on evolution, forecasting, scheduling, risk analysis focusing on integration, to optimize the allocation of resources and information interoperability. Aimed at intelligent functional design, it should adapt to the generation of flood control scheduling scheme scientifically and fast. Besides, in order to take real-time flood control as the core, it should analyzes the different characteristics of the region and consider the requirement of real-time computing, with the construction of professional computing model. The three- dimensional simulation is used to realize the spatial scene display of the scheduling results, and show the regional flood control situation, scheduling simulation and flood loss assessment in a vivid way.

The system should be developed based on GIS technology, professional business model and business support platform technology, which provides fast analysis and judgment of the situation of flood control. Meanwhile it should provide various hydrology and hydrodynamics flood model in the computing service, to provide flood control, scheduling scheme optimization, flood simulation visualization and results management services for user. To Achieve the combination scheduling and single node scheduling of reservoir and river channel, manual scheduling node by node, a variety of scheduling results could be compared and analysed for better one, and carry out comprehensive management of flood forecasting and dispatching results. When floods occur, it can quickly collect and transport of water-rainfall regime, the labor, drought disaster information, and make a prediction about the trend and forecast, flood control scheduling scheme being put forward by the analysis. This scheme is the key to minimize disaster losses, to provide technical support for flood control decision-making. The system requires large amount of data, large amount of calculation analysis and short response time, that business flow and data flow should be analysed with solution.

4. Framework design of flood control dispatching system

The framework of flood control dispatching system is divided into six levels: infrastructure layer, data resource layer, business support layer, basic model layer, business application layer and function display layer [7].

4.1. Infrastructure layer
This layer is the material base of each business subsystem, including network environment, hardware server, storage device, security equipment, etc.

4.2. Data resource layer
The data resource layer includes the public data, special social and economic topics, water conservancy project, real-time hydrological information, thematic information, calculation model of flood characteristic information, flood damage information, flood control command reference information, spatial information, 3-D model information and so on. All kinds of data enter the corresponding subject database through ETL method, respectively. The data model and database application in flood control is created, via various views of application theme above the data model, to meet the demand of all kinds of data in the flood control system processing and information platform publication.

4.3. Business support layer
This layer provides unified technical framework and operating environment for flood control dispatching system. Meanwhile it provides universal application service and integrated service for system construction, as well as operation platform for resource integration and information sharing. According to the needs of the business, the flood control system calls for the business support platform file services, computing services, general data services, video services, access services, search / reporting services, real-time data and KPI services, map services, integrated services, mobile portal service, interface service, model service, database service, log service, distributed real-time computing services, batch computing services, to achieve the functional modules of comprehensive inquiry, water situation forecasting, flood control situation analysis, flood control dispatching, comprehensive analysis of decision making, achievement display and application, mobile terminal application, etc.

4.4. Basic model layer
Flood control system is designed for real-time flood control management capabilities to provide scheduling scheme generation, scheduling scheme simulation and effect analysis in one system, so that the system must be equipped with the corresponding calculation model, mainly includes confluence model, flood evolution model and cascade reservoir group scheduling model, etc.

4.5. Business application layer
According to the analysis of flood control and the business process, the design of business application layer covers the forecast, flood routing, river basin flood control situation analysis, flood control scheduling plan, proposed scheduling for flood control effect analysis, the proposed scheduling for flood risk assessment, flood control scheduling scheme generation module, the module support each other, complement each other, completed in flood control and the management of the business, as shown in figure 1.

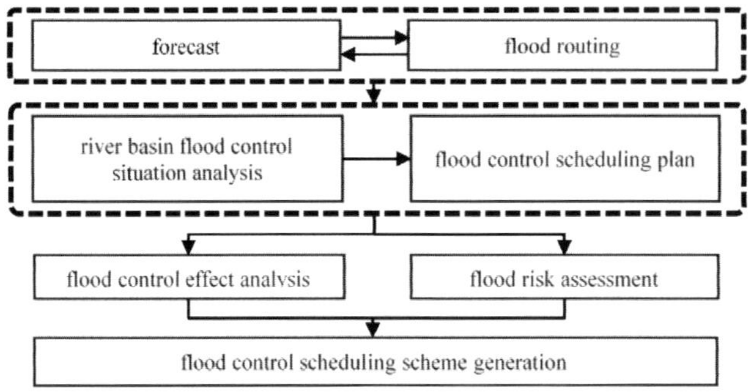

Figure 1. Service application layer module

4.6. Function display layer

On the basis of in-depth analysis of flood control and business requirements, the model based on professional technology integrates the use of online transaction processing technology, component technology, geographic information system (GIS), decision support system (DSS) and other high and new technology. Through the combination of flood control and special business, we can construct the advanced, scientific, efficient and practical flood control scheduling system, the flood control scheduling management and consultation with the daily work of operation interface and its function display module.

5. Organic integration of the whole system

Flood control operation system is composed of various business data, different business functions, business model and structure of the interface. To establish the system, it is necessary to integrate these modules into an organic wholeness. According to this idea, the option interface design for the flood control decision support system of Yangtze River basin (FCDSS-YR) is demonstrated in figure 2-3.

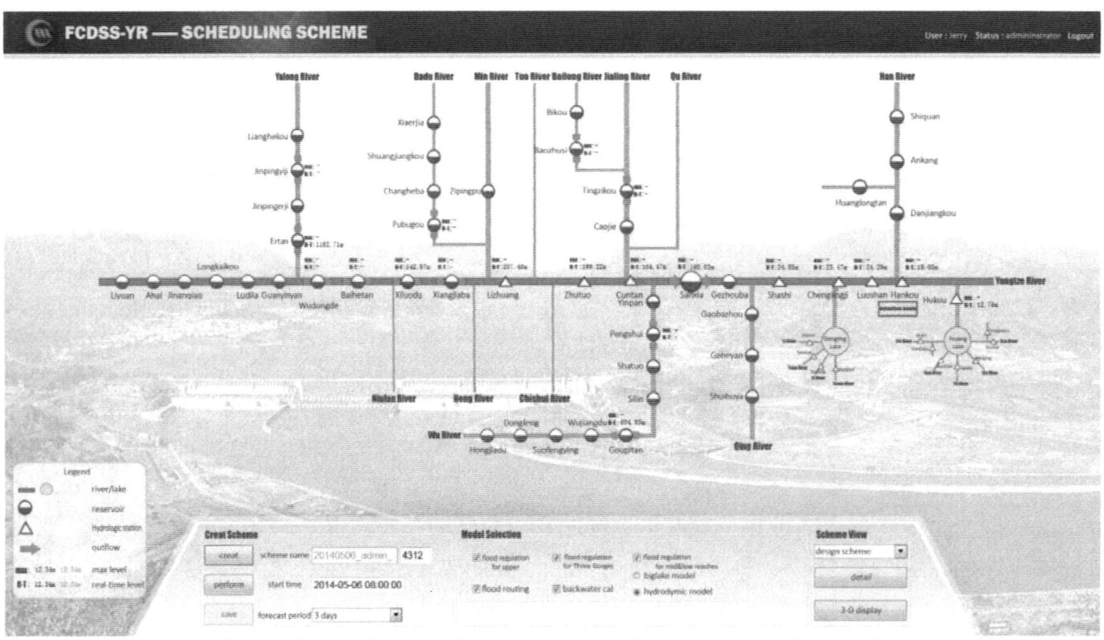

Figure 2. Interface of dispatching calculation in the FCDSS-YR

Figure 3. 3-D display interface of submergence in the FC-YR

In FCDSS-YR, it has implemented the integration of data, function, model and interface, which will be gone into details one by one.

For the data integration, it is essential to solve the problem of data sharing to make the data widely used in the system. Meanwhile, it contains multi-type data in the flood control system, such as water-rainfall data, hydraulic engineering data and model calculation data. Since these data source may be heterogeneous and be distributed in different regions, it is necessary to collect and integrate the multi-source data; and then converge to information data center through the joint transformation and copying technology.

As to the function integration, this system attempts to integrate the functions of hydrological forecast, flood propagation, flood control operation, flood risk analysis, information query and other business functions through the SOA based ESB (Enterprise Service Bus) technology. For the common functions of flood control system, such as information query, model calculation and file service, they can be provided by the enterprise service bus. Especially, the query of result can be fed back to the business function module for further calculation, analysis and display through the bus service.

As to the model integration, kinds of models such as hydrological forecast model, flood regulation model, calculation of backwater model, flood propagation model of downstream, risk assessment model are in modularity to realize above functions. It is worthy to point that these models are conjunctly integrated into the business support platform, which provides a unified computing service.

For the integrated interface, although the interface of each function module in the flood control scheduling system is not the same, we keep the unified style in the interface design. Meanwhile, the universal UI is addressed for different module function, such as user login interface, report export, authority management, information query interface and so on. Moreover, these interfaces are wholly integrated to business support platform.

The completion of FCDSS-YR can facilitate to comprehensively master the water-rainfall information and engineering disaster. What's more, it can accurately calculate the rainstorm and flood forecasting, quickly make various feasible dispatching solutions, and evaluate the scheduling schemes overall. With the aid of the FCDSS-YR, the group decision can achieve the scientific operation schemes as far as possible. Therefore, it can not only ensure the safety of flood control projects and the minimum loss of flood disaster, but also can give full play to the effect of flood control engineering, so as to improve the flood control capacity for the crucial area and cities in the Yangtze River.

6. Conclusion

The flood control operation of the Yangtze River Basin involves the livelihoods of the people on both sides of the river, and it is of great significance to decide whether the dispatching is correct or not. The design of flood control system integrates the functions of hydrological forecasting, flood routing and control, flood risk analysis. It can strengthen the interaction and share massive information resources, improve the flood control decision support technology process, enrich the simulation of flood control operation and the expression means, as well as to improve the speed of decision-making, the level and quality to a new the level. Besides, the system can effectively improve the regional flood risk analysis and flood control ability for the key area in the Yangtze River Basin through the flood emergency transfer capability of three-dimensional display and flood. With the aid of the system, the flood control and disaster reduction of decision for administrative leadership to provide visual display has enhanced flood control and management of the Yangtze River basin level, while the efficiency and quality of the flood control work has been greatly improved. Overall, it will greatly reduce the economic loss, enhance the guarantee of sustainable economic-social development, and protect the people's lives and social stability.

Acknowledgments

This work is supported by the National Key Research and Development Program of China (Item Nos. 2016YFC0402207). The authors also appreciate the insightful comments and suggestions from anonymous reviewers.

References

[1] LIU H, LIU Y, XiONG Z, et al. Study and application of urban flood risk map information management system based on SOA[J]. Journal of Software, 2015, 10(2): 180-190.
[2] Ahmad S, Simonovic S P. An intelligent decision support system for management of floods[J]. Water Resources Management, 2006, 20(3): 391-410.
[3] Ahmad S, Simonovic S P. System dynamics modeling of reservoir operations for flood management[J]. Journal of Computing in Civil Engineering, 2000, 14(3): 190-198.
[4] Choudhury P. Reservoir flood control operation model incorporating multiple uncontrolled water flows[J]. Lakes & Reservoirs: Research & Management, 2010, 15(2): 153-163.
[5] LI Anqiang, ZHANG Jianyun, ZHONG Zhiyu, etc. Study on joint flood control operation for leading reservoirs in the upper Changjiang River [J]. Journal of Hydraulic Engineering, 2013, 44(1):59-66. (in Chinese)
[6] LI Shufei, CHEN Shuli, HE Xiaocong. Key Technologies in United Flood Control Operation System of Lower Jinsha Cascade Reservoirs and Three Gorges Project[J]. Hydropower and New Energy, 2014(2):1-5. (in Chinese)
[7] YU Shan, LUO Bin, ZHANG Hengfei. Design of decision support system for flood control operation in Yangtze River Basin [J]. Yangtze River, 2015, 46(21):5-7. (in Chinese)

IOP Conf. Series: Materials Science and Engineering **366** (2018) 012008 doi:10.1088/1757-899X/366/1/012008

Study on Flood Resources Utilization Dispatching Method of the Three Gorges Reservoir Based on the Demand of Power Generation

He Xiaocong[1], Yan Binjun[2] and Yan Lingzhi[1]

[1]Changjiang Survey, Planning, design and Research CO., LTD., Wuhan 430010, China;
[2]State Grid Hubei Electric Power Company Limited, Wuhan 430077, China)

E-mail (He Xiaocong): hexiaocong@cjwsjy.com.cn

Abstract: In order to give full play to the power generation benefits of the Three Gorges Project, this paper based on the concept of floodwater utilization, studied the flood control modes of the Three Gorges Reservoir in flood season by combining the hydrological forecast in the middle and upper reaches of the Yangtze River. The benefit and risk of flood resource utilization was compared. The results show that through the implementation of flood control and utilization scheduling of the Three Gorges Reservoir in flood season, under the controllable of flood risk, the power generating capacity of the Three Gorges Reservoir in flood season can be increased by 933 million kW · h, and power generation overflow decreased 4.1 billion m^3 per year, which can effectively raising the power generation benefit of the Three Gorges Reservoir and providing important technical support for the scientific dispatch of the Three Gorges Reservoir.

1 Introduction

Water can benefit, also can become a catastrophe. With the social-economic development and improvement of living standards of people, the demand for water for production, domestic utilization and ecological protection continues to increase [1]. The shortage of water resources has become the major contradiction in the water issue in China. After years of construction of water conservancy projects, the improvement of hydrological and meteorological forecasting, as well as the accumulation of experience in flood control and drought relief, the people's ability to control floods has been continuously enhanced. The understanding of floods has been changed from passive defence to proper use while defensive. The utilization of water resources in flood season has become a great potential for water resources utilization [2].

The Three Gorges Reservoir located at the middle of the Yangtze River in China. It is a key backbone project for harnessing the Yangtze River and for the development and utilization of water resources [3]. The normal water storage level is 175m, the flood control limit level is 145m [4]. The reservoir has flood control capacity of 22.25 billion m^3 and beneficial storage capacity of 16.5 billion m^3, has a huge capacity of water storage and transportation, flood control, power generation, shipping and other comprehensive utilization tasks. Since the Three Gorges Reservoir was put into operation, it has provided a large amount of clean energy for the country, effectively promoted the economic development and created favourable conditions for the national energy conservation and emission reduction. With the increasing demand for electricity and the gradual implementation of energy conservation and emission reduction measures, there is a higher demand for implementing optimal reservoir operations to improve the efficiency of power generation. Utilizing flood resources without affecting the flood control safety is an effective measure to improve the efficiency of the comprehensive utilization of water resources for Three Gorges Reservoir.

2 Dispatching mode of Three Gorges Reservoir

2.1 Conventional dispatching method

Flood control is the primary task of the Three Gorges Reservoir in flood season [5]. Its major flood control targets are Zhicheng and Chenglingji in the middle reaches of the Yangtze River, with safety flow of 56700m^3/s and 60000m^3/s respectively [6]. The dispatching schedule of the Three Gorges Reservoir in flood season adopts a graded compensation schedule, which divides the 22.5 billion m^3 flood control storage capacity into three parts from the bottom up. The first part has a storage capacity of 5.65 billion m^3 and the corresponding water level is from 145m to 155m, for flood control in Zhicheng and Chenglingji, to control the maximum flow at Chenglingji station not exceeding 60000m^3/s as well as the maximum flow at Zhicheng station not exceeding 56700m^3/s. The second part of the storage capacity is 12.85 billion m^3, the corresponding water level is from 155m to 171.32m, only for flood control in Zhicheng, to control the maximum flow at Zhicheng station not exceeding 56700m^3/s. The third part of the storage capacity is 3.65 billion m^3, the corresponding water level is from 171.32m to 175m, reserved for extraordinarily large flood control at Jinjiang river reach, to control the maximum flow at Zhicheng station not exceeding 80000m^3/s.

It identified the existing flood control rules of the Three Gorges Reservoir as a conventional dispatching method for not considering the utilization of flood resources. The conventional dispatching method is shown as follows:

a. When the Three Gorges reservoir water level is equal to 145m,

$$q_{s,t} = \min(p_{s,t}, sp_{z,t}, sp_{c,t}) \tag{1}$$

$$sp_{z,t} = 56700 - q_{z,t} \tag{2}$$

$$sp_{c,t} = 60000 - q_{c,t+2} \tag{3}$$

Where, $p_{s,t}$ is outflow of the Three Gorges reservoir when time t; $sp_{z,t}$ is outflow of the Three Gorges reservoir when guarantee the safety of Zhicheng when time t; $q_{z,t}$ is interval flow between the Three Gorges Reservoir and Zhicheng when time t; $sp_{c,t}$ is outflow of the Three Gorges reservoir when guarantee the safety of Chenglingji when time t; $q_{c,t+2}$ is interval flow between the Three Gorges Reservoir and Chenglingji two days later than time t.

b. When the Three Gorges reservoir water level is between 145m to 155m,

$$q_{s,t} = \min(sp_{z,t}, sp_{c,t}) \tag{4}$$

c. When the Three Gorges reservoir water level is between 155m to 171.32m,

$$q_{s,t} = sp_{z,t} \tag{5}$$

d. When the Three Gorges Reservoir water level is higher than 171.32m,

$$q_{s,t} = 80000 - q_{z,t} \tag{6}$$

Normally, $sp_{z,t}$ and $sp_{c,t}$ are larger than the full power generation flow, which is around 30000m^3/s. So when the inflow of the Three Gorges Reservoir is larger than 30000m^3/s, the water resource exceed 30000m^3/s will be abandoned and the utilization of flood resources will not be sufficient. The distribution of runoff in the Yangtze River is uneven and the water resources in the flood season is much more abundant than dry season. Table 1 shows the daily discharge of Yichang Hydrological Station (the representative hydrological station of the Three Gorges reservoir, and after constructed the data is the inflow of the Three Gorges reservoir) from June to September from 1881 to 2013.

Table 1. Different discharge level percentage at Yichang Hydrological Station

Inflow	<20000	20000~30000	30000~40000	40000~50000	≥60000
Percentage	32.9%	38.2%	19.6%	6.9%	2.4%

It can be seen from the table that the average daily flow of more than 30000m^3/s accounts for 28.9% of the days, which exceeds the full power generation flow of the Three Gorges Reservoir and must be abandoned without power generation, resulting in the inability to fully utilize the water resources.

2.2 Dispatching method of flood resource utilization

The implementation of flood resource utilization under the premise of not affecting the flood control safety is one of the effective measures to improve the comprehensive utilization efficiency of water resources in the Three Gorges Reservoir. In recent years, there had been many practices on the utilization of flood resources [7]. Floodwater resources are generally used through the allocation of water volume within the river basin and inter-basin water diversion, water storage and power generation, water diversion and recharge, ecological water and groundwater supplement. For the Three Gorges Reservoir, it can make use of the floodwater resources during the flood season by generating electricity [8].

The Three Gorges reservoir flood resources utilization method introduced in this paper was when the inflow is larger than the full power generation flow of 30000m^3/s, and less than $sp_{z,t}$ and $sp_{c,t}$ in the next few days by hydrological forecasting, part of the flood control capacity may be utilized to reserve inflows larger than 30000m^3/s, and increase the outflow when the inflow of the Three Gorges Reservoir is less than 30000m^3/s, which would reduce the amount of abandoned water and increase power generation benefits.

Take the inflow from July 1st to July 22nd in 2013 for example, Zhicheng and Chenglingji had flood control damage until July 21st. For the conventional scheduling, the outflow before July 21th equalled to the inflow of the Three Gorges Reservoir, the water resource would be abounded when the inflow exceeded 30000m^3/s, which called power generation overflow. The inflow, outflow, power flow and power generation overflow according to the conventional scheduling was shown in Figure 1.

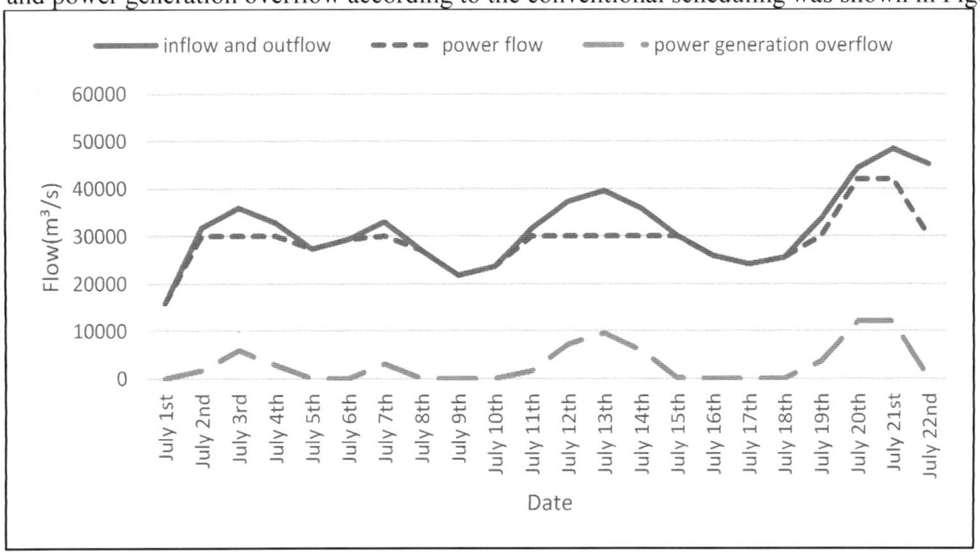

Figure. 1 The flow according to the conventional scheduling

But for flood resource utilization dispatching method, the flood control capacity could be used for regulating power generation overflow, when the inflow exceeded 30000m^3/s, it reduced outflow to 30000m^3/s and storage water, and the water level went up. When the inflow was less than 30000m^3/s, it increased outflow to 30000m^3/s, and the water level went down. The inflow, outflow, power flow and power generation overflow according to the conventional scheduling was shown in Figure 2.

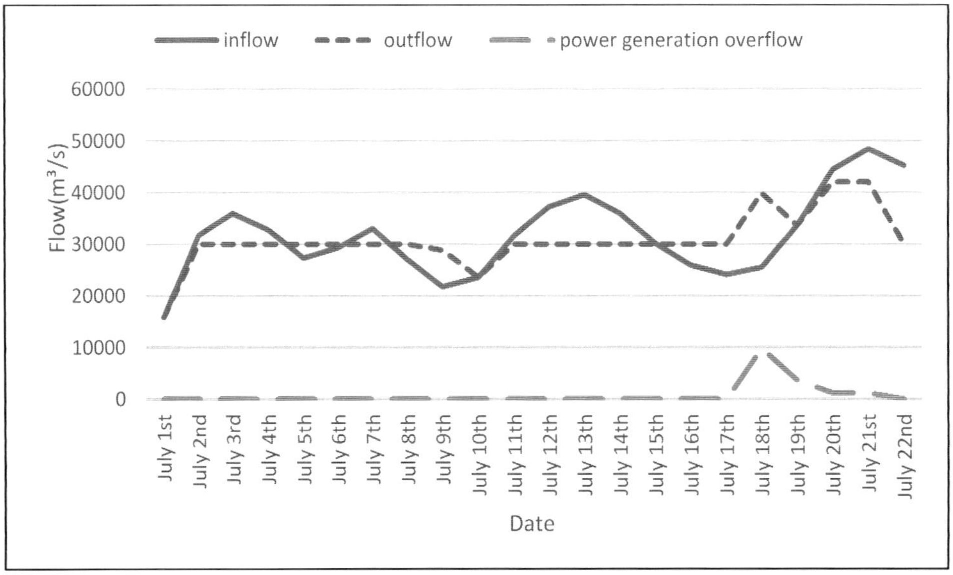

Figure. 2 The flow according to dispatching method of flood resource utilization

The water level process in the two kinds of scheduling of the Three Gorges reservoir is shown in Figure 3.

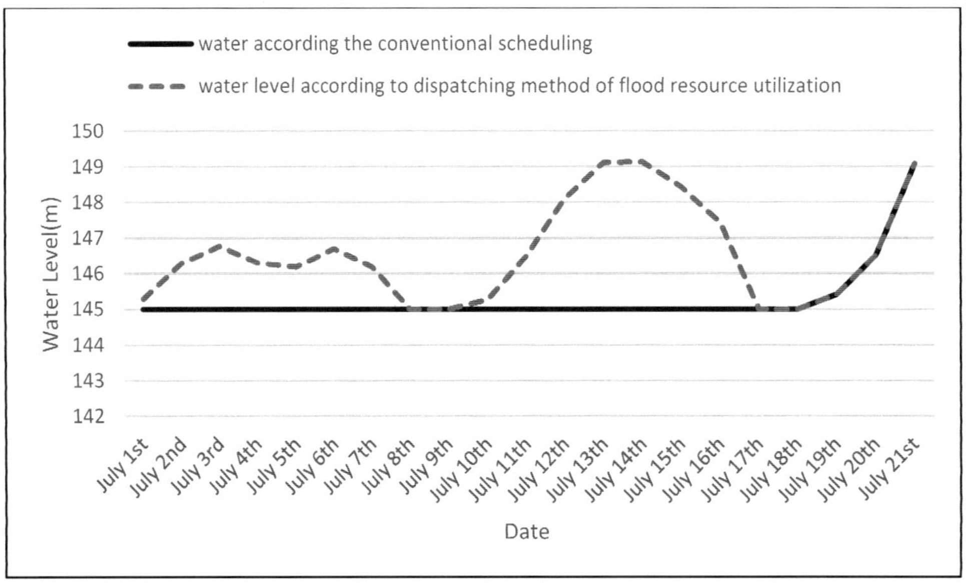

Figure. 3 The water level process in the two kinds of scheduling

The maximum flood protection capacity for flood resource utilization can be calculated by the following formula,

$$V_{max} = \sum_{t=1}^{N}\left(\min\left(sp_{z,t}, sp_{c,t}\right) - q_{s,t}\right) \cdot \Delta t \tag{7}$$

Where V_{max} is the maximum flood protection capacity for flood resource utilization, and Δt is the time step, for example when the time step is daily, then the Δt is equal to 86400s.

3 Power generation benefits and flood control risk analysis

According to the current hydrological forecast level in the Three Gorges Reservoir dispatching, the Δt is 5 days. According to flood control status of Zhicheng and Chenglingji, the safety outflow of the Three Gorges reservoir was taken as 42000m^3/s.

The actual discharge process of Yichang Hydrological Station in flood season from 1881 to 2013 was used to simulate the calculation. The power generation capacity, the amount of power generation overflow and the flood control effect by the two methods were shown in Table 2.

Table 2. The power generation and flood control risk by the two methods

method	average power generation every year(billion kW · h)	average power generation overflow every year(billion m^3)	average outflow exceed 42000m^3/s every year((billion m^3)	The number of years when the maximum level is higher than 155m of the Three Gorges Reservoir
Conventional Dispatching Method	43.52	24.54	1.15	15
Dispatching method of flood resource utilization	44.46	20.38	1.15	15

It can be seen from the table that according to the proposed flood resource utilization mode, the power generation is larger than that of the conventional dispatching method, and the amount of power generation overflow is reduced. The power generation increased by 3.25% from 43.52 billion kW · h to 44.46 kW · h during flood season per year, and the spillage decreased by 21.19% from 24.54 billion m^3 to 20.38 billion m^3 during flood season per year. At the same time the outflow exceeding 42000m^3/s and the years with the maximum water level exceeding 155m did not increase, which shows that the proposed flood resource utilization mode would not increase flood control risk by pre-discharging stored water to reduce the reservoir water level to its flood control level 145m before the next flood.

4 Conclusions

Aiming at the problem that conventional dispatching mode of Three Gorges reservoir in flood season cannot make full use of flood resources, this paper presents a method that utilizes part of flood control capacity to integrate flood control with water resources utilization based on flood forecasting. The proposed pre-discharge measure could reduce the flood risk of using flood control capacity. The results show that this method could help the reservoir benefit more from power generation in flood season.

Acknowledgement

This work is supported by the National Key R&D Program of China (Item Nos. 2016YFC0402203).

References

[1] Dottori F, Salamon P, Bianchi A, et al. Development and evaluation of a framework for global flood hazard mapping[J]. Advances in Water Resources, 2016, 94:87-102.
[2] Gober P, Wheater H S. Debates—Perspectives on socio - hydrology: Modeling flood risk as a public policy problem[J]. Water Resources Research, 2015, 51(6): 4782–4788.
[3] Chang J, Guo A, Du H, et al. Floodwater utilization for cascade reservoirs based on dynamic control of seasonal flood control limit levels[J]. Environmental Earth Sciences, 2017, 76(6):260.
[4] Liu P, Li L, Guo S, et al. Optimal design of seasonal flood limited water levels and its application for the Three Gorges Reservoir[J]. Journal of Hydrology, 2015, 527:1045-1053.
[5] Zhong P, Kong Y, Wang X, et al. Study on method for estimation of dynamic control bounds of flood limited water level in cascade reservoirs[J]. Journal of Hydroelectric Engineering, 2014, 33(5):36-43.
[6] Merz B, Vorogushyn S, Lall U, et al. Charting unknown waters—On the role of surprise in flood risk assessment and management[J]. Water Resources Research, 2015, 51(8):6399-6416.
[7] Nabangchang O, Allaire M, Leangcharoen P, et al. Economic costs incurred by households in the 2011 Greater Bangkok flood[J]. Water Resources Research, 2015, 51(1):58-77.
[8] Dai L, Mao J, Wang Y, et al. Optimal operation of the Three Gorges Reservoir subject to the ecological water level of Dongting Lake[J]. Environmental Earth Sciences, 2016, 75(14):1-14.

IOP Conf. Series: Materials Science and Engineering **366** (2018) 012009 doi:10.1088/1757-899X/366/1/012009

A Research on Multi-objective Generation Optimization Scheduling of the TGC Reservoirs Considering Ecological and Shipping Benefits

Hu Xiangyang[1], Wang Xuemin[1], Chen Fang[2] and Yu Shan[1]

[1] Changjiang Survey, Planning, design and Research CO., LTD., Wuhan 430010, China

[2] Bureau of Hydrology, Changjiang Water Resources commission, Wuhan 430010, China

E-mail (Wang Xuemin):wang_xuemin@foxmail.com or 1198780078@qq.com

Abstract. Due to the increasing energy demand and the lack of comprehensive utilization of water resources, this paper has established a generation-shipping-ecological optimization (GSEO) model to Balance the benefits between shipping, power generation and ecology. then, the Three Gorges Cascade (TGC) is taken as the research object, and the efficiency evaluation methods of shipping and ecological benefits are proposed. Finally, the actual inflows of the abundant, average and dry years are studied as the samples to solve the GSEO model. The simulation results indicate that the model and the benefit evaluation methods proposed in this paper are feasible and effective.

1. Introduction

Along with China population growth and rapid economic development, energy demand is increasing at the same time. As a pure and clean energy, hydropower has the advantages of renewable, pollution-free, low operating cost and easy for peak load regulation, which is widely favoured by all countries in the world. The hydropower station is also carrying out other comprehensive benefits while generating electrical energy. How to consider multiple benefits in reservoir scheduling and improve the operation level of reservoir is the target of this paper. For this purpose, this paper will take the Three Gorges Cascade (TGC) in China as the research object, consider other benefits in the generation scheduling, and realize the efficient utilization of water resources.

As the "golden waterway" of China, the Yangtze River runs through China's eastern and western regions, and its great capacity is irreplaceable by other means of transportation. As a control center of the middle and lower reaches of the Yangtze River, the TGC has an important strategic position for the regulation of navigation in the middle and lower reaches. Therefore, it is necessary to study the shipping schedule of the TGC. The operation of the TGC will inevitably pose a threat to the ecosystem of the river basin, therefore, how to effectively reduce the adverse effects on the ecological habitat of the basins while making full use of the large-scale cascade, is also the difficult problem to be solved. This paper will take the TGC in China as the research object, consider the ecological and shipping benefits in the generation scheduling, and realize the efficient utilization of water resources.

At present, there are few researches on multi-objective scheduling of shipping. In terms of ecological benefit evaluation, the commonly used methods are IFIM[1] method, RVA[2] method,

Tennant[3] method and so on. In terms of power generation, Kumar[4], Wu[5] and Hui[6] have in-depth research, and the simulation results are good. The multi-objective optimization algorithm has been widely used in the reservoir dispatching[7][8], and the algorithm is rich in variety and improvement. On the basis of existing research and the current situation of the TGC dispatching, this paper proposes the method of considering the shipping benefit in the reservoir dispatching by simulating the operation process of the actual ship lock. Then, an ecological benefits calculation method is introduced to establish a multi-objective generation scheduling model considering ecological and shipping objectives. A multi-objective optimization algorithm is used to solve the model, and finally the TGC operation schemes with generation, shipping and ecological benefits are given.

2. Multi-objective power generation optimization model

The contradictions between the three benefits of shipping, power generation and ecology are obvious during the Storage period. Therefore, the paper has established a generation-shipping-ecological optimization (GSEO) model.

2.1. Objective function

In this paper, the GSEO model includes three scheduling objectives of generation, shipping and ecology, and the mathematical expression of the objective function is shown below

(1) Shipping benefit: the total accumulative retention weight of goods (TAG) in the waterway is the evaluation standard for shipping

$$\min f_1 = \min \sum_{t=0}^{T} PL(D_t, Q_t) \tag{1}$$

(2) Generation benefit: the total power generation of cascade power station

$$\max f_2 = \max \sum_{i=1}^{Num} \sum_{t=1}^{T} K_i Q_{i,t}^f H_{i,t}^f \Delta t = \max \sum_{i=1}^{Num} \sum_{t=1}^{T} N_{i,t} \Delta t \tag{2}$$

(3) The ecological benefits are evaluated by the average habitat area (See section 4 for details of WUA method) of the Chinese sturgeon

$$\max f_3 = \max \frac{1}{T} \sum_{t=0}^{T} \mathrm{WUA}(Q_t) \tag{3}$$

where $PL(D_t, Q_t)$ is the TAG in tth period, Num is the number of the power stations in the cascade, T is the total number of the periods, K_i is the output coefficient of the ith power station, $Q_{i,t}^f$ 、 $H_{i,t}^f$ and $N_{i,t}$ are the generation flow, average water head and average output respectively, Δt is the time length of a period, $\mathrm{WUA}(Q_t)$ is the average habitat area calculated by the WUA method.

2.2. Model constraints

In order to ensure the safe and stable operation of the cascade power station, the relevant dispatching rules of cascade hydropower stations are referenced. The dispatch constraints of the cascade hydropower stations mainly includes water level constraints, water balance constraints, power output constraints and discharged flow constraints.

(1) Water level constraints

$$Z_{i,t}^{\min} \le Z_{i,t} \le Z_{i,t}^{\max} \tag{4}$$

where $Z_{i,t}$ is the water level of the i-th power station at the t-th period; $Z_{i,t}^{\min}$ and $Z_{i,t}^{\max}$ respectively stand for the upper limit and lower limit of water level.

(2) Power output constraints

$$N_{i,t}^{\min} \le N_{i,t} \le N_{i,t}^{\max} \tag{5}$$

where $N_{i,t}$ is the power output of the i-th power station during the t-th period; $N_{i,t}^{\min}$ and $N_{i,t}^{\max}$ respectively stand for the upper limit and lower limit of power output.

(3) Flow constraint

$$Q_{i,t}^{\min} \le Q_{i,t}^{X} \le Q_{i,t}^{\max} \tag{6}$$

where $Q_{i,t}^{\min}$ is the discharge of the i-th power station during the t-th period; $Q_{i,t}^{\min}$ and $Q_{i,t}^{\max}$ respectively stand for the upper limit and lower limit of outflow.

(4) Water balance constraints

$$V_{i,t+1} = V_{i,t} + (I_{i,t} - Q_{i,t}^{X}) \times \Delta t \tag{7}$$

$$I_{G,t} = Q_{S,t-\tau}^{X} + q_t \tag{8}$$

where $V_{i,t}$ is the capacity of the i-th power station of the t-th period; $I_{i,t}$ is the inflow of the i-th power station; τ is the flow time; $I_{G,t}$ is the inflow of the power station in the lower reaches; $Q_{S,t-\tau}^{X}$ is the discharge flow of the power station in the upper reaches of the ($t-\tau$)-th period; q_t is the inflow of the river between these two power stations.

(5) Limitation of shipping capacity

$$D_t^{s} \le D_t^{\max} \tag{9}$$

Where D_t^s is the amount of goods through the lock in the tth period; $D_t^{\max}$ is the capacity of shipping in the tth period.

3. Shipping benefit evaluation

In this paper, the research object is the TGC, so when discussing the shipping benefit, the default object is the TGC.

The TGC is a comprehensive hydro project built on the main stream of the Yangtze river, its main tasks are flood control, power generation and shipping. The TGC is composed of the Three Gorges Dam and the Gezhouba Dam, the Gezhouba Dam is 38km downstream. The TGC has a double-line ship lock. The construction of the ship lock was started on April 17, 1994, and on June 16, 2003, it entered the commissioning phase and passed the acceptance inspection on July 8, 2004. As at December 25, 2013, the ship lock had operated 94558 times, passed 59.30 million vessels and 645 million tons of goods, undertook a important transport mission.

For the TGC, The most influential factor for shipping is the discharge flow of the Three Gorges Dam, and the can also directly reflect the operation effect of the dispatch. Therefore, the discharge flow process is used to analyse the shipping benefits.

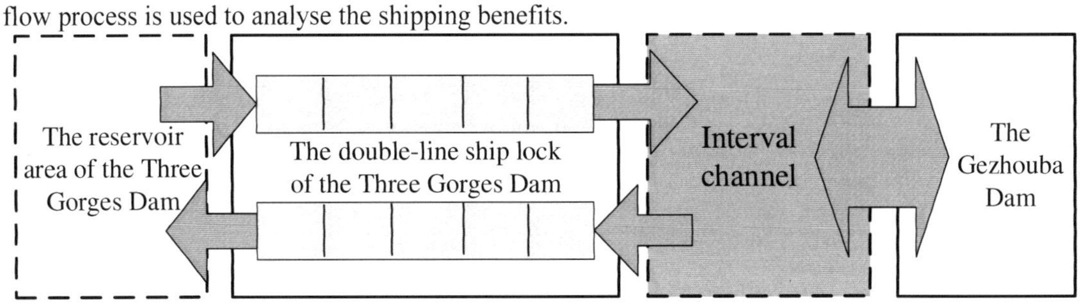

Figure 1. Schematic diagram of navigation structure of the TGC

The navigation structure of the TGC is shown in figure. 1. The navigation ability is up to the water flow of the interval channel, which is determined by the discharge flow of the Three Gorges Dam. Ships of different tonnage follow different rules, which are show in Table 1.

Table 1 The navigation limit flows of the interval channel (m^3/s)

Ship power (Tons)	the highest discharge going up	the highest discharge going down	Ship power	the highest discharge going up	the highest discharge going down
> 4000	45000	45000	1001 to 2000	30000	35000
3001 to 4000	40000	45000	501 to 1000	25000	30000
2001 to 3000	35000	40000	< 500	25000	25000

The way to calculate the TAG is shown in Fig.2.

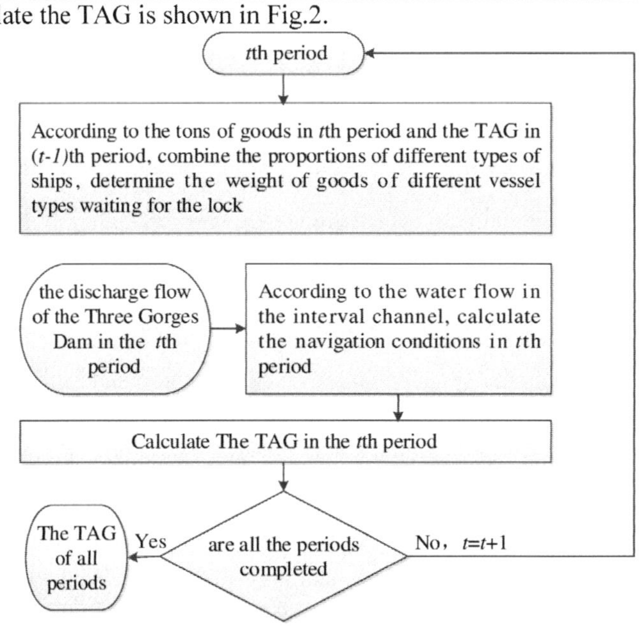

Figure 2. Schematic diagram of TAG calculation

4. Ecological benefit evaluation

This paper introduces the WUA[9] method to evaluate the ecological benefit. WUA is a biological physical habitat simulation method, this paper uses this method to evaluate the suitability of target species habitat under different water flows. The Chinese sturgeon is a rare species living in the downstream channel of the TGC, and The research results of its habitat and breeding conditions are abundant. Therefore, this paper chooses Chinese sturgeon as the target species of ecological evaluation. The WUA evaluation curve used in this paper is from the literature[10], the relationship between the discharge flow of the TGC and the suitable habitat area of the Chinese sturgeon is shown in Fig.3.

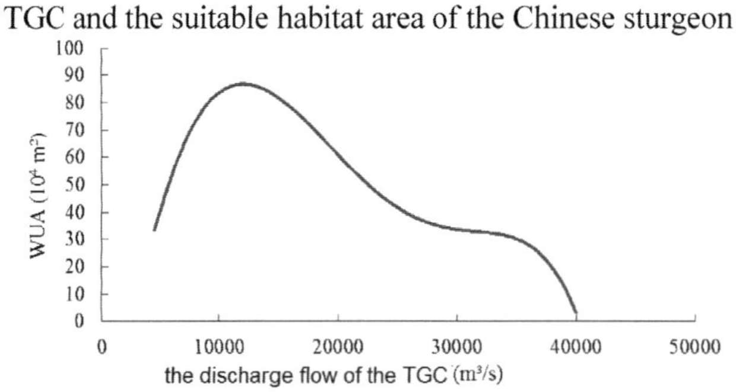

Figure 3. The WUA evaluation curve of the Chinese sturgeon

5. Case study

In this paper, a multi-objective optimization algorithm is introduced, and the above shipping and ecological evaluation methods are applied to study the GSEO model.

5.1. Optimization algorithm

This paper uses a kind of modified multi-objective differential evolution algorithm (MODE) [11] to solve the GSEO model proposed. The flow chart of the algorithm is shown in Fig.4.

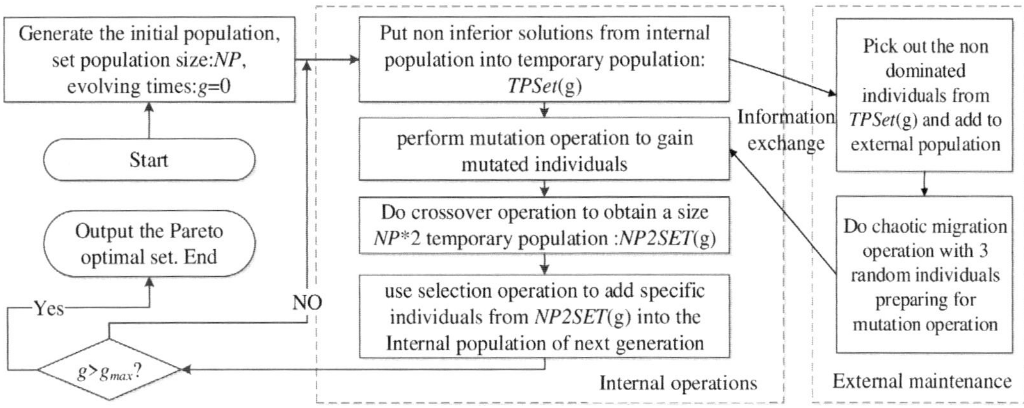

Figure 4. Flow chart of multi-objective difference algorithm [11]

5.2. Simulation results and analysis

The start time of the water storage schedule is September 10, which lasts for 80 days. The abundant year is 1907, the average year is 2014, the dry year is 2006. So the inflows of the year 2006, 2014 and 1907 are as the study cases, and the simulation results are shown in fig.5, 6 and 7.

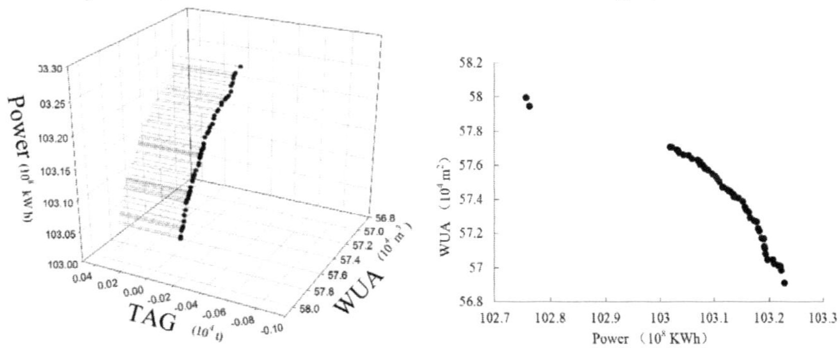

Figure 5. Optimization scheduling solution set of year 2006

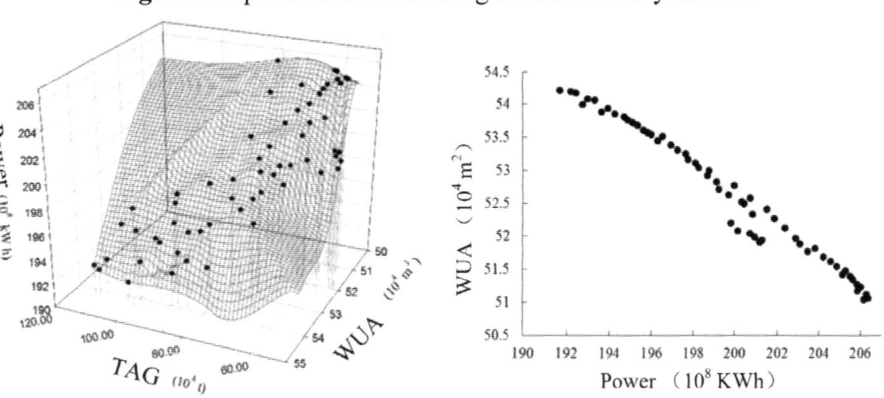

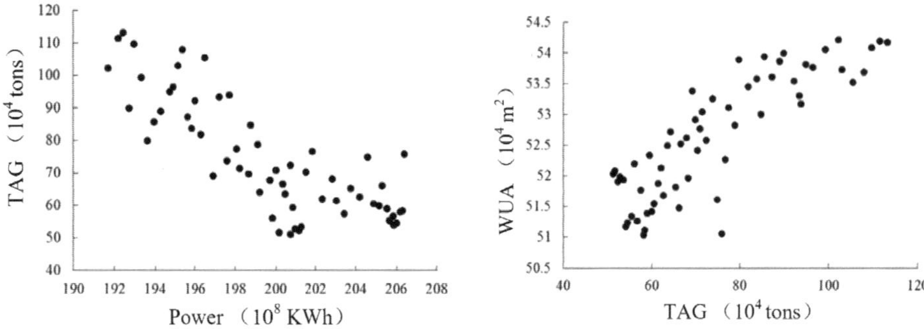

Figure 6. Optimization scheduling solution set of year 2014

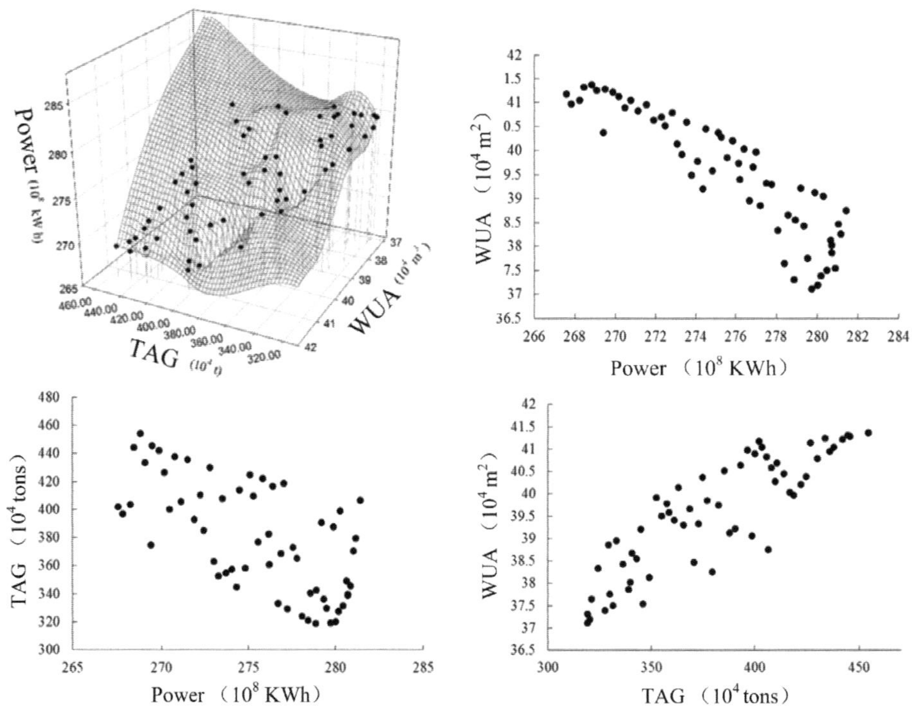

Figure 7. Optimization scheduling solution set of year 1907

As can be seen from the fig.5, the TAGs of all the schemes in the solution set of year 2006 are 0, this is due to the small amount of water in the storage period of the year 2006. In order to achieve the purpose of full impoundment, the discharge flow should be small, so that the flow of the interval channel has no impact on shipping. At the same time, it can be seen that there is an obvious antagonistic relationship between the generation and the ecological benefit without considering the shipping. From fig.6 and 7, it can be seen that the smaller the water inflow is, the more obvious the antagonistic relationship between power generation and ecology benefits are, so a greater ecological benefit would require more water consumption, which would lead to reduction of the power generation. Because the inflow in early storage period is still large, the operation mode is similar to that of the flood period, so as to improve the shipping benefit by decreasing the discharge during the period with small water inflow. However, in the later period of the storage period, the water inflow decreases, and the discharge flow has no longer impeded the shipping benefits.

6. Conclusions

This paper take the TGC as the research object, through the analysis of actual operation data, the flow in the interval channel is taken as the reflection of shipping benefit, which can be calculated based on the simulation of the actual ship lock operation. WUA method was introduced to evaluate the ecological benefits of the Chinese sturgeon habitat in the lower reaches of the TGC. Then, a kind of modified multi-objective differential evolution algorithm is applied to solve the GSEO model. Finally, the actual inflows of the three years are studied. It can be concluded that the GSEO model and the shipping and ecological evaluation methods proposed in this paper are feasible and effective.

Acknowledgments
This work is supported by the National Key Research and Development Program of China (Item Nos. 2016YFC0402209). The authors also appreciate the insightful comments and suggestions from anonymous reviewers.

References
[1] Bovee, K. D. A guide to stream habitat analysis using the instream flow incremental methodology[J]. 1982.
[2] Richter, B. D.,Baumgartner, J. V.,Powell, J.,Braun, D. P. A Method for Assessing Hydrologic Alteration within Ecosystems[J]. Conservation Biology, 1996, 10 (4): 1163–1174.
[3] Tennant, D. L. Instream Flow Regimens for Fish, Wildlife, Recreation and Related Environmental Resources[J]. Fisheries, 2011: 6-10.
[4] Kumar D. N., Reddy M. J. Ant Colony Optimization for Multi-Purpose Reservoir Operation. Water Resources Management, 2006, 20(6): 879-898.
[5] Wu J, Guo Z, Wu F. Short-term multi-objective optimization scheduling for cascaded hydroelectric plants with dynamic generation flow limit based on EMA and DEA[J]. International Journal of Electrical Power & Energy Systems, 2014, 57(5):189-197.
[6] Hui Q, Zhou J, Lu Y, et al. Multi-objective differential evolution with adaptive Cauchy mutation for short-term multi-objective optimal hydro-thermal scheduling[J]. Energy Conversion & Management, 2010, 51(4):788-794.
[7] Nazari-Heris M, Mohammadi-Ivatloo B, Gharehpetian G B. Short-term scheduling of hydro-based power plants considering application of heuristic algorithms: A comprehensive review[J]. Renewable & Sustainable Energy Reviews, 2017, 74:116-129.
[8] Xie M, Zhou J, Li C, et al. Daily Generation Scheduling of Cascade Hydro Plants Considering Peak Shaving Constraints[J]. Journal of Water Resources Planning & Management, 2015, 142(4):04015072.
[9] Bovee, K. D.,Lamb, B. L.,Bartholow, J. M.,Stalnaker, C. B.,Taylor, J. Stream habitat analysis using the instream flow incremental methodology[R]. DTIC Document, 1998.
[10] Zhao Yue. Study on Modeling Theory and Methodology of Reservoir Operation for River Ecosystem Restoration [D]. Huazhong University of Science and Technology, 2014.
[11] Wang X, Zhou J, Wang C. Ecology and Water Supply Based Multi-objective Optimal Dispatch Model and Its Case Study in Yangtze basin[J]. International Journal of Modeling & Optimization, 2015, 5(3):241-245.

Operation Mode of Danjiangkou Reservoir under Water Diversion Conditions of South-to-North Water Diversion Middle Route Project

Shuo Ouyang[1], Hui Qin[2*], Jun Shao[1], Rui Zhang[3] and Minglong Dai[1]

[1]Bureau of Hydrology, Changjiang Water Resources Commission, Wuhan 430010;
[2] School of Hydropower and Information Engineering, Huazhong University of Science and Technology, Wuhan, 430074, People's Republic of China;
[3]Changjiang Institute of Survey, Planning, Design and Research, Changjiang Water Resources Commission, Wuhan 430010

E-mail (Hui Qin): hqin@hust.edu.cn

Abstract: In order to explore operation scheme of downstream controlling reservoir in water source area for optimized allocation of inter-basin water resources, this paper selected the south-to-north water diversion middle route project in Hanjiang River basin as an instance. The optimized dispatching mode of Danjiangkou Reservoir is built with multiple dispatching objective functions and different dispatching scenarios set, such as maximum water supply during water supply period, minimum discharge, minimum abandoned water and maximum power generation. The simulation results indicate that the model and methods proposed in this paper are feasible and effective.

1. Introduction

For more than half a century, domestic and foreign experts and scholars propose many inter-basin water diversion project planning, decision-making models and methods of management and operation, which could be divided into two categories: firstly, after simplifying complex inter-basin water diversion system through various methods, single mathematical programming model or simulation model is adopted for planning, management and operation decision-making research on inter-basin water diversion projects[1]. Secondly, large-scale system optimized decision model and method are directly adopted. Planning management decision-making research on inter-basin water diversion projects is conducted by setting various large-scale system hierarchical structure models firstly and then applying solution method combined with multiple mathematical programming or simulation techniques (including self-optimization simulation technique) [2]. In recent years, along with continuous development and perfecting of decision-making support system and expert system[3], fuzzy mathematics [4, 5], cooperative game theory [6, 7] and other new theories[8, 9] and methods, people begin to explore application possibility of these theories and methods in planning management decision-making research on inter-basin water diversion project.

However, along with building of inter-basin water diversion projects, the inflow from upstream is obviously reduced, and the hydrological characteristics of river network are vary greatly. Currently, domestic and foreign research mainly focuses on the analysis of impact of inter-basin water diversion on downstream hydrology[10], water quality[11, 12], ecology and salinization[13] of water sources, and less quantitatively focuses on the effect of downstream controlling reservoir of water source in inter-basin water diversion. As the inter-basin water diversion with the most work quantities since new China was found, the south-to-north water diversion project (SNWDP) diverts a part of abundant water resources from Yangtze River basin to North China and northwest districts to relieve water resources shortage there [14, 15].

Therefore, in order to explore the operation scheme of controlling reservoir at downstream water source area specific to the optimized allocation of inter-basin water resources, this paper selected the Middle Route Project (MRP) of SNWDP as an instance, set multiple dispatching objective functions and different dispatching scenarios such as maximum water supply during water supply period, minimum discharge, minimum abandoned water and maximum power generation, and built the optimized dispatching mode of Danjiangkou Reservoir. Optimized dispatching of Dangjiangkou Reservoir was simulated by adopting natural runoff of Danjiangkou Reservoir from 1956 to 2010.

The structure of this paper is as follows: Chapter 2 describes the reservoir dispatching model, Chapter 3 discusses the water diversion scheme of MRP, Chapter 4 gives instance analysis, and Chapter 5 makes a summary.

2. Reservoir dispatching model

This paper selected four objective functions, i.e. maximum power generation, minimum abandoned water, maximum water supply during water supply period and minimum discharge, suitable for different scenario stages, and established the controlling reservoir regulation and storage model specific to upstream inter-basin water diversion. The generalized figure (Fig. 1) of the model is as follows:

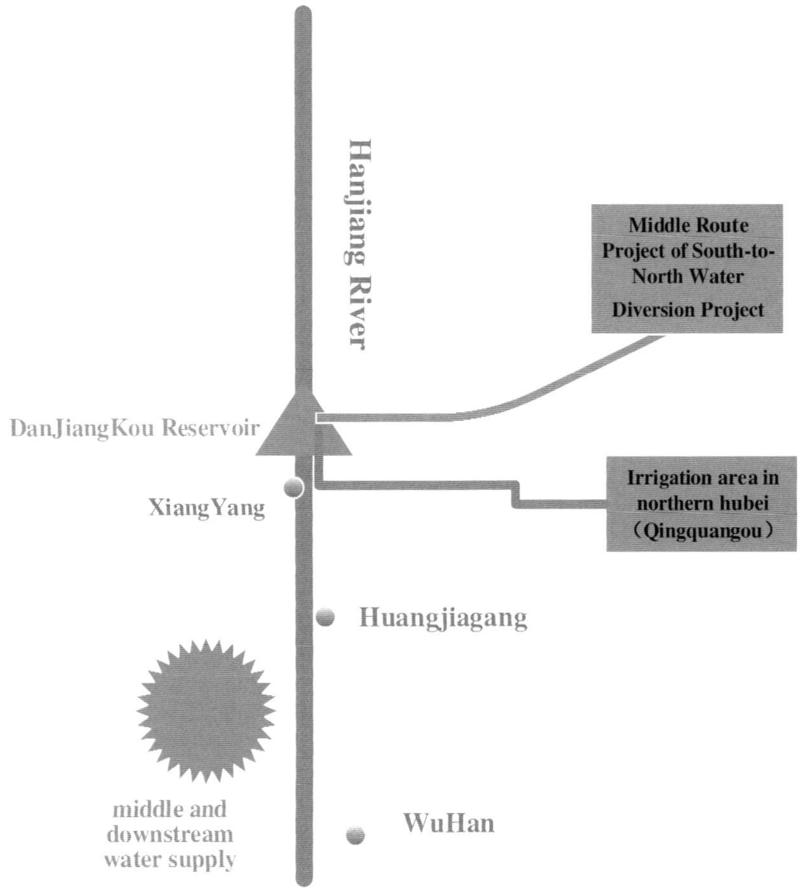

Figure 1 The geographical locations and topological structures of the Middle Route Project of South-to-North Water Diversion Project and Danjiangkou hydraulic complex project

IOP Conf. Series: Materials Science and Engineering **366** (2018) 012010 doi:10.1088/1757-899X/366/1/012010

2.1 Objective functions

(1) Minimum abandoned water (Obj1)

$$\min W_d = \min\{\sum_{t=1}^{T} Q_{t,d} \cdot \Delta t\} \tag{1}$$

Where, $Q_{t,d}$ represents abandoned water flow of reservoir during period t, and W_d represents abandoned water volume.

(2) Maximum water supply during water supply period (Obj2)

$$\max W_S = \max\{\sum_{t=1}^{T_S} Q_t \cdot \Delta t\} \tag{2}$$

Where, W_S represents water supply during water supply period, Q_t represents discharge of reservoir during period t, and T_S represents water supply dispatching period.

(3) Minimum discharge (Obj3)

$$\max Q_{f\min} = \max\{\min(Q_{t,f}, t \subset T)\} \tag{3}$$

(4) Maximum power generation (Obj4)

$$\max P = \max\{\sum_{t=1}^{T} N_t \cdot \Delta t\} \tag{4}$$

Where, N_t represents output of reservoir during period t.

2.2 Constraint conditions

Constraint conditions of the above-mentioned model are as follows:

(1) Upper and lower constraint of reservoir capacity (water level):

$$Z_{t,\min} \le Z_t \le Z_{t,\max} \tag{5}$$

Where, $Z_{t,\min}$, $Z_{t,\max}$ and Z_t respectively represent allowable minimum water level, maximum water level and water level in current period of reservoir during period t.

(2) Reservoir water balance constraint:

$$V_t = V_{t-1} + (I_t - Q_t - S_t)\Delta t \tag{6}$$

Where, V_t, I_t and Q_t respectively represent reservoir capacity, reservoir inflow and reservoir outflow during period t of dispatching; and S_t represents water diversion flow of upstream inter-basin water diversion project.

(3) Reservoir discharge capacity constraint

$$Q_{\min} \le Q_{out,t} \le Q_{\max} \tag{7}$$

Where, $Q_{out,t}$ represents discharge of reservoir during period t of dispatching, and $Q_{\max}$ and $Q_{\min}$ respectively represents maximum and minimum discharge capacity.

(4) Reservoir outflow constraint:

$$\left|Q_{out,t} - Q_{out,t-1}\right| \le \Delta Q \tag{8}$$

Where, $Q_{out,t}$ represents reservoir outflow during period t of dispatching, and ΔQ represents maximum amplitude of daily discharge corresponding to the reservoir.

(5) Flood control dispatching rules requirements and basin flood control standards for each reservoir of cascade reservoirs are conformed to.

3. Water division scheme for MRP

MRP diverts water from Danjiangkou Reservoir in Hanjiang River, a tributary of Yangtze River, channelizes and delivers water along plains of Funiu Mountain and Taihang Mountains to the end point, Beijing. Water supply range of MRP mainly includes more than 20 large and medium-sized middle and west cities in Tang-Bai River Plain and Huang-Huai-Hai Plain. MRP can relieve water resources crisis in Beijing, Tianjin and North China and increase water supply for living and industry of Beijing, Tianjin and cities of Henan, Hebei along the route and promote economic development in central China region.

Main monitoring cross-sections of main canal of MRP includes Taocha, Diao River, Fangcheng Caodun River, north bank of Yellow River, north bank of Zhang River, Gangtou Tunnel, north Juma River, Wangqingtuo (Hebei-Tianjin border). The key indicators are as shown in Table 1.

Table 1 Design flow and design level of main monitoring cross-sections

Name of monitoring cross-section	Location	Flow indicator(m^3/s)		Water level indicator
		Design flow	Increased flow	Design level (m)
Taocha	Behind Taocha sluice	350	420	147.38
Diao River	Diao River flume inlet regulating sluicc	350	420	146.80
Fangcheng Caodun River	Caodun River inlet regulating sluice	330	400	136.04
North bank of Yellow River	Yellow River-crossing outlet regulating sluice	265	320	111.93
North bank of Zhang River	Zhang River outlet regulating sluice	235	265	91.87
Gangtou Tunnel	Gangtou Tunnel inlet regulating sluice	125	150	65.99
North Juma River	North Juma River underground canal inlet regulating sluice	50	60	60.73
Wangqingtuo (Hebei-Tianjin border)	Wangqingtuo junction chamber inlet sluice	50	60	5.11

4. Instance analysis

4.1 Dispatching operation mode of Danjiangkou Reservoir

After Danjiangkou dam is heightened, the main task of comprehensive utilization is flood control, water supply (including irrigation), power generation, navigation, etc. Because *Danjiangkou Reservoir Dispatching Regulation* has not been complied, this paper proposed the *Flood Control and Water Supply Schemes for Danjiangkou Reservoir* according to the *Water Dispatching Scheme for Middle Route Project Phase I of South-to-North Water Diversion Project* and *Water Allocation Scheme in Hanjiang River Basin* reviewed and approved by the State Council in 2014, under the conditions of meeting water demand by Hangjiang River middle and downstream, following inflow and water levels of Danjiangkou Reservoir and combined with water diversion quantity required by areas receiving water, and conducted partition dispatching according to water levels of Danjiangkou Reservoir. The operating water level of Danjiangkou Reservoir: normal pool level is 170 m; limited water level in flood season is 160 m in Summer (from June 21 to August 20), and is 163.5 m in Autumn (from September 1 to September 30); dead water level is 150 m; and limiting drawdown water level is 145 m.

The specific operation and dispatching schemes of Danjiangkou Reservoir:

(1) Flood control dispatching

The reservoir water level is gradually reduced from May 1, and reduced to 160 m on June 20. From June 21 to August 20, 160 m is controlled as initial dispatch stage of flood control (similar to 145 m operation mode of Three Gorges Reservoir), 160 m water is transited to 163.5 m from August 21 to September 1. 163.5 m is controlled as initial dispatch stage of flood control from September 1 to

September 30, and the reservoir can gradually store water to normal pool level 170 m from October 1 to October 10. During flood period from June 20 to October 10, discharge of reservoir shall be not more than 21,000 m^3/s so as to ensure flood control safety at downstream area.

(2) Water supply dispatching

Water supply objects of Danjiangkou Reservoir include: i) Cities and irrigating areas located in middle and lower Hanjiang River (let-down discharge of reservoir); ii) water diversion of Qingquangou; iii) water diversion for MRP (diversion of Taocha canal head). Where: the designed diversion discharge volume is 350 m^3/s and the increased water volume is 420 m^3/s for Taocha canal head diversion; and the designed water diversion volume of Qingquangou is 100 m^3/s. According to the dispatching principles and dispatching graph regulations, the water is supplied to the water-receiving areas of MRP Phase I in guaranteeing the diversion requirements of the middle and downstream of Han River and Qingquangou.

The dispatching lines include: i) normal pool level line (NPLline); ii) flood control dispatching line (FCline); iii) reduced water supply line 1 (RWSline1); iv) reduced water supply line 2 (RWSline2); v) limited water supply line (LWSline); vi) limiting drawdown water level (LDWline). The six water supply dispatching lines divide the Danjiangkou Reservoir dispatching chart into 5 zones. The water level of the six dispatching lines at the beginning of the month is shown in Table 2.

Table 2 Water level for six dispatching lines of Dangjiangkou Reservoir **Unit: m**

S/N	Line	May	Jun	Jul	Aug	Sep	Oct	Nov	Dec	Jan	Feb	Mar	Apr
1	NPLline	170	170	170	170	170	170	170	170	170	170	170	170
2	FCline	170	164	160	160	163.5	163.5	170	170	170	170	170	170
3	RWSline1	156	156	156	159	161.4	163.5	168	168	168	168	165	160.5
4	RWSline2	155	155	155	155	156	161	163	163	163	161	158	155
5	LWSline	148	148	150	152	152	152	152	152	152	152	152	151
6	LDWline	145	145	145	145	145	145	145	145	145	145	145	145

4.2 Comparison analysis of dispatching results

Based on the above proposed Danjiangkou Reservoir dispatching scheme, 1956-2010 historical runoff data of ten-day natural storage runoff were selected. The DP algorithm and reservoir dispatching model designed in Chapter 2 were applied. Three dispatching cases were set, i.e. initial 157 m (case 1), without consideration of water diversion of phase 1 of MRP after the Danjiangkou dam was heightened (case 2), and in consideration of water diversion of phase 1 of MRP (case 3), to develop the optimized dispatching scheme with different objective functions. The comparison analysis of water resources allocation efficiency in the basin was carried out. See Table 3 for the simulated dispatching results of three cases.

Table 3 water supply dispatching results of Danjiangkou Reservoir in three cases

Case	Object	FSR (%)	MFR (%)	MGE ($\times10^8$kWh)	*MAW* ($\times10^8m^3$)	MSW ($\times10^8m^3$)	MWD ($\times10^8m^3$)	
							TC	QQG
Case 1	Obj4	89	99.83	51.05	49.20	**382.30**	0	3.46
	Obj1	15	97.36	54.27	41.18	**382.28**	0	3.48
Case 2	Obj2	11	96.23	45.92	82.83	**382.56**	0	3.20
	Obj3	15	97.37	42.76	126.40	**382.28**	0	3.48
	Obj1	11	96.54	41.97	22.59	**289.49**	92.81	3.46
Case 3	Obj2	0	94.39	39.11	58.20	**317.91**	64.75	3.10
	Obj3	11	96.52	34.23	72.61	**282.20**	100.11	3.45

The above table shows that in case 1 fill storage ratio (FSR) is 89%, mean annual full ratio (MFR) is 99.83%, mean annual quantity of water supply for downstream (MSW) is 38.23 billion m^3, mean annual abandoned water (MAW) is 4.92 billion m^3, and mean annual generated energy (MGE)is 5.105 billion kWh; in case 2 the abandoned water is minimum, the water supply capacity is the largest, and the minimum discharge is the highest. In the three optimized dispatching schemes, FSR is 15%, 11%

and 15% respectively, MFR is 97.36%, 96.23% and 97.37% respectively, MSW is 38.228, 38.256 and 38.228 billion m^3 respectively, MAW is 4.118, 8.283 and 12.64 billion m^3 respectively, and MGE is 5.427, 4.592 and 4.276 billion kWh respectively. In case 3, the abandoned water is minimum, the water supply capacity is the largest, and the minimum discharge is the highest. In the three optimized dispatching schemes, FSR is 11%, 0 and 11% respectively, MFR is 96.54%, 94.39% and 96.52% respectively, MSW is 28.949, 31.791 and 28.220 billion m^3 respectively, MAW is 2.259, 5.820 and 7.261 billion m^3 respectively, and MGE is 4.197, 3.911 and 3.423 billion kWh respectively. Compared to case 1 and case 2, case 3 reduces FSR by 78-89% and 4-11% respectively, MFR by 3.3-5.5% and 0.853-1.84% respectively, MSW by 6.4-10 billion m^3, and MGE by 0.9-1.7 and 0.6-1.3 billion kWh respectively.

On the other hand, MWD effectively alleviates the problem of water shortage in the north, and MAW is reduced by 0.9-2.7 and 1.8-5.4 billion m^3, which improves the utilization of water resources in the basin.

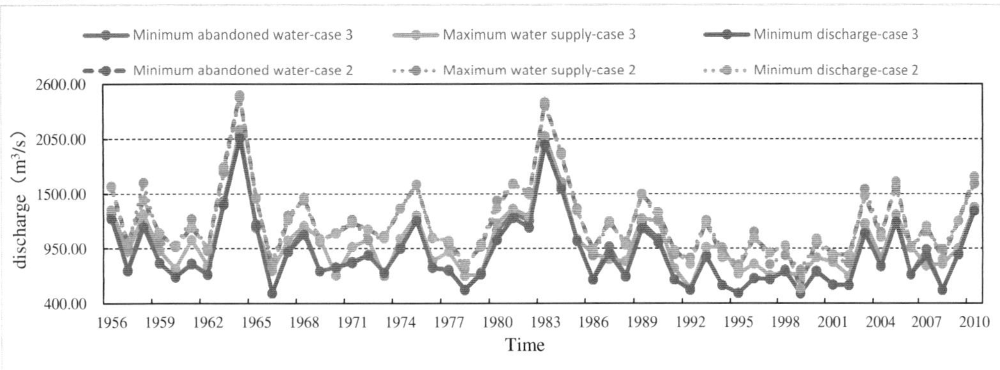

Figure 2 Discharge process of Danjiangkou Reservoir

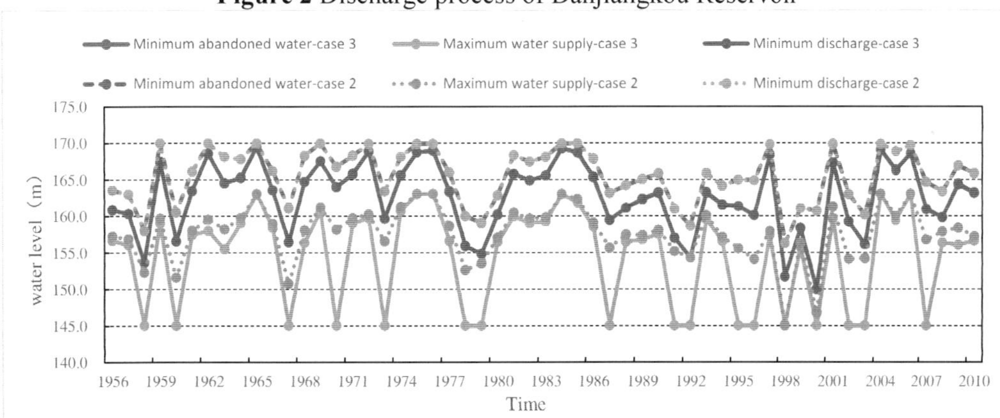

Figure 3 Water level process of Danjiangkou Reservoir

As can be seen from Table 3, Fig. 2 and Fig.3, due to the serious problem of water shortage, phase 1 of MRP has the large water diversion, which has a significant impact on the overall efficiency of water supply, water storage and power generation. In consideration of phase 1 of MRP water diversion conditions, MGE and FSR are less affected in the minor task, reduced by about 4% and 5% respectively.

5. Conclusion

In order to explore the downstream controlling reservoir operation scheme specific to optimal allocation of inter-basin water resources, MRP was selected as an instance to propose the Danjiangkou

Reservoir flood control and water supply scheme. This paper studies the operation mode of downstream controlling reservoir specific to optimal allocation of inter-basin water resources, and quantitatively analyzes the relationship and law of function between inter-basin water diversion, controlling water conservancy project at water source and the downstream reach of water source, which provides theoretical and data support for the development of rational operation scheme of downstream controlling water conservancy project at inter-basin water diversion water source.

Acknowledgments

This work was financially Supported by Hubei Provincial Natural Science Foundation of China (Grant No. 2016CFB391), supported by the Youth Funds of National Natural Science of China (Project No. 51609007), and supported by the National Key R&D Program of China (Item Nos. 2016YFC0402201). Special thanks are given to the anonymous reviewers and editors for their constructive comments.

References

[1]. Tu, M.-Y., N.-S. Hsu, and W.W.-G. Yeh, *Optimization of reservoir management and operation with hedging rules.* Journal of Water Resources Planning and Management, 2003. **129**(2): p. 86-97.

[2]. Jafarzadegan, K., A. Abed-Elmdoust, and R. Kerachian, *A stochastic model for optimal operation of inter-basin water allocation systems: a case study.* Stochastic Environmental Research and Risk Assessment, 2013. **28**(6): p. 1343-1358.

[3]. Ballestero, E., *Inter-Basin Water Transfer Public Agreements: A Decision Approach to Quantity and Price.* Water Resources Management, 2004. **18**(1): p. 75-88.

[4]. Xuesen, L., et al., *Consideration of trends in evaluating inter-basin water transfer alternatives within a fuzzy decision making framework.* Water resources management, 2009. **23**(15): p. 3207-3220.

[5]. Dariane, A.B., M.M. Javadianzadeh, and L.D. James, *Developing an Efficient Auto-Calibration Algorithm for HEC-HMS Program.* Water Resources Management, 2016. **30**(6): p. 1923-1937.

[6]. Zhang, C., et al., *A Negotiation-Based Multi-Objective, Multi-Party Decision-Making Model for Inter-Basin Water Transfer Scheme Optimization.* Water Resources Management, 2012. **26**(14): p. 4029-4038.

[7]. Wang, X., et al., *Bayesian network-based risk assessment for hazmat transportation on the Middle Route of the South-to-North Water Transfer Project in China.* Stochastic Environmental Research and Risk Assessment, 2015: p. 1-17.

[8]. Matete, M. and R. Hassan, *An ecological economics framework for assessing environmental flows: the case of inter-basin water transfers in Lesotho.* Global and Planetary Change, 2005. **47**(2–4): p. 193-200.

[9]. Matete, M. and R. Hassan, *Integrated ecological economics accounting approach to evaluation of inter-basin water transfers: An application to the Lesotho Highlands Water Project.* Ecological Economics, 2006. **60**(1): p. 246-259.

[10]. Li, Y., et al., *Reservoir Operation with Combined Natural Inflow and Controlled Inflow through Interbasin Transfer: Biliu Reservoir in Northeastern China.* Journal of Water Resources Planning and Management, 2015: p. 05015009.

[11]. Tigkas, D., V. Christelis, and G. Tsakiris, *Comparative Study of Evolutionary Algorithms for the Automatic Calibration of the Medbasin-D Conceptual Hydrological Model.* Environmental Processes, 2016: p. 1-16.

[12]. Andersson, J., et al., *Key factors for improving large-scale hydrological model performance.* European Water 2015. **49**: p. 77-88,.

[13]. Zhang, Q., *The South-to-North Water Transfer Project of China: Environmental Implications and Monitoring Strategy1.* JAWRA Journal of the American Water Resources Association, 2009. **45**(5): p. 1238-1247.

[14]. Liu, C. and H. Zheng, *South-to-north water transfer schemes for China.* International Journal of Water Resources Development, 2002. **18**(3): p. 453-471.

[15]. Peng, Y., et al., *Optimization Operation Model Coupled with Improving Water-Transfer Rules and Hedging Rules for Inter-Basin Water Transfer-Supply Systems.* Water Resources Management, 2015. **29**(10): p. 3787-3806.

IOP Conf. Series: Materials Science and Engineering **366** (2018) 012011 doi:10.1088/1757-899X/366/1/012011

Feasibility Study of a PV Hydro Hybrid System, With Photovoltaic Panels on Floating Structures

G Vasco[1,2*], **J S Silva**[1] **and A Beluco**[1]

[1] Departmemnt of Hydrodynamic and Hydrology, Institute of Hydraulic Research (IPH), Federal University of Rio Grande do Sul (UFRGS), Av. Bento Gonçalves, 9500 CEP: 91501-970 - Porto Alegre - RS – Brazil.
[2] Faculty of Environmental Engineering and Natural Resources, Zambeze University (UniZambeze), Chimoio – Mozambique.

* Correspondig author: **G Vasco** - gvasco23@gmail.com

Abstract: This paper intends to contribute, presenting a feasibility study for the implementation of a PV hydro hybrid system in Laranjeiras dam, in southern Brazil, with photovoltaic panels on floating structures. The study was carried out using software HOMER, which exported 8 760 annual values of solar radiation for the city of Três Coroas, and the power to be generated by the hydroelectric plant was determined from 8 760 hourly values of the Canastra turbinated flow, which were made available by CEEE. The study determined that the operation of the hybrid system would be optimal, by installing a hydroelectric plant at the foot of the dam, with 1 497 kW of installed capacity, operating with photovoltaic panels on the dam's water mirror, with a minimum capacity of 180 kW installed, and with 400 kW as a limit to purchase and sale energy to grid, to meet consumer demand up to 40 MWh/d. This combination would result in an initial cost of US $ 3 984 885 / kW and an energy cost of US $ 0.026 / kWh.

1. Introduction

With this serious energy crisis, the vast majority of people in the world are living below the poverty line, and especially in the rural areas [1]. Energy is central to nearly every major challenge and opportunity the world faces today, so, access to energy for all is essential [2].

Nevertheless, energy is the dominant contributor to climate change, accounting for around 60% of total global greenhouse gas emissions [2]. This fact is very worrying because the share of renewable energy in the world energy matrix was not more than 20% by 2015 [3]. That is why global interest and investments in renewable energy sources has increased considerably in recent years. These technologies have the advantage of using locally available resources, reducing dependence on external energy sources. Another good advantage of this energy is to meet rural energy demand, reduce the pollution of the environment, respond to climate change, and promote the development of the local business with small factory [4].

However, most renewable energy technologies suffer from an intermittent characteristic due to the diurnal and seasonal patterns of the natural resources needed for power generation [3]. In this context, hybrid systems based on complementary energy resources can be used as a tool for management of renewable resources [5]. A possible complementarity between energy resources can help reduce the risks of the hybrid system fails during periods of reduced energy availability and may also contribute to the installed capacities are lower. Methodologically, this paper was carried out using software HOMER, aiming to contribute presenting a feasibility study of a PV hydro hybrid system to be installed in Laranjeiras dam, in southern Brazil

2. Simulations with HOMER, the Micropower Optimization Model

HOMER [6] is a software for optimization of hybrid systems on micro and small scale. It was originally developed by U.S National Renewable Energy Laboratory (NREL) and is available for universal access in its Legacy version. HOMER simulates a system for power generation over the time period of 25 years at intervals of 60 minutes, presenting the results for a period of one year [7] [8].

In this paper, PV modules with costs between 4000 US$/kW and 2000 US$/kW were considered as suggested by [9]. The installation of photovoltaic panels on floating structures, as suggested by [10] [11], raise the cost by 30%. The PV system has a useful life of 12.5 years, and was considered 80% of initial cost to the cost of switching the photovoltaic system at the end of the 12.5 years' service life, due to the possibility of recovering 20% of the PV system, with the sale of the system that will be in operation. The annual O&M cost for the photovoltaic system to be adopted will be 5% of the installation cost of the hybrid system considered as an optimal solution, which results in a value of 3200 US$/kW.

The engine room of the hydropower plant represented by the "Hydro" button will be located at the foot of the dam, with a useful life of 25 years. The hydroelectric cost was considered as a sensitivity variable with values of 750.000.00 US$/kW, 900.000.00 US$/kW and 1050.00 US$/kW, as recommended by [1]. In addition, was considered 80% of initial cost to the cost of switching the hydroelectric plant, due to the possibility of recovering 20% of the value of the hydroelectric plant with the sale of the equipment that was in operation. The annual cost of O&M for the hydroelectric plant has a typical value of 4% of the installation cost according [1]. In this context, the power supplied by the grid will obey the table of values determined by [12], with a cost of 0.100 US$/kWh outside peak hours and 0.500 US$/kW at peak hours, while the sale occurs with values of 0.080 US$/kW outside peak hours and 0.250 US$/kW at peak hours.

Figure 1 shows the mean incident solar radiation in a horizontal plane for each month, the deviations around these averages and maximum and minimum values are shown. The maximum insolation occurs in January, while the minimum occurs in June. Moreover, figure 2 shows the hourly variation of the flow rate with an average of 9 171 L/s, where the maximum flow values occur in September and October, while the minimum occurs in March.

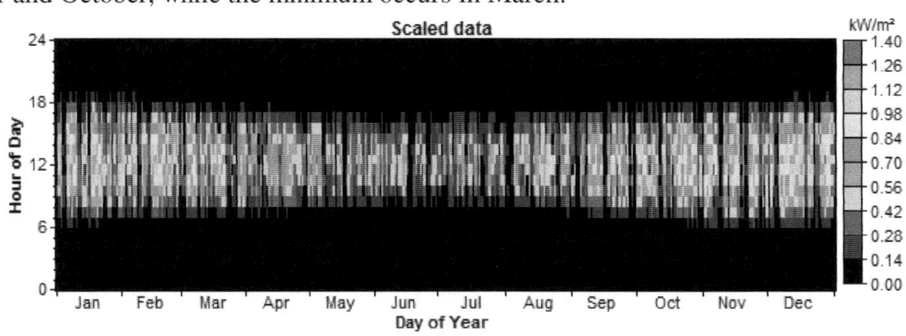

Figure 1. Time values of the solar radiation incident on a horizontal plane, obtained with HOMER.

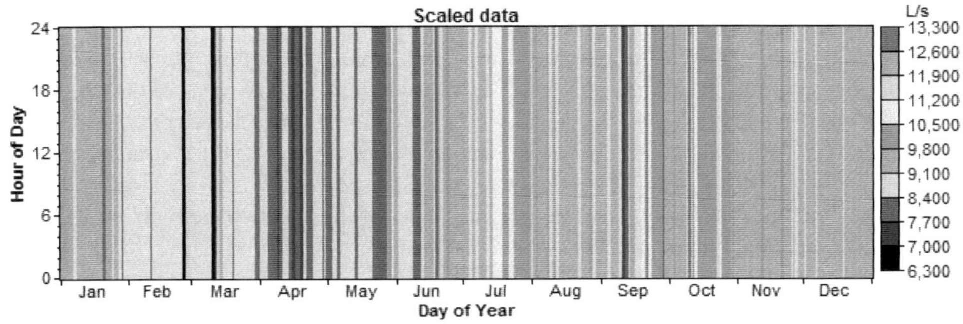

Figure 2. Hourly flow series obtained, through 8 760 hours per year, for the year 2015.

3. Results and Discussion

The simulations results of the proposed hybrid system (figure 3), connected to the grid to sell surplus energy and to buy energy in case of insufficiency, indicate some optimal solutions, as shown in figure 4, which indicates the monthly averages of electricity generation per year. The purchase and sale of energy o the grid in the national interconnected system has been taken into account in all cases, with less dependence on the grid in October, and the participation of photovoltaic solar energy within the system under study is much lower in July, but for the present study it was considered as a solution a system that does not depend on the energy coming from grid.

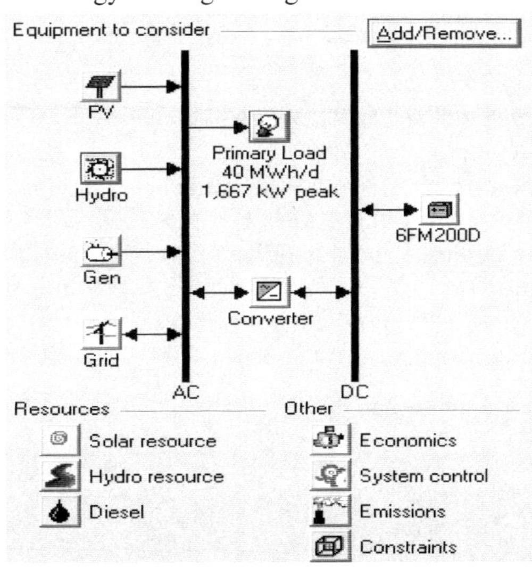

Figure 3. Schematic diagram of the hybrid system proposed in this paper.

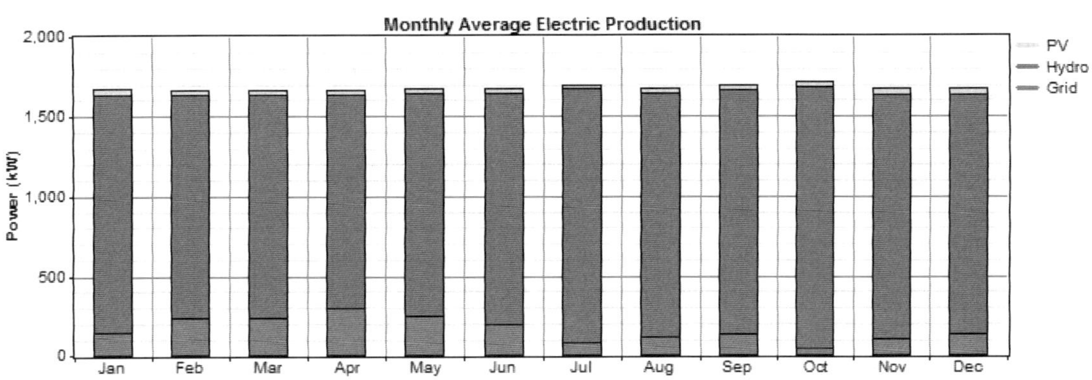

Figure 4. Simulations results, with monthly averages of electricity generation per year.

In case of low incidence of solar radiation and/or low flow rate, the solution to buy energy from the grid should be preferred. On the other hand, in cases of high incidence of solar radiation, multiples of 30 kW for the photovoltaic system become viable options to supply the load, eliminating the option of storage of energy in batteries, which was not recommended in the viable solution presented by HOMER.

Figure 5 shows the optimization space for the initial cost of the hydroelectric plant as a function of the demand of the consuming loads, for the system of Figure 3, with a maximum cost of PVs, acquired for 3.200 US$/kW, with an average flow of 9 171 L/s, and a power limit to purchase and sale to the grid equal to 400 kW to meet consumer loads up to 40 MWh/d. The results showed two solutions: the connection to the grid and hydropower plant, with or without photovoltaic modules.

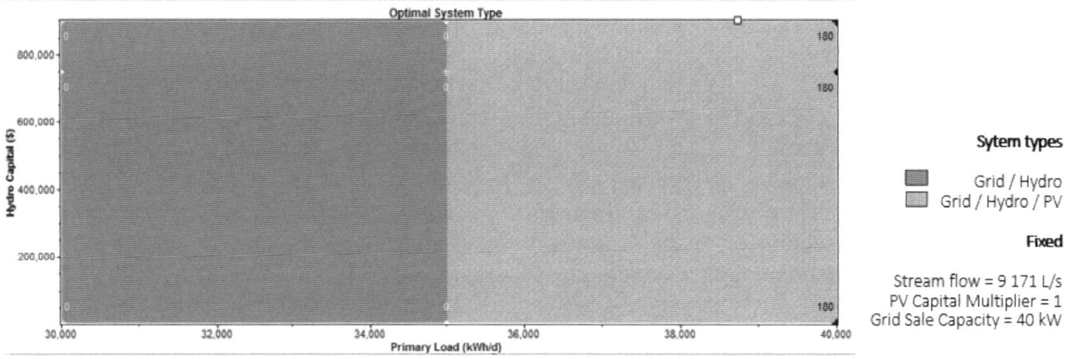

Figure 5. Results for the simulation of the complete system of figure 3 connected from the grid, with maximum cost of PVs

Figure 6 shows the optimization space for the initial cost of the hydroelectric plant as a function of the cost multiplier (0.25) of photovoltaic panels purchased at 3.200 US$/kW, and a flow rate of 9 171 L/s, with a power limit for purchase and sale of the network equal to 400 kW, to avoid failures to meet the demand of the consumer loads. The results indicated two solutions: the connection to the grid and hydropower plant, with or without photovoltaic modules. As shown in Figure 6, lowering to 25% of the cost of PVs increases the number of solutions including PVs with powers up to 540 kW.

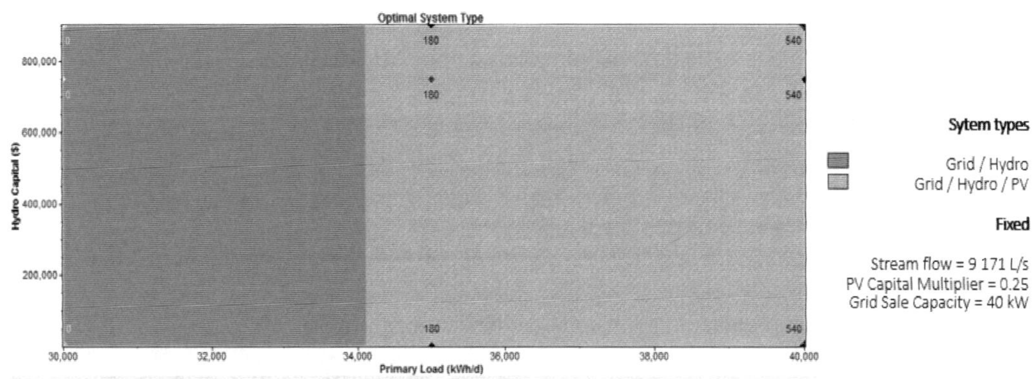

Figure 6. Results for the simulation of the complete system of figure 3 connected from the grid, with PV capital multiplier equal to 0.25

Figure 7 shows the optimization space for the cost of the hydroelectric plant as a function of the cost multiplier (0.15) of photovoltaic panels purchased at 3200 US $/kW, an average flow rate of 9 171 L/s, with 400 kW as a limit to purchase and sale energy to the grid. The results showed all optimization space indicating solutions containing PVs.

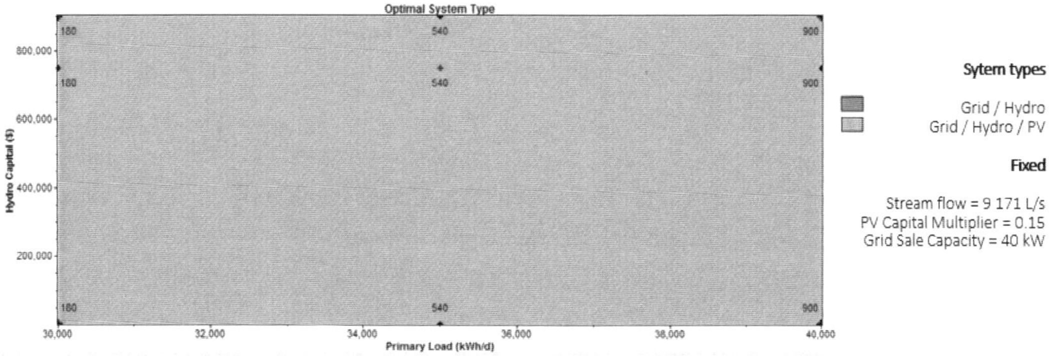

Figure 7. Results for the simulation of the complete system of figure 3 connected from the grid, with PV capital multiplier equal to 0.15

4. Final remarks

This paper presented the main results of the feasibility study, which is the first step in the design of the hybrid system to be implemented in Laranjeiras dam, in southern Brazil to generate energy. The feasible solution, recommended the installation of photovoltaic panels of 180 kW capacity, added to the hydropower with 20 m of height, an average flow of 9 171 L/s, to generate 1 497 kW and with 400 kW as a limit to purchase and sale energy to the grid, to supply stably 40 MWh/d. This combination will results in an initial cost of 3 984 885 US$/kW and an energy cost of 0.026 US$/kWh, which is within the range defined by [13].

Photovoltaic modules allow a greater supply of energy during the daytime period, but these will require overnight power supplies, which can be obtained from the hydropower and/or from the supply obtained in the interconnected national system. As mentioned by [14], these results should serve as a horizon for policy makers to encourage the use of renewable energies, specifically solar energy, with a governmental incentive of 15% and 25% of the total cost of acquisition of the photovoltaic panels.

However, other studies can assess whether the use of reversible reservoirs and historical series from other years can increase these extra supplies, as well as study the possibility of inserting another energy resource to maximize the use of available energy resources in the region.

Acknowledgments

This paper was developed, as a part of advanced research activities on water resources management, renewable energies and environmental sustainability, developed at the Institute of Hydraulic Research (IPH) of Federal University of Rio Grande do Sul (UFRGS) – Brazil. The main author would like to acknowledge the PEC-PG, of CAPES/CNPq/MRE-Brazil, for the scholarship granted to pursue successfully his master degree in water resources and environmental sanitation.

References

[1] IRENA. International Renewable Energy Agency. (2012) Mozambique RRA Preliminary Findings. [Available at: http://sdg.iisd.org/news/irena-publishes-renewables-readiness-assessment-of-mozambique/?rdr=energy-l.iisd.org]. Access: 14 Jul. 2017
[2] UN. United Nations. (2015) Sustainable Development Goals. [Available at: http://www.un.org/sustainabledevelopment/energy/].
[3] CANALES, Fausto A.; BELUCO, Alexandre; MENDES, Carlos André B. Modelling a hydropower plant with reservoir with the micro power optimization model (HOMER). *International Journal of Sustainable Energy*, v. 36, 1-14p. 2015
[4] HAMMAR, L.; EHNBERG, J.; MAVUME, A.; BOAVENTURA, Cuamba, C. and MOLANDER, S. (2012). Renewable Ocean Energy in the Western Indian Ocean. *Renewable and Sustainable Energy Reviews*, 16, 4938-4950.

[5] BELUCO, Alexandre; DE SOUSA, Paulo Kroeff; LIVI, Flávio Pohlman; CAUX, Johann. Energetic Complementarity with Hydropower and the Possibility of Storage in Batteries and Water Reservoirs. In: SORENSEN, Bent. Solar Energy Storage. 1. ed. Amsterdam, Netherlands: Elsevier, 2015, 155-188pp.

[6] Homer Energy, L. L. C. (2014) Software HOMER, Version 2.68 Beta; the Micropower Optimization Model. [Available at: www.homerenergy.com]

[7] LILIENTHAL, P. D.; LAMBERT, T. W.; and GILMAN, P. (2004) Computer Modeling of Renewable Power Systems. In: Cleveland, C. J., Ed., *Encyclopedia of Energy*, Elsevier, Amsterdam, 633-647. [DOI: http://dx.doi.org/10.1016/b0-12-176480-x/00522-2].

[8] LAMBERT, T. W., GILMAN, P. and LILIENTHAL, P. D. (2005) Micropower System Modeling with Homer. In: Farret, F. A. and Simões, M. G., Eds., *Integration of Alternative Sources of Energy*, John Wiley & Sons, Hoboken, 379-418. [DOI: http://dx.doi.org/10.1002/0471755621.ch15].

[9] TEIXEIRA, L. E.; CAUX, J.; BELUCO, A.; BERTOLDO, I.; LOUZADA, J. A. S.; EIFLER, R. C. Feasibility Study of a Hydro PV Hybrid System Operating at a Dam for Water Supply in Southern Brazil. *Journal of Power and Energy Engineering*, n.3, 2015. 70-83pp. DOI: http://dx.doi.org/10.4236/jpee.2015.39006

[10] GISBERT, C. M. F.; GONZÁLVEZ, J. J. F.; SANTAFÉ, M. R.; GISBERT, P. S. F.; ROMERO, F. J. S. and SOLER, J. B. T. (2013) A New Photovoltaic Floating Cover System for Water Reservoirs. *Renewable Energy*, 60, 63-70. [DOI: http://dx.doi.org/10.1016/j.renene.2013.04.007]

[11] SANTAFÉ, M. R.; GSBERT, P. S. F.; ROMERO, F. J. S.; SOLER, J. B. T., GONZÁLVEZ, J. J. F. and GISBERT, C. M. F. (2014) Implementation of a Photovoltaic Floating Cover for Irrigation Reservoirs. *Journal of Cleaner Production*, 66, 568-570. [DOI: http://dx.doi.org/10.1016/j.jclepro.2013.11.006]

[12] ANEEL. Agência Nacional de Energia Elétrica. (2008) Atlas de energia elétrica do Brasil, 3ª Ed. ANEEL, Brasília. [Available at: http://www.aneel.gov.br/arquivos/PDF/atlas3ed.pdf].

[13] IRENA. International Renewable Energy Agency. (2012) Renewable energy technologies: cost analysis series - Hydropower. [Available at: <RENEWABLE ENERGY TECHNOLOGIES: COST ANALYSIS SERIES>].

[14] SILVA, Jones S.; CARDOSO, A. R.; BELUCO, Alexandre. (2012) Consequences of Reducing the Cost of PV Modules on a PV Wind Diesel Hybrid System with Limited Sizing Components. *International Journal of Photoenergy*, v. 2012, 1-7pp. [DOI: http://dx.doi.org/10.1155/2012/384153]

Prediction Techniques for Accelerating the Convergence of Alternating Solution Method during the Power System Dynamic Simulation Using SDC Integration Algorithm

Hanyang Dai*, Xinli Song, Tao Liu, Zhida Suand Xia Li

State Key Laboratory of Power Grid Safety and Energy Conservation (China Electric Power Research Institute), Beijing 100192, China.

*840519184@qq.com

Abstract. The electric power of generator P_e and bus voltage of network V have a strong nonlinear coupling relationship with other variables, which results in bad effects on the convergence procedure for the alternating solution method during the power system dynamic simulation using the SDC integration algorithm. Deep research was conducted into this problem. The convergence procedure for P_e and V was illustrated during the simulation, and prediction techniques by applying the Lagrange interpolation to construct a better iterative initial value with smaller error for them was proposed. Based on the SDC simulation method, the proposed method was tested on an IEEE standard 3-generator 9-bus system on the platform of MATLAB, and results show that the prediction techniques performs quite well for the acceleration of convergence procedure for P_e and $\boldsymbol{V}$ and simulation efficiency of SDC integration algorithm.

1. INTRODUCTION

The Spectral Deferred Correction method, also known as SDC for short, is a kind of new integration algorithm with high precision proposed by Alok Dutt, Leslie Greengard, and Vladimir Rokhlin in 2000[1]. The algorithm first conducts an interpolation in one integration step using the Gussian polynomials and integrations between adjacent interpolation nodes with the alternating solution method(ASM)[2] by applying the low-order integration algorithms to make sure of the numerical stability, and then implements a spectral integration using the spectral integration matrix[3] which is a high-order integration scheme for all the interpolation nodes to correct the integration results of the low-order integration, as a result, the final integration converges with high precision by the iterative correction[4]. Theoretical research claims that the SDC integration algorithm has an arbitrary convergence precision with the capability of large simulation step when solving the ordinary differential equations and it's been applied widely to the simulations of various practical engineering problems[1], such as the ordinary differential equations as well as the differential algebraic equations.

Our research group applied the SDC integration algorithm in the simulation of the power system dynamics, and obtained a promising performance with good numerical stability and high precision of simulation capability. But we also revealed a problem during the integration of ASM for each interpolation interval that the electric power of generator P_e and bus voltage of network $\boldsymbol{V}$ affect the convergence procedure of the ASM due to the strong nonlinear coupling relations with other variables. Therefore, researches about improving the convergence procedure for ASM should be conducted.

Literature research has summarized properties for the network algebraic equations in the ASM and reported optimized linear solvers, such as UMFPACK[5], MKL, SuperLU[6], SuperNodal, ILU[7], and Krylov methods, to increase the simulation speed of ASM with outstanding performances. But research on how to provide better iterative initial values for the strong nonlinear coupling variables is not related. This paper illustrates the convergence procedure for P_e and $\boldsymbol{V}$ during ASM, and then constructs prediction techniques for them based on the Lagrange interpolation, aiming to accelerate the convergence procedure of ASM and to improve the efficiency of SDC integration algorithm.

2. DYNAMIC SIMULATION WITH SDC INTEGRATION ALGORITHM

The mathematical model of the power system dynamic simulation is set of differential and algebraic equations (DAE) with the differential equations describing the dynamics of the equipment and the algebraic ones indicating the network constraints, as showed in (1) with x, y, t represent algebraic variables, differential variables, and the simulation time[8].

$$\begin{cases} \frac{dy}{dt} = \boldsymbol{f}(t, \boldsymbol{x}, \boldsymbol{y}) \\ \boldsymbol{g}(\boldsymbol{x}, \boldsymbol{y}) = \boldsymbol{0} \end{cases} \quad (1)$$

The dynamic simulation of power system dynamic simulation is applying numeric integration algorithms to solve the DAE equations.

2.1. *SDC integration algorithm*

Taking the trapezoidal rule as the low-order integration algorithm, we implemented the following differential scheme to solve the differential equations in ASM while applying the SDC integration algorithm to simulate the dynamic of power system.

$$\widetilde{\boldsymbol{y}}_{i+1}^{k+1} = \widetilde{\boldsymbol{y}}_{i}^{k+1} + \frac{1}{2}\Delta t_i\left[\boldsymbol{f}\left(t_i, \widetilde{y}_i^{k+1}\right) + \boldsymbol{f}\left(t_{i+1}, \widetilde{\boldsymbol{y}}_{i+1}^{k+1}\right)\right] + \boldsymbol{I}_i^{i+1}\left(\widetilde{\boldsymbol{y}}^k\right) - \frac{1}{2}\Delta t_i\left[\boldsymbol{f}\left(t_i, \widetilde{y}_i^{k}\right) + \boldsymbol{f}\left(t_{i+1}, \widetilde{\boldsymbol{y}}_{i+1}^{k}\right)\right] \quad (2)$$

where $i = 0,1,2 \ldots \ldots, n-1$ with n represents the interpolation nodes dividing the simulation step into n intervals, t_i is the time for node i, $\Delta t_i = t_{i+1} - t_i$, k is the iterative counts, and $\widetilde{y}$ is the numerical calculation of $\boldsymbol{y}$. $I_i^{i+1}\left(\widetilde{\boldsymbol{y}}^k\right)$ is calculated by the history value though the spectral integration[1].

It's clear that there is an additional high-order correction item compared to the traditional differential scheme, making (2) being an integration scheme with high-order truncation error.

2.2. *Iterative procedure for ASM*

2.2.1. *Nonlinearity*

The ASM integrates the interval Δt_i by solving the differential equations one by one using (2) while fixing $\boldsymbol{x}$ variables unchanged, and the solving the network algebraic equations while fixing $\boldsymbol{y}$ variables unchanged. Iterations are required until all the variables converge to the desired precision. The iteration is actually a kind of fixed point iteration with super-linear convergence [9]. But there are variables converge slower than others due to the strong nonlinear coupling as following.

1) Electric power of generator P_e. As the most important linkage between the electrical dynamics and mechanical dynamics, P_e is coupled with various variables in the generator system. It's decided by the bus voltage $\boldsymbol{V}$ and input current $\boldsymbol{I}$, and then fed back to rotor motion.

2) Bus voltage of network V. Transformation between d-q coordinates and x-y coordinates is required during the generator interface equation, because the bus voltage in generator is represented in d-q coordinates while the network algebraic equations need them in x-y. The transformation matrix is constructed by trigonometric function of power angle δ, making V variables strong nonlinear coupled.

2.2.2. *Influence on convergence*

Here we present the 3-generator 9-bus system as shown in figure 1 to illustrate the influence of the nonlinear coupling on the procedure of convergence for ASM.

The mathematical model for the generators in the system is 6-order detailed sub-transient model. A three-phase ground fault occurs to the transmission line between B4 and B6 with a ground impedance of 0.1+0.1j at t=1.0s and disappears at t=1.1s. Execute the simulation on the platform of MATLAB using the SDC integration algorithm with termination time of 20s.

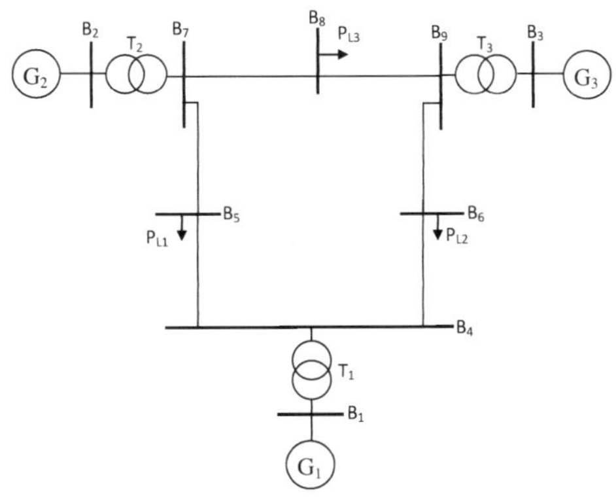

Figure 1. Structure of 3-gen 9-bus system

Figure 2 shows the dynamic response of electric power for G1. And figure 3 illustrates the convergence procedure for different variables with err-dw, err-cur, and err-pacc representing power angle δ, $\boldsymbol{V}$, and P_e, respectively. We can see that δ converges to the prescribed error 10^{-10} quite well at the 4th iteration, while P_e and $\boldsymbol{V}$ converges at the 6th iteration due to a larger iterative initial error.

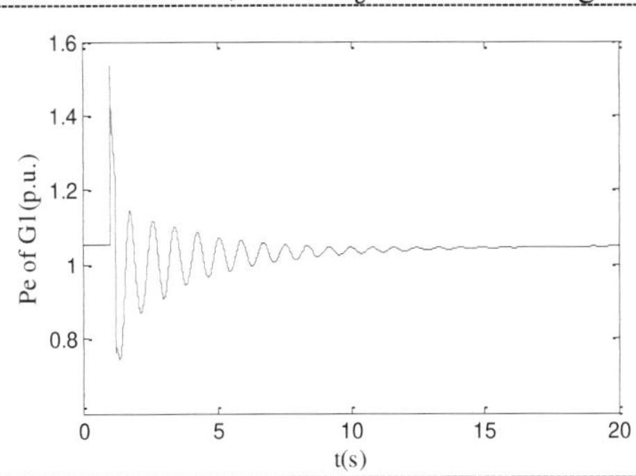

Figure 2. Electrical power P_e of G1

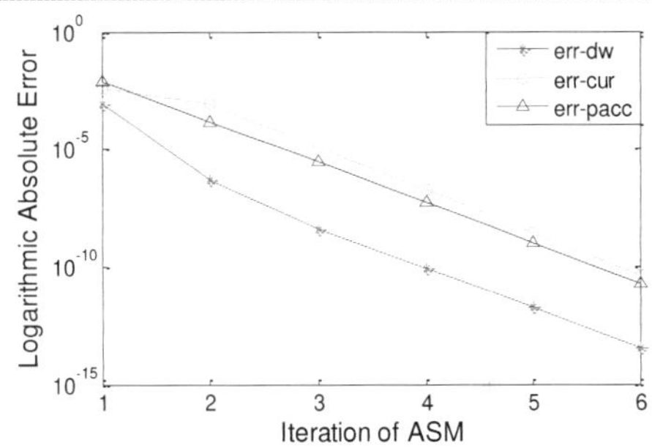

Figure 3. Convergent procedure for different variables

Seeing that all the variables have a nearly the same error decaying rate, we can accelerate the convergence procedure by constructing a better iterative initial value for them.

3. PREDICTION TECHNIQUES

The iterative initial value adopts the results at the previous step during the traditional simulation, and the error becomes larger under a larger step size. Here we consider the prediction techniques to construct a better iterative initial value using the Lagrange interpolation[10] for P_e and $\boldsymbol{V}$.

3.1. Lagrange interpolation prediction

Assuming function $y(x)$ satisfy the Lipschitz continuity condition, we can construct an interpolation polynomial $\mathcal{L}(x)$ by $n+1$ interpolation nodes

$$\mathcal{L}(x) := \sum_{j=0}^{n}\left[\ell_j(x) * y(x_j)\right] \tag{3}$$

Where $\ell_0(x), \ell_1(x), \ldots, \ell_n(x)$ are the basis for each nodes $x_0, x_1, \ldots, x_n$ defined by

$$\ell_j(x) = \prod_{0\le k\le n, k\ne j}\frac{x-x_k}{x_j-x_k} \tag{4}$$

Then (4) has a truncation error of order n.

Generally, P_e and $\boldsymbol{V}$ satisfy the Lipschitz continuity condition except when a fault occurs, therefore, we can construct the Lagrange interpolations for them.

3.2. Prediction for P_e

Let the previous k step value noted as $\left(t_{n-k+1}, P_e(t_{n-k+1})\right),\ldots,(t_n, P_e(t_n))$ where t_j represents the time for each node, $(n-k+1) \leq j \leq n$, the Lagrange basis is defined by

$$\ell_j(t) = \prod_{0\leq m\leq k-1, m\neq j} \frac{t-t_{n-m}}{t_j-t_{n-m}} \tag{5}$$

and the interpolation polynomial for P_e is

$$\mathcal{L}^{P_e}(t) := \sum_{j=(n-k+1)}^{n} \left[\ell_j(t) * P_e(t_j)\right] \tag{6}$$

therefore, the prediction value for P_e at t_{n+1} is

$$P_e^0(t_{n+1}) = \mathcal{L}^{P_e}(t_{n+1}) \tag{7}$$

Equation (7) provides a better iterative initial value for P_e during the practical simulation.

Extract the first 4 steps after fault in figure 2 to test the prediction technique by comparing the predicted value P_e^0 with the exact value $P_e(t_{n+1})$ mark by red '*', in figure 3. Set the following 2 kinds of simulation scenarios:

Lagrange-2: let *k=2* in (5), namely adopt 2 historical nodes to construct the interpolation, mark the prediction result with '+' in figure 3.

Lagrange-3: let *k=3* in (5), namely adopt 3 historical nodes to construct the interpolation, mark the prediction result with green 'o' in figure 3.

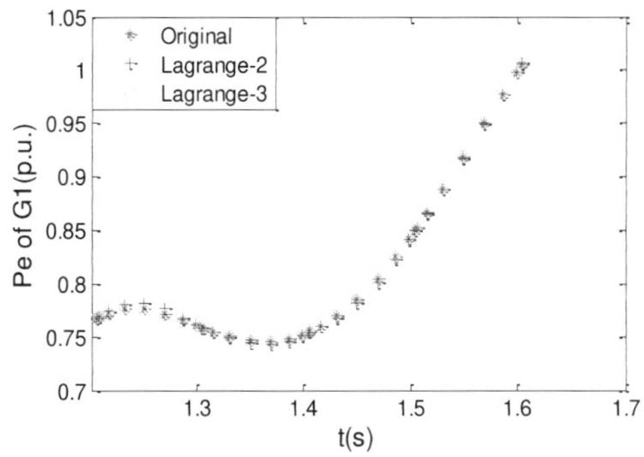

Figure 4. Prediction for P_e with Lagrange Quadrature

We have the following observation from figure 4.

(1) The initial error is large when P_e changes rapidly for the case of adopting the value of previous step as the iterative initial value (noted as original strategy).

(2) The prediction technique of Lagrange-2 strategy doesn't fit well and even worse than original strategy when the P_e curve has a strong nonlinearity.

(3) The prediction technique of Lagrange-3 strategy has an excellent performance for all steps because it's actually a quadratic fitting with a higher truncation error of order 2, ensuring the prediction of a relatively high precision.

3.3. Prediction for V

Apply the similar procedure to construct the interpolation prediction for bus voltage V_x, V_y, the corresponding interpolation polynomials are

$$\mathcal{L}^{V_x}(t) := \sum_{j=(n-k+1)}^{n} \left[\ell_j(t) * V_x(t_j)\right] \tag{8}$$

$$\mathcal{L}^{V_y}(t) := \sum_{j=(n-k+1)}^{n} \left[\ell_j(t) * V_y(t_j)\right] \tag{9}$$

And the prediction values for V_x, V_y at t_{n+1} becomes

$$V_x^0(t_{n+1}) = \mathcal{L}^{V_x}(t_{n+1}) \tag{10}$$

$$V_y^0(t_{n+1}) = \mathcal{L}^{V_y}(t_{n+1}) \tag{11}$$

Following the configuration and testing method in section 2.2, the prediction curves for V_x, V_y are illustrated in figure 5, separately.

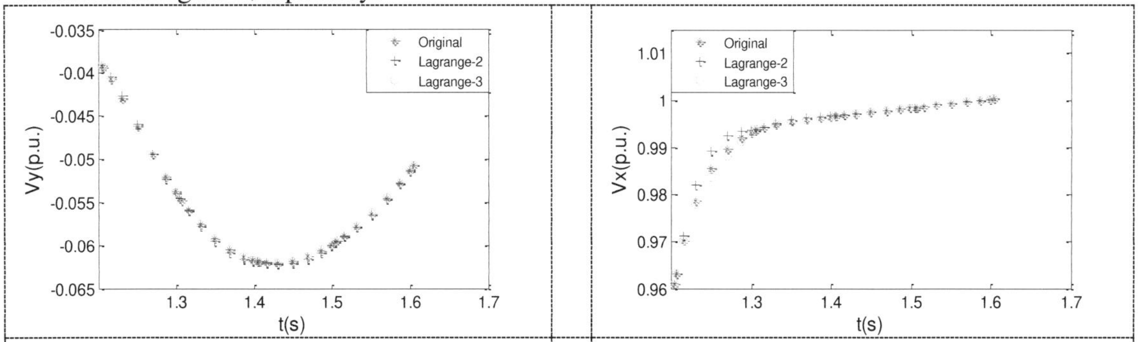

(a) Component V_y (b) Component V_x

Figure 5. Prediction for voltage $\boldsymbol{V}$ with Lagrange Quadrature

We have the following observation from figure 5.

(1) The initial error is large when V_x and V_y change rapidly for the case of adopting the value of previous step as the iterative initial value (noted as original strategy).

(2) The prediction technique of Lagrange-2 strategy doesn't fit well and even worse than original strategy when V_x or V_y curve has a strong nonlinearity.

(3) The prediction technique of Lagrange-3 strategy has an excellent performance for all steps because it's actually a quadratic fitting with a higher truncation error of order 2, ensuring the prediction of a relatively high precision. It's very clear in figure 5(b) that Lagrange-3 has a much better performance than Lagrange-2.

4. NUMERICAL RESULTS

Applying the constructed iterative initial values for P_e and $\boldsymbol{V}$, we can make a research on the acceleration for the ASM procedure in this section.

Based on figure 3, figure 6 illustrates the convergence procedure for different variables noted as '-pred' after applying the prediction techniques. We can see that the error decaying rate for different variables remains the same, but the initial error during the iteration becomes smaller. As a result, all the variables converge to the prescribed error of 10^{-10} at the 5th iteration, making the ASM iteration terminates ahead of time.

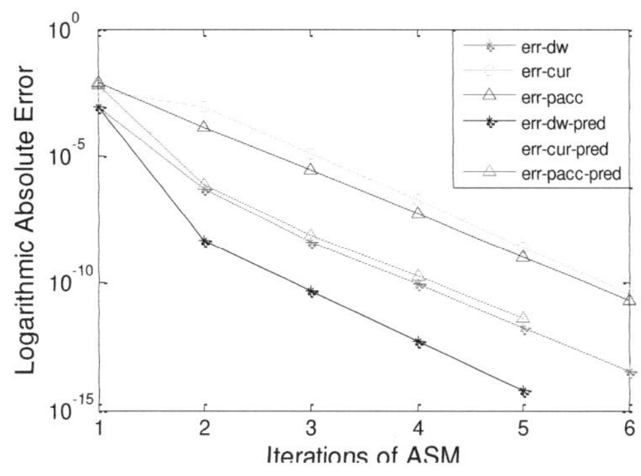

Figure 6. Convergent procedure for different variables after prediction

The total iterations before prediction for each SDC simulation step in the first 4 steps are also listed in table 1. As a comparison, the iterations after prediction are also listed where 'P' and 'V' represents applying the prediction for and ***V***, separately, and 'P+V' represents applying the prediction for and ***V*** simultaneously.

The data suggests that:

(1) the prediction for P_e alone doesn't have any obvious affection to the iteration of ASM;

(2) the prediction for V_x, V_y are very effective that iterations has been decreased by 73 times for the 4 SDC steps which means 22.8% increase for the ASM efficiency.

(3) iterations have further decreased when applying both prediction techniques, and the increased efficiency for ASM is raised up to 25.4%.

Table 1. Number of iterations of ASM in one SDC step before prediction and after prediction

SDC step	Before Prediction	After prediction		
		P	**V**	**V+P**
1	65	65	61	61
2	84	84	66	64
3	84	84	56	53
4	80	80	63	59
Summation	316	315	243	235

As a result, we can conclude that the proposed prediction techniques for the strong nonlinear coupled variables P_e and ***V*** are very effective, it decreases the iterations for ASM and accelerates the simulation efficiency of SDC integration algorithm.

5. CONCLUSION

The paper conducts a research into the simulation efficiency of ASM iteration for SDC integration algorithm. During the ASM iteration, P_e and ***V*** have relatively larger iterative initial errors than others because of their strong nonlinear coupling with other variables, which makes ASM iteration a slow procedure. The paper illustrates the convergence procedure for different variables during simulation for the standard 3-generator and 9-bus system using SDC integration algorithm, and then proposes prediction techniques to construct better iterative initial values with smaller error for P_e and ***V*** based on the Lagrange interpolation. Numerical results show that the prediction techniques can accelerate the simulation efficiency effectively by decreasing the iterations of ASM. The proposed method has practical applications and can also be applied to other integration algorithms when the ASM is adopted.

References

[1] Dutt A, Greengard L, Rokhlin V. Spectral Deferred Correction Methods for Ordinary Differential Equations. Bit Numerical Mathematics, 2000, 40(2):241-266. differential equations." BIT Numerical Mathematics 40.2 (2000): 241-266.

[2] Xinli, Song, et al. "New mixed integral algorithm for unified dynamic power system simulations of transient, medium-term and long-term stabilities." Power Systems Conference and Exposition (PSCE), 2011 IEEE/PES. IEEE, 2011.

[3] Gottlieb, David, and Steven A. Orszag. Numerical analysis of spectral methods: theory and applications. Society for Industrial and Applied Mathematics, 1977.

[4] Jingfang, Huang, Jia, Jun, and Minion Michael. "Accelerating the convergence of spectral deferred correction methods." Journal of Computational Physics 214.2 (2006): 633-656.

[5] Khaitan, Siddhartha Kumar, James D. McCalley, and Qiming Chen. "Multifrontal solver for online power system time-domain simulation." IEEE Transactions on Power Systems 23.4 (2008): 1727-1737.

[6] Li, Xiaoye S., and James W. Demmel. "SuperLU_DIST: A scalable distributed-memory sparse direct solver for unsymmetric linear systems." ACM Transactions on Mathematical Software (TOMS) 29.2 (2003): 110-140.

[7] Chuan Fu. High-speed extended-term time-domain simulation for online cascading analysis of power system. Iowa State University, 2011.
[8] P. Kundur, N. J. Balu, and M. G. Lauby, Power system stability and control. Vol. 7. New York: McGraw-hill, 1994.
[9] Machowski J, Bialek J W, Bumby J R. Power System Dynamics. Stability and Control. West Sussex, UK:John Wiley, 2006.
[10] Crow, Mariesa L. Computational methods for electric power systems. CRC Press, 2015.

Dynamic islanding strategy regarding sustainable on-load capacity of photovoltaic/energy-storage hybrid system in microgrid

Bo-Da Chen[1,a] **and Yong-Jun Zhang**[1]

[1] South China University of Technology, 381 Wushan Road, Tianhe District, Guangzhou, P.R.China

[a] E-mail of corresponding author: 919046909@qq.com

Abstract. To measure sustainable loading capacity of photovoltaic/energy-storage system, indexes for sustainable loading power margin and energy margin are defined. With this basis, a dynamic islanding strategy for distribution system with photovoltaic/energy-storage system is proposed. The reliability assessment algorithm of the distribution network is improved based on Monte Carlo Simulation, by considering the fluctuation of PV output and load demand. Finally, it conducts a reliability assessment for modified RBTS Bus6, and results of simulation show that the proposed strategy can ensure continuous and stable operation of important loads within Micro-Grid, and enhance the reliability of distribution network.

1. Introduction

In order to make distributed generation (DG) play a better role in its power supply, standard IEEE1547.4-2011 encourages conscious islanding operation. It is pointed out that planning islanding should take dynamic characteristics, switching strategy of micro-power supply and load into consideration [1].

Currently, there are lots of researches focuses on the division strategy of islanding within the microgrid globally. It is proposed an off-island planning based on DG capacity, location and network structure in [2, 3], but it is not always applicable in the microgrid with strong power output and load demand volatility. Wang Y considered importance of load in microgrid, classified islanding by pre-determined load reduction strategy under island mode [4]. It is proposed a multi-objective islanding optimization model for distribution network with potential power supply of microgrid in [5].

However, above literatures are based on a certain moment when distribution network fails, takes output power and load status of DG as a certain amount, with this benchmark to divide static island search, ignores the dynamic random changes in power supply and demand in the island operation. Their static islanding strategy is hard to guarantee sustainable and stable supply of important loads during the fault period. Therefore, the reliability assessment based on the expected operation status of islanding strategy needs to be discussed.

In this paper, it firstly proposes concept and evaluation index of sustainable on-load capacity of photovoltaic/energy-storage hybrid system. Secondly, dynamic islanding model of micro-grid is proposed with considering sustainable on-load capacity of photovoltaic/energy-storage hybrid system, which makes up the lack of static islanding. Finally, it conducts a reliability assessment for distribution network with different islanding strategy under microgrid, by using reliability evaluation algorithm that takes account of the PV output and the load demand volatility.

2. Component model

2.1. Probabilistic Model of Load
In order to work out timing of PV output, this paper adopts the photovoltaic model of Reference [6], which can reflect characteristics of timing and randomness of PV output under different weather types in the target area.

2.2. Probabilistic Model of PV
Distribution network load has time-varying and randomness. Therefore, result of load forecasting is actually a random variable, and normal distribution is usually used to approximate reflect load uncertainty. In this paper, load stochastic model in Reference [7] is used.

2.3. Storage Energy Constraint Model
With considering charging and discharging efficiency of storage energy, the maximum output power of storage energy and the capacity limitation, the charging and discharging process should meet the constraints mentioned in Reference [8].

3. Sustainable on-load capacity of photovoltaic/energy-storage hybrid system

3.1. Photovoltaic reliable power output
The uncertainty of PV system output makes people only pay attention to their power value. In fact, PV system has some capacity value simultaneously. In this paper, concept of photovoltaic credible output is proposed. The credible output ($P_{\alpha}(t)|_{\Delta t}$) of photovoltaic in Δt period is the least output level that PV can reach with certain probability (confidence) α in Δt period after time t. For example, if α = 90%, Δt = 15min, so photovoltaic credible output power is $P_{90\%}(t)|_{15\text{min}}$, which means the least PV output level can reach with 90% confidence in 15min is $P_{90\%}(t)|_{15\text{min}}$.

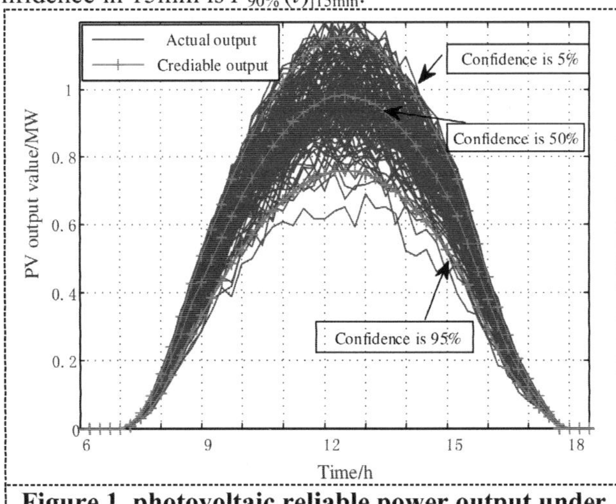

Figure 1. photovoltaic reliable power output under different confidential level

During actual operation, a certain prediction error will be taken into consideration in the PV prediction to determine the PV's credible output. This paper uses above PV model, to get the PV credibility curve under different weather types through random sampling simulation method. Specific calculation steps are as follows:
(1) Establish a PV power generation timing model under different weather types.

(2) Generate Ne daily timing samples of PV output by sampling.
(3) Work out credible curve of PV power output by calculating Credible value for PV power output ($P_\alpha(t)|_{\Delta t}$) in different times of the day under different weather types. The schematic diagram of the PV credible output under different confidence levels is shown in Figure 1, among them, the sampling interval on the curve is 15 minutes, therefore, $P_\alpha(t)|_{\Delta t}$ can be calculated by the following formula:

$$P_\alpha(t)|_{\Delta t} = \left\{ P_\beta \mid \mathrm{P_r}\left[\underset{i\in N_e}{P_i}(t:t+\Delta t) > P_\beta \right] \geq \alpha \right\} \tag{1}$$

In the formula, $\mathrm{P_r}[\cdot]$ indicates the probability value that the inequality constraint holds.

Please note in the following, the credible output of PV at a certain moment all refers to the confidence output when Δt is 15mins, so that $P_\alpha(t)|_{15\text{min}}$ is abbreviated as $P_\alpha(t)$.

3.2. Assessment Indicators

In order to evaluate the continuous power supply capability of load under distribution network with optical storage in the fault condition, this paper presents evaluation index of the sustainable load capacity of optical storage system. Among them, sustainable load capacity of optical storage system means: the capability that optical storage system can provide uninterrupted power supply to load on island within a certain period, which includes both sustainable power supply and energy supply, details are as followed:

(1) Continuous loading power margin (M_p): the load supply margin that optical storage system can achieve when power balance in isolated island can be met during the fault period, which is calculated as follows:

$$M_\mathrm{p}(t) = \min\left(1 - \frac{\sum_{i\in\mathrm{K}}(1+\beta)P_{Li}(\tau)}{\sum_{i\in\mathrm{K}}(P_{\alpha i}(\tau) + P_i^{\mathrm{dmax}})}\right) \times 100\% \quad \tau \in [t, t_{\mathrm{end}}] \tag{2}$$

In which, P_{Li} is the predicted demand power of the ith load, β is load forecast error, $P_{\alpha i}$ is the credible output in the ith PV under the confidence α, P_i^{dmax} is the maximum discharge power of the ith energy storage, and t_{end} is predicted fault end moment, while K is the range of micro-grid islanding.

(2) Continuously loading energy margin (M_w): the load supply margin that optical storage system can achieve when energy balance in isolated island can be satisfied during the fault period, the formula is as follows:

$$\begin{cases} A(t) = E_{\mathrm{re}} - \delta\int_t^{t_{\mathrm{end}}} \sum_{i\in K}((1+\beta)P_{Li}(\tau) - P_{\alpha i}(\tau))d\tau \\ M_\mathrm{w}(t) = \left(\frac{A(t)}{\sum_{i\in\mathrm{K}} E_{\min i}} - 1\right) \times 100\%, \quad \tau \in [t, t_{\mathrm{end}}] \end{cases} \tag{3}$$

In which, $A(t)$ is the remaining storage capacity after $[t, t_{\text{end}}]$ period under the loading range. $E_{\min i}$ is the minimum allowable remaining capacity of the ith energy storage. E_{re} is the sum of the initial capacity of each energy storage device, δ is the charge-discharge efficiency.

$M_\mathrm{p}(t)$ and $M_\mathrm{w}(t)$ reflect the continuous loading capacity of island optical storage system in $[t, t_{\text{end}}]$ period from the perspective of power balance and energy adequacy respectively. The larger the value is, the larger the load increased margin brought by optical storage system will be, otherwise will be smaller.

3.3. Dynamic islanding strategy

To ensure the sustained and stable power supply to important users, this paper proposes a dynamic islanding strategy, which including the strategy of optimization and reconstruction on a long time scale and the strategy of load-shedding on a short time scale. The optimization and reconstruction strategy is to work out islanding partitioning model and optimize islanding partitioning plan with a certain time interval T_z, while load-shedding strategy refers to priority to remove the less important load in island real-time operation, if the active power imbalance exceeds a certain limit. The proposed strategy uses evaluation index of sustainable load capacity to assess current increased margin for optical storage system, and to incorporate more load into island under the condition that island can run stably and continuously. Its mathematics model is as below:

$$\max L = \max \sum_{i \in K} \lambda_i L_i$$
$$\text{s.t.} \begin{cases} M_{\mathrm{P}}(t) \geq 0 \\ M_{\mathrm{W}}(t) \geq 0 \\ \text{Meet the energy storage constraints mentioned in Section 2.3} \end{cases} \tag{4}$$

In which, if the value of L_i is 1 or 0, it means whether the load point i is included in the island operation plan load set K. λ_i is load importance index, the greater the value is, the more important it is, the more priority to incorporated into K.

4. Distribution network Reliability Assessment Considering Load and Power Fluctuation

To cover the volatility and randomness of PV and load demand during islanding operation in the reliability assessment, this paper improves the reliability assessment algorithm based on Monte Carlo. The specific algorithm flow and method are based on the traditional reliability Evaluation method [4]. For island load operation, the reliability assessment should base on actual operation situation, now only presents assessment process of island operation load as follows:

(1) Island data initialization: setting predict fault repair time as T, confidence level of PV credible output as α, then work out PV credible output curve and load forecast curve within [t_{occ}, t_{end}] period, t_{occ} is the moment when breakdown occur, T=t_{end} - t_{occ}. Setting dynamical division time interval as T_z. Set n as1, n means the nth dynamic partition island during breakdown period. K_n is the nth island load set, initial K_0 is empty set.

(2) Islanding scheme Determination: According to the PV credible output curve and load forecasting curve, the islanding partition model is used to determine the islanding scheme load set K_n.

(3) Simulation operation: use PV and load model to work out PV output and load demand curve with considering stochastic fluctuation, and then simulate islanding operation under constraint according to islanding scheme load set K_n. If there is insufficient power supply, power balance will be satisfied by cutting load point with less importance, thereby to accumulate times and duration of power outage, and insufficient amount of power supply at each load point during the process, and then to update K_n.

Judge whether the time nT_z is greater than or equal to the repair time T, if yes, then re-sampling the faulty component to continue this year's simulation, if not, then make n=n+1, t=t_{occ}+nT_z and then return to step 2.

5. Case Study

5.1. Example parameters

This paper selects modified RBTS Bus6 F4 feeder system as the simulation example as shown in Figure 2.This article assumes that the order of importance of the loads in the microgrid is 8, 7, 6...2 and 1, that is to say, the priority order of load access silos is LP19>LP11>LP20>LP12>LP21>LP13> LP22>LP23. Component reliability parameters refers to Reference [9].The installation capacity of photovoltaic is 2MW, the rated capacity of energy storage is 2MW·h, the rated power of energy storage is 0.5MW.

The load data in references [10,11] are used as the annual demand forecast curve. The load forecasting curve is used as the average load. The load model described in Section 1.2 is used, where the variance of load fluctuations is taken as 8% of the load average at the corresponding time. The prediction error of the load demand forecast β in the evaluation model of sustainable on-load capacity of photovoltaic energy-storage hybrid system is set to be 3σ.

PV Output Data are generated randomly according to the type of weather selected in section 1.1. The credible output curves under different weather types are obtained as described in Section 2.1 in this paper, where the confidence $\alpha = 95\%$.

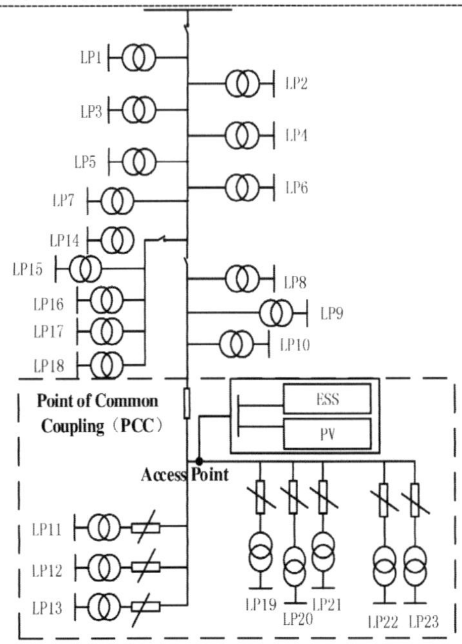

Figure 2. Modified RBTS Bus6 F4 feeder system

5.2. *Comparison of different islanding strategies*

This paper evaluates the reliability of the following three islanding strategies for microgrids.

- Strategy 1: Do not consider islanding in microgrids.
- Strategy 2: Consider the operation of islanding and use the islanding method of literature [4], that is, only use static power balance as the basis for dynamic islanding.
- Strategy 3: Consider islanding and use the dynamic islanding strategy regarding sustainable on-load capacity of photovoltaic energy-storage hybrid system, T_z =1h.

Table 1. Reliability indices of distribution system

Reliability indices	Strategy 1	Strategy 2	Strategy 3
SAIFI	1.0382	0.9512	0.8541
SAIDI	8.3296	7.4241	7.4238
ASAI	0.99904	0.99915	0.99916
EENS	56.5767	53.1402	54.4198
SAIFI of microgrid	1.7897	1.4622	1.0961
Energy depletion times	/	7856	0
Number of secondary faults	/	41599	17

Table 1 shows the results of reliability evaluation of distribution system. The failure rate and the average power outage time of typical load point under three kinds of strategies are shown in Figure 3. Among them, LP9 and LP10 is non-island load while others are microgrid loads.

Below conclusions are from table 1 and figure 3:

(1) Strategy 3 has better reliability than Strategy 2. This is mainly because islanding strategy 2 based on static power balance does not take into account the uncertainty of the PV and load during the fault,

resulting in repeated resection and incorporation of load points and depletion of stored energy, which increases the failure rate of the microgrid load.

(2) As shown in Figure 3, Strategy 3 can reduce the failure rate and average power failure time of important loads to a greater extent. In the process of islanding, the number of energy storage exhaustion is 0 and the number of secondary faults also much lower than strategy 2.

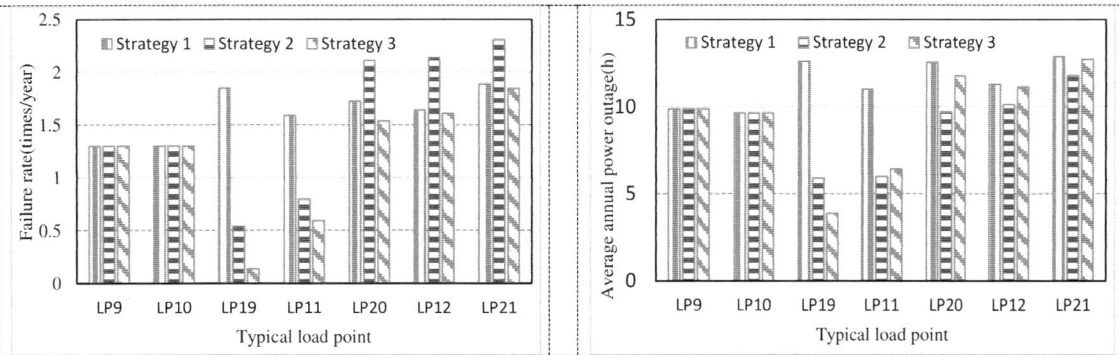

(a) Failure rate (b) Average power outage time

Figure 3. Index comparison of typical load point

6. Conclusion

In this paper, it considers the uncertainty of PV output and load demand in the micro-grid, and proposes a dynamic islanding model based on the sustainable on-load capacity of photovoltaic energy-storage hybrid system. Simulation results show that the proposed strategy can take uncertainty of PV output and load into account during the breakdown period and ensure that the important loads in the microgrid will maintain stable power supply and improve overall power supply reliability of the system.

Acknowledgment

This work was supported by Natural Science Foundation of Guangdong Province, China (2017A030313304).

References

[1] IEEE Std 1547.4-2011 2011 *S. The Institute of Electrical and Electronics Engineers*

[2] Barsali S, Ceraolo M, and Pelacchi P 2002 *C. Power Engineering Society Winter Meeting* **2** 789-794

[3] Marei M I, El-Saadany E F, and Salama M M A 2004 *J. IEEE Transactions on Power Delivery* **19(3)** 1384-92.

[4] Wang Y, Wan L Y and Hu B 2014 *J. Power System Technology* **38(9)** 2379-85.

[5] Zhang X X, Niu H N and Zhao J X 2016 *J. Electric Power Automation Equipment* **36(11)** 51-58.

[6] Li Z K, Zhou W J and Wang J M 2016 *J. Automation of Electric Power Systems* **40(14)** 58-64.

[7] Sui B Y, Hou K and Jia H J 2016 *J. Power System Technology* **2016(43)** 3696-3750.

[8] LI C 2015 *D. China Electric Power Research Institute.*

[9] A Allan R N, Billinton R and Sjarief I 1991 *J. IEEE Transactions on Power Systems* **6(2)** 813-820.

[10] Billinton R and Jonnavithula S 2002 *J. IEEE Transactions on Power Systems* **11(4)** 1670-76.

[11] Alkuhayli A A, Raghavan S and Chowdhury B H 2012 *C. North American Power Symposium* **2012** 1-6.

IOP Conf. Series: Materials Science and Engineering **366** (2018) 012014 doi:10.1088/1757-899X/366/1/012014

Photovoltaic Single-Phase Grid-Connected Inverter Based on Voltage and Reactive Power Support

Jin Huang[1,a], Ziqiang Wu[1], Zhihao Dong[1] and Yang Zhu[1]

[1] School of Automation, Wuhan University of Technology, Wuhan, China

[a] Corresponding author's e-mail: 15827183459@163.com

Abstract. In this paper, a kind of PV grid-connected inverter suitable for low voltage ride through is proposed. In order to alleviate the voltage drop at the power grid access point during the fault, the photovoltaic inverter needs to provide a certain amount of reactive power support to the power grid. The orthogonal quantities in the time domain are transformed into direct quantities by the dq rotating orthogonal coordinate system, so that decoupling control of active power and reactive power can be achieved. In order to achieve the smooth switching of the working mode, the parameter design of the PID controller needs to be analyzed according to the controlled object model. The simulation results show that the photovoltaic grid-connected inverter under the decoupling control can realize the reactive power support to the voltage drop at the grid-connected point in time.

1. Introduction

In recent years, the increase of the high-power impact load continuously threatens the stability of the grid voltage. The latest "Photovoltaic Power Station Access Power System Technical Regulations": Large and medium-sized grid-connected PV systems must have a certain low voltage ride through to ensure that the equipment can operate safely and stably for a period of time after the grid falls [1].

When voltage instability occurs at the PV inverter interconnection point, the grid-connected system needs to provide a certain amount of inductive or capacitive reactive power to prevent the collapse of the local grid voltage[2]. Since reactive power can not be transmitted over long distances, a static reactive power compensation device (SVC) is often used to make up for the reactive power shortage [3]. However, these devices undoubtedly increase the system's cost and complexity. In this paper, the energy transfer of photovoltaic system under normal conditions and the reactive power regulation of grid-connected inverter under fault condition are realized by vector decomposition of rotating coordinate system.

2. Reactive power compensation

GB/29321-2012"Photovoltaic Plant Reactive Power Compensation Technical Specification" stipulates that the photovoltaic power station should make full use of grid-connected inverter reactive power capacity and its regulatory capacity to configure adequate dynamic reactive power compensation capacity. So as to ensure the photovoltaic inverter in the specified node voltage range safe and reliable, continuous and stable operation.

According to the domestic industry on the specific requirements of reactive power voltage support, single-phase grid inverter reactive power compensation is usually determined by the grid voltage drop depth [4]. As shown in Figure 1.

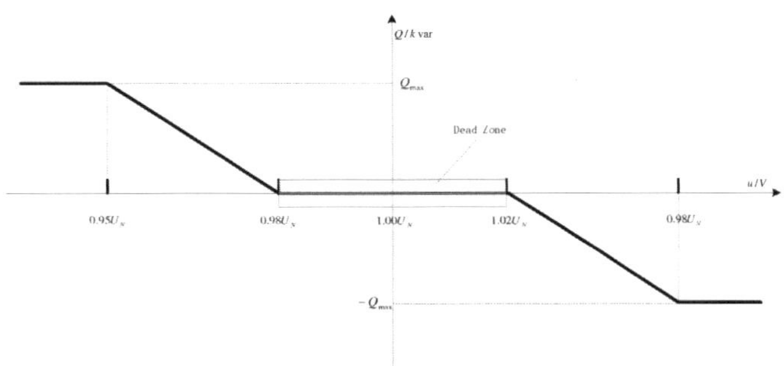

Figure 1. reactive compensation value

In the current working state, the controller uses the sensor to collect and process the voltage of the grid connection point in real time. When the busbar access point voltage is in the range of $0.98U_N\sim1.02U_N$ dead zone, the PV inverter works in constant power factor control mode without reactive power compensation. When busbar access point voltage is $0.95U_N\sim0.98U_N$ or when $1.02U_N\sim1.05U_N$, Calculated as follows:

$$Q=\begin{cases} Q_{max} & ,(u\le 0.95U_N) \\ Q_{max}+\frac{\partial Q}{\partial U}\left(u-U_{ref(low)}\right) & ,(0.95U_N<u<0.98U_N) \\ 0 & ,(0.98U_N<u<1.02U_N) \\ -Q_{max}+\frac{\partial Q}{\partial U}\left(u-U_{ref(high)}\right) & ,(1.02U_N<u<1.05U_N) \\ -Q_{max} & ,(u\ge 1.05U_N) \end{cases} \tag{1}$$

3. Photovoltaic grid system

3.1 Design of Photovoltaic Grid Connection Inverter

The system equivalent model of the single-phase grid-connected inverter is shown in the figure 2[7]. The DC voltage is modulated by a full-bridge circuit into a set of unequal-width pulse signals U_H. After the reactor filters out some of the high-frequency harmonics, it is connected to the power grid through an isolated step-up transformer. Where L includes the inductance value of the filter inductance and the leakage inductance of the isolation transformer. R is determined by the inductor internal resistance, the transformer copper loss, and the line impedance loss.

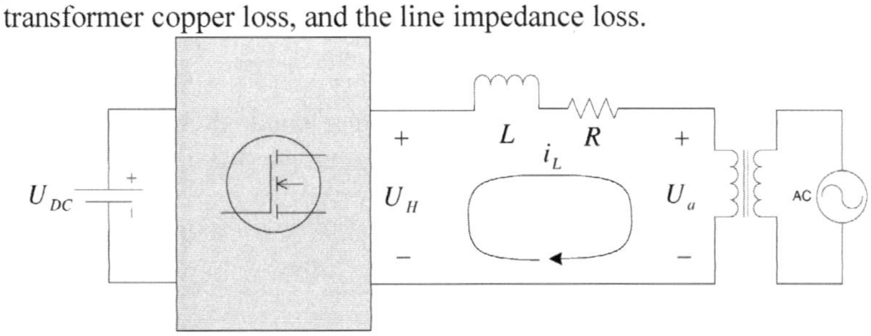

Figure 2. Single-phase grid-connected inverter system equivalent model

In the above system equivalent model, the time-domain equation of the loop current in the stationary coordinate system is listed according to Kirchhoff's voltage theorem as follows:

$$V_H=Ri_L+L\frac{di_L}{dt}+V_a \tag{2}$$

Where V_H is the single-phase full bridge output voltage, i_L is the inductor loop current, and V_a is the equivalent grid voltage coupled to the primary side of the isolation transformer. For computational convenience, these variables are transformed from a stationary two-phase orthogonal coordinate system αβ to a rotating orthogonal coordinate system dq using a Park transformation matrix. The principle of transformation is to produce the same total combined current [8]. As shown in the following figure 2.

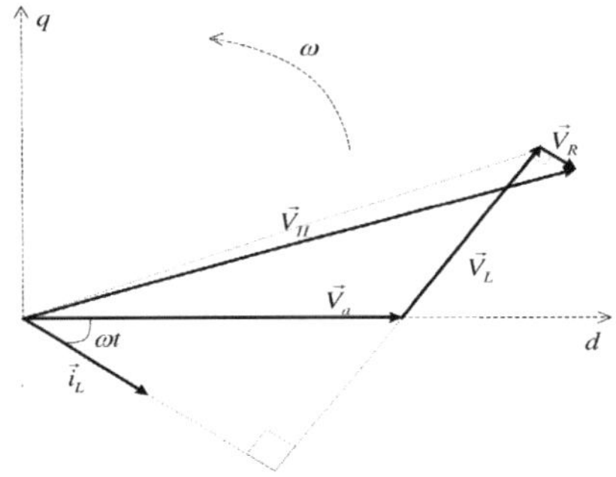

Figure 3. rotating coordinate system of the voltage and current vector diagram

The voltage and current vectors in the dq coordinate system are plotted in Figure 3. The vector components of each voltage and current in the rotating coordinate system can be obtained by decomposing the vectors along the coordinate system rotated by the angular frequency ω of the industrial frequency. Calculated as follows:

$$V_H = \begin{bmatrix} \cos(\omega t) & -\sin(\omega t) \\ \sin(\omega t) & \cos(\omega t) \end{bmatrix} \begin{bmatrix} V_{H(d)} \\ V_{H(q)} \end{bmatrix} \tag{3}$$

$$i_L = \begin{bmatrix} \cos(\omega t) & -\sin(\omega t) \\ \sin(\omega t) & \cos(\omega t) \end{bmatrix} \begin{bmatrix} i_{L(d)} \\ i_{L(q)} \end{bmatrix} \tag{4}$$

$$V_a = \begin{bmatrix} \cos(\omega t) & -\sin(\omega t) \\ \sin(\omega t) & \cos(\omega t) \end{bmatrix} \begin{bmatrix} V_{a(d)} \\ 0 \end{bmatrix} \tag{5}$$

Where $V_{H(d)}$ is the component of V_H on the d axis and $V_{H(q)}$ is the component of V_H on the q-axis. $i_{L(d)}$ is the component of i_L on the d axis, and $i_{L(q)}$ is the component of i_L on the q-axis . $V_{a(d)}$ is the component of V_a on the d-axis. Considering that the direction of V_a vector is consistent with the d-axis, its component on the q-axis is zero.

The above formula (3) (4) (5) into the formula (2), available:

$$V_{H(d)} = Ri_{L(d)} + L\frac{di_{L(d)}}{dt} - \omega L i_{L(q)} + V_{a(d)}$$
$$V_{H(q)} = Ri_{L(q)} + L\frac{di_{L(q)}}{dt} + \omega L i_{L(d)} \tag{6}$$

When the circuit is in constant power mode, by adjusting the single-phase full-bridge voltage component $V_{H(q)}$, you can change the size of the equivalent inductor current $i_{L(q)}$ in the same direction with the grid current to control the PV inverter on the ac side Transmission of active voltage of grid voltage. When the circuit is in constant reactive power mode, by adjusting the inverter bridge voltage component $V_{H(d)}$, you can change the size of the equivalent inductor current $i_{L(d)}$ component perpendicular to the direction of the grid voltage to control the photovoltaic inverter absorption or send reactive power.

3.2 PID control in a rotating coordinate system

The above equation (7) (8) is normalized. The vector equation of the controlled object in the rotating two-phase orthogonal coordinate system is as follows:

$$\vec{V}_{H(s)} = (R + j\omega L)\vec{i}_{L(s)} + \vec{V}_{a(s)} \tag{7}$$

By deducting (9) above, the transfer function of PV inverter in the frequency domain is as follows:

$$G_{H(s)} = \frac{\vec{V}_{H(s)} - \vec{V}_{a(s)}}{\vec{i}_{L(s)}} = \frac{1}{R+LS} \tag{8}$$

IOP Conf. Series: Materials Science and Engineering **366** (2018) 012014 doi:10.1088/1757-899X/366/1/012014

PWM pulse width modulator can be seen as a high-gain proportion of links. Often the delay in software operation leads to a delay in its response [9]. The delay time T_{pwm} is related to the switching period, which is typically on the order of 0.01ms. The transfer function of the PWM pulse width modulator is as follows:

$$G_{pwm} = \frac{K_{pwm}}{1+sT_{pwm}} \tag{9}$$

K_{pwm} is the amplification factor of PWM pulse width modulator in the formula (11), $K_{pwm} = 220$ in this example. T_{pwm} is the delay time of the PWM pulse width modulator. Since the switching frequency is 20kHz in this project, $T_{pwm} = 5 \times 10^{-5}$. Photovoltaic single-phase grid inverter closed-loop control diagram is shown in the Figure 4:

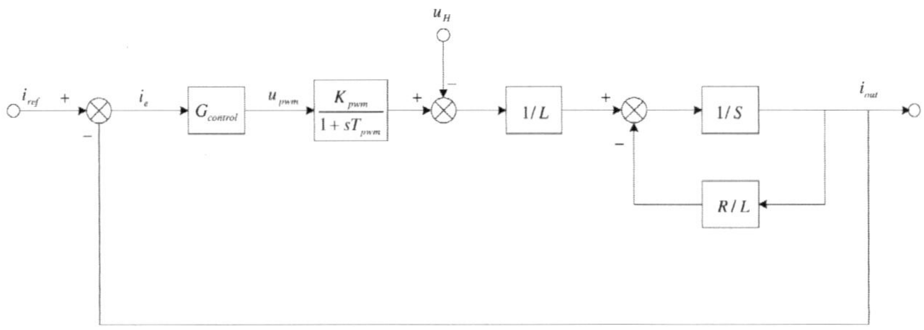

Figure 4. Photovoltaic single-phase inverter with closed-loop control block diagram

As can be seen from Figure 5 above, the open-loop transfer function of the control loop is:

$$\begin{aligned} G_{op} &= G_{control}(s)\, G_{pwm}(s)\, G_H(s) \\ &= G_{control}(s) \frac{K_{pwm}}{1+sT_{pwm}} \frac{1/R}{1+(L/R)S} \end{aligned} \tag{10}$$

When designing a closed-loop controller, the dynamic performance of the system should be improved as much as possible to reduce the overshoot and adjustment time during the follow-up process. Typical Type I systems operating in underdamped conditions greatly reduce the risk of single-phase inverter currents exceeding allowable values during dynamic response, ensuring safe and reliable operation of the device [10]. The transfer function of the current controller is as follows:

$$G_{control}(s) = K_p + \frac{K_i}{s} = \frac{K_p(\tau s+1)}{\tau s} \tag{11}$$

Where $\tau = K_p/K_i$, K_p is a proportional constant, K_i is an integral constant. Use the $(\tau s + 1)$ in the regulator to eliminate the inertial component $[1 + (L/R)S]$ of the larger time constant in the controlled object. The corrected open-loop transfer function of the system is as follows [11]:

$$\begin{aligned} G_{op} &= \frac{K_p(\tau s+1)}{\tau s} \frac{K_{pwm}}{1+sT_{pwm}} \frac{1/R}{1+(L/R)S} \\ &= \frac{K}{s(1+Ts)} \end{aligned} \tag{12}$$

Where $T = T_{pwm}$, $K = (K_i K_{pwm})/R$, $\tau = K_p/K_i = L/R$. The logarithmic frequency characteristics of the open loop transfer function is shown in the Figure 5:

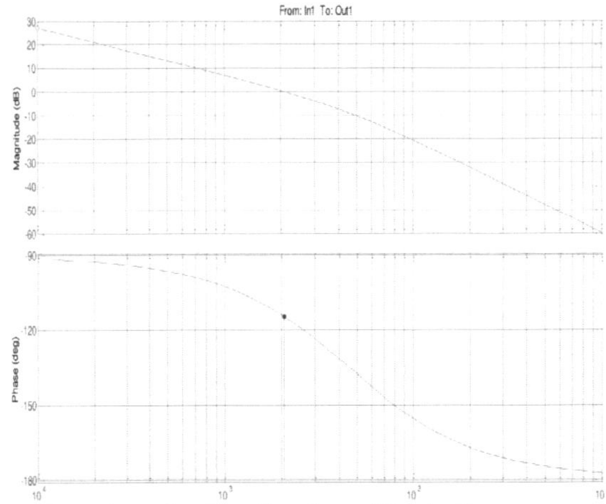

Figure 5. open-loop transfer function logarithmic frequency characteristics

In the log amplitude frequency characteristic, the intersection of the curve and the horizontal axis is the cutoff frequency. For a typical I-type system, the greater the cut-off frequency, the better the dynamic performance, but the smaller the phase margin is, the poorer the stability of the system. According to the system dynamic performance index, using Siemens "best tuning" method. The parameters are as follows:

$$\xi = 0.707,\ KT = 0.5 \tag{13}$$

From the above formula (14) (15), calculated:

$$K_p = 0.036,\ K_i = 2.27 \tag{14}$$

4. Simulation analysis

In this paper, according to the principle of photovoltaic single-phase grid, it is proposed to design a 6kw inverter to convert 380V DC bus voltage into 220V AC power. In order to optimize the power distribution for optimal purposes, a more flexible inverter grid control strategy is obtained. Specific design indicators are shown in the Table 1:

Table 1. photovoltaic grid inverter design parameters

Input voltage	380V±10V
output voltage	220V
Output voltage ripple	<3%
Maximum output power	6kw
Switching frequency	20kHz

Using Matlab / Simulink to build single phase inverter simulation model [12]. As shown in FIG. 6, the measurement module collects the instantaneous value of the grid voltage and current in real time, and the phase angle of the grid voltage is calculated by the PLL (Phase Locked Loop). The feedback signal is compared with the given target current value after passing through the dq coordinate transformation matrix . The PI controller is used to realize the zero-static error correction of the controlled object. Finally, the anti-dq coordinate transformation matrix is used to obtain the PWM pulse-width modulated fundamental signal.

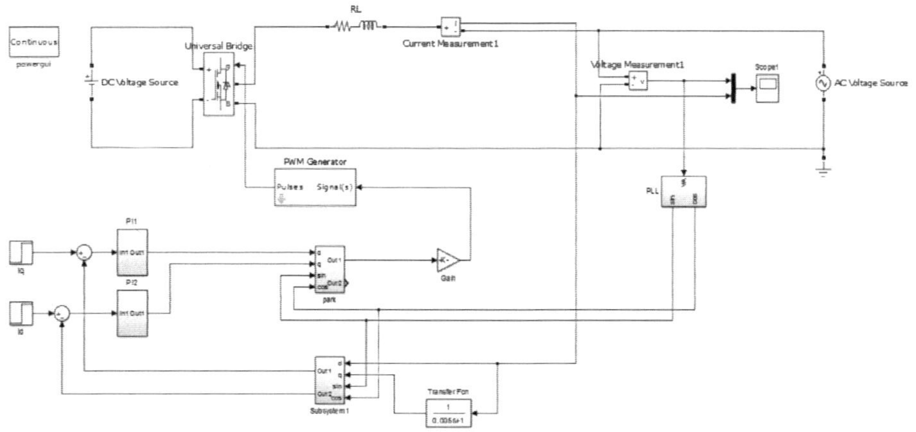

Figure 6. Photovoltaic grid-connected inverter simulation model

In Figure 6, the use of univeral bridge module to achieve full-bridge main circuit functions. The simulation time is 0.1s. By adjusting the values of the i_d and i_q step signal modules, the effective values of the active and reactive power current components are changed. Scope oscilloscope output waveform as shown in Figure 8 below, the horizontal axis is the time axis, the blue waveform represents the grid voltage, the red waveform represents the filter inductor current. The target value of the active current component is 36A. The simulation result on the oscilloscope shows that the effective value of the inductor current is 35.4A.

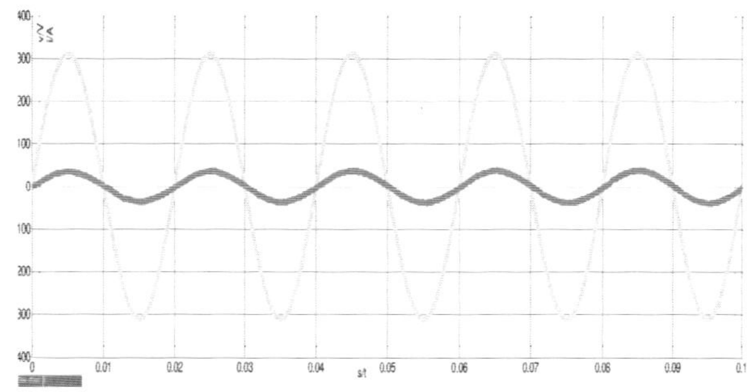

(A)

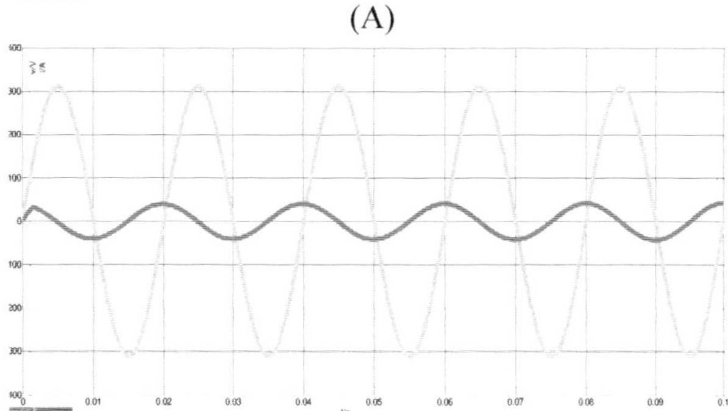

(B)

Figure 7. **(A)** Inductance current and grid voltage waveform in constant power mode
(B) Inductance current and grid voltage waveform in constant reactive power mode

As can be seen from the simulation results in FIG. 7A, the filter inductor current and the grid voltage are in the same direction to achieve constant power mode operation. From FIG. 7B, it can be seen that the current phase of the filter inductor leads the grid voltage by 90 degrees, and the PV inverter operates at constant in reactive power mode.

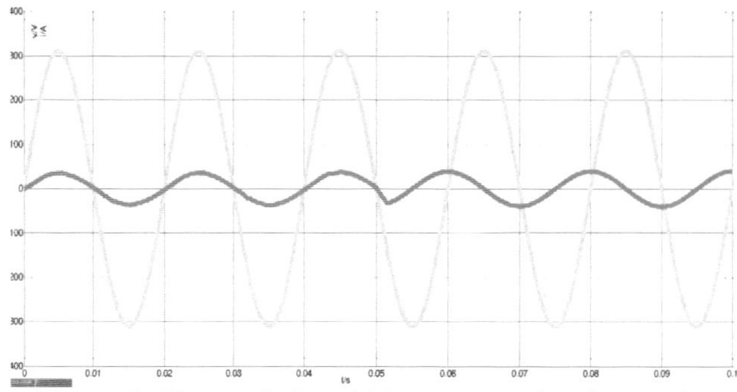

Figure 8. Photovoltaic grid inverter mode of operation

In the simulation, the jump time of the step signal generator module is used to simulate the controller response when the grid fault occurs. The switching time is 0.05s. It can be seen from the simulation results of the oscilloscope in Figure 8 that the fundamental current value of the filter inductor can reach the steady state value in time and the working state of the photovoltaic inverter is switched from the constant power mode to the constant reactive power mode. K_p, K_i parameter design rationality.

5. Conclusion

In the field of photovoltaic in new energy sources, this paper draws the following conclusions for photovoltaic grid-connected inverters: (1) The single-phase grid-connected photovoltaic inverter based on reactive voltage support can realize the low voltage crossing of grid-connected equipment and can to alleviate the voltage drop at the grid-connected point under the fault condition. (2) The PI control in the rotating coordinate system can realize the decoupling control of grid-connected power quickly and stably. The simulation model built by Matlab / Simulink proves that the system has good dynamic and static performance.

References

[1] ZHANG Haiyu, LIU Chuang, CHAO Qin, et al. Research on relay protection issues of grid-connected photovoltaic system with LVRT ability[J]. Power System Protection and Control, 2015, 43(3): 53-60.

[2] K. &C. Mamandur, R. D. Chenoweth. Optimal Control of Reactive Power Flow for Improvements in Voltage Profiles and for Real Power Loss Minimization[J], IEEE Trans On PAS, 1981, 100(7): 3185～3194.

[3] Harmonics Characteristic Parameter, Methods of Study ,Estimates of Existing Values in Network ,WG 36-O,Electra No.77,1982.

[4] Antonios Marinopoulos, Fabio Papandrea, Muhamad Reza, et al. Grid Integration Aspects of Large Solar PV Installation: LVRT Capability and Reactive Power/ Voltage Support [A].IEEE Trondheim Power Tech [c]. 2011:1-8

[5] S. B. Kjaer, J. K. Pedersen, and F. Blaabjerg, "A review of single-phase grid-connected inverters for photovoltaic modules," IEEE Trans. Ind. Appl., vol. 41, no. 5, pp. 1292–1306, Sep./Oct. 2005.

[6] J. P. Benner and L. Kazmerski, "Photovoltaics gaining greater visibility," IEEE Spectr., vol. 29, no. 9, pp. 34–42, Sep. 1999.

[7] ZHANG Chenghui, YE Ying, CHEN Allan, et a1. Research on grid-connected photovoltaic inverter based on output current control [J]. Transactions of China Electrotechnical Society, 2007, 22(8): 41- 45.

[8] ROWE C N, SUMMERS T J, BETZ R E, et al. Implementing the virtual output impedance concept in a three phase system utilising cascaded pi controllers in the dq rotating reference

frame for microgrid inverter control[C]//15th European Conference on Power Electronics and Applications(EPE). Lille, France, 2013: 1～10.
[9] A. Capel, J. Jalade, J. C. Marpinard, and M. Valentin, "Large signal dynamic stability analysis of synchronized current controlled modulators. Application to sine-wave high power inverters," in 1982 IEEE PESC Rec., pp. 101–109.
[10] Zames, G. Feedback and optimal sensitivity: model reference transformations, multiplicative seminorms and approximate Inverse. IEEE Trans. Automat. Contr. 198l, V01. AC-26: 301-320
[11] M. N. El-Gamal and G. W. Roberts, "Very high-frequency log-domain bandpass filters," IEEE Trans. On Circuits and Systems II, vol. 45, no. 9, pp. 1188-1198, Sept. 1998.
[12] Henriksson D, Cervin A, Aren K E. TrueTime: Real-time control system simulation with Matlab/Simulink [A].Proceedings of the Nordic Matlab conference [C]. Copenhagen Denmark, 2003.

Short-Term Local Prediction of Wind Power Based on Singular Spectrum Analysis and Self-Organizing Maps

T Y Ji[1], L Liu[1], D Y Hong[1], M S Li[1] and Q H Wu[1]

[1] School of Electric Power Engineering, South China University of Technology, Guangzhou, 510640, China

Corresponding author: Dr. M. S. Li. Email: mengshili@scut.edu.cn.

Abstract. Along with the increasing penetration of wind power into power systems, more accurate forecast of wind power becomes more and more important for real-time scheduling and operation. This paper proposes a novel model for short-term wind power forecast based on singular spectrum analysis (SSA) and self-organizing maps (SOM). In order to deal with the impact of high volatility of the original time series, SSA is utilized to extract the mean trend from the original time series. After that, SOM is applied to select the similar segments from mean trend, which are then employed in local prediction by support vector regression (SVR). Simulation studies are conducted on real wind power time series, and the final results indicate that the proposed model is more accurate and stable than other models.

1. Introduction

In recent years, wind energy as a non-pollution, promising type of energy has been the fastest growing renewable energy technology [1]. Obviously, with the increasing popularity of wind power inputting into power systems, the intermittency and randomness of wind have brought great challenge to the safe and stable operation of power systems [2]. Therefore, more accurate and stable forecast is required to guarantee the stability and security of power system operation without increasing the operating cost [3].

Many research works have been focusing on improving the accuracy of wind power forecast, and several forecast models have been proposed, which can be approximately classified into two categories: physical models and statistical models [4]. Physical models are usually referred to as meteorological prediction of wind power, which is related to the numerical approximation of the models that describe the state of atmosphere [5]. Unlike physical models, statistical models are established by discovering the relationship of historical data, whose advantages for wind power forecast are the simplicity of model construction, reduced computational requirement, and better performance than physical models in short-term forecast [6]. Therefore, in order to provide a more accurate guidance for power system real-time operation, a short-term forecast model belonging to the statistical models is proposed in this paper.

In order to gain more accurate forecast results, it is crucial to reduce the impact of strong volatility of wind power. To achieve this, a number of methods have been applied to extract the mean trend from the original time series. Empirical mode decomposition (EMD) [7], which is a typical approach among them, decomposes the original time series into multiple intrinsic mode functions (IMFs) and considers the first IMF or the first several IMFs as noise [8]. However, this practice is based on experience without theoretical guidance. In this paper, SSA, which is a powerful technique for nonlinear time series analysis [9] [10], is applied to extract the mean trend from the original time series. The main advantage of SSA compared to other methods is that each component obtained by SSA has its own contribution rate, which makes the decomposition more meaningful.

According to the previous research, local prediction outperforms global prediction [11]. Thus, the SOM [12] [13] is used to select the similar segments of the forecast segment. The SOM is an unsupervised competitive learning algorithm which has been generally applied to vector quantization, data visualization tasks and clustering, thus, it is adopted in this research.

In this paper, a novel forecast model based on SSA and SOM for short-term wind power prediction is proposed. SSA is utilized to extract the mean trend from the original time series. After that, SOM is used to select the similar segments from the mean trend to participate in local prediction. Finally, the similar segments are trained by SVR [14].

2. Methodologies

2.1. Singular Spectrum Analysis (SSA)

SSA is an effective way to study the meaningful features of nonlinear time series. It extracts the mean trend from the original time series. The standard SSA is detailed in the following steps.

Step 1: Embedding. Given an original time series $\mathbf{y} = [y_1\ y_2\ \cdots y_N]$ of length N and a parameter L as the embedding dimension of SSA, the original time series $\mathbf{y}$ is transformed to L-lagged vectors $\mathbf{x}_1, \mathbf{x}_2 \cdots \mathbf{x}_K, K = N - L + 1$, and $\mathbf{x}_i = [y_i\ y_{i+1}\ \cdots\ y_{i+L-1}]^{\mathrm{T}}, i = 1, 2, \cdots, K$ is the expression of each L-lagged. The trajectory matrix $\mathbf{x} \in \mathbb{R}^{L\times K}$ is expressed as

$$\mathbf{X} = [\mathbf{x}_1\ \mathbf{x}_2\ \cdots\ \mathbf{x}_K] = \begin{bmatrix} y_1 & y_2 & \cdots & y_K \\ y_2 & y_3 & \cdots & y_{K+1} \\ \vdots & \vdots & \cdots & \vdots \\ y_L & y_{L+1} & \cdots & y_N \end{bmatrix}. \quad (1)$$

Step 2: Singular values decomposition (SVD). d eigenvalues which meet the condition $\lambda_1 \ge \lambda_2 \ge \cdots \ge \lambda_d \ge 0$ .are produced by computing the eigenvalues of the covariance matrix $\mathbf{S} = \mathbf{X}\mathbf{X}^{\mathrm{T}}$. $\mathbf{u}_1, \mathbf{u}_2, \cdots, \mathbf{u}_d$ are the corresponding orthogonal eigenvectors. Then the trajectory matrix $\mathbf{X}$ is decomposed by SVD into

$$\begin{gathered} \mathbf{X} = \mathbf{e}_1 + \mathbf{e}_2 + \cdots + \mathbf{e}_d \\ \mathbf{e}_i = \sqrt{\lambda_i}\mathbf{u}_i\mathbf{v}_i^{\mathrm{T}}, \mathbf{v}_i = \mathbf{X}^{\mathrm{T}}\mathbf{u}_i\sqrt{\lambda_i} \\ d = \min\{L, K\}, i = 1, 2, \cdots, d \end{gathered} \quad (2)$$

where the vectors $\mathbf{v}_1, \mathbf{v}_2, \cdots, \mathbf{v}_d$ are regarded as the principle components of trajectory matrix $\mathbf{X}$. Therefore, the collection $(\sqrt{\lambda_i}, \mathbf{u}_i, \mathbf{v}_i)$ is defined as the ith triple feature vector of matrix $\mathbf{X}$.

Step 3: Grouping. The indices $\{1, 2, \cdots, d\}$ are divided into m independent groups $I_1, I_2, \cdots, I_m$ and each group $I_j = \{i_{j_1}, \cdots, i_{j_p}\}$, $j = 1, 2, \cdots, m$ is denoted as a group of indices. The trajectory matrix $\mathbf{X}$ is expressed as

$$\mathbf{X} = \mathbf{E}_{I_1} + \mathbf{E}_{I_2} + \cdots + \mathbf{E}_{I_m} \quad (3)$$

$$\mathbf{E}_{I_j} = \mathbf{e}_{i_{j1}} + \mathbf{e}_{i_{j2}} + \cdots + \mathbf{e}_{i_{jp}} \quad (4)$$

where the contribution rate of $\mathbf{E}_{I_j}$ is $\sum_{i\in I_j} \lambda_i / \sum_{i=1}^{d} \lambda_i$. In this paper, the original time series is decomposed into two components. Hence, we have $\mathbf{X} = \mathbf{E}_{I_1} + \mathbf{E}_{I_2}$.

Step 4: Diagonal averaging. Each matrix $\mathbf{E}_{I_j} \in \mathbb{R}^{L\times K}$ is converted to a time series by the following ways. Let $z_{il}, i=1,2,\cdots,L, l=1,2,\cdots,K$ be the elements of the matrix with $\mathbf{E}_{I_j}$. Set $L^*=\min\{L,K\}$, $K^*=\max\{L,K\}$, and $N=K+L-1$. $z_{il}^*=z_{il}$ when $L<K$; otherwise, $z_{il}^*=z_{li}$. The matrix $\mathbf{E}_{I_j}$ is transformed into a time series $\left[y_1^{(j)}\ y_2^{(j)} \cdots y_N^{(j)}\right]$ via the following equation

$$y_k^{(j)}=\begin{cases}\frac{1}{k}\sum_{q=1}^{k+1} z_{q,k-q+2}^*, 1\le k\le L^* \\ \frac{1}{L^*}\sum_{q=1}^{L^*} z_{q,k-q+2}^*, L^*\le k\le K^* \\ \frac{1}{N-K+2}\sum_{q=k-K^*+2}^{N-K^*+1} z_{q,k-q+2}^*, K^*<k\le N\end{cases} \quad (5)$$

In this paper, the diagonal averaging is applied to $\mathbf{E}_{I_1}$ and $\mathbf{E}_{I_2}$, respectively, and the two corresponding time series are $\mathbf{y}^{(1)}=\left[y_1^{(1)}\ y_2^{(1)} \cdots y_N^{(1)}\right]$ and $\mathbf{y}^{(2)}=\left[y_1^{(2)}\ y_2^{(2)} \cdots y_N^{(2)}\right]$ according to (5). Thus, the original time series $\mathbf{y}$ is expressed by $\mathbf{y}=\mathbf{y}^{(1)}+\mathbf{y}^{(2)}$, where $\mathbf{y}^{(1)}$ is the mean trend and $\mathbf{y}^{(2)}$ is treated as noise.

An example is given in Figure 1, where the mean trend is extracted from a wind power time series collected from Elia, the transmission system operator (TSO) of Belguium [15], and the contribution rate of each eigenvector is demonstrated in Table 1.

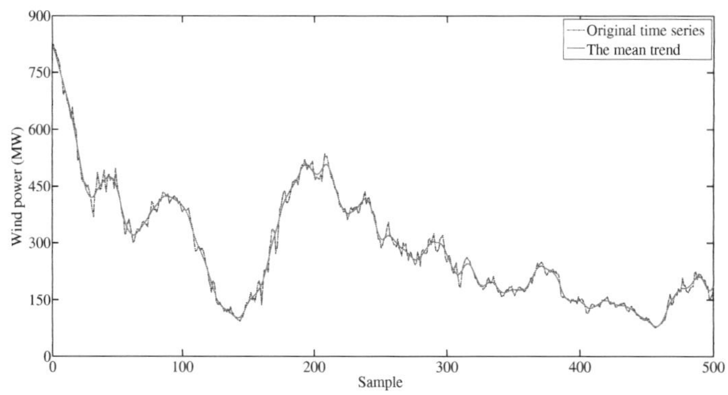

Figure 1. A time series of wind speed from the Elia database and the mean trend extracted by SSA

Table 1. The contribution rate of each eigenvector in embedding dimension L=20

Contribution rate (%)				
(1)	(2)	(3)	(4)	(5)-(20)
98.438	1.153	0.194	0.074	0.141

2.2. *Self-Organizing Maps (SOM)*

Considering the continuity characteristic of meteorology, wind power at a certain time instant is closely related to data collected in a short period of time before. Thus, instead of using all the sample segments to train the model, it is more effective to use only the segments that bear the same trend of the forecast segment. The process of SOM is explained in 4 steps as follows.

Step 1: Initialization. The mean trend $\mathbf{y}^{(1)}$ is reconstructed into a higher-dimensional phase space with the embedding dimension of s and the time constant τ. It can be reconstructed into $N-(s-1)\tau$ segments and each segment is expressed as

$$\mathbf{y}_i^{(1)}=\left[y_i^{(1)}\ y_{i+\tau}^{(1)} \cdots y_{i+(s-1)\tau}^{(1)}\right] \quad (6)$$

IOP Conf. Series: Materials Science and Engineering **366** (2018) 012015 doi:10.1088/1757-899X/366/1/012015

where $i = 1, 2, \cdots, N-(s-1)\tau$.

Initialize the parameters connection weights vector $\mathbf{w}_j(0) = [w_{j1}(0)\ w_{j2}(0)\ \cdots w_{js}(0)],\ j = 1,\ \cdots, D$ of D neurons, learning rate η_0 and radius of the neighborhood σ_0 with small random values.

Step 2: Competition. For each input pattern, the neurons compute their respective values of a discriminant function, which provides the basis for competition. The particular neuron with the smallest value of the discriminant function is the winner. In this paper, the discriminant function is the squared Euclidean distance between the mean trend $\mathbf{y}_i^{(1)}$ and the weight vector $\mathbf{w}_j(k), k = 1, 2\cdots, T$, for each neuron j

$$d_j(\mathbf{y}_i^{(1)}) = \| \mathbf{y}_i^{(1)} - \mathbf{w}_j(k) \| \tag{7}$$

Step 3: Cooperation. The winner neuron determines the spatial location of a topological neighborhood neurons, thereby providing the basis for cooperation among neighboring neurons. The topological neighborhood $T_{j,I(\mathbf{y}_i^{(1)})}$ is defined as

$$T_{j,I(\mathbf{y}_i^{(1)})} = \exp\left(-S^2_{j,I(\mathbf{y}_i^{(1)})} \big/ 2\sigma^2\right) \tag{8}$$

$$\sigma(k) = \sigma_0 \exp(-k / T),\ \ k = 1, 2, \cdots T \tag{9}$$

where $I(\mathbf{y}_i^{(1)})$ is the index of the winner neuron, $S_{j,I(\mathbf{y}_i^{1})}$ is the distance between the connection $\mathbf{w}_j(k)$ and $\mathbf{w}_{I(\mathbf{y}_i^{(1)})}(k)$, $\sigma(k)$ is radius of the neighbourhood at step iteration k.

Step 4: Adaptation. The weight vector $\mathbf{w}_j(k)$ of the current winner neuron and the learning rate $\eta(k)$ are simultaneously adjusted according to the following learning rule

$$\Delta w_{ji} = \eta(k) \cdot T_{j,I(x)}(k) \cdot (y_i^{(1)} - w_{ji}) \tag{10}$$

$$\eta(k) = \eta_0 \exp(-k / T),\ \ k = 1, 2, \cdots T \tag{11}$$

An instruction on how to implement the proposed model is shown in the framework of Figure 2.

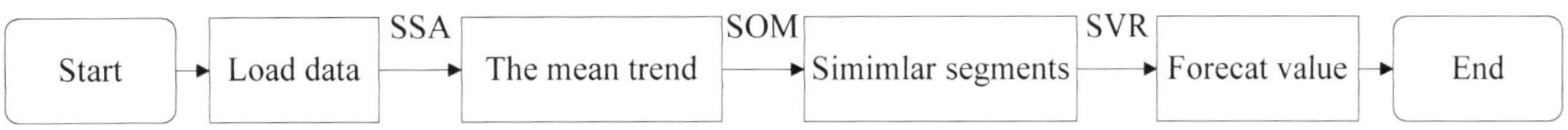

Figure 2. The Framework of the proposed model

3. Simulation Studies

3.1. Performance qualification

In order to assess the performance of the proposed model, two commonly used criteria are employed to measure the errors: normalized mean absolute error (NMAE) and normalized root mean squared error (NRMSE) [16]. The NMAE indicates how close the forecast results are to the actual values, and the NRMSE evaluates the standard deviation between the forecast results and the actual values, which reflects the stability of the forecast model. The smaller the NMAE/NRMSE is, the more accurate and stable the forecast results are. They are defined by:

$$\text{NMAE} = \frac{1}{N} \sum_{i=1}^{N} \frac{|y_i - \hat{y}_i|}{Y} \times 100\% \tag{12}$$

$$\text{NRMSE} = \frac{1}{Y} \sqrt{\frac{1}{N-1} \sum_{i=1}^{N} (y_i - \hat{y}_i)^2} \times 100\% \tag{13}$$

where N is the size of the validation data set, $\hat{y}_i$ is the forecast value, y_i is the actual value, Y is the installed capacity of the wind farm at the forecasting moment.

3.2. Forecast results and discussion

To evaluate the performance of the SSA-SOM-SVR model, comparison studies are carried out on the persistence (Per.), SVR and SSA-SVR models for 1~20 look-ahead steps, and the performances have been illustrated in Figure 3. It is obvious that the SSA-SOM-SVR model achieves the smallest NMAEs and NRMSEs. Furthermore, NMAE and NRMSE are improved by 12.38% and 13.90% in 20 look-ahead steps, respectively, by the SSA-SOM-SVR model over the Per. model. To make a more comprehensive study, the numerical results of 4, 12, 20 look-ahead steps are listed in Table 2. As wind may have various natural characteristics in different seasons, the forecast is carried out in March, June, September and December, respectively. It is obvious that the proposed model gains the smallest NMAEs and NRMSEs for all situations, which indicates that the proposed model outperforms the other ones in both accuracy and stability.

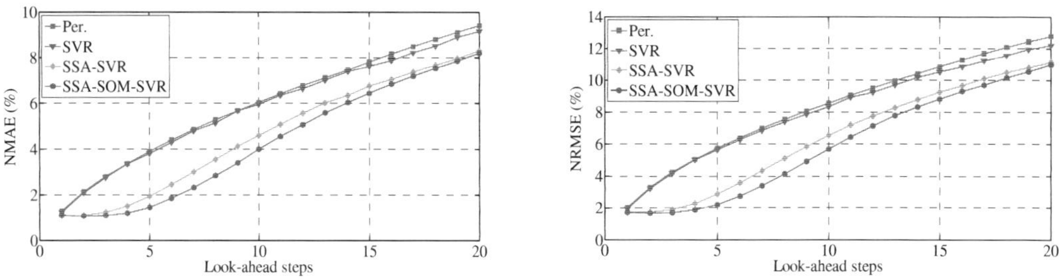

(a) (b)

Figure 3. The performance of the four models on the Elia data set in June (a) NMAEs (b) NRMSEs

Table 2. Performance evaluation of the four forecast models on the Elia data set (%)

		month	Per	SVR	SSA-SVR	SSA-SOM-SVR
4-step	NMAE	3	2.6018	2.5433	1.3322	**1.0318**
		6	3.3608	3.2890	1.6240	**1.2129**
		9	2.2986	2.2146	1.0423	**0.7316**
		12	3.3782	3.3599	1.6283	**1.1074**
	NRMSE	3	3.6154	3.5836	1.8750	**1.4994**
		6	5.0544	4.9797	2.4484	**1.8661**
		9	3.3571	3.2318	1.5565	**1.1504**
		12	4.7742	4.6830	2.3031	**1.6866**
12-step	NMAE	3	5.1307	5.1524	4.2830	**3.8897**
		6	6.7770	6.6244	5.6338	**5.1062**
		9	4.8966	4.7201	3.9833	**3.4471**
		12	7.3559	7.2613	6.2025	**5.1865**
	NRMSE	3	6.9956	6.9042	5.8747	**5.2941**
		6	9.5244	9.2134	7.8885	**7.1293**
		9	6.8697	6.5273	5.5086	**4.7814**
		12	10.2746	9.6897	8.2331	**7.0041**
20-step	NMAE	3	7.0272	6.9700	6.2935	**6.1695**
		6	9.4153	9.1108	8.3938	**8.2494**
		9	6.8036	6.7274	6.1369	**5.7239**
		12	10.6998	10.8987	9.7743	**8.9914**
	NRMSE	3	9.5660	9.1924	8.3764	**8.2779**
		6	12.7807	12.2040	11.2540	**11.0038**
		9	9.3852	8.9187	8.1482	**7.7337**
		12	14.7701	14.3054	12.8965	**12.0250**

4. Conclusion

This paper proposes a forecast model based on SSA, SOM and SVR for short-term wind power prediction. The SSA is employed to extract the mean trend from the original time series. Afterwards, SOM is used to select similar segments to perform local prediction. In order to evaluate the accuracy of the proposed model and highlight the effect of SSA and SOM methods, simulation studies are conducted on the Per., SVR and SSA-SVR models for the purpose of comparison, and the final results prove that the application of SSA and SOM improve the accuracy and stability.

Acknowledgments

This work was supported by Guangdong Innovative Research Team Program (No. 201001N0104744201) and the State Key Program of National Natural Science of China (No. 51437006).

References

[1] Zhang J, Mingjian Cui, Bri-Mathias Hodge, et al. Ramp forecasting performance from improved short-term wind power forecasting over multiple spatial and temporal scales[J]. Energy, 2017, 122:528-541.
[2] Wu J L, Ji T Y, Li M S, et al. Multi-step wind power forecast based on similar segments extracted by mathematical morphology[C].Power and Energy Engineering Conference. IEEE, 2015:1-6.
[3] Zhang C, Wei H, Zhao X, et al. A Gaussian process regression based hybrid approach for short-term wind speed prediction[J]. Energy Conversion & Management, 2016, 126:1084-1092.
[4] Foley A M, Leahy P G, Marvuglia A, et al. Current methods and advances in forecasting of wind power generation[J]. Renewable Energy, 2012, 37(1):1-8.
[5] Liu H, Shi J, Erdem E. Prediction of wind speed time series using modified Taylor Kriging method[J]. Energy, 2010, 35(12):4870-4879.
[6] Akçay H, Filik T, Yan J. Short-term wind speed forecasting by spectral analysis from long-term observations with missing values[J]. Applied Energy, 2017, 191:653-662.
[7] Guo Z, Zhao W, Lu H, et al. Multi-step forecasting for wind speed using a modified EMD-based artificial neural network model[J]. Renewable Energy, 2012, 37(1):241-249.
[8] Boudraa A O, Cexus J C. EMD-Based Signal Filtering[J]. IEEE Transactions on Instrumentation & Measurement, 2007, 56(6):2196-2202.
[9] Zhang X, Wang J, Zhang K. Short-term electric load forecasting based on singular spectrum analysis and support vector machine optimized by Cuckoo search algorithm[J]. Electric Power Systems Research, 2017, 146:270-285.
[10] Zhang W, Su Z, Zhang H, et al. Hybrid Wind Speed Forecasting Model Study Based on SSA and Intelligent Optimized Algorithm[J]. Abstract and Applied Analysis,2014,(2014-4-3), 2014, 2014(1):1-14.
[11] Lau K W, Wu Q H. Local prediction of non-linear time series using support vector regression[J]. Pattern Recognition, 2008, 41(5):1539-1547.
[12] Liu J, Djurdjanovic D. Topology Preservation and Cooperative Learning in Identification of Multiple Model Systems[J]. IEEE Transactions on Neural Networks, 2008, 19(12):2065-72.
[13] Júnior A H S, Barreto G A, Corona F. Regional models: A new approach for nonlinear system identification via clustering of the self-organizing map[J]. Neurocomputing, 2015, 147:31-46.
[14] Zhu L, Wu Q H, Li M S, et al. Support vector regression-based short-term wind power prediction with false neighbours filtered[C]. International Conference on Renewable Energy Research and Applications. IEEE, 2014:740-744.
[15] Elia. Wind power data [online]. Available:http://www.elia.be/en/griddata/power-generation/wind-power.
[16] Bludszuweit H, Dominguez-Navarro J A, Llombart A. Statistical Analysis of Wind Power Forecast Error[J]. IEEE Transactions on Power Systems, 2008, 23(3):983-991.

IOP Conf. Series: Materials Science and Engineering **366** (2018) 012016 doi:10.1088/1757-899X/366/1/012016

A Two-Stage Distributionally Robust Coordinated Dispatch for Integrated Electricity and Natural-gas Energy Systems Considering Uncertainty of Wind Power

XingquanXIAO[1], Yue SHUI[2,a], Hongjun GAO[2], Junyong LIU[2], QianCHEN[1], Ting LI[1], Ying LIU[1] and XuetongOUYANG[1]

[1]Sichuan Electric Power Corporation Power Economic Research Institute, Chengdu 610000, China;
[2]College of Electrical Engineering and Information Technology, Sichuan University, Chengdu 610065, China

[a] Corresponding author: 178332804@qq.com

Abstract. Application of gas-fired units and power-to-gas devices promotes coupling between power system and natural gas system, and hence provides an alternative solution for the wind power integration. Also, the traditional stochastic optimization and robust optimization methods may result in uneconomic or conservative decisions in dealing with the uncertain problems. Thus, a data-driven based distributionally robust dispatch model for the integrated electricity and natural gas system is proposed in this paper, which is constructed in a two-stage optimization fashion. The day-ahead total cost for the integrated system is regarded as the optimization objective in the first stage where the forecast wind power information is also taken into consideration while the output adjustment of thermal generation units and the supply regulation of natural gas source are included in the real-time dispatch (second stage in this study). Besides, the norm-1 and norm-inf are simultaneously combined to constrain the confidence set of wind power probability distribution. Then the model is solved by the Column-and-Constraint Generation algorithm. Finally, numerical results verify the effectiveness of the proposed model.

1. Introduction

In order to improve the efficiency of energy, the "Energy Internet" concept has been proposed recently which aims to promote coupling among different networks [1]. Compared with other energy, natural gas keeps a rising integration in electricity system, owing to low cost, low carbon emission, so it is vital to research the coupling between power system and natural gas system. Moreover, the dispatch of the integrated system is regarded as an efficient way to promote the use of wind power.

Currently, the gas-fired unit servers as a linkage between the electricity and natural gas system. The reference [2] introduced a carbon trading mechanism to study the economic operation of the integrated system. The reference [3] made the optimal operation decisions of the integrated system considering security of the electricity and natural gas networks. However, the gas-fired units only convert the natural gas into electricity, extensive researches consider the Power-To-Gas (P2G) devices which could convert the energy in a contrary direction and thus make the energy flow in the integrated system run in a closed loop [4-6]. In [7], the author conducted a dispatch model for integrated electricity-gas-heating system with P2G devices, and discussed the accommodation of the wind power.

In [8], the authors proposed a two-stage economic dispatch model for integrated electricity and natural gas system. However, the uncertainty of the wind power was not considered in the previous research.

Generally, stochastic optimization and robust optimization methods were used in dealing with the uncertainties of the wind power. In [9], an optimization method based on the scenario of wind power output is proposed to make the day-ahead dispatch of the integrated system. Paper [10] conduct a robust optimization dispatch model of the electricity and natural gas system considered the security of integrated system and the dynamic characteristics of the natural gas system. However, the stochastic optimization method needs to obtain the accurate probability distribution function[11] while the robust optimization may result in over-conservative decisions[12]. Based on the known problems, recent researches combine the two existing methods and propose a distributionally robust optimization (DRO)[13-15] methods to deal with the uncertainties. In this paper, a data-driven distributionally robust optimization dispatch model based on the historical wind power output data is proposed. In the model, norm-1 and norm-inf constraints are simultaneously included to constrain the confidence set of wind power probability distribution in order to find the optimal solution under the worst probability distribution. The paper [16] and [17] apply the DRO method into the reactive optimization problem and the unit commitment respectively, but these researches only consider the norm-1 or norm-inf constraint which could make the decisions more uneconomic. We apply the DRO method which considers the norm-1 and norm-inf simultaneously to the integrated electricity and natural gas system coupled with gas-fired unit and P2G devices for the first time.

2. Integrated electricity and natural gas system modeling

The proposed model of the integrated system can be illustrated in Fig. 1. The system includes electricity transmission system, gas supply system, electricity load, gas load, gas-fired units and P2G devices. The electricity and natural gas system are closely interconnected by the gas-fired units and P2G devices which can be treated as energy converters among these two energy networks. As we can see, the wind power can be accommodated by the decrease output of gas-fired units and the increase usage of P2G devices.

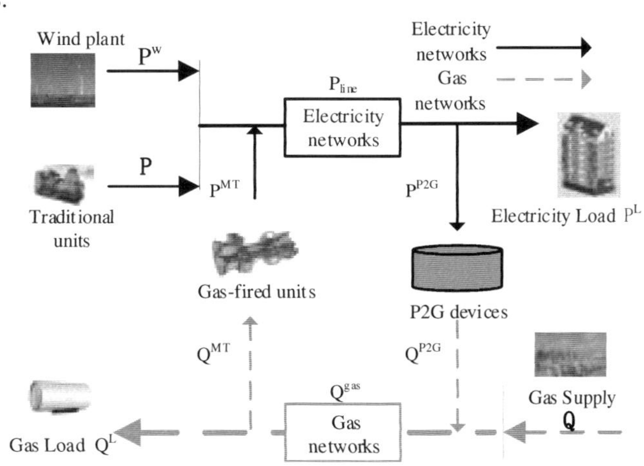

Figure 1. The structure of the integrated system

2.1. Objective Function

The integrated systems set the day-ahead dispatch cost and the real-time adjustment cost as the optimization targets which is represented in (1):

$$\min \sum_{t=1}^{T}[\sum_{i=1}^{N_G} S_{Ti} I_{i,t}(1-I_{i,t-1}) + S_{Di} I_{i,t-1}(1-I_{i,t}) + (a_i P_{i,t}^2 + b_i P_{i,t} + c_i I_{i,t})]$$
$$+\sum_{t=1}^{T}[\sum_{i=1}^{N_{GT}} S_{Ti}^{gt} I_{i,t}^{gt}(1-I_{i,t-1}^{gt}) + S_{Di}^{gt} I_{i,t-1}^{gt}(1-I_{i,t}^{gt}) + (a_i^{gt}(P_{i,t}^{gt})^2 + b_i^{gt} P_{i,t}^{gt} + c_i^{gt} I_{i,t}^{gt})] + \sum_{t=1}^{T}\sum_{i=1}^{N_W} \delta(P_{i,t}^{w0} - P_{i,t}^{w}) \quad (1)$$
$$+ \max_{\{p_k\}\in\Omega} \sum_{k=1}^{K} p_k[\min \sum_{t=1}^{T}(\sum_{i=1}^{N_G} c_{i,t}\Delta P_{i,t,k}) + \sum_{t=1}^{T}(\sum_{i=1}^{N_{MT}} c_{i,t}^{gt}\Delta P_{i,t,k}^{gt}) + \sum_{t=1}^{T}(\sum_{i=1}^{N_W} c_{i,t}^{wind}(P_{i,t,k}^{w0} - P_{i,t,k}^{w}))]$$

The first part of (1) represents the day-ahead dispatch cost of the integrated system, which includes the start-up and shut-down cost, the generation cost of traditional units and gas-fired units, and the curtailment cost of wind power, where T represents the total scheduling time; N_G represents the number of the traditional units; S_{Ti}、S_{Di} represents the start-up and shut-down cost of traditional units; $I_{i,t}$ represents the status indicator of traditional units; a_i、b_i、c_i is the cost coefficient; $P_{i,t}$ is the power output of traditional units; N_{GT} represents the number of gas-fired units; S_{Ti}^{gt} 、S_{Di}^{gt} represents the startup and shutdown cost of gas-fired units; $I_{i,t}^{gt}$ represents the status indicator of gas-fired units; a_i^{gt} 、b_i^{gt} 、c_i^{gt} is the cost coefficient of gas-fired units; $P_{i,t}^{gt}$ is the power output of gas-fired units; N_W is the number of wind plant; δ is the penalty coefficient of wind power curtailment; $P_{i,t}^{w0}$ and $P_{i,t}^{w}$ represent the forecast output and the actual output of wind power, respectively.

The second part of (1) represents the real-timeadjustment cost of the integrated system, which includes the adjustment cost of the power unit and gas supplier. Based on the historical data, this paper construct K scenarios to describe possible wind power output realizations. The basic empirical probability in each given scenario k can correspondingly be obtained, which is denoted as p_k^0 ; p_k represents the probability of each scenario; $\Delta P_{i,t,k}$ and $\Delta P_{i,t,k}^{gt}$ represent the adjustment of traditional units and gas-fired units respectively; $P_{i,t,k}^{w0}$ and $P_{i,t,k}^{w}$ represent the forecast output and the actual output of wind power in scenario k; $c_{i,t}$ 、$c_{i,t}^{gt}$ and $c_{i,t}^{wind}$ represent the adjustment penalty coefficient in real-time dispatch.

2.2. Day-ahead Dispatch Constraints

In the day-ahead dispatch, for each period t, the electricity, natural gas system and the coupling devices constraints are expressed as follows:

2.2.1 Power System Constraints

1) Power Balance

$$\sum_{i=1}^{N_G} P_{i,t} + \sum_{i=1}^{N_W} P_{i,t}^{w} + \sum_{i=1}^{N_{GT}} P_{i,t}^{gt} = P_t^{L} + \sum_{i=1}^{N_{P2G}} P_{i,t}^{P2G} \quad (2)$$

where P_t^{L} is the total electricity load; N_{P2G} is the number of P2G devices; $P_{i,t}^{P2G}$ is the power usage of P2G devices.

3) Wind power output Constraints

$$0 \le P_{i,t}^{w} \le P_{i,t}^{w0} \quad (3)$$

2.2.2 Natural Gas System Constraints

1) Node Pressure Constraints

In the gas networks, each gas node has the upper and lower pressure limit,which is denoted as follows:

$$\pi_{i,\min} \le \pi_{i,t} \le \pi_{i,\max} \quad (4)$$

where $\pi_{i,t}$ is the pressure of the gas node; $\pi_{i,\min}$ and $\pi_{i,\max}$ is the minimum and maximum of the node pressure respectively.

2) Gas Balance Constraints

The steady-state natural gas injection at each node is equal to flow extracted from the node. The (5) ensures the nodal balance at the natural gas transmission system:

$$Q_{i,t}^{\text{gs}} + Q_{i,t}^{\text{P2G}} - Q_{i,t}^{\text{gt}} - Q_{i,t}^{\text{L}} = Q_{ij,t}^{\text{pipe}} \tag{5}$$

where $Q_{i,t}^{\text{gs}}$ 、 $Q_{i,t}^{\text{P2G}}$ 、 $Q_{i,t}^{\text{gt}}$ is the gas injection of the corresponding devices; $Q_{ij,t}^{\text{pipe}}$ is the gas flow of the gas pipeline.

3) Flow Conservation Constraints

Natural gas is delivered to customers via pipelines. The pipelines include passive pipeline and active pipeline. In this paper, we consider the gas flow between nodes *i* and *j* is a quadratic function of the pressure at the two end nodes:

$$Q_{ij,t}^{\text{pipe}} = \text{sgn}(\pi_i, \pi_j) \cdot C_{ij} \sqrt{|\pi_i^2 - \pi_j^2|} \tag{6}$$

where C_{ij} is the pipeline constant that depends on the temperature,length, friction and natural gas compositions.

$$\text{sgn}(\pi_i, \pi_j) = \begin{cases} 1, \pi_i \geq \pi_j \\ -1, \pi_j \geq \pi_i \end{cases} \tag{7}$$

4) Gas Supplier Constraints

The gas is usually supplied by the gas well, and the gas supplier has the limitation which is denoted as follows:

$$Q_{i,\min}^{\text{gs}} \leq Q_{i,t}^{\text{gs}} \leq Q_{i,\max}^{\text{gs}} \tag{8}$$

2.2.3 Coupling Devices Constraints

The integrated system is coupled by gas-fired units and P2G devices.

1) Gas-fired Units Constraints

$$Q_{i,\min} \leq Q_{i,t}^{\text{gt}} \leq Q_{i,\max} \tag{9}$$

$$Q_{i,t}^{\text{gt}} = P_{i,t}^{\text{gt}} / \eta \tag{10}$$

where η is the efficiency of gas-fired units, which equals to 40%.

2) P2G Constraints

$$Q_{i,t}^{\text{P2G}} = \eta^{\text{P2G}} \cdot P_{i,t}^{\text{P2G}} \tag{11}$$

$$P_{i,\min}^{\text{P2G}} \leq P_{i,t}^{\text{P2G}} \leq P_{i,\max}^{\text{P2G}} \tag{12}$$

where η^{P2G} is the convert efficiency of P2G devices, which equals to 80%.

2.3. Real-time Dispatch Constraints

The generation units and gas suppliers have to adjust their outputs when the wind power is revealed in the real-time dispatch model. The constraint in each scenario *k* is denoted as follows:

2.3.1 Power System Constraints

1) Power Balance

$$\begin{aligned} &\sum_{i=1}^{\text{N}_\text{G}} (P_{i,t} + \Delta P_{i,t,k}) + \sum_{i=1}^{\text{N}_\text{W}} P_{i,t,k}^{\text{w}} + \sum_{i=1}^{\text{N}_\text{gt}} (P_{i,t}^{\text{gt}} + \Delta P_{i,t,k}^{\text{gt}}) \\ &= P_t^{\text{L}} + \sum_{i=1}^{\text{N}_\text{P2G}} P_{i,t}^{\text{P2G}} \end{aligned} \tag{13}$$

3) Wind power output Constraints

$$0 \leq P_{i,t,k}^{\text{w}} \leq P_{i,t,k}^{\text{w0}} \tag{14}$$

2.3.2 Natural Gas System Constraints

1) Node Pressure Constraints

$$\pi_{i,\min} \leq \pi_{i,t} + \Delta\pi_{i,t,k} \leq \pi_{i,\max} \tag{15}$$

where $\Delta\pi_{i,t,k}$ is the pressure change of the gas node.

2) Gas Balance Constraints

$$(Q_{i,t}^{\text{gs}}+\Delta Q_{i,t,k}^{\text{gs}})+(Q_{i,t}^{\text{P2G}}+\Delta Q_{i,t,k}^{\text{P2G}})-(Q_{i,t}^{\text{gt}}+\Delta Q_{i,t,k}^{\text{gt}}) \\ -Q_{i,t}^{\text{L}}=Q_{ij,t}^{\text{pipe}}+\Delta Q_{ij,t,k}^{\text{pipe}} \tag{16}$$

where $\Delta Q_{i,t,k}^{\text{gs}}$ 、 $\Delta Q_{i,t,k}^{\text{gt}}$ 、 $\Delta Q_{ij,t,k}^{\text{pipe}}$ is the gas injection change of the corresponding devices.

3) Flow Conservation Constraints

$$Q_{ij,t}^{\text{pipe}}+\Delta Q_{ij,t,k}^{\text{pipe}}= \\ \text{sgn}(\pi_i,\pi_j)\cdot C_{ij}\sqrt{\left|(\pi_i+\Delta\pi_{i,k})^2-(\pi_j+\Delta\pi_{j,k})^2\right|} \tag{17}$$

4) Gas Supplier Constraints

$$Q_{i,\min}^{\text{gs}}\le Q_{i,t}^{\text{gs}}+\Delta Q_{i,t,k}^{\text{gs}}\le Q_{i,\max}^{\text{gs}} \tag{18}$$

where $\Delta Q_{i,t}^{\text{gs}}$ is the gas injection change of the gas supplier.

2.3.3 Coupling Devices Constraints. There is no adjustment power of P2G devices, so we only consider the regulation of the gas-fired units in the real-time dispatch:

$$Q_{i,\min}\le Q_{i,t}^{\text{gt}}+\Delta Q_{i,t,k}^{\text{gt}}\le Q_{i,\max} \tag{19}$$

$$Q_{i,t}^{\text{gt}}+\Delta Q_{i,t,k}^{\text{gt}}=(P_{i,t}^{\text{gt}}+\Delta P_{i,t,k}^{\text{gt}})/\eta \tag{20}$$

In the two-stage of the model, we also consider the constraints of the traditional generation units which include ramping rate, minimum/maximum power output, and minimum ON/OFF time. Besides, the reserve rate of integrated system which is used to handle the uncertainties is also considered (which is set to 10% of the wind power forecast value). In this paper, we apply the DC power flow model which is proposed in paper [18].

3. SolutionAlgorithm

3.1. Linearization of the Norm Constraints

In the two-stage DRO model, we propose to use the norm-1 and norm-inf simultaneously to constrain the probability distribution of the wind power in order to find optimal solution under the worst distribution. The constraints can be expressed by follows:

$$\Omega=\left\{\{p_k\}\left|\begin{array}{l}p_k\ge 0,k=1,...,\text{K}\\ \sum_{k=1}^{\text{K}}p_k=1\\ \sum_{K=1}^{K}\left|p_k-p_k^0\right|\le\theta_1\\ \max\limits_{1\le k\le K}\left|p_k-p_k^0\right|\le\theta_\infty\end{array}\right.\right\} \tag{21}$$

where θ_1 and θ_∞ are norm-1 and norm-inf tolerance value, which becomes smaller when more statistic historical data are provided. Supposing K scenarios from M historical samples.

3.2. Solution of The Two-Stage DRO Model

The two-stage DRO model is a three-level min-max-min optimization problem, which is usually solved by Benders decomposition algorithm or CCG algorithm [16]. The CCG algorithm divides the model into two problems that are the master problem (MP) and the sub-problem (SP), The iteration would stop until the difference of optimization results between MP and SP meet the predefined tolerance requirement.In this model, the master problem aims to find the optimal robust first-stage solution under some given finite worst-case probability distributions (possible realizations obtained from sub-problem). The MP provides a lower bound for model.

$$\text{(MP)} \quad \min_{x\in X, y_0\in Y(x,\xi_0), y_k^m\in Y(x,\xi_k), L} \quad a^T x + b^T y_0 + c^T \xi_0 + L \tag{22a}$$

$$L \geq \sum_{k=1}^{K} p_k^m (b^T y_k^m + c^T \xi_k), \forall m = 1, \cdots, n \tag{22b}$$

The equation (22b) is the second part of (1).

The sub-problem gives an upper bound for model by optimizing the worst probability distribution after first-stage variables x^* are given. The SP can be described in (23):

$$\text{(SP)} \quad L(x^*) = \max_{\{p_k\}\in\Omega} \sum_{k=1}^{K} p_k \min_{y_k\in Y(x^*,\xi_k)} (b^T y_k + c^T \xi_k) \tag{23}$$

The inner min model can be decoupled into K small independent problems that are suitable to be solved in parallel and be expressed into (24):

$$L(x^*) = \max_{\{p_k\}\in\Omega} \sum_{k=1}^{K} f(x^*, \xi_k) p_k \tag{24}$$

4. Numerical Simulation

4.1. System Description

In this part, we use acase study consisting of a 6-bus natural gas system and the IEEE39 power system to test the proposed model. The 6-bus natural gas system is illustrated in Fig. 2. The specific data of the 6-bus system can be seen in [7]. In the IEEE39 system,the gas-fired units are connected to nodes 30-32. The traditional generation units are connected to nodes 33-39. Three wind plants are connected to nodes 37-39.The electricity load and gas load are shown in Fig.3.

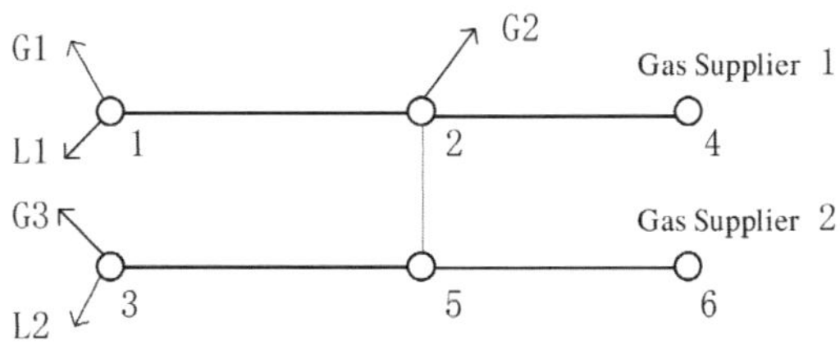

Figure 2.6-node natural gas system

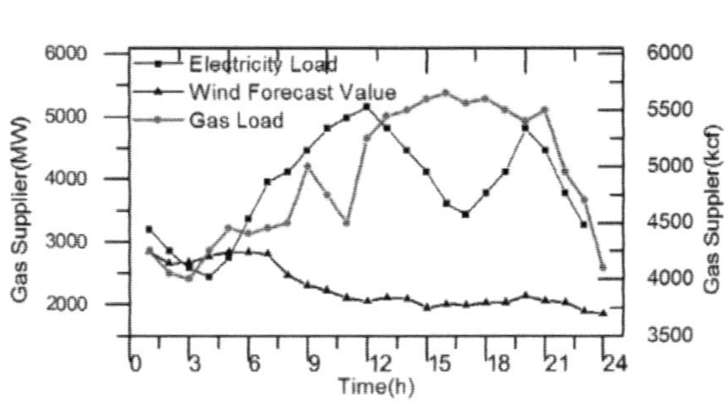

Figure 3. The daily curves of Electricity load, Gas load and wind power

4.2. Results Comparisons

The DRO model proposed in this paper is used to obtain the worst probability distribution wind power. We give the following comparisons to verify the efficiency of the model.

4.2.1 Results Comparisons under different number of historical data. We set different number of historical data to compare the results, which are shown in Table 1.

Table 1. Effects of Historical Data

Historical Data	norm-1	Proposed model	norm-∞	Stochastic model
100	2.3538	2.2995	2.3190	1.1422
500	1.6071	1.5052	1.6003	1.1422
1000	1.5411	1.4419	1.5311	1.1422
2000	1.5185	1.3073	1.4073	1.1422
5000	1.4603	1.2783	1.3620	1.1422
10000	1.1446	1.1431	1.1445	1.1422

As we can see in Table. 1, when the number of historical data increases, the total cost of the integrated system decreases, that is because when the historical data number increases, the tolerance value decreases so the conservatism of the model decreases. Also,we can conclude that under the same number of historical data, the model proposed in this paper is more economic than the others. Comparing with the stochastic optimization method, the proposed model becomes more economic when the number of historical data increases, and the results are much closer to stochastic method. So we choose 1000 historical data in the following calculation in order to guarantee the conservatism of the model.

4.2.2 Results Comparisons under different confidence set. In this part, we set different confidence set value to analysis the results, which are shown in Table.2.

Table 2. Results under different Confidence set

α_1 \ α_∞	0.5	0.9	0.99
0.2	1.4412	1.4414	1.4420
0.5	1.4419	1.4427	1.4429
0.9	1.4419	1.4429	1.4431

We can conclude from Table.2that when the confidence set increases, the total cost of the integrated system increases simultaneously. The reason is that when the confidence set increases, the uncertainty of the wind power output increases as well, the integrated system has to adjust more units and gas supplier to deal with the uncertainty which means more cost.

Moreover, we set theα_∞as 0.99, the range of α_1 as [0.2, 0.9] to compare with the model only considering the norm-∞.The results are shown in Table.3.

Table 3. Comparison between comprehensive norm and ∞-norm

α_1	Cost($\$\times10^6$)	
	Proposed model	norm-∞
0.2	1.4420	1.4436
0.5	1.4429	1.4436
0.99	1.4431	1.4436

Then we set$\alpha_1$as 0.5,the range of α_∞ as [0.5, 0.99] to compare with the model only considering the norm-1.The results are shown in Table.4.

Table 4. Comparison between comprehensive norm and 1-norm

α_∞	Cost($\$\times10^6$)	
	Proposed model	norm-1
0.5	1.4419	1.5411
0.8	1.4427	1.5411
0.99	1.4429	1.5411

4.2.3 Results Comparisons under different methods. We also analyze the optimization results among traditional stochastic optimization (SO), robust optimization (RO), and distributionally robust optimization methods under 1,000,000 test probability distributions generated by Monte Carlo simulation, as shown in Table.5.

Table 5. The results comparison between different algorithms

α_∞	DRO($\$\times10^6$)			SO($\$\times10^6$)			RO($\$\times10^6$)		
	First Stage Cost	Test Distributions	Probability	First Stage Cost Mean	Test Distributions	Probability	First Stage Cost	Test Distributions	Probability
		Mean	Max		Mean	Max		Mean	Max
0.5	1.4419	0.6194	0.9568	1.1422	1.1453	2.6665	7.1693	1.4383	4.4768
0.9	1.4427	0.9541	2.0977	1.1422	1.1453	2.6665	7.1693	1.4383	4.4768
0.99	1.4429	1.1225	2.5659	1.1422	1.1453	2.6665	7.1693	1.4383	4.4768

We can conclude from the Table.8 that the DRO method can obtain a balanced dispatch solution in robustness and economic perspectives compared with the traditional SO and RO methods. The first stage results of the DRO method is smaller than that of the RO and larger than that of SO. Under the second-stage simulation results, the DRO shows a better expected performance in both averaged and worst-case probability distribution scenarios compared with the SO and RO methods.

5. Conclusions

From the simulation results above, we can draw the following conclusions:1)By simultaneously considering norm-1 and norm-inf, the proposed model exhibits an economic advantage with the lowest total cost. The total cost becomes larger with the increase of the amount of available historical data.2)

The decisions are more conservative and the total cost increases when the confidence level settings α_1 and α_∞ increase.3) Compared with SO and RO, the DRO approach can obtain a balanced dispatch solution in robustness and economic perspectives. The first-stage cost of DRO is smaller than that of RO and larger than that of SO. Also, DRO shows a better expected performance in both averaged and worst-case probability distribution scenarios.

Acknowledgments
This work was supported in part by State Grid Corporation of China under Grant SGSCJY00JHJS201700009, and in part supported by the Fundamental Research Funds for the central Universities under Grant YJ201750.

References

[1] Dong Zhaoyang, Zhao Junhua, Wen Fushuan. From Smart Grid to Energy Internet: Basic Concept and Research Framework. Automation of Electric Power Systems, 2014, 38(15):1-11
[2] Wei Zhinong, Zhang Side, Sun Guoqiang. Carbon Trading Based Low-carbon Economic Operation for Integrated Electricity and Natural Gas Energy System. Automation of Electric Power Systems, 2016, 40(15):9-16
[3] Zheng J H, Wu Q H, Jing Z X. Coordinated scheduling strategy to optimize conflicting benefits for daily operation of integrated electricity and gas networks. Applied Energy, 2016, 192.
[4] Wei Zhinong, Zhang Side, Sun Guoqiang. Power-to-Gas Considered Peak Load Shifting Research for Integrated Electricity and Natural-Gas Energy Systems. Proceedings of the CSEE, 2017, 37(16):4601-4609
[5] Simonis B, Newborough M, Simonis B. Sizing and operating power-to-gas systems to absorb excess renewable electricity. International Journal of Hydrogen Energy, 2017, 42(34).
[6] Estermann T, Newborough M, Sterner M. Power-to-gas systems for absorbing excess solar power in electricity distribution networks. International Journal of Hydrogen Energy, 2016, 41(32):13950-13959.
[7] Li Yang, Liu Weijia, Zhao Junhua. Optimal Dispatch of Combined Electricity-Gas-Heat Energy Systems with Pow-er-to-Gas Devices and Benefit Analysis of Wind Power Accommodation. Power System Technology，2016, 40(12):3680-3688
[8] Li G, Zhang R, Jiang T. Security-constrained bi-level economic dispatch model for integrated natural gas and electricity systems considering wind power and power-to-gas process. Applied Energy, 2017.
[9] Alabdulwahab A, Abusorrah A, Zhang X. Coordination of Interdependent Natural Gas and Electricity Infrastructures for Firming the Variability of Wind Energy in Stochastic Day-Ahead Scheduling. IEEE Transactions on Sustainable Energy, 2015, 6(2):606-615
[10] He C, Wu L, Liu T. Robust Co-Optimization Scheduling of Electricity and Natural Gas Systems via ADMM. IEEE Transactions on Sustainable Energy, 2017, 8(2):658-670.
[11] Liu Yixin, Guo Li, Wang Chenshan. Economic Dispatch of Microgrid Based on Two Stage Robust Optimization. Proceedings of the CSEE, to be published.
[12] Zhou Renjun, Min Xiongbang, Tong Xiaojiao. Distri-butional Robust Optimization under Moment Uncertainty of Environment and Economic Dispatch for Power System. Proceedings of the CSEE, 2015,35(13):3248-3256.
[13] Chen Y, Wei W, Liu F. Distributionally robust hydro-thermal-wind economic dispatch. Applied Energy, 2016, 173:511-519.
[14] Bian Q, Xin H, Wang Z. Distributionally Robust Solution to the Reserve Scheduling Problem With Partial Information of Wind Power. IEEE Transactions on Power Systems, 2015, 30(5):2822-2823.
[15] Delage E, Ye Y. Distributionally Robust Optimization under Moment Uncertainty with Application to Data-Driven Problems. Operations Research, 2010, 58(3):595-612.
[16] Ding T, Yang Q, Yang Y. A Data-Driven Stochastic Reactive Power Optimization Considering Uncertainties in Active Distribution Networks and Decomposition Method. IEEE Transactions on Smart Grid, 2017, to be published.
[17] Zhao C, Guan Y. Data-Driven Stochastic Unit Commitment for Integrating Wind Generation. IEEE Transactions on Power Systems, 2016, 31(4):2587-2596.
[18] GaoHongjun, Liu Junyong, Wei Zhenbo. The Security-Constrained Dispatching Model With Wind Generator Based on the Extreme Scenario Set Optimization. Power System Technology，2013, 37(6):1590-1595

IOP Conf. Series: Materials Science and Engineering **366** (2018) 012017 doi:10.1088/1757-899X/366/1/012017

A Demand Response Scheme for Wind Integrated Power System Based on Mechanism Design Theory

Xueliang Li[1], Donglei Sun[1], Yumin Zhang[2a], Xueshan Han[2], Mingqiang Wang[2], Si Yang[1], Wenbo Li[3] and Xin Wei[1]

[1] Economic & Technology Research Institute of State Grid Shandong Electric Power Company, Jinan, China
[2] Key Laboratory of Power System Intelligent Dispatch and Control of Ministry of Education, Shandong University, Jinan, 250061, China
[3.] State Grid Shandong Electric Power Research Institute Jinan, China

[a] Corresponding author: zymcth@126.com

Abstract. The increasing integration of high penetration of wind power into power systems have created great challenges for stable operation of the power system. It is important to design a reasonable demand response scheme (DRS) to make decision the optimal load relief, specially, for the wind power suppliers (WPS) and their customers (WPSC). Therefore, this paper proposes a demand response scheme for wind integrated power system based on mechanism design theory. Firstly, we derive the outage cost function and incentive function modeled by the two variables: the customers' type parameter and the location value. Secondly, incorporate customers' type parameter and locational flexibility into the DRS, and based on game theory, a demand response scheme with incentive compatible constraint and participation constraint is designed to maximize the benefit both the wind power suppliers and their customers. The implication of demand response scheme is to design an incentive structure that encourages customers to sign up for the right scheme and reveal their true value of load reduction. Finally, the eight-bus system example can demonstrate the effectiveness of the proposed method.

1. Introduction

With the large-scale penetration of wind generations in power grid and the increasing load demand from customers, power systems are becoming increasingly time-varying stochastic at both demand and generation sides. Although wind generations can reduce the stress between supply and demand to some extent, their inherent volatility, uncertainty and intermittent [1] will make its quality as a power product is inferior to conventional energy and further influence the safe and economic operation of power system. The development of the power system is faced with two challenges, one is the exacerbation of issues of system security and stability caused by renewable energy integration and market competition, the other is the urgency for efficient use of energy which we must face due to the exhaustion and pollution of fossil energy. Therefore, it has become the strategy for the power system development to promote implementation of various demand response (DR) measures [2].

As an important means of demand side management (DSM), interruptible load (IL) provides an effective way to solve the above problems [3]. By signing a contract in advance, users' loads can be interrupted according to the contract during the peak periods, users can get some certain compensation or price discounts [4]. IL increases the reserve power of system and improved the reliability of power system, which enable it has a good prospect of application [5]. The present research of DR mainly focuses on the peak clipping, load shifting [6], reserve services [7], congestion management [8], improving power system reliability [9], increasing operating benefits [10], etc. Research on the mechanism design of DR participating in power grid operation still should be improved. In [11], an incentive-compatible mechanism is designed to encourage consumers to reveal their true type parameters, which could ensure the effective implementation of DR programs. In [12], DR compensation mechanism is constructed from the perspective of day-ahead scheduling. [13] designs a compensation mechanism in which IL is considered as the reserve power of system. [14] considers the uncertainty of distributed generation (wind power), constructs a market-based mechanism to control and coordinate the interconnection between distributed generation and power grid. All show that the critical issue is how to design DRS for the wind power suppliers and their customers to alleviate stress on the power grid through load curtailment.

Therefore, on the basis of fully analyzing and grasping the characteristics of market participants (supply side and demand side). In this paper, a rational demand response scheme (DRS) for wind integrated power system is designed which considers the interests of the demand side and generating side. Firstly, we derive the outage cost function and incentive function for different type of customers at critical locations. Therefore, the customer type parameter, as a parameter to characterize the customer performance in DRS, and the locational value parameter, as a parameter to characterize importance of critical location, are two significant variables in these studies. Then, based on game theory, a demand response scheme with incentive compatible constraint and participation constraint is designed to makes sure that the wind power supplier benefit is maximized and their customers take participate in the DRS voluntarily since they are given a satisfactory compensate fee to do so. The idea behind demand response scheme is to design an incentive structure that encourages customers to sign up for the right scheme and reveal their true value of load reduction. The eight-bus system example can illustrate the effectiveness of the proposed method.

2. Demand response scheme design for wind power supplier

The outage cost of the customer to reduce their load demand depends on the curtailment amount of customers and customers' preference for curtailment. The assumption is that outage cost $c(\theta,x)$ of the customer θ for curtailing x MW is:

$$c(\theta,x)=K_1x^2+K_2x-K_2x\theta \tag{1}$$

where θ is used to characterize customers' preference for curtailment probabilistically, and we assume this "preference parameter" θ possesses a uniform probability distribution $f(\theta)$ in the interval [0,1]. It sorts the customers from the least willing to the most willing to curtail load, and let $F(\theta)=\int_0^{\theta} f(\theta)\, d\theta$. K_1 and K_2 are the quadratic and linear term coefficients of the outage cost function, respectively. These parameters are private customer information that is unknown to the supply-side. They are all positive, and they are estimated using the existing data. For the sake of simplicity, this paper assumes that K_1 =0.5, and K_2 =1 [15]. In our previous work [16-17], we give a simple example to illustrate the derivation process of the assumed values of K_1 , K_2 and all the θ .

Customer is rational enough, if the customers do not receive any compensation for their demand reduction, they will not choose to participate DR program. It means that customers will self-select the amount of demand curtailment x MW based on their monetary compensation y offered by energy supplier in order to maximize their utility naturally. The customers' benefit $u_1(x,y,\theta)$ is defined as the amount of monetary compensation they received from energy supplier minus their outage cost due to the curtailment.

$$u_1(x,y,\theta)=y-c(x,\theta) \tag{2}$$

Wind power suppliers' benefit function u_2 generated by the load curtailment of customer j can be expressed as

$$u_2 = (\lambda_w + \lambda_s)x - y = \lambda x - y \tag{3}$$

where λ is spot market price, $\lambda = \lambda_w + \lambda_s$. λ_w is the on-grid tariff of wind power. λ_s is the price subsidies for wind power supply per MWh provided by Chinese government in order to promote the penetration of wind generation. Usually, the electricity subsidy is 0.25-0.3 \$/kWh. Here, this article takes $\lambda_s = 0.25\,\$/\,\mathrm{KWh}$.

The aim of designing DR compensation mechanism is to make optimal compensation fee scheme for a customer about some amount MW of load curtailment, so as to reduce hidden danger of security in power grid. It is actually a mathematical optimization problem, which should meet the objective of maximizing the total utility for DR. The objective function is

$$\max_{X(.),Y(.)} E_\theta u_0(X(\theta,\lambda_w), Y(X(\theta,\lambda_w)), \lambda_w) \tag{4}$$

Both energy supplier and customer are rational. Energy supplier will only provide compensation for customers at more critical locations due to it can reduce hidden danger of security in power grid by curtailing some amount of load. Customers will not curtail load unless there is a positive incentive for them. Thus both them subject to the individual rational constraint (IRC):

$$u_1(X(\theta,\lambda_w), Y(X(\theta,\lambda_w)), \theta) \ge 0 \tag{5}$$

And the incentive compatibility constraint (ICC) [16].

$$u_1(X(\theta,\lambda_w), Y(\theta,\lambda_w), \theta) \ge u_1(X(\tilde{\theta},\lambda_w), Y(X(\tilde{\theta},\lambda_w)), \theta) \tag{6}$$

Constraint (6) illustrates that if a customer θ_j reports his information $\tilde{\theta}_j$ wrongly, his benefit will not maximum, it is not wise for the customer to lie. Hence, (6) can prevent customer from lying about θ, and it ensures customer to select their special compensation scheme according to their true willing.

Wind power suppliers can obtain the amount of voluntary reduction of load by each user through questionnaires or surveys, and determine the outage cost for each customer according to the incentive compatible contract. This maximization problem are expressed through the mechanism design and display principle using the distribution function parameterization method to obtain the contract expression of the continuous variable.

$$x(\theta) = \begin{cases} 0 & if\, 0 \le \theta < \frac{7}{8} - \frac{\lambda_w}{2} \\ 2\theta + \lambda_w - \frac{7}{4} & if\, \frac{7}{8} - \frac{\lambda_w}{2} \le \theta < 1 \end{cases} \tag{7}$$

$$y(\theta) = \begin{cases} 0 \quad if\, 0 \le \theta < \frac{7}{8} - \frac{\lambda_w}{2} \\ \theta^2 - \frac{3}{2}\theta + 2\theta\lambda_w + \frac{3}{4}\lambda_w^{\ 2} - \frac{13}{8}\lambda_w + \frac{35}{64} \\ \quad if\, \frac{7}{8} - \frac{\lambda_w}{2} \le \theta \le 1 \end{cases} \tag{8}$$

Therefore, the proposed mechanism design depends on the two important parameters: the local value λ_w and the customer preference parameter θ .

3. Simulation Results

The numerical experiments are carried on the eight-bus system to demonstrate the validity and effectiveness of the proposed model. The eight-bus system with 2 thermal units, 6 load notes, and 9 branches are shown in Figure 1. All the parameters can be found in [16]. We use wind units instead of thermal units. The total load capacity of the eight-bus system is 315MW, assuming that the maximum output of each wind power unit is 2.5MW, and the total amount of wind power installed in the system is 140. So the total wind output is 350MW. Due to the uncertainties and intermittency of wind power, and the impact of wind speed and geographical location, the output efficiency of wind power is hard to reach 100%, so the utilization rate of wind power is not very high. The wind power profile and the load profile, as well as the difference between them are shown in Figure 2. The example involves three stages to illustrate the proposed optimal demand response scheme can benefit both the interests of the demand side and wind generating side.

Stage 1: Sensitivity analysis to determine the most valuable loads location

Stage 2: The Optimal DRS for wind power suppliers (WPS) and their customers (WPSC).

Stage 3: Compare the DRS for thermal energy suppliers (TES) and their customers (TESC) [15-16], as well as WPS and WPSC under different scenarios

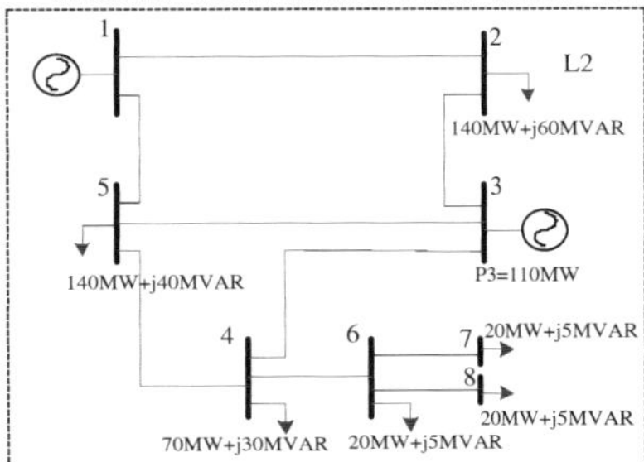

Figure 1. Line diagram of eight-bus system.

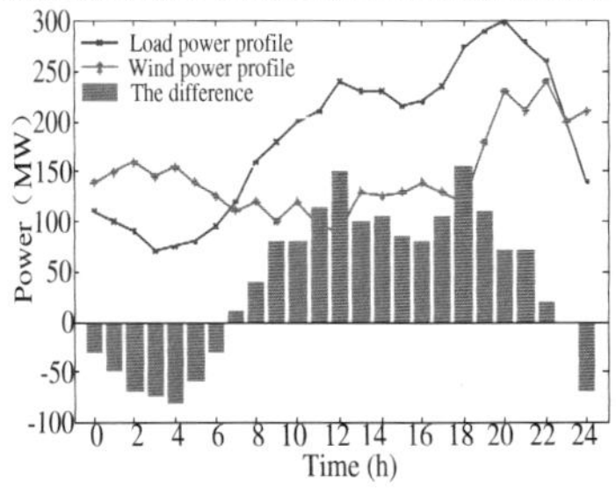

Figure 2. Wind power profile and load profile.

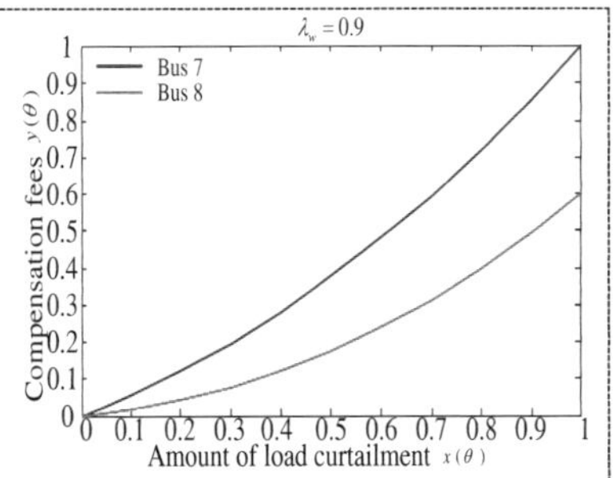

Figure 3. The DRS design for customers at bus 7 and at bus 8.

3.1. Stage 1 Analysis

The aim of sensitivity analysis is to find the most vulnerable load nodes and arranging all of them. The sensitivity of the loading margin to voltage collapse with respect to a change in each load is shown in Table 1. It can be noted that the most valuable loads buses are 7 and 8. This means that the location 7 and location 8 are the most helpful in case of wind power shortage, line congestion, voltage collapse, etc. Based on these ranks, the electrical market operation will make decision the quantitative impact of loads location information on the reliability of the system. Then electrical market operation design the specific mechanism for different location customers to curtail some amount of load when required.

3.2. Stage 2 Analysis

After the relative value of all load locations is determined, another critical customer attributes, customer private information are list in Table 2. It can be seen that the bus 7 and bus 8 have the same location value (equivalent sensitivity), but have different DRS for them due to they have different customer private information. Figure 3 illustrates the optimal mechanism design. As shown in Figure 3, the customers at bus7 and at bus 8 will sign up for the different demand response scheme even though they are at equally sensitive locations. This is because the customer at bus 7 has a higher marginal cost for load reduction than the customer at bus 8.

In the following, we will discuss how do the two critical parameter (λ_w and θ) impact the design of DRS.

As shown in Table 2, the bus 4, bus 6, bus 7 and bus 8 have the same location value $\lambda_w = 0.9$. According to (7) and (8), when fix $\lambda_w = 0.9$, the formulations of load curtailment function and the compensation function as:

$$x(\theta) = \begin{cases} 0 & 0 \le \theta < 0.425 \\ 2\theta - 0.85 & 0.425 \le \theta \le 1 \end{cases} \tag{9}$$

$$y(\theta) = \begin{cases} 0 & 0 \le \theta < 0.425 \\ \theta^2 + 0.3\theta - 0.308125 & 0.425 \le \theta \le 1 \end{cases} \tag{10}$$

Table 1. Sensitivity of the loading margin to voltage collapse with respect to each load.

Load Bus	Sensitivity (MW/MW)
2	-0.03
4	-0.89
5	-0,12
6	-0.48
7	-1.73
8	-1.73
Load marginal =36.18MW	

Table 2. Customer attributes.

Customer	Locational Value(λ)	Customer Private Information(θ)
Bus 2	0.4	0.7
Bus 4	0.9	0.8
Bus 5	0.8	0.7
Bus 6	0.9	0.8
Bus 7	0.9	0.5
Bus 8	0.9	0.9

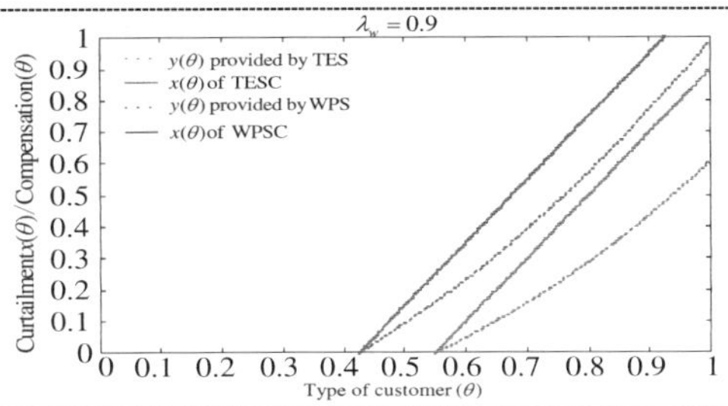

Figure 4. The DRS for customers when fix the locational value $\lambda_w = 0.9$.

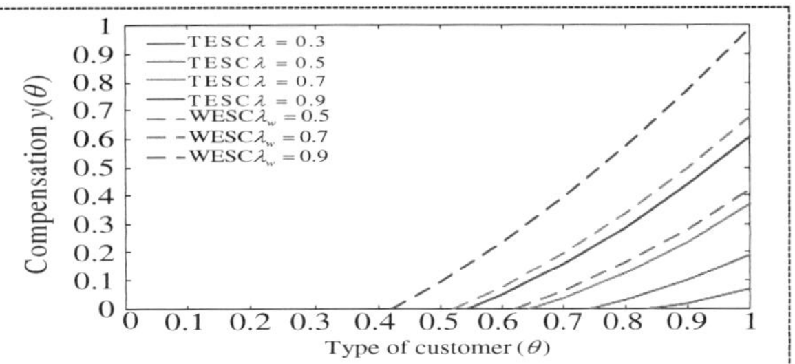

Figure 5. The compensation fee under different locational value.

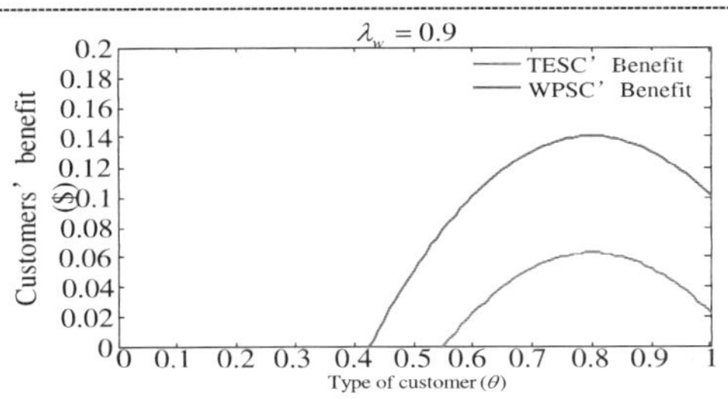

Figure 6. The customer' benefit of WPSC and TESC when their true preference parameter $\theta = 0.8$, and locational value $\lambda_w = 0.9$

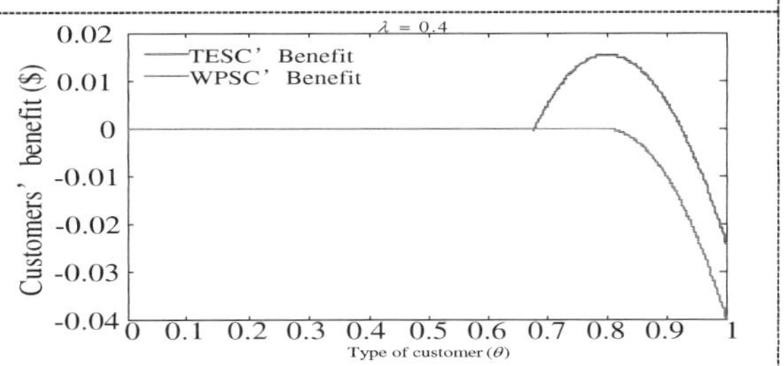

Figure 7. The customer' benefit of WPSC and TESC when their true preference parameter $\theta = 0.8$, and locational value $\lambda_w = 0.9$

Both the load curtailment $x(\theta)$ and the compensation fee $y(\theta)$ are the function of the customer preference parameter θ . Figure 4 design the specific DRS for customers with the locational value $\lambda_w = 0.9$. Figure 5 shows that the number of customers participating DR project varies with the value of λ_w.

The role of the customer preference parameter θ is obvious, because it determines the type of customer (which in turn determines the cost of their load curtailment). However, the role of local value λ_w may be subtle. Since the local value λ_w is a parameters of economic analysis from engineering, which indicates that the customer's location is one of the most important aspects to design the optimal DRS. Some locations will be more expensive to transmit electricity than others. It makes sense for electricity market operation to design the specific DRS for customers at expensive locations.

Figure 6 shows the customer' benefit of WPSC and TESC in bus 4 and bus 6, respectively. Their true preference parameter $\theta = 0.8$, and locational value $\lambda_w = 0.9$. As shown in Figure 6, obviously, only the user report his true value, his interest is the biggest of all. According to the formula (9) and (10), the calculation:

Load curtailment: $x(\theta)=7.50\text{MW}$; Compensation fee: $y(\theta)=\$5718.75$; Customers' benefit: $V_1(\theta,x,y)$ =\$1406.25.

Customer is rational, in order to obtain more compensation fee, he may lie. If the customer with his true value $\theta=0.8$ sign up the scheme for the customer with $\theta=0.9$, then their contract will be:

Load curtailment: $x(\theta)=9.50\text{MW}$; Compensation fee: $y(\theta)=\$7718.75$; Customers' benefit: $V_1(\theta,x,y)$ =\$1306.25.

It is easy to find out that the benefit of the customer who lies is reduced, so the customer is best to tell their true value and sign up for a scheme that is specifically designed for him.

In addition, compare with the DRS for TES and TESC, the DRS for WPS and WPSC not only increase the benefit both WPS and their customers, but also encourages more customers to sign up for the right scheme and reveal their true value of load reduction. All the merits improve the utilization efficiency of wind power and promote the accommodation of wind generation.

According to (7) and (8), take bus 2 as an example, when fix $\lambda_w=0.4$, the formulations of load curtailment function and the compensation function as:

$$x(\theta)=\begin{cases}0 & 0\le\theta<0.425\\ 2\theta-0.85 & 0.425\le\theta\le1\end{cases} \tag{11}$$

$$y(\theta)=\begin{cases}0 & 0\le\theta<0.425\\ \theta^2+0.3\theta-0.308125 & 0.425\le\theta\le1\end{cases} \tag{12}$$

As mentioned above, the different customers have different preference parameter θ. The sensitivity is also different at different load location. Thus the amount of load curtailment and compensation fee are different. As shown in Figure 7, the thermal energy suppliers' customer signing up for the DRS at bus 2 will not benefit. Their benefit is zero or negative according to formula (11) and (12). However, if the wind power suppliers' customers to sign up for the DRS, the customers in the range of $\theta\in[0.675,0.925]$ can still benefit. The benefit of customers at bus 2 and interval of $\theta\in[0,0.675]$ is zero, and it is negative in the interval of $\theta\in[0.925,1]$. Through the comparison analysis, wind power suppliers provide a DRS for the customers at the more expensive location, it also can make some customers benefit. This illustrates that the DRS for wind power suppliers has more extensively applicable scope, which may bring more benefits to customers. On the on hand, it encourage more customer to sign up for the right scheme and reveal their true value of load reduction. On the other hand, it promotes the accommodation of high penetration of wind generation, which achieves a win-win result.

Table 3 lists the optimal DRS results for wind and thermal power.

Table 3. A optimal result of DRS.

Customer	Thermal		Wind generation	
	$x(\theta)$ / (MW)	$y(\theta)$ / (\$)	$x(\theta)$ / (MW)	$y(\theta)$ / (\$)
Bus 2	0.00	0.00	0.5	168.75
Bus 4	5.0	2875	7.5	5718.75
Bus 5	2.00	900	4.5	2868.75
Bus 6	5.00	2875	7.5	5718.75
Bus 7	0.00	0.00	1.5	918.75
Bus 8	7.00	4375	9.5	7718.75
Optimal Load Relief=19 MW			Optimal Load Relief=31 MW	
Increase in Margin=24.34 MW			Increase in Margin=29.86 MW	

Table 4. Non-optimal result of DRS when fix $\lambda=0.7$.

Customer	Thermal		Wind generation	
	$x(\theta)$ / (MW)	$y(\theta)$ / (\$)	$x(\theta)$ / (MW)	$y(\theta)$ / (\$)
Bus 2	1.00	375.00	3.5	1968.75
Bus 4	3.00	1275.00	5.5	3368.75
Bus 5	1.00	375.00	3.5	1968.75
Bus 6	3.00	1275.00	5.5	3368.75
Bus 7	0.00	0.00	0.00	0.00
Bus 8	5.00	2375.00	7.5	4968.75
Optimal Load Relief=13 MW			Optimal Load Relief=25.5 MW	
Increase in Margin=15.91 MW			Increase in Margin=21.035 MW	

As shown in Table 3, under the same DRS mechanism, the penetration of wind generation integrated into power grid increase the amount of customers participating in demand response projects. The amount of load curtailment has increased by 38.7%, and the load margin increased by 18.5%. Which will improve the safety margin of the system and alleviate the pressure of transmission line more effectively during the peak and the failure period, thus improve the safety level of the system.

3.3. Stage 3 Analysis

In order to emphasize the importance of the two parameters: The local value λ_w and the customer preference parameter θ. Table 4 gives the non-optimal results of DRS for WPSC and TESC when fixed $\lambda_w = 0.7$, respectively. It can be seen that the amount of load curtailment for WPSC is more than it for TESC, and the result is also a smaller increase in the loading margin (LM).

Table 5 gives the non-optimal results of DRS for WPSC and TESC when fixed $\theta = 0.7$, respectively. It can be seen that the amount of load curtailment for WPSC is more than it for TESC, and the result is also a smaller increase in the LM. The DRS for WPSC has an obvious advantage in terms of both the amount of load curtailment and the increase in load margins better than DRS for TESC.

Table 5. Non-optimal result of DRS when fix $\theta = 0.7$.

Customer	Thermal		Wind generation	
	$x(\theta)$ / (MW)	$y(\theta)$ / ($)	$x(\theta)$ / (MW)	$y(\theta)$ / ($)
Bus 2	0.00	0.00	0.5MW	168.75
Bus 4	3.00	1575.00	5.5MW	391 8.75
Bus 5	2.00	900.00	4.5MW	2868.75
Bus 6	3.00	1575.00	5.5MW	391 8.75
Bus 7	3.00	1575.00	5.5MW	391 8.75
Bus 8	3.00	1575.00	5.5MW	391 8.75
Optimal Load Relief=14MW			Optimal Load Relief=27MW	
Increase in Margin=14.73MW			Increase in Margin=27.12 MW	

We carry on the comprehensive comparison to the three scenarios, as shown in Table 6. It is concluded that only when the optimal DRS enables the WPS, TES and WPSC, TESC to get maximum benefit. Namely, only the customer tell their true private information and sign up for the specific DRS, their benefit is the most. The amount of load curtailment is most and the load margin increase is also the largest. Which reduces transmission line overload, ensure the safety of the system, thus avoid the occurrence of accidental events such as voltage collapse. It also proves the absolute advantage of the penetration of wind power.

Table 6. The Comparison results under different scenarios.

	Scenarios	C(MW)	LM(MW)	SP($)	CP($)
Thermal	Optimal	19	24.34	5875.00	2775.00
	$\lambda_w = 0.7$	13	15.91	3425.00	1125.00
	$\theta = 0.7$	14	14.73	5200.00	1000.00
Wind	Optimal	31	29.86	11837.5	12387.5
	$\lambda_w = 0.7$	25.5	21.035	10181.25	3531.25
	$\theta = 0.7$	27	27.12	11637.5	8887.5

4. Conclusions

This paper proposes a demand response scheme for wind integrated power system based on mechanism design theory. Under the scenario of power shortage, we carry on the compare the DRS for WPSC with TESC. In the perspective of utility theory, the DRS for wind power suppliers has more extensively applicable scope, which may bring more benefits to customers. On the on hand, it encourage more customer to sign up for the right scheme and reveal their true value of load reduction. On the other hand, it promotes the accommodation of high penetration of wind generation. In addition, the government subsidy policy for wind power supplier, further increase the utilization efficiency of wind power, which achieve a win-win result. From the perspective of stable and reliable operation of the system. During peak periods, when the system is suffering from peak load or emergencies, customers can curtail the load actively or controllably according the DRS' requirements to alleviate the stress brought on the system by events and contingencies.

Acknowledgment
This paper is supported by the project of Economic & Technology research Institute, State Grid Shandong Electric Power Company, and National Natural Science Foundation of China (51407106).

References

[1] Zhao C Y, Wang J H, Watson J P, Guan Y P. Multi-Stage Robust Unit Commitment Considering Wind and Demand Response Uncertainties. *IEEE Trans. on Power Syst.*, 2013, 28, 3, pp. 2708-2717.
[2] Jason W, Black. Integrating Demand into the U.S. Electric Power System: Technical, Economic, and Regulatory Frameworks for Responsive Load. *Massachusetts Institute of Technology*, Boston, 2005.
[3] Singh S N, 0stergaard J. Use of Demand Response in Electricity Markets: An Overview and Key Issues. *International Conference on the European Energy Marke*t, Madrid, 2010, pp. 1-6.
[4] Quan S S, Cao J, Chen X H. Research on Interruptible Load Management in Power Market. *2014 China International Conference on Electricity Distribution (CICED)*, Shenzhen, 2014, pp. 1127-1130.
[5] Billinton R, Lakhanpal D. Impacts of Demand-side Management on Reliability Cost/Reliability Worth Analysis. *IEE Proceedings-Generation, Transmission and Distribution*, 1996, 143, 3, pp. 225-231.
[6] Kirschen D S. Demand-side View of Electricity Markets. *IEEE Trans. on Power Syst*, 2003, 18, 2, pp. 520-527.
[7] Xia L M, Gooi H B, Bai J. Probabilistic Spinning Reserves with Interruptible Loads. *Power Engineering Society General Meeting*, 2004, 1, pp. 146-152.
[8] Yu C W, Zhang S, Chung T S, Wong K P. Modeling and Evaluation of Interruptible Loads in Electricity MarketS," *IEE Proceedings-Generation, Transmission and Distribution*. 2005, 152, 5, pp.581-588.
[9] Borsvert R N, Cappers P A, Neenan B. The Benefits of Customer Participation in Wholesale Electricity Markets. *The Electricity Journal*, 2002, 15, 3, pp. 41-51.
[10] Fahrioglu M, Alvarado F L. Using Utility Information to Calibrate Customer Demand Management Behavior Models. *IEEE Trans. on Power Syst.*, 2001, 16, 2, pp. 317-322.
[11] Fahrioglu M, Alvarado F L. Designing Incentive Compatible Contracts for Effective Demand Management. *IEEE Trans. on Power Syst.*, 2000, 15, 4, pp. 1255-1260.
[12] Li H Y, Li Y Z, Li ZY. A Multi-period Energy Acquisition Model for a Distribution Company with Distributed Generation and Interruptible Load. *IEEE Trans. on Power Syst.*, 2007, 22, 2, pp.588-596.
[13] Bai J, Gooi H B, Xia L M, Strbac G. Venkatesh B. A Probabilistic Reserve Market Incorporating Interruptible Load. *IEEE Trans. on Power Syst.*,2006, 2, 3, pp. 1079-1087.
[14] Bitar E Y, Rajagopal R, Khargonekar P P. Bring Wind Energy to Market. *IEEE Trans. on Power Syst.*, 2012, 27, pp.1225-1235.
[15] Jian X H, Zhang L, Miao X F, Zhang Y M, Han X S. Designing interruptible load management scheme based on customer performance using mechanism design theory. *Electrical Power and*

Energy Systems, 2018, 95, pp.476–489.

[16] Zhang Y M, Han X S, Zhang L, Xu B, et al. Integrated generation–consumption dispatch based on compensation mechanism considering demand response behavior. *J. Mod. Power Syst. Clean Energy*, 2017, pp.1-17. (In press)

[17] Zhang Y M, Han X S, Xu B, Zhang L, Sun D L, et al. Unit commitment considering demand response behavior based on utility maximization. *2017 IEEE Conference on Energy Internet and Energy System Integration (EI2)* , Beijing, 2017, pp.1-6.

The Research on the Braking System of the Small Danielle Wind Generators

Na LI[1], Xiaoliang WANG[2], Zhenhua ZHU[1], Ying LIN[1] and Jie HUI[2]

[1] State Grid Shandong Electric Power Research Institute Jinan, China;
[2] State Grid Shandong Electric Power Maintenance Company

273580859@qq.com

Abstract. By researching on the mathematical model of small danielle wind generators, it is easily to find that reasonable load connection and disconnection or changing the load characteristics could control the rotational speed and output power of wind generators. Using the controlling strategy combined with switching braking resistor and three-phase short-circuit braking, the rotational speed of wind generator could be controlled accurately in the scenes of sustainable strong wind or the wind speed continuously increasing. At the same time, users' cost could be ensured. A lot of field experiments had proved the feasibility and practicality of the controlling strategy. The research of this paper would promote the application of small danielle wind generators.

1. Introduction

Along with the economic development and environment deterioration, the energy problem has become a problem urgent to be solved. Developing environment friendly and increasing the utilization of renewable energy had become increasingly important subject.

Wind energy was a typical renewable energy with large content, which had been paid more attention for generating. The wind energy was 10 times of the hydropower in the world and the wind energy distributed in almost all the countries and regions. The energy which wind energy could be transformed into power was 53 billion kW•h[1]. As the important part of wind generating system, small wind generators had the advantages of low cost, little environmental influence and convenient installation. Because of the complex topography, large population and residential dispersion in our country, the traditional power grid could not supply power to special areas. Developing small wind generating system could make up the defects of traditional grid[2]. Although the small wind generating technology had a lot of advancement, the safety protection problem under large wind speed scene had not been solved. So the small wind generating and its' grid-connected technology should be improved furthermore.

Danielle wind generators had the characteristics such as simple structure, high tip-speed ratio, and high utilization coefficient of wind energy. Danielle wind generators usually used fixed pitch generating model[3-4]. Restricted by the volume, mechanical structure and cost, general mechanical braking structure could not apply to the danielle wind generators. It was important for reliability operation of the danielle wind generating system to research unloading equipment to control the rotational speed under large wind speed. This paper established the mathematical model of danielle wind generators and provided the controlling strategy combined with switching braking resistor and three-phase short-circuit braking. A lot of field experiments verified that the controlling strategy could accurately control the rotational speed of wind generator, which had feasibility and practicality for engineering application. The research of this paper could apply for small wind generators controlling to improve operating reliability.

2. Mathematical model

2.1 The typical power supply system of danielle wind generators

The small danielle wind generators usually used in off-network scenes and could supply power to loads directly. The typical power supply model of danielle wind generators was composed by wind generator, controller, inverter and load. The model of the system was shown as figure 1. The controller was the key part of the system. The controller could do MTTP controlling strategy, battery controlling strategy and switching controlling strategy of braking resistor considering comprehensively by the output of wind generator, load power demand and the status of charging and discharging for battery.

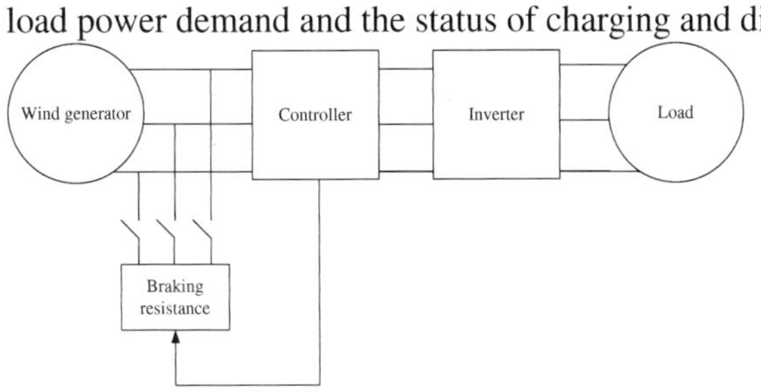

Figure 1. The typical power supply model of danielle wind generators

2.2 Mathematical model

The energy for danielle wind generator accepted by wind could be expressed as formula (1).

$$P_{\text{in}}=\frac{1}{2}\rho AV^{3} \quad (1)$$

In formula (1), ρ -the air density;

A -the sweeping wind area of the wind generator;

V -the wind speed.

The output power of danielle wind generators could expressed as formula (2).

$$P_{\text{out}}=C_P\cdot\frac{1}{2}\rho AV^{3} \quad (2)$$

In formula (2),C_p meaned the utilization coefficient of wind energy. C_p was related to the function of velocity ratio.

$$\lambda=\frac{R\omega}{V} \quad (3)$$

R - radius of the wind generator;

ω - angular velocity of wind generator.

C_p would change when λchanged. The relationship of C_p *and* λ was showed as figure2. C_p would increase firstly and then decrease along with λ increasing and the maximum was close to 0.4.

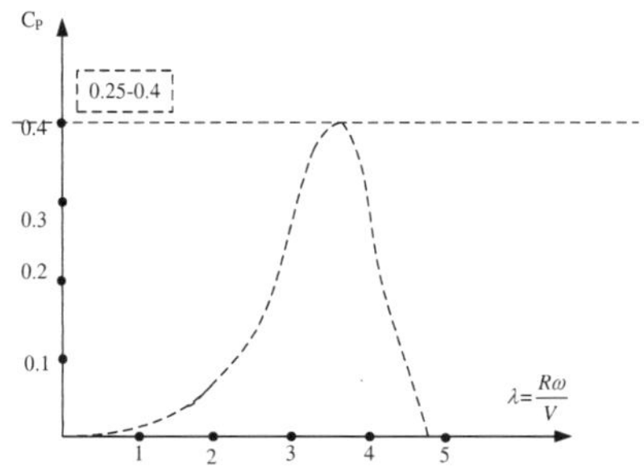

Figure 2. The relationship between Cp and λ

According to formula (1) and (2), the output power of danielle wind generators was effected by C_p. According to formula (3), reasonable switching loads or changing the load characteristics could control C_p accurately and control the output power of the wind generators. So, switching braking resistance could control the rotational speed and output power of wind generators.

3. The design of controlling strategy

Literature [5] provided to control the rotational speed of wind generators using braking resistance. The braking resistance could be adjusted according to the detection result of wind speed so as to control the rotational speed of wind generator. Literature [6] provided that although the small wind generators could not use variable propeller pitch control, it was feasible to improve the fault ride-through capability by serial braking resistance. A lot of field experiments had shown that when encountering large wind or wind speed sustainable increasing during braking procedure, the small danielle wind generators could not realize expected braking effect only using switching braking resistance to control wind generator rotational speed. The related parameters of this braking strategy were shown as table 1.

Table 1. The related parameters of small danielle wind generating braking system

parameters	number
Rated power	300W
Starting wind	2m/s
Rated wind	9-11m/s
Braking Rotational speed	600r/min

The braking resistance used the three-phase Y connection mode and every phase resistance was 1Ω and had 100Ω cement resistance.

In the process of experiments, when the rotational speed of wind generators reached 600r/min, put the braking resistance into operation and sustainable increasing wind speed. The detection results of the wind generators rotational speed and voltage and current were shown as table 2.

Table 2. The test results of braking process

Wind speed(m/s)	Rotational speed(r)	Line voltage(v)	Phase current(A)
11.40	596	24.5	10
11.40	0	0.01	0.2
12.87	0	1.8	0.96
14.69	88	2.85	1.73

16.90	139	4.5	2.75
18.85	186	6	3.7
19.89	210	6.7	4
21.32	252	8.12	4.85
22.10	360	11.85	6.7
23.14	405	13	7.22

According to table 2, when the rotational speed reached 600r/min, the controller got into controlling state and switched the braking resistance. At this time, the rotational speed of wind generators would decrease rapidly. But, when the wind speed continued to increase or the wind speed reached over 23.14m/s, the rotational speed of wind generators could not be controlled as expected. So the simple controlling strategy only by switching braking resistance could not control the rotational speed of wind generator accurately.

Literature [7] and [8] had provided to use NTC resistance to be braking resistance. By the heat detection produced in the braking process, it was easily to adjust the resistance size. The controlling strategy could reduce the controlling difficulty and could realize three-phase short circuit braking when the heat produced was more than heat output. In the state the rotational speed of wind generators could be controlled accurately. The use of NTC would be restricted by the environment and its' using frequency was about 700. In the regions where wind energy was rich, frequent braking control would endanger the safety operation of wind generator. In addition, the NTC cost of large power was high and this method could not satisfy the economic demand of small danielle wind generators.

By theory research and a lot of field experiments, this paper provided an improved braking control strategy. The flow chart was shown as figure 3.

From figure 3 we could know the whole controlling process. When the rotational speed reached the braking rotational speed, the braking resistance could be switched into automatically. At the same time, in this status:

(1) If the wind speed could not increase continuously, the rotational speed could be controlled only by switching braking resistance.

(2)If the rotational speed of wind generator would continue to increase after switching the braking resistance, the second braking step could be carried out. When the voltage and current were in safe range, use three-phase short circuit braking to control the rotational speed.

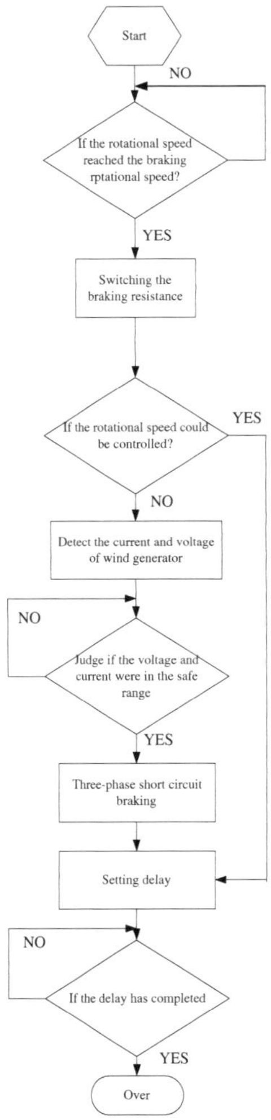

Figure 3. The flow chart of the controlling strategy

The reasonable delay was set in the above two situations. When the wind speed reached safe speed, the brake control could be relieved and the whole braking process was completed. The controlling strategy provided by this paper avoided the braking failure caused by large wind speed or burned out the wind generators caused by short circuit current.

4. Test results

According to the power supply model of danielle wind generator and the improved controlling strategy, a lot of field tests had done to verify the feasibility and accuracy of the controlling strategy. The test results were shown as table 3.

Table 3. The test results of the improved controlling strategy

Wind speed（m/s）	Rotational speed（r）	Line voltage（v）	Phase current（A）
11.25	589	24.4	10
11.25	0	0.01	0.2
12.87	0	1.8	0.96
14.69	88.5	2.85	1.73
14.69	0	0.01	0.2

19.89	0	0.01	0.2
21.32	0	0.02	0.2
23.40	0	0.02	0.2

According to the test results, when the rotational speed reached 600r/min, the controller got into controlling state and switched the braking resistance. At this time, the rotational speed of wind generator would decrease rapidly. The wind speed continued to increase and the rotational speed of wind generators could not be controlled by braking resistance. When the phase current of wind generator reached close to 1.5A, use the three-phase short circuit braking method. As table 3 showed, test results showed that when the wind speed was 14.69m/s, the three-phase short circuit braking method was put into use. Compared with table 2, in the process of wind speed increasing continuously, the improved controlling strategy had better controlling effect and the wind generators could be controlled precisely.

5. Conclusion

This paper used unloading brake method to realize the rotational speed control of small danielle wind generators. The rotational speed could be controlled precisely in the status of wind speed increasing continuously and ensured the economic request of users. Field tests had verified the feasibility and accuracy of the improved controlling strategy. The research had high practical value.

References

[1] Miao W, Zhiqiang Z,Cuiping Z, et al. 2010 Design of variable pitch constant speed system. *Electrical Technology* number 4 pp 62-67.
[2] Xinliang W, Wanzhu G, Jian Z, et al 2014 Discussion of a practical 2D plane calibration approach of non-measured digital camera. *Journal of Green Technology* number 8 pp 289-294.
[3] Zhenyu Z, Jingru C, Tongguang W, et al 2017 Effect analysis of disturbance airflow on pneumatic attack angle of darrieus-type wind turbine. *Acta Energy Solars Sinica* vol 38 number 2 pp 444-449.
[4] Jianlin L, Zhigang G,Xunbo F, et al. 2007 The status and trend of wind power technology in the world and suggestion for future development in China. *Power Supply Ttechnology and Application* vol 10 number 9 pp 42~47.
[5] Freitas W, Morelato A, Xu W. 2004 Improvement of induction Generator stability using braking resistors. *Power Systems, IEEE Transactions* on vol 19, number.2 pp 1247- 49.
[6] Causebrook A, Atkinson D, Jack AG. 2007 Fault ride-through of large wind farms using series dynamic rraking resistors. *Power Systems IEEE Transactions on* vol 22 number 3 pp 966-975.
[7] Guoliang Y, Ruibin G, Chunjiang Z. 2009 Design of braking and protection circuit on vertical axis wind turbine system. *Power Electronics* vol 43 number 10 pp 53-59.
[8] Matsui Y, Sugawara A, Sato S, et al 2007 Braking circuit of small wind turbine using NTC thermistor under natural wind condition. *Power Electronics and Drive Systems* vol 12 number 34 pp 910-915.

Wind farm's clustering equivalent method based on mixed-copula function considering wind speed correlation of complex terrain

Wang Chenggen[1], Zhou Qian[1] and Bu Jing[2]

[1]Power Grid Technology Department State Grid Jiangsu Electric Power Company Research Institute, Nanjing, 211100, China
[2]School of Automation, Nanjing University of Science and Technology, Nanjing, 210094, China

Email: 535462103@qq.com

Abstract. Large-scale grid connected wind farms are the main development trend of wind power in China. In the planning and design stage of wind farms, a simplified equivalent model is often used in wind farm equivalents. It makes the wind farm simply equivalent to a single wind turbine, neglecting the change of wind speed correlation caused by the terrain effect. This makes a large error in estimating wind power. In order to solve this problem, a method based on wind speed correlation is proposed to cluster the wind farm of complex terrain. By using the mixed Copula function with tail characteristics, a wind field correlation model with complex terrain features is built. Based on that, EM function is used to identify the Copula function parameters, and the equivalent clustering method of wind farms is built. Finally, the simulation analysis shows that the complex terrain wind farm clustering equivalent method considering wind speed correlation is verified.

1. Introduction

As an intermittent energy, the randomness and volatility of wind power lead to large errors in its prediction. Meanwhile, because of the existence of Meteorology inertia within a certain area, the wind speed of different wind turbines has a significant correlation. This correlation will be converted to the correlation between wind power[1], which makes the overall characteristics of the wind farm different from the characteristic of the single wind turbine[2], which will lead to a large error in the prediction of the output of the wind farm.

In the equivalent modeling of wind farm, the wind speed of the same type of wind turbine is also different because of the difference of the spatial factors inside the wind farm. The wind speed correlation caused by the wake effect and the topography and geomorphology leads to a large degree of correlation between wind farms. Based on the correlation characteristics of wind speed, the wind turbines are equivalent to simplify the calculation. Many researches have been studied on the multi-machine equivalence or single-machine equivalence in wind farms. References [3, 4] improve the prediction accuracy by improving the prediction method of wind power, but the prediction error can only be controlled at 10%-20%, and the prediction accuracy cannot meet the power balance accurately. Reference [5] optimizes the same row of wind turbines equivalents as one machine on the basis of this. Although the optimization method simplifies the wind farm model, the difference of wind speed distribution of the actual wind farm has great influence on the output of the wind turbine, and there is a large computational error for large wind farms. According to Reference [6], a general statistical rule of cross correlation between adjacent units is obtained based on actual wind farm's data for common rectangular and one wind turbine arrangement modes in wind farms. Other traditional probabilistic clustering methods are based on the K-MEANS algorithm, and the state variable matrix of the wind turbine is used as a classification criterion for clustering. Due to the fluctuation of wind speed, the change of classification and partition is greatly increased due to the fluctuation of wind speed. At the same time, the correlation of wind speed is not considered.

In order to solve this problem, this paper presents a method of clustering equal value of wind speed correlation of wind farm considering the influence of complex terrain. It uses a Mixed-Copula function considering the tail effect. Then based on the wind speed sample data, it uses EM method of mixed Copula function parameter identification. Finally, it uses WAsP software to construct simulation model for algorithm verification.

2. Wind speed correlation model based on Copula theory

2.1 Wind speed correlation and distribution characteristics

The probability distribution of wind speed can be described by the Weibull distribution [7], which expression is shown as (1):

$$P(v) = \frac{k}{v}(\frac{v}{c})^{k-1}\exp[-(\frac{v}{c})^{k}] \tag{1}$$

Where v represents wind speed, k and c represents the shape and scale parameters, which could be calculated by measured data of wind speed.

Tail characteristics are used to describe the degree of interaction between variables when extremely small probability events occur, and the upper and lower tail variables are used to indicate the correlation characteristics of variables tend to be minimum and maximum. Thus the tail characteristics can be used to analyze the correlation between the samples when the wind speed is minimization and maximum.

When the wind speed is close to the minimum and maximum, it is possible to analyze whether the two variables are asymptotically independent or independent by introducing tail characteristics, and it is also a theoretical support for wind turbine convergence based on wind speed. The tail characteristics are common in the field of financial analysis to describe the degree of interaction between variables when extremely small probability events occur[8]. The correlation features of the minimum and maximum variables are represented by the upper and lower tails respectively.

The joint distribution function of random vector (X, Y) is F, and the edge distribution functions are F_1 and F_2 respectively, and the tail correlation is expressed by the limit. The characteristics of the upper and lower tails are as follows:

$$\lim_{u\to 1^{-1}} P(F_1(X) > u \mid F_2(Y) > u) = \lambda_U \tag{2}$$

$$\lim_{u\to 0^{+}} P(F_1(X) < u \mid F_2(Y) < u) = \lambda_L \tag{3}$$

Where, λ_U and λ_L represent the upper and tail correlation coefficients and the lower tail correlation coefficients respectively, respectively, which are represented by the generating element of the Copula function, which are expressed as follows:

$$\lambda_L = \lim_{u\to 0^{+}} \frac{\varphi^{[-1]}(2\varphi(u))}{u} = \lim_{x\to\infty} \frac{\varphi^{[-1]}(2x)}{\varphi^{[-1]}(x)} \tag{4}$$

$$\lambda_U = \lim_{u\to 1^-}\frac{1-\varphi^{[-1]}(2\varphi(u))}{1-u} = 2-\lim_{x\to\infty}\frac{1-\varphi^{[-1]}(2x)}{1-\varphi^{[-1]}(x)} \tag{5}$$

2.2 Copula correlation evaluation index considering wind speed correlation

In order to facilitate describing nonlinear and non normal distribution of random variables, and considering the tail characteristics of correlation distribution, two rank correlation coefficient expressions are introduced. They are Kendall rank correlation coefficient and Spearman rank correlation coefficient.

1) Kendall rank correlation coefficient

In order to represent the nonlinear parameters and non normal distribution, we use τ to represent the Kendall rank correlation coefficient, u, V represent two random variables to determine the correlation. The expression of Kendall rank correlation coefficient is as follows:

$$\tau = 4\int_0^1\int_0^1 C(u,v)dC(u,v) - 1 \tag{6}$$

Where, in the range of τ is [-1,1]. Less than zero of negative correlation degree, greater than zero to represent the positive correlation.

2) Spearman rank correlation coefficient

The expression of the Spearman rank correlation coefficient is similar to the linear correlation coefficient. The specific difference is that one variable is the rank of the sample and the other is the coordinate of the sample. As to the correlation of the expression variables, it is clear that the Spearman rank correlation is more widely applicable and is no longer limited to the nonparametric test of the overall distribution. The expression of the Spearman rank correlation coefficient is as follows:

$$\rho = 12\int_0^1\int_0^1 uvdC(u,v) - 3 \tag{7}$$

Where, ρ is the same Spearman rank correlation coefficient, u and v represent two random variables to determine the correlation.

2.3 Mixed-Copula function model of wind speed correlation with tail characteristics

According to the difference of the fluctuation size, the wind speed fluctuation range is divided into three levels which are high, medium and low [9]. Among them, the high and low wind speed level are shown by the tail characteristics, and the wind speed level in the middle is expressed with the symmetry. In the Copula function, the Gumbel-Copula function and the Clayton-Copula function can be used to characterize the tail features, and the Frank-Copula function is selected to characterize the symmetry. Symmetry can well reflect the distribution characteristics of the middle level of the wind speed, but there is a great difference between the wind speed's tail characteristics. So single Copula function cannot reflect this characteristic synthetically. Therefore, the Mixed-Copula function is selected to fit the wind speed. The three Copula functions are defined as follows [10]:

1) Gumbel-Copula function

The Gumbel-Copula function is more sensitive to the change of the upper end to deviate from the farther and larger data of the center. The function expression is shown as follows:

$$C_2(u,v,\theta_i) = \exp\left\{-\left[(-lnu)^{\theta_i} + (-lnv)^{\theta_i}\right]^{\frac{1}{\theta_i}}\right\} \tag{8}$$

The corresponding generating element is $(-\ln u)^{\vartheta}$

2) Frank-Copula function

The Frank-Copula function can reflect the characteristics of the symmetrical tail. That is, the fitting effect is better for the speed segment of the stroke. The function expression is shown as follows:

$$C_1(u,v,\theta_i) = -\frac{1}{\theta_i}\left[1+\frac{(e^{-\theta_i u}-1)(e^{-\theta_i v}-1)}{e^{-\theta_i}-1}\right] \tag{9}$$

The corresponding generating element is $-ln\frac{(e^{-\theta u}-1)}{(e^{-\theta}-1)}$.

3) Clayton-Copula function

It is very sensitive to the change of the lower tail. So it is better to deviate from the farther and larger data from the center. The function expression is shown as follows.

$$C_3(u,v,\theta_i) = \max\left[(u^{-\theta_i}+v^{-\theta_i}-1)^{-\frac{1}{\theta_i}},0\right] \tag{10}$$

The corresponding generating element is $\frac{1}{\theta}(t^{-\theta}-1)$.

The Mixed-Copula function expression to construct wind speed model considering the tail characteristics is as shown in (11):

$$\begin{aligned} C_M(u,v) &= \sum_i^K \lambda_i C_i(u,v,\theta_i) \\ &= \lambda_1 C_F(u,v,\theta_1)+\lambda_2 C_G(u,v,\theta_2)+\lambda_3 C_C(u,v,\theta_3) \end{aligned} \tag{11}$$

Where, C_i represents different Copula function, λ_i represents weight coefficient of Frank-Copula function, Gumbel-Copula and Clayton-Copula function in the Mixed-Copula function; θ_i represents the correlation coefficient between random variables; u and v represent uniform distribution between the two random variables within [0,1].

3. Wind turbine clustering method considering wind speed correlation

The specific flow of the wind turbine clustering method considering wind speed correlation is shown as Figure 1. Firstly, we use historical wind speed data collected by wind-measuring points in the wind farm to construct Mixed-Copula function as Equ.(11). Secondly, we use EM (expectation maximization) method to estimate parameters of Mixed-Copula function. And then we use inverse transformation of its probability density function to fit wind speed. Finally, we use their rank correlation coefficient to judge the degree of relevance and construct wind farm clustering equivalent model.

In this Mixed-Copula function, λ_i and θ_i are the parameters to be estimated. We use EM method which is the non-hierarchical clustering algorithm to estimate. The correlation coefficient of weight coefficient reflects the characteristics that each tail of Copula function and reflect the relationship of the node velocity, joint probability the distribution of final node velocity of two adjacent columns.

And we can use the rank correlation test on the wind speed of any two wind turbine installation points in the wind farm, the strong correlation is determined by the confidence level. If the correlation of two machines is strong correlation, they could be equivalent. The output of the equivalent unit can be obtained by the conversion of the capacity and wind speed of these two machines.

4. Simulation analysis

The wind farm model was established by WAsP software as an example, the wind farm has 24 Bonus 2MW wind turbines of SIEMENS with height of 60m which are the same types. The arrangement of these 24 wind machines is as shown in Figure 2. Where, the red line represents the contour, blue line and the gray line represents the roughness change. And its wind's rose map is shown in Figure 3.

We use WAsP software to generate wind speed prediction data considering wind speed correlation, which fully considers the wake effect caused by wind turbine arrangement, the surface roughness and the influence of shelter. Table 1 lists a row of four matches' related data.

IOP Conf. Series: Materials Science and Engineering **366** (2018) 012019 doi:10.1088/1757-899X/366/1/012019

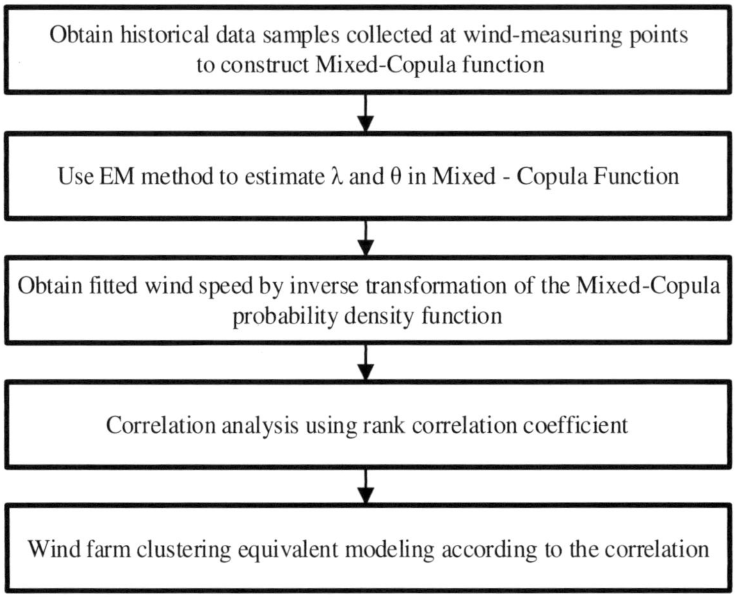

Figure 1. Flow chart of wind turbine clustering method considering wind speed correlation

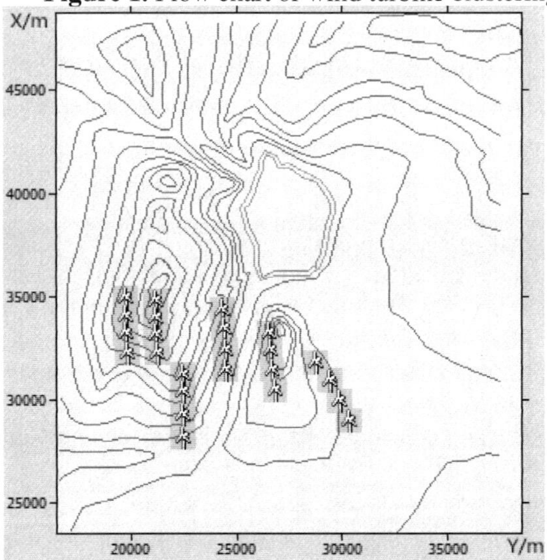

Figure 2. Distribution of Wind Turbines in a complex terrain

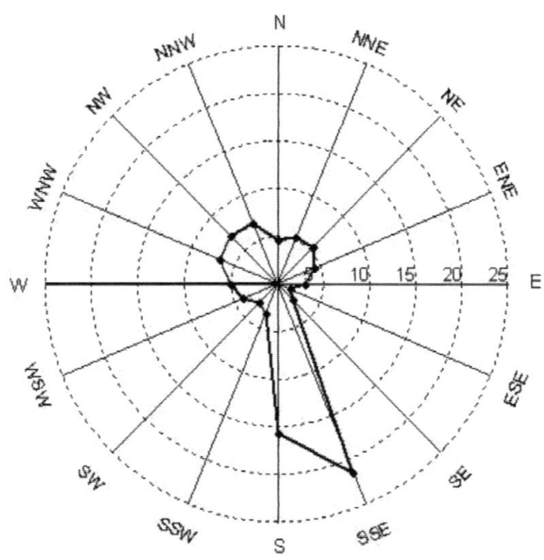

Figure 3. Wind direction distribution roses of the wind farm

TABLE 1. Wind speed prediction data considering wind speed correlation

Position	Coordinate[m]	Wake loss[%]	Wind speed[m/s] Average	Wind speed[m/s] Extreme
1	(24298.8,33871.5)	1.60	7.36	10.88
2	(24342.9,32943.9)	1.87	7.25	10.71
3	(24387.1,31972.2)	2.23	6.88	10.26
4	(24431.3,31000.5)	1.82	7.07	10.50

First, according to the historical data of wind speed, we estimate the two parameter of Weibull, which are the shape and the scale parameter. Through these two parameters, we can generate the wind speed distribution scatter diagram fitting Weibull distribution, as shown in Figure 4. And then according to the historical data of wind speed, we use Mixed-Copula function to construct their distribution, which is shown in Figure 5. Through fitting results, we can see that by constructing the

Mixed-Copula function to fit the joint probability distribution of wind speed, the general distribution characteristic is similar to the Weibull shown as Figure 4. The Weibull fitting curve is fitted by the Mixed-Copula function as shown in Figure 6.

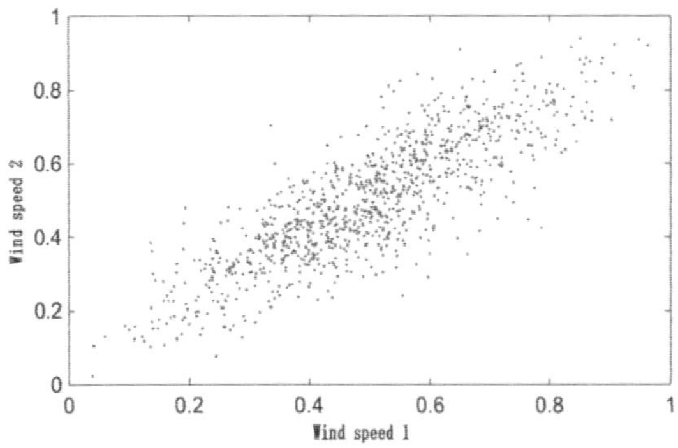

Figure 4. Scatter plot of Binary wind velocity meets Weibull distribution

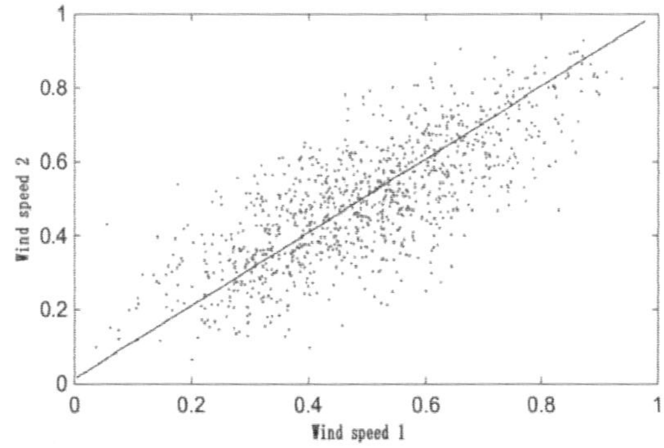

Figure 5. Scatter plot of Binary wind velocity meets Mixed-Copula distribution

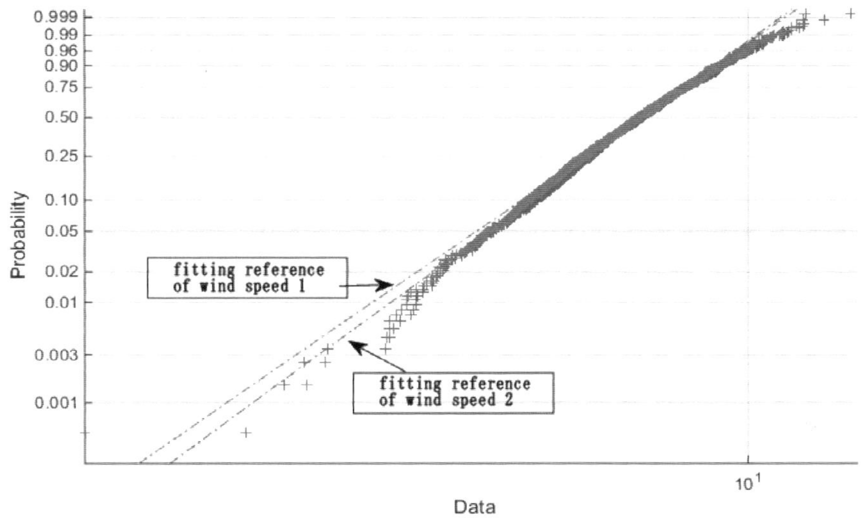

Figure 6. Mixed-Copula function meets the wind velocity

Using the above method to estimate the parameters of λ and θ, the results of which are shown in Table 2. Taking the relevant parameters of the first and second wind turbines as an example, according to the expression, we can make correlation analysis on the wind speed samples of any two wind turbines. A total of 276 sets, and then the equivalent level of wind farms is over 0.99, and the results are aggregated as shown in Table 3.

According to the correlation results, the wind farm is polymerized. After that, the number of the machines changed from 24 to 12 equivalent units. In the above experimental environment, based on the wind speed data obtained from the example and the results of the correlation polymerization, the output of the wind farm considering the correlation of wind speed is calculated. According to each machine using the MPPT control, the traditional single row equivalent simulation and equivalent method considering wind speed correlation clustering results were compared, found that the active power output size differences, through the comparison the cluster correlation of wind speed equivalence than single row equivalent more detailed model.

Figure 7 shows the comparison of the output of wind farm calculated by different equivalent methods. Table 4 shows the accurate value in Figure 7. Through this figure, we can clearly see that the output of which the non-equivalence state is additive, and the clustering equivalent represents 12 machines. The single row equivalent represents 6 machines. The Figure 7 can be seen both in average and extreme wind speed, relative aggregated than single row equivalent active power output of the wind farm output closer to the MPPT model, so the clustering method according to the correlation of wind speed is more effective than the traditional single row equivalent.

TABLE 2. Mixed-Copula function related parameters

Paraments	Gumbel-Copula	Clayton-Copula	Frank-Copula
λ	0.339	0.357	0.304
θ	1.524	7.236	9.146

TABLE 3. The wind turbines polymerization results by the rank correlation coefficient

	Wind turbines' combination	Correlation level
Group 1	(4,14,19,23)	0.996
Group 2	(11,13,20)	0.994
Group 3	(7,10,17)	0.995
Group 4	(1,22)	0.997
Group 5	(3,5)	0.998
Group 6	(6,12)	0.997
Group 7	(8,9)	0.991
Group 8	(18,21)	0.992
Not clustering	16, 24, 2, 15	

TABLE 4. Output comparison between average wind and extreme wind speed

Wind Speed	Active power (MW)		
	Single row equivalence (6 machines)	Clustering equivalence (12 machines)	Non-equivalence (24 machines)
Average	11.343	13.950	16.326
Extreme	32.649	41.333	46.888

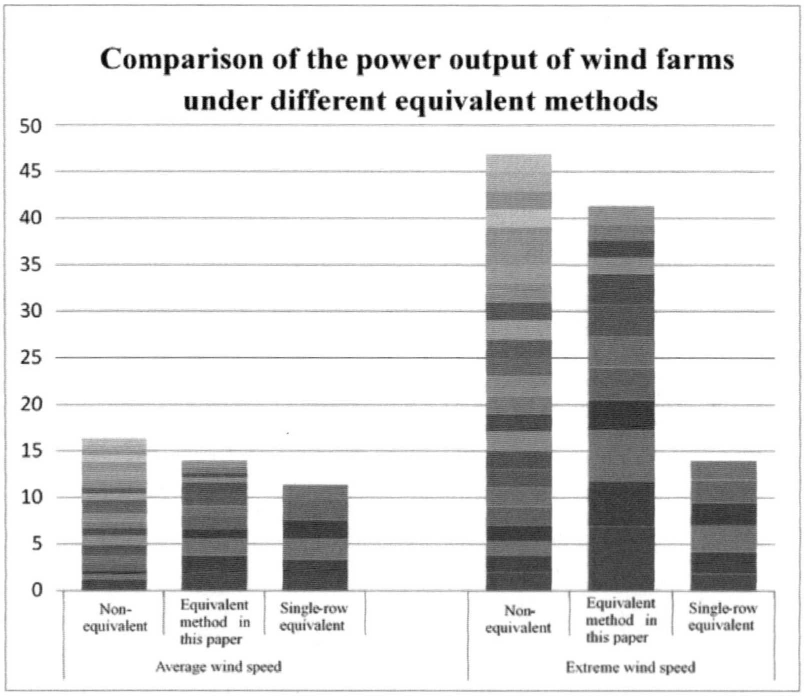

Figure 7. Output comparison between different ways of equivalences

5. Conclusion

In this paper, a method based on wind speed correlation is proposed to cluster the wind farm of complex terrain. The Copula theory is introduced to analyze the correlation of wind farms. The Mixed-Copula function is used to achieve equivalent polymerization by using the correlation of wind speed.

This method is used to test the hypothesis by rank correlation coefficient, to judge the correlation degree between them, and to make equivalent output with higher value, so as to achieve equivalent simplification of wind farms. The wind speed and extreme wind speed under the same climate environment are compared with the traditional single row equivalent method, the unequal value detailed model and the wind farm output of this method. Simulation validates the effectiveness of this method.

Acknowledgment

This work is supported by the Science and Technology Project of State Grid Jiangsu Electric Power Company (J2017046).

References

[1] Papaefthymiou G, Schavemaker P H, Sluis L V D, et al. Integration of stochastic generation in power systems[J]. *International Journal of Electrical Power & Energy Systems*, 2006, 28(9):655-667.

[2] Wu Yuwei, Shi Bin, Zhu Xueqiong, et al. Research on the correlation of wind farm output power based on M-Copula theory [J] .*Electronic Applications*, 2015 (6): 72-78.

[3] Ji Feng, Cai Xingguo, Wang Jun.Analysis of Wind Power Correlation Based on Hybrid Copula Function [J]. *Automation of Electric Power Systems*, 2014, 38 (2): 1-5..

[4] Wang Wei. Analysis on Factors and Routes of Financing of New Energy Industry in China [D]. *China University of Mining and Technology*, 2014.

[5] Zhang Li, Liu Junyong, Liu Youbo, et al. An optimal rotational reserve capacity under wind speed dependence [J]. *Power System Technology*, 2014, 38 (12): 3412-3417.

[6] Yang Jingyi, Li Zheng. Analysis on output correlation of fan with different arrangement [J]. *China Industrial Control Computer*, 2014 (8): 153-155.

[7] Ge Jiangbei, Zhou Ming, Li Gengyin, et al. A review on modeling of large-scale wind farms [J]. *Power System Protection and Control*, 2013 (17): 146-153 (in Chinese).

[8] Louie H. Evaluation of bivariate Archimedean and elliptical copulas to model wind power dependency structures[J]. *Wind Energy*, 2014, 17(2): 225–240.

[9] Liu Zhuo, Li Yanqing, Qi Tingting. The study on power system stability and stable operation under wind convergence effect [C] .*Yunan Electric Power Technology Forum*. 2015.

[10] Cai Fei. Establishment of Wind Speed Dependent Structure Modeling in Wind Farms with Correlation and Its Impact on Transmission and Distribution System Reliability [D]. *Shanghai Jiaotong University*, 2014.

IOP Conf. Series: Materials Science and Engineering **366** (2018) 012020 doi:10.1088/1757-899X/366/1/012020

Research on Coordinated Development Strategy of Northeast Power Grid and National Energy Grid

J K Zhang[1], X Ma[1], P C Ren[1], C S Li[1], J C Liu[1], S H Zhang[2] and S J Lin[2]

[1] Northeast branch of State Grid company, Shenyang Hunnan Camp No. 1 North Street, Shenyang 110180, China
[2] School of Electric and Electronic Engineering, North China Electric Power University, Beijing 102206, China

[3] 13718899557@163.com

Abstract. In the new era, the Northeast power grid is facing a series of opportunities and challenges. In this paper, the strategy of "complementary", "transmission" and "mutual benefit" coordinated development of Northeast China Power Grid and national energy grid is formulated based on the relevant national policy planning; Considering the difference of unique geographical location, natural resources, human resources, technology and other factors of the northeast and other areas; taking into account the development situation and future development needs of all regions, formed a coordinated development strategy layout, covering the unified planning, new energy development, peaking capacity construction, power substitution and talent support. It is beneficial to improve the efficiency of the national energy allocation and utilization, and to solve the practical problems faced by the power grid, so as to realize the coordinated development of the Northeast Power Grid and the national energy grid.

1. Introduction

Under the new environment, the rapid development of China's economy, the continuous progress of technology, the expansion of the scope of cooperation and the new round of electricity market reform have brought a series of opportunities and challenges for the development of Northeast Power Grid. Judging the role orientation of Northeast Power Grid in the future energy grid reasonably and scientifically is the foundation to ensure the sustainable development of Northeast Power Grid. According to the Northeast power grid's development conditions and external challenges and opportunities, the Northeast power grid is positioned as the future power hub in Northeast Asia Energy Internet. And the coordinated development strategy of Northeast Power Grid and national energy grid is put forward, which can provide reference for the future planning of Northeast power grid, and ensure the power grid science and sustainable development.

2. Power grid positioning of Northeast China Power Grid

2.1. Geographical advantage brings up the hub position of Northeast Power Grid

From the domestic point of view, the northeast is located in the northernmost part of China, the southwest connects to the North China, and the south is adjacent to the Bo Sea and the Yellow Sea; from an international point of view, the northeast is located in the central region of Northeast Asia, adjacent to North Korea, Russia and Mongolia, and across the sea from Japan and Korea. Therefore, Northeast China will become the hub of interconnection between China's power grid and Northeast Asia Power Grid.

2.2. Energy situation promotes northeast power grid to pivot positioning transformation

With the growing depletion of fossil energy, the development of renewable energy has become the focus of development. Northeast China has a large amount of wind resources and light resources, which can be used as renewable energy sources in the future [1]. According to the seasonal variation of renewable energy, the remaining renewable energy power generation is transported to the surrounding areas or neighbouring countries through UHV transmission lines, and when renewable energy output is insufficient, it can receive electricity from other countries.

2.3. Unbalance of power supply and demand makes northeast power grid develop to hub

In terms of electricity supply and demand, the power supply in the Northeast Power Grid is concentrated in the north with the load concentrated in the south, presenting the pattern of sending electricity from the west to the east and sending it to the north.

In recent years, the northeast region load growth slowly, and the scale of clean energy development is gradually expanding, the scale of local consumption ability is insufficient, leading to widespread abandoned wind, nuclear phenomena. In addition, due to the randomness of renewable energy output, it may not be able to meet the demand of local and electric power delivery, In the future, the northeast can be connected with the Northeast Asia grid to meet the power demand of itself and other parts of the country.

2.4. "One Belt and One Road" policy to promote the formation of the Northeast Power Grid hub location

"One Belt and One Road" series of policy clear program of cooperation in Northeast China and the eastern region, and support the eastern region in Northeast China to jointly promote the Mongolia and Russia economic corridor construction, carried out jointly for open cooperation in Northeast Asia, which provided significant opportunities for the interconnection of Northeast China Power Grid and Northeast power grid [2].

3. Constructing concentric circles of coordinated development

The coordinated development strategy originated in Northeast China, to expand the domestic Chinese and other regions, and then extended to the Northeast Asian countries, the Northeast Power Grid and the national energy network influence the coordinated development of "concentric circles" such as water waves ripple across.

3.1. "Complementary" coordinated development

The Northeast Power Grid concentrated in the north, the load concentrated in the south, showing the power transmission from east to west and North to the South. In addition, the total energy resources in Northeast China are rich, but unevenly distributed. And the provinces in the region have obvious differences in economic development, technological innovation, institutional mechanisms, talent reserves and so on. Therefore, for the protection of the northeast region of power balance, to optimize the allocation of various energy resources, and to increase network efficiency, should further optimize the internal power structure of three provinces and one district, and realize the coordinated development of the complementary between the northeast provinces.

3.2. "Transfer type" coordinated development

At the national level, due to geographical constraints, the northeast is only directly connected with the North China. Northeast China and Northwest China are rich in energy resources, but are lack of consumptive ability, and the development is relatively backward [3]; Beijing Tianjin Hebei region has a large number of talents, but the problem of environmental pollution is serious; in East China, the economic development is faster, the technology level is higher, and the management mechanism is perfect. Therefore, in order to realize the comprehensive and coordinated development of Northeast China Power Grid and national energy grid, it is necessary to give full play to the unique advantages of each region and to achieve coordinated development across the country.

3.3. "Mutually beneficial" Coordinated development

In Northeast Asia, Russia and Mongolia have abundant energy resources, but their consumptive ability is limited [4]. Japan and South Korea's demand for energy resources is relatively large, while their own energy resources are relatively scarce. And Northeast China has a good market potential and geographical advantages. Due to the time difference and climate difference, the load peak periods of different countries in Northeast Asia are different. Therefore, Northeast China should give full play to the role of the hub to achieve interoperability with Russia, Mongolia, South Korea, and Japan. Give full play to the advantages of all countries, cooperation and sharing, so as to achieve mutually beneficial coordinated development between the two sides.

4. Strategic focus of coordinated development

4.1. In the unified planning: "power grid load" coordination

4.1.1. Coordinated planning between power grids. According to the idea of national unified planning, considering the power supply and load situation of each district, the planning of power grid and the planning of liaison line among regions are formulated [5]. To further expand the construction of inter regional power grid interconnection, to break the balance of power with the principle of "balancing the provinces, districts and districts", and to improve the inter provincial power grid inter regulating capacity [6].

4.1.2. Coordinated planning between multiple power sources. The key to the coordinated development of multiple sources in Northeast China is the peak load regulation of the system. It is necessary to reasonably increase the proportion of peak load power supply and flexible regulation power supply in the power installation, and solve the problem of system peaking caused by the large-scale development of the wind power TV University. Therefore, we should speed up the construction of pumped storage power station and gas station, to improve the renewable energy consumptive capacity of Northeast Power Grid.

4.1.3. Coordinated planning of power grid and power supply. In order to ensure the optimal allocation of energy resources, it is necessary to ensure the scale and speed of power construction match the grid construction. In view of the situation in Northeast China, the wind power planning can be integrated into the unified planning of energy and electricity, and the speed of wind power project construction and grid construction should be synchronized to achieve the overall planning and coordinated development of wind power and power grid.

4.2. In the development of new energy: "renewable energy centralized power generation and distributed generation coordination and complementation"

4.2.1. Vigorously develop renewable energy distributed generation. The northeast is rich in solar and wind energy resources. Distributed photovoltaic power generation and distributed wind power generation should be developed according to local conditions to provide useful supplement for power supply of large power grid [7], which can not only meet the heating demand in some remote areas, saving the cost of construction of power grids, but also can realize the local consumption of solar and wind resources, so as to achieve the comprehensive and coordinated development of the Power Grid.

4.2.2. Construct the intelligent micro grid based on distributed generation system. Smart microgrid can meet the access and consumption of Distributed Renewable Energy of high penetration, can meet the special requirements of the power quality and reliability of users. Therefore, the Northeast power grid should do a good job in supporting the construction of intelligent micro grid, combine the production and consumption of distributed energy to supply energy directly to users, and the remaining power should be integrated into the power grid through smart micro grid, which can effectively improve the reliability and flexibility of power system, and reduce power loss.

4.3. In the aspect of peaking capacity construction, "Measures should be taken to improve peak shaving capacity from Power side load side of grid side"

4.3.1. Construct Pumped Storage Power Station to peak shaving. The pumped storage power station has dual functions of peak shaving and valley filling. The peak shaving range is large and environmental friendly and economical, which can ensure the power grid to meet the real-time power demand of the user side. At the same time, the stability of the output is guaranteed by the lower power generation cost and the environmental burden. In Northeast China the construction of large pumped storage power stations should be accelerated.

4.3.2. Construct gas fired power plant and import flexible thermal power technology to peak shaving. Gas power generation has good regulation ability and peak shaving performance. Facing the severe peak shaving problem, the northeast region should develop a certain scale of gas-fired power generation, mainly as peaking power supply. In addition, it is necessary to build coal power unit peaking capacity improvement project, to learn from the experience of other regions and foreign countries [8], and to speed up thermal storage reform of thermal power units and flexibility transformation of pure condensing unit in Northeast China so as to speed up the thermal power response and deep peak shaving capacity.

4.3.3. Develop energy storage technology and thermal power generation peak shaving system. In the future, Northeast China power grid should accelerate the development of energy storage battery, cold storage, heat storage and other energy storage methods, adhere to the centralized and distributed energy storage as a whole, and improve the energy density of the Northeast power grid energy storage device through the application of large capacity, long life, low cost, high safety new energy storage technology to meet the large-scale energy storage demand of power system, and to improve the peak load capacity of Northeast Power Grid. In addition, vigorously develop solar thermal power industry, give full play to its advantages of heat storage and energy storage, controllable output and adjustable output, to improve the absorptive capacity of wind power.

4.3.4. Strengthen power demand side management and tap demand side response to peak shaving capacity. The Northeast power grid should adopt a comprehensive and interactive demand side management approach. The demand side management should be established in the market to improve the price incentive policy to ease the tension situation of Northeast Power Grid peaking. At the same time, guiding users to change the way of electricity consumption, develop the valley power market, realize peak load shifting, improve the load rate of the grid, and ensure the continuous growth of power supply throughout the year.

4.4. In terms of electric energy substitution, "substituting electricity for coal, substituting electricity for oil, electricity comes from afar."

In the operating area of Northeast Power Grid, it started the substitution work of electric energy in an all-round way. In response to the national electric energy alternative strategy and the advantages of convenient, safe, clean and efficient electric power, the Company started from the areas of heating, industrial production, transportation, service and agriculture, Actively advocating a new mode of energy consumption "substituting electricity with coal, replacing oil with electricity, and electricity from afar", the proportion of electricity consumed in terminal energy consumption has been continuously raised.

4.4.1. Replacing coal with electricity. Replacing coal with electricity in terminal consumption to reduce the environmental damage caused by coal use. In the city central heating, commercial, industrial and agricultural production field, vigorously promote the heat pump, electric heating, electric boiler, double storage alternative energy technology to promote electricity instead of direct consumption of coal, and guide the whole society to change the habit of useful energy [9].

4.4.2. Replacing oil with electricity. Speeding up commercial application of electric vehicles, to promote port power, airport operation vehicle oil to electricity according to local conditions. fossil fuel vehicles will be gradually replaced with electric vehicles. Accelerate the construction of intelligent transportation system, fully realize the "electricity instead of oil", so as to promote the efficient use of energy in the transportation industry and reduce environmental pollution.

4.4.3. Electricity comes from afar. Construct UHV network, the thermal power, wind power and solar power in the West and north of the northeast will be transported to the southern load center of China to achieve large-scale and comprehensive energy substitution in Northeast China.

4.5. In terms of talent support: "adapt to the networking needs, improve the talent training mechanism."

4.5.1. Establish the talent training mechanism in Northeast Power Grid. According to the transformation from northeast power grid to northeast Asia power hub, the paper analyzes in depth the trend of talent demand in the future power grid, formulates a new training mechanism for talents, adapts to the future development of networking in Northeast Asia, and reserves talents for operation and management of Northeast Power Grid.

4.5.2. Establish the Northeast Power Grid think tank. It is necessary to set up the Northeast Power Grid and the National and Northeast Asia Energy Internet Strategy Research Fund to guide the study of major issues in Mongolia, Russia, Japan and South Korea to establish a high-end think tank for energy internet in Northeast Asia. In view of the Northeast Power Grid and the National Energy Network, Northeast Asia Internet development regularly released research report to ensure the coordinated development of the Northeast Power Grid and the National Energy Network.

5. Conclusion

This paper positioned Northeast power grid as the hub of power network in Northeast Asia in the future, and put forward the "concentric circle" macro strategy of coordinated development between Northeast Power Grid and national energy grid, which will realize the "complementary", "Transfer type" and "mutually beneficial" coordinated development of the Northeast China Power Grid and the national energy grid. At the same time, it put forward the coordinated development strategic layout, which covers five aspects: unified planning, new energy development, peak regulation capacity construction, power substitution and talent support, provided a reference for the coordinated development of Northeast China Power Grid and national energy grid, which is of great significance to the sustainable development of power grid Science.

Acknowledgment

2017 National Grid company Management Consulting Project: Research on coordinated development strategy of Northeast China Power Grid and national energy network under new normal conditions.

References

[1] Kuiyan Z 2002 *J. Chem. Manage.* **1** 42-43
[2] Xianliang X 2015 J. *Int. Trade* **11** 26-33
[3] Qiang Z, Ningbo W and Shien H 2017 *J. Power Syst. Prot. Control* **45** 146-154
[4] Yan L 2017 *D.* Research on Sino Russian energy cooperation since twenty-first Century *Hebei Normal U.* p 6-7
[5] Fei F, Puchun R, Chunshan L and et al. 2016 *J. Jilin Electr. Power* **44** 1-4
[6] The national development and Reform Commission, the National Energy Board issued the "13th Five-Year" "power development planning" (2016-2020)
[7] Hua L 2010 *J. Sec. Int. Renew. Energ. Equip. Technol.*
[8] Quan L, Tianyou C, Haixia W and et al. 2014 *J. Automat. Electr. Power Syst.* **38** 34-41
[9] Jing Z and Jianzhong Z 2015 *J. Power Demand Manage.* **1** 41-44

Demand Analysis of Wireless Sensor Networks for New Energy Micro-grid

DI Zhuo[1] , LI Chunlong[2],YU zhipeng[3,*], CHEN Yang[3] and YUAN Jianan[3]

[1] State Grid Liaoning Electric Power Co., Ltd. information communication branch, Liaoning 110000, China
[2] Global Energy Interconnection Research Institute co.,ltd., Nanjing 210000, China
[3] North China Electric Power University, Beijing 102206, China

*E-mail: 171743566@qq.com

Abstract. Wireless sensor networks are self-organizing networks composed of a large number of energy-limited sensor nodes. Due to the factors such as distributed power, energy storage equipment, power load density, operation mode and geo-environment in the microgrid region of the new energy grid, each device in the new energy microgrid has multiple regions and diversified forms, So it is difficult to carry out a unified real-time monitoring. Therefore, this paper designs to integrate the detailed operation status of the independent new energy power supply network into the existing station monitoring system through the wireless sensor network to implement unified energy management. According to the real-time acquisition ,immediately to take appropriate measures to greatly improve work efficiency.

Keywords: new energy microgrid,sensor network, ZigBee,Information Physics fusion system

1. Introduction

According to factors such as distributed power, energy storage equipment, density of electricity load, operation mode and geographical environment of new energy microgrid, each equipment in new energy microgrid has different distribution forms: (1) Multi-regionalization: the new energy micro-grid is not only distributed in a relatively dense population, the power load is relatively concentrated geographical area, also located in remote areas, so the new energy micro-grid has the characteristics of geographical diversification. (2) Diversification of forms[1]: The forms of new energy microgrids are diversified due to the new energy access forms of new energy microgrids, differences in energy storage equipment and load types, and differences in the capacity of distributed power sources.(3) Spacing Randomization: Taking into account the different influences of power load density and environmental factors in different regions and regions where new energy micro-grid power supply, energy storage and electric equipment are distributed in the new energy micro-grid, equipment selection, installation and geography[2]. The position is different, so the new energy micro-grid equipment presents a multi-spacing and distribution of random distribution characteristics. It is necessary to take into account factors such as the number of power supplies, energy storage and electrical equipment distributed in the new energy microgrid, the operating environment and the uneven geographical distribution, and comprehensively construct a reasonable topology to meet the requirements of the new energy microgrid communication network .

2. Analysis of Wireless Sensor Information Transmission Methods

In the new energy microgrid region, the sensor nodes can utilize short-range wireless communication for information transmission. At present, short-range wireless communication network technologies include Bluetooth, Wi-Fi, WirelessUSB, and ZigBee[3]. Bluetooth is a short distance communication technology, which stipulates the communication speed of 1Mbps and the communication range of which is within 20m, adopts the 2.4GHz frequency band wireless communication technology and only supports point-to-point and point-to-multipoint communication, which restricts its application in the development of wireless sensor networks. Wi-Fi is IEEE defined wireless network technology, the maximum transmission rate is 11Mbps, supporting for data connections below 100 meters, the bandwidth performance is good but it also consume a lot of power . WirelessUSB (WUSB) is a wireless technology based on the star topology[4]. The maximum bandwidth is 480Mbps, and do not exceed more than 10 meters. ZigBee technology is a close-range ,low-power, low-rate, and low-cost two-way communication technology. The transmission rate is 10 ~ 250KB / s, which can work in the 2.4GHz, 868MHz, 965MHz three bands.The frequency 2.4GHz is the global free band. Communication distance can reach more than 400 meters. Signal communication distance could amplify to more than one kilometer[5]. Meanwhile ZigBee technology has strong network expansion ability .Because ZigBee technology has the features such as safe and reliable, short delay, large network capacity, low cost and low power consumption, good network topology capability, large effective range and flexible working frequency band. considerting comprehensively network cost, communication distance and Convenience of network combination, ZigBee wireless network communication technology can be used to achieve new energy micro-network monitoring system for information transmission.

3. Wireless sensor network optimization of CyberPhysical System for new energy micro-grid

3.1 Overall design

The independent power grid remote monitoring system is mainly composed of the collection terminal (data acquisition node), concentrator (aggregation node), the main station (data processing center) and the system which constituted of the efficient and reliable data transmission and communication mode that connects these devices to each other. Independent new energy power supply network unified management of energy in Figure 1.

As the hardware distribution structure of the network is closely related to the network topology, in the design process, the topology of the network remote monitoring network should be determined first. According to the distribution characteristics of power network, its obvious tree-like characteristics can be seen[6]. Therefore, Zigbee network topology selection tree topology used in this paper. The data acquisition node and sink node must use full-featured device FFD. Each acquisition node in the network can play the role of coordinator to form a wireless sensor network communication lines along the power network.

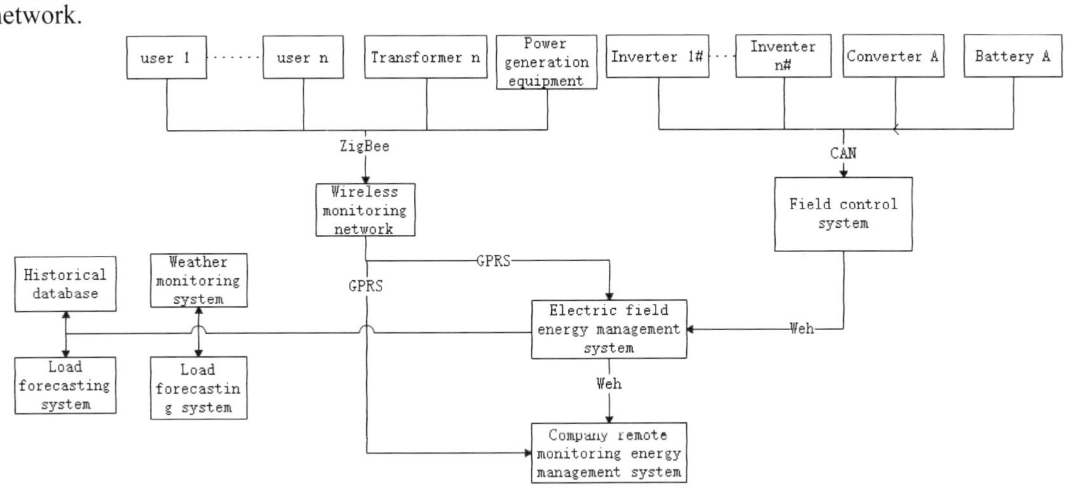

Figure 1 independent new energy supply network management system diagram

In order to transmit the information received by the sink node to the data center in real time, we chose the GPRS technology which is currently relatively mature [7]. ZigBee network transmission distance is relatively limited, if the data monitoring center distance from the collection area is very far , the use of wireless sensor network routing method will be collected data sent to the data center, both from a cost point of view or the difficulty of system implementation ,it is not worth. Therefore, the collected data is sent by the convergence node to the data center through the GPRS device point. Finally, the data center will draw the real-time data of the scene according to the data sent from each place. The schematic diagram of the whole system is shown in Figure 2.

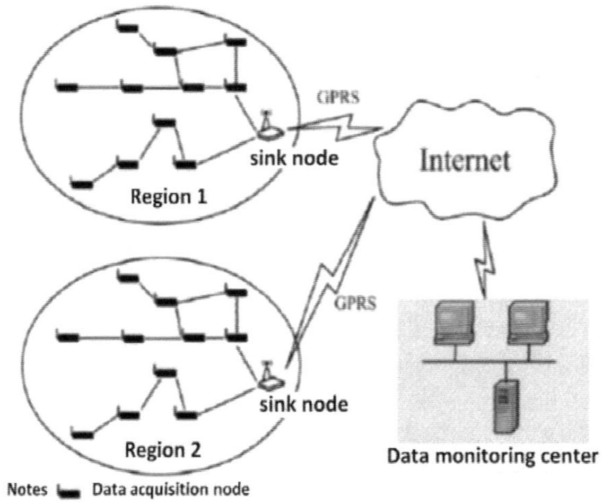

Figure 2 Zigbee-based remote monitoring system structure diagram

3.2 Components of wireless sensor network based monitoring system

This monitoring system designed for the 3 parts: Zigbee Self-network, GPRS network, data monitoring center. Zigbee Self-network can realize data acquisition, storage and communication in complex area. GPRS network can realize data communication between aggregation node and data monitoring center. data monitoring center is equipped with database and microprocessor to store all the data, Analyze, process, and make decisions.

Data collection node: The collection node is set at any position of the power grid and transmits the power quality parameters of the location,such as voltage and current,and also surrounding data such as temperature, humidity and brightness to the regional joint equipment immediately through the Zigbee network. Each node has its own ip address, so the data center can find the data relative location.

Aggregation node: A summary of all the data in a certain area in the Zigbee network. Only one aggregation node in a region of the wireless monitoring system . aggregation nodes collect all information form all the collecting nodes and transit nodes , then stores the data. If necessary, the data can be uploaded to the data monitoring center through the GPRS network, and it can also receive data and transmit it to data acquisition node which from data monitoring Center.

The data center receives the data uploaded by the joint node then through further processing and storage by the computer. Through the configuration software to build a data center management platform for all data analysis, and dispaly. managers can make observations by discussing the data to make decisions.

4. Sensor Network Rate, Communication Quality and Life Cycle Performance Analysis

In view of the multi-regionalization, diversification of forms and randomization of new energy microgrids, the wireless sensor networks are used to monitor the environment. collected data flows are evenly distributed in time and space and it is very stable. collected data used for monitoring new energy network equipment.To ensure a high communication rate and high reliability of communication quality, meanwhile the reliable operation of the sensor network is essential. With the development of science and technology, the communication speed and reliability of the sensor network have been greatly improved , which has basically satisfied the monitoring requirements for the equipment in the new energy micro-grid. The reliable operation of the sensor network mainly refers to the network can maintain a longer life cycle, and the case when no energy exhaustion it can not work. Its working life is defined as the period from the start of the network to the death of the first node. Since WSNs are characterized by such features as no-center ,self-organizing, dynamic network topology, multi-hop routing, limited energy and bandwidth limitation , large network size . However nodes use the same transmit power, multi-hop relay communication and many- Traffic characteristics, making the entire network serious uneven distribution of network traffic. The nodes close to the receiving station consume energy first due to excessive traffic, causing network partitioning and destroying the connectivity of the network, resulting in a serious decline of the network service life . Due to the large number of nodes and the working environment, the sensor network can not supplement the battery energy. Limitation node battery energy will severely restrict the working life of the wireless sensor network and bring a high network maintenance costs, the sensor network must extend the service life, to ensure reliable monitoring of the new energy micro-grid in order to maintain its stable operation.

4.1 Life cycle extension method based on route optimization

In the new energy micro-grid environment, the sensor nodes according to the types of information collected can be divided into fixed nodes and mobile nodes. The fixed node collects the information of a device in the microgrid, and the mobile node is used for the safety of the site operation.

Wireless sensor networks collect information in real time through a large number of sensor nodes distributed over a large area, and then process the information through an embedded system. Finally, the sensed information is transmitted to the user terminal through the wireless communication so that the user Take full control of and respond to the situation in the monitored area. As the key technology in wireless sensor network, routing technology is mainly used to select the path with better performance between the source node and the destination node. The routing information is transmitted from the source node to the destination node along the optimized path , to complete the data communication between nodes. each node's energy is limited, which requires WSN routing protocols and algorithms with efficient energy utilization, and the existing routing protocol has been unable to meet the requirement of wireless sensor network .especially in reducing the energy consumption, effectively using the node energy and prolonging the lifetime of the network, it shows the congenital deficiencies. Therefore, it is necessary to design a reasonable and efficient routing protocol to prolong the life cycle of the wireless sensor network,in order to enhance aspect of timeliness and reliability, etc.

4.2 Life Extension Method Based on Wireless Charging

For the new energy microgrid in a remote environment, once the battery is exhausted, the node will no longer have the ability to work and can not monitor important equipment in the microgrid, so a wireless chargeable sensor network with energy harvesting capability has emerged. In a wireless chargeable sensor network, a node can collect a radio-frequency injection signal in the environment to charge itself. Wireless charging technology original from the wireless power transmission technology, which uses electromagnetic resonance or electromagnetic jurisdiction to transmit electrical energy from the power side of through wireless transmission to the electrical equipment, so it can achieve the energy transfer easily. Due to the economic and security restrictions, the wireless charging device can not be completely installed on the monitoring site. Therefore, the wireless charging device can be charged by a carrier / robot which is automatically operated in a running environment. When the node has been deployed in a remote area In the wild, energy may be supplemented by UAVs. Because of the periodicity of the power supply of the wireless charging device, it is not possible to supply power to the node in real time. Therefore, the routing algorithm should be improved to adapt to the chargeable wireless sensor network and prolong the life of each node in the network to ensure the reliable operation of the sensor network.

4.3 Function expansion analysis

New energy microgrid is the distributed power generation access and the user terminal, in order to ensure the safety of electricity users and to avoid the microgrid information island, so it is necessity to monitor power quality continuous at real-time and control power quality. microgrid contains a large number of Distributed renewable energy generation systems, and it has great connection with the natural environment ,it output power will change with the factors like sun, wind speed, temperature and other external changes in the objective . The information physics fusion system realizes the real-time and dynamic information control and information service of objects based on the environment perception and the interconnection of objects and objects through the deep interaction and fusion of the information system and the physical system, and can be implemented in a safe, reliable and efficient manner Real-time way to detect or control physical entities. Application of CPS technology can be real-time processing of a large number of different types of data sent from different sensor, real-time control of electrical physical devices. Compared to other components of the power system running for many years, micro-grid is more conducive to the use of new technologies and technological advantages. In response to the trend of the development of smart grid in the future, information fusion system technology can be introduced into the construction of the microgrid, and technologies such as environment awareness, embedded computing, network communication and network control are integrated to serve the construction of a new energy microgrid Operation, making the micro-grid electrical equipment system has abilities with computing, communications, precise control, remote coordination and autonomy. the establishment of new energy micro-network CPS, in order to achieve human interaction with electrical equipment.

5. Conclusion

In this paper, the design of the independent new energy power supply network in the detailed operation of the state through the wireless sensor network into the existing station monitoring system to implement a unified energy management, management staff in the data center can grasp data and surrounding environment information Such as the voltage, current, power factor, power output of the power generation equipment and the power consumption of the user terminal from the power network nodes .Meanwhile it may analyze and decide the data of a period of time, such as the peak current at morning and night, different regions Fluctuations in the use of electricity, accident inquiries, prediction and avoidance.staff can maintain power network more convenient and economical. When there is an accident on the power network, the power grid remote monitoring system can provide the function of fixed point or two-point positioning to find out the accident accurately and quickly .at the same time, according to the data Judge the type of accident, and immediately to take appropriate measures to greatly improve work efficiency.

Acknowledgment

National network company science and technology project:Research and Application on key technology of Rechargeable Wireless Sensor Network

References

[1] Sean Dieter Tebje Kelly, gender Kumar Suryadevara, and Subhas Chandra Mukhopadhyay. Towards the implementation of Io T for environmental condition monitoring in homes[J]. IEEE Sensors Council,2013,13(10):3846 -3853.

[2] Michael Angelo A Pedrasa, Ted DSpooner, Iain FMac Gill. Coordinated Scheduling of Residential Distributed Energy Resources to Optimize Smart Home Energy Services[J]. IEEE Transaction on Smart Grid,2010,1(2):134-143.

[3] Dae-Man Han, Jae-Hyun Lim. Design and Implementation of Smart Home Energy Management Systems based on Zig Bee[J]. IEEE Transactions on Consumer Electronics,2010,56(3): 1417-1425.

[4] Rim Missaoui, Hussein Joumaa, Stephane Ploix, Seddik Bacha. Managing energy smart homes

according to energy prices: Analysis of a building energy management system[J]. Energy Buildings, 2014, vol. 71: 155–167.
[5] Keigo Nomura, Fumiaki Sato.A Performance Study of Zig Bee Network under Wi-Fi Interference. 2014 17th Internation Conference on Network-Based Information Systems [C]. Salerno. 2014. IEEE,2014:201-207.
[6] Chauvenet C, Etheve G, Sedjai M, et al. G3-PLC based IoT sensor networks for SmartGrid[C]// IEEE International Symposium on Power Line Communications and ITS Applications. IEEE, 2017.
[7] Lieven Tytgat, Opher Yaron, Spfie Pollin, Ingrid Moerman, Piet Demeester. Analysis and Experimental Verification of Frequency-Based Interference Avoidance Mechanisms in IEEE 802.15.4[J].IEEE/ACM Transactions on Networking,2015,23(2): 369-382.

Design of Wireless Sensor Network Routing for Renewable Energy Microgrid

LIU Guanke[1] , ZHANG Yanling[2], LUO Fang[2,*] and YUAN Jianan[2]

[1] Dongguan Power Supply Bureau of Guangdong Power Grid Co., Ltd.,Guangdong 523000, China
[2] North China Electric Power University, Beijing 102206, China

*E-mail: 1156037316@qq.com

Abstract. Wireless sensor network (WSN) is a self-organizing network composed of a large number of energy limited sensor nodes. As the wireless sensor network in new energy micro grid environment may be close to the node receiving station due to the amount of traffic and the first run out of energy, resulting in shorter life cycle, part of the micro network environment for many sensor nodes, working environment and other factors can not be timely replacement of batteries, the reliability of sensor network is difficult to guarantee. Therefore, this paper optimizes the design of LEACH routing algorithm. The simulation results show that the improved routing algorithm has obvious advantages in prolonging the lifetime of sensor networks.
Key words. renewable energy microgrid; sensor network; clustering; mobile node

1. Introduction

With the increasing penetration of renewable energy, renewable energy is becoming more and more difficult to deal with the challenges of network consolidation, energy consumption, the existence of grid-connected, low energy efficiency of long-distance transmission, high loss, low utilization of equipment, poor system capacity investment marginal efficiency and so on. Under normal circumstances, renewable energy microgrid run with large power grid. When a large grid failure is detected, it will separate from the big power grid timely and began to run independently[1]. Whether the renewable energy microgrid is running with large power grid or running independently. It is important to control the distributed power and energy storage equipment to ensure system stability. Therefore, we need real-time monitoring of photovoltaic arrays, wind turbines, batteries, inverters, controllers and other equipment timely to ensure timely, accurate, all-round reflect the operation of independent power supply network.[2]

Wireless sensor network combine sensor technology, embedded computing technology, distributed information processing technology and communication technology. It can perceive and collect the information of the monitored object collaboratively, and then And then transmit within the area and upload to the background management system. Therefore,wireless sensor network is widely used in national defense and military, environmental monitoring, urban management, health care, industrial control, anti-terrorism and many other areas. As the wireless sensor network has the characteristics of no center self-organization, dynamic network topology, multi-hop routing, limited energy and bandwidth,large network size and the nodes use the same transmit power, multi-hop relay communication and many-to-one traffic, so that the traffic distribution in the whole network is seriously uneven.The nodes close to the receiving station is the first to run out of energy due to excessive traffic, resulting in network partition and disruption of network connectivity, then the service life of network will be a serious decline[3]. And the micro network environment for many sensor nodes, working environment and other factors can not supplement the power of battery timely, limited node energy will severely restrict the working life of wireless sensor network and increase the cost of network maintenance. It is necessary to extend the life of the sensor network to ensure reliable monitoring of renewable energy microgrid to maintain its stable operation[4]. Therefore, based on the routing algorithm of a variety of wireless sensor network, this paper optimizes the design of LEACH routing algorithm,extend the life cycle of the wireless sensor network,and realize real-time monitoring,optimize scheduling, rapid maintenance and integrated management in microgrid.

2. Analysis of Typical Routing Protocols in Wireless Sensor Network

At present, the design of wireless sensor network routing algorithm is divided into two types according to its networking and routing:plane type and hierarchical type. Among them, the planar routing algorithms are flooding, SPIN (sensor protocol for information elder negotiation), SAR (sequential assignment routing), GEAR algorithm and Direct Diffusion algorithm. Typical hierarchical routing algorithms include low power adaptive clustering routing algorithm (TEACH), TEEN (threshold sensitive energy efficient sensor network protocol) and multi-layer clustering algorithm[5-8]. The following is a brief introduction of flooding algorithm, GEAR algorithm, directional diffusion algorithm and low power adaptive clustering routing algorithm.

The multiple parameters of several typical routing algorithms are compared, as shown in Table 1.

Table 1 Typical Routing Algorithm Performance Comparison

parameter	Flooding	LEACH	Direct Diffusion	GEAR
Route optimization capabilities	no	no	yes	no
Network life cycle	Short	Long	Long	Long
Route robustness	Good	Better	Good	Better
Resource availability	Bad	Good	Better	Good
Data aggregation	No	Yes	Yes	No
Data / query cache	No	Yes	Yes	No
Metadata description	No	No	Yes	Yes
Data transmission mode	Continuous	Continuous	Query driver	Query driver
Path Selection	Multi-hop	single hop	Multi-hop	Multi-hop

The flood routing protocol directly forwards the packets of the message to all neighboring nodes in the form of broadcasts. The Flooding protocol is the oldest and classic flood routing protocol. Its research goal is how to effectively solve the message "implosion" and "overlap" and "resource blind spot" problem; Hierarchical routing protocol efficiently transports sensor data by dividing the sensor nodes into different subnets, LEACH is the first hierarchical routing protocol proposed in a wireless sensor network; The factors that influence the formation of subnets and the communication of sub - network nodes are the hotspots of future research; Data-centric routing protocols are based on data queries and rely on data naming. Directed Diffusion is a milestone in the development of data-centric routing protocols. Efficient standard data naming mechanism and the efficient transmission of query information is the direction of future research; The location-based routing protocol utilizes the location information of the nodes for data transmission. GEAR is a location-based routing protocol that takes full account of energy efficiency[9]. The key to the study is how to intelligently utilize the location information of the nodes to achieve the goal of energy efficiency.

3. Wireless Sensor Network Mobile Node Routing Optimization Scheme

In view of the environment and business characteristics of the new energy micronetwork, this paper designs the improvement scheme of wireless sensor network mobile node cluster routing. The wireless sensor network of design includes five stages, namely cluster heads, cluster heads, node cluster, stable communication and cluster head rotation.

3.1 Clustering

Clustering is accomplished on the basis of the following assumptions: First, all the nodes in the monitoring area are known for the location information of the nodes themselves. Second, all the nodes in the monitoring area are known to the location information of the monitored range. According to the location information of each node in the monitoring range and the communication radius of the node, the monitoring range can be divided into several clusters. The monitoring nodes in each cluster decide the cluster head according to the rules mentioned below. The cluster head is responsible for collecting the cluster The data collected by the node is transmitted directly after data fusion or transmitted to the Sink node via other cluster head nodes. As shown in the figure 1, the wireless sensor play distribution area is divided into several virtual clusters. The following figure shows a schematic representation of the virtual clusters in the area.

where the wireless sensor network is located. In the monitoring area, there are 16 square areas with a length of a, and the sensor nodes in each square area form a virtual The nodes in the cluster are responsible for collecting the data and sending them to the cluster head. In order to ensure that any node in the cluster can send data to the cluster head node, the relationship between the length of the virtual cluster and the communication radius R of the sensor node is: $a=\frac{2}{\sqrt{2}}R$ .This setting ensures that common node data can be sent to cluster heads.The node cluster flow chart is shown in Figure2.

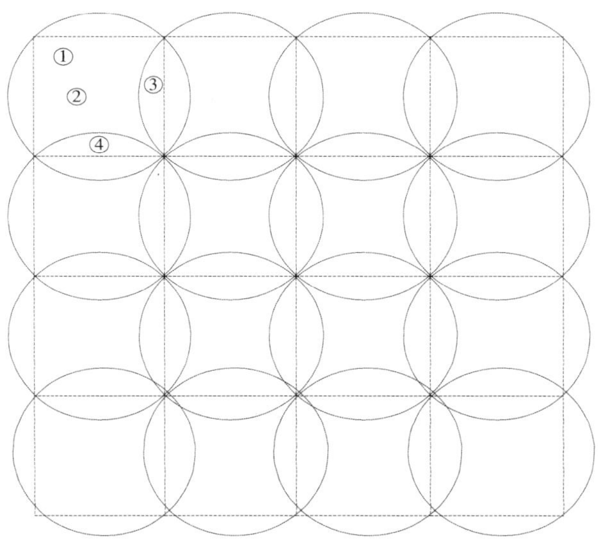

Figure 1 Node clustering diagram

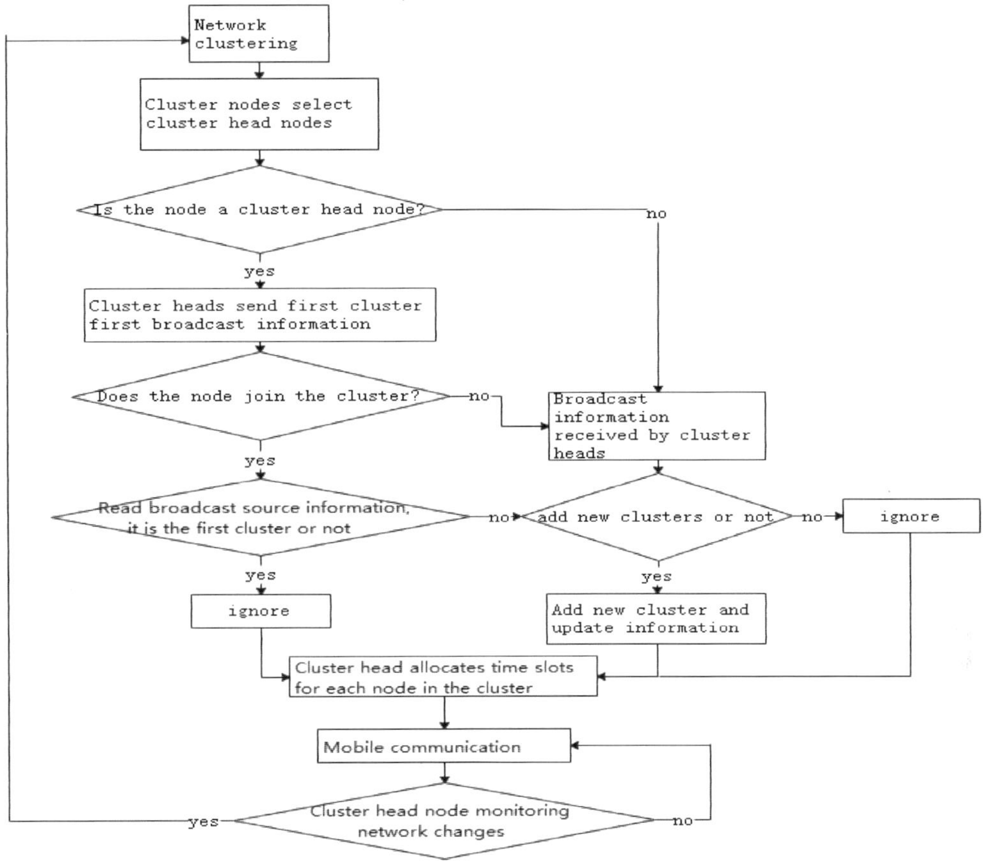

Figure2 Node clustering flow chart

3.2 cluster head

In order to select the node with the highest stability in the cluster as the cluster head, we give the concept of the stability of the node. The stability of the node is related to the residual energy of the node and the relative velocity of the node. We introduce the weight index to represent the contribution of the residual energy of the node and the relative velocity of the node to the stability of the node. Node stability is defined as:

$$S = w_1 \times E + w_2 \times R \tag{1}$$

In formula (1), $w_1 > w_2, w_1 + w_2 = 1$.

The cluster head node is responsible for collecting the data sent by the ordinary node in the cluster, and the data is fused to the sink node (Sink Node). Therefore, the energy consumption of the cluster head node is faster. In order to ensure the energy balance between nodes and maximize the survival time of the extension of the network, the cluster head node need to be replaced regularly. In the process of cluster head replacement, we consider two factors: first, the residual energy E of the nodes, the more the remaining energy of the node, the greater the probability that it is selected; and the second, the node In the wireless sensor network, the nodes are moving. When the cluster head node is running, it is possible to enter the range of other virtual clusters, resulting in the distance between the nodes in the cluster and the increase of the energy consumption , The worse the situation caused by the loss of data, so we consider the node movement speed as a factor, R take a negative value, the greater the node movement speed R, the lower the probability of being selected as the cluster head node.

The node residual energy E is defined as:

$$E = \frac{Energy\_residual}{Energy\_average} \tag{2}$$

In the formula (2), the $Energy\_residual$ is residual energy of the node $Energy\_average$ is the average energy of the nodes in the cluster.

The node motion velocity R is defined as:

$$R = \frac{V_C}{V_{MAX}} \tag{3}$$

In equation (3), V_C is the current velocity of the sensor node, V_{MAX} is the maximum velocity of the sensor node.

In the above figure 2, the overlapping of the signals is generated in the adjacent clusters. The nodes in the overlapping area can receive the broadcast information sent by the cluster heads. After receiving the broadcast message sent by the first cluster head, If the node is able to receive broadcast messages from other cluster heads, it will compare the distance between the node and the cluster head. If the new cluster head is closer to the node, the node will join the new cluster. Otherwise, the cluster will be ignored. Head broadcast message.

4. Simulation analysis

In this paper, we use matlab to compare the performance of the proposed routing protocol with LEACH routing protocol.Firstly, the network lifetime of the two algorithms is simulated and analyzed. The lifetime of a wireless sensor network is defined as the time from the beginning of the simulation to the failure of the first node in the network. If the remaining energy of the node is less than 5%, the node is considered to be exhausted.

Simulation parameter setting, such as table 2. The monitoring area of the wireless sensor network is set to 200m × 200m square area. 400 wireless sensor nodes are randomly arranged in the monitoring area, and the Sink node of the network is set in the upper right position, that is(200,0) .The initialized energy of all nodes in the simulated wireless sensor network is the same, set to 1J. All nodes in the wireless sensor network can move, the maximum speed V_{MAX} is 2m / s, the actual movement of the node V_S takes a random value between $0 \sim V_{MAX}$.

Table 2 Simulation parameter settings

Parameter	Setting parameters
The size of the sensing area	200×200
Number of sensing nodes	400
Send energy consumption	δ*Dist3
the initial energy of a sink node	5J
The initial energy of an ordinary node	1J
Broadcast energy consumption	0.0001
δ	0.0001*(1/(25^3))

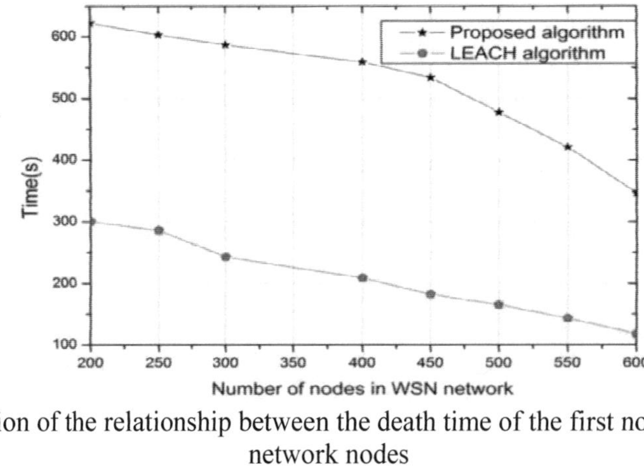

Figure 3 Continuation of the relationship between the death time of the first node and the number of network nodes

Figure 3 shows the relationship between the death time of the first node and the number of nodes in the network. The figure shows the time corresponding to the energy consumption of the wireless sensor network at the first node. The size of the network is different, and the time at which the first node is exhausted is different. Figure 2 shows the time at which the network's first node dies when the network size is from 200 to 600 nodes. As can be seen from the figure, the larger the network size, the earlier the death of the first node. By using the algorithm proposed in this paper, we can delay the death time of the first node of the network to a certain extent and improve the life cycle of the network.

5. Conclusion

Based on the analysis of the routing algorithm of traditional wireless sensor networks, this paper proposes an improved mobile node clustering based on clustering, cluster node selection cluster head, node into cluster, stable communication and cluster head rotation based on Leach algorithm algorithm. The simulation results show that this algorithm has obvious advantages in prolonging the life cycle of wireless sensor network and provide a favorable foundation for the reliable operation of wireless sensor network.

Acknowledgment

China Souther Grid Corp science and technology project: Research on Intelligent wearable power field operation safety protection technology

References

[1] YICK J, MUKHERJEE B, GHOSAL D. Wireless sensor network survey[J]. Computer Networks, 2008, 52(4): 2292-2330.

[2] CHOUHAN S, BOSE R, BALAKRISHNAN M. Integrated energy analysis of error correcting codes and modulation for energy efficient wireless sensor nodes[J]. IEEE Transactions on Wireless Communica- tions, 2009, 8(10): 5348-5355.

[3] KALIS A, KANATAS A G, EFTHYMOGLOU G P. A co-operative beamforming solution for eliminating multi-hop communications in wireless sensor networks[J]. IEEE Journal on Selected Areas in Com- munications, 2010, 28(7): 1055-1062.

[4] ABOUEI J, PLATANIOTIS K N, PASUPATHY S. Green modula- tions in energy-constrained wireless sensor networks[J]. IET Commu- nications, 2011, 5(2): 240-251.

[5] ZHOU Z, ZHOU S, CUI S, et al. Energy-efficient cooperative com- munication in a clustered wireless sensor network[J]. IEEE Transac- tions on Vehicular Technology, 2008, 57(6): 3618-3628.

[6] SIMMER MAN S,OSBOR NE J,JIAN HUANG. Eden: simplified management of atypical high-performance computing jobs[J]. Computing in Science & Engineering,2013,15(6) : 46 — 54.

[7] SINGH A K,DZIUR ZANSKI P,INDR USIAK L S. Market-inspired dynamic resource allocation in many-core high perform-ance computing systems ［C］ // In High Performance Computing & Simulation (HPCS) ,2015 International Conference,2015(20 — 24) : 413 — 420.

[8] TAO C,FR ANCHETTI F. A multi-core high performance computing framework for probabilistic solutions of distributionsystems[C] // In Power and Energy Society General Meeting,IEEE,2012(1 — 6) : 22-26.

[9] Mohammed Zaki Hasan,Hussain Al-Rizzo,Melih Günay. Lifetime maximization by partitioning approach in wireless sensor networks[J]. EURASIP Journal on Wireless Communications and Networking,2017,2017(1):.

IOP Conf. Series: Materials Science and Engineering **366** (2018) 012023 doi:10.1088/1757-899X/366/1/012023

Design of Rechargeable Wireless Sensor Network for new energy micro grid

HUANG Hui[1] ,LI Chunlong[2] ,WANG Jinshuai[3] ,TU Xin[4] ,YUAN Jianan[5]

[1] Global energy Internet Research Institute, Nanjing 210000, China
[2] Global energy Internet Research Institute, Nanjing 210000, China
[3] North China Electric Power University, Beijing 102206, China
[4] North China Electric Power University, Beijing 102206, China
[5] North China Electric Power University, Beijing 102206, China

E-mail: 1053349427@qq.com

Abstract. In traditional wireless sensor networks, the node battery capacity is very limited, which seriously restricts the sensor network long time monitoring operation. Therefore, how to improve the life cycle of WSN has always been the focus of research. In order to ensure the normal operation of WSN as long as possible, the energy of nodes must be utilized to the maximum extent.in this paper, Based on the wireless transmission technology, the mobile wireless charging equipment, energy supplement of rechargeable sensor nodes deployed in the complex environment, solve the network energy bottleneck problem essentially, effectively reduce the node failure rate, greatly improves the network lifetime, has broad prospects and research value.

1 Introduction[1]

In general, the power of the sensor node is provided by the disposable battery. In order to ensure the normal operation of the WSN for as long as possible, the energy of the node must be utilized to the maximum extent. Most of the research is devoted to formulating a reasonable energy allocation and management strategy , But this approach can only extend the life cycle of the network as much as possible and can not fundamentally maintain the lasting and normal work of the network[1]. Where the environment allows, the battery can be artificially changed, but for special applications such as environmental monitoring systems, field expedition, etc., this regular replacement has become difficult to achieve. And drawing natural energy from the environment does not guarantee continuity and controllability[2].

On the basis of wireless transmission technology, the use of mobile charging devices to supplement the rechargeable energy sensor nodes deployed in complex environments is gradually becoming a hot issue in recent years[3]. The research on charging wireless sensor networks mainly focuses on charging strategies. Wireless charging devices are expensive and their mobile costs are quite large. Therefore, it is necessary to develop a reasonable and effective charging strategy and node energy supplement scheme.

Rechargeable sensor node with all the advantages of traditional sensor nodes, such as low power

National network company science and technology project:Research and Application on key technology of Rechargeable Wireless Sensor Network consumption, micro-size, easy deployment, scalable, ad hoc network, robustness and other characteristics, but also has the following potential: battery life is strong, at Rechargeable sensor network, the sensor node energy from the surrounding environment to obtain their own, and these energy sources are usually renewable. There are many sources of energy[4]. The node can be based on the specific characteristics of the environment and the implementation of the work to choose the appropriate Energy collection module. low maintenance cost. energy source is controllable and in the rechargeable sensor network, energy charge chess strategy can be optimized by controlling the energy source to the network. good service quality. environmental damage is small and the energy collected by rechargeable sensor nodes usually comes from the surrounding nature, which belongs to renewable green energy, so it is more environmentally friendly.

2 Charging mode of rechargeable sensor node

Wireless charging technology is developed based on wireless power transmission technology, is not by physical (wire, cable) connect the electric power transmission in a wireless way directly or indirectly out, charging the electronic equipment or wireless sensor nodes for target[5]. At this stage, the diversification of wireless energy transmission technology makes wireless charging have different charging methods. According to the principle of wireless charging technology, there are mainly three charging technologies: inductive coupling technology, electromagnetic radiation technology and magnetic coupling resonance technology [6-8]. These three kinds of wireless charging technology have their own advantages and disadvantages to meet different application requirements, the specific comparison in Table.1.

Table 1 Comparison of three wireless charging technology

Wireless charging technology	Transmission power	Transmission distance	Advantage	Shortcoming
Inductance coupling technology	A few to a few hundred watts	3-4m	Achieve simple, in the very close distance has a higher transfer rate	Effective transmission distance is short
Electromagnetic radiation technology	Highest 100mW	Up to 10m	Can maintain a high transmission rate over long distances	Transmission between the two sides can not be obstructed by obstacles, the equipment is very large
Magnetic Coupling Resonance Technology	A few kilowatts	Less than 1m	Maintain a higher transmission rate, without alignment, allowing blocking, without environmental impact	Transmission efficiency decreases sharply with increasing distance, efficient transmission distance is very short

3 Rechargeable Sensor Network Design for New Energy Microgrid

3.1 Mobile charging base station model

In the periodic charging model, the Wireless Charging Vehicle (WCV) initially moves from the service station (ie, the maintenance station) to the network and transmits the sensor nodes wirelessly at certain fixed points for energy supplement. After all the sensor nodes in the network have been given an energy supplement, the WCV returns to the service station for its own energy supply. The interval at which WCV starts charging twice from the service station is an energy replenishment period, denoted by T.

In an energy replenishment cycle T, WCV has three states: moving state, charging state, rest state. Rest state refers to WCV service station for its own energy supply and maintenance of rest, we can use (3-1) to said T:

$$T = T_P + T_{vac} + \sum_{i \in N} T_i \tag{3-1}$$

The total time that WCV consumes movement in one charging cycle is T, the charging time of any node i means the idle time staying at the service station to be adjusted, and the optimization goal of the periodic charging model is to maximize the totalizing time of MC Replenishment period ratio: T_{vac}/T.

In the periodic charging model, according to whether the mobility of the base station or not, it can be further divided into a fixed base station periodic charging model and a mobile base station periodic charging model. Mobile base station model is the base station and charger together on a mobile device, the implementation of the charging task.

3.2 Fixed charging base station model

The location of the base station remains unchanged throughout the charging process. The sensing data generated by the sensor nodes are transmitted to the base station through direct or multi-hop transmission. The nodes in the monitoring process are generated per unit time, and the data flow received and transmitted remains constant. The MC periodically traverses the nodes for point-to-point charging. The MC is charged only when it is as close as possible to the node. Each sensor node charging receive power is U, any sensor node i unit time to receive and generate data volume is $\sum_{k \in N}^{k \neq i} f_{ki}$ and G_i, sent to the remaining nodes and base station data volume is $\sum_{j \in N}^{j \neq i} f_{ij}$ and f_{iB}, follow the flow conservation constraint:

$$\sum_{k \in N}^{k \neq i} f_{ki} + G_i = \sum_{j \in N}^{j \neq i} f_{ij} + f_{iB}, i \in N \tag{3-2}$$

Then node i energy consumption per unit time r is:

$$r_i = \rho \sum_{k \in N}^{k \neq i} f_{ki} + \sum_{j \in N}^{j \neq i} C_{ij} f_{ij} + C_{iB} f_{iB}, i \in N \tag{3-3}$$

Where ρ denotes the energy consumption of the receiving unit data, C_{ij} is the energy consumption of sending unit data from node i to node j, and C_{iB} is the energy consumption of transmitting unit data directly to the base station (C_{ij} and C_{iB} are all distance related). The maximum battery capacity of ordinary nodes are E_{MAX}, node failure when the power is lower than E_{MIN}. All nodes in the network can work normally, then the remaining energy $e_i(t) > E_{\min}$ of the node at any moment t, namely:

$$E_{\max} - (T - T_i), r_i > E_{\min}, i \in N \tag{3-4}$$

In order to maintain the continuity and periodicity of node work, it is necessary to meet the node energy consumption equal to the amount of charge in each cycle in a cycle:

$$T \cdot r_i = T_i \cdot U, i \in N \tag{3-5}$$

WCV to get the maximum body-building time in one charging cycle means that the WCV moves all the nodes in the network as far as possible and the WCV reaches all nodes when all the nodes are still valid and can be charged more energy to reduce the frequency of charging. Shi proved that maximizing the station idle time of mobile chargers is to minimize the movement time, and thus the shortest path traversing all nodes is actually the Traveling Salesman Problem (TSP), so (3-1) can be transformed into:

$$T = T_{tsp} + T_{vac} + \sum_{i \in N} T_i \tag{3-6}$$

After the movement path of WCV is determined, the periodic charging problem that maximizes the station time ratio can be transformed into a linear programming problem by introducing auxiliary variables and relaxing the constraints, and the approximate optimal solution is obtained by using the CPLEX solving tool solution.

Define the maximum distance between sensor node and WCV when the received power is not less than the threshold value δ is the chargeable range D_{ij}. For node i, the received power U is a distance-related parameter:

$$U_i = \begin{cases} \mu(D_i) \cdot U_{\max}, if D_i(\rho) \le D_\delta \\ 0, if D_i(\rho) > D_\delta \end{cases}, 0 < \mu(D_i) < 1 \tag{3-7}$$

Where $U_{\max}$ is the maximum charge output power of WCV, $\mu(D_i)$ is the charge transfer efficiency, which is determined by the distance between WCV and node i. Based on this charging model, the entire network area is divided into regular hexagonal cells of side length D centered on the base station, as shown in Figure1.

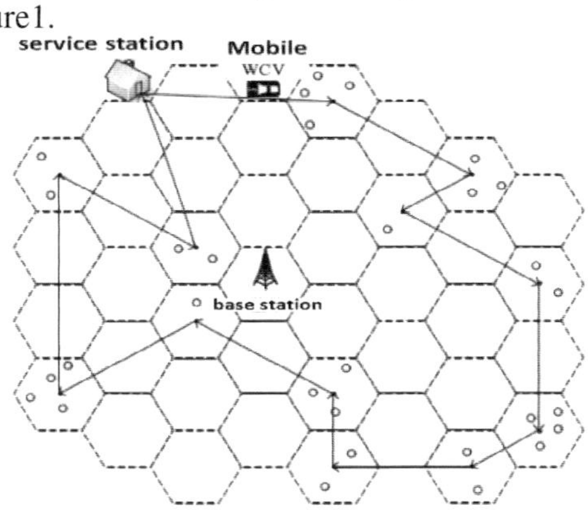

Figure1 Periodically traversing the cell's point-to-multipoint charging

4 Simulation results

The matlab simulation software is used to simulate and compare the performance of LBRWRSN protocol and LEACH routing protocol proposed in this paper. In the simulation, the monitoring area of the wireless sensor network is set to a square area of 200m × 200m. 400 wireless sensor nodes are randomly arranged in the monitoring area. The Sink node of the network is set in the upper right corner position (200,200).Within the monitoring range, 4 charging stations are laid at (50, 50), (150, 50), (50150), (150150), the charging station is fixed, and can not move, as shown in Table 2.

Table 2 Simulation parameters setting

parameter	Set up	parameter	Set up
Sensing area size	200×200	ε_{fs}	10Pj/bit/m2
Number of sensor nodes	400	ε_{mp}	0.0013Pj/bit/m4
Initial energy of sink nodes	5J	α	0.2
Initial energy of ordinary node	1J	β	2.0
E_{elec}	50nJ/bit	R	10m

The initialization energy of all nodes in the simulation wireless sensor network is the same, set as 1J. All nodes in WSN can move, the maximum moving speed VMAX is 2m / s and the actual moving speed of nodes VT is a random value between 0 ~ VMAX. The length of the data packet between nodes is 5 kbit, the length of the control packet is 150 bits, the charging radius of the charging station is 4 meters, and the energy consumption of node data fusion is 4 nJ / bit.

Figure 2 shows the relationship between the number of network survivors and the number of network runs. The figure shows the change curve of the number of surviving nodes in a wireless sensor network when the network size is 400 nodes. The proposed algorithm LBRWRSN can effectively delay the emergence of the first node in the network and ensure that there are enough surviving nodes in the network to complete the related monitoring tasks. The improvement of LBRWRSN performance is due to the introduction of a charging station to charge the nearby sensor nodes, and the algorithm of node communication radius adjustment and improved cluster head election mechanism are introduced.

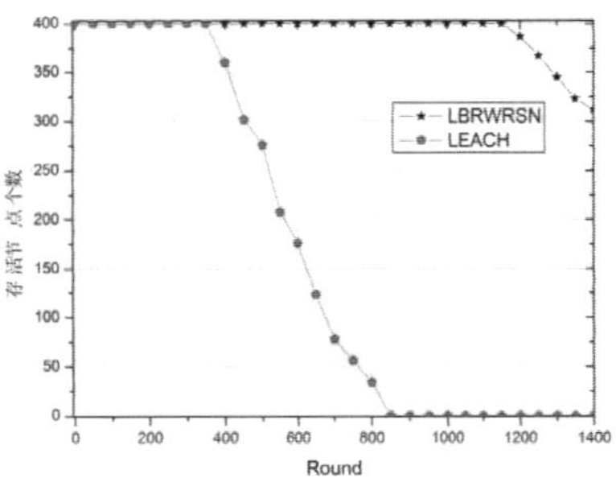

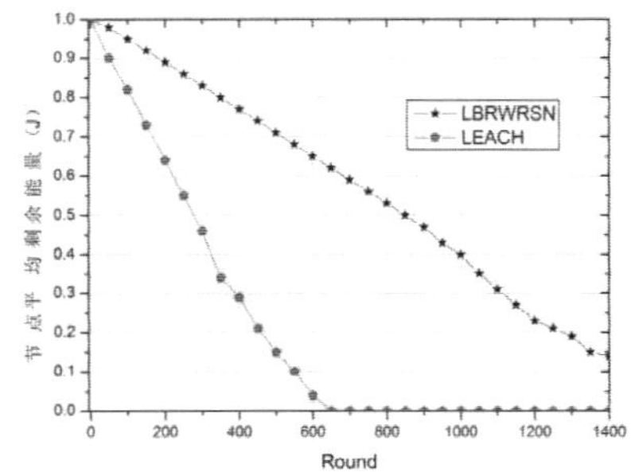

Figure 2 The number of network survivors **Figure 3** Average energy consumption

Figure 3 shows the relationship between nodes average energy consumption and network running times. The figure shows the average remaining energy curve of nodes during the operation of wireless sensor network. As can be seen from the figure, by using the LBRWRSN algorithm proposed in this paper, the average energy consumption of its nodes is reduced to a certain extent relative to the LEACH algorithm, which can effectively improve the life cycle of the network.

5 Conclusion

In this paper, based on the wireless transmission technology, a mobile charging device is used to add wireless energy to rechargeable sensor nodes deployed in a complex environment. Different charging models are set up for different situations to analyze. Finally, the simulation results show that the introduction of charging station pairs The charging of nearby sensor nodes has obvious advantages in prolonging the working time of the wireless sensor network and providing a favorable foundation for the reliable operation of the wireless sensor network.

Reference

[1] EPRI. Technical and system requirements of advanced distribution automation[R]. Palo Alto,CA：EPRI,2004.
[2] Singh A K, Sood Y R, Singh H, et al. Smartgrid: An Introduction[J]. International Journal of Advanced Computer Research, 2013, 3(13).
[3] United States Department of Energy. The smart grid: An introduction[J]. 2011.
[4] European commission. European technology platform smart grids：vision and strategy for Europe’s electricity networks of the future [EB/OL] 2008.10
[5] Smart grid working group. Challenge and opportunity：charting a new energy future，appendix A：working group reports[R]. USA：Energy Future Coalition，2003.
[6] Lu B, Habetler T G, Harley R G, et al. Energy evaluation goes wireless[J]. Industry Applications Magazine IEEE, 2007, 13(2):17-23.
[7] Leon R A, Vittal V, Manimaran G. Application of Sensor Network for Secure Electric Energy Infrastructure[J]. Power Delivery IEEE Transactions on, 2007, 22(2):1021-1028.
[8] Gungor V C, Lu B, Hancke G P. Opportunities and Challenges of Wireless Sensor Networks in Smart Grid[J]. IEEE Transactions on Industrial Electronics, 2010, 57(10):3557-3564.

Performance Improvement of BPLC Based on Quantum Algorithm For New Energy Micro-grid

YING Zegui[1] ,LIU Junjie[2] ,HU Yonghong[3],ZHANG Chao[4] and PAN Ke[5]

[1] National grid Sichuan electric power company skills training center, Sichuan 610041,China
[2] North China Electric Power University, Beijing 102206, China
[3] National grid Sichuan electric power company skills training center, Sichuan 610041,China
[4] National grid Sichuan electric power company skills training center, Sichuan 610041,China
[5] North China Electric Power University, Beijing 102206, China

E-mail: ybpanke1992@163.com

Abstract: The channel environment of the new energy microgrid is worse than the traditional power line. On the one hand, the grid connection and switching of a large number of energy terminals in the new energy microgrid will change the impedance characteristics of the power grid and thus affect the transmission characteristics of the power line carrier communication. At the same time, the new energy terminal grid-connected devices make extensive use of power electronic devices, making the noise of the power grid more complicated and greatly different from the noise distribution in the traditional power distribution network, which makes it difficult to detect the carrier communication signal. This article broadband PLC system uses OFDM technology as an effective way to overcome the multipath fading modulation. Although broadband powerline OFDM systems represent the direction of broadband PLC technology, the communications capabilities and quality of service currently provided by the technology are far from broadband telecommunications service standards and require further optimization.

1 Introduction

At present, the research on power line carrier communication technology is basically carried out for the traditional distribution network. On the one hand, a large number of energy terminal equipment in the new energy micro-grid will cause the impedance characteristics of the power grid to change and thus affect the transmission of power line carrier communication[1] Characteristics; at the same time, the new energy terminal grid connected equipment uses a large number of power electronic devices that makes the power grid noise is more complex, there is a great difference between the noise and the traditional distribution network in the distribution, impulsive noise occupies a larger proportion of the noise, which makes it difficult to detect the carrier communication signal.

State Grid Sichuan electric power company:Research on power line and wireless heterogeneous communication technology for new energy micro grid.

Broadband power line communication, referred to as BPLC, is a communication technology that uses medium and low voltage distribution networks as a communication medium to implement integrated services such as data, voice and video transmission. Generally, the communication speed is above 1Mbps, and the frequency range is 1-40MHz. Among the core technologies of broadband PLC, there are orthogonal frequency division multiplexing (OFDM), discrete multitone DMT and spread spectrum modulation, among which OFDM modulation is the most representative, and its application is the most[1-2]. With the increasing speed of smart grid services for PLC communication, the current technology development trend is to use OFDM mode to realize power line carrier communication.

2 Analysis of subcarrier allocation scheme for broadband power line OFDM

The OFDM system divides the actual channel into several subcarrier channels, in order to improve the transmission efficiency of the OFDM system, a dynamic resource allocation mode may be adopted to select different modulation modes and allocate different powers for each subcarrier channel according to its own channel conditions, excellent channel selects high modulation mode, poor channel selects low modulation mode, shut down a few very poor channel[3]. With the advent of quantum computing theory, Quantum Genetic Algorithm (QGA) increases the advantages that traditional genetic algorithms do not have, using quantum bits to represent information and expanding the range of population search through the randomness of quantum bit measurement , through the direction of evolution to quickly approximate the optimal solution[4]. Compared with the traditional genetic algorithm, it has the characteristics of less population, less iterative evolution and less likely to fall into local extremum, therefore, this paper uses quantum genetic algorithm to optimize the allocation of broadband PLC subcarriers[5].

On the basis of quantum computation, quantum genetic algorithm expresses the probability amplitude of quantum bits as the encoding applied to chromosomes so that one chromosome can express the superposition of multiple states and realize the update of chromosomes by using quantum logic gates, so as to achieve the goal of optimization. Specific steps include: quantum bit coding, quantum state measurement, individual evaluation, individual evolution and so on.

3 Quantum Genetic Algorithm

On the basis of quantum computation, quantum genetic algorithm expresses the probability amplitude of quantum bits as the encoding applied to chromosomes so that one chromosome can express the superposition of multiple states and realize the update of chromosomes by using quantum logic gates, so as to achieve the goal of optimization.

(1) Quantum bit code: In genetic algorithm, the chromosome is expressed by definite value, and in quantum genetic algorithm, the chromosome is expressed by quantum bit (qubit). Unlike classical bits, qubits can be in superposition state[6]. A qubit can be represented as:

$$\varphi = \alpha|0\rangle + \beta|1\rangle, \alpha^2 + \beta^2 = 1 \tag{1}$$

Where $|0\rangle$ and $|1\rangle$ denote two different quantum states, and α and β denote the probabilities of the two quantum states, respectively, so a qubit can store and express the information of two states simultaneously. If the gene contains multiple states, you need to convert to binary. Thus a gene that contains 2^K states can be expressed as:

$$q = \begin{bmatrix} \alpha_1 & \alpha_2 & \cdots & \alpha_K \\ \beta_1 & \beta_2 & \cdots & \beta_K \end{bmatrix} \tag{2}$$

(2) Quantum state measurement: The measurement of quantum states is a measurement of a quantum superposition state by a random variable of 0~1, which causes it to collapse from a probability state to a specific state. The specific process is: randomly produce a number between 0 to 1, if it is greater than the square of the probability amplitude, then the measurement results take 1, otherwise the value of 0. Therefore, by measuring, the qubit of quantum states becomes real valued bits. It can be seen that the measurement of Quantum States has certain randomness, so it can expand the range of population search and reduce the population quantity to a certain extent[7].

(3) Individual evaluation: Individual evaluation and conventional genetic algorithm. That is to say, an appropriate fitness function is set up to evaluate the merits of the current individual.
(4) Individual evolution: Quantum evolutionary operations have two strategies, one is the use of conventional cross evolution operation, and the other is the use of quantum rotation gate strategy. In the quantum theory, the transfer between states is realized through the quantum door transformation matrix, the research shows that the rotation angle of the quantum revolving door can also characterize the evolutionary operation, so that it is convenient to add the information of the best individuals to the evolution, accelerate the convergence of the algorithm[8-9]. Quantum gate evolutionary strategy is used to change the probability of each ground state through the quantum gate acting on the ground state of quantum superposition state or entangled state in order to maintain the diversity of the population and objectively simplify the algorithm itself. Therefore, the evolutionary update method of quantum gates is particularly critical, which is directly related to the performance of the algorithm. According to the characteristics of quantum genetic algorithm, quantum rotation gate U is chosen to evolve and update, quantum rotation gate is defined as follows:

$$U(\theta) = \begin{bmatrix} \cos\theta & -\sin\theta \\ \sin\theta & \cos\theta \end{bmatrix} \tag{3}$$

In the formula, θ is the angle of rotation evolution, and the probability amplitude becomes the quantum evolution after the quantum gate evolution:

$$\left[\alpha^*, \beta^*\right]^T = U(\theta_i) \cdot \left[\alpha_i, \beta_i\right]^T \tag{4}$$

In the formula, $[\alpha_i, \beta_i]^T$ is the probability amplitude of the ith gene in the individual; $\left[\alpha_i^*, \beta_i^*\right]^T$ is the probability amplitude of the ith gene updated by the quantum rotation gate, θ_i is the rotation angle of the quantum gate,

$$\theta_i = \Delta\theta \cdot f(\alpha_i, \beta_i) \tag{5}$$

In the formula, $\Delta\theta$ is the size of the rotation angle to control the convergence speed of the algorithm; $f(\alpha_i, \beta_i)$ is the direction of the rotation angle to determine the search direction of the algorithm. $\Delta\theta$ must be reasonable value, if too large, prone to premature phenomenon; the other hand, if too small, the algorithm search grid is small, too slow, or even at a standstill. $\Delta\theta$ is defined as a variable related to evolutionary algebra:

$$\Delta\theta = 10 \cdot \exp(-q / n_c)$$

The formula is q evolution algebra, and n_c is a constant according to the complexity of optimization problem. In this way, the step size of the search can be adaptively changed.

4 Broadband Powerline Communication Simulation Based on Quantum Algorithm for Subcarrier Allocation

The subcarrier allocation firstly initializes the population, then performs quantum state measurement and individual evaluation, and finally realizes the evolution. The specific algorithm flow is shown in Figure 1. In order to prove the effectiveness of the algorithm in the power line channel environment, according to the modeling method of the channel characteristics of the broadband power line OFDM communication system, 8 user channel models are established, and the channel parameter settings are shown in table 1.

Table 1 channel simulation parameters

Channel number	Multipath number	Impulse response duration	Initial delay (μs)	Maximum range (head diameter)	Minimum range (end diameter)	Amplitude variable
1	20	2	1	0.003	0.0003	0.9
2	20	1.5	0.5	0.002	0.0002	0.8

3	10	1	0.5	0.02	0.003	0.8
4	10	1	0.5	0.02	0.003	0.7
5	7	0.7	0.3	0.03	0.004	0.7
6	7	0.7	0.3	0.03	0.004	0.6
7	5	0.5	0.2	0.05	0.005	0.5
8	5	0.5	0.2	0.05	0.005	0.5

Figure 2 shows the transmission characteristics of 8 channels, the selection of 8 channels takes into account the channel conditions under different multipath numbers and fading, and the channel differences among users are obvious. Table 1 shows the other parameters required for the simulation.

It can be seen that in the power line OFDM system, the number of subcarriers is very large, so the sub carrier clustering method is adopted to reduce the complexity of resource allocation. The 10 subcarriers are divided into a cluster, as a resource allocation granularity, so the system only 128 sub carrier clusters to be allocated, greatly reducing the complexity of the allocation of computing. The modulation mode uses M-QAM modulation, and the maximum modulation mode is 1024-QAM. The power spectral density of the signal is selected as -60dBm/Hz according to the HPAV standard.

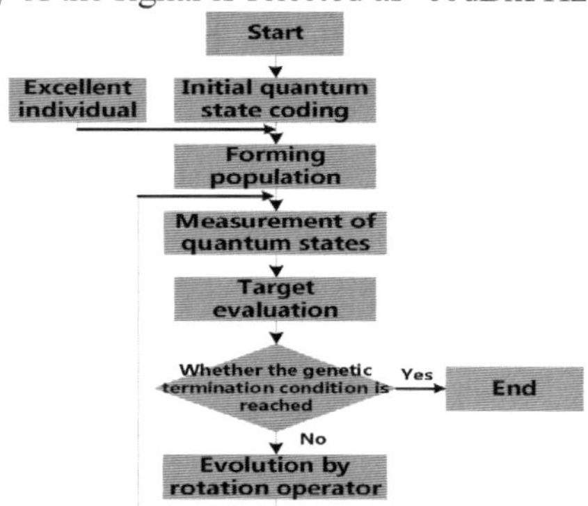

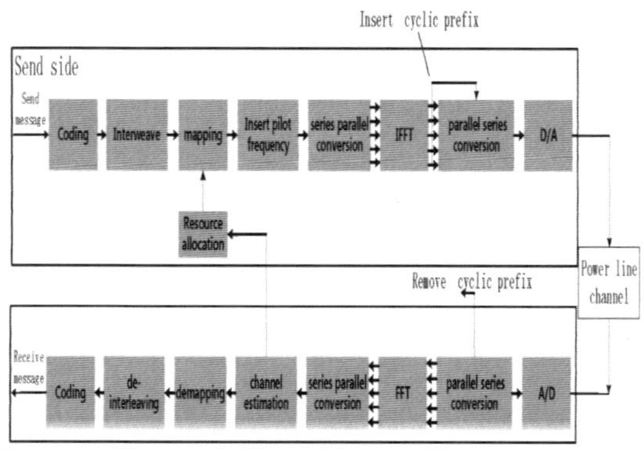

Figure1 Resource allocation algorithm **Figure 2** OFDM based PLC system

Table 2 is the simulation parameter of the power line system OFDM, in which the number of subcarriers is modified appropriately on the reference HPAV standard, selected as 1280, occupying the frequency band from 1 to 30MHz, the main purpose is to facilitate calculation. Multiuser channel transmission characteristic curve as shown in Figure 3.

Table 2 simulation parameters for resource allocation

Parameter name	Parameter value
Number of subcarriers	1280
Subcarrier cluster	128
Number of users	1，2，4，6，8
Band range	1~30MHz
Error rate	10^{-6}
Power density spectrum of colored background noise	-120+40*exp（-f/0.7）dBm/Hz
Signal power spectral density PSD	-60dBm/Hz

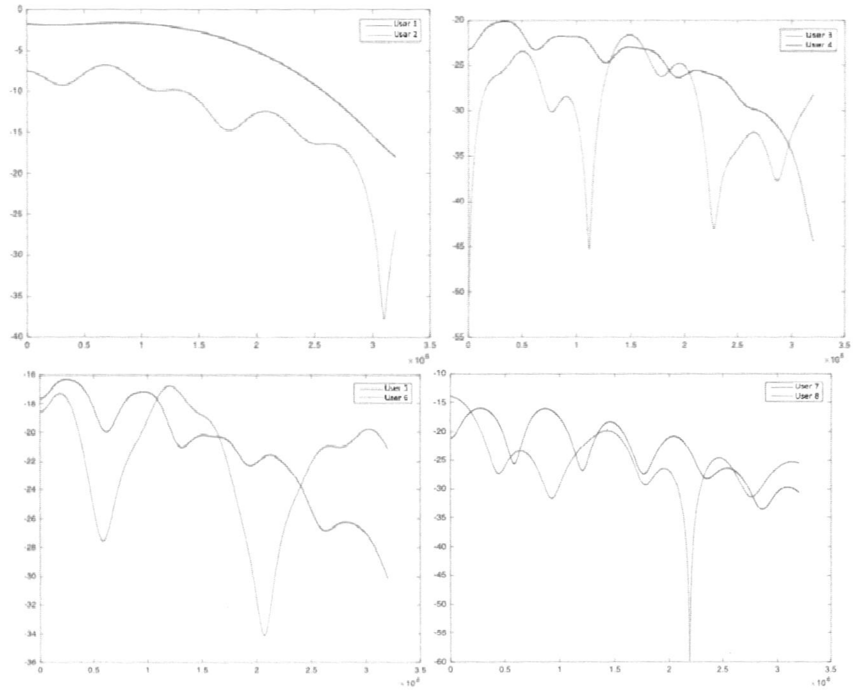

Figure 3 Multi-user channel transmission characteristics

5 Conclusion

According to the features of the power line channel under the new energy micro-grid environment, the signal transmission model of PLC is established. Due to the shortage of spectrum, the noise interference is serious in some frequency ranges, to avoid these frequency bands and achieve higher signal transmission rate, it is necessary to use the frequency aggregation technology to broaden the frequency band used; at the same time, it is necessary to analyze the acquisition of various types of channel parameters of the PLC and create favorable conditions for improving power line communication performance through networking.

Reference

[1] Zheng Jianhong, Lin Huan, Li Xiang. Limited Feedback Precoding Method for Broadband MIMO Power Line System [J/OL]. Application Research of Computers,2018,(05):1-5(2017-08-18).http://kns.cnki.net/kcms/detail/51.1196.TP.20170818.1703.066.html.

[2] Lu Wenbing, Zhang Hui, Zhao Xiongwen, Suo Chaonan. Influence of Network Parameters on Low Voltage Broadband Power Line Channel [J]. Transactions of China Electrotechnical Society,2016,31(S1):221-229.

[3] Huang Yudong. Adaptive resource allocation algorithm for low voltage power line based on OFDM [D]. Hunan University,2016.

[4] Li Zhe. Design and Implementation of ODMA-WiFi Broadband Access Layer [D]. Hebei University,2015.

[5] Zhang Xiaofeng, Sui Guifang, Zheng Ran, Li Zhinong, Yang Guowei. An Improved Quantum Rotation Gate Quantum Genetic Algorithm [J]. Computer Engineering,2013,39(04):234-238.

[6] Zhou Jianping, Han Lin, Wen Buying. Application of Improved Quantum Genetic Algorithm in Transmission Network Planning [J].Power System Protection and Control,2012,40(19):90-95.

[7] Liang Changyong, Bai Hua, Cai Meiju, Lu Wenxing. Research Progress in Quantum Genetic Algorithm [J]. Application Research of Computers,2012,29(07):2401-2405.

[8] Li Huangqiang. Research on Resource Allocation in Broadband Power Line OFDM Communication System [D]. WuHan University,2010.
[9] Cun Qiaoping. Reactive Power Optimization of Power System based on Quantum Genetic Algorithm [D]. Southwest Jiao Tong University,2008.

Research on the Optimization Plan Method based on Electric Vehicles' Scheduling Strategy in Active Distribution Network with Wind/Photovoltaic/Energy Storage Hybrid Distribution System

LIU Yali[1], ZHAO Xin[1], LI Guodong[1], LIU Yujun[2] and YIN Hongyuan[2]

[1] Electric Power Research Institute of State Grid Tianjin Electric Power Company, Tianjin 300384, China
[2] School of Electrical Engineering, Southeast University, Nanjing 210096, China

liuyujun0713@126.com

Abstract: The rapid development of electric vehicles (EVs) and distributed generations (DGs) has brought new challenges to the safe and economical operation of power system. This paper mainly works on how to optimize the management of plug-in EVs in the distribution system with fixed DGs, in order to achieve the economic and stability in power grid. Firstly, the management characteristic of EV is discussed and classified to obtain the access probability model. Secondly, being in distribution network investors' shoes and considering the constraints of distribution network, DG, ESS and EV, the economic management of EV optimization bi-level model is established with the objective function of maximum annual income and minimum fluctuation of output power of equivalent load. Then the genetic algorithm(GA) is used to solve this model. Finally, the experimental results of 33-bus system with DGs and EVs verifies the effectiveness of the proposed method, and the influence of different penetration of EV on the annual income of distribution network is also analysed to provide a new idea for the future development of distribution network.

1. Introduction

Recent years, with the environmental pollution and energy shortage problems become increasingly serious, clean energy including wind and solar and the zero carbon emissions of EVs get more attention in the distribution network. However, with the higher penetration of DGs and the large popularity of EVs, their uncertainty will have a serious impact on the safe and economical operation of the distribution network [1-3].

Energy storage system (ESS) can quickly adjust power flow. It can store energy or release the excess power, which is the key to adjust network's operation state[4-5]. But it is difficult to configuration on a large scale because of the expensive price[6]. Because of the characteristics of EVs' battery, EV is not only the load in power system, but also can be as a mobile energy storage device. Under the control strategy, EV can use V2G technology undertake the task of smooth power fluctuation of load in network with traditional ESS[7-8]. Therefore, the reasonable guidance and configuration of the plug-in EV will directly affect the operation and management level and economy of the distribution network, which is the key in future development.

In [9], a stochastic economic scheduling model was constructed, which took the uncertainly of EV and wind power output into account. And the expectation expression of wind turbine and EV in V2G state were derived. [10] established the EVs and wind power co-scheduling model in multi-time scale with the aim of smoothing the system load fluctuation. However, only the charging mode is considered in this model, and V2G technology is not involved. [11-12] proposed a multi-objective optimization model to decrease the fluctuation caused by DG by using different algorithms respectively. Through dynamically adjusting the charging and discharging power of EV to match the demand of system. But both of them changed EV into ESS directly, which didn't consider the uncertainty and randomness of plug-in EV in actual situation.

This paper is in distribution network investors' shoes and considers the constraints of distribution network, DG, ESS and EV, proposes the economic management of EV optimization bi-level model. Objective of the model is maximum annual income and minimum power fluctuation of equivalent load. The GA is used to solve the algorithm.

2. Dispatchable model of electric vehicle

2.1. The status classification of electric vehicle

According to the purpose, EV can be divided into two types including public and private. According to the mode connection to grid, EV can be divided into three types including slow charge, fast charge and replacing battery. This paper mainly focuses on the private EV with mode of slow charge. Statistics show that about 90% time of private cars are in a state of suspension during one day [13]. Excluding some of the EVs may be in the state of unconnected to grid, the remaining EVs will connect into grid to charge power according to their need without guide incentives. From the grid point of view, these charging vehicles constitute a disordered load.

However, the more attractive research is the large size of storage resource provided by EVs. Through incentive strategies, EVs users can management their charge action to meet the need of grid[14]. As long as the user is willing to access the EV to the grid and follow the grid's dispatch, EV can fully realize the energy storage function.

Here the following assumptions are made for the control of dispatchable EVs:

(1)The dispatchable EVs connected into the same node will be management by the aggregator[15];

(2)The EVs that have registered into dispatchable will automatically been identified, and obey the management of grid;

(3)The charge and discharge function is achieved by DC/AC converter of aggregator and EV battery.

2.2. Access probability analysis of dispatchable EVs

To analyze the dispatchable EVs, it is necessary to master the access probability at any time in typical day. In this paper, the Monte-Carlo simulation method is used to study the probability model. By simulating it, the access probability result is shown in Figure.1.

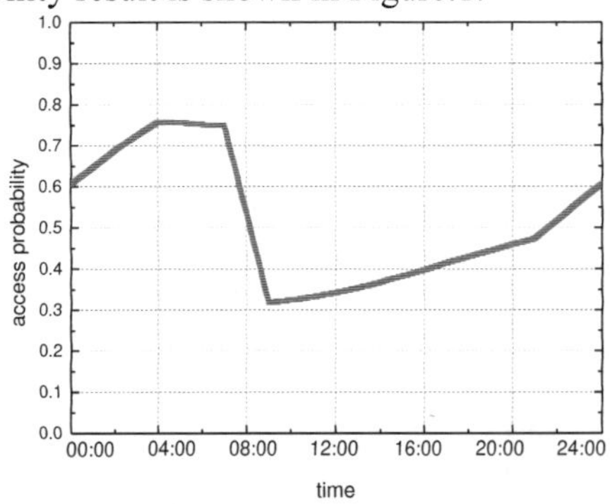

Figure 1. access probability of EV

2.3. Analysis of influencing factors of dispatchable EVs

At present stage, managers mainly use the way of price compensation to guide EVs to participate in the scheduling. Considering the big difference of EVs' usage during the day and night, and in different times their reaction are also vary greatly. Reasonable subsidy price at different times can guide the users to make more favorable choice for the operation of the grid. However, [16] pointed that the frequent changes in the price will lose the guiding role. So this paper set the subsidy price into two time periods to consider:6:00—20:00 and 20:00—6:00 of the next day. Using different mathematical formulas to fit the relationship between subsidy price and EV number[17-18], the expression is as follows:

$$n=\begin{cases} 0.013\ln(x)N & 6:00—20:00 \\ \dfrac{1}{10+57e^{-0.01x}}N & 0:00—6:00,20:00-24:00 \end{cases} \tag{1}$$

Where, x is subsidy price. N is the total number of plug-in EVs. n is the total number of dispatchable EVs in the corresponding periods.

3. Optimization model of EV scheduling strategy

3.1. Objective function

The outer optimization is mainly from the perspective of the distribution network investment manager, taking into account the part of the profit part and the cost part, analyze the economic benefits brought about by the dispatchable EVs. The objective function for outer optimization is as follows, and all costs are converted into one-year calculations:

$$\max\ F_{total}=f_{pur}+f_{save}+f_{loss}+f_{sale}-f_{cons}-f_{main}-f_{V2G} \tag{2}$$

Where, f_{pur} is the saved purchasing cost after the DG connected to network; f_{save} is the saved expansion investment; f_{loss} is saved network loss; f_{sale} is the profit from sale power to the EVs which don't participate in dispatch; f_{cons} is the total investment; f_{main} is the total operation and maintenance cost; f_{V2G} is the total subsidy cost for dispatchable EVs.

The target of inner optimization is mainly smoothing the fluctuation of equivalent load. The function is as follows:

$$\min\ F_{flu}=\sum_{i=1}^{T-M+1}\sum_{t=i}^{i+M-1}(P_l(t)-P_{DG}(t)-P_{ESS}(t)+P_{EVl}(t)-P_{av}(i))^2 \tag{3}$$

Where, T is the number of sampling data in the period; M is the number of sampling data in one time window; $P_l(t)$ is the load power at time t; $P_{DG}(t)$ is output power of DG at time t; $P_{ESS}(t)$ is the total power of dispatchable EVs and ESS at time t; $P_{EVl}(t)$ is the EVs load at time t; $P_{av}(i)$ is the mean value of the equivalent load during this time window.

3.2. Constraints

- Power flow constrains: The state of distributed network should always satisfy the power flow formulas.
- The constrain of voltages: The newly-allocated DGs will change the voltages. While the voltages should always follow the constrains as

$$U_N(1-\varepsilon_1)\le U_i\le U_N(1+\varepsilon_2) \tag{4}$$

Where, U_i is the voltages of the nodes; ε_1 and ε_2 is the allowed deviation of voltages.

- The constrain of DGs: The output of DGs has its own maximum limits and is limited by the real-time weather condition.

$$\begin{cases} 0\le P_{DG}\le P_{DG\max} \\ Q_{DG}\to 0 \end{cases} \tag{5}$$

Where, P_{DGmax} is the maximum output of DGs. Assume that DGs generate no inactive power.

- The constrain of DESS: All the DESS should meet the constrains about SOC, charging and discharging power[19].

$$\begin{cases} SOC_{\min} \le SOC_t \le SOC_{\max} \\ 0 \le \left|P_{\text{DESS},t}\right| \le P_{\text{DESS,max}} \end{cases} \tag{6}$$

Where, $SOC_{\max}$ and $SOC_{\min}$ is the upper and lower limits of the SOC of each distributed energy storage; $P_{\text{DESS,max}}$ is the maximum power of each DESS; SOC_t and $P_{\text{DESS,t}}$ is SOC and power of each distributed energy storage at time t.

- The constrain of EV: All the EVs should meet the constrains about charging and discharging power.

$$0 \le \left|P_{\text{EV},t}\right| \le P_{\text{EV,max}} \tag{7}$$

Where, $P_{\text{EV,max}}$ is the maximum power of each EV; P_{EV},t is power of each electric vehicle at time t.

3.3. Solution and algorithm

GA is a randomized search method based on the evolution of biological evolution, which is suitable for solving the mathematic problem of incentive strategy in this paper.

3.4. Model and parameter settings

In this paper, IEEE 33 nodes distribution network is selected as the test system. The topology of its structure is shown in Figure.2. Set node 19 and 31 connect to wind generation(WG), node 20 and 32 connect to photovoltaic(PV). The rated power of them are all 100kW.And the output data of WG and PV is selected from the typical day in a certain area, the sampling period is 15min. Their output power curves are shown in Figure.3. The load data of the typical day refers to [20]. Capacity lithium-ion battery is used in ESS which has 92.74 percent efficiency.

It is assumed that the total number of EVs in this network is 400, and the corresponding node of DG is connected to 100 electric vehicles, the maximum charging and discharging power of EVs is 3kW, and the discharge efficiency is 95%.At the same time, the scheduling priority of the EV is higher than the ESS.

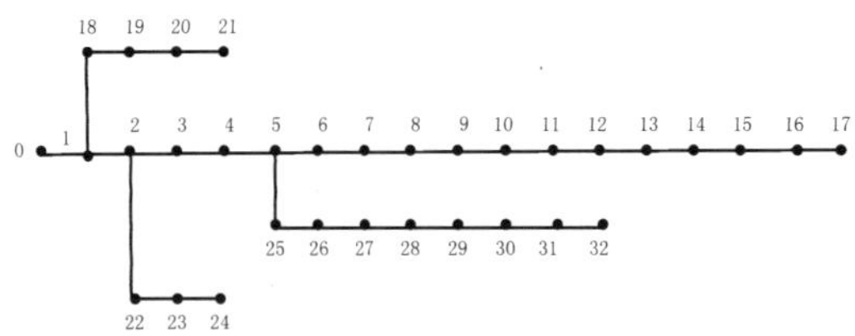

Figure 2. IEEE 33-bus system

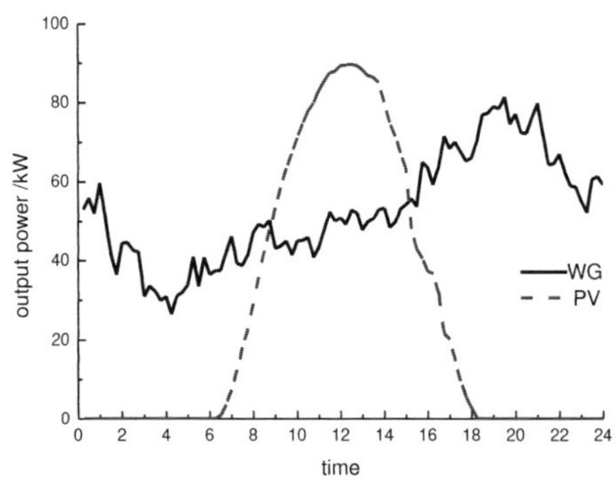

Figure 3. Curves of WG and PV output power in typical day

3.5. Results and analysis

The optimization of the EV management strategy is carried out in the network with wind/photovoltaic/energy storage hybrid system. And the GA has evolved to the 14th generation convergence through the method mentioned in this paper. The optimal configuration scheme of dispatchable EVs subsidy price is shown in Table 1. The subsidy price means that each dispatchable EV can get the corresponding subsidy when they plug-in grid one hour.

Table 1. The optimal configuration scheme of subsidy price

Time interval	6:00—20:00	20:00—6:00
subsidy price(yuan/h)	2.51	1.32
Upper limit of EV	3	6

In this scheme, the optimized strategy of the time-sharing subsidy price is adopted to control the EVs subject to power grid dispatch to make the maximum annual profit, about 353,828 yuan. And the access situation of dispatchable EVs in one node can be seen in Figure.4.

In the process of smooth the fluctuation of equivalent load, due to the dispatchable EVs can't guarantee meeting the demand of grid all the time in one day. So it's necessary to use traditional ESS to auxiliary dispatchable EVs in the shortage period. The rated power of ESS scheme is shown in Table.2. And the output power change of the joint system is shown in Figure.5. We can see it clearly that the effective reduction of ESS configuration with the EVs' participation.

Table 2. The optimal configuration scheme of ESS capacity

Node	19/31	20/32
Rated power/(kW)	7.03	14.42

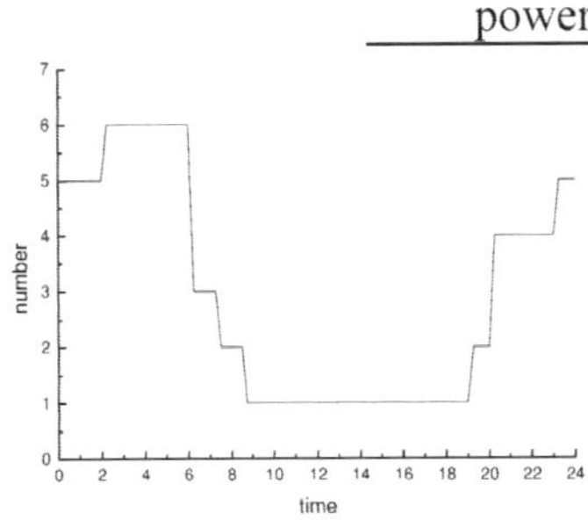

Figure 4. The change state of dispatchable EVs power

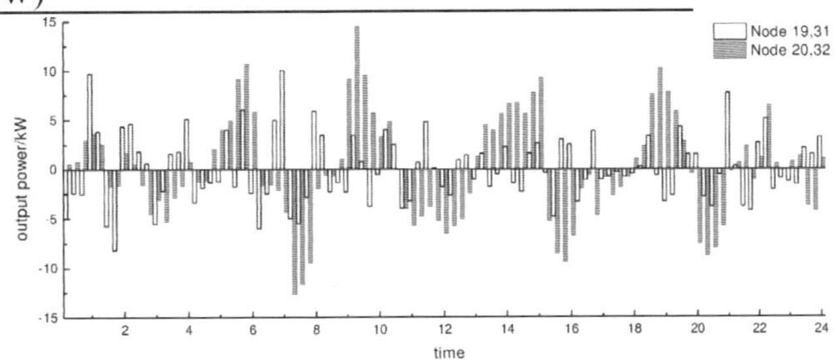

Figure 5. The change state of joint system output

To study the variation of annual income of distribution network with the number of EVs. It can be found that with the increase of EVs, the overall income is on the rise. But to one certain point, the income stop increasing and even falling fast as shown in Figure.6. Analysis of this phenomenon, when the rated power connected into grid of DGs is fixed, it will bring much disorder load with the huge number of EVs to the grid which can't be balanced by the dispatchable EVs. So the income is published by the algorithm. Also we can get the maximum number of EVs is 480 in this system.

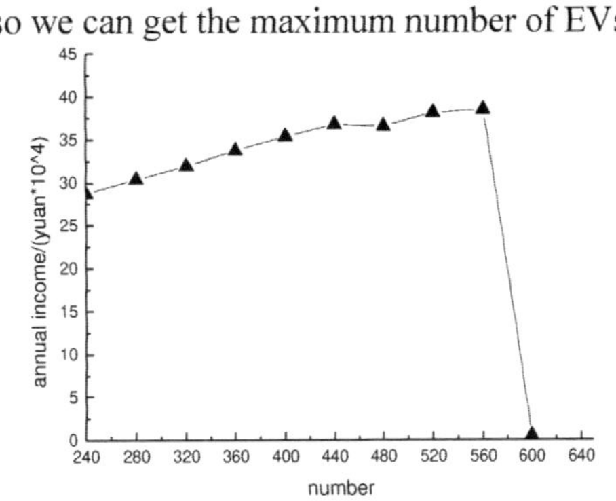

Figure 6. The curve of annual income varies with the number of EVs

4. Conclusions

In this paper, the study of the optimization of EVs scheduling strategy in the background of distribution network with wind/photovoltaic/energy storage hybrid system is studied.

The article from the perspective of power distribution network investment managers, in order to improve the annual income and decrease the fluctuation of equivalent load as the objective function, establishing the bi-level optimization model to seek the optimal scheduling strategy of dispatchable EVs subsidy price. The proposed model of is solved by the GA with good searching ability and convergence. The results of the example analysis show that the proposed model method can be considered as a more reasonable configuration scheme considering economy and power grid security.

Acknowledgments

This paper is funded by Science and Technology Project of State Grid Tianjin Electric Power Company (kj17-1-02).

References

[1] El-Khattam W, Salama M M A. Distributed generation technologies, definitions and benefits[J]. Electric Power Systems Research, 2004, 71(2): 119-128.

[2] Kempton W,Letendre S . Electric vehicles as a new power source for electric utilities[J] . Transportation Research : Part D,1997,2(3): 157-175 .

[3] Fernandez L P , Roman T G S , Cossent R , et al. Assessment of the impact of plug-in electric vehicles on distribution networks[J]. IEEE Transactions on Power Systems,2011,26(1): 206-213.

[4] ESMAILI A, NOVAKOVIC B, NASIRI A, et al. A hybrid system of Li-ion capacitors and flow battery for dynamic wind energy support[J]. IEEE Trans on Industry Applications, 2013,49(4): 1649-1657.

[5] Shen X, Shahidehpour M, Han Y, et al. Expansion Planning of Active Distribution Networks With Centralized and Distributed Energy Storage Systems[J]. IEEE Transactions on Sustainable Energy, 2016, 8(1):126-134.

[6] Xiao Hao, PEI Wei, YANG Yanhong, et al. Energy Storage Capacity Optimization for Microgrid Considering Battery Life and Economic Operation[J]. High Voltage Engineering, 2015, 41(10):3256-3265.

[7] PENG Xiangang, LIN Lixiang, LIU Yi, et al. Multi-objective optimal allocation of distributed generation considering uncertainties of plug-in electric vehicles and renewable energy sources[J]. Power System Technology,2015, 39(8): 2188-2194.

[8] Lu Linrong,Wen Fushuan,Xue Yusheng,et al. Economic analysis of ancillary service provision by plug-in electric vehicles[J]. Automation of Electric Power Systems,2013 37(14): 43-49.

[9] Zhao Junhua,Wen Fushuan,Xue Yusheng,et al. Power system stochastic dispatch considering uncertain outputs from plug-in electric vehicles and wind generators[J] Automation of Electric Power Systems,2010,34(20): 22-29.

[10] Yu Dayang,Song Shuguang,Zhang Bo,et al. Synergistic dispatch of PEVs charging and wind power in Chinese regional power grids[J] Automation of Electric Power Systems,2011,35(14): 24-29.

[11] Bao Yan,Jiang Jiuchun,Zhang Weige,et al. Model and control strategy of electric vehicle mobile energy storage system[J]. Automation of Electric Power Systems,2012,36(22): 36-43.

[12] Li Huiling,Bai Xiaomin,Tan Wen,et al. Coordination control for grid-connection of plug-in hybrid electric vehicles and distributed generation[J] . Power System Technology,2013,37(8): 2008-2015.

[13] Kempton W,Letendre S . Electric vehicles as a new power source for electric utilities[J]. Transportation Research: Part D, 1997, 2(3): 157-175.
[14] Yang Xiaodong, Zhang Youbing, Zhao Bo. Automated Demand Response Method for Electric Vehicles Charging and Discharging to Achieve Supply-demand Coordinated Optimization[J]. Proceedings of the CSEE, 2017, 37(1):120-129.
[15] Han S , Sezaki K. Development of an optimal vehicle-to-grid aggregator for frequency regulation[J]. IEEE Trans on Smart Grid, 2010, 1(10):65-72.
[16] Dong Zhaoyang,Zhao Junhua,Wen Fushuan,et al. From smart grid to energy internet: basic concept and research framework[J]. Automation of Electric Power Systems. 2014, 38(15):1-11.
[17] Xu Zhiwei, Hu Zechun, Song Yonghua, et al. Coordinated charging strategy for PEV charging stations based on dynamic time-of-use tariffs[J]. Proceedings of the CSEE,2014,34(22):3638-3646.
[18] Sun Xiaoming, Wang Wei, Su Su,et al. Coordinated charging strategy for electric vehicles based on Time-ofuse Price[J]. Automation of Electric Power Systems, 2013, 37(1):191-195.
[19] Wang Y, Tan K T, Peng X Y, et al. Coordinated Control of Distributed Energy-Storage Systems for Voltage Regulation in Distribution Networks[J]. IEEE Transactions on Power Delivery, 2016, 31(3):1132-1141.
[20] Li Peng. Wind/PV Contain Storage System Economic Operation Study based on Power Demand Response Mechanism [J]. Electrical technology, 2015(3):57-60.

IOP Conf. Series: Materials Science and Engineering **366** (2018) 012026 doi:10.1088/1757-899X/366/1/012026

Opportunistic Maintenance Model for Wind Turbine Based on Reliability Constraint

Hongbo Li[1], Pei Li [1], Nannan Gao[1], Lin Niu[1], Yanyan Zhang[1], Yongbin Bao[2] and Wenqi Xu[2]

[1] State Grid of China Technology College, Jinan 250002, China;

[2] Yushu Electric Power Supply Company of National Power Grid,Yushu 815000, China.

13305318311@163.com

Abstract: Preventive maintenance of wind turbine can effectively improve the reliability of wind turbine. This paper firstly set single component reliability of wind turbine as constraint condition, single component minimum unit time maintenance cost as objective function to obtain optimal maintenance cycle and maintenance times. On this basis, repairing component which meet certain conditions, then build opportunistic maintenance model of wind turbine. Lastly, set the minimum total cost under opportunistic maintenance as objective function, the wind turbine availability as constraint condition, then solve this model to obtain optimal threshold of opportunistic maintenance and minimum total maintenance cost using genetic algorithm. The example analysis shows that the model can effectively save the total maintenance cost. This study has some reference significance for the maintenance department to make the maintenance plan.

1. Introduction

As a clean renewable energy, the development and utilization of wind energy has been paid more and more attention by various countries, and the installed capacity of wind power is also increasing year by year[1-2]. As the key equipment of the wind power system, the wind turbine is complicated and runs in the field for a long time. The working environment is bad, and it is difficult to install and repair. In order to make the wind farm with large income, the wind turbine must meet certain reliability to reduce the loss of the system. The reasonable maintenance plan not only can improve the reliability of the wind turbine, but also can effectively reduce the maintenance cost of the wind turbine.

In recent years, the maintenance of wind turbines has been highly concerned by the enterprises and the academia. For example, document [3] focuses on the maintenance of wind turbine blades. Based on the damage degree and degradation data of blade of wind turbine, the optimization of blade maintenance strategy is realized by simulation in this paper. In document [4], the semi Markov model was used to study the maintenance decision of the gear box of the wind turbine. In document [5], the system failure model and delayed maintenance model are used to study the problem of maintenance and optimization of the key components of the wind turbine. The problem in the above literature is to optimize the maintenance of the single component of the wind turbine. Such a time based single component preventive maintenance strategy is lack of coordination and coordination for each component's maintenance, which will cause expensive overhaul costs and high downtime losses[6].

The wind turbine is composed of key parts, such as gear box, spindle, generator set and impeller. It is a typical multi component series repairable system. It is pointed out that there is correlation between components of multi-component system in document [7-8]. Batch maintenance, group maintenance and opportunistic maintenance can effectively reduce maintenance costs. The opportunistic maintenance model is proposed by Berg.M[9]. The model defines the maintenance and replacement of other components together if the other parts exceed the given threshold when a component is overhauled in the system. It is proposed an opportunistic maintenance model based on reliability of wind turbines in document [10]. This model optimizes the total maintenance cost with the opportunistic maintenance reliability as the optimization variable in order to obtain the minimum maintenance cost. It is presented a state-opportunity maintenance model for wind turbines in document [11-12]. The model is based on the system state indicator. When a component of a unit is overhauled, if the status indicator of other components exceeds its opportunistic maintenance threshold, it will be overhauled.

On the basis of the above documents, the opportunity maintenance model of wind turbine based on the requirement of reliability is proposed in this paper. The work done is as follows: First, the life distribution function of each component is obtained according to the historical data. And with the reliability of each component as a constraint condition, the best repair period and maintenance times of each component are obtained. Secondly, according to the opportunistic maintenance strategy, we define the opportunistic maintenance threshold of each component, and solve the expression of total overhaul cost of wind turbine under opportunistic maintenance strategy. Finally, we use genetic algorithm to optimize the model to get the best opportunistic maintenance threshold and minimum total maintenance cost, and compare it with the single component maintenance model to verify the validity of the established model.

2. The opportunistic maintenance strategy of wind turbine based on reliability

2.1 Principle of opportunistic maintenance

The maintenance methods of the wind turbines involved in this article mainly include the following four kinds,Unwanted corrective maintenance, the minimum maintenance mode is adopted, and the failure rate of the component is not changed after overhaul; Preventive maintenance and opportunistic maintenance using incomplete opportunistic maintenance method、 Parts replacement, Part 1 is replaced during the second preventive maintenance, replacement to restore the parts as new.

The wind turbine is a system composed of gear boxes, bearings, generators, and blades, and the reliability function of the parts of the unit is monotonically decreasing with time. The principle of opportunistic maintenance for wind turbines based on reliability is as follows: When the reliability of a component in the unit reaches the reliability of preventive maintenance, preventive maintenance is carried out. This opportunity can be used to repair the other parts of the unit in advance, so as to improve the reliability of the unit, and reduce the loss of downtime and the cost of maintenance.

3. An Opportunistic Maintenance Model for Wind Turbines

3.1 Model Hypothesis

In order to simplify the study of the model and the main problems, the following assumptions are made in the construction of the model:

(1) All the components of the wind turbine are all new in the initial time, and the distribution of the faults of each component is independent of each other, and all of them obey the Weibull distribution.

(2) When an unintended fault occurs in a wind turbine, it is immediately overhauled and the maintenance time cannot be ignored.

(3) The wind turbine works continuously, and the failure of a single component will cause the whole unit to stop.

3.2 Analysis on the cost of opportunity maintenance for wind turbines

The maintenance cost of the wind turbine consists of two parts, a part of the direct overhaul cost of the unit, and the other part of the cost of the outage loss due to the maintenance. The direct overhaul cost of any part k in the unit can be divided into preventive maintenance cost C_{pk}, the replacement cost C_{rk} of the component and the cost C_{fk} of the unintended failure maintenance of the component. The cost of outage loss for wind turbines includes C_{pd} for maintenance and maintenance loss and C_{fd} for unintended failure maintenance. The failure of any part of the wind turbine will cause the shutdown of the whole unit. C_d is used to indicate the cost of shutdown loss per unit time, and the above maintenance costs are analyzed respectively.

(1) Direct Overhaul Cost

After the completion of the $n-1$ prevention and maintenance from the time t_{n-1}, the wind turbine will complete the n preventive maintenance after the time t_n, and the direct maintenance cost of the component k is:

$$C_{p(k,n)}=\begin{cases} C_{fk}\int_0^{(t_n-t_{n-1})}\lambda_{nk}(t)dt & Y(k,t_n)=0 \\ C_{pk}+C_{fk}\int_0^{(t_n-t_{n-1})}\lambda_{nk}(t)dt & Y(k,t_n)=1 \\ C_{rk}+C_{fk}\int_0^{(t_n-t_{n-1})}\lambda_{nk}(t)dt & Y(k,t_n)=2 \end{cases} \tag{1}$$

In the formula, $Y(k,t_n)$ means the maintenance mode adopted by the component k at the time t_n of maintenance of the wind turbine. 0 indicates no repair for the component k, 1 means the preventive maintenance of the component k, and 2 for the preventive replacement of the components.

If the wind turbine has carried out N preventive maintenance during the limited operation period $[0,T]$, the direct maintenance cost of the component k in the cycle is as follows:

$$C_k=\sum_{n=1}^{N}C_{p(k,n)} \tag{2}$$

Assuming that the wind turbine is composed of S components, the direct overhaul costs of all components in $[0,T]$ are:

$$C_m=\sum_{k=1}^{S}\sum_{n=1}^{N}C_{p(k,n)} \tag{3}$$

(2) Cost of shutdown for preventive maintenance

The loss cost C_{pd} of the shutdown for the preventive maintenance of the wind turbine is proportional to the time T_{pd} of the shutdown time, and the C_{pd} is proportional to the shutdown time. And T_{pd} is related to the number of tasks per preventive maintenance. When the wind turbine carries on the n

preventive maintenance, the longest part of the maintenance time should be selected as the stop time. As shown below:

$$T_{pdn} = \max_{1\le k\le s}(t_{pkt_n}, t_{rkt_n}) \tag{4}$$

In the formula, S is the unit number of units; t_{pkt_n} indicates the time required for the preventive maintenance of the component k at the time of t_n ; t_{rkt_n} represents the time required for the replacement of the component k at the time of t_n . The cost of the preventive maintenance and shutdown for the wind turbine in the limited period of $[0,T]$ is as follows:

$$C_{pd} = C_d T_{pd} = C_d \sum_{n=1}^{N} T_{pdn} \tag{5}$$

In the formula, C_d represents the cost of outage loss per unit time of a wind turbine.

3.3 Solution of opportunity maintenance model

(1)We can obtain the shape parameter β_k 、 Scale parameter η_k 、 Replacement cost C_{rk} 、 The cost of preventive maintenance C_{pk} 、 The cost of unintended fault maintenance t_{pk} 、 The time required for a preventive maintenance t_{fk} Time required for an unwanted troubleshooting t_{fk} and the failure rate function of the component k after each completion of the preventive maintenance $\lambda_{(i+1)k}(t) = b_{ik}(t + a_{ik}T_{ik})$ by statistics and analysis of the historical data of the wind turbine.

(2) According to the method in the second section, the optimal maintenance interval T_{ik} of the component k and the best number of preventive maintenance times N_k are obtained.

(3) $t_n = \min(t_{(1,n)}, t_{(2,n)}, \cdots t_{(k,n)} \cdots t_{(s,n)})$ as the time for the n maintenance of the unit, Where $t_{(k,n)}$ indicates the time for the unit k to do a n times repair.

(4) Set the opportunistic maintenance threshold of the unit ΔR_k . Then compare $R_{kn} - R_{km}$ and ΔR_k .If $R_{kn} - R_{km} > \Delta R_k$, The part k does not require an opportunistic maintenance at the time of t_n ;If $0 < R_{kn} - R_{km} \le \Delta R_k$, The part k require an opportunistic maintenance at the time of t_n ,meanwhile, The number of preventive maintenance of component k changed to $n_k + 1$;If $0 < R_{kn} - R_{km} \le \Delta R_k$ and $n_k = N_k + 1$,a preventive replacement of the component k is required.

(5) The time required for the repair of the component k at the time t_n can be expressed as:

$$\tau(k, n_k) = \begin{cases} 0 & Y(k,t_n) = 0 \\ t_{pkt_n} & Y(k,t_n) = 1 \\ t_{rkt_n} & Y(k,t_n) = 2 \end{cases} \tag{6}$$

In the formula, t_{pkt_n} indicates the time required for a preventive maintenance of the component k ; t_{rkt_n} indicates the time required for a preventive replacement of the component k . Choose the time of maintenance for the longest time required for maintenance, as the stopping time of the unit for this maintenance, that is $T_{pdn} = \max_{1\le k\le s}(\tau(k, n_k))$.

(6) The time for the next preventive maintenance of the component k after the n times preventive maintenance of the wind turbine can be expressed as:

$$t_{(k,(n+1))} = \begin{cases} t_{(k,n)} + T_{pdn} & Y(k,t_n) = 0 \\ t_n + T_{pdn} + T_{(n+1)k} & Y(k,t_n) = 1,2 \end{cases} \tag{7}$$

(7)The model established in this paper is a multi-variable and nonlinear global optimization problem, and the MATLAB genetic toolbox can be called to solve the model. The specific process of genetic algorithm is as follows[13-14]:

Step 1: a population $P_0 = \left(\overline{\Delta R(1)}, \overline{\Delta R(2)} \cdots \overline{\Delta R(n)}\right)$ of a number of initial individuals is formed using a computer, in which $\overline{\Delta R(n)}$ represents a feasible solution of the model.

Step 2: The values of each individual's C and A are calculated, and the individuals who are not satisfied with the reliability requirements are eliminated.

Step 3: Select the sorting operation for the population. The most adaptive individuals are retained to the next generation, and the lowest adaptive individuals are eliminated, and the remaining individuals are chosen according to the roulette algorithm.

Step 4: The crossover and mutation operation of the population was carried out according to the crossover probability P_n and the mutation probability P_m, producing a new generation of population and returning to step 2.

Step 5: The optimal $\overline{\Delta R}$ is output when the population iterates enough iterations to terminate the calculation.

4. analyses of example

4.1 Determination of model parameters

To verify the reasonableness of the model built in this paper. Choose a wind turbine, The 4 key parts of the gear box, the main bearing, the generator and the impeller are analyzed in the opportunistic maintenance strategy. Through the statistical analysis of the fault data of large wind farms, it is found that the fault time of the 4 key components all obey Weibull distribution. The Weibull distribution parameters of each component and other parameters are shown in Table 1. To make the components of the age reduction factor and failure rate increase factor same. that is: $a_k = a = 0.1$, $b_k = b = 1.1$. The study cycle of the overhaul of the wind turbine is $[0,365]$ days.

Table 1. Maintenance parameters of critical components of wind turbine

Part Number	β	η	C_{pk} / Million Yuan	C_{fk} / Million Yuan	C_r / Million Yuan	t_p /Days	t_{fk} / Days	t_{rk} / Days
1	9.1604	1207.9	3.8	6.6	16.7	2	4	2
2	7.3317	1189.1	1.5	3.1	15.6	2	3	2
3	12.3972	1204.6	2.5	5.6	19.4	4	5	4
4	10.5149	1134.8	2.8	4.8	12.5	4	6	3

The reliability requirements for no fault operation of the 4 key components are given in JB/T10396. As shown in Table 2, when the reliability of the key components of the wind turbine is lower than the data shown in Table 2, preventive maintenance is carried out.

Table 2. Preventative maintenance reliability requirements of critical components of wind turbine

Part	Gear Box	Main Bearing	Generator	Impeller
Reliability	0.95	0.94	0.95	0.96

4.2 The solution of the interval of single part preventive maintenance and the number of overhaul times

Take the parameters in the 4.1 section into the second section model, and the model is solved by MATLAB software programming. Table 3 is the change table for the maintenance costs per unit time of each component with the number of times of maintenance. The maintenance cost per unit time in the table is only calculated to $N_k + 1$ times.

Table 3. the relation of each component unit time maintenance cost versus maintenance times

Part	Maintenance cost for unit time of each component Yuan/Days					Optimal number of preventive maintenance
	1	2	3	4	5	
1	212.3	206.8	197.6	216.5	-	3
2	223.8	207.1	201.2	197.4	204.6	4
3	478.5	465.2	482.8	-	-	2
4	115.7	104.8	102.9	112.3	-	3

From the above table, it can be seen that the maintenance cost per unit time of a single unit decreases first and then rises with the increase of the number of maintenance. Therefore, there is a best maintenance times make the maintenance cost of single part unit time minimum. The best preventive maintenance cycle in its life cycle is listed in Table 4. It can be seen from table 4 that the preventive maintenance cycle of each component is gradually shortened, which is in conformity with the actual situation. Along with the increasing age of the parts should repair more frequently on the parts, to ensure its reliability.

Table 4. Single component optimal maintenance interval

Part	Maintenance cycle of each component /Days				
	T_{1k}	T_{2k}	T_{3k}	T_{4k}	T_{5k}
1	120.1	114.5	106.5	98.7	-
2	95.4	76.2	70.8	67.6	55.4
3	165.4	154.2	123.6	-	-
4	89.7	78.4	72.5	68.9	-

4.3 Solution of an opportunistic maintenance model for wind turbines

The GA toolbox in MATLAB is called to solve the opportunity maintenance model of the wind turbine. The initial population size is set to 300, the maximum number of iterations $d=100$, the cross probability $P_m=0.8$, the mutation probability $P_n=0.02$, and the variable precision of 0.01, The other parameters use the system default value, After 100 iterations, the optimal total maintenance cost is 254 thousand yuan, An optimal set of opportunity maintenance thresholds corresponding to them is $\overline{\Delta R}=(0.198,0.124,0.167,0.114)$.

5. Conclusions

From the perspective of operational reliability and economy of wind turbines, an opportunity maintenance model for wind turbines is built in this paper, and applies genetic algorithm to solve this model to get the best opportunity maintenance threshold and minimum total maintenance cost. On this basis, the single component maintenance model which does not consider the opportunity overhaul is compared. The results show that the opportunistic maintenance model saves about 1/3 of the total maintenance cost compared with the single component maintenance model. The model provides a new idea for the maintenance and maintenance of the wind turbine, and has a certain application prospect. It is necessary to point out that this article is aimed at the establishment and solution of the opportunity maintenance model. It is still relatively lacking in the accumulation of basic data. If the model is to be used in the formulation of the actual maintenance plan, a large amount of basic data is needed to accumulate.

Reference

[1] ZHANG Yingchao, GUO Xiaojie, YE Xiaoling, et al. An integrated forecasting method of short-term wind power[J]. Power System Protection and Control, 2016, 44(7): 90-95.

[2] LIAO Zhigang, HE Shien, DONG Xinzhou, et al. Research on the system protection for enhancing the accommodation and transmission of large scale wind power[J]. Power System

Protection and Control, 2015, 43(22): 41-46.

[3] BESNARD F, BERTLING F. An approach for condition-baesd maintenance optimization applied to wind turbine blades[J]. IEEE Transactions on Sustainable Energy, 2010, 1(2): 77-83.

[4] SU Chun, ZHOU Xiaoquan. Condition-based maintenance optimization for wind turbines based on semi-Markov decision process[J]. Journal of Mechanical Engineering, 2012, 48(2): 44-49.

[5] NIELSEN J J, SORENSEN J D. On risk-baesd operation and maintenance of offshore wind turbine components[J]. Reliability Engineering and System Safety, 2011, 96(1): 218-229.

[6] Xiao Yunqi, WANG Kunpeng, He Guanju, et al. Fuzzy comprehensive evaluation for operation condition of large-scale wind turbines based on trend predication[J]. Proceeding of the CSEE, 2014, 34(13): 2132-2139.

[7] WANG Hongzhou. A survey of maintenance policies of deteriorating systems[J]. European Journal of Operational Research, 2002, 139(3): 469-489.

[8] HORENBEEK A V, PINTELON L. A dynamic predictive maintenance policy for complex multi-component systems[J]. Reliability Engineering &System Safety, 2013, 120(1): 39-50.

[9] Nicolai, RP, Dekker R. Optimal maintenance of multi-component systems: a review[R]. Rotterdam: Econometric Institute Report, 2008.

[10] ZHAO Hongshan, ZHANG Lupeng. Reliability Preventive Opportunistic Maintenance Strategy for wind Turbines Based on Reliability[J]. Proceedings of the CSEE, 2014, 34(5): 3777-3783.

[11] ZHAO Hongshan, ZHANG Jianping, GAO Duo, et al. A Condition Based Opportunistic Maintenance Strategy for Wind Turbine[J]. Proceeding of the CSEE, 2015, 35(15): 3851-3858.

[12] ZHAO Hongshan, ZHANG Jianping, CHENG Liangliang,et al. A Condition Based Opportunistic Maintenance Strategy for Wind Turbine Under Imperfect Maintenance[J]. Proceeding of the CSEE, 2016, 36(3): 701-706.

[13] XU Yan, ZHI Jing. Optimal configuration based on improved adaptive genetic alorithm[J]. Power System Protection and Control , 2015, 43(2): 55-62.

[14] YAN Xiaojuan, GONG Renxi, ZHANG Qianfeng. Application of optimization SVM baesd on improved genetic algorithm in short-term wind speed prediction[J]. Power System Protection and Control , 2016, 44(9): 38-42.

IOP Conf. Series: Materials Science and Engineering **366** (2018) 012027 doi:10.1088/1757-899X/366/1/012027

Investigation on Overvoltage caused by Connecting Shunt Reactors with Vacuum Circuit Breakers in Offshore Wind Farms

W. T. Zhou[1], Y. L. Xin[1], W. H. Tang[1], J. J. Huang[1] and Q. H. Wu[1]

[1] School of Electric Power Engineering, South China University of Technology, Guangzhou, 510641, China.

Corresponding author: Prof. W. H. Tang, E-mail: wenhutang@scut.edu.cn.

Abstract. Nowadays, the transient overvoltage occurs in offshore wind farms has aroused widespread concern. To improve the existing offshore wind farm model's accuracy in high frequency conditions, this paper proposes a high frequency model of the vacuum circuit breaker as well as a shunt reactor model considering the influence of stray capacitance. After that, a model of a typical offshore wind farm is constructed for the purpose to analyze the overvoltage occurring at bus bars and wind turbine transformers in the process of connecting a shunt reactor. Simulations show that connecting a shunt reactor in offshore wind farms cause overvoltage of high peak and high steepness compared than with those in onshore wind farms.

1. Introduction

Operational analysis of offshore wind farms is one of hotspots in China [1] for several reasons. First, the topology of offshore wind farms is different from onshore wind farms. Moreover, long and capacitive submarine cables are used more frequently in offshore wind farms. As a result, the charging power in offshore wind farms is greater. Therefore, it is necessary to employ shunt reactors and other devices to absorb the charging power in offshore wind farms. Because the submarine cable impedance is relatively small and the reactor inductance is relatively large, the amplitude and the steepness of overvoltage caused by VCB when switching shunt reactors in offshore wind farms are high [2]. These overvoltages may threaten the operation stability of offshore wind farms, so it is necessary to study the formation occurrence of such transients.

Vacuum circuit breaker is a widely used switching device in offshore wind farms for its advantages of strong arc extinguishing ability, high reliability, long service life, suitability for frequent operation and no fire hazard [3]. Reference [4] introduced the re-ignition and the prestrike phenomena of vacuum circuit breaker in the process of switching operation, and built a simplified model of this process by establishing formulas for dielectric recovery strength and high frequency quenching characteristics based on the experimental data of a 40.5 kV system in [5]. Reference [6] summarized the forming and developing mechanism for three-phase multiple re-ignition. Reference [7] established a simplified offshore wind farm model based on Power Systems Computer Aided Design (PSCAD), and reference [8] proposed a high frequency model of transformer considering terminal capacitances. Reference [8, 9] analyzed internal overvoltages occur in different operation conditions considering the position of transformers, the number of feeders and so on in offshore wind farms.

Overall, there have been abundant researches on the analysis of the overvoltage caused by VCB in the process of switching shunt reactors in traditional power system, while related researches are rare in scenarios for offshore wind farms. In this paper, the switching process of shunt reactors using a vacuum circuit breaker is simulated. In order to study the switching transient overvoltages in accordance with actual working conditions, the high frequency equipment models are built to replace power frequency models which usually used for the steady-state analysis of electric power system in offshore wind farms, namely, VCB and parallel reactors. To study the electromagnetic properties of these overvoltages, PSCAD, an electromagnetic transient simulation software widely used for solving differential equation in the time domain, is employed.

2. Model development for an offshore wind farm

Figure 1 shows a model of an offshore wind farm in Guangdong Province of China. In this model, a shunt reactor switched by a vacuum circuit breaker (VCB_2) is connected to the 35 kV bus. An ideal power supply of 110 kV is employed as the external power grid of the offshore wind farm and is connected to the offshore wind farm through a 35 kV/ 0.69 kV three-phase step-up transformer. The model of the overhead line between the ideal power supply and the step-up transformer is of the π type. Submarine cables are simulated using the frequency dependent phase model in PSCAD/EMTDC component library, because this model has a high accuracy in the wide band analysis considering the frequency dependence of component parameters and line losses. It is suitable for the simulation of the transient process during pre-strikes [7]. According to [9, 10], the duration of the transient process caused by switching a shunt reactor is short, during which the state of wind generators can be treated as unchanged. Therefore, in this study, inductances of 1.2 mH are used to represent the wind power generators. Transformers connected to the wind generators are simulated by the high-frequency transformer transient model considering the influence of stray capacitance in reference to [8].

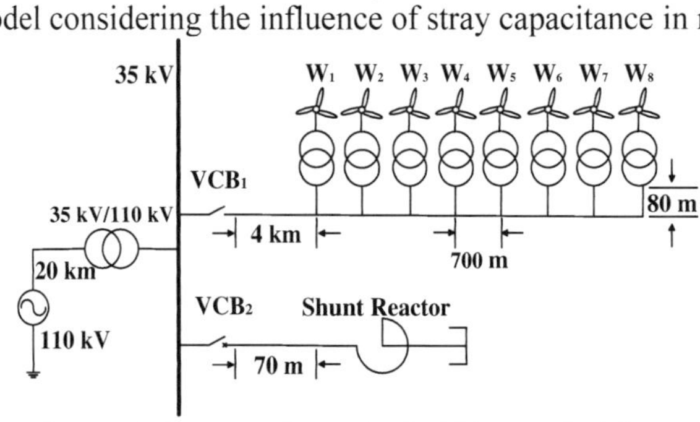

Figure 1. Layout of a typical offshore wind farm.

2.1. Vacuum circuit breaker model

Based on the vacuum switch model proposed in [11], a modified vacuum circuit breaker model considering basic characteristics of vacuum switching is employed in this study. This model, which is built in PSCAD/EMTDC, consists of an ideal circuit breaker and a parallel branch made up of a resistor, an inductor and an capacitor. Furthermore, the prestrike process is simulated by using a customized module to control switch opening and closing.

This model divides a closing transient process into four stages: first prestrike, repeated pretrikes, high frequency arc quenching and physical close, which are presetented by *fp, rp, ro* and *fc* respectively. The flowchart in Figure 2 shows the control logic of this vacuum circuit breaker model.

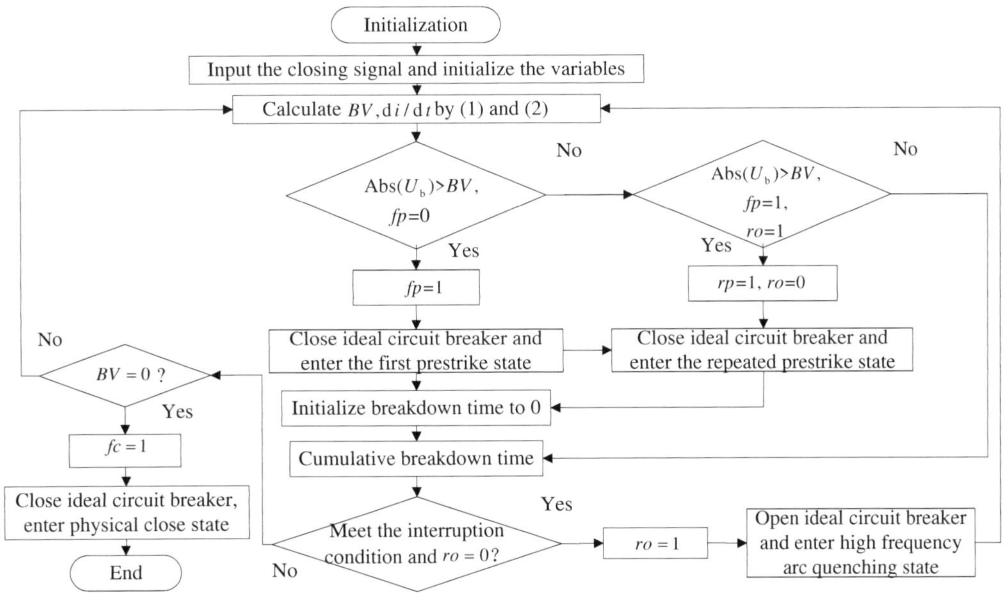

Figure 2. Control logic of the vacuum circuit breaker model.

2.1.1. The first prestrike stage. When the model receives a switch-in signal and the absolute value of voltage across the vacuum circuit breaker is more than the dielectric strength, the system gets into the first prestrike stage and the module sets *fp* to 1 to simulate the prestrike phenomenon. Before the prestrike happens, the frequency of the voltage across vacuum circuit breaker is 50 Hz and thus, the first prestrike also behaves as the frequency of 50 Hz. The dielectric strength in the vacuum circuit breaker is the key factor to determine whether the breaker enters the first prestrike. The prestrike frequency, duration of prestrike current and recovery voltage after prestrike interruption are regarded as constant, which actually decrease over time beause the dielectric strength of the vacuum circuit breaker decreases as the distance between contacts decreases [12]. Reference [13] proposed an approximate fitting method to calculate the dielectric strength based on test results of a specific type vacuum circuit breaker, which can't be used in other system. Different from the above mentioned method, general first order polynomial was employed for the calculation of the dielectric strength in [5]:

$$BV = BV_{\text{limit}} - A(t - t_{\text{close}}) - B \tag{1}$$

where BV denotes the dielectric strength, BV_{imit} represents the maximum dielectric strength, A denotes the closing speed, B is the dielectric constant constant, t is the current time and t_{close} is the time when the contacts starts to close.

2.1.2. High frequency quenching stage. After the first prestrike, the high frequency current which behaves in the form of high frequency arcs, appears between the contacts of the breaker. The interruption condition are:

- The duration of prestrike current is over the limit value or the absolute value of the high frequency arcs is less than the value of the high frequency chopping current;
- The high frequency arc across the vacuum circuit breaker is at the zero crossing point, whose changing rate is less than the high frequency quenching capability.

When the transient process gets to this stage, the custom control module sets *ro* to 1 and swiches-off the ideal circuit breaker, and then the voltage across the breaker restores gradually. The general formula for calculating the high frequency quenching capability is proposed in [5]:

$$\mathrm{d}i/\mathrm{d}t = C(t - t_{\text{close}}) + D \tag{2}$$

where C is a constant in proportion to the arc extinguishing ability, D denotes restorability of the arc when the breaker begins to close.

2.1.3. The repeated prestrike stage. When the transient process enters the high frequency arc quenching recovery stage and the voltage across the vacuum circuit breaker is higher than the dielectric strength again, the custom module sets *rp* to 1 and the transient process turns into the repeated prestrike stage. Different from the first prestrike stage, in this stage, the voltage across the breaker has a relatively higher frequency rather than 50 Hz. The prestrike value is different from those mentioned in section 2.1.1, while the extinguishing conditions of the arc are the same as mentioned in section 2.1.2.

2.1.4. The physical close stage. During the transient process the high frequency quenching stage and the repeat prestrike stage repeat alternately. When in the prestrike stage, the dielectric strength is zero and the contacts of the breaker is closed, *fc* is set to 1 and the transient process enters the physical closing stage.

Figure 3 shows a comparison of the measured and simulation curves of the current and voltage across a vacuum circuit breaker during the closing transient prosess. In Figure 3, U_L is measured at the load side and I_L is the current across the vacuum circuit breaker, U_{L1} and I_{L1} are simulated results respect to U_L and I_L. In Figure 3, the first prestrikes occurs at point a and after that the frequency of prestrikes changes from 50 Hz to higher frequencies. At that time, the voltage across the breaker drops to 0 and the high frequency current and voltage at the load side increase rapidly. Then, at point b, the high frequency quenching stage begins. The voltage across the breaker gradually increases while the voltage at the load side decreases until the repeated prestrikes stage begins at point c. The position of points a_1, b_1, c_1 and points a, b, c, are similar and thus verifies that the proposed four-stages vacuum breaker model is accurate in the high frequency switching transient process.

2.2. Shunt reactor model

The simulation system employs a shunt reactor with a rated voltage of 35 kV, the capacity of 3330 kVar and the reactance value of 106 mH. The neutral point of the shunt reator is float. When the vacuum circuit breaker is interrupted, the peak voltage between the contacts is calculated by:

$$U_m = \sqrt{\frac{L_S}{C_S} I_j^2 + U_j^2} \tag{3}$$

where U_m is the peak value of the overvoltage, U_j is the peak value of power supply phase voltage, I_j is the peak value of chopping current, L_s is the reactor inductance and C_s is the stray capacitance.

Because the stray capacitance of the shunt reactor is rather small, it has a great influence upon the amplitude of the interrupting overvoltage according to Equation (3). The traditional shunt reactor model in PSCAD only considers the three-phase inductance value and neutralpoint-to-ground capacitance, but does not take the stray capacitance such as interphase capacitance and phase-to-ground capacitance into account. In this study, a high frequency model of the reactor considering the stray capacitance is built based on the model proposed in [14]. A switching test circuit is used to compare the traditional reactor model with the high frequency model proposed in this study. In Figure 4, the simulated closing transient overvoltage amplitude obtained by using the high frequnecy model oscillates more significantly than that obtained by using the traditional model, which reflects the effect of the stray capacitance of the shunt reactor on the amplitudes of overvoltage. Therefore, the developed high frequency model is more accurate than the current model in transient simulations, and it has great significance for the study of transient overvoltage simulation in large systems.

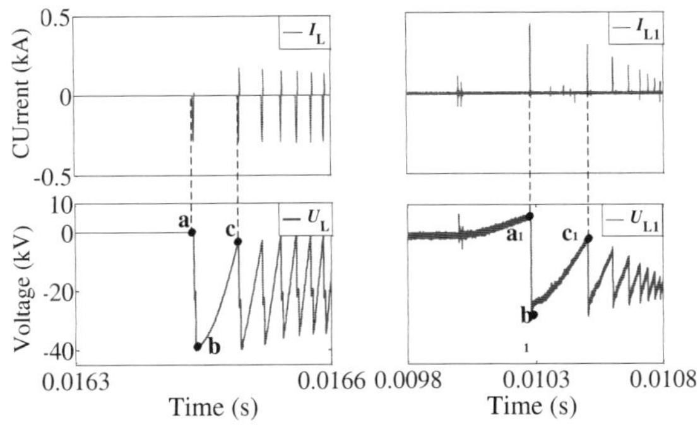

(a) simulation results (b) measured results

Figure 3. A comparison of simulation and measured results of vacuum circuit breaker

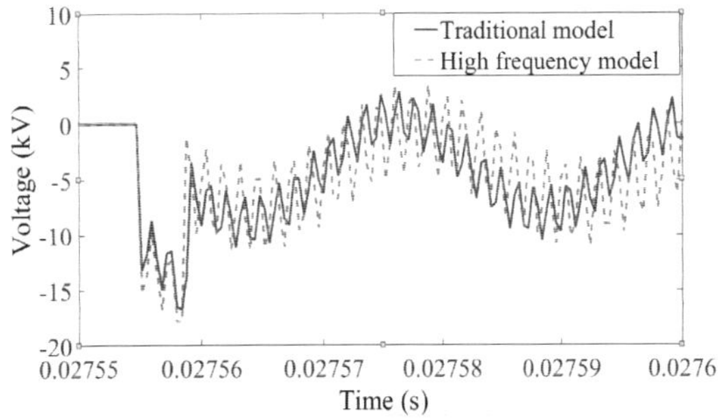

Figure 4. A comparison of simulations by the and high frequency model of shunt reactor.

3. Simulation results and discussion

Based on the offshore wind farm model mentioned above, the overvoltages is simulated in the process of closing VCB_2 when VCB_1 is already closed. The simulation results in comparison to the results of an onshore wind farm is listed in Table 1. In the model of the onshore wind farm the submarine cable model is replaced by the overhead lines of π type, and other conditions are the same as these in the offshore wind farm.

Table 1. Simultated overvoltages in offshore wind farms and onshore wind farms without protection

Equipment	Offshore wind farm		Onshore wind farm	
	Peak voltage (p.u.)	du/dt kV/μ s	Peak voltage (p.u.)	du/dt kV/μ s
Shunt reactor	6.100	116.30	1.160	27.30
35 kV bus	2.510	47.83	1.158	18.10
Transformer of W_1	2.130	20.32	1.155	17.89

Table 1. is the worst case of the closing overvoltage, which happens when the closing phase angle is near 90 degrees. Table 1 shows that overvoltage peak value and steepness of reactor terminal in an offshore wind farms are 5.2 times and 4.2 times larger than in the onshore wind farms. In addition, the overvoltage at the junction of the bus and transformer of W_1 in offshore wind farms is near 2 times greater than that in onshore wind farms. So the overvoltage of swiching-in shunt reactors in offshore wind farms is higher than that in onshore wind farm. The operation overvoltage appears not only at shunt reactors and VCB, but also at the bus and the transformers. The submarine cables are capacitive and the reactors are nonlinear inductive load. At the beginning, VCB_2 received a signal to switch on and the voltage exceeds the dielectric strength in as discussed in section 2.1.1. The energy stored in submarine cable capacitors flows rapidly to the reactor across the 35 kV bus and causes a large current and overvoltage in the system. And then, the quenching condition is satisfied and then VCB_2 opens and the 35 kV bus charges the submarine cable again. The reactor continues charging its stray capacitances as well as the submarine cable capacitors until the next breakdown occurs. As a result, the repeated prestrike of the vacuum circuit breaker causes high frequency oscillations in the system. In the process of repeated prestrike, the continuous energy exchange between inductances of the reactor and capacitances of the submarine cables leads to the increase of amplitude of high frequency oscillation. In onshore wind farms, transmission lines are usually inductive and the energy exchange in this system is mainly between inductances, and therefore the voltage amplitude and steepness are significantly smaller.

4. Conclusion

This paper investigates the overvoltage when connecting reactors in offshore wind farms. A vacuum circuit breaker model considering the arc characteristics, and a high-frequency shunt reactor model introducing stray capacitance are proposed and they are proved to be more accurate than the current models, which can be applied in the transient analysis of power systems. Long submarine cables with a large capacity in offshore wind farms store a lot of energy before they close, which cause overvoltages at the closing moment. The simulation results have shown that in offshore wind farms the connecting process of shunt reactor causes overvoltage with higher amplitude and steepness at shunt reactors, VCB, bus and wind turbine transformers than those of onshore wind farms.

Acknowledgments
This work was supported by the Natural Science Foundation of China (No. 51477054).

References

[1] Q. Sang, Y. He, X. D. Feng, et al. Modeling and Simulation of Combined Repair Oriented Operation and Maintenance for Offshore Wind Farms[J]. Automation of Electric Power Systems, 2016, 40(20): 83-91.
[2] Z. Liu, J. Wang, S. Xiu, et al. Development of high-voltage VCB in China[J]. IEEE Transactions on Plasma Science, 2007, 35(4): 856-865.
[3] Q. Zhou, Y. Cheng, X. Bian, et al. Analysis of restrike overvoltage of circuit breakers in offshore wind farms[J]. IEEE Transactions on Applied Superconductivity, 2016, 26(7): 1-5.
[4] Y. G. Guan, Q. W. Tang, W. D. Liu. Overvoltage mechanism of switching off shunt reactors for 40.5 kV VCB[J]. Proceedings of the CSEE, 2012, 32(33): 124-132.
[5] N. Du, Y. G. Guan, J. S. Zhang, et al. Field experiments of 40.5 kV vacuum circuit breakers switching off shunt reactors[J]. Journal of Tsinghua University, 201 0, 50(4): 517-520.
[6] Q. Zhang, H. Yuan, Y. Liu, et al. The operating transient process analysis and synchronous switching strategies research of vacuum breaker[C]// International Conference on Genetic & Evolutionary Computing. IEEE, 2010: 142-145.
[7] B. Liu, W. H. Tang, X. D. Chen. Modeling of transient overvoltages in wind power plants[C]// 2014 IEEE PES Asia-Pacific Power and Energy Engineering Conference (APPEEC). IEEE, 2014: 1-6.
[8] Y. L. Xin, B. Liu, W. H. Tang, et al. Modeling and mitigation for high frequency switching transients due to energization in offshore wind farms[J]. Energies, 2016, 9(12): 1044-1059.
[9] J. D. Wang, G. J. Li, H. Qin. Simulation of switching over-voltages in the collector networks of offshore wind farm[J]. Automation of Electric Power Systems, 2010(2): 104-107.
[10] L. Liljestrand, A. Sannino, H. Breder, et al. Transients in collection grids of large offshore wind parks[J]. Wind Energy, 2008, 11(1): 45-61.
[11] J. Helmer, M. Lindmayer. Mathematical modeling of the high frequency behavior of vacuum interrupters and comparison with measured transients in power systems[C]// Proceedings of 17th International Symposium on Discharges and Electrical Insulation in Vacuum. IEEE, 1996: 323-331.
[12] Y. Z. An, X. S. Wen, T. T. Zhang. Overvoltage of shunt reactor Caused by prestrike of SF_6 breaker and protection measures[J]. High Voltage Engineering. 2013, 39(1): 75-80.
[13] Z. H. Zhang, G. Wei, Q.Yang, et al. Study of pre-strike overvoltage caused by 12 kV vacuum circuit breaker switching shunt reactors[J]. High Voltage Apparatus, 2016, 52(3): 1-6.
[14] L. Huang, J. J. Niu, Y. Hou. Modeling of overvoltage for 24 kV vacuum circuit breaker switching off shunt reactors[J]. Electrical & Energy Management Technology, 2016(11): 19-27.

IOP Conf. Series: Materials Science and Engineering **366** (2018) 012028 doi:10.1088/1757-899X/366/1/012028

Behaviour of an Integrated EV Based on Renewable Energy Linked to the Distribution Grid: An Educational Tool

Husam Qutaishat[1], Qudaih Yaser[1], Hassan Mighdadi[1] and Uglanov Dmitriy Aleksandrovich[2]

[1]American University of Madaba (AUM), Jordan
[2]Samara National Research University, Samara, Russia

Email: y.qudah@aum.edu.jo

Abstract. Electric Vehicles (EVs) are one of the renewable energy sources that can bring significant positive impacts to power system during their charging and discharging operations. According to this research an educational tool has been created to show students/users the Electric Vehicle's behavior when integrated to the distribution grid. This research helps students/users to develop new algorithms for renewable energy and energy storage applications. A Matlab/Simulink module has been utilized to implement the integrated systems with simple, easy and flexible Graphical User Interface (GUI) for experimentation purposes. Furthermore, the developed educational tool aims to show that EVs can be efficiently charged by renewable energy sources connected to the grid with reduced charging costs. In addition, users can understand how those results can be sustainable for further studies and development. Unlike commercial wise tools, such a tool would add a promising methodology for the research methods and educational systems due to availability and user friendly approach.

1. Introduction

The global electricity supply sector accounts for the release to the atmosphere of over 7700 million tons of carbon dioxide annually [1]. So, increasing the efficiency of electric utilities and integrating modern and sustainable renewable energy resources to electrical grids will have the potential to decrease the carbon dioxide emissions to low levels. On the other side, transportations based on oil cause high amounts of carbon dioxide emissions too. Thus, technologies such as Electric Vehicles (EV) can bring significant improvements and solutions on both electric grids and transportation sectors [2, 3]. Clearly these new technologies require more research efforts in order to improve their sustainability with electric grids and roads infrastructure. Meanwhile, smart grids became the main title behind the modern construction of the modern systems integrating all renewable sources and communications [4]. Furthermore, Energy Management System (EMS) and Buildings Energy Management Systems (BEMS) are important parts of the new systems infrastructure.

As a matter of fact, Distributed Generation (DG), loads and storage units are forming the power grid. By controlling, monitoring and managing the grid's operations, a smart power grid system is implemented. To achieve this, an extensive communication and networking infrastructure and system in deeded The interplay between the power networks and communication networks has been studied at several occasions [5-6] Due to these improvements and needs, educational tools must be found to help researchers, engineers and students about this topic and make them able to understand the complex technical issues. In this research an interactive educational tool has been utilized to implement the behavior of integrated EVs based on Renewable Energy Sources (RES) linked to the electrical distribution grid. An important approach of this research is to provide simple, easy and flexible GUI for experimentation purposes using Matlab/Simulink.

This paper is constructed as follows: Section one introduces the topic, section two explains the model configuration, section three illustrates a sample results and discussion results and section four concludes.

2. System Configuration

2.1. Distribution System (Microgrid)

The targeted system in this study is in the form of the latest designs of distribution systems which is micro-grid. In fact, due to deregulation implemented at many countries across the globe, micro-grids became the main target for electrical engineers to develop new projects and improve the performance of the distribution grid. Similarly, a 33kv, 50 Hz and about 10 MW load system includes all elements such as bus bars, transformers and renewable sources has been chosen in Matlab/Simulink environment for educational purpose as shown in Fig. 1.

In the figure, Diesel generator is the main source for the stable energy which is responsible about keeping the system reliable with its minimum generation cooperated with the storage systems. The storage system in the form of Electric Vehicles is integrated to the system based on special management system coordinating between the renewable and conventional resources. The GUI design will provide the users with enough parameters and structures to practice the management system for all the elements of the grid.

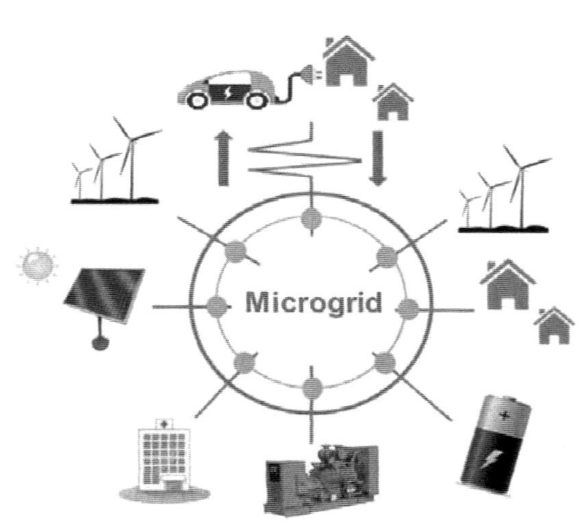

Figure 1. System Configuration

2.2. Renewable Energy

a) Wind farm:

The wind farm model is a model built using Simulink to represent simplified (DFIG) Doubly Fed Induction Generator based wind turbines to produce electrical power as a function of wind speed given as (1).

$$P_w = \frac{\rho}{2} C_P(\gamma, \beta) A v_w^3 \quad (1)$$

Where ρ =1.229 kg /m^3 is the air density, A is the area swept by the turbine blades, γ is the Tip-Speed-Ratio (TSR), β is the pitch angle.[7]

Equation (1) shows that wind energy variation is caused by wind speed fluctuations, as wind speed increases as the output power of the wind farm will increase and when it reaches its nominal speed, wind turbines will reflect the nominal output power of the wind farm.

The wind farm will shut down, only if wind speed increases above its maximum rated value and it will keep shut down until wind speed returns again to its nominal value or below it.

b) PV System
The PV system model is a model built using Simulink to represent a PV farm that produces electrical energy based on three factors: The area of the PV farm, Solar panels efficiency and the irradiance profile through the day. The relationship between these three factors and the output power of the PV farm is given as (2).

$$P_{pv} = AEI \quad (2)$$

Where A is the area of the PV farm, E is the solar panels efficiency per m2 and I is a variable represent the irradiance profile through the day per watt/ m^2.

Where A is the area of the PV farm, E is the solar panels efficiency per m2 and I is a variable represent the irradiance profile through the day per watt/ m2.

2.3. Electric Vehicles
Electric vehicles models are models built using Simulink to represent random plugging profiles of electric vehicles with two behaviours as storage systems connected to the grid. The first behaviour is the charging behaviour where power flows from the grid towards EVs batteries and the second behaviour is the regulation behaviour where power flows from the EVs batteries towards the grid.

State of Charge (SOC) and the grid frequency are the main conditions that control the battery behaviour of every EV has been modelled. The state of charge (SOC) of EV (n) at instant (t) is given by $n, \leq 1$, and the SOC dynamics are described by the simplified model as (3).

$$X_{n,t} + 1 = X_{n,t} + \frac{\alpha_n}{\beta_n} u_{n,t} \quad (3)$$

Where $X_{n,t}$ the State of Charge of EV (n) at instant (t), α_n is the charging efficiency of EV (n), β_n is the battery capacity of EV (n) and $u_{n,t}$ is the charging rate of EV (n) at instant (t).[8]

EVs batteries will remain only in the charging mode as long as their SOC is below 0.9, if their SOC exceeds 0.9 then the EVs batteries can operate on both charging and regulation modes.

The regulation mode operation of EVs batteries is controlled through PI controller. The PI controller controls the amount of power that can be flow from the EVs batteries towards the grid at a certain time. The rotor speed of the diesel generator has been taken as a feedback for the PI controller. When the rotor speed of the diesel generator decreases the diesel generator has to provide more power to the grid as a swing bus. The PI controller role is to detect the decrease of the rotor speed to provide an equivalent amount of power to the grid from EVs batteries until the rotor speed back again to a previously determined reference, as shown in Fig. 2

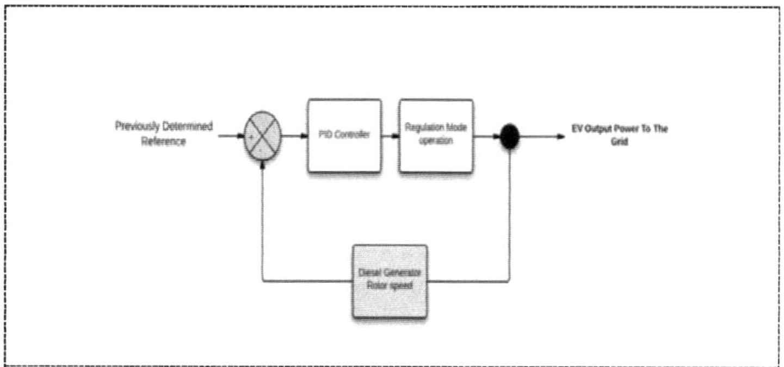

Figure 2. Control diagram of the regulation mode of EVs

3. Graphical User Interface (GUI)

The presented GUI was constructed as follows: six Inputs forms and four Output Graphs as shown in Fig. 3

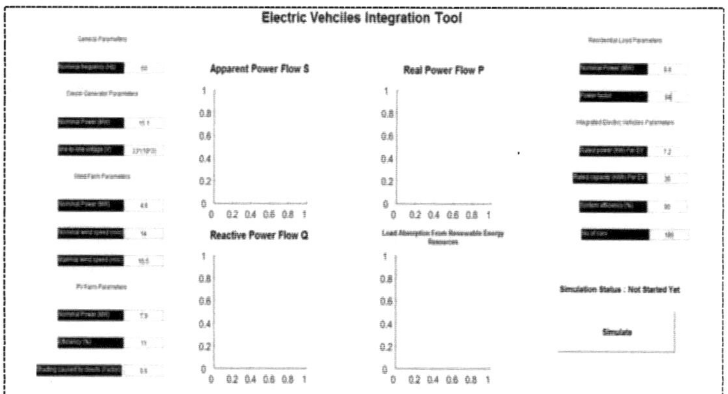

Figure 3. Main GUI Configuration

The input forms are as follows:

3.1 Diesel/grid parameters

Parameters shown in Fig. 4 were chosen to implement a similar Jordanian medium voltage distribution grid

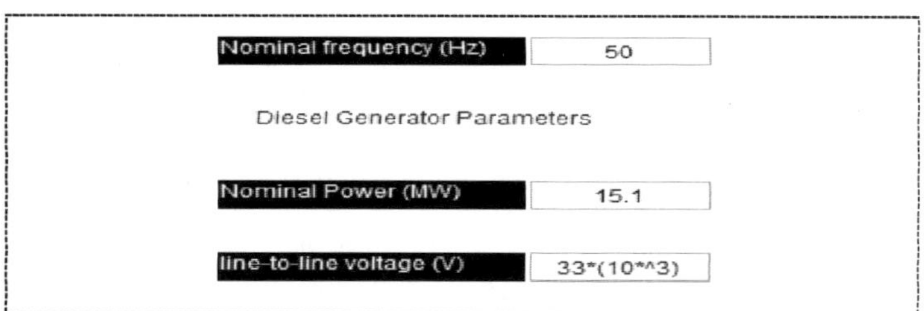

Figure 4. Diesel/grid parameters

3.2 Wind farm parameters

Parameters shown in Fig. 5 are implementing a wind farm connected to the medium voltage distribution grid. The wind farm can be isolated from the system if the maximal wind speed is below the nominal wind speed. Parameters shown in Fig. 6 were chosen to implement a similar Jordanian medium voltage distribution grid.

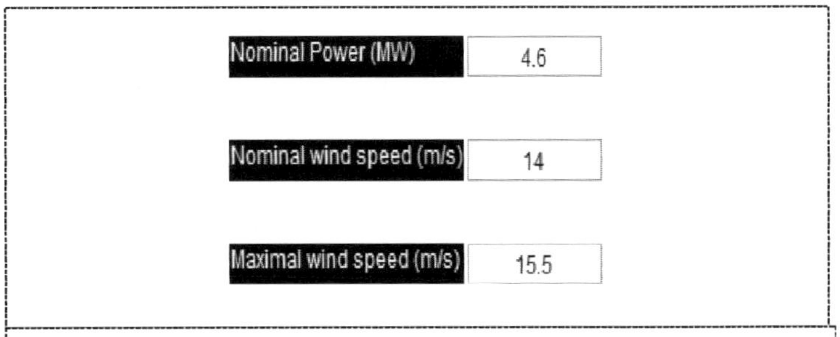

Figure 5. Wind farm parameters

3.3 PV farm Parameters

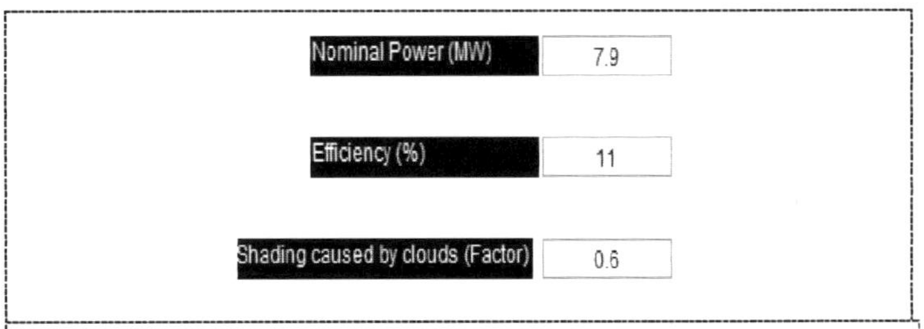

Figure 6. PV farm Parameters

3.4 EV Parameters

Parameters shown in Fig. 7 are illustrating randomly distributed EVs connected to the low voltage distribution grid. The total rated power of EVs connected to the grid is 1.340 MW [9-10].

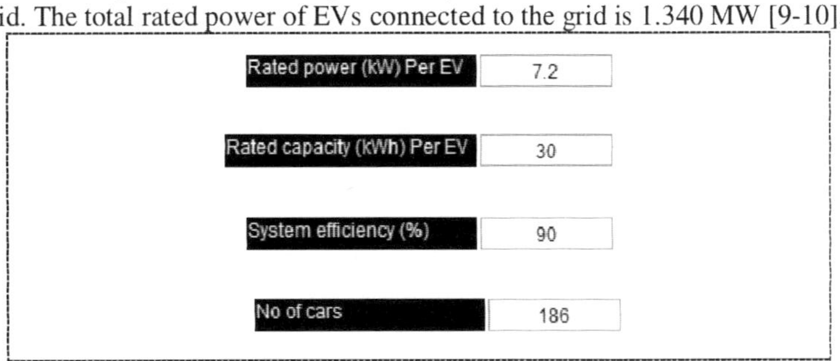

Figure 7. EVs Parameters

3.5 Residential Load Parameters

Parameters shown in Fig. 8 are representing the residential loads connected to the low voltage distribution grid.

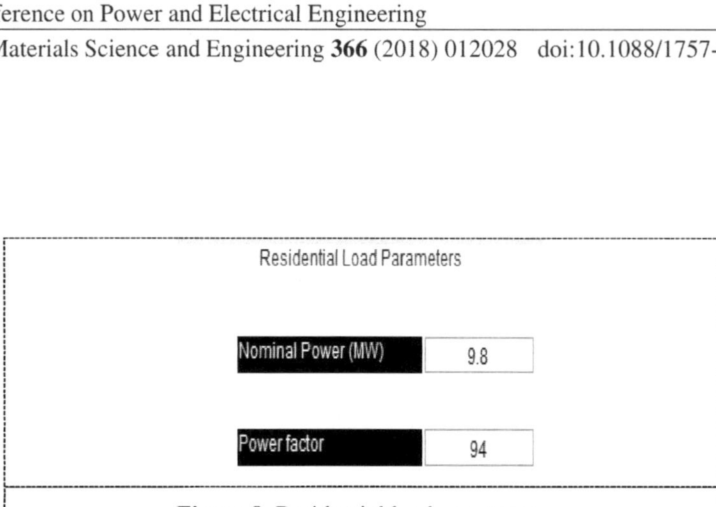

Figure 8. Residential load parameters

On the other hand, output graphs are as follows: The first three graphs represent apparent, real and reactive power flow from the diesel generator towards the load, while the last graph represents load absorption from renewable energy resources as shown in fig. 3

4. Result And Discussion

The result is an example of using the GUI. However, an execution file can be created in a further step to simplify the process and to have a comprehensive coverage of the subject. According to Fig.9, the performance of the system in the presence of the electric vehicles is improving in the term of reducing the generation from the diesel unit. In this research, the integration of EVs considering their State of Charge (SOC) in addition to all charging and discharging configuration. Data have been taken following the commercially popular available cars in the Jordanian market [9-10].

In addition, the reduced amount of energy from the diesel unit is already stored in the EVs which results in increasing the sharing amount of energy produced by the renewable energy sources as shown in Fig.9

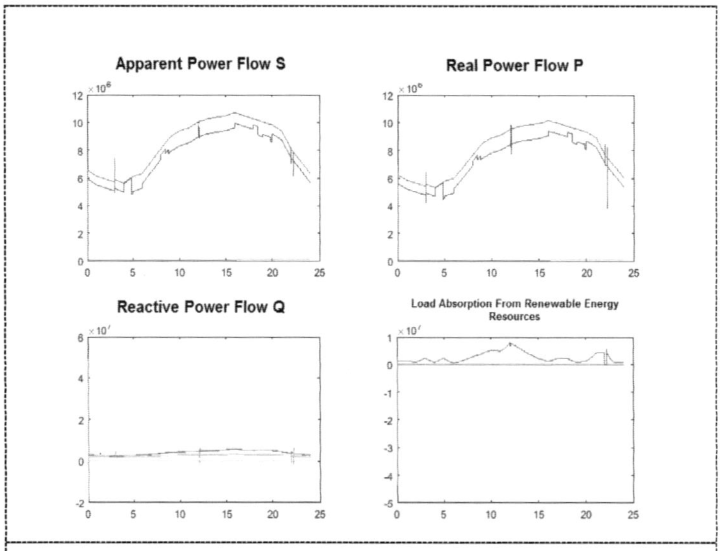

Figure 9. System performance monitoring

5. Conclusion

This study develops and describes an educational tool capable in a simple way to illustrate the utilization of storage systems in the form of electric vehicles in a medium size distribution grid. The grid is somehow smart as the renewable energy and the two way communication configuration are important part of the system structure. The study guaranties an important educational platform for students and researchers in the area of micro grids and opens the door for a future improvement in the latest applications for energy related issues. Execution file will be created with the ability to develop an online wider GUI as a future work plan.

References

[1] Ralph E.H. Simsa,Hans-Holger Rognerb,Ken Gregoryc (2003) 'Carbon emission and mitigation cost comparisons between fossil fuel, nuclear and renewable energy resources for electricity generation', Energy Policy , (31), pp. 1315–1326.

[2] G. Magdy1,2*, G. Shabib2, Adel A. Elbaset3, Thongchart Kerdphol1, Yaser Qudaih1, Yasunori Mitani1 (2017) 'A New Coordinated Fuzzy-PID Controller for Power System Considering Electric Vehicles', Energy and Power Engineering, (9), pp. 425-435.

[3] Hossam Gabbar, Smart Energy Grid Engineering, Book, 1st Edition, Academic Press, 2017.

[4] Sood VK, Fischer D, Eklund JM, Brown T. Developing a communication infrastructure for the smart grid. InElectrical Power & Energy Conference (EPEC), 2009 IEEE 2009 Oct 22 (pp. 1-7). IEEE.

[5] Yan Y, Qian Y, Sharif H, Tipper D. A survey on smart grid communication infrastructures: Motivations, requirements and challenges. IEEE communications surveys & tutorials. 2013 Mar;15(1):5-20.

[6] Reddy KS, Kumar M, Mallick TK, Sharon H, Lokeswaran S. A review of Integration, Control, Communication and Metering (ICCM) of renewable energy based smart grid. Renewable and Sustainable Energy Reviews. 2014 Oct 31;38:180-92.

[7] T. Luu ; A. Abedini ; A. Nasiri (2008) 'Power smoothing of doubly fed induction generator wind turbines', Industrial Electronics, 2008. IECON 2008. 34th Annual Conference of IEEE.

[8] Zhongjing Ma, Duncan S. Callaway, and Ian A. Hiskens (2013) 'Decentralized Charging Control of Large Populations of Plug-in Electric Vehicles', IEEE TRANSACTIONS ON CONTROL SYSTEMS TECHNOLOGY, 21(1)

[9] https://www.nissanusa.com/electric-cars/leaf/charging-range/charging/ avaliable on 15/Dec./2017

[10] Siemens VersiCharge Hard-Wired (VC30GRYHW) : Fast Charging, Easy Installation, Flexible Control, Award Winning, UL Listed, J1772 Compatibility, 14ft Cable, Hard-Wired

IOP Conf. Series: Materials Science and Engineering **366** (2018) 012029 doi:10.1088/1757-899X/366/1/012029

An improved design of damped AC test system for partial discharge measurement in distribution power cables

He Weisheng[1], Shen Xielin[2], Guo Jianzhao[2], Wang Qinghua[2], Lu Yuxin[1] and Li Hongjie[1]

[1] State Key Laboratory of Electrical Insulation and Power Equipment, School of Electrical Engineering, Xi'an Jiaotong University, Xi'an, China
[2] State Grid Quanzhou Electrical Power Company, Quanzhou, China

E-mail: hwsxjtu@126.com

Abstract. This paper presents an improved technique to generate damped AC voltage for on-site PD detection on distribution power cables. The proposed technique is based on conventional frequency-tuned series resonant system and two additional high-speed switches. In order to reduce the size and weight of the proposed damped AC test system, electric simulation by Simulink and heat simulation by ANSYS are carried out to determine the system parameters. High-voltage test on two 10kV XLPE cable segments was conducted to validate the proposed technique.

1. Introduction

Power cables are of great importance in power distribution system[1]. However, cable insulation is deteriorated by various issues such as imperfections insulation with voids or bubbles formed in the process of manufacturing, protrusions in cable joints introduced in the process of joint fabrication[2]. These defects would not directly lead to insulation breakdown but cause partial discharge (PD) phenomenon. Therefore, partial discharge measurement is an effective method in cable insulation condition detection. Over the past decades, several techniques have been proposed and applied to on-site partial discharge measurement for estimating insulation condition of distribution power cables. According to the type of high voltage applied to the cable, DC voltage, AC voltage, very low frequency (VLF) voltage and damped AC (DAC) voltage are the common used technique. DC voltage is a technique for cable testing with a long application history. However, it is harmful to XLPE cable which is widely used at present[3]. For this reason, DC voltage has been removed from on-site testing on XLPE cable. Frequency-tuned series resonant technique has been proposed to meet the power requirement of capacitive load. However, the electronic devices used in the frequency-tuned system introduce severe noise, which is not suitable for PD measurement, into the testing circuit. VLF system generates a very low frequency square-wave or sine-wave voltage, which is usually 0.01~1 Hz, on the cable. Several researches have proved that PD behaviors under VLF voltage are quite different from those under 50 Hz AC voltage[4]. DAC test system has gained more and more interests in recent years for its effectiveness in on-site PD measurement. Conventional DAC test system contains high voltage switches which cost too much and introduce unstable issues. As a consequence, it is essential to design a cost-effective test system for PD detection of power cables.

This paper introduce an improved damped AC voltage test system which is based on the frequency-tuned series resonant technique. The proposed technique is suitable for on-site PD measurement of distribution cable due to its effectiveness and portability. The following chapters in the paper would demonstrate how to improve conventional frequency-tuned series resonant technique to conduct PD detection and how to reduce size and weight of the system to meet the on-site testing requirements. Besides, laboratory test on a 10 kV XLPE cable has been carried out to validate the proposed technique.

2. Technical theories

As mentioned above, for the convenience of on-site cable test, several test methods have been put forward and applied[5]. To achieve the aim to design a cost-effective system for on-site cable test, this paper has combined the advantages of two common techniques and the technical theories are presented as following.

2.1. Conventional DAC technique

DAC test technique has been brought out for decades and put into application all over the world in recent years[6][7]. The basic operation principle of conventional DAC system is to charge the cable to test voltage and discharge the cable via a high-voltage switch. A conventional DAC test system generally consists of a high-voltage DC supply, a PD-free inductor and a high-voltage switch, as shown in figure 1. When the cable is charged to test voltage, the high-voltage switch is closed and a DAC voltage is generated on the cable via the circuit consists of the cable, inductor and the high-voltage switch. During the DAC voltage stage, PD signals can be detected and analysed to evaluate the condition of the test cable insulation and locate the defects induce the PD signals. Unlike conventional DC voltage test, no 'steady state' DC conditions occur in the insulation of the test cable during the energizing stage under DAC voltage[8]. Besides, DAC test system is portable in test site for DC charge process reduces the power requirement.

During last few years, researches have demonstrated that the influence of different voltage shape and frequency applied on the cable is negligible in practice[9]. That is to say that the DAC voltage is suitable for on-site cable testing and behave well in PD detection and location.

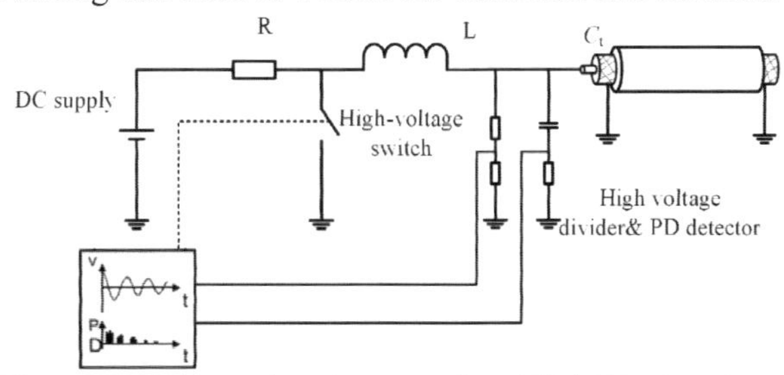

Figure 1. Schematic of conventional DAC test system.

2.2. Frequency-tuned series resonant technique

Series resonant technique has been put into application for a long time and can be classified into two types, frequency-tune and inductance-tuned. The frequency-tuned type is used more frequently than the other type for the weight-power-ratio of frequency-tuned series resonant system is about 3~5 times than the inductance-tuned one[8]. Therefore, this paper only takes frequency-tuned series resonant technique into consideration.

The frequency-tuned series resonant system generally consist of the following parts and correspondingly shown in figure 2.

a. control and sample unit
b. variable frequency power source (frequency-tuned power source)
c. exciting transformer
d. resonant inductor

e. cable
f. high-voltage divider

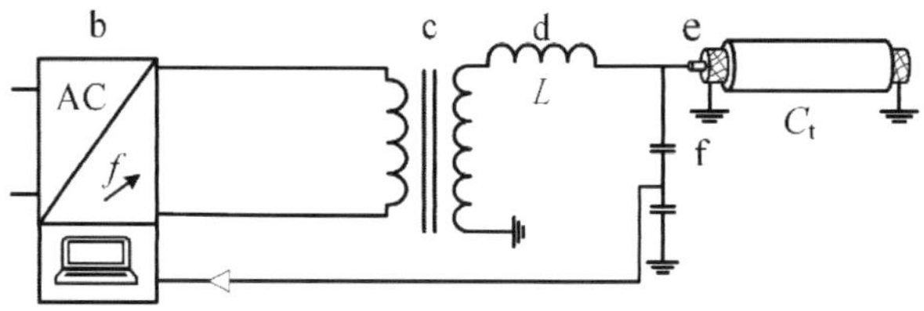

Figure 2. Schematic of frequency-tuned resonant test system.

When the frequency of the variable frequency source is tuned to equal to the natural frequency (resonant frequency) of the cable-inductor circuit, it will generate a sine-wave voltage several tens of times than the output of the exciting transformer on the cable[10]. Under this resonant condition, only active power loss exists in the system so that the frequency-tuned resonant system is portable in test site. Usually, the frequency of the system ranges from 30~300Hz to meet the requirement of different cable lengths[11][12]. Besides, it has been verified and accepted that the frequency of the voltage stress in this range is similar to the operation frequency.

2.3. *Improved system for DAC test*

As described above, DAC system and frequency-tuned series resonant system are both suitable and effective for on-site cable test. Frequency-tuned series resonant system is widely used in the field of withstand test with the development of power electronic devices[13]. For on-site cable tests, frequency-tuned series resonant system is often used to conduct withstand test not diagnostic test. The general reason for this phenomenon is due to the power electronic devices introduce numerous switching noise which brings disadvantage to PD detection. As for DAC system, the solid-state high-voltage switch is relatively expensive. This paper proposes an improved technique which combines the advantages of both DAC technique and frequency-tuned series resonant system, i.e. the proposed system generate a DAC voltage based on frequency-tuned series resonant technique. Therefore, the proposed system can be denoted as AC-DAC system. The schematic of this technique is depicted in figure 3.

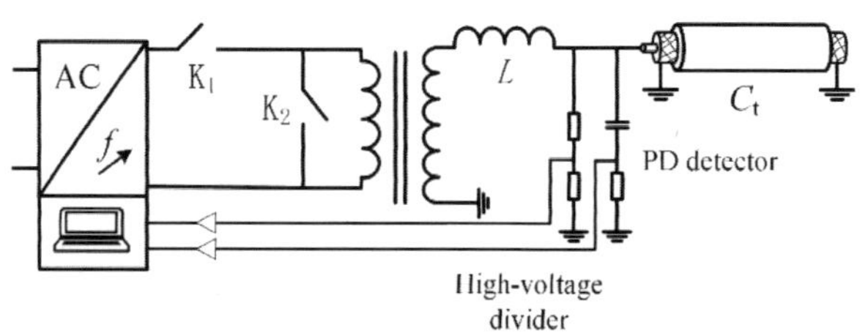

Figure 3. Schematic of AC-DAC system.

Comparing to conventional frequency-tuned series resonant system shown in figure 2, AC-DAC system mainly owns two more parts, switch K_1 and K_2 with the rated voltage of a few hundreds of volts. The basic operation principle of this AC-DAC test system is described as following.

a. When the system starts to work, K_1 is closed and K_2 is opened. Under this condition, the system could be regarded as a conventional frequency-tuned series resonant system.
b. Tune the frequency of the variable frequency source output to the natural frequency of the circuit. Generally, the natural frequency could be estimated by

$$f = \frac{1}{2\pi\sqrt{LC_t}} \tag{1}$$

c. Increase the output of the variable frequency source to boost the voltage on the cable to predetermined test level.
d. Open K_1 and close K_2 when the voltage on the cable reaches the peak value, the cable will discharge via the cable-inductor-transformer circuit. Due to the existence of the resistance in the inductor and the winding in transformer, the voltage generated on the cable will be in the form of damped AC voltage.

As the operation process shown above, the AC-DAC technique can be divided into two stages, resonant AC stage and DAC stage. During the resonant AC stage, the system contains a lot of switching noise from the variable frequency source. With the actions of K_1 and K_2, switching noise has been eliminated from the PD detection unit. Therefore, during the DAC stage, it is suitable for PD detection and data processing.

To validate the proposed technique, simulation of the AC-DAC system is carried out by Simulink software, as shown in figure 4. The variable frequency source is based on single-phase rectifier and corresponding inverter. The zoomed detail of the output voltage on the cable from the simulation result depicted in figure 4(b) has clearly demonstrate the resonant AC stage and DAC stage.

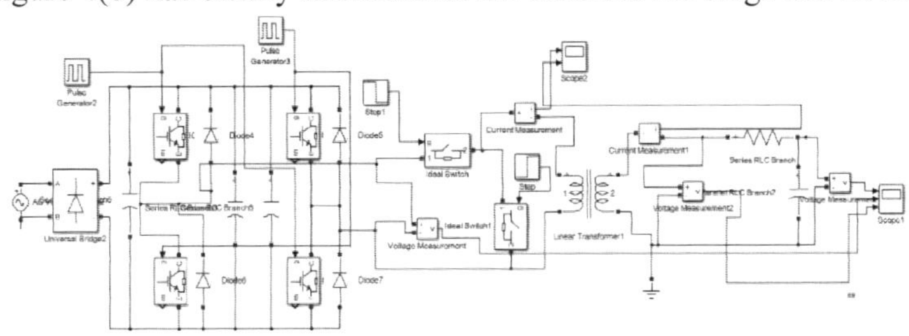

(a) Schematic diagram of the simulation.

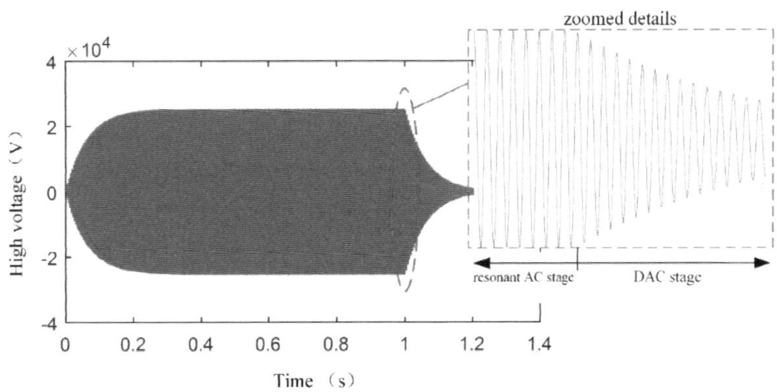

(b) Output of the simulation.

Figure 4. Simulation of the AC-DAC system.

3. Power design of the proposed technique

For the portability of the system in test site, the AC-DAC system should be small in size and light in weight. Therefore, accurate power design of the system is of importance. The main parts that contribute to the total weight of the system are the exciting transformer and resonant inductor. Thus, special attention should be paid to the issues that could reduce the weight of the transformer and the inductor.

Firstly, set the maximum load capacitance as 0.5μF which generally could represent a cable with the length of 2~2.5 km[14]. The inductor used in conventional frequency-tuned series resonant is usually in the range of several tens of Henry with iron core[15]. However, the inductor used in DAC test system is usually in the range of 0.7~1 H with air core. In order to improve the quality factor and reduce the weight of the proposed system, air-cored inductor with the value of 1H is a better alternative in this paper. The current flows in the resonant circuit can be evaluated by the following equation,

$$I_s = \sqrt{\frac{C}{L}}U \tag{2}$$

According to IEEE guide, the voltage applied on the cable when conducting diagnostic test should be twice as the operation voltage U_0[12]. For the commonly used 8.7/10 kV XLPE distribution cable, the maximum current in the resonant circuit can be calculated as

$$I_{s\max} = \sqrt{\frac{0.5\times10^{-6}}{1}}\times2\times8.7\times10^3 = 12.3\text{ A} \tag{3}$$

The voltage ratio of the exciting transformer can be chosen as 1:4 according to a simple simulation by Simulink software. Under this condition, the current flows in the primary winding of the exciting transformer can be approximately evaluated as 49.2 A.

Heat issues of the inductor and transformer are the key aspects that influence the power and weight of the system. Conventional frequency-tuned resonant system uses iron-core inductor for it maintains in the condition of maximum current for about 60 min. However, the proposed AC-DAC system employs an air-core inductor which leads to high current and low thermal capacity. The AC-DAC system works in discontinuous mode to solve this problem. Denote the process that the system generate a resonant AC voltage and transits to the DAC voltage as one DAC shot. One DAC shot totally takes about 1.5 seconds and then it takes about 1~2 seconds to conduct data transmission. During the data transmission time, the whole system gets natural cooling. Denote the sum of one DAC shot time and one natural cooling time as a heating cycle. The AC-DAC discontinuous heating mode is further explained in figure 5.

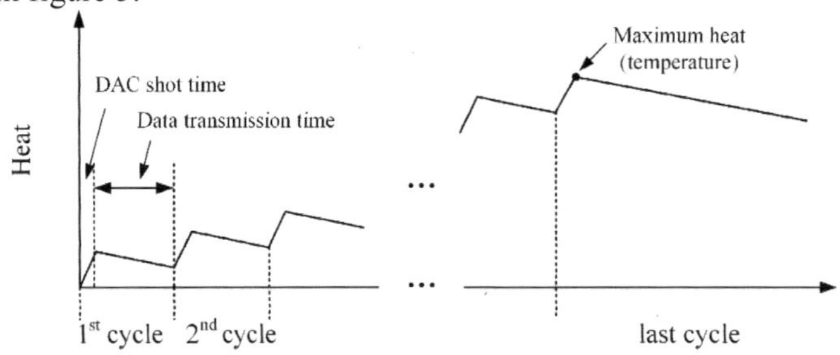

Figure 5. The heating mode of the AC-DAC system.

To evaluate the temperature condition of the AC-DAC system, a finite-element model was established by ANSYS software. A 3 kVA dry-type transformer was modelled with a double-E type iron core, depicted in figure 6(a). The air-core inductor was modelled as 4 winding layers with epoxy resin embedded. Apply the current obtained above to the simulation model and set the total operation time as 30 minutes, the results are shown in figure 6(b) and (c).

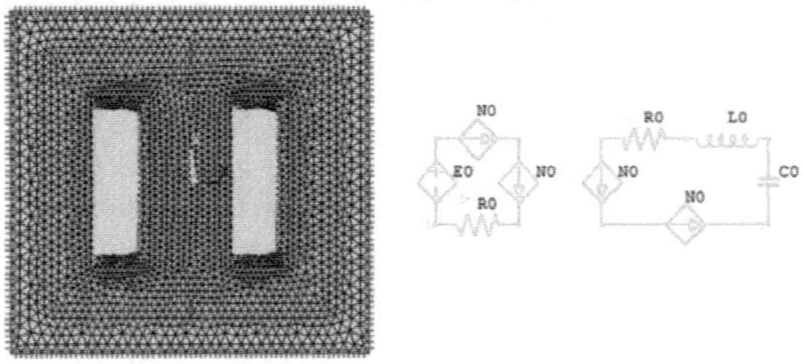

(a) Finite-element model of the transformer.

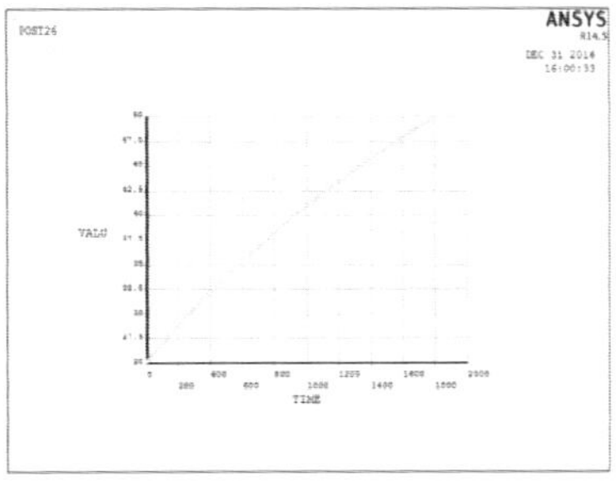

(b) Temperature rise of the transformer.

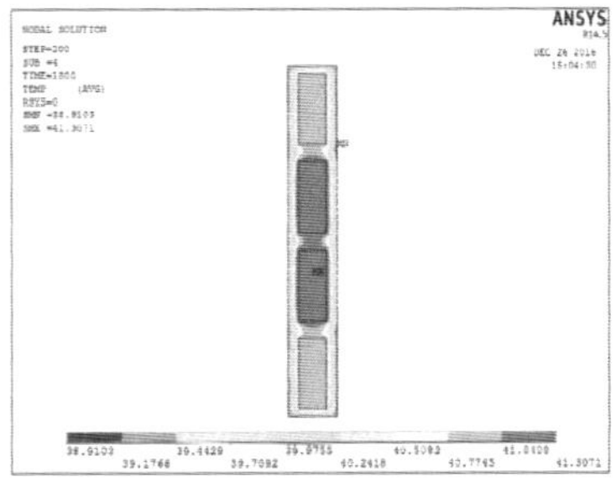

(c) Temperature profile graph of the inductor.

Figure 6. Simulation model and results.

The simulation results show that the maximum temperature of transformer rise from 25℃ (the pre-setting initial temperature) to 50.4 ℃ and the maximum temperature of inductor rise from 25℃ to 41.3 ℃. The simulation result of temperature rise is within the range of critical warning value of this kind of apparatus. The power parameters of the AC-DAC system designed in this paper is thus validated.

4. Laboratory test

An AC-DAC test system was manufactured on the basis of the proposed technique and validated parameters, shown in figure 7. The weight of the air-core inductor and the transformer are 35kg and 23kg, respectively. Therefore, it is suitable for transport and operate in the test site.

As shown in figure 7, a laboratory test on two 10 kV XLPE cable segments with the total length of 200m was carried out to further verify the proposed technique. An artificial insulation defect was made at the connection point of two cable segments.

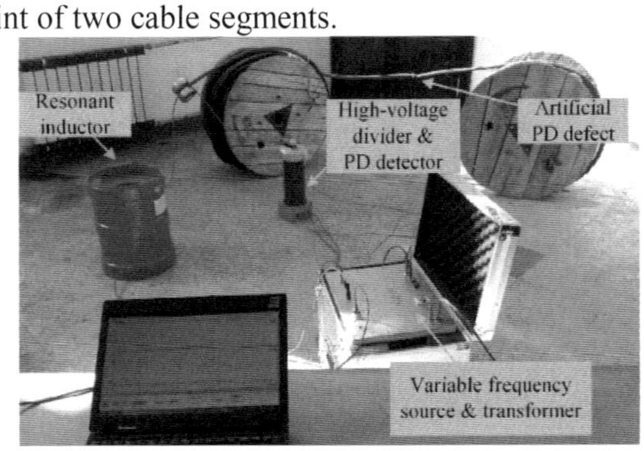

Figure 7. Setup of the laboratory test.

The data detected at 16 kV (peak value) is shown in figure 8(a) and it is clear that PD data under the DAC stage can be used for PD source location. PD location can be identified according to time domain reflectometry technique and the result is drawn in figure 8(b). It is clear that the PD source is located at the position 101.8 m from the detection point. The result is consistent with the actual PD location in the layout of the experiment network.

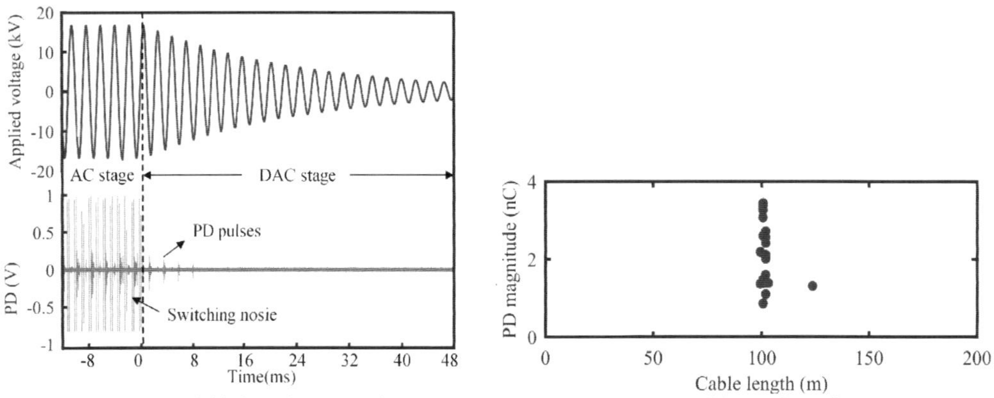

(a) Detected high voltage and PD. . (b) PD location map.

Figure 8. Laboratory test results.

5. Conclusion

The proposed system for PD detection using DAC voltage based on improved frequency-tuned series resonant technique has proven to be effective. the proposed technique shows economic benefit by adding two extra relatively low-voltage switches into conventional frequency-tuned resonant system rather than the high-voltage solid state switch in conventional DAC system. Through electric and heat simulation and design, the units employed in the proposed system also show advantages in on-site transportation and operation. A high-voltage experiment on two 10 kV XLPE cable segments was conducted and the results further validate the proposed technique.

References

[1] Elfaraskoury A, Mokhtar M, Mehanna M and Ossma G 2012 Conventional and un-conventional partial discharge detection methods in high voltage XLPE cable accessories *Advances in Electrical Engineering Systems* 1 pp 170-175.

[2] Mashikian M S and Szarkowski A 2006 Medium voltage cable defects revealed by off-line partial discharge testing at power frequency *Electrical Insulation Magazine* 22 pp 24-32

[3] Hauschild W 2013 Critical review of voltages applied for quality-acceptance and diagnostic field tests on high-voltage and extra-high-voltage cable systems *Electrical Insulation Magazine* 29 pp16-25

[4] Jespers S, Vanossi M, Green M, Lanz B 2011 the use of after-installation commissioning tests to assure MV power cable systems meet manufacturers' standards 8th Int. Conf. on Insulated Power Cables (Versailles)

[5] Gulski E, Cichecki P, Wester F, Smit J.J, Bodega R, Hermans T, Seitz P, Quak B and Vries F 2008 On-site testing and PD diagnosis of high voltage power cables. *IEEE Transactions on Dielectrics & Electrical Insulation* 15 pp 1691-1700.

[6] Gulski E, Rakowska A, Siod¿A K, Jongen R, Minassian R, Cichecki P, Parciak J and Smit J.J 2014 On-site testing and diagnosis of transmission power cables up to 230 kv using damped ac voltages *IEEE Electrical Insulation Magazine* 30 pp 27-38

[7] Cichecki, P, Gulski, E, Smit, J.J, Chmura, L and Jongen, R 2010 On-site diagnosis of XLPE transmission power cables with Damped AC technique. *Int. Symp. on Electrical Insulation* (*San Diego*) pp 1-5

[8] Kubler B and Hauschild W 2004 What to do after the warning of an on-line insulation diagnostic system? Off-line PD monitoring with HVAC testing! *Power Systems Conference and Exposition* (New York) 111 pp 822-828

[9] Wester F J, Guilski E and Smit J J 2007 Detection of partial discharges at different ac voltage stresses in power cables *IEEE Electrical Insulation Magazine* 23 pp 28-43

[10] Gockenbach E and Hauschild W 2000 The selection of the frequency range for high-voltage

on-site testing of extruded insulation cable systems *IEEE Electrical Insulation Magazine* 16 pp 11-16.
[11] IEC 60840, Third Edition, 2004-04 power cables and their accessories for rated voltages above 30kv up to 150kv- test methods and requirements
[12] IEEE 400, 2001 IEEE guide for field testing and evaluation of the insulation of shielded cable Systems
[13] Schufft W, Coors P, Hauschild W and Spiegelberg J 2002 Frequency-tuned resonant test systems for on-site testing and diagnostics of extruded cables *11th Int. Symp. on High Voltage Engineering* 5 pp 335-339
[14] Coors P and Schierig S 2008 HV AC Testing of Super-Long Cables *Int. Symp. on Electrical Insulation* pp 636-640
[15] Hauschild W 1997 Frequency-tuned resonant test systems for HV on-site testing of XLPE cables and SF6 insulated apparatus *5th Int. Conf. on Properties and Applications of Dielectric Materials* (*South Korea*) pp 1151-1158

Compact 30kV 0.1 Hz cosine square wave testing system for PD detection on XLPE cables

Shuo Chen[1], Xielin Shen[2], Jianzhao Guo[2] and Qinghua Wang[2]

[1]Xi'an Jiaotong University, No.28 Xianning West Road, Xi'an 710049, Shaanxi Province, China

[2]State Grid Quanzhou Electrical Power Company, Quanzhou, Fujian Province, China

E-mail: x321cs@163.com

Abstract. To provide a more convenient and efficient method for detecting partial discharge (PD) on XLPE cables, this paper describes a 0.1 Hz cosine-square wave testing system, which can output peak voltage up to 30 kV. Due to L-C series resonance principle, the wave of discharge process is in cosine form which can be equivalent to AC voltage. Therefore, PD detection can be conducted under the rising and falling edges. The vital component of the proposed system is high-voltage semiconductor switch which consists of several series connected IGBTs. Based on the idea of compactness, a prototype system is integrated in the lab. Withstand test and PD detection test are also carried out to validate the proposed system. The experiment result shows that the system is effective to detect PD signal in power cables and has certain development prospects on insulation condition assessment of XLPE cables.

1. Introduction

The applied quantity of power cables increases year by year, as power demand is rising, which proposes high standard for safe operation of cables. XLPE cables have superiority in fine heat-resistance, mechanical behaviour, installation and preventive maintenance. However, XLPE cables usually get damaged during production progress and cable laying[1-2]. There are two main methods for power cable insulation testing, namely, online test and offline test[3]. Offline test has the features of high precision and easy operation, and thus it's widely used in engineering. Depending on the difference of excitation source, offline test mainly includes power frequency AC voltage test, DC voltage test, oscillating wave test and VLF (very low frequency) voltage test etc[4-9]. Power cables often have large equivalent capacitance due to the long laying distance, and it is not convenient to engineer while using AC voltage test, because they have to adopt high capacity and massive transformer. Furthermore, it'll develop additional electric field after DC voltage withstand test[10-12]. Therefore, VLF voltage test and oscillating wave test are mainly used in the cable insulation test.

Over the years, VLF cosine square wave is the develop priority in the power industry. It can conduct DC voltage withstand test for cable by using the part of square wave, and lastly we can obtain leakage current and dielectric loss reflecting the whole insulation status of cables. In addition, the rising and falling edges similar to AC voltage can carry out partial discharge detection[13-14]. This paper presents compact 0.1 Hz cosine square wave testing system designed for on-site cable insulation test, and it can output 30 kV peak voltage and has the advantage of compactness and portability. Several XLPE cables have been tested by using the system, and experiments results show that this system have excellent performance in the field of cable insulation test.

2. Basic principle of system

2.1. Componentf system

As depicted in the Figure 1, the main components are described as follows:

a. Positive high-voltage source, namely, voltage doubling rectifier circuit.

b. Discharge unit consisted of bipolar high voltage semiconductor switch.

c. Detection unit including high voltage divider and PD detection unit.

d. Capacitive load.

In the Figure 1, S_r is the relay controlling on-off of electric supply, and T_1 is step-up transformer, and R_1 is limiting resistor. S_p, S_1 and S_2 are high-voltage switch, and L_1 is air-core reactor. C_x is equivalent capacitance, and C is auxiliary capacitance.

High-voltage semiconductor switch is the vital component of system, and switch has to meet the demand of high speed and high withstand voltage. High-voltage switch proposed consists of several series-connected IGBTs.

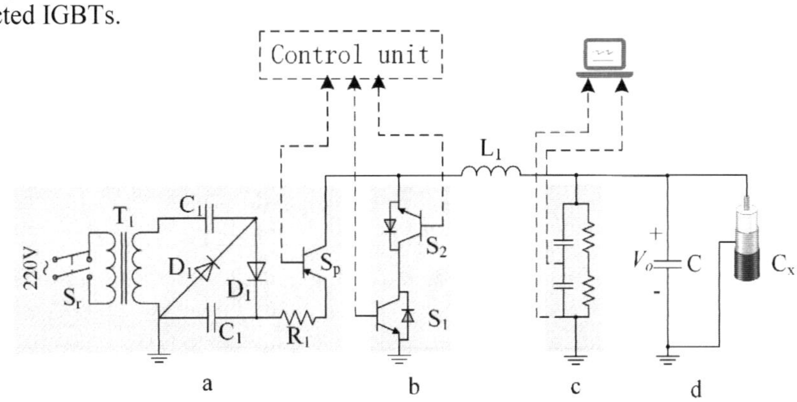

Figure 1. Topology model of 0.1Hz testing system

2.2. Working process of system

In working condition, cable as capacitive load is charged to a default voltage through positive DC source firstly, and it's in the part of square wave. Thus, load begins to discharge via inductance and load capacitor, and voltage polarity of load reverses. It's worth mentioning that energy of load is recycled under the discharge process. The typical waveform is shown in Figure 2. Specific process is depicted as follows:

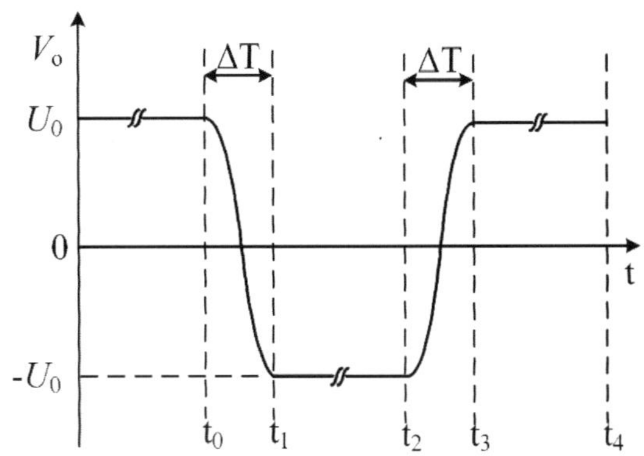

Figure 2. Typical waveform of 0.1Hz cosine square wave

A. Phase $[t_0\text{-}t_1]$:

Before t_0, cable is charged to a positive level, and at t_0, S_r and S_p are turned off making DC source stop power supply. Then, S_1 is turned on and S_2 is still in off mode. Load capacitor begins to discharge via reactor L_1, forming into a L-C resonance loop. Voltage polarity of load turns to reverse, and the waveform of V_o shows cosine wave form. Mathematical expression is depicted as below:

$$V_o = \frac{U_0}{A} e^{-\delta(t-t_0)} \cos[\omega(t-t_0)+\beta] \tag{1}$$

Where,

$$A = \left(1 - \frac{R_l^2 (C_x + C)}{4L_1}\right)^{1/2} \tag{2}$$

$$\delta = \frac{R_l}{2L_1} \tag{3}$$

$$\omega = \left(\frac{1}{L_1(C_x+C)} - \left(\frac{R_l}{2L_1}\right)^2\right)^{1/2} \tag{4}$$

$$\beta = \arccos(A) \tag{5}$$

Where U_0 is the default voltage, and R_l is equivalent resistance of the air-core reactor. Value of ΔT is related to L_1, C and C_x, and mathematical expression is depicted as below:

$$\Delta T = \frac{\pi}{\omega} = \pi\left(\frac{1}{L_1(C_x+C)} - (\frac{R_l}{2L_1})^2\right)^{-1/2} \tag{6}$$

Usually, R_l is so small that it can be neglected, and then expression can be simplified as below:

$$\Delta T \approx \pi\left(\frac{1}{L_1(C_x+C)}\right)^{-1/2} = \pi\left(L_1(C_x+C)\right)^{1/2} \tag{7}$$

B. Phase $[t_1\text{-}t_2]$:
In fact, a small energy loss occurs in the cable when cable is in the process of discharge, therefore peak voltage at t_1 of cable is smaller than voltage at t_0. From t_1 to t_2, switches S_1 and S_2 are both turned off, and S_p and S_r are still in off mode. Time period of this process is 5 seconds.

C. Phase $[t_2\text{-}t_3]$:
Similar to first step, at t_2, S_p and S_r are still in off mode. And S_2 is turned on, but S_1 is still in off mode. At the moment, cable begins to discharge via reactor L_1, forming into a L-C resonance loop. Finally voltage polarity of load turns to reverse, and the waveform of V_o shows cosine wave form. Mathematical expression is depicted as below:

$$V_o = -\frac{KU_0}{A} e^{-\delta(t-t_2)} \cos[\omega(t-t_2)+\beta],\ 0<K<1 \tag{8}$$

Where, K is loss coefficient due to energy loss.

D. Phase $[t_3\text{-}t_4]$:
At t_3, voltage polarity of V_o is positive. In order to replenish energy loss, S_r and S_p are turned on, but S_1 and S_2 are turned off. Therefore, DC source works following that cable begins to be charged. And time period of this process is also 5 seconds.

3. System design

3.1. High-voltage switch

According to explanation above, the testing system need three high-voltage switches, and switches require higher operating speed, higher withstand voltage and smaller size. There is comparison of characteristics among three electronic power switches as depicted in Table 1.

Table 1 shows that IGBT and MOSFET both have low driving power and high operating speed, but MOSFET has higher saturation voltage. Therefore, IGBT is chosen as vital component of high-voltage switch. Taking various factors into consideration, high-voltage semiconductor switch designed is based on several series connected IGBTs. The whole switch unit consists of IGBT, drive circuit, isolation power supply system and signal control system.

Table 1. Comparison of characteristics among three electronic power.

	IGBT	MOSFET	BJT
Driving mode	Voltage	Voltage	Current
Operating speed	ns	ns	μs
Driving power	Low	Low	High
Saturation voltage	Low	High	Low

When several IGBTs are connected in series, question about series even-voltage on different IGBTs has to be taken into account. Respective IGBTs have different static parameter and dynamic parameter. Besides, delay of trigger signal and exist of stray capacitance will have effect on even-voltage on IGBTs. To solve the problem, each IGBT is in parallel with high-voltage resistor calculated to be 25 MΩ. When IGBT operates, there is transient pulse voltage, which will affect normal operation of IGBT, therefore each IGBT is also in parallel with transient voltage suppressors (TVS). TVS has fast response time that is less than 1.0 ps from 0 volts to breakdown voltage, and can completely satisfy the requirement. For integration of system, high-voltage switch is designed in circle outline as shown in Figure 3. A high-voltage switch board consists of eight IGBT modules, which are connected in series.

Figure 3. Physical map of high voltage switch

Reference potential difference of respective drive circuit is up to a few thousand volts, therefore supply power of drive circuit must be isolated respectively. Flyback switching power supply has advantage of stable operation and good isolation. Besides, it is easily realizable to have multiplexed output. A multi-output signal flyback switching power supply is proposed in this paper. High-frequency transformer is placed on the centre of switch board. And there are eight outputs in the secondary side of transformer. As shown in Figure 4, the primary side of transformer is controlled by MOSFET, forming into main circuit of power supply with relevant driving circuit.

IOP Conf. Series: Materials Science and Engineering **366** (2018) 012030 doi:10.1088/1757-899X/366/1/012030

Figure 4. Physical map of flyback switching power supply

3.2. Control unit

From Figure 1, we can see that system has S_p, S_1 and S_2 totaling three high-voltage switches, and one relay. In order to ensure electrical isolation, control unit needs three ways of optical signal to control on-off of high-voltage switches and two ways of optical signal to control on-off of isolation power supply system. For switches, control signal is firstly generated by FPGA control board, and thus it's transferred to multichannel control board by optical fiber, and multichannel control signal are also transferred to corresponding IGBT modules by optical fiber at last.

For power supply system, it will generate much noise interference. In order to detect accurately partial discharge in cable, power supply system must be shut up in the process of polarity reversal. Similarly, two ways of optical signal are generated by FPGA control board, and thus they will be transferred to power supply system by optical fiber. Besides, for relay, control signal is generated by FPGA control board in the mode of electrical signal, and it can drive the relay corresponding with audion.

3.3. Physical design

In the final design, the diameter of high-voltage switch is 520 mm, each weighed 3kg. The charge unit consists of two high-voltage switches, and sixteen IGBTs rated for 4 kV are chosen in these switches. The discharge unit consists of four high-voltage switches, and thirty-two IGBTs rated for 3 kV are chosen in these ones. The vertical distance between each switch board is about 80 mm, assuming safe enough insulation distance.

Finally, HV source, charge unit, detection unit and air-core reactor are put together in one epoxy cylinder, and the inductance of reactor is 1 H. And discharge unit and control unit are integrated in another epoxy cylinder, which is connected with the former cylinder by flexible cable.

4. High voltage test

After system integration, PD detection test for cable was conducted in the laboratory. Before the experiment, for system's reliability and security, withstand voltage test was carried out using oil filled capacitor instead of XLPE cable as load. As can be seen in Figure 5, the peak value of output voltage is up to 30 kV. In the test, the value of capacitor is 0.1 μf, and according to expression (7), the time of rising edge and falling edge can be calculated.

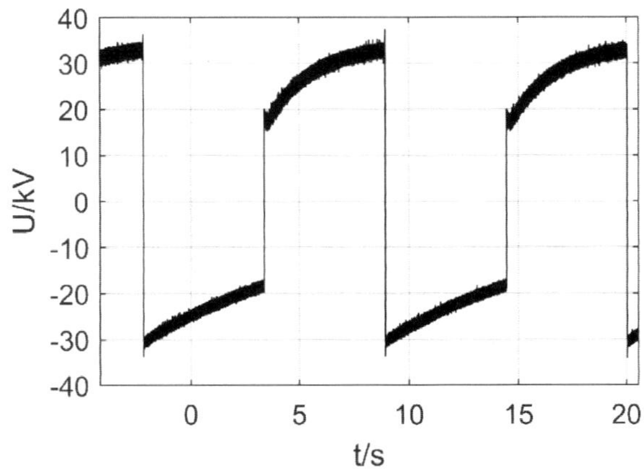

Figure 5. Output waveform of system

For PD detection test, a 200 m single core 10 kV XLPE cable was chosen as test object, which was connected by two 100 m XLPE cables. Besides, a defect was made artificially in the intermediate joint of this cable, which could cause partial discharge under high voltage. As depicted in Figure 6, a needle was stuck into insulating layer of intermediate joint, and the depth was about 2.5 mm.

Using 0.1Hz cosine square wave testing system, PD detection on the XLPE proposed was conducted in the laboratory. Experiment results are shown in Figure 7 and Figure 8. From results, we can clearly see that PD signal was accurately detected by the testing system.

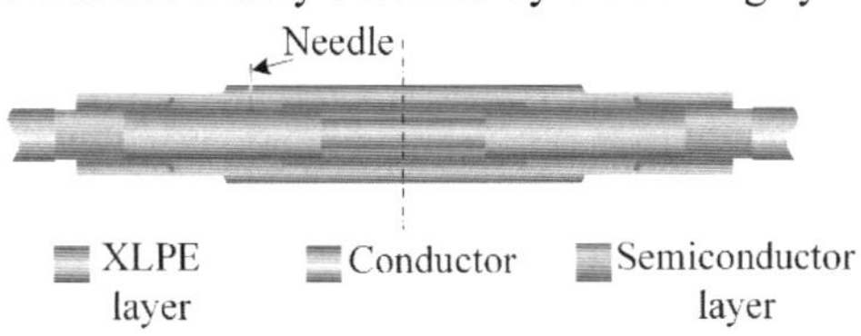

Figure 6. Defect in the intermediate joint of this cable

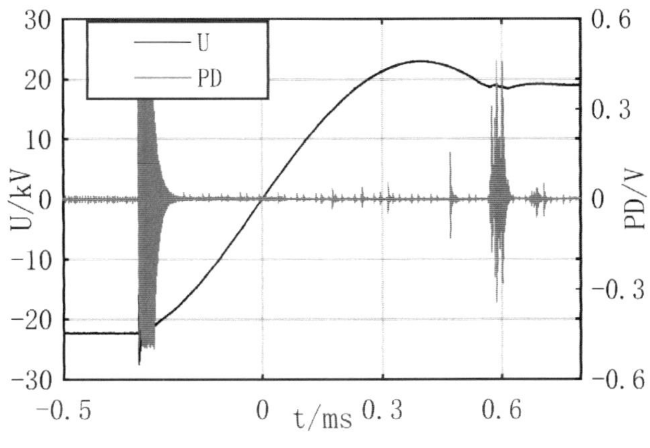

Figure 7. PD detection on cable under the rising edge

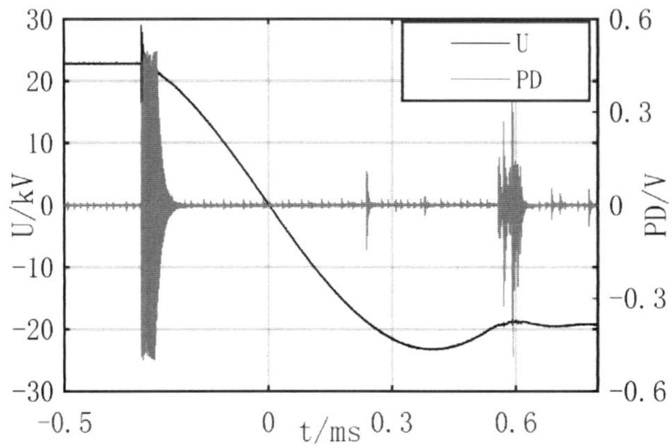

Figure 8. PD detection on cable under the falling edge

5. Conclusion

A compact testing system for PD detection on power cable is presented in this paper. Benefiting from high-voltage semiconductor switch proposed, it can output 0.1 Hz cosine square voltage rated for 30 kV. Through lab tests, system has been proven to be effective on PD detection. Because energy loss is much small under the process of polarity reverse, DC source has low power and small size. The system proposed is more suitable for on-site cable test than conventional AC system. In fact, this system also can locate defect point by using TDR and it will be discussed in the future.

References

[1] Tanaka T, Okamoto T, Hozumi N and Suzuki K 1996 Interfacial improvement of XLPE cable insulation at reduced thickness *IEEE Transactions on Dielectrics & Electrical Insulation* pp 345–350

[2] Sarathi R, Nandini A and Tanaka T 2012 Understanding electrical treeing phenomena in XLPE cable insulation under harmonic AC voltages adopting UHF technique *IEEE Transactions on Dielectrics & Electrical Insulation* pp 903–909

[3] Ahmed N and Srinivas N 2001 Experience gained with on-line partial discharge testing in power cable system *Transmission and Distribution Conference and Exposition* vol 2 (Atlanta Ga) pp 859–864

[4] Siregar Mutiara Sofia 2017 Partial discharge patterns on cross-linked polyethylene DC power cables *Power Engineering and Renewable Energy IEEE*

[5] Wong, Belle Valerie, T Yuan and T Neier 2016 Practical experience using VLF Tan Delta and Partial Discharge measurement in medium voltage cable *International Conference on Condition Monitoring and Diagnosis IEEE* pp 106–109

[6] Kim, Dong sub, Y Cho and S M Kim 2014 A study on three dimensional assessment of the aging condition of polymeric medium voltage cables applying very low frequency (VLF) tan δ diagnostic *IEEE Transactions on Dielectrics & Electrical Insulation* pp 940–947

[7] Takahashi and Toshihiro 2016 AC and DC partial discharge measurements on defective cables *IEEE Electrical Insulation Conference* pp 375–378

[8] Bach, Robert, S Sulk and C Walter 2016 Suitable voltage levels for on-site tests of MV and HV-cables with Damped AC to detect service endangering defects *IEEE International Conference on Dielectrics*

[9] Zhijuan Chen 2010 Test and study of 10kV XLPE power cables' typical defects based on Oscillating Wave Test System pp 1–4

[10] Wang Yajun 2014 The partial discharge characteristic of typical XLPE cable insulation defects under damped oscillating voltage *Electrical Insulation Conference* pp 290–293

[11] Oyegoke B 2003 Selectivity of damped AC (DAC) and VLF voltages in after-laying tests of extruded MV cable systems *IEEE Transactions on Dielectrics & Electrical Insulation* pp 874–882
[12] Xue Rong, Zhang Long and Huang Zhiwei 2015 Comparative Study on Partial Discharge Characteristics of XLPE Cable under Damped Oscillating Wave Voltage and AC Volatge *Insulation Materials* pp 43–48
[13] Salathe Daniel 2017 New methods for offline PD diagnosis on MV cable systems *Power System Conference*
[14] Tang Shihu 2014 Research and Design on 0.1Hz VLF Cosine Square Wave System (Southwest Jiaotong University)

The Study of Thin Film Capacitor on Loss and Current Distribution

Tianyi Guan[1,2,a], Feng Zhao[1,3,4],Tao Fan [1,3,4]and Xuhui Wen [1,3,4]

[1]University of Chinese Academy ofSciences ,Beijing,China
[2] Institute of Electrical Engineering ,Chinese Academy of Sciences, Beijing, China
[3]Key Laboratory of Power Electronics and Electrical Drive, Chinese Academy of Sciences, Beijing ,China
[4]Collaborative Innovation Center of Electric Vehicles in Beijing, Beijing 100190,China

[a]Corresponding author: guantianyi@mail.iee.ac.cn

Abstract. The corestructure of inverter DC-link bus film capacitor is too complex, it is more urgent to build the simulation model oncapacitor core loss. Thus,in this paper,the current distributionand the loss distribution of thin film capacitor metal plate were analyzed, and then the mathematical expressions with current distribution and loss distribution were derived on the capacitorplate. On this basis, the concept of equivalent resistivity is put forward, and thelossmodel of equivalent capacitor core was established. By this method, the simulation model with capacitor core was simplified and verified by simulation. At the same time, the structure model of thin film capacitor core was established, and the influence of different core terminals and core shape were analyzed in thefilm capacitor.

1. Introduction

The capacitor plays an important role in the EV traction system, it is very important to choose the type of capacitor. Compared with the electrolytic capacitor, metalized film capacitor with high current capacity can withstand reverse voltage and long service life, no pollution and other advantages of acid [1], which is more suitable for using in a vehicle traction system. However, the film capacitor heating problem is the main reason for its service life, but want to understand the temperature of each part on the film capacitor, it is necessary to accurately know the loss of each part with the film capacitor. In view of the above problems, some papers concerned about the loss calculation in the film capacitor [2-3], but only the results of the loss calculation are given and the simulation is not verified. At the same time, the shape and the number of terminals have a great influence on the current distribution andthe core loss distribution in the film capacitor, so the question should be analyzed.

In this paper, this thin film capacitor core will be divided into nine parts according to the internal structure. The current distribution and the loss distribution are analyzed, and the mathematical expressions are derived on the film capacitor plate .According to the concept of equivalent resistivity, the simulation model of thin film capacitor isestablished. The capacitor model is used to simulate the loss with the 87.8uf film capacitor. At the same time, the equivalent structure model of thin film capacitor core is established. The model can analyze the influence of capacitor terminal and capacitor core shape on the current distribution and loss distribution. Capacitors with lower height, larger diameter and multiple terminals have higher life expectancy.

2. Current distribution and loss distribution of thin film capacitor plates

2.1. Theoretical analysis

A common equivalent circuit model of thin film capacitor is shown in figure 1.

Figure 1. The equivalent circuit model of thin film capacitor[4-5]

R_{ESR} is the equivalent resistance of the film capacitor, and $R_{ESL}$is the equivalent inductance of the film capacitor.The capacitor loss can be expressed as,

$$P_{cap} = I_{crms}^2 \times R_{ESR} \tag{1}$$

I_{crms} is the RMS of thin film capacitor,this formula can only calculate the overall loss of the film capacitor, however, the loss of each part on the film capacitor cannot be obtained, it is necessary to understand the internal structure of the film capacitor.

The internal structure of thin film capacitor core mainly includes four parts in figure 2: thesprayed layer, the metal plate layer,the margin area and the dielectric layer. The current distribution of the film capacitor can be obtained through the film capacitor structure.

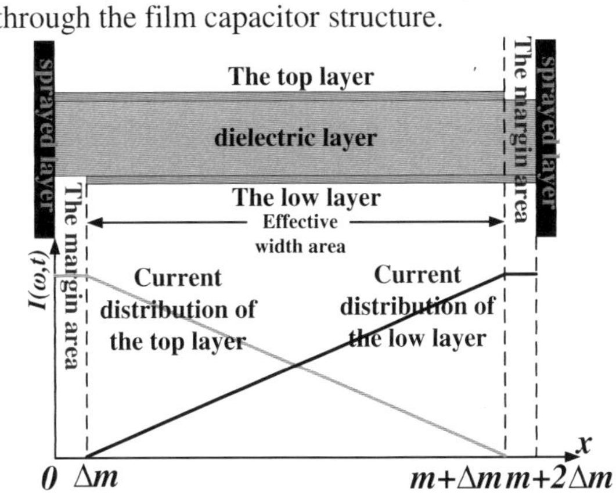

Figure2. The current distribution of thin film capacitor plate

Δm is the margin width ,m is the effective width of the metal plate, I is current density in the margin area, L is the plate length,$\rho$is the plate resistivity, d is the metal plate thickness.

The current density can be expressed in $[0,\Delta m]$ region as,

$$\begin{cases} I_{top} = I(\omega,t) \\ I_{low} = 0 \end{cases} \tag{2}$$

The current density can be expressed in $[\Delta m,m+\Delta m]$ region as,

$$\begin{cases} I_{top} = \left(-\dfrac{I(\omega,t)}{m} x + I(\omega,t)\left(1+\dfrac{\Delta m}{m}\right)\right) \\ I_{low} = \left(\dfrac{I(\omega,t)}{m} x - I(\omega,t)\dfrac{\Delta m}{m} \right) \end{cases} \tag{3}$$

The current density can be expressed as in $[m+\Delta m,m+2\Delta m]$ region as,

$$\begin{cases} I_{top} = 0 \\ I_{low} = I(\omega,t) \end{cases} \tag{4}$$

Themetal plate formula can be obtained at the dx width, the plate loss can be expressed in $[0,\Delta m]$ region as,

$$\begin{cases} d(P_{top}) = \dfrac{\rho}{L \times d} \times I^2(\omega,t)dx \\ d(P_{low}) = 0 \end{cases} \tag{5}$$

The plate loss can be expressed in [$\Delta m, m+\Delta m$] region as,

$$\begin{cases} d(P_{top}) = \dfrac{\rho}{Ld} \times \left(-\dfrac{I(\omega,t)}{m} x + I(\omega,t)\left(1+\dfrac{\Delta m}{m}\right) \right)^2 dx \\ d(P_{low}) = \dfrac{\rho}{Ld} \times \left(\dfrac{I(\omega,t)}{m} x - I(\omega,t)\dfrac{\Delta m}{m} \right)^2 dx \end{cases} \tag{6}$$

The plate loss can be expressed in [$m+\Delta m, m+2\Delta m$] region as,

$$\begin{cases} d(P_{top}) = 0 \\ d(P_{low}) = \dfrac{\rho}{L \times d} \times I^2(\omega,t)dx \end{cases} \tag{7}$$

After the loss expression of the film capacitance is obtained, the equivalent resistivity of each part can be further calculated, which greatly simplifies the establish process of the simulation model on the film capacitor.

2.2. Deduction of equivalent resistivity

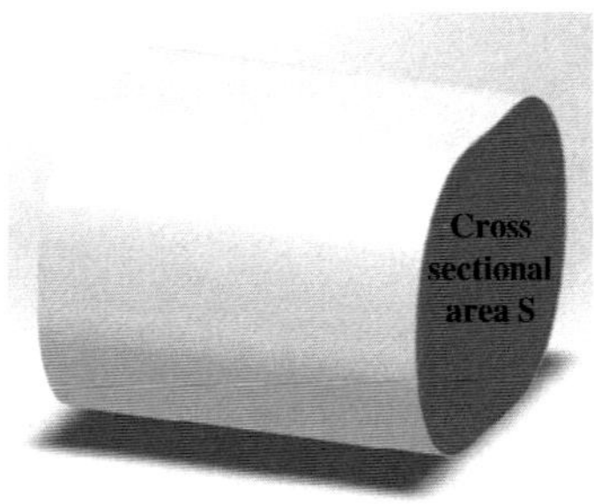

Figure3. The shape of the film capacitor core

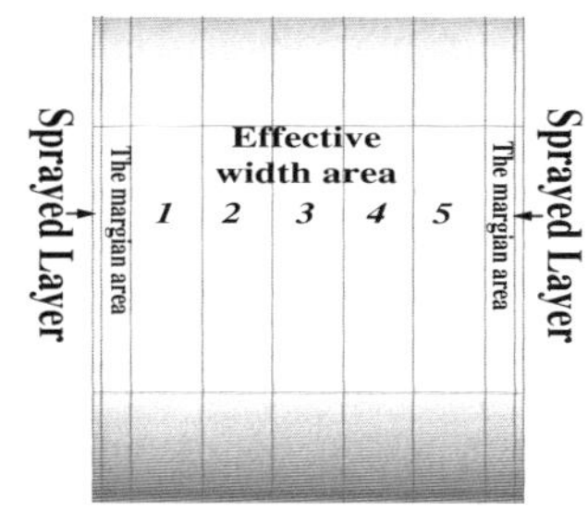

Figure4. The different regions of capacitor equivalent loss model

The equivalent core loss model can be established according to the equal loss principle.From the formula (2) (3) (4), it can be considered that the current in the equivalent model is equal to the sum of the current of the top and low plates.

It can be obtained by making the total loss of the top and low plates equal to the equivalent model losson the marginarea.

$$d(P_{top}) + d(P_{low}) = \frac{\rho}{L \times d} \times I^2(\omega,t)dx = \frac{\rho_1}{S} I^2(\omega,t)dx \\ x \in [0,m] \cup [m+\Delta m \ , \ m+2\Delta m] \tag{8}$$

then

$$\rho_1 = \frac{\rho S}{L \times d} \tag{9}$$

S isthe cross sectional area of the core, as shown in figure 3.ρ_1 is the equivalent resistivity of the margin area.

It can be obtained by making the total loss of the top and low plates equal to the equivalent model loss on the effective width area.

$$d(P_{up})+d(P_{low})=\left(2x^2+Ax+B\right)\frac{I^2(\omega,t)\rho}{m^2Ld}dx=\frac{I^2(\omega,t)\rho_2}{S}dx \quad x\in\left[m,m+\Delta m\right] \tag{10}$$

then

$$\rho_2=\left(2x^2+Ax+B\right)\times\frac{S\rho}{m^2Ld} \tag{11}$$

ρ_2 is the equivalent resistivity of the effective width area, A is $-2m-4\Delta m$, B is $(m+\Delta m)^2+\Delta m^2$.According to (11), the effective width area resistivity of the equivalent model is the quadratic function about x.Therefore, the effective width area is divided into five parts along the axial direction, and the equivalent resistivity value is calculated. Then, the distribution of the effective width area can be fitted by the simulation data.

The equivalent resistivity of five parts can be expressed as:

$$\rho_{2i}=\frac{\int_{m_i}^{m_{i+1}}\left(2x^2+Ax+B\right)\times\frac{S\rho}{m^2Ld}dx}{\frac{m}{5}} \quad i=1\cdots5 \tag{12}$$

ρ_{2i} is the equivalent resistivity in i part of the effective width area.

$$\begin{cases} m_1=\Delta m \\ m_{i+1}=m_i+\frac{m}{5} \end{cases} \quad i=1\cdots5 \tag{13}$$

Through the above analysis, the capacitor can be divided into nine parts in figure 4, and the different parts have corresponding equivalent resistivity.

2.3. Simulation results

87.8uFcapacitor core parameters are shown in Table 1.

Table 1. The Capacitor core parameter

	Material	Width (mm)	Resistivity (Ω·m)	Thickness (nm)	Length (m)
Sprayed layer	*SnZn-2+SZSC-3*	0.55	$8.307\cdot10^{-8}$	—	—
Metal plate	*Zn+Al*	25	$7\cdot10^{-8}$	2.08	550.35
Margin area（*mm*）		Effective width area（*mm*）		Cross section area（m^2）	
2		23		$\pi(24.05\cdot10^{-3})^2$	

The equivalent resistivity of each partscan be obtained according tothe capacitor parameters.

$$\begin{aligned} &\rho_1=1.11\times10^{-4}\Omega\cdot m &&\rho_{21}=2.01161\times10^{-6}\Omega\cdot m \\ &\rho_{22}=1.42759\times10^{-6}\Omega\cdot m &&\rho_{23}=1.23299\times10^{-6}\Omega\cdot m \\ &\rho_{24}=1.42759\times10^{-6}\Omega\cdot m &&\rho_{25}=2.01161\times10^{-6}\Omega\cdot m \end{aligned} \tag{14}$$

The model shown in figure 5 is added with a current excitation source of 10A and 10kHz.As seen in figure.6 and figure.7, the loss of the terminal end in the spray layer is the largest, indicating the most serious heating here. The loss in the effective width region is decreasing along the radial direction from the capacitor terminal side.Generally, the loss distribution of the effective width area along the radial is related to terminals shape, so the radial loss distribution has no regularities.

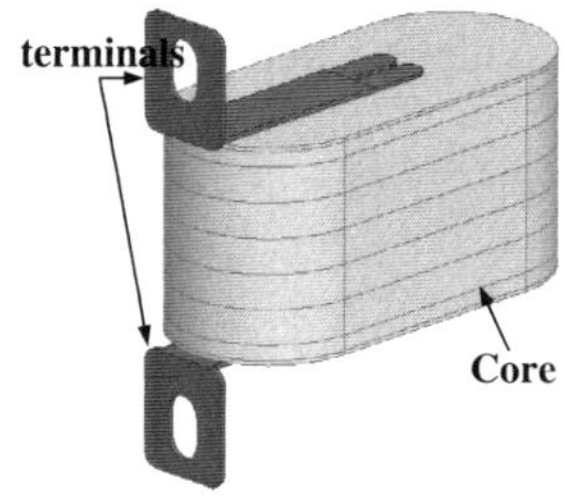

Figure5. The simulation model of film capacitor core

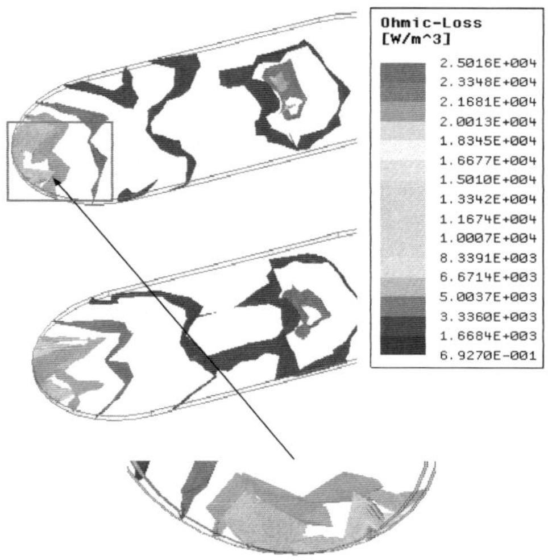

Figure6. The loss nephogram of the spraying layer

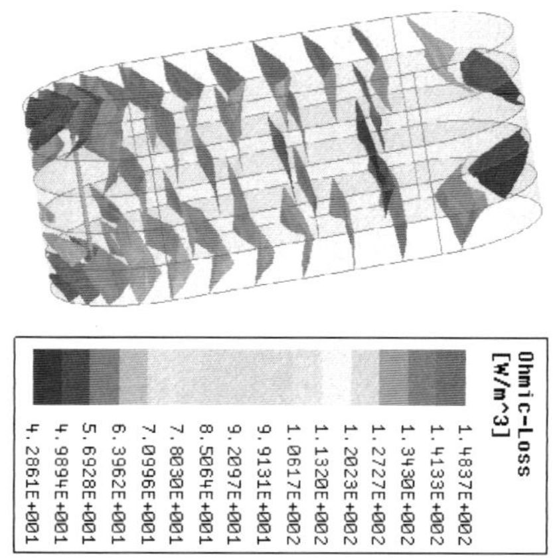

Figure 7. The loss nephogram of effective width area

The simulation results show that the loss of the sprayed layer is much larger than that of the effective width area.

In order to observe more intuitively the loss distribution of the capacitorin the axial direction, a loss path can be added to the capacitor model, as shown in figure 8.

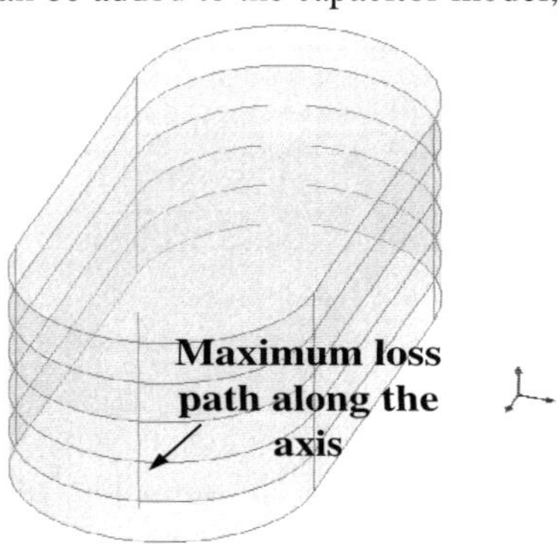

Figure 8. Themodel loss path

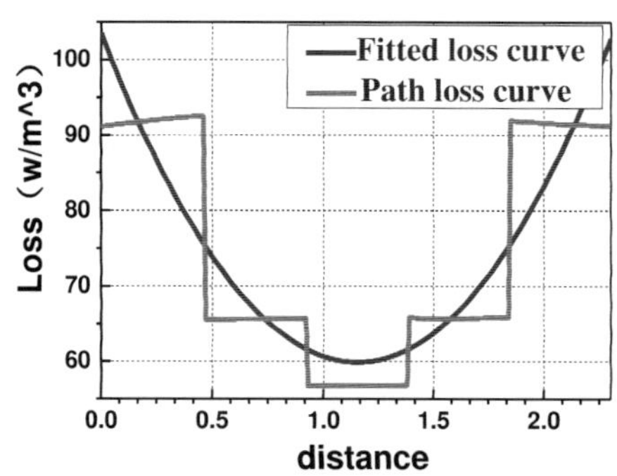

Figure 9. Thepath loss curve and fitted loss curve

It can be seen from the path loss curve in figure. 9 that the five parts of the equivalent model is symmetrically distributed, and the curve is fitted to the quadratic curve as the loss distribution of the actual core.

3. The influence of size and terminals on the capacitor core

IOP Conf. Series: Materials Science and Engineering **366** (2018) 012031 doi:10.1088/1757-899X/366/1/012031

3.1. Model establishment

The thickness of the winding layer is a few nanometers, and the winding layer reaches thousands of layers.theentitycapacitor model cannot be established, so the capacitor model is made up of a few layersto investigate the effect of different factors on the current and loss inside the core. At the same time, in order to simplify the analysis, the two winding layer can be equivalent to the cylinder with close contact.

The capacitance of a film capacitor can be expressed as,

$$C = 2\frac{\varepsilon_1 \varepsilon_0 wl}{d} \tag{15}$$

C is the capacitance value, d is the film thickness, ϵ_1 is the dielectric constant of film, ϵ_0 is the dielectric constant of vacuum, w is the effective width, l is the winding length of capacitor core. In order to analyze the influence of different sizes on the current distribution, it is necessary to ensure that the capacitance value of different capacitors are equal, the product wl should be equal.

Three different dimension capacitor core models are established in figure 10 and figure 11, w_i is the effective width of *No. i* core model.l_i is the winding length of *No. I* core model.n_{ij} is the diameter of *No.j* layer in *No.i* core m-odel.

The *No.1* core model is made up of four metal layers.

$$l_1 = \pi(n_{12} + n_{14}) = \pi(0.51 + 1.03) = 4.84mm \tag{16}$$

$$l_1 \times w_1 = 4.84 \times 8 = 38.7mm^2 \tag{17}$$

The *No.2* core model is made up of six metal layers.

$$l_2 = \pi(n_{22} + n_{24} + n_{26}) = 9.7mm \tag{18}$$

$$l_2 \times w_2 = 9.7 \times 4 = 38.8mm^2 \tag{19}$$

The *No.3* core model is made up of eight metal layers.

$$l_3 = \pi(n_{32} + n_{34} + n_{36} + n_{38}) = 16.21mm \tag{20}$$

$$l_3 \times w_3 = 16.21 \times 2.4 = 38.9mm^2 \tag{21}$$

The product wl are equal basically, the capacitance of three capacitor models are equal.

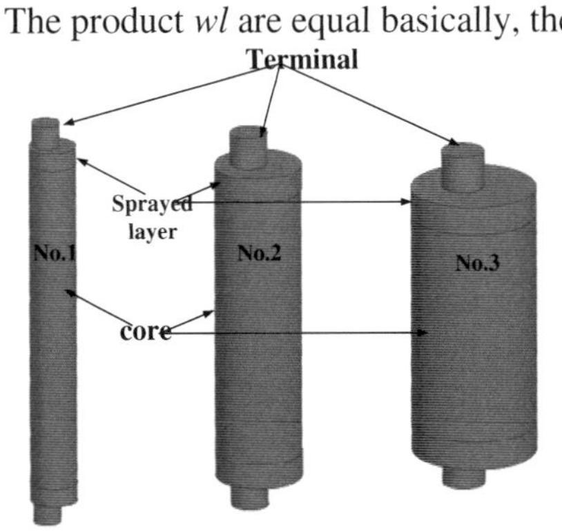

Figure 10. The three capacitor core model

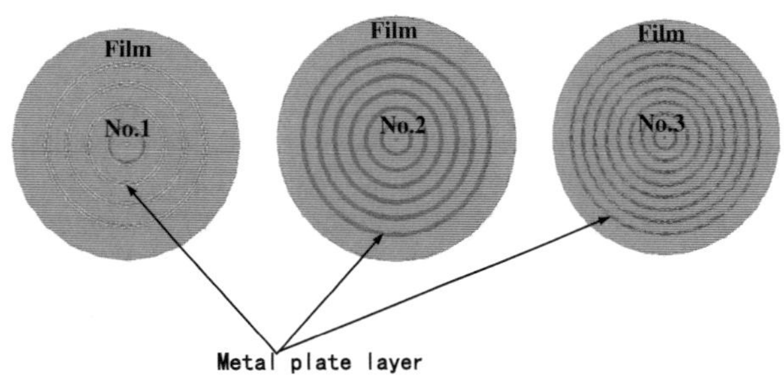

Figure 11.capacitor film and metal plate layer

3.2. Simulation result on different size capacitors

The excitation used in the simulation is a current source with 10A and 10kHz.Capacitor model materials are shown in Table 2.

Table 2. Capacitor model material

Metal plate layer	Al
Terminal	Cu
Thin film	polythylene
Sprayed layer	Zn

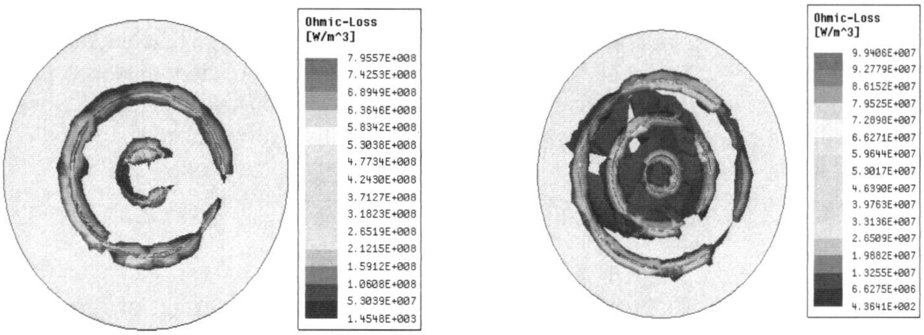

(a) No.1sprayed layer loss (b) No.2sprayed layer loss

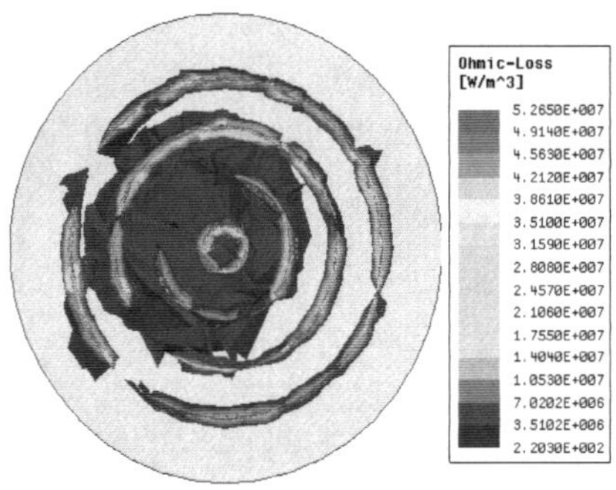

(c) No.3sprayed layer loss

Figure12. The loss nephogram of the sprayed layer

*3.2.1. Loss on sprayed layer.*The location ofmaximumloss in the sprayed layer is shown infigure12.*No.1*capacitor sprayed layer maximum loss is 7.9e+008w/m^3. *No.2*capacitor sprayed layermaximum loss is 9.9e+007w/m^3. *No.3*capacitor sprayed layer maximum loss is 5.2e+007w/m^3.The maximum loss area of the *No.1* capacitor sprayed layer is the largest in the three capacitorsmodel.

3.2.2. Loss on metal plate layer. P_{ij} is the metal plate loss distribution on *No.j* layer, *No.i* capacitor core model. The metal plate layer losses are compared in different capacitors. In order to research the maximum loss relationship in different metal layers, A scatter diagram can be drawn.

It can be concluded that the capacitor with lower height and larger diameter can produce less loss in the capacitor, can reduce the heat conduction path and can reduce the heat inside core under identical capacitance circumstance from figure13.

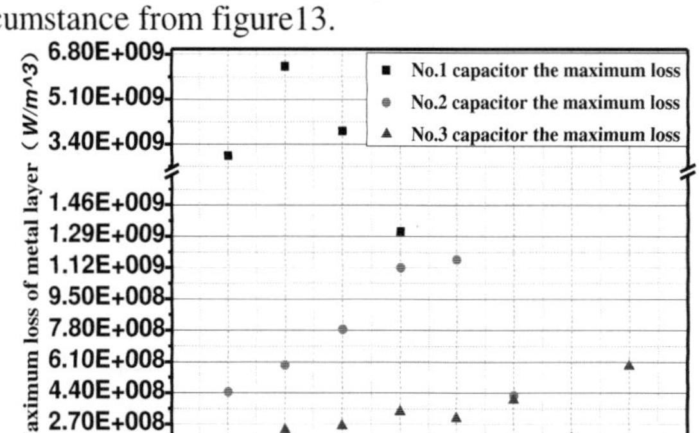

Figure 13. The maximum loss of the different capacitorson metal plate layers

IOP Conf. Series: Materials Science and Engineering **366** (2018) 012031 doi:10.1088/1757-899X/366/1/012031

3.3. Simulation result on different terminal capacitors

In order to research the terminaleffect on the capacitor loss distribution, the core model was established in figure 14, in which only the number of terminals changed, while the material and size of the three capacitor cores did not change. Capacitance values are equalon the three capacitor core models.

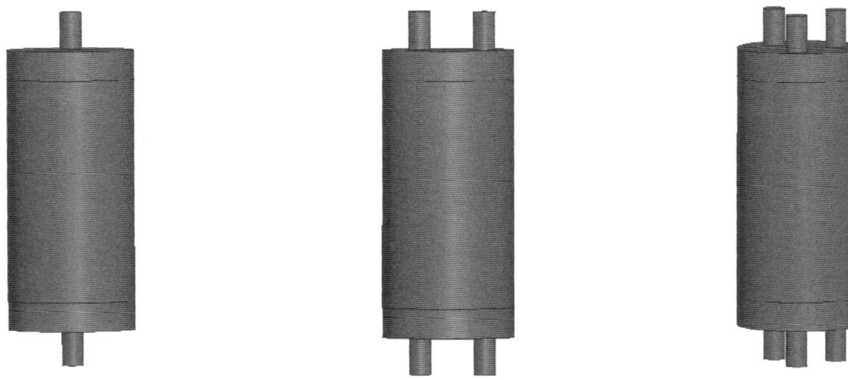

（a）No.1 capacitor model （b）No.2 capacitor model （c）No.3 capacitor model

Figure 14. The Capacitor core models with different terminals

The excitation used in the simulation is a current source with 10A and 10kHz.

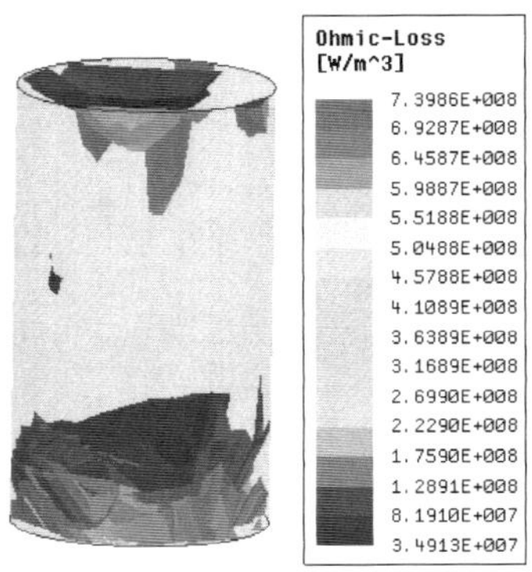

（a）Terminal loss nephogram of No.1 capacitor

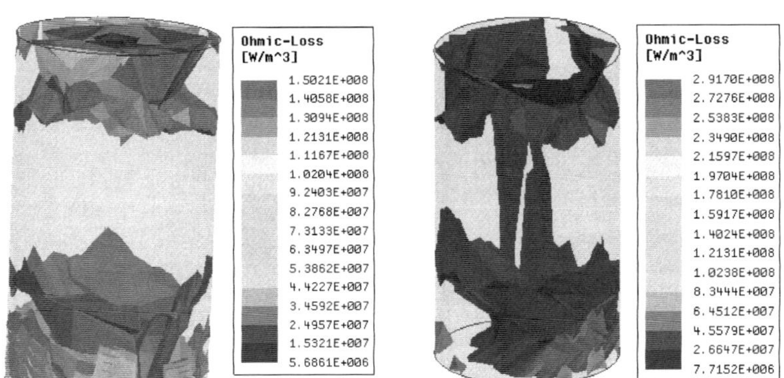

（b）Terminal loss nephogram of No.2 capacitor

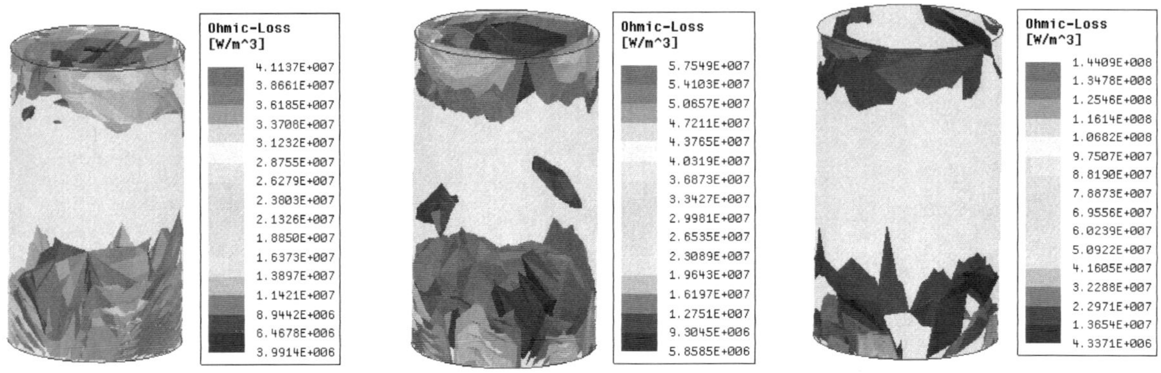

（c） Terminal loss nephogram of No.3 capacitor

Figure 15. Loss distribution nephogram of different capacitor terminals

*3.3.1. Terminal loss.*In figure15,the maximum loss of *No.1* capacitor terminal is 7.39e+008w/m^3,Themaximu m loss of *No.2* capacitor terminals are 2.91e+008w/m^3, The maximum loss of *No.3* capacitor terminals are 2.91e+008w/m^3.With the number of terminals increases, the maximum loss of capacitor terminal can be reduced, the local overheat of terminal is reduced.

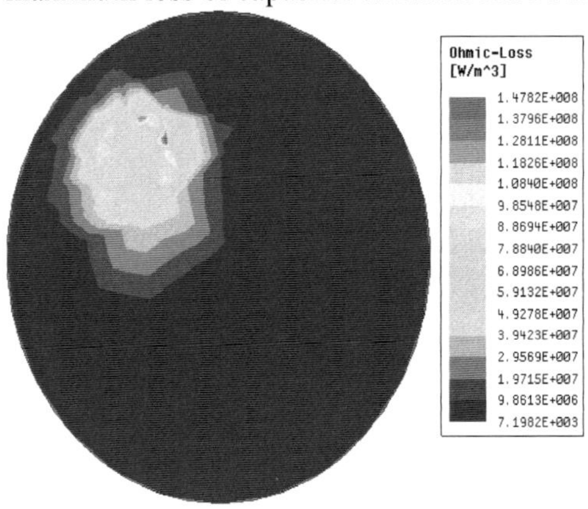

(a) No.1sprayed layer loss

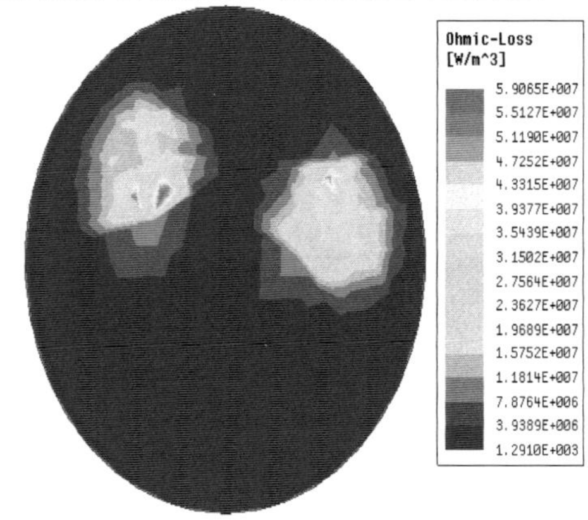

(b) No.2 sprayed layer loss

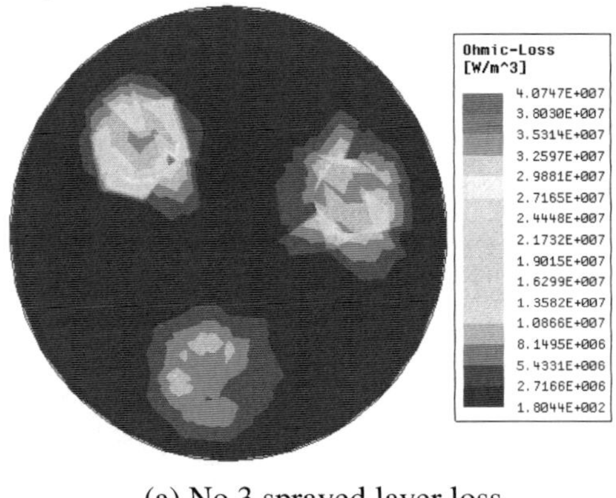

=

(a) No.3 sprayed layer loss

Figure 16. The loss nephogram of the sprayed layer

3.3.2. Loss on sprayedlayer. In figure16,The maximum loss of *No.1* capacitor sprayed layer is 1.4782e+008w/m^3, the maximum loss of *No.2* sprayed layeris 1.4782e+008w/m^3, the maximum loss of *No.3* capacitorsprayed layer are 1.4782+008w/m^3.The simulation result show that incrementalterminalquantities can reduce the maximum loss on sprayed layer, which makes the loss distribution more uniform, and the local overheatingare reduced on sprayed layer.

*3.3.3. Loss on metal plate layer.*In order to research the influence of terminal on the film metal layer more intuitively, Scatter diagram can be drawn from the maximum loss of the metal layer with different capacitance models.
It can be obtained from figure17 that the maximum loss value is 8.9e+009*w/m*on the second layer with the *No.1*capacitor model[3],which is obviously higher than the maximum loss of other metal layers. It can be concluded that when the number of terminal is one, the current distribution on the metal plate layer will be more inhomogeneous,which will lead to obvious hot spot in the capacitor.

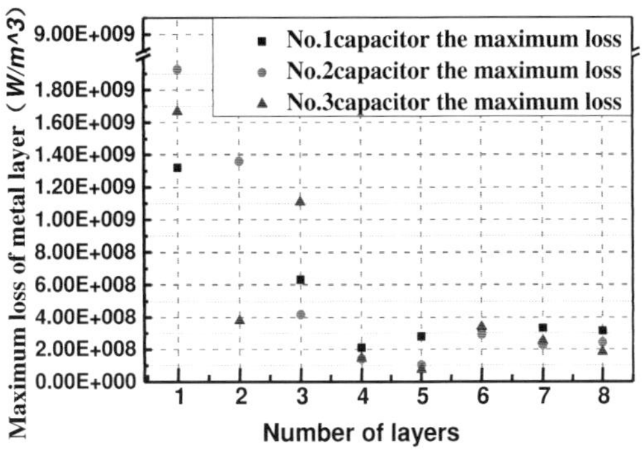

Figure 17. The maximum loss of the different capacitor on metal plate layers

4. Conclusion

In this paper, a method establishing the loss model is proposed on thin film capacitor, First of all, The concept of equivalent resistivity is put forward, which solves the problem on establishing film capacitor loss model. At the same time, the loss distribution of different film capacitor cores is compared by simulation, The simulation result show that the capacitor core with largerdiameter and lower heighthas less loss, multi-terminal capacitor can make core loss distribution be more uniform and can reduce hot spots.

Acknowledgments

This work was supported byNational Key R&D Program of China,and the project number is 2016YFB100700.

References

[1] R.P.Deshpande2015 *Capacitors*(India: McGraw-Hill Education)

[2] ZijianWang, FeiYan, Zheng Hua, Lingna Qi, ZhijianHou and ZhiniuXu2016 Geometric optimization of self-healing power capacitor with consideration of multiple factors*Journal of Power Sources* pp 147-157

[3] S.Qin, J.Ho, M.Rabuffi and G.Borelli, T.R.Jow 2011 Implication of the anisotropic thermal conducitvity of capacitor windings*IEEE Electrical Insulation Magazine* vol.27,no.1,February pp 7-18

[4] Shanshan Qin and Steven A.Boggs 2010 Design of longitudinal multisection foil-film Capacitors *IEEE Transactions on Dielectrics and Electrical Insulation* vol.17,no.6,December pp 1884-1887

[5] Robert W.Brown 2007 Distributed Circuit Modeling of Multilayer Capacitor Parameters Related to the Metal Film Layer *IEEE Transactions on Componentsand PackingTechnologies* vol.30 no.4,December pp 764-772

IOP Conf. Series: Materials Science and Engineering **366** (2018) 012032 doi:10.1088/1757-899X/366/1/012032

Research on the Improved Reliability Evaluation Method of Active Distribution System

Na LI[1], Xiaoliang WANG[2], Zhenhua ZHU[1], Ying LIN[1], Ming LI[1] and Jie HUI[2]

[1]State Grid Shandong Electric Power Research Institute, Jinan, China;
[2]State Grid Shandong Electric Power Maintenance Company, Jinan, China

273580859@qq.com

Abstract. The reliability of active distribution system is important for power system operation and users convenient. In order to evaluate the reliability of distribution system quickly and effectively, this paper established the output model of photovoltaic generation system and wind turbine generators considering more random factors. Then considering that the calculating complexity bring by above model and Monte Carlo method, this paper provided network simplification method based on feeder partition. Then this paper designed parallel calculation project for reliability evaluation to improve calculating efficiency. At last, simulation example verified the effectiveness of active distribution system reliability evaluation.

1. Introduction

Active distribution system is one of the distribution systems which can increase renewable energy and distributed energy. The features of distribution system will affect the reliability of the distribution system[1], such as AC and DC distribution form, the kinds of distributed generators(DG), the sustainable generating ability and demand response. The reliability evaluation method of traditional power system cannot adapt to the new features, so it is important for researching on reliability evaluation of active distribution system.

Reference [2]established the reliability model of DG combined with stored energy, and realized evaluating the reliability of microgrid distributed system based on sequential Monte Carlo simulation. The model of DG and loads in this paper were too simple for application. Reference[3-5]built the reliability model of microgrids generating system based on the factors such as correlation of DG output, forced outage and nonlinear output. Monte Carlo simulation verified the effectiveness and generality of the model, but the complex model increased the computational complexity. Reference[6] built the coordinative optimization model of load- stored energy device considering electricity price and used the model combined with Monte Carlo simulation method. The reliability evaluation results of this paper approached the practical microgrids operation, but the calculating complexity would increase correspondingly. Reference[7]established a short-term interrupt model of main feeder and distributor considering the affection of protection system to microgirds reliability, and used non-sequential operating state to simulate reliability evaluation. Above all, simulation method such as Monte Carlo could simulate the characteristics of loads and devices flexibly, so it has strong adaptability for evaluating active distribution system. But along with the increasing of distribution system nodes and kinds, numbers, the simulation method for reliability evaluation will treat huge data calculation and the time calculation needing will increase rapidly. Parallel calculation is an effective method to solve the time wasting problem. Parallel calculation has been used in some fields such as power flow calculation, safety analysis, risk assessment[8]. The parallel calculation has more adaptability for simulation reliability evaluation.

This paper established the output model of photovoltaic generation system and wind turbine generators considering more random factors. Considering that the calculating complexity bring by above model and Monte Carlo method, this paper provided network simplification method based on feeder partition. Then this paper designed parallel calculation project for reliability evaluation to improve calculating efficiency. At last, simulation example verified the effectiveness of active distribution system reliability evaluation.

2. The improved model of DG

2.1The output model of photovoltaic power generation system

The output of photovoltaic power generation system depended on many factors. For convenience, the steady power of PV modules could be affected by external conditions such as irradiation, environmental temperature and rated capacity[9]. The output power could be expressed as formula (1).

$$P_{PV} = P_{STC}\frac{S}{S_{STC}}[1+K(T-T_{rate})] \tag{1}$$

In formula (1), S means irradiation. P_{STC} means the rated power of normal condition. S_{STC} means the irradiation of normal condition. k means the temperature coefficient. T means the working temperature of solar panels. T_{rate} means the normal temperature.

According to reference [9], considering the decreasing power operation caused by failure of photovoltaic array, the output model of m*n PV matrix could be expressed as formula (2).

$$P_{VT}(t) = P_{STC}(t)\frac{S_{std}(t)+\Delta S(t)}{S_{STC}}[1+k(T_{std}(t)+\Delta T(t)-T_{rate})] \tag{2}$$

In formula (2), $\Delta S(t)$ means the difference between irradiation and the standard radioactivity curve. $\Delta T(t)$ means the difference between temperature and the standard temperature curve.

2.2 The random output model of wind turbine generators

The wind speed modelling is the key factor for getting the output of wind power. Considering the minimum wind speed limit in the same time period, this paper used third order Willbur distribution as the wind speed model.

$$f_t(v) = \begin{cases} \frac{b}{a}(v-c(t))^{b-1}e^{-(\frac{v-c(t)}{a})^b} & v > c(t) \\ 0 & v \le c(t) \end{cases} \tag{3}$$

In formula (3), v means the wind speed. a means the scale parameter which affects the average wind speed. b means the shape parameter which affects the distribution curve. $c(t)$ means the position parameter, $c(t)<\min v(t)$.

The wind speed parameters could be obtained by least squares method, method of moments estimator and maximum likelihood estimation method[9]. The v(t) could be obtained by inverse transformation.

Combined with wind speed model, the output power model could be obtained on the basis of wind turbines operating model, which is expressed as formula (4).

$$P_W = \begin{cases} 0 & v \le v_{ci} \cup v \ge v_{c0} \\ P_R\frac{v-v_{ci}}{v_R-v_{ci}} & v_{ci} < v \le v_R \\ P_R & v_R < v < v_{c0} \end{cases} \tag{4}$$

In formula (4), v_{ci}, v_{co}, $v_R$mean the cut in wind speed, the cut out wind speed and rated wind speed. P_R means the rated power of wind turbine generators.

3. The feeder partition method

The traditional reliability method of distribution network should enumerate all the devices including distribution transformers, overhead lines and cables. The equipment number of distribution network is large, analysis for every equipment will cost large calculation, so it is important to simplify the system.

The network simplification of this paper uses switchgear equipment as boundary to divide distribution network and at the same time, all the sub-distribution systems could be regarded as zone nodes. A complex distribution was shown as figure 1(a), which had 2 branch feeders and 3 connected switches. As figure 1 showed,F1 was the bus bar of substation. B1 was circuit breaker and S1~S4 were sectional switches. Lp1~Lp7 were loads nodes. N/O1~N/O3 were loop switches. The devices in the dotted line frames were sub-distribution systems. Figure 1(b) was the simplification of figure 1(a). All the sub-distribution systems could be regarded as zone nodes. By simplification, the number of devices was reduced, so the reliability evaluation efficiency could be improved greatly.

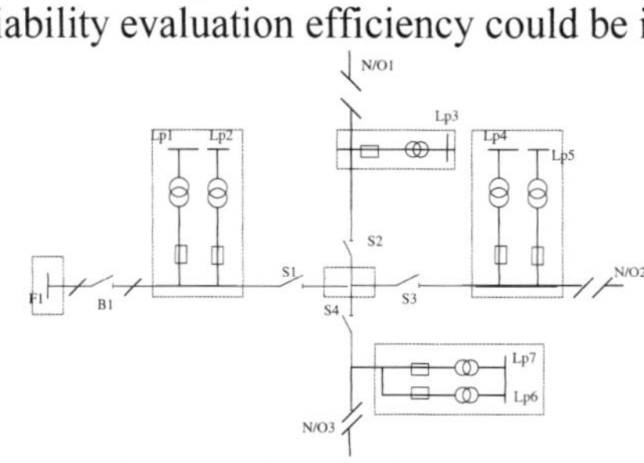

(a) the complex distribution system

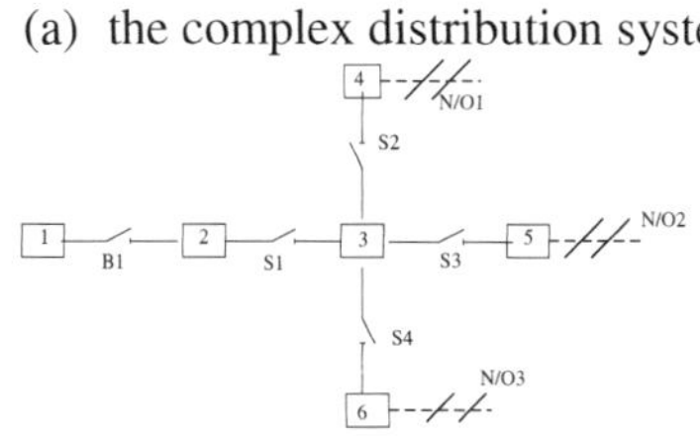

(b) the simplified distribution system

Figure 1. The complex distribution system and its' simplification system

4. The reliability evaluation of distribution network based on parallel calculation

4.1 Sequential Monte Carlo simulation

Sequential Monte Carlo method could construct the state duration sequence and realize multistate simulation of active distribution network by system decision.

1) The construction of state sequence Monte Carlo simulation

Network devices(such as lines, transformers, converters and disconnectors)reliability evaluation model usually used two-state models, which ignored transient failures. The device operating time and repairing time in failure state obeyed exponential distribution. The construction of state sequence included two steps.

Step 1: Hypothesis that all the devices are in the operating state. The non-failure time and failure repairing time could be obtained by sample according to failure rate and repairing time. Considering the active distribution system island operating state caused by public distribution system failure, Assuming that the failure rate λ_s and repairing rate μ_s were constants, then

$$\lambda_{eq} = \lambda_S + \lambda_D \tag{5}$$

$$\mu_{eq} = \frac{\lambda_{eq}\mu_S\mu_D}{\lambda_S\lambda_D + \lambda_D\mu_S + \lambda_S\mu_D} \tag{6}$$

In formula (5) and (6),λ_D and $\mu_D$mean the failure rate and repairing rate of grid connected switches.

Step 2: According to the non-failure operating time and repairing time of devices, construct the system state duration sequence $S_{sd}=[S_k,d_k]^T$,$K\epsilon[1,N_{sys}]$,N_{sys} is the system state numbers. S_k is the state combination of kth devices. d_k is the duration time.

2) System state judgment

For a system operating state (s_k,d_k),the judgment process was shown as below.

Step 1: Loads nodes. The loads nodes of active distribution system could be divided into connecting-grid loads, interrupted loads nodes and island loads nodes. The reliability of connecting-grid loads was ensured by public distribution system. Interrupted loads had non-connected pathway with DG and connecting-grid loads, so the power supply would be interrupted. The power supply of island loads was supported by DG and stored energy.

Step 2: Island operating topology identification. In order to improve the applicability of every topology structure in public distribution system and island operating running condition, this paper designed a limited priority aggregation (LPA) method to realize operating topology judgment. LPA method could choose the island nodes by aggregating the network nodes according connected domain.

Step 3: Simulating loads. Sampling simulation could get the output of loads according to the reliability models of DG and loads.

Step 4: State updating of the system. Update the system state of current hour including island controlling and SOC devices.

4.2 The design of parallel calculation scheme

In order to improve calculating effectiveness, traditional Monte Carlo method designed the parallel calculation strategy using fixed cycle times. Every circulation is independent. The calculating result was collected by main server. The traditional parallel calculation would encounter redundancy. In order to improve the calculation effectiveness to a large extent, this paper provided the dynamic parallel calculation method. When the sampling sample X* was sure, the calculation of the random variables was independent and had little influences. So the method can improve calculation effectiveness further more.

Different from the traditional parallel calculation with fixed cycle times, this paper made the variance convergence as the judgment basis. The main server formed the final sample matrix using sampling method and allocated the sample value of X^* to all the sub-servers for calculation.

In the process of calculation, assumption that the numbers of servers is p and the ID of every server is 0 to (p-1). The ID of main server is 0 and other ID numbers are servers for parallel calculation. I means initializing working. T means one Monte Carlo task. C means convergence controlling. S means stopping the simulation. F means calculating and outputting the reliability indexes.

As the figure 2 showed, after the initializing working I0, the main server will send the random quantity of sample X*. Receiving the calculating task, the sub-servers complete their calculating tasks and send results to the main server. The main calculating steps are shown as below.

(1) Initializing working.

(2) The main server formed the final sample matrix X^* using sampling method and send calculating tasks to sub-servers.

(3) Sub-servers selected a system state from sample matrix X^* and marked the random quantity avoiding repeating calculation. Then topology analysis and power flow calculation would be completed according to the state system. After the state analysis, a next random quantity would be gotten for the next calculation until all the random quantity calculated.

(4) All the sub-servers send their calculating results to main server. The main server read the calculating results and judged if the calculation reached the convergence precision. If the calculation satisfied the convergence conditionthen send stopping command $S_0$to sub-servers and all the sub-servers would stop calculation and send the residual result to main server. At the same time the main server carried out command F_0 and generated the reliability indexes. Otherwise, the calculation would jump to step (2) and continue the calculation until the ending conditions are satisfied.

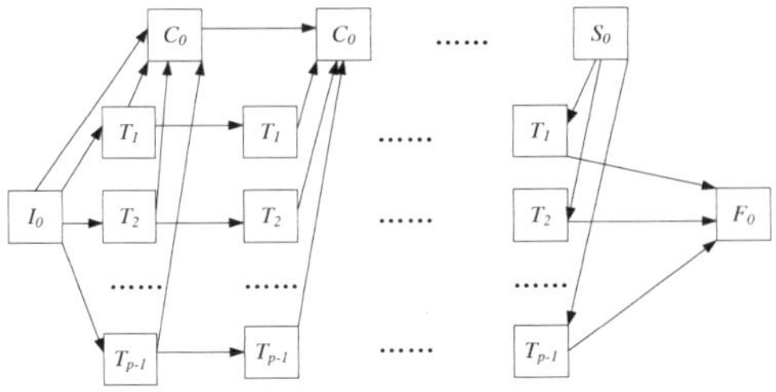

Figure 2. The schematic diagram of parallel calculation

5. Simulation example analysis

5.1 The test system and its' parameters

This paper selected two feeders of RDTS Bus2 system[10]and allocated it into an active distribution system, which is the test system of this paper and the structure was shown as figure 3.The test system had a wind turbine generator with 1MW capacity, 2 photovoltaic power generation systems with 0.8MWcapacity and 2 stored energy devices with 0.36MW capacity. The maximum charging and discharging power was 0.12MW.The average loads power of all the nodes was 4.936MW. The operating environment parameters were obtained by a certain area of Shandong. The loss cost coefficient of all the energy stored devices was same.

The device failure rate, repairing rate and the user number of loads nodes could be obtained by reference [10]. The failure rate and repairing time of power were shown as table 1. The reliability evaluation indexes of active distribution system were system average interruption duration index (SAIDI), expected energy not supplied (EENS) and system average interruption frequency index (SAIFI).

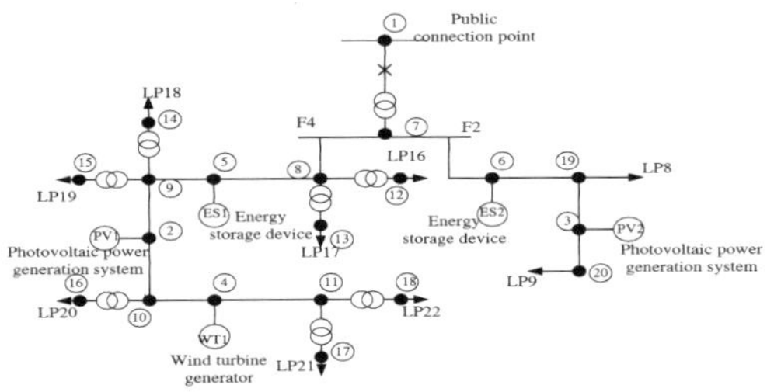

Figure 3. Test system for reliability evaluation

Table 1. Reliability data of power sources

Power sources/devices	Failure rate（time • a^{-1}）	Average repairing time/h
Public distribution system	2.00	4.5
Photovoltaic array	0.05	5.0
Photovoltaic converter	0.12	70.0
Wind turbine generator	0.25	60.0
Energy storage device	0.12	20.0

5.2 The calculation results and analysis

The simulation time was selected as 350a, the serial calculation and parallel calculation were both done after network simplification. The reliability evaluation results were shown as table 2.

Table 2. The comparison of different evaluation methods

model	EENS/	SAIDI/	SAIFI/	Calculatingtime/

	((MW·h)·a-^{1})	(h·a^{-1})	(time·(subscriber·a)$^{-1}$)	s
Serial method with non-simplification	50.8903	11.2976	1.7827	2039.799
Serial calculation after network simplification	49.2986	11.5285	1.7267	1796.271
Parallel calculation	50.2831	12.9532	1.7936	238.219

According to the comparison of table 2, the network simplification method based on feeder partition could improve calculating efficiency and keep the accuracy invariant. Using the parallel calculation method, the calculating efficiency could be improved greatly. At the same time calculating accuracy could be kept stable. In this paper, when the number of workers was 16, the parallel calculation efficiency could be improved 86.75% compared with serial calculation. The example analysis verified the practicality of the method proposed by this paper.

6. Conclusion

This paper established the output model of DG and provided the network simplification method of feeder partition. Then this paper provided the active distribution reliability evaluation method based on parallel Monte Carlo simulation. At last, the simulation example verified the effectiveness and practicality the method of this paper, which could be used in complex distribution system reliability evaluation.

References

[1] Kroposki B, Lasseter R, Ise T, et al. 2008 Making microgrids work. *IEEE Power and Energy Magazine* vol 6 mumber 3 pp40-53
[2] Liangl HS, Cheng L and Liu S 2011 Monte Carlo simulation based reliability evaluation of distribution system containing microgrids. *Power System Technology*2011vol 35 mumber10 pp76-81
[3] Luo Y. 2013 *Research on the models and methods of reliability evaluation in distribution system*(Guangzhou :South China University of Technology)
[4] Luo Yi, Wang L and Wang G 2014 Reliability assessment of microgrid generation system considering output correlation of distributed generators. *Automation of Electric Power Systems*vol 38 number 12 pp34-38
[5] Bie C H, Zang P, Li G F, et al 2012 Reliability evaluation of active distribution system including microgrids. *IEEE Trans on Power Systems*vol 27 number 4 pp2342-50
[6] Bie C H, Li G F and Xie H P 2014 Reliability evaluation of microgrids considering coordinative optimization of loads and storage devices. *Transactions of China Electrotechnical Society* vol 29 number 2 pp 65-73
[7] Xu X F, Mitra J, Wang T T, et al 2014 Evaluation of operational reliability of a microgrid using a short-term outage model. *IEEE Trans on Power Systems* vol 29 number 5 pp 2238-47
[8] Hu B, Xie K G and Cao K 2011 Parallel CMRES techniques for solving newton power flow Of large scale power systems on Beowulf cluster. *Transactions of China Electrotechnical Society* vol 26 number 4 pp 145-152
[9] Guo XY, Xie K G, Hu B, et al 2013 A time-interval based probabilistic production simulation of power system with grid-connected photovoltaic generation. *Power System Technology*vol 37 number 6 pp 1499-1505
[10] Allan R N, Billinton R, Sjarief I, et al 1991 A reliability test system for educational purposes-basic disteibution system data and results. *IEEE Trans on Power Systems*vol 6 number 2 pp 813-20

IOP Conf. Series: Materials Science and Engineering **366** (2018) 012033 doi:10.1088/1757-899X/366/1/012033

The impact of quick charge on power quality of high-voltage grid

ZHAO Li-shuang[1] and YUAN Hui-mei[1]

[1]Capital Normal University, Haidian District, Beijing 100048, China

Email: lszhaolx@163.com

Abstract. With the continuous development of electric vehicles, a large number of electric vehicles have connected to the grid. This impacts the security and economic operation of power systems deeply. When large-scale electric vehicles are charging, it will bring the problem of electricity load growth and cause the peak-valley difference of power grid. The paper analyzes the impact of electric vehicles quick charge on power quality. According to Monte Carlo method, the daily load curve of quick charging mode in Beijing in 2030 is set up. Newton Raphson method and forward-backward substitution method are used to analyze the voltage deviation and power loss of each node in the typical power system.

1. Introduction

Electric vehicles charging mode is divided into conventional charging mode, quick charging mode and mechanical charging mode [1]. Due to the quick charging mode has characteristics like charging time is short, long battery life, without memory and so on [2], in this paper, we study the quick charging mode. Electric vehicles quick charge will produce very big power load and high harmonic in a short time, which can produce great influence to power grid electric energy quality. The factors that affect the quality of the electric energy include the fluctuation and the flicker of voltage, the voltage deviation, frequency deviation, harmonics and three-phase voltage unbalance [3]. Electric vehicles will produce a lot of harmonic while charging. Harmonic pollution will have a negative impact on the power quality of the power supply system [4]. Therefore, it is significant to study the effect of power quality on electric vehicles quick charge and propose the corresponding strategies.

2. Electric vehicles charging load model and realization method

2.1. *Total charge power calculation model*

The total charge load curve is the sum of each electric vehicle charging load curve. The charging load is calculated with the day as the unit, the time is accurate to every minute, a total of 1440 minutes per day. The total charge power for the first i minutes is the sum of the charging load of all the electric vehicles which are being charging at the first i time. Equation (1) for the total charge power formula:

$$L_i = \sum_{n=1}^{N} P_{n,i} \quad i=1,2,\ldots\ldots 1440 \tag{1}$$

Among them: L_i is the total charge power for the first i minutes; N is the sum of the electric vehicles; $P_{n,i}$ is the charging power of the first N vehicle in the first i minutes.

2.2. *Load model for electric vehicles quick charge*

In the quick charging mode, the power of the charger is very large. The charging current is much higher than the normal current, which can improve the battery SOC in a very short period of time. The number of electric vehicles that set one day quick charging mode is N, S_1 is the initial SOC, which obeys $N(\mu_1, \sigma_1^2)$; S_2 is the end charge SOC, which obeys $N(\mu_2, \sigma_2^2)$, S1<S2≤1. Set current charging with 1C, charging power is P, which obeys the uniform distribution in the 20~30kW range. It is assumed that the quick charge time of the electric vehicles is uniform distribution, and the distribution probability function can be obtained by the equation (2).

$$f_t(x) = \begin{cases} \alpha, x \in T_a \\ \beta, x \in T_b \\ \gamma, x \in T_c \end{cases} \tag{2}$$

Among them: *Ta*, *Tb* and *Tc* represent the different time periods.

Electric vehicles charging duration can be expressed by equation (3) [5].

$$T = \frac{(s_2 - s_1)C}{P} \tag{3}$$

2.3. *Method for realizing the quick charge load model of electric vehicles*

According to the quick charge load model of the electric vehicles, the daily load curve of quick charge for electric vehicle can be obtained by Monte Carlo simulation method.

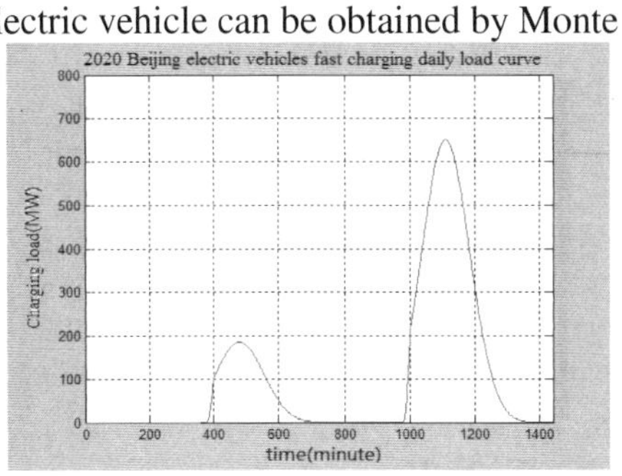

Figure 1. 2020 Beijing electric vehicles quick charge daily load curve.

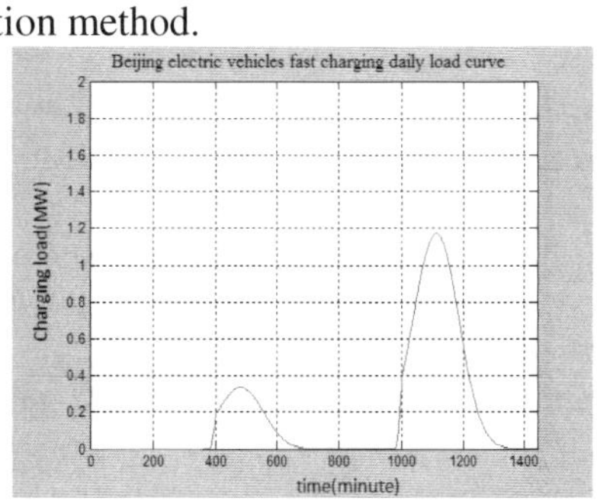

Figure 2. 2030 Beijing electric vehicles quick charge daily load curve.

From the figure 1 and 2 we can learn that the daily load of the two peak values have a big difference between 2020 and 2030. The value of charging load can affect the running state of power system, and its power quality. Therefore, it is very important to control the charging load in a certain range and reduce the effect on the power quality.

3. The effect of quick charge on power quality of high-voltage grid

3.1. *Voltage deviation of high voltage power network during quick charge*

Using the Typical network topology data of 10 machines 39 nodes. We calculate the voltage deviation when the electric vehicles are quick charged according to the Newton Raphson power flow calculation method. By analyzing the voltage deviation, we can study on the effect of electric vehicles on power quality while they are being quick charged [6].

The network topology structure of 10 machines 39 nodes includes 10 generators, 39 nodes, 12 transformers, 34 lines. The system reference power is 100MVA, the reference voltage is 345kV. Figure 3 is a network topology structure connection diagram of 10 machines 39 nodes.

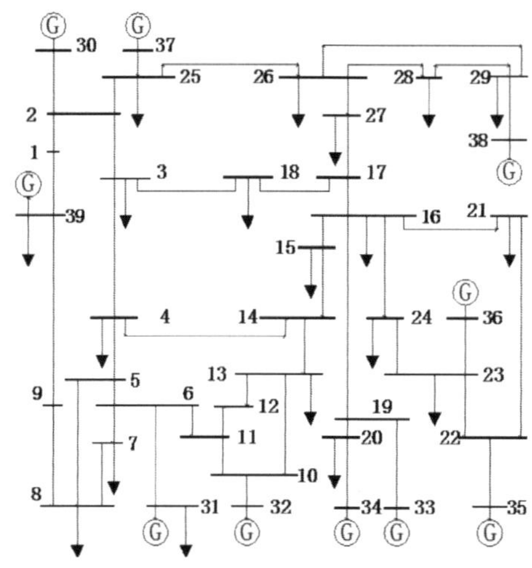

Figure 3. Network topology structure of 10 machines 39 nodes.

When the reactive power load is added to the system nodes, the system can get the voltage value of each node in the actual operation by the Newton Raphson method for power flow. Through the calculation formula of the voltage deviation, the voltage deviation of each node is obtained eventually [7]. When the electric vehicles quick charging daily load curve data of 2030 is added to the system nodes, we can calculate the voltage deviation of the system. In 2030, the daily load peak was 1172MW, and the power load was added to the single node of the system according to the 10%, 20%, 30%, 40% of the permeability respectively for calculating the voltage deviation of the system. After adding the power load, the voltage deviation variation of 6 nodes in the system is obvious. We select the 6 feature nodes to analyze the voltage deviation.

From figure 4, it can be seen that the voltage deviation of the nodes after adding the 40% of the power load is bigger than that with no power load. This can be known, the user charging behavior will affect the size of the power load, while the power load will affect the index of the voltage deviation in power quality. Therefore, the reasonable range of the power load to the nodes of the grid can make the normal operation of the network. If exceeding a reasonable range, the voltage deviation in the power grid will be greater and the power quality will also be seriously affected.

The above is the result of voltage deviation by adding the power load to the single node. In real life, electric vehicles charging will select multiple nodes at the same time in the power grid. Then we will select two nodes to add the 10%, 20%, 30% and 40% of the power load, and study their voltage deviation variation [8].

After according to the voltage deviation variation of the single feature node adding power load, we select any two nodes in the feature node.

Voltage deviation is an important index in power quality. According to the provisions mentioned above concerning the voltage deviation standard for power supply (GB/T 12325-2003), the voltage deviation of the supply voltage higher than 35kV shouldn't exceed 10%. The reference voltage of the power grid is 345kV. It can be seen from figure 5, the deviation of the voltage deviation from the load without power load is increasing with the increase of the power load. After any two feature nodes adding 40% of the power load, most of the value of the voltage deviation exceed the standard range of voltage deviation, reaching 11%~12%. The voltage deviation exceeding the standard range will increase the loss of line in the grid system and increase the power consumption of the grid. It's also has a bad impact on the power grid [9].

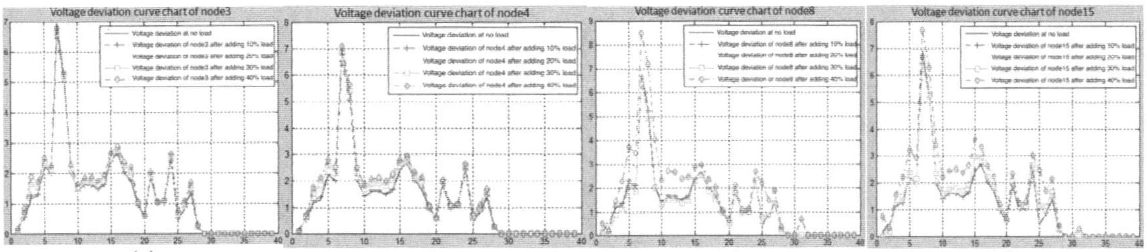

(a) (b) (c) (d)

Figure 4. Voltage deviation of single feature node comparison chart.

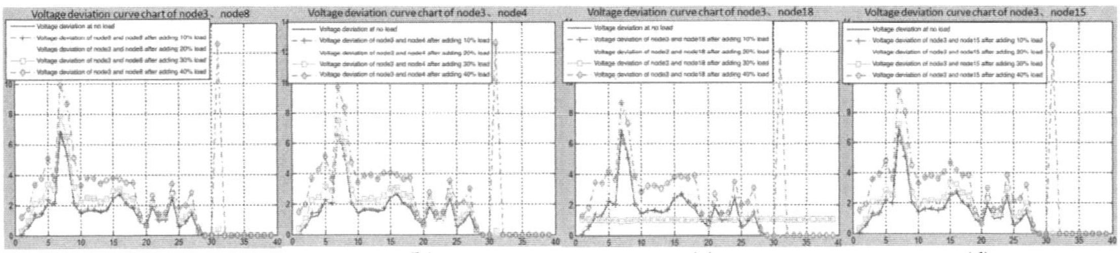

(a) (b) (c) (d)

Figure 5. Voltage deviation of two feature nodes comparison chart.

3.2. *The influence of voltage deviation on power quality in high voltage power grid*

The large voltage deviation will affect the quality of power directly, which shows the increase of line loss and the decrease of power factor in power grid. The reactive power loss and the active power loss of the power grid increase with the low voltage operation, which increases the cost of power consumption and affects the economic operation of the power grid. In addition, voltage deviation is too large will make the operation of the power system is not stable and easy to cause damage to electrical equipment easily [10]. In this paper, reactive power loss, active power loss and reactive power factor of the power grid are analyzed to study the influence of voltage deviation on the power quality of the electric vehicles charging at multiple nodes simultaneously.

From the table 1 we can see that, after two feature nodes are added at 30% and 40% of the power load, the reactive power loss is higher than that are added at 10% and 20% of the power load. Therefore, the increase of voltage deviation will directly affect the reactive power loss of the power grid. After adding 40% of power load, the reactive power loss rate is 4%~6% in Table 1, the loss of reactive power is larger. The line loss rate in the power grid is mainly the reactive power loss. Therefore, it can be explained that the voltage deviation will affect the line loss rate of the power grid, thus affecting the power quality [11].

The cause of the voltage deviation is reactive power compensation is excessive, the transmission distance is too long, excessive overloading and lack of reactive power [12], the reactive power shortage is major cause of the voltage deviation. Because of the lack of reactive power is generated due to the reactive power factor is low. The following will mainly analyze the reactive power factor of the two feature nodes after adding power load and study the reasons for the shortage of reactive power.

From table 2 indicates that part of the power factor is reduced to 0.8 after two feather nodes of the grid system adding to 30% and 40% power load. The low power factor of power system causes insufficient reactive power.

We charge with the power load of 10%, 20%, 30% and 40% on the single node or two feature nodes. It can be seen that the voltage deviation is within the standard range after the power load is added to the single feature node. As the power load increases, the deviation of voltage deviation from the original voltage deviation is getting larger and larger. Therefore, we should add the reasonable power load at the nodes, avoiding the voltage deviation exceeding limit. After two feature nodes are added to 20%, 30% and 40% of the power load, part of the voltage deviation is greater than 10%. Lead to voltage deviation exceeding the limit, resulting in line loss and the reactive power shortage. After node 4 and node 15 are added 20% of the power load at the same time, the voltage deviation exceeds the limit. We need to avoid adding too much power load on these two nodes at the same time. When electric vehicles quick charge in the two nodes, the number of electric vehicle charging needs to be restricted, avoid grid voltage deviation is too large, which can affect the power quality.

Table 1. The power loss after two nodes added the power load.

node	10% load reactive power loss	10% load active power loss	20% load reactive power loss	20% load active power loss	30% load reactive power loss	30% load active power loss	40% load reactive power loss	40% load active power loss
node 3，node 4	0.489	0.484	1.188	0.520	2.419	0.587	6.604	0.832
node 3，node 8	0.520	0.486	2.389	0.582	2.344	0.581	6.166	0.800
node 3，node 15	0.431	0.480	2.349	0.582	2.364	0.587	6.781	0.851
node 3，node 18	0.451	0.481	2.362	0.582	2.439	0.590	5.512	0.768
node 3，node 27	0.418	0.473	2.275	0.569	2.355	0.576	4.266	0.690
node 4，node 8	0.591	0.488	2.389	0.582	2.735	0.597	4.762	0.716
node 4，node 15	0.485	0.482	4.038	0.674	2.581	0.593	4.619	0.713
node 4，node 18	0.487	0.482	1.206	0.517	5.582	0.772	4.391	0.695
node 4，node 27	0.457	0.474	1.158	0.505	5.431	0.758	4.336	0.686
node 8，node 15	0.517	0.483	1.248	0.519	2.524	0.588	4.440	0.699
node 8，node 18	0.519	0.483	2.345	0.577	5.286	0.750	4.192	0.681
node 8，node 27	0.491	0.476	2.293	0.568	2.371	0.569	4.159	0.672
node 15，node 18	0.442	0.479	1.184	0.517	5.736	0.787	4.643	0.720
node 15，node 27	0.415	0.471	2.433	0.583	2.535	0.590	4.659	0.723
node 18，node 27	0.431	0.472	1.176	0.509	2.562	0.590	4.678	0.722

Table 2. Reactive power factor after two nodes added power load.

node	10% no load Power factor	20% no load Power factor	30% no load Power factor	40% no load Power factor
node 3，node 4	0.9	0.9	0.9	0.9
node 3，node 8	0.9	0.9	0.9	0.8
node 3，node 15	0.9	0.9	0.9	0.8
node 3，node 18	0.9	0.9	0.9	0.8
node 3，node 27	0.9	0.9	0.9	0.9
node 4，node 8	0.9	0.9	0.9	0.8
node 4，node 15	0.9	0.8	0.8	0.8
node 4，node 18	0.9	0.9	0.8	0.8

node 4，node 27	0.9	0.9	0.9	0.9
node 8，node 15	0.9	0.9	0.9	0.8
node 8，node 18	0.9	0.9	0.8	0.8
node 8，node 27	0.9	0.9	0.9	0.8
node 15，node 18	0.9	0.9	0.9	0.9
node 15，node 27	0.9	0.9	0.9	0.9
node 18，node 27	0.9	0.9	0.9	0.9

4. Conclusion

Based on the daily load curve of quick charge for electric vehicles in Beijing in 2030, this paper analyzes the influence of the electric vehicles quick charge on voltage deviation of high voltage power grid. Through the typical network topology data of 10 machines 39 nodes, using Newton Raphson method analysis the voltage deviation and loss, when a large number of electric vehicles quick charge on the high voltage power grid. And study its effect on power quality. Quick charging mode can generate a large charge power in a short time, which can meet the electric vehicles in a short time full of electricity requirements. However, it also causes the increase of voltage deviation, increases the line loss in the power grid system, and increases power consumption of the power grid. At the same time, it has a negative impact on the power grid, affecting the reliability of the distribution network operation. The calculation method of this paper can predict the future electric vehicles ownership. And the load model can also be used to study the impact of future electric vehicles charging on the power grid, which has certain practical value.

Reference

[1] Hao Juan, Li Qiang and Yue Jianhua 2010 The charging mode of electric vehicle charging station . Inner Mongolia Electric Power 28 7–9
[2] Kang Jiguang, Lin Wei, Cheng Danming and Xu Fan 2009 Mode and charging station construction of electric vehicle charging *Power Demand Side Management* **11** 64–66
[3] Shi Haiying Influencing factors and measures of power quality *New Technique* **15** 77–78
[4] Huang Shaofang electric car charger (station) Research on harmonic problem 2008 *Beijing Jiaotong University*
[5] Zhang Shigang and Meng Qinglin 2011 A summary of power flow calculation in distribution system *Tianjin Electric Power Technology* **2** 1–5
[6] Hu Hao, Yan Yingmin and Chen Yongli 2012 Power flow calculation based on MATLAB *Application World* **31** 55–59
[7] Zhang Shigang and Meng Qinglin 2011 A summary of power flow calculation in distribution system *Tianjin Electric Power Technology* **2** 1–5
[8] Zhang Jianxiang, Zhang Bingliang and Li Bingqiang 2013 Modeling method of electric vehicle charging demand load model *Conceptual Design* **32** 44–48
[9] XU Aiguo and XIE Shaojun 2009 A multipulse-structure-based bidirectional PWM converter for high-power applications *IEEE Trans on Power Electric* 1233–42
[10] Qin Wenhong, Zhu Wenguang and Li Qingpeng 2007 Decision making of 500kV substation site selection based on fuzzy comprehensive judgment *Journal of Nanchang Institute Technology* **71** 46–49
[11] Ma Jian 2012 Influence factors and control methods of power quality *Research and Development* **19** 40–41
[12] Cheng Haozhong 2007 Lecture second on power quality, voltage deviation and frequency deviation *Special Lecture* 55–59

Assessment Index System for Optimal D5000 Alarm Data Platform

Xinjie Zhang and Jian Wang

State Grid Tianjin Chengxi Power Supply Company, Tianjin, China

Email: funnyzxj@qq.com

Abstract. Nowadays, alarm information collected by dispatching and control systems is increasing substantially and the problem of frequent and false information happens occasionally, which may influence the operation status analysis of power grid and the following work of big data mining and cloud-computing. Thus this paper proposed an optimal D5000 alarm data platform, using the intelligent techniques like alarm restraining, information folding, etc. Besides, assessment index system was also established to analyse the optimized effect of the alarm data processing system, adopting the fishbone diagram and improved Grey incidence approach. The practical results show that the optimal platform is practical and valuable in following aspects: eliminating redundant and interference information, reducing the man-made missing or judging problem of urgent and important information, improving the big data analysis and management, and so on.

1. Introduction

Under the integrated mode of dispatching and control, real-time alarm information increases sharply, equipment abnormal data sent frequently and mistakenly and plenty of simultaneous interference information produced during power grid fault, which pose great challenge to the normal equipment supervision work [1-2].

Faced with this, Tianjin Chengxi Power Supply Company introduced an advanced smart grid dispatching and control system (hereinafter referred to as D5000 systems) in 2014. As a big data platform, D5000 systems are of great use in collecting, storing and processing massive real-time state data [3-5], bringing a new concept for solution of big alarm data information analysis and management.

This article proposed an optimal D5000 alarm data platform, which adopts the intelligent techniques, including uploading-time delay, single-point or interval restraining, automatic alarm folding, etc., to solve relevant problems in alarm information processing of substation equipment [6-8]. Assessment index system was also built to analyse the optimized effect of the alarm data processing system, compared with ON2000 systems. The fishbone diagram and improved Grey incidence approach are applied to simultaneously consider subjectivity and objectivity [9-11].

2. D5000 alarm service

2.1. *Alarm service function*

Supervision alarm refers to the alarm terms uploaded to the alarm window after the alarm information processing, which is also the key attention object of supervision operation, including operation state change, future system prediction, equipment supervision and control, dispatcher operation records, etc. [12-14].

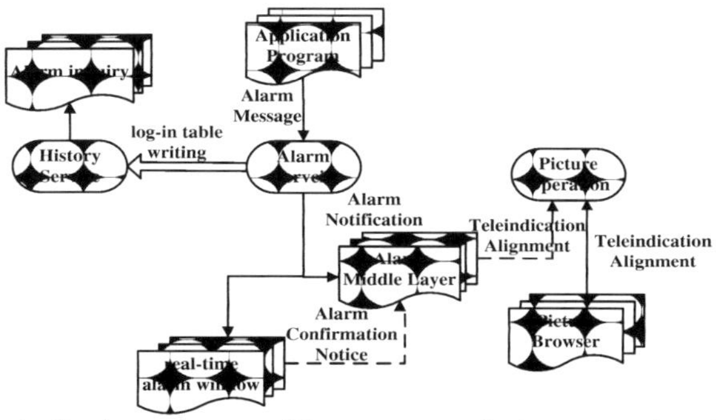

Figure 1. Flow diagram of alarm service function and real-time alarm window display.

Figure 1 displays the specific process of alarm service function in D5000 systems. The alarm service serves as a central nervous control system which is capable of receiving alarm source information (including sequence of event SOE, protection action and change signal, etc.) sent by application programs (SCADA, PAS, AGC, etc.), assigning tasks to the connected alarm clients, and notifying them to adopt corresponding alarm behaviors.

2.2. *Alarm information classification*

According to the requirement of State Grid Enterprise Standard "Q/GDW 11398-2015 Substation Equipment Supervision Information Specification" [15], five alarm levels of accident, abnormal, out-of-limit, change and notification are classified by the effect on the power grid and the equipment and displayed in the real-time alarm window by different colour and category. See Figure 2.

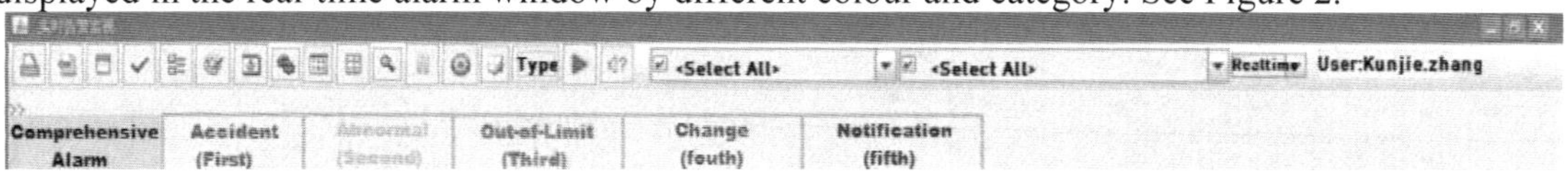

Figure 2. Real-time alarm window display of supervision information.

The daily work of supervision personnel is to monitor the real-time alarm information in the "comprehensive alarm" window, which displays the accident, alarm, out-of-limit and change information, and adopt corresponding treatment measures according to the alarm category. By checking the alarm type, workload of supervision personnel can be reduced effectively.

2.3. *Alarm filtering function*

In the process that D5000 alarm information is received in the middle layer via the alarm service, the information can be processed through the background file configuration and self-defined filtering function.

Table 1. Custom filtering condition and meaning of alarm information.

Filtering Condition	Filtering meanings
st filtering	The substation state attribute is non-supervised state in the real-time database
if_sum_group filtering	"Uploading to alarm-window" is not selected in the alarm definition
warn_type filtering	Alarm information of such certain type is not configured in "detail.ini" file
restrain_flag filtering	In case of alarm restraining or filtering due to other reasons, the "alarm_server" program assigns values (the value larger than 0 indicates restraining)
occur_time filtering	Alarm time is 10mins later or 10s earlier than the scada host

Table 1 just lists several filtering conditions and the corresponding meanings. By setting the specific filtering condition of alarm data, redundant and interference information could be effectively filtered, which helps to reduce supervision risk of massive alarm information within a short time, release the occupied system resources and improve the system operation efficiency.

3. D5000 alarm optimal processing

Based on D5000 alarm service, the optimal data platform adopted several optimized measures for alarm information processing, including uploading-time delay, single-point or interval restraining, automatic alarm folding, etc. The uploading-time delay and single-point or interval restraining can be achieved by setting the filtering parameter restrain_flag, as shown in Table 1, while automatic alarm folding is achieved by further modifying conditions in the configuration file.

3.1. Uploading-time delay

The alarm information should be delayed several seconds to upload to real-time alarm window when the signals acting frequently and then resetting immediately. Figure 3 shows the specific time-delay processing flow. Supervision personnel judge whether the received alarm information needs time-delay setting; then judge whether the resetting information is received in the time-delay period. When the time-delay for alarm information is set and the resetting information is received within the time-delay period, the information can be filtered, which is favourable to monitor the alarm information in real-time and treatment it timely.

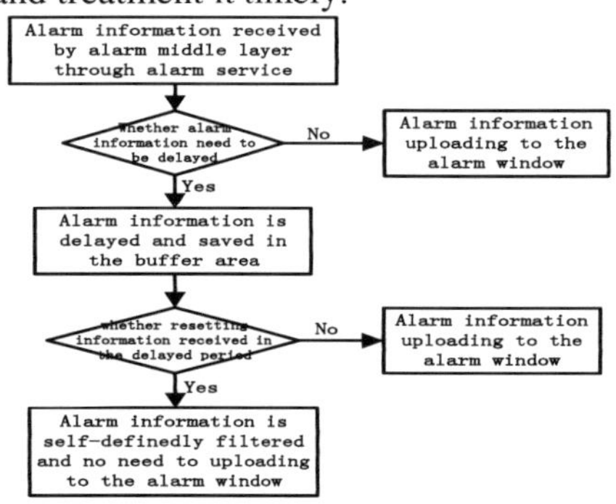

Figure 3. Flow diagram of window-uploading time delay for alarm information.

In actual application, there is mainly the following alarm information, which can be processed by setting the uploading-time delay:

1) The associated signals produced during equipment operation, which are usually instantaneous information. Generally a 10s time-delay processing method is adopted;

2) The voltage, current and other telemetry out-of-limit data. A 30s time-delay processing is used to shield the out-of-limit information which can be reset within the stipulated delay period;

3) Abnormal simultaneously or out of service of one communication channel can be filtered. Only when both 101 and 104channels are disconnected, the information should be displayed in real-time alarm window;

4) During telecontrol of some substations, the tap switchgear may be short to 0 in the lifting process due to equipment cause. Such information should be filtered by a 10s time-delay.

3.2. Single-point or interval restraining

There is mainly 2 alarm information restraining methods in D5000 systems:

1) Equipment listing processing, mainly used in secondary equipment overhaul and test to shield a large amount of debugging signals produced within a short time and avoid the fault judgment of the equipment true condition.

2) Single-point, interval and even substation alarm restraining, mainly targeted to frequent information caused by abnormal equipment operation to avoid the related signals frequently uploaded to the alarm window and release the occupied alarm window space.

IOP Conf. Series: Materials Science and Engineering **366** (2018) 012034 doi:10.1088/1757-899X/366/1/012034

Table 2 displays the data structure of supervision information, i.e. the real-time non-recovery action record table. The above 2 restraining methods can be achieved by separately configuring the alarm parameter reserved_2 and restrain_flag to display results required by users in the alarm window.

Table 2. The Non-recovery action record structure of alarm information.

```
typedefstructreal_warn_rec_new_scada
// Structure of action non-resetting table
{
   long id;   /* Keywords */
   long occur_time;   /* Occurrence time */
   ......
   int restrain_flag;  /* Alarm restraining tab */
   int if_fg;  /* Whether to reset */
   ......
   int reserved_2; /* Tag type */
   long dev_id;   /* Signal id */
}REAL_WARN_REC_NEW_SCADA;
```

Equipment listing property is shown by reserved_2, and the value is the tag ID. In case reserved_2 ≤0, the information is not listed or filtered, otherwise, find out whether the tag ID has the alarm restraining property, filter it if any.

Single-point and interval alarm restraining just focus on the value of restrain_flag. When 1, 2 and 3 was separately assigned to restrain_flag, it indicates single-point, interval and substation restraining respectively.

The restrained or listed alarm information should be shown in the corresponding alarm information summary list, including substation name, information name, restraining or listing time, etc. Supervision in the normal state will be recovered automatically after restraining relieving and alarm information also displayed in the “comprehensive alarm” window.

3.3. Alarm information folding

Single-point or interval alarm restraining method still needs supervision personnel to perform manual judgment and restraining operation on the uploaded information. Therefore, a more intelligent and automatic folding method is proposed to effectively solve the problem of frequent alarm information by folding the redundant repeated information within a period, classified into two types: real-time alarm information folding and designated alarm information folding.

Only the newest folded alarm information is displayed in the alarm window, including folding tab, newest alarm time of the folded alarm, folding time, etc., as shown in Figure 4. Meanwhile, the information will keep normal uploading sequence, without affecting the time sequence analysis of the supervision information.

Figure4. The alarm information folding display in real-time alarm window.

Information folding calculation is performed by the alarm folding rule set in the configuration file to judge whether folding information is required. If the alarm information only acts and is reset for one or two times within 5min, it will be displayed normally; if the alarm information acts and is reset for no less than three times, it will be provided with the folding tab and uploaded to the alarm window.

Table3. The Non-recovery action record structure of alarm information.

Type	Name	Description
long	dev_id	Signal id

long	occur_time	Signal action time
int	flod_type	Folding type
int	flod_time	Minimum interval of alarm-folding
int	flod_num	Minimum folding times of alarm interval

Table 3 shows part of basic configuration information about the alarm-folding rule in D5000 systems. By acquiring minimum interval and minimum folding times as the comparison conditions for folding calculation, alarm-folding analysis is performed.

The folded alarm information can be displayed by right clicking the option "Inquire alarm folding detailed information" to inquire corresponding alarm information from the background shared memory by searching for signal id, folding times, start time, etc.

4. Assessment index system

Assessment index system was established to analyse the optimized effect in D5000 systems and the fishbone diagram was used to brainstorm all the possibilities that could cause the effect. Besides, an improved Grey incidence approach is applied to evaluate the performance of big data processing in D5000 systems compared with that in ON2000 systems.

4.1. Index system analysis

Establishing an index system should follow the five principles, including systematic, scientific, independence, adaptability and comparability, and ensure that the indexes have clear meanings and could calculate easily.

This paper adopt the fishbone diagram to analyse the factors causing the optimized effect, in which the various causes are grouped into categories and cascade from the main categories, flowing towards the effect, forming what resembles a fishbone appearance [9].

Figure 5 shows the main factors influencing on big data analysis and management. Considering three categories of D5000 systems, supervision personnel and optimized methods, the index system was built up, specific indexes are shown in Figure 5.

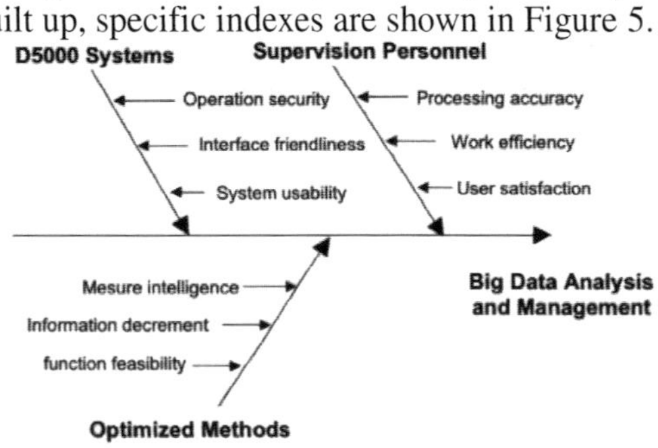

Figure 5. The alarm information folding display in real-time alarm window.

4.2. Index weight calculation

The assessment index weight is an important link in comprehensive evaluation. Firstly, Delphi method is adopted to set the index weight and the improved Grey incidence approach is chosen to reduce artificial factors, which makes subjectivity and objectivity simultaneously considered [10]. Specific steps are as follows:

1) Determine the evaluation indicators and employ experts to make the experience weights. See Equation 1, the matrix with n indexes and m experts.

$$A = \left[A_1, A_2, \cdots, A_n\right]^T$$
$$A = \begin{bmatrix} a_{11} & a_{12} & \cdots & a_{1m} \\ a_{21} & a_{22} & \cdots & a_{2m} \\ \cdots & \cdots & \cdots & \cdots \\ a_{n1} & a_{n2} & \cdots & a_{nm} \end{bmatrix} \quad (1)$$

2) Determine the reference sequence A_0, selecting maximum weight values from the empirical judgment matrix for each index, as shown in Equation 2.

$$A_0 = \left(a_{01}, a_{02}, \text{L}\ , a_{0m}\right) \quad (2)$$

3) Calculate the distance between all the index sequence $A_1, A_2, \cdots, A_n$ and the reference sequence A_0 as shown in Equation 3.

$$D_{0i} = \sum_{k=1}^{m}\left(a_{0k} - a_{ik}\right)^2 \quad (3)$$

4) Obtain the weights of each index and do the normalization. See Equation 4 and 5.

$$\omega_i = \frac{1}{1 + D_{0i}} \quad (4)$$

$$\bar{\omega}_i = \frac{\omega_i}{\sum_{i=1}^{n} \omega_i} \quad (5)$$

With the opinions of 10 experts, the weights of each index are finally determined, as shown in Table 4.

4.3. Index score and result

According to the experts' opinions, this paper adopted Delphi method again to obtain the discrete index score standard, and did the curve fitting with "Curving fitting tool" of Matlab. The index weight and score function are shown in Table 4.

Table 4. Index weight and score function of influencing factors

Index Name	Index Weight	Index Score Function
Operation security	0.1124	y=0.0007x^3-0.093x^2+4.51x
Interface friendliness	0.0857	y=0.0054x^2+0.46x
System usability	0.1011	y=-0.011x^2+2.1x
Processing accuracy	0.1429	y=x
Work efficiency	0.1254	y=0.0054x^2+0.46x
User satisfaction	0.0955	y=-0.011x^2+2.1x
Measure intelligence	0.1314	y=0.0007x^3-0.093x^2+4.51x
Information decrement	0.1124	y=-0.024x^2+2.97x
Function feasibility	0.0932	y=-0.024x^2+2.97x

According to the score function given in Table 4, the score for each index can be calculated and the overall score can also be obtained, which comparing the practicability of optimized measures in D5000 systems with that in ON2000 systems, as shown in Table 5.

Table 5. Evaluation results of index system

Evaluation result	D5000 systems	ON2000 systems
Operation security	92.46	77.63

Interface friendliness	78.16	47.04
System usability	95.63	88.64
Processing accuracy	88.03	65.12
Work efficiency	71.36	36.50
User satisfaction	96.48	90.68
Measure intelligence	96.75	79.62
Information decrement	86.99	39.15
Function feasibility	91.65	85.05
Comprehensive score	88.53	67.03

5. Conclusion

To analyse the optimized processing effect in the D5000 alarm data platform, this paper introduced several optimized techniques including uploading-time delay, single-point or interval restraining, alarm information folding, etc., and established the assessment index system and evaluation method, using the fishbone diagram and improved Grey incidence approach.

According to the data in Table 5, all the indexes were improved in D5000 systems, especially the three indexes of interface friendliness, work efficiency and information decrement, whose scores are separately promoted to 78.16, 71.36 and 86.99, compared with 47.04, 36.50 and 39.15 in ON2000 systems. Besides, the comprehensive score of D5000 systems was 88.53, which has a 21.5 improvement compared to that in ON2000 systems, indicating obviously the intelligence and feasibility of optimal D5000 alarm data platform.

The practical results show that the optimal alarm data platform is practical and valuable since redundant and interference information was eliminated effectively, supervision information amount and the man-made missing or judging problems were both reduced and big alarm data analysis and management was significantly improved. Meanwhile, it provided reference for online application work of D5000 systems; thereby achieve great generalization and reference significance.

References

[1] C. Fan, Y. Ni, R. Dou, et al 2012 Interpretation of relevant specifications of integrated supervision and control system in smart substations *J. Automation of Electric Power Systems* **36** 1-5

[2] L. Yang, H. Wang, X. Shen et al 2014 Information classification and optimization management of technical support *J. Zhejiang Electric Power* **8** 50-52

[3] G. Zhang2016 Application of big data in the field of smart grid *J. Power System and Clean Energy* **32** 114-117

[4] K. Wu, W. Liu, Y. Li, et al 2015 Cloud-computing based power big data analysis technology and its application *J. Electric Power* **48** 111-116

[5] G. Li2014 Research on the application of D5000 system for intelligent power network dispatch master station *D.* Baoding: North China Electric Power University

[6] W. Wang, M. Shi, J. Song et al 2015 Optimization treatment of alarm information in centralized monitoring mode of substation *J. Hebei Electric Power* **34** 23-24,27

[7] X. Zhou, Z. Chen, L. Tao et al 2016 Study of index presentation application based on grid dispatch and control big data platform *J. Electric Power Information and Communication Technology* **12** 15-19

[8] X. Zhang, J. Wang 2017 Optimized processing of monitoring information based on smart grid dispatching and control system D5000 *J. Electric Automation* **39** 30-33, 68

[9] Z. Han, Z. Huang, S. Ge, et al A comprehensive evaluation system of urban distribution network *J. Power System Technology* **36** 95-99

[10] X. Zhang, S. Ge, H. Liu, et al 2014 Comprehensive assessment system and method of smart distribution grid *J. Power System Technology* **38** 40-46

IOP Conf. Series: Materials Science and Engineering **366** (2018) 012034 doi:10.1088/1757-899X/366/1/012034

[11] P. Lü, J. Zhao, D. Li, et al 2015 An assessment index system for power grid operation status and corresponding synthetic assessment method *J. Power System Technology* **39** 2245-2252
[12] S. Sun, W. Jiang, H. Peng, et al 2011 Construction of the intelligent alarm system for regional power grids *J. China Southern Power Grid Technology* **5** 73-76
[13] X. Shan, Z. Dai, Z. Zhang, et al 2015 Research on and application of integrated smart alarm based on smart grid dispatching and control systems *J. Automation of Electric Power Systems* **39** 65-72
[14] F. Jin, H. Wang, G. Fan, et al 2015 Design of centralized substation monitoring functions for smart grid dispatching and control systems *J. Automation of Electric Power Systems* **39** 241-247
[15] 2015 Q/GDW11398-2015 Information specification for supervision and control of substation devices *S* China

IOP Conf. Series: Materials Science and Engineering **366** (2018) 012035 doi:10.1088/1757-899X/366/1/012035

The Evaluation of Ultimate Capacity for Distributed Photovoltaic Access Distribution Network Considering Uncertainty

Jun-Huang Zhou[1,a] and Yong-Jun Zhang[1]

[1] South China University of Technology, 381 Wushan Road, Tianhe District, Guangzhou, P.R.China

[a] E-mail of corresponding author: 272757783@qq.com

Abstract. Aiming at the problem of the increasing voltage over limit risk by a large number of distributed PV access network, firstly by analyzing the volatility and time-series property of the PV output, a probability model of PV output based on the clearance model is established in this paper. Then the risk assessment index of system voltage exceeding the limit based on the probability and severity of node voltage at each time interval is established. Based on this, using the hybrid approximation to solve the ultimate capacity of distributed PV access network is proposed. Finally, a typical IEEE 33-node power distribution system is taken as an example to analyze the risk of system voltage under different load characteristics, load levels and line types. From these three aspects, the ultimate capacity of distributed PV access network is evaluated.

1. Introduction

A large number of distributed photovoltaic (DPV) accesses to distribution network (DN), making the original single-source radiation DN into an active distribution network, which also changes the power flow and voltage distribution of DN and increases the risk of voltage over-limit (VOL) [1-4]. However, due to the uncertainty of PV output and load, it is difficult to measure their impacts on the voltage of DN accurately by certain power flow assessment method, which is mainly related to the location and capacity of the PV. Therefore, how to assess the voltage over limit risk of DN caused by PV access and determine the ultimate capacity of PV will be an urgent problem to be solved.

At present, lots of references have studied the influence of DPV access on the voltage of DN [5-8]. Reference [5] proposed using Thevenin's equivalent model based on DG to analyze the effect of DG access on voltage stability of the system. Reference [6] proposed using sensitivity analysis method to analyze the influence of DG output and access location on the system voltage. Reference [7] established the probability model of DPV power generation system based on Beta distribution of solar radiation and analyzed the influence law of the steady-state voltage distribution and voltage amplitude fluctuation of DN after PV access. Reference [8] proposed solving the problem of VOL in DN by PV power supply active/reactive power integrated control scheme. The above references all analyze the impact of PV access on DN voltage from the perspective of static power flow. However, in fact, PV output has uncertainties and time-series property, that is to say, the risk of system VOL caused by the fluctuation of PV output during different periods of the day is not the same. Therefore, when assessing the risk of VOL caused by PV access, the model considering the uncertainty and time-series of load and PV output simultaneously is more practical.

In this paper, based on the analysis of the PV output characteristics, a probability model of PV output based on the clearance model is established. The risk assessment index of VOL in DN considering the uncertainty and time-series property of PV output and the ultimate capacity of PV access network are proposed. Taking the IEEE33-node standard power distribution system as an example, this paper evaluates the ultimate capacity of DPV from three aspects of load characteristic, load level and line type.

2. Probabilistic model of PV and load

2.1. Time-series model of PV output based on clearance theory

Related researches show that there is a strong positive correlation between PV output and solar irradiance. But the intensity of sunlight varies greatly with different weather types and seasons. Therefore, in order to show the differences of PV output due to different solar irradiance in different weather types, the eigenvector is constructed in this paper according to the indexes that describe the average output and volatility of PV, and then the generalized weather types are classified by cluster analysis. Then, the time-series models of PV output under different weather types are established.

The PV output contains both deterministic and uncertain components, the deterministic property mainly reflects the time-series characteristics of PV output due to the sunrise and sunset of the sun, while the uncertainty mainly reflects the PV output due to cloud changes fluctuation characteristics. This paper refer to the analysis method of reference [9], the relative output at each moment breaks down into the day's benchmark output and fluctuation coefficient at each moment. Finally, for PV output benchmark data and fluctuation coefficient data of different weather types, the least square method is used to fit the PDF of the PV benchmark output and the fluctuation coefficient under different weather types. The fitting function is：

$$f(x)=\sum_{i=1}^{n}\alpha_i\frac{1}{\sqrt{2\pi}\sigma_i}e^{-\frac{1}{2\sigma_i^2}(x-\mu_i)^2} \tag{1}$$

In (1), α_i, σ_i and μ_i are fitting parameters whose values are related to the type of weather and the type of fitting variables.

2.2. Probabilistic model of load

The errors of load forecasting and load measuring are inevitable. Therefore, its active power P_L and reactive power Q_L should be described as random variables. Many researches demonstrate that these two random variables obey normal distribution function.

3. The evaluation of ultimate capacity for DPV access distribution network

PV access generally raises the voltage of DN, resulting in the risk of VOL. And the PV output is generally larger in sunny days and the risk of VOL in DN will be higher. Therefore, the following analysis uses the probability model of PV output in sunny days as an example to analyze the ultimate penetration (ultimate capacity) of PV access. The PV penetration mentioned in this article is the ratio of the total installed capacity of DPV in the DN to the maximum power supply capacity of the feeder.

3.1. System voltage over-limit risk assessment index

Risk index not only need to reflect the possibility of fault occurred in the system, but also need to reflect the severity of the fault occurred. In this paper, we use the voltage probability distribution model to quantify the probability, use the severity function to quantify the severity and establish the system VOL risk assessment based on the voltage probability distribution and severity function of each node in each period. Therefore, the definition of the system VOL risk assessment $R_{\mathrm{vs\_max}}$ is the maximum VOL risk of all nodes in the system during the operation period (generally 24h), that is：

$$R_{\mathrm{vs\_max}}=\max\{R_{\mathrm{vs}}(1),R_{\mathrm{vs}}(2),...,R_{\mathrm{vs}}(t),...,R_{\mathrm{vs}}(24)\} \tag{2}$$

Where $R_{\mathrm{vs}}(t)$ is the sum of VOL risk for all nodes at time t and its calculation method is as follows：

$$R_{vs}(t)=\sum_i\int_0^{V_{min}} f(V_i^t)U_{dl}(V_i^t)dV_i^t+\sum_i\int_{V_{max}}^{\infty} f(V_i^t)U_{ul}(V_i^t)dV_i^t \tag{3}$$

In (3), V_{max} and V_{min} are upper limit and lower limit of the node voltage respectively; $f(V_i^t)$ is the probability density function of node i voltage at time t; $U(V_i^t)$ is the function of the severity of the VOL as follows:

$$U_{dl}(V_i^t)=\frac{e^{V_{min}-V_i^t}-1}{e-1} \qquad U_{ul}(V_i^t)=\frac{e^{V_i^t-V_{max}}-1}{e-1} \tag{4}$$

3.2. *The calculation of the assessment index*

In this paper, R_{vs} is calculated by the method of probabilistic power flow calculation based on cumulant method [10], in which the probabilistic power flow model uses the linearization model [11] of the AC power flow and the Polar-Newton-Raphson method is used as the power flow calculation method.

3.3. *Assessment process*

In order to ensure the safe operation of the power grid and improve the power quality of the consumer, if given the limiting value R_{vs_limit} of the risk of the DN VOL in actual operation, we can use the hybrid approximation method to obtain the ultimate capacity H_{limt} of the DPV access DN. The main steps of the assessment process are as follows and the process is shown in Figure 1.

1) Entering the original data, building simulation network model;

2) Collecting the historical data of PV output and load in the target area to construct the probabilistic model of load and PV output;

3) Given the initial PV penetration rate H;

4) Calculating R_{vs_max}: calculate the risk assessment index of DN VOL at a given PV penetration rate according to the calculation method of risk assessment index of VOL in Section 3.1 and 3.2;

5) Judging whether R_{vs} is greater than R_{vs_limit}, if not, let H=H*(k+1), where k is the penetration rate adjustment coefficient and 0<k<1, and return to step 4; if yes, go to step 6;

6) Reducing the penetration rate H and penetration rate adjustment coefficient k: let H = H*(1-k) and k = k / 2;

7) Judging whether k is less than the accuracy e, if yes then return to step 4; if no, then output the ultimate capacity H_{limt} of PV.

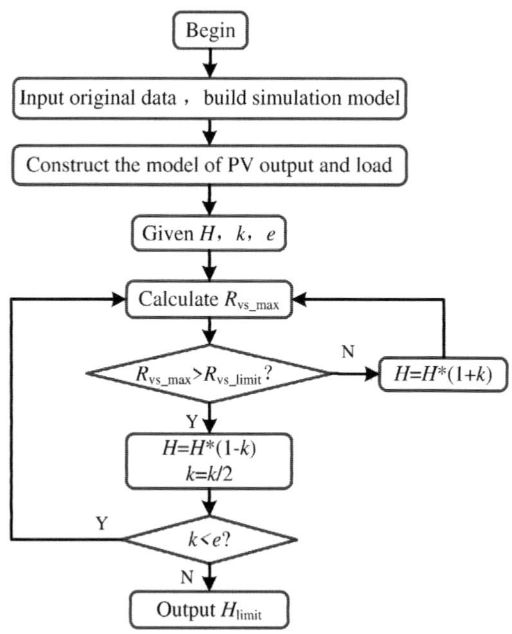

Figure 1. Assessment process for H_{limt}

4. Case study

The IEEE 33-node distribution system is chosen for simulation. As the main line, Node 2 to Node 19 selects the 10kV line model LGJ-240, which is commonly used in China currently. Node 1 is a balance bus, its voltage is set to 1.05pu. The system equivalent reactance Xs is 0.32Ω and DPV connected to Node 19 equivalently and centrally at the end of system. The improved network model is shown in Figure 2.

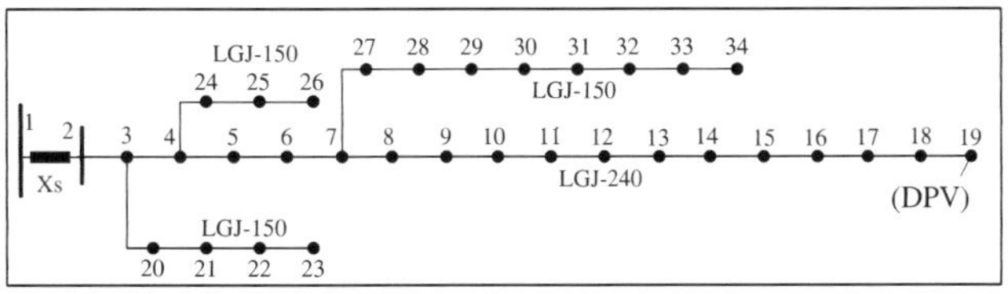

Figure 2. DPV access distribution grid structure

The main line of main simulation is LGJ-240. According to the current carrying capacity, its maximum power supply capacity P_{Lmax} is 9.509MW. Because the DN is a single radiation network, its upper limit of PV penetration rate is 100%, only need to ensure that the current of main line is not overloaded.

4.1. *The ultimate penetration rate of PV under different load characteristics*

To study the influence of PV access on voltage level under different penetration rates, based on the DN structure shown in Fig.2, we assume that load types of all nodes are industrial load and the peak value of total load is 3.715MW, (that is, the maximum load rate of line is 39.07%).The assessment index of VOL risk R_{vs_max} in DN is calculated under different line lengths and different PV penetration rates, as shown in Figure 3. The load characteristics of different load types are different. So the load types of nodes are set as agricultural load, commercial load and resident load respectively. The maximum risk of VOL of system node under different line lengths and PV penetration rates is studied under different load types. The PV penetration rate is obtained under different lengths of main line when $R_{vs\_limit} = 0.005$, as shown in Fig.4(a).

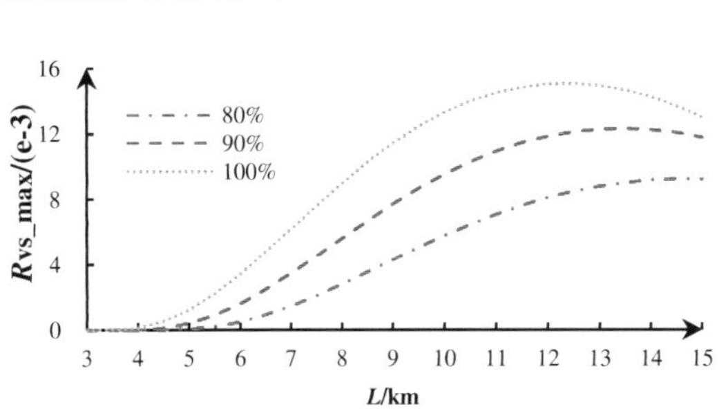

Figure 3. Curves of voltage over limit risk

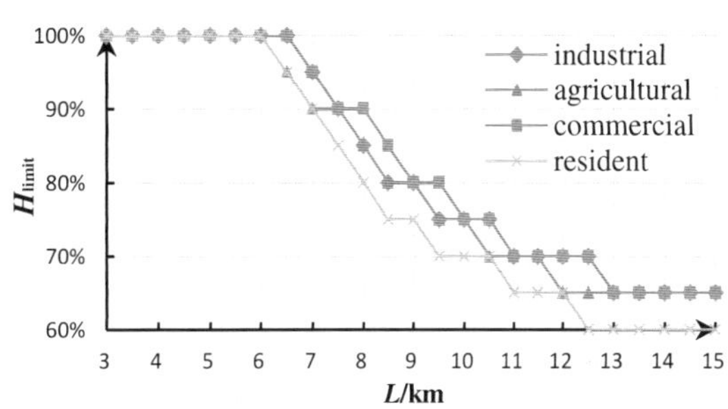

Figure 4(a). Under different load characteristics and R_{vs_limit}=0.005

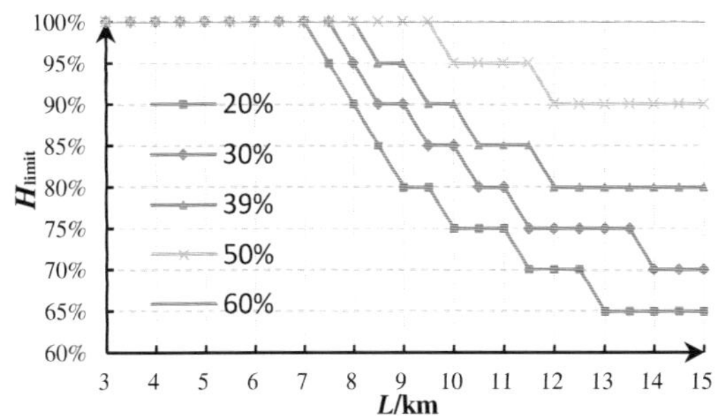

Figure 4(b). Under different load levels and R_{vs_limit}=0.01

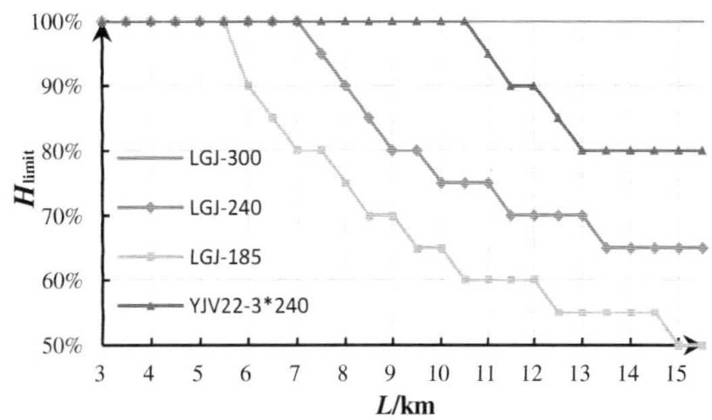

Figure 4(c). Under different line types and R_{vs_limit}=0.005

Figure 4. The relationship between H_{limt} and L under different conditions

From Fig.3 and Fig.4(a), we can see that: 1) When the penetration rate is less than 80%, the risk of VOL in DN will increase with the increase of PV penetration rate and main line length. When the penetration rate is more than 80% and the length of the overhead line is too long, the reactive power loss of the line will increase significantly if the direction of the active and reactive power flow in the line is opposite, which leads to lower voltage at some nodes in the front part of the line. Therefore, under the condition of large PV penetration rate, the risk of voltage over-upper-limit will slightly decrease as the length of the main line increases. 2) When the line length does not exceed 6km, the ultimate penetration rate of PV access is mainly limited by the line current carrying capacity, and its ultimate penetration rate is 100%. 3) When the line length is more than 6km and $R_{vs\_limit} = 0.005$, the accessible PV capacity in DN with the characteristics of resident load is the smallest under different load types, which is mainly the residents load curve and PV output curve does not match well, leading to the maximum risk of VOL when in the same penetration rate. When $R_{vs\_limit} = 0.01$, the maximum over-limit probability constraint is relaxed, and the ultimate PV penetration rate increases under a certain length, but the overall trend remains unchanged.

4.2. The ultimate penetration rate of PV under different load levels

Based on the grid structure and line parameters in Fig.2, assuming that all nodes are industrial load types, their peak loads are set at 1.9, 2.9, 3.7, 4.8 and 5.7MW respectively, that is, the corresponding maximum load rates of the line are respectively 20%, 30%, 39%, 50% and 60% respectively. The maximum risk of VOL of the system node under different line lengths and different PV penetration rate is studied, and the PV penetration rates under different main line lengths are obtained when $R_{vs\_limit} = 0.01$, the result is shown in Fig.4(b).

The figure shows that: 1) When the line length does not exceed 7km, the PV access ultimate penetration rate is mainly limited by the line safety current, and its ultimate penetration rate is 100%. When the line length is more than 7km but not more than 15km, the PV access ultimate penetration rate is affected by the limit of the risk of VOL, and remains above 60%. 2) Under the same line length, the penetration rate of PV access increases with the increase of line load rate.

4.3. The ultimate penetration rate of PV under different line types

Based on the grid structure in Fig.2, assuming that all nodes are industrial loads and the peak of total load is 3.7MW, we study the maximum risk of VOL of the system node under different line lengths and different PV penetration rates, and the PV penetration rates under different line types are obtained when $R_{vs\_limit} = 0.005$, the result is shown in Fig.4(c).

The figure shows that: 1) For overhead lines with the length between 5.5km and 15km, the ultimate penetration rate of PV access increases as the wire diameter of the line increases. 2) Under the same line length and the same diameter conditions, the cable line can access larger penetration rate of PV than overhead line.

5. Conclusions

In this paper, a time-series model of PV output is established, and an evaluation method of ultimate capacity for DPV access DN considering uncertainty based on the risk index of voltage over-limit is proposed. The simulations show that:

(1) When the other conditions are the same, the distribution network with the majority of industrial load has the highest receiving capacity of photovoltaic power. Therefore, it is a reasonable choice to install DPV on the roof of a factory and realize local consumption.

(2) When the line length and load type are the same, the penetration rate of photovoltaic access increases with the increase of the line load level. Therefore, the DPV should be given priority to access to the heavy-load medium voltage feeder.

(3) In general, the ultimate penetration rate of PV access can reach 100% when the length of medium-voltage line does not exceed 5km, and the ultimate penetration rate of photovoltaic access can reach at least 50% or so when the line length is more than 5km and less than 15km.

Acknowledgment

This work was supported by Natural Science Foundation of Guangdong Province, China (2017A030313304).

References

[1] Ding M, Wang W S, Wang X L, Song Y T, Chen, D Z and Sun M 2014 *J. Proceedings of the Csee* **34** 1-14.
[2] Chen J D, Qi L Z, Sun M Y and Xue Y D 2016 *J. Power System Protection and Control* **44** 129-134.
[3] Huang W, Liu J, Wei H K and Zhang Z H 2015 *J. Power System Protection and Control* **3** 22-28.
[4] Chen Q X, Zhao X Y and Gan D H 2017 *J. Protection & Control of Modern Power Systems* **2** 29.
[5] Ma C, Li C L, Chen Y Y and Zhang Y F 2014 *J. Renewable Energy Resources* **32** 424-428.
[6] Li B, Liu T Q and Li X Y 2009 *J. Power System Technology* **33** 84-88.
[7] Chen X, Yang Y Y, Zhang Y J and Ye L H 2015 *J. Journal of South China University of Technology (Natural Science Edition)* **43** 112-118.
[8] Li Q R and Zhang J C 2015 *J. Automation of Electric Power Systems* **39** 117-123.
[9] Li C 2015 *D. China Electric Power Research Institute*.
[10] Wu W, Wang K, Li G, Jiang X and Wang Z 2016 *J. Iet Generation Transmission & Distribution* **10** 1703-1709.
[11] Shi D Y, Cai D F, Chen J F, Duan X Z and Li H J 2012 *J. Proceedings of the Csee* **32** 104-113.

IOP Conf. Series: Materials Science and Engineering **366** (2018) 012036 doi:10.1088/1757-899X/366/1/012036

A coordinated voltage control scheme for power system with wind farm integration based on EID compensation

F Liu[1], Z Xu[1], B Hu[2] and W D Zhang[3]

[1] School of Information Science and Engineering, Central South University, Changsha 410083, China
[2] Dept. of Power Grid Comprehensive Development Research, State Grid Energy Research Institute, Beijing 102209, China
[3] Hangzhou Water Holding Group Co., LTD, Hangzhou 310009, China

[1]csuliufang@csu.edu.cn; [2]hubo@sgeri.sgcc.com.cn

Abstract. Rapid increase of wind power installation brings new issues and challenges for grid security operation and one of the most important key issues is reactive power and voltage control capability of wind farms. This paper proposes a coordinated voltage control strategy for power system with wind farm based on equivalent input disturbances (EID) compensation. Firstly, the voltage control model of power system with wind farm based on doubly fed induction generator (DFIG) is established. The power system adopts reactive power compensation device for voltage control, and the wind farm connected to the grid adopts frequency converter of DFIG to adjust reactive power output. They can coordinate each other to realize voltage stability control of the wind farm connected to the grid. Moreover, when the system is subjected to load disturbance or wind speed disturbance, the proposed method based on EID compensation also can keep the voltage stability of power system with wind farm. Finally, the simulation results show the feasibility and effectiveness of the proposed voltage coordinated control method in voltage regulation.

1. Introduction

The voltage stability of wind farm grid connected power system is affected by the active power and reactive power output of wind farm. At present, a variety of voltage coordinated control of wind farms and power systems have been proposed. A terminal voltage control of a doubly-fed induction generator (DFIG) driven by a pitch controlled wind turbine for the supply of autonomous system without any auxiliary source is proposed in [1]. The influence of d axis rotor current on the reactive power output of fixed rotor in DFIG under five different operation modes is investigated in [2], in which an iterative algorithm of rotor current under different operating modes is proposed, which can adjust the reactive power and voltage of the system. The reactive power supply in wind farm includes wind turbine and other reactive power compensation device. When the power system voltage fluctuate or drop, reactive power capacity of wind turbines will first be used in voltage regulation. When it can not meet the reactive demand of the power grid, the reactive voltage compensation equipment, such as on-load tap changer (OLTC), Static Var Compensator (SVC), will be used to realize voltage control. Voltage fluctuations occur frequently in wind farms as a result of wind power perturbation, while reactive power control of various device is always uncoordinated and often has a time lag when the system operates in the traditional control mode that only uses the current data for decision-making. In

[3], a method for coordinated control of the reactive power of wind turbine generators (WTGs) and static var generators is proposed based on model predictive control (MPC). In [4], it proposes an optimal tracking secondary voltage control scheme applied to double fed induction generator-based wind generators. More, coordinated with the voltage control means of power system, the wind farm can support system voltage quickly by regulating reactive power output, which is benefit to strengthen the voltage control capability and improve the voltage stability of power system [5-8]. By coordinating the internal wind turbine and other reactive power compensation devices, the reactive power regulation capability of the wind farm can be better utilized, but the control strategy is more complex, and the influence factors such as time delay are introduced, and the control effect is affected by the algorithm.

In order to meet the demand of wind farm and the grad for reactive power, a coordinated reactive power and voltage control strategy of doubly fed induction generator (DFIG) based on EID compensation is proposed in this paper. The proposed method with EID compensation does not need accurate model information and has an inherent robustness against various disturbance and internal/external uncertainties. The plant, based on doubly fed induction generator (DFIG) have fast reactive and voltage regulation capability. It is helpful to reduce adverse effects and ensure the stable operation of the wind farm by regulating rationally. The study is mainly related to improve the voltage stability of wind power integrated system. By regulated the wind power generation and allocating the reactive power rationally, it can take advantage of power control capabilities of the wind farm to participate in the coordinated voltage control of power system.

2. Voltage control model

Figure.1 is a block diagram of the voltage control transfer function of a power system with wind farm. Where $\frac{K_1}{\tau_1 s+1}$ is the reactive power compensation device of the power system; $\frac{K_2}{\tau_2 s+1}$ is the transmission line transfer function, in which τ_2 is 0; $\frac{K_3}{\tau_3 s+1}$ is the voltage link, and the gain K_3 is 1 desirable for simulation; Q_s and Q_{qr} respectively indicate reactive power compensation and reactive power output of double-fed induction generator. Q_d represents the system's reactive power fluctuation.

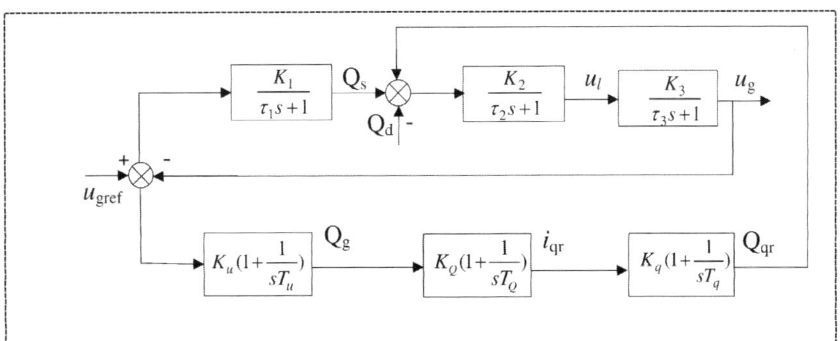

Figure 1. Transfer function diagram of voltage control system.

The state space expression of the voltage control system is as follows:

$$\begin{cases} \dot{x}(t) = Ax(t) + Bu(t) + B_d w(t) \\ y(t) = Cx(t) \end{cases} \tag{1}$$

Where $x(t) = [K_u u_{gref} - K_u u_g - Q_g \;\; K_Q Q_g - i_{qr} \;\; K_q i_{qr} - Q_{qr} \;\; Q_s \;\; u_g]^{\mathrm{T}}$, $u(t) = u_{gref}$, $w(t) = Q_d$, $y(t) = u_g$.

3. Design of controller based on the compensation of EID

This section designs a voltage controller based on the compensation of EID to construct a control system for power system (1).

Based on EID compensation the power system (1) can be written as [9]

$$\begin{cases} \dot{x}(t) = Ax(t) + B[u(t) + w_e(t)] \\ y(t) = Cx(t) \end{cases} \tag{2}$$

3.1. Estimation of EID

It is obvious that the original plant expressed by (1) is controllable, observable and there is no zero on *jw*-axis. The structure diagram of the EID is shown in Figure 2, where K_p and K_R are state feedback gain, and

$$B^+ := (B^T B)^{-1} B^T \tag{3}$$

To obtain an EID compensation with high precision and a high tracking performance, the following IM and GSO observer are used to reproduce the state of the plat:

$$\text{IM: } \dot{x}_R(t) = A_R x_R(t) + B_R[\Delta y_{ref}(t) - y(t)] \tag{4}$$

$$\text{GSO:} \begin{cases} \dot{z}(t) = \Phi z(t) + \Psi y(t) + \Gamma u_f(t) \\ \hat{x}(t) = T^{-1} z(t) \end{cases} \tag{5}$$

Where Φ, Ψ and Γ are the appropriate matrices to be determined; and $\hat{x}(t)$ is the reconstruction state of the original state $x(t)$.

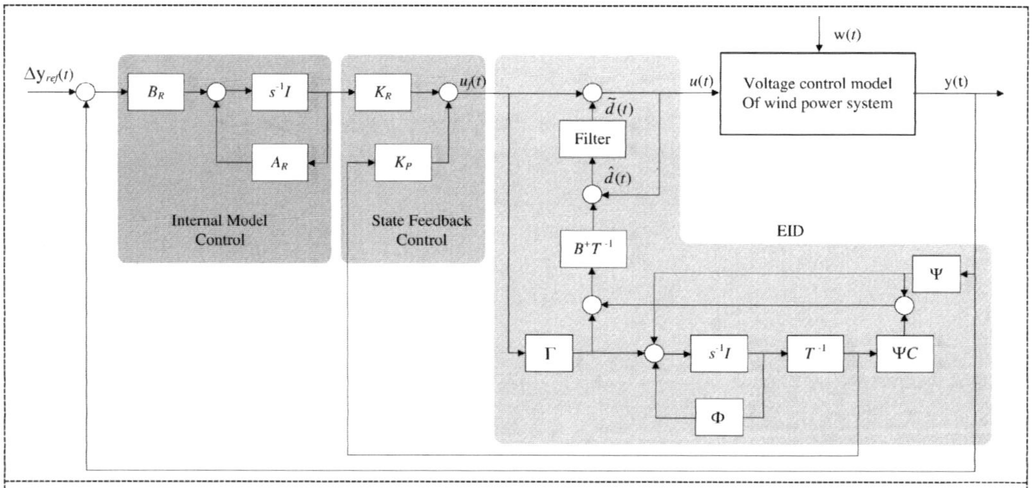

Figure 2. Voltage control structure of wind power system.

3.2. Robust Controller Design

To guarantee the stability of the whole close-loop control power system, first we consider the system without exogenous signals since exogenous signals do not influence the system stability. So letting $\Delta f_{ref}(t)=0$, $w(t)=0$, the power system model can be expressed as

$$\begin{cases} \dot{x}(t) = Ax(t) + Bu(t) \\ y(t) = Cx(t) \end{cases} \tag{6}$$

Thus

$$\Delta\dot{x}(t) = T^{-1}\Phi T \Delta x(t) + B\tilde{w}(t) \tag{7}$$

and

$$\hat{w}(t) = B^+(T^{-1}\Phi T - A)\Delta x(t) + \tilde{w}(t) \tag{8}$$

IOP Conf. Series: Materials Science and Engineering **366** (2018) 012036 doi:10.1088/1757-899X/366/1/012036

Where $\hat{w}(t)$ is an estimation of the actual load variation, $\tilde{w}(t)$ is the filtered disturbance estimation and $\Delta x(t) = \hat{x}(t) - x(t)$.

From (7) and (8), the transfer function from $\tilde{w}(t)$ to $\hat{w}(t)$ can be expressed as

$$\begin{aligned} G_w(s) &= B^+(T^{-1}\Phi T - A)(sI - T^{-1}\Phi T)^{-1}B + 1 \\ &= B^+(sI - A)T^{-1}(sI - \Phi)^{-1}TB \end{aligned} \tag{9}$$

According to the small gain theory, when the H_∞ norm of the closed-loop systems satisfies

$$\|G_w(s)F(s)\| \le 1 \tag{10}$$

Then, the closed-loop system is stable. Here, F(s) is a low-pass filter.

To design the controller, we construct an augmented system that contains the power system and IM, that is

$$\dot{\bar{x}}(t) = \bar{A}\bar{x}(t) + \bar{B}u(t) + \bar{B}_R \Delta f_{ref}(t) \tag{11}$$

Where $\bar{x}(t) = \begin{bmatrix} x(t) \\ x_R(t) \end{bmatrix}$, $\bar{A} = \begin{bmatrix} A & 0 \\ -B_R C & A_R \end{bmatrix}$, $\bar{B} = \begin{bmatrix} B \\ 0 \end{bmatrix}$, $\bar{B}_R = \begin{bmatrix} 0 \\ B_R \end{bmatrix}$.

We set $\Delta y_{ref}=0$, by employing the liner-quadratic-regular method [10] to determine the state-feedback gain.

$$J_K = \int_0^\infty [\bar{x}^T(t)Q_K x(t) + R_K u(t)]dt \tag{12}$$

Where Q_k and R_k are weighting matrices.

The optimal state-feedback control gains are determined by minimizing J_k.

$$[K_P \quad K_R] = -R_K^{-1}\bar{B}^T P \tag{13}$$

Where the matrix P is a solution of the following Riccati equation:

$$P\begin{bmatrix} A & 0 \\ -B_R C & A_R \end{bmatrix} + \begin{bmatrix} A & 0 \\ -B_R C & A_R \end{bmatrix}^T P + Q_K - P\begin{bmatrix} B \\ 0 \end{bmatrix} R_K^{-1} \begin{bmatrix} B \\ 0 \end{bmatrix}^T P = 0 \tag{14}$$

4. Case study

The power system with doubly fed wind farm is selected as the object of study. The parameters of the controlled object are expressed as follows in Table 1:

Table 1. Parameters of the power system with wind farms

K_1 (pu)	K_2 (pu)	K_3 (pu)	K_u (pu)	K_q (pu)	K_Q (pu)	T_u (s)	T_Q (s)	T_q (s)	τ_1 (s)	τ_2 (s)	τ_3 (s)
1.0	12	1.0	1.0	1.8	2.0	0.2	0.16	0.3	0.001	0	0.05

Figure 3 shows the dynamic response of a power system with wind farm when the system is subjected to the external reactive disturbance Q_d. It can be seen that with the load variation shown in Figure 3(a), the system with traditional controller has serious voltage variations, and the voltage deviation Δu_g are varied at [-0.25 0.25], as shown in Figure 3(b). However, when using the proposed voltage controller with EID compensation, as shown in Figure 3(b), the voltage deviations are varied at [-0.15 0.15]. The proposed voltage control method with EID compensation can better reduce the voltage fluctuation range of the system, and make the voltage fluctuation of the system more smooth. Compared with the traditional PI controller, it has obvious advantages.

IOP Conf. Series: Materials Science and Engineering **366** (2018) 012036 doi:10.1088/1757-899X/366/1/012036

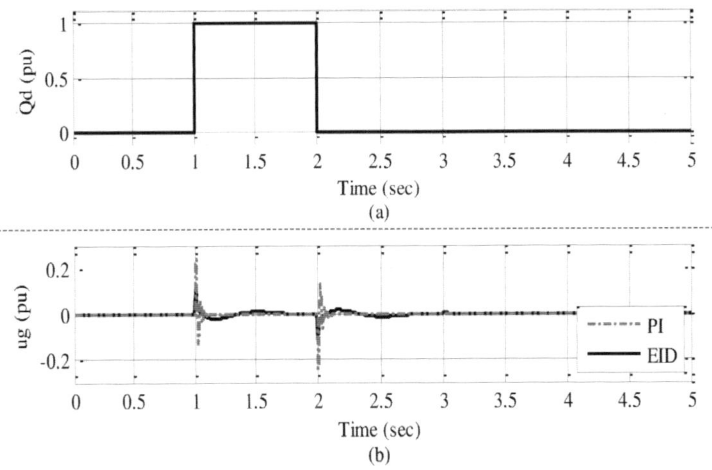

Figure 3. Dynamic response of voltage control system of wind farm with load disturbance, (a) Load disturbance; (b) Voltage fluctuation under different control strategies.

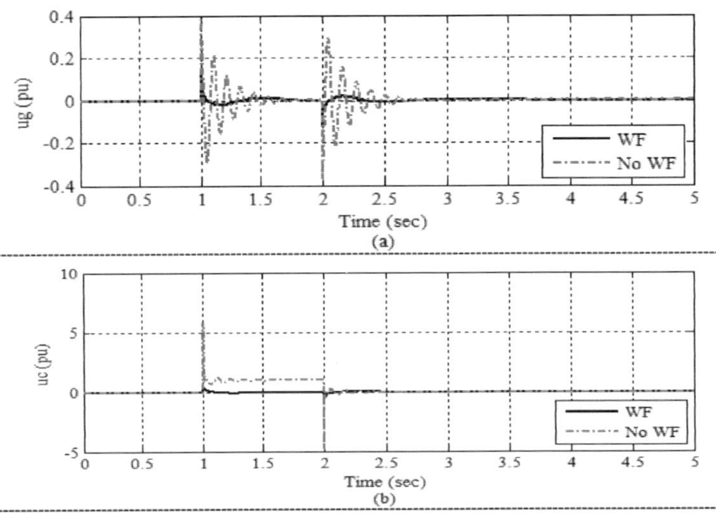

Figure 4. Dynamic response of voltage control system under different participation degree of wind farm, (a) Voltage fluctuation; (b) Control inputs for active interference suppression strategies.

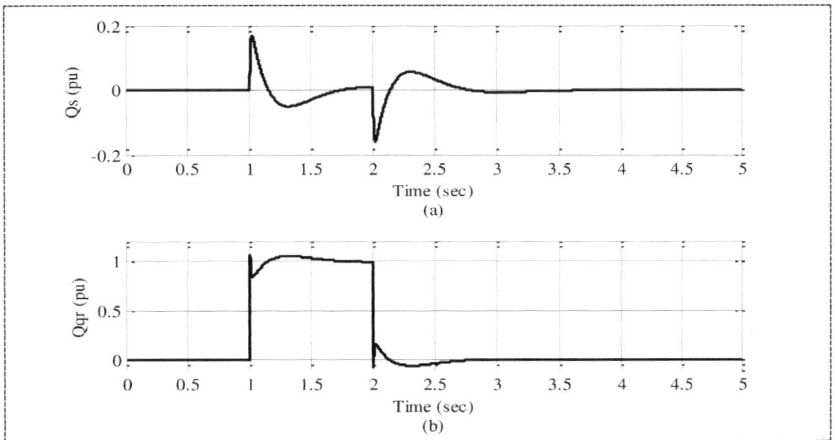

Figure 5. Reactive power response of voltage control system with wind farm under load disturbance, (a) Reactive power compensation device; (b) Double-fed induction generator.

In order to verify that the wind farm has a positive effect on the system voltage stability, Figure 4 uses "NO, WF" and "WF" to indicate that the wind farm is not involved in voltage control and wind farm participation in conventional power system voltage control. When the wind farm actively participates in the system voltage control, the reactive power output of the reactive power compensation unit and the DFIG wind turbine is shown in Figure 5(a) and (b). The simulation results show that double-fed induction generator bear most of the reactive power compensation needed for voltage regulation, and reactive power compensation devices serve as supplementary means to further smooth the system voltage fluctuation.

5. Conclusions

In this paper, a controller method with EID compensation is proposed for voltage regulation of DFIG wind farm grid-connected system. The voltage regulation model of wind power system based on DFIG is established. The controller structure and basic design principle are introduced. Since the internal/external disturbances are transformed into EID and one compensator is designed for EID, the proposed method has strong robustness on load variations and wind power variation. Then the paper examines the capability of the proposed control approaches for improvement of voltage regulation under load disturbance or wind speed disturbance. Finally, the implementation of coordinated voltage control strategy for wind power grid-connected system is presented. Simulation result confirm the better dynamic performance of the proposed control method than the traditional one and show the feasibility and effectiveness of the proposed voltage coordinated control method based on EID compensation in voltage regulation.

Acknowledgments

This work was supported in part by the National Natural Science Foundation of China under Grant 61673398, in part by the Key S&T Special Project of Hunan Province of China under Grants 2017RS3006 and in part by the Innovation Innovation-Driven Project of Central South University under Grants 2016CX013.

References

[1] Abdoune F, Aouzellag D and Ghedamsi K 2016 Terminal voltage build-up and control of a DFIG based stand-alone wind energy conversion system *Renewable Energy*. **97** 468-480

[2] Weng Y T and Hsu Y Y 2016 Reactive power control strategy for a wind farm with DFIG *Renewable Energy*. **94** 383-390

[3] Fengda X, Qinglai G and Hongbin S 2015 Automatic Voltage Control of Wind Farms Based on Model Predictive Control Theory *Automation of Electric Power Systems.* **7** 59-67

[4] Moursi M E, Joos G and Abbey C 2007 High-performance voltage control scheme for windpark integration *IET Renewable Power Generation.* **1** 151-159

[5] Mohammed E M, Geza J and Chad A 2008 A secondary voltage control strategy for transmission level interconnection of wind generation *IEEE Transactions on Power Electronics.* **23** 1178-1190

[6] Hee-Sang K, Gi-Gab Y and Won-Pyo H 2007 Active use of DFIG-Based Variable-Speed Wind-Turbine for Voltage Regulation at a Remote Location *IEEE Transaction on Power System.* **22** 1916-1925

[7] Kayikci M and Milanovic JV 2007 Reactive power control strategies for DFIG-based plant *IEEE Transactions on Energy Conversion* **22** 389-396

[8] Yongqiang L, Xueguang Z and Dianguo X 2007 Reactive Power Analysis and Control of Doubly Fed Induction Generator Wind Farm *Proceeding of the CSEE* **27** 77-82

[9] Jinhua S, Mingxing F, Yasuhiro O, Hiroshi H and Min Wu 2007 Improving disturbance-rejection performance based on an equivalent-input-disturbance approach *IEEE Transactions on Industrial Electronics* **55** 380-389

[10] Anderson B D O and Moore J B 1990 *Optimal control: linear quadratic method* (NJ: Prentice-Hall)

A Single-phase Grounding Fault Judgment Method Based on Steady-state Current Characteristics in Distribution

Xu Yuhang [1], Li Tongning[1], Lin Yibin,[1] Lu Chaoteng[1], Lu Huiji[1] and Zhang Wei[2]

[1] State Grid Xiamen Electric Power Supply Company, 361000 Xiamen, Fujian, P. R. China
[2] Jicheng Electronics Co., Ltd, 250100, Jinan, Shandong, P. R. China

E-mail:zhangweiieslab@163.com

Abstract. In order to solve the problem of single-phase grounding fault judgment in non-solid-earthed distribution network, the power flow in non-solid-earthed distribution is analyzed. The steady-state current value in the upstream fault phase will be abruptly changed, and the current in the downstream fault phase will keep the load current constant. The steady-state current values in other non-faulty phases and non-faulty lines do not change abruptly. A single-phase grounding fault judgment method based on steady-state current characteristics is proposed. The power supply path real-time current matrix, the power supply path history current matrix, the power supply path mutation current matrix and the power supply path mutation current logical matrix are established. The problem of the single-phase grounding fault judgment is transformed into the matrix operation problem. The detailed calculation formulas are given. The simulation results show that the method proposed in this paper has a good adaptability to the permanent grounding fault and it is worth popularizing.

1. Introduction

FA (Feeder Automation), as an important means to improve the reliability of power supply, has been widely deployed in the current DMS (Distribution Management System), and it plays an important role in distribution fault location. However, at present, the feeder automation system can only deal with the short-circuit fault location, but there is no better way to deal with the single-phase grounding fault judgment in non-solid-earthed distribution network [1-3].

The neutral grounding mode of 10kV distribution network in China mainly includes two kinds of neutral unearthed and neutral earthed via Peterson coil. When the single-phase grounding fault occurs, these grounding modes can limit the short-circuit current to a smaller value. But because of the step voltage, it often leads to human and livestock casualties. If the neutral grounding mode is changed into neutral earthed through small resistance, when the single-phase grounding fault occurs, the substation outlet circuit breaker will trip. Although this method can isolate the fault, the reliability of power supply is seriously reduced due to frequent tripping of the outlet switch. Therefore, the non-solid-earthed operation mode is desirable in most areas of our country, but it also needs to be removed as soon as possible [4-6].

Passive method and active method are two commonly used methods for single-phase grounding fault location in non-solid-earthed distribution [7-8]. However, the installation of special terminal equipment is required for both the active and passive methods to meet the special sampling requirements. The current distribution network has installed a large number of sampling terminal equipment, such as FTU(Feeder Terminal Unit), DTU(Distribution Terminal Unit), fault indicator, etc., most of these devices do not have a special sampling and determination requirements. If a special sampling decision device is installed, most of the equipment needs to be installed under the condition of power failure, and the reliability of power supply is further reduced. Therefore, the best possible use of existing equipment and sampling data to solve the problem of single-phase grounding fault judgment, try not to install special equipment, to avoid duplication of waste of resources, these practices can be better to build a conservation-oriented society.

Based on the characteristics of single-phase grounding fault location, this paper analyzes the power flow in non-solid-earthed distribution, and presents a single-phase grounding fault judgment method based on steady-state current characteristics.

2. Steady-state characteristic analysis of single-phase grounding fault

In the non-solid-earthed distribution, whether the grounding transformer is grounded via Peterson coil or not, its influence on distribution network is the same, which is to reduce the fault short-circuit current. Figure 1 is the distribution of capacitance current and load current in the non-solid-earthed distribution network. Wherein, the G1 is the equivalent power supply of the system, #1~#4 is the bus, and the ABC are the three phases of bus #3.TRA1 is Y/Y/D type three winding transformer, and rated voltage is 110kV/35kV/10kV. TRA2 is Y/Y type two winding transformer, rated voltage is 10kV/400V, and it can be used as grounding transformer. Feeder1 and Feeder2 are two feeders. Load1 and Load2 are equivalent loads. FP is the fault point.

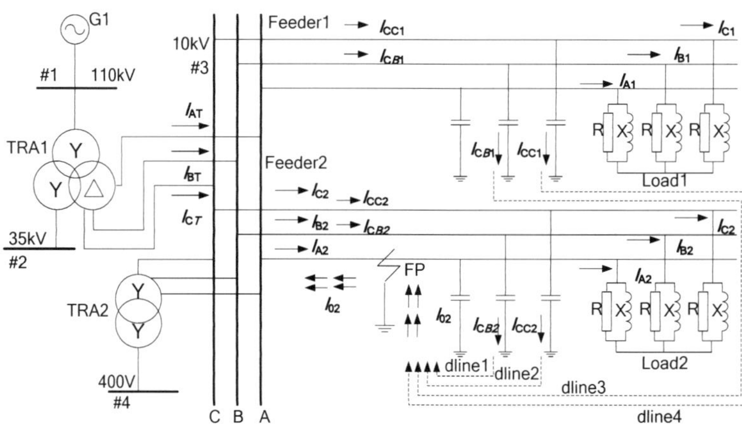

Figure 1. The non-solid-earthed distribution power flow.

When the A phase of the feeder Feeder2 is single-phase grounding fault at the FP point, the B phase and C phase capacitive current phasors $\boldsymbol{I}_{CB1}$, $\boldsymbol{I}_{CC1}$ and $\boldsymbol{I}_{CB2}$, $\boldsymbol{I}_{CC2}$ in the non-fault line Feeder1 and the fault line Feeder2 flow from the bus to the feeder and flow to the fault point. As shown in figure dline1~dline4. TRA1 transformer low side because it is triangular wiring, so the winding between the ground capacitance is small, negligible, and the capacitor current between the windings have been incorporated into the load current, as shown phasors $\boldsymbol{I}_{AT}$, $\boldsymbol{I}_{BT}$, $\boldsymbol{I}_{CT}$ in Figure 1. TRA2 ground transformer due to neutral point is not grounded, so the capacitance current is also negligible. The total capacitance current flowing through the fault point is the zero sequence current $\boldsymbol{I}_{02}$, which is the sum of the four capacitor currents whose direction flows from the fault point to the bus.

In the feeder Feeder1 and Feeder2, the load current of each phase is denoted by $\boldsymbol{I}_{A1}$, $\boldsymbol{I}_{B1}$, $\boldsymbol{I}_{C1}$, $\boldsymbol{I}_{A2}$, $\boldsymbol{I}_{B2}$, $\boldsymbol{I}_{C2}$ respectively, as shown in the figure. When the fault occurs, due to symmetry, the load current of each phase does not change.

As can be seen from Figure 1, the current $\boldsymbol{I}_{AL2}$ flowing through the fault phase A in the upstream region of the fault point FP is a superposition of the load current $\boldsymbol{I}_{A2}$ and the zero sequence current $\boldsymbol{I}_{02}$. And the current flowing through the downstream of the fault point is only the load current $\boldsymbol{I}_{A2}$. The current flowing through the BC phase of the fault line is the same as before and after the fault, and the

IOP Conf. Series: Materials Science and Engineering **366** (2018) 012037 doi:10.1088/1757-899X/366/1/012037

value is the load current. The current of ABC three-phase flow in non fault line Feeder1 has no change before and after the fault, and its value is the load current.

For cable lines or cable and overhead hybrid lines, the line capacitance current will be much larger than the overhead line capacitance current, and the substation 10kV feeder is generally more than a dozen, in addition feeder branch more effective length. Therefore, in the city or suburban distribution network with cable line and cable overhead hybrid line, the ground current flowing through the fault point will not be neglected relative to the load current when the single-phase grounding fault occurs.

3. A single-phase grounding fault judgment method based on steady state current characteristics

3.1. Establishment of fault judgment model

According to the above analysis, when the single-phase grounding fault occurs, only the steady-state current of the fault phase load superimposed on the fault feeder upstream to the substation switch will change, while the steady-state current of the other paths will not change. Therefore, the problem of fault judgment is to find the power supply path from the substation switch to the fault point.

Figure 2 is the distribution network typical feeder; # 1 is the substation bus; S1 is substation outlet switch; A~K are sectional switches which is installed with FTU, DTU, fault indicator and other acquisition devices. DT1 ~ DT5 are distribution transformers.

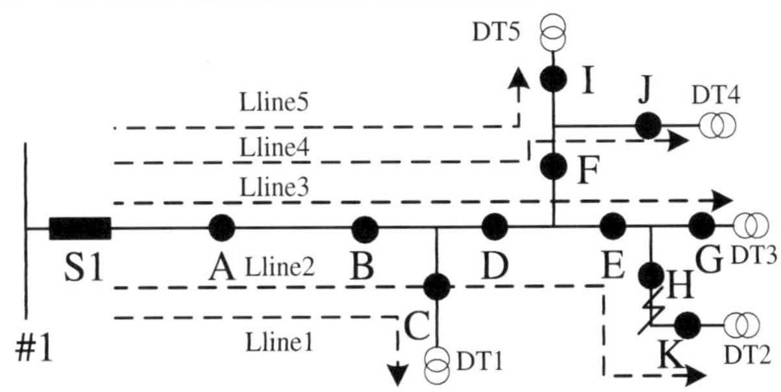

Figure 2. The distribution network typical feeder.

As can be seen from Figure 2, there are 5 power supply paths from substation outlet switch S1 to terminal distribution transformer DT1~DT5, which are respectively Lline1~Lline5, as shown by the dashed line in Figure 2. The single-phase grounding fault will also take place on one of the 5 power supply paths. When a single-phase grounding fault occurs, the steady-state current value of one of the five power supply paths will change, and the changed path is the fault path.

Starting from the distribution transformer, it is traced back in the opposite direction of the flow direction until the substation outlet switch, which is defined as a set of power supply path switches. Each of the power supply path switches is generated a power supply path sequence table according to the order of the current flows. The power supply path sequence table of the power supply path switch in Figure 2 is shown in Table 1.

Table 1. The power supply path sequence table

Path	Path1	Path2	Path3	Path4	Path5
Switch	S1,A,B,C	S1,A,B,D,E,H,K	S1,A,B,D,E,G	S1,A,B,D,F,J	S1,A,B,D,F,I

The power supply path real-time current matrix is established to describe the real-time operation state of a distribution feeder A, B and C three-phase. The A phase power supply path real-time current matrix RCM_A is described as follows:

$$RCM_A = \begin{bmatrix} ra_{12} & ra_{22} & \cdots & ra_{1j} & \cdots & ra_{1m} \\ ra_{21} & ra_{22} & \cdots & ra_{2j} & \cdots & ra_{2m} \\ \vdots & \vdots & \vdots & \vdots & \vdots & \vdots \\ ra_{i1} & ra_{i2} & \cdots & ra_{ij} & \cdots & ra_{im} \\ \vdots & \vdots & \vdots & \vdots & \vdots & \vdots \\ ra_{n1} & ra_{n2} & \cdots & ra_{nj} & \cdots & ra_{nm} \end{bmatrix} \quad (1)$$

Wherein, i=1,2, n, n is the number of feeder power supply paths , that is, the number of distribution transformers; j=1,2, ..., m, m is the number of switches included in the maximum power supply path; ra_{ij} is the real-time current value of the j switch in the i power supply path. If the switch does not exist, it is represented as -1. RCM_B and RCM_C are used to describe the real-time current matrix of B and C phase power supply path. The specific method is similar to RCM_A, and no longer is described.

The power supply path history current matrix is established to describe the history operation state of a distribution feeder A, B and C three-phase. The A phase power supply path history current matrix HCM_A is described as follows:

$$HCM_A = \begin{bmatrix} ha_{12} & ha_{22} & \cdots & ha_{1j} & \cdots & ha_{1m} \\ ha_{21} & ha_{22} & \cdots & ha_{2j} & \cdots & ha_{2m} \\ \vdots & \vdots & \vdots & \vdots & \vdots & \vdots \\ ha_{i1} & ha_{i2} & \cdots & ha_{ij} & \cdots & ha_{im} \\ \vdots & \vdots & \vdots & \vdots & \vdots & \vdots \\ ha_{n1} & ha_{n2} & \cdots & ha_{nj} & \cdots & ha_{nm} \end{bmatrix} \quad (2)$$

Wherein, ha_{ij} is the certain historical moment current value of the j switch in the i power supply path. If the switch does not exist, it is represented as -1. The historical moment can be set according to the terminal type. If the sampling devices are high-precision real-time devices with short sampling interval such as FTU and DTU, the historical moment may be set to 2 to 5 minutes before the current moment. If the sampling devices are sampling devices with a longer sampling interval such as fault indicator, the historical moment may be set as 5 to 10 minutes before the current moment. If the sampling devices are mixed configurations of FTU, DTU and fault indicator, the historical moment can be set as 5 to 10 minutes before the current moment, so as to ensure that all data are sent to DMS. HCM_B and HCM_C are used to describe the history current matrix of B and C phase power supply path. The specific method is similar to HCM_A, and no longer is described.

3.2. A single-phase grounding fault judgment method

When the grounding fault occurs, the substation bus will be generated zero sequence voltage. In this case, the steady-state current mutation caused by the fault can be calculated by using the power supply path real-time current matrix after the fault and the power supply path history current matrix before the fault, and the detailed description is as follows:

$$ACM_A = RCM_A - HCM_A \quad (3)$$

$$ACLM_A = \begin{bmatrix} la_{12} & la_{22} & \cdots & la_{1j} & \cdots & la_{1m} \\ la_{21} & la_{22} & \cdots & la_{2j} & \cdots & la_{2m} \\ \vdots & \vdots & \vdots & \vdots & \vdots & \vdots \\ la_{i1} & la_{i2} & \cdots & la_{ij} & \cdots & la_{im} \\ \vdots & \vdots & \vdots & \vdots & \vdots & \vdots \\ la_{n1} & la_{n2} & \cdots & la_{nj} & \cdots & la_{nm} \end{bmatrix} \qquad la_{ij} = \begin{cases} 1 & aa_{ij} \geq \alpha Ma \\ 0 & aa_{ij} < \alpha Ma \end{cases} \quad (4)$$

Wherein, ACM_A is A-phase power supply path mutation current matrix, and it is also n row m column matrix. For the convenience of calculation, the mutation of the A phase current is expressed by the power supply path mutation current logic matrix $ACLM_A$. la_{ij} is the mutation current logic value of the j switch in the i power supply path. aa_{ij} is the i row and j column element of the A-phase power supply path mutation current matrix ACM_A. Ma is the capacitance current threshold, which can be determined according to the maximum steady-state current change of all the substations after a grounding fault within one year of the substation. α is the threshold coefficient, the value range is (0,1).

If the grounding fault occurs in phase A, the power supply path mutation current logic matrix $ACLM_A$ is a matrix of 0 and 1. The path with the most mutation current switch in all the power supply paths is the fault path, that is, the power supply path represented by the row with the most logic value 1 in all the rows in the matrix is the fault path. The detailed description is as follows:

$$RF_A = Max\left[sla_1, sla_2, \cdots, sla_i, sla_n\right] \qquad sla_i = \sum_{j=1}^{m} la_{ij} \tag{5}$$

Where, sla_i is the sum of all the elements in line i. RF_A is the judgment result of A-phase grounding fault, if its value is sla_k and not 0, it means that the ground fault point is after the sla_k switch of A-phase k (k = 1,2, ..., n) power supply path. In summary, the specific steps to determine the grounding fault are as follows:

Step 1: According to substation bus voltage information to determine whether a grounding fault occurs. If the grounding fault occurs then go to step 2, if the ground fault does not occur then terminate;

Step 2: Select any substation feeder, and generate A-phase, B-phase and C-phase power supply path mutation current logic matrix according to the formula (1) ~ (4).

Step 3: The results of the grounding fault judgment of the three phases are calculated respectively according to the formula (5). If judgment result value is sla_k and not 0, it means that the ground fault point is after the sla_k switch of A-phase k power supply path. If the judgment result value is 0, it indicates that there is no grounding fault on the feeder.

Step 4: Repeat step 2 to 3 until all substations have been fully analyzed, and then all the grounding fault feeders and grounding fault points can be determined.

4. Case analysis

As shown in Figure 2, it is assumed that a permanent fault of A-phase occurs between the switches H and K. Through the PSCAD simulation system, the three-phase simulation current waveform of substation outlet switch S1 is shown in Figure 3. The horizontal axis is time, the unit is s; the longitudinal axis is current, and the unit is A.

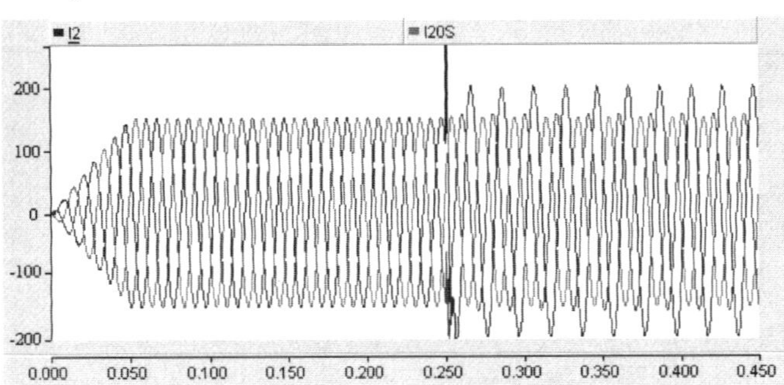

Figure 3. The three-phase current simulation waveform

The real-time measured data RD of each switch after the fault occurred and the historical measured data HD of 5 minutes before the fault occurred are shown in Table 2.

Table 2. The switch measurement value before and after fault (A)

	S1	A	B	C	D	E	F	G	H	I	J	K
RD	101.6	101.6	101.6	23.3	78.3	57.3	21.0	22.3	35.0	7.9	13.1	27.8
HD	94.4	94.4	94.4	24.1	70.3	49.9	20.6	22.1	27.8	7.9	12.7	27.8

According to the formula (1)~(4), *Ma* is 8, α is 0.5, then the A-phase power supply path mutation current logic matrix $ACLM_A$ is:

$$ACM_A = \begin{bmatrix} 7.2 & 7.2 & 7.2 & -0.8 & 0 & 0 & 0 \\ 7.2 & 7.2 & 7.2 & 8 & 7.4 & 7.2 & 0 \\ 7.2 & 7.2 & 7.2 & 8 & 7.4 & 0.2 & 0 \\ 7.2 & 7.2 & 7.2 & 8 & 0.4 & 0.4 & 0 \\ 7.2 & 7.2 & 7.2 & 8 & 0.4 & 0 & 0 \end{bmatrix} \quad ACLM_A = \begin{bmatrix} 1 & 1 & 1 & 0 & 0 & 0 & 0 \\ 1 & 1 & 1 & 1 & 1 & 1 & 0 \\ 1 & 1 & 1 & 1 & 1 & 0 & 0 \\ 1 & 1 & 1 & 1 & 0 & 0 & 0 \\ 1 & 1 & 1 & 1 & 0 & 0 & 0 \end{bmatrix} \quad RF_A = Max[3,6,5,4,4]=6$$

The analysis process of phase B and phase C is similar and will not be repeated here. Therefore, the grounding fault occurs after the 6th switching device of the A-phase 2nd power supply path. According to Table 1, the fault occurs between the H switch and the K switch.

5. Conclusion

According to the above analysis, it can be seen that in the non-solid-earthed distribution network, when the single-phase grounding fault occurs, the current in the upstream fault phase is abruptly changed due to the superposition of the normal load current and the capacitance current, and the current in the downstream fault phase will keep the load current constant. For other non-fault phase and non-fault feeder current in the upstream or downstream of the fault point, the steady-state current values are load current without mutation.

A single-phase grounding fault judgment method based on steady-state current characteristics is proposed. The power supply path real-time current matrix, the power supply path history current matrix, the power supply path mutation current matrix and the power supply path mutation current logical matrix are established. The problem of the single-phase grounding fault judgment is transformed into the matrix operation problem. The simulation results show that the method proposed in this paper has a good adaptability to the permanent grounding fault and it is worth popularizing.

References

[1] Senger E.C. Manassero G. Jr.Goldemberg 2005 *Automated fault location system for primary distribution networks*(American: IEEE Transactions on Power Delivery) chapter 2.

[2] Cliff J Williams 2009 *A 15 Year Roadmap to Advanced Distribution Automation* (International Seminar On New Technology And Applications Of Distribution Automation).

[3] Ni Guang-kui Yang Yi-han Bao Hai 2009 *Design of New Fault Indicator of Earthling Fault* (China: Modern Electric Power) chapter 6 pp 35-38.

[4] CHEN Bobo,QU Weifeng, YANG Hongyu. 2016 *Research on single phase grounding arc model and line selection for neutral ineffectively grounding system* (China:Power System Protection and Control), chapter 16 pp 1-7.

[5] LAI Ping, ZHOU Xiangling, QIU Dan.2015 *Research on transient-current frequency analysis and faulty line detecting method in indirectly grounding power system*(China: Power System Protection and Control), chapter 4 pp 51-57.

[6] ZHANG Wei 2013 *An intelligent distributed feeder automation fault judgment* (China: Power System Protection and Control) chapter 5 pp 108-113.

[7] YanFeng,YangQixun,QiZheng. 2004 *Study on fault location for distribution network based on travelling wave theory*(China: Proceedings of the CSEE), chapter 9 pp 38-43.

[8] XUE Yongduan,LI Juan,XU Bingyin. 2015 *Transient Equivalent Circuit and Transient Analysis of Single-phase Earth Fault in Arc Suppression Coil Grounded System*(China: Proceedings of the CSEE), chapter 22 pp 5703-5714.

IOP Conf. Series: Materials Science and Engineering **366** (2018) 012038 doi:10.1088/1757-899X/366/1/012038

Research on Purchasing Strategy of 110 kV Cable Accessories based on Life Cycle Cost Management

Yue Li[1,2], Bengang Wei[1,2] and Honglei Li[2]

[1] Shanghai SEPRI Power Technology Co., LTD, 200437 Shanghai, China
[2] State Grid Shanghai Electric Power Research Institute, 200437 Shanghai, China

ivymax5261@qq.com; wbgsj@126.com; lihonglei@sh.sgcc.com.cn

Abstract. Cable accessories, as an important part of cable system, we often only focus on its initial acquisition costs and ignore the late costs in its life cycle. In order to maximize the economic benefits and to play the use of cable accessories, we need to introduce an equipment life cycle cost management (LCC) for 110kV cable accessories. In this paper, a life cycle cost management of 110kV cable accessories is studied, and an evaluation model of cable accessories purchase based on LCC is established. The LCC theory is applied to the bidding and purchasing of cable accessories, which can make the purchased and managed reasonably, and save the social resources effectively.

1. Introduction

Life Cycle Cost (LCC) manages the asset management in the whole life cycle of planning, project approval, design, operation and retirement to find the best solution to achieve the overall optimization throughout the various stages. Its concept originated in the Swedish railway system in 1904 [1]. In 1996, the International Electrotechnical Commission (IEC) issued the international standard "credibility management" (IEC60300-3-3) [2], to further promote the application of LCC. In 2004, the International Conference on Large Power Grid (CIGRE) proposed the use of LCC equipment management, to encourage manufacturers to provide products LCC report.

At present, life cycle cost management of electric power equipment is mainly applied to the main transformer, circuit breaker and other equipment in the world. The literature [3] introduced a life cycle cost method covering a complete 110 kV overhead transmission grid with air insulated substations which can be used effectively for comparing different network topologies and equipment types. In [4], a life cycle cost method was used for analysing the various operational scenarios of maintenance power transmission lines. A principle calculation on the power transformer was showed in [5] which can determine the economic life cycle from economic aspect of asset management.

Chinese electric power enterprises explored its own LCC method with the characteristics of Chines electric power. The literature [6] showed a more comprehensive transformer life cycle cost model with environmental cost and preventive test cost. In [7], a new power transformers maintenance decision making model was showed to solve the large financial losses to electric enterprises based on life cycle cost analysis. The literature [8] introduced the LCC method and application of the 2010 Shanghai Expo 500kV underground substation GIS procurement, which became the first successful use of LCC method of bidding and purchasing.

The LCC management technology of main transformers and circuit breakers is becoming more and more mature in China. However, cable accessories, as an important part of cable system assets, lack the applicable LCC management method to control their economy and reliability. In order to fill the vacancy of 110kV cable accessories in LCC management, this paper presents a modified LCC management model for cable accessories, which is applied to the bidding and purchasing of cable accessories, and the application case is used.

2. Research on Evaluation Model of LCC Purchasing for 110kV Cable Accessories

At present, the LCC model of power equipment widely used IEC60300-3-3 application guide. Cable accessories as an important part of the cable line is the weakest link of whole cable line insulation, coupled with its complex construction process and special underground laying layout, we should amend the evaluation model according to the actual operation of the situation in LCC evaluation. The corresponding amendments can be expressed as:

$$LCC = CI + COM + CF + CD \tag{2-1}$$

LCC : Life Cycle Cost
CI : Investment Costs
COM : Operation and Maintenance Costs
CF : Failure Costs
CD : Disposal Costs

2.1. The Cost of Inputs (CI)

Electricity enterprises can learn the composition of quotation from equipment manufacturers by CI which including cable accessories price and installation costs. Cable accessories price includes the purchase price of equipment, factory test inspection fees, special tools and first time spare parts price, after-sales service fees, supplier-shipping costs, etc. The input cost CI can be expressed as:

$$CI = C_{sj} + C_{az} \tag{2-2}$$

In the formula, C_{sj} indicates the price of the cable accessories, C_{az} indicates the cable attachment installation fee.

Cable accessories are usually packaged to purchase. The calculation of different equipment classification can help power enterprises to understand the structure of input cost longitudinally. Therefore, C_{sj} also expressed as the price of cable terminal fittings, insulated joints, coaxial cables, direct grounding boxes and other complete sets of cable accessories. It should be noted that the values of C_{sj} calculated in two ways should be equal.

The above methods are used to decompose the price from the two dimensions of the cost nature and equipment classification, which can intuitively show the composition and proportion of price, and the quotation of "foaming price" is eliminated to the maximum extent. Purpose.

2.2. Cost of Maintenance (COM)

Remote manipulation such as on-line monitoring is usually used for daily inspections of cable accessories, so the daily inspection costs are negligible. At the same time, cable connector and terminal in the event of a fault or defect, usually directly replace the replacement. Therefore, the cable accessories operating and maintenance costs can be simplified according as COM, which can be expressed as:

$$COM = C_{jb} + C_{ew} \tag{2-3}$$

In the formula, C_{jb} indicates the basic maintenance costs, and C_{ew} indicates additional maintenance costs. Let A be the best life cycle of the 110kV cable accessory, $C_{jb}(t)$ for the basic operation maintenance cost of the year ***t***, and ***i*** for the discount rate, then the basic maintenance costs converted to the initial purchase can be calculated using the formula (2-4).

$$C_{jb} = \sum_{t=1}^{A} \frac{C_{jb}(t)}{(1+i)^t} \tag{2-4}$$

Similarly, conversion to the initial purchase of the additional maintenance costs can be used to calculate (2-5).

$$C_{ew} = \sum_{t=1}^{A} \frac{C_{ew}(t)}{(1+i)^t} \tag{2-5}$$

2.3. The Cost of Failure (CF)

The failure costs for 110kV cable accessories include failure and maintenance losses, which can be expressed as:

$$CF = C_{gj} + C_{gs} \tag{2-6}$$

In the formula, C_{gj} indicates the troubleshooting fee, C_{gs} indicates the cost of the failure. C_{gj} can be expressed as:

$$C_{gj} = \alpha \cdot C_{ar} \tag{2-7}$$

In the formula, α indicates the equipment failure rate guaranteed by the equipment manufacturer C_{ar} indicates the average repair cost of 110kV cable accessories failure, which is provided by the power company.

C_{gs} can be expressed as:

$$C_{gs} = \alpha \cdot C_{at} \cdot P_{rt} \cdot MTTR \tag{2-8}$$

In this formula, C_{at} indicates 110kV average transmission price, P_{rt} indicates rated conveying capacity, $MTTR$ indicates the repair recovery time, provided by the power companies.

2.4. The Cost of Disuse (CD)

As the scrap recovery costs of the cable accessories are too small to be ignored, so the cost of disuse can be expressed as:

$$CD = C_{cc} \tag{2-9}$$

In the formula, C_{cc} indicates the removal costs of cable accessories.

2.5. Life Cycle Cost Evaluation Model of 110kV Cable Accessory

According to the above analysis, the LCC calculation model of the 110kV cable accessories can be sorted into:

$$LCC = C_{sj} + C_{az} + \sum_{t=1}^{A} \frac{C_{jb}(t)}{(1+i)^t} + \sum_{t=1}^{A} \frac{C_{ew}(t)}{(1+i)^t} + \alpha \cdot C_{ar} + \alpha \cdot C_{at} \cdot P_{rt} \cdot MTTR + C_{cc} \tag{2-10}$$

The LCC evaluation model Breakdown Structure can be shown in Figure 2-1.

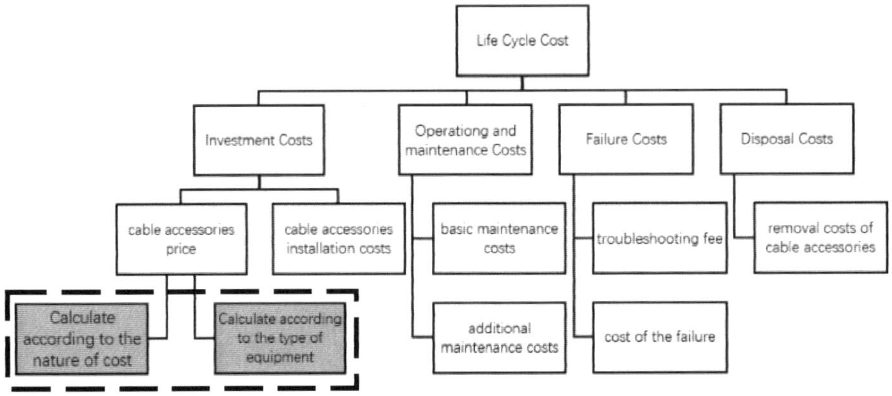

Figure2-1. The Life Cycle Cost Breakdown Structure of 110kV Cable Accessories.

This model can be used not only for the bidding and purchasing, but also for providing the theoretical basis to optimize the life-cycle cost for the 110kV cable accessories.

3. Research on Bidding and Purchasing Strategy of LCC Evaluation Model Based on 110kV Cable Accessories

After finished the bidding evaluation of technical and business conditions, electric power company can choose a number of top equipment manufacturers to conduct the LCC evaluation. Firstly, before the equipment manufacturers provide the LCC calculation data, the electric power company needs to convene the equipment manufacturers to convene the preface, detailing the filling way of calculation data and the needing attention. After that, the equipment manufacturers shall complete the LCC evaluation report according to the actual situation and then perform the LCC verification by the electric power company to ensure that the LCC calculation data provided by the equipment manufacturer is real. Finally, based on the preliminary verification and clarification, we can get the final LCC calculation results and evaluation by summarizing the cost of power enterprises and equipment manufacturers, which are the main foundations to evaluate the bidding.

3.1. LCC Evaluation Report

The LCC evaluation report written by equipment manufacturers should focus on the measures, which are used to reduce the failure rate, repair and maintenance costs in design, manufacture, material selection and testing. In this report, electric power company can learn the experiences and effect in reducing the LCC costs from manufacturers.

3.2. LCC Verification Method

The LCC calculation data and LCC report should be verified after picked up. According to the "grasp the big release" principle, the verification of LCC calculation data mainly consider the following 4 points:

- Verify the initial cost of input: according to the initial cost of input provided by the equipment manufacturer, we can use contrastive method to verify the rationality. The price of calculated in terms of cost nature and equipment classification shall be equal.
- Verify the basic operation and maintenance costs: the basic operation and maintenance costs can refer to the operation and maintenance technical standards of local power grid and verify by contrastive method and experiences preliminarily.
- Verify the additional maintenance costs: The additional maintenance costs usually refers to other maintenance costs other than periodic maintenance requirements, such as special commissioning costs, condition monitoring equipment costs, etc. We can verify the additional maintenance costs by actual situation. For equipment with distinct differences, the manufacturer can clarifies if necessary.

- Verify 110kV cable accessories failure probability: the power company can be provided by the equipment manufacturers in recent years, the failure rate guarantee value and the relevant failure statistics and power companies in the region where the power grid for several years of failure to compare the situation, for the statistics. To determine part of the further research, clarify, eliminate the dead ends of statistics.

3.3. The Punishment when LCC Verification Fail to Performance Assurance Assessment

When LCC calculation data provided by the equipment manufacturer is verified to be inaccurate or excessive from the actual value, the bid shall be annulled, furthermore, this equipment manufacturer may automatically be listed in the credit blacklist. If the equipment during the warranty period cannot meet the commitment at the purchase time and failure or defect, the equipment manufacturer should repair it within specified time, restore power transmission, take responsibility for the accident, and bear the costs of maintenance, replacement and actual loss. If the equipment outside the warranty period happen to failure or make an extra loss, the equipment manufacturer should compensate for the direct loss and bear the corresponding expenses.

4. Applications

This section selects three 110kV cable accessories bidding plans for LCC calculation. Among them, set the depreciation rate of $\boldsymbol{i}$ is 5%, the best life of cable accessories $\boldsymbol{A}$ for 30 years to calculate.

4.1. LCC analysis

4.1.1. The Cost of inputs (CI)

The input cost CI is mainly composed of the quotation and the final value of the CI converted to the best life of the equipment is shown in Table 4-1:

Table 4-1. Input Costs of Three Plans. (Unit: RMB)

	Plan A	Plan B	Plan C
Cable accessories manufacturing cost	50000	45000	65000
Factory test inspection fee	800	1000	1500
Special tools and initial spare parts cost	1000	2000	2000
After-sales service fee	2000	1500	2000
Supplier shipping cost	800	1000	1500
Taxes and fee	5000	4500	6500
Other fee	1000	500	2500
110kV cable terminal price	24000	18000	24000
110kV insulated connector price	24000	24000	45000
Coaxial cable price	1000	500	1200
Ground wire price	500	500	1100
Direct grounding box price	3000	4000	3000
Cross-connect protection earth box price	4000	4000	2500
With protective earth box price	3500	4000	3000
Return line price	600	500	1200
Cable attachment installation fee	2000	5000	2500
CI present value	62600	60500	83500
CI final value	270553.59	261477.51	360882.19

In table 4-1, the C_{sj} calculated by cost nature and equipment classification (gray shading part) in the three plans are equal, LCC calculation is valid.

4.1.2. The Cost of inputs (CI)
The value of COM for conversion to the best life of the equipment is shown in Table 4-2.

Table 4-2. Operation and Maintenance Costs of Three Plans. (Unit: RMB)

	Plan A	Plan B	Plan C
Regular operation and maintenance costs in second year	30000	30000	30000
Additional maintenance costs	10000	5000	5000
COM value in second year	40000	35000	35000
COM present value	36281.18	31746.03	31746.03
COM final value	156805.17	137204.52	137204.52

4.1.3. Cost of Failure (CF)
Table 4-3 shows the calculation results of the CF final value of the best life of the equipment.

Table 4-3. Failure Costs for Three Plans. (Unit: RMB)

	Plan A	Plan B	Plan C
Failure rate	0.5%	0.1%	0%
Average cost of repair	140000	140000	140000
110kV average transmission price (RMB / kW.h)	0.6	0.6	0.6
Repair time (h)	3	3	3
Rated conveying capacity (kW)	50000	50000	50000
CF second year value	115000	115000	0
CF present value	104308.39	20861.68	0
CF final value	450814.85	90162.97	0

4.1.4. Cost of disuse (CD)
The value of the depreciation cost is the final value, as shown in Table 4-4.

Table 4-4. Waste Costs for Three Scenarios CD. (Unit: RMB)

	Plan A	Plan B	Plan C
Cable accessories removal costs	8000	5000	3000
CD final value	8000	5000	3000

4.2. LCC analysis
Based on the above calculation results, we can get LCC calculation results shown in Table 4-5 and Figure 4-1:

Table 4-5. 110kV cable accessories LCC cost deconstruction comparison. (Unit: RMB)

	CI	COM	CF	CD	Total
Plan A	270553.59	156805.17	450814.85	8000	1313532.37
Plan B	261477.51	137204.52	90162.97	5000	892527.04
Plan C	360882.19	137204.52	0	3000	999173.42

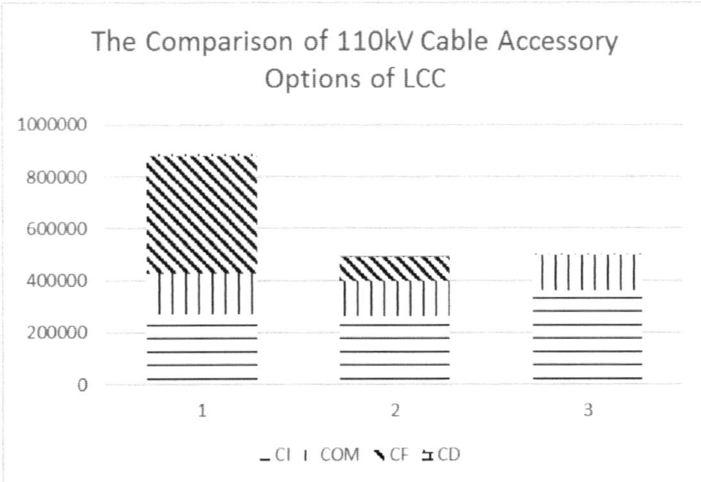

Figure4-1. The Comparison of 110kV Cable Accessories Options of LCC.

From Table 4-5 and Figure 4-1, it can be concluded that LCC final value of plan A is the highest, and the difference between plan B and plan C is small, so the safety should be considered. In this three plans, the CI of program C is the highest, and the three plans are comparatively comparable in terms of COM. The difference between three plans is reflected in the cost of failure. Plan A is guaranteed by the high values of failure rate lead to a significant increase in CF, and the reliability of the equipment is low. Moreover, the failure rate guarantee value of plan C is 0, no CF is generated. If the plan reports the failure rate guarantee value verified by the real effective, we can determine that the program C equipment safety performance is much more higher.

In summary, it can be seen that the cost performance of plan B is higher than that of the other two plans. Although the CI of plan C is slightly higher than the cost of the plan B, the lower COM and higher reliability and security results show some advantages. It is suggested that the electric power company should fully verify the authenticity of COM and the failure rate guaranteed value in bidding process, consider the proportion of COM and CF, and reduce the cost of labour and equipment.

5. Conclusion

The LCC management of the power system is based on the reliability of the equipment to make the life of the lowest cost of ownership management. In the procurement of equipment, we should consider not only the purchase price of equipment, but also the cost throughout the life cycle. Considering all the costs incurred, and seek the best solution, find a balance between appropriate availability and full cost, then find the smallest solution.

In this paper, the traditional LCC evaluation model is amended to establish a LCC evaluation model suitable for 110kV cable accessories, and applied it to the bidding and purchasing of 110kV cable accessories, which is practical significance. At the same time, three different plans are used to analyze the application cases, CI, COM, CF and CD are analysed. The LCC method is not only used to realize the unity of economy and reliability in equipment procurement, but also to deeply promote the power system of LCC technology and management applications.

References

[1] H. Akselsson and B. Burstrom. *Life cycle cost pocurement of Swedish State Railways' high-speed train X2000*[J]. Proceedings of the Institution of Mechanical Engineers, Part F: Journal of Rail and Rapid Transit, 1994,208: 51 -59

[2] International Electrotechnical Commission, *Life Cycle Costing* [Z]. International Standard 60300-3-3, 2004

[3] I. Jeromin, G. Balzer, J. Backes, R. Huber. *Life cycle cost analysis of transmission and distribution systems*, 20th International Conference and Exhibition on Electricity Distribution (CIRED 2009)

[4] Ł. Dzierżanowski, B. Ruszczak and M. Tomaszewski, *Frequency of power line failures in Life Cycle Cost analysis*, 2013 International Symposium on Electrodynamic and Mechatronic Systems (SELM), Opole-Zawiercie, 2013, pp. 55-56.

[5] S. H. Lee, A. K. Lee and J. O. Kim, *Determining economic life cycle for power transformer based on life cycle cost analysis*, 2012 IEEE International Power Modulator and High Voltage Conference (IPMHVC), San Diego, CA, 2012, pp. 604-607.

[6] Q. Guo-Hua, R. Zheng, S. Lei, Z. Bo, X. Jian-Gang and Z. Xiang-Ling, *A new life cycle cost model of power transformer and its comprehensive sensitivity analysis*, 2014 International Conference on Power System Technology, Chengdu, 2014, pp. 1342-1348.

[7] Y. Zhang, J. Jiao, Y. Yang, X. Li, T. Zhou and X. Huang, *A new maintenance decision making model based on life cycle cost analysis for power transformers*, 2016 IEEE International Conference on High Voltage Engineering and Application (ICHVE), Chengdu, 2016, pp. 1-4.

[8] YAO Ming, CHEN Hong-bing, LI Li-hua, HAN Tian-xiang, GAO Kai. *Research of LCC method and its applications to GIS purchases in 500kV underground substations* [J]. Electric Power, 2008, (08):32-34.

Comparison of all-side risk and down-side risk constrained probabilistic dispatch models considering wind power

T. Y. Ji[1], D. Y. Hong[1],M. S. Li[1]and Q. H. Wu[1]

[1] School of Electric Power Engineering, South ChinaUniversity of Technology, Guangzhou, 510641, China.

Corresponding author:Dr. M. S. Li, E-mail: mengshili@scut.edu.cn.

Abstract. This paper, for the first time, compares the probabilistic optimal power flow (POPF) models targetingdifferent objectives -- minimizing the expected generation cost (EGC) and all-side risk (AR) or down side risk (DR)simultaneously. In the POPF models, wind power generation isconsidered in dispatching, the uncertainty of which can be explained by building a model to describe the stochasticcharacteristicof the forecast error. The distribution of the forecast error can be obtained by sampling from the historical error time series.The POPFsaresolvedusinggroup search optimizer with multiple producers (GSOMP), and a decision-making methodTOPSIS is used to select the most suitablesolution. Simulation studies are conducted on a modified IEEE 118-bus power system with wind farms integrated, and the results show that the EGC-DR modelfind better solution compared tothe EGC-AR model.

1. Introduction

With the increasing amount of wind power generation used in power systems, the uncertaintyofthe system dispatch ability is increasing [1]. To forecast wind power accurately is thereforeextremely important because it enables better unit commitment and dispatch [2]. To consider the stochasticcharacteristic of wind power, probabilistic optimal power flow (POPF) is calculated to determine the dispatch strategy.

The wind power forecast models can be classified into deterministic and probabilisticforecast. Due to the deterministicprediction result is a point value, it is called point forecastas well. Efforts have been made to improve the accuracy in point forecast [3, 4].Reference [3] proposes a mean trend detector to decompose the wind power data and a local predictor based on mathematical morphology to predict the fluctuation component accurately.Reference [4] predicts the wind speed first and then gains the wind power forecast result through the fitted speed-power curve. The result of probabilistic prediction turns out to be a series of intervals with their probabilities [5],or the possible values corresponding to their probability distributions [6].In reference [5], the conditional copula function is suggested to describe the relationship among the historical wind power internals.In [6], the error series is forecasted by support vector machine (SVM) to correct the original SVM forecast value.

In order to improve the accuracy of the model, the probability distribution of wind power forecast error is obtained from the kernel density estimation of the time series of historical error, instead of assuming itsprobabilistic distribution function.

As wind power is widely and increasingly used in power systems, system operators pay moreattention to the influence of wind power penetration to the safety and economy of power systemoperation. There are a number of studies focusing on weakening the negative impact of wind power forecasterror on power system operation [7, 8].

Considering the stochasticcharacteristic of wind power, POPF is proposed and found to be an effectivetool in power system operation [9]. The probabilistic optimal dispatch model usually has multi-objectives of minimizing the expected generation cost (EGC) [10] and risk simultaneously, where the risk is usually manifested by thestandard deviation of generation cost in an uncertainenvironment [11, 12]. The risk index, standard deviation, was proposed by Markowitz to dealwith the portfolio optimization problem [11]. However, in thiscase, the variance, which is the so-called all-side risk (AR) in this paper,treats the gain (the up-side risk) and the potentialloss (the down-side risk) as equal, i.e., it weighs the gains that investors like the same tothe potential losses that investors dislike[11]. As a matter of fact,investors concern more about potential loss. Inthis case, it is no longer appropriate to use the variance as the measure. Therefore,the down-side risk (DR), which is usually measured by the lower semi-variance, was proposed by Roy to emphasize thepotential loss [13], which has nowbecome the industry standard for risk management andis regarded as a better measure of investment risk, comparedwith the variance [13].

To solve the multi-objectiveoptimization function, group search optimizer with multipleproducers (GSOMP) isemployed. A decision-making method, the techniquefor order preference similar to an ideal solution (TOPSIS)[14],is used to select the optimal solution betweenthe EGC and risk. Moreover, the differences between the dispatch strategies aiming at EGC-AR minimization and EGC-DR minimization are analyzed.

2. The modified IEEE 118-bus test system

Fig. 1 showsa modified IEEE 118-bus test system, where a wind power farmis addedto bus 7. The wind power data is obtained from the Elia database [15], which is the total wind power of Belgium collected once per hour. Other parameters remain the same as in the traditional IEEE 118-bus test system.

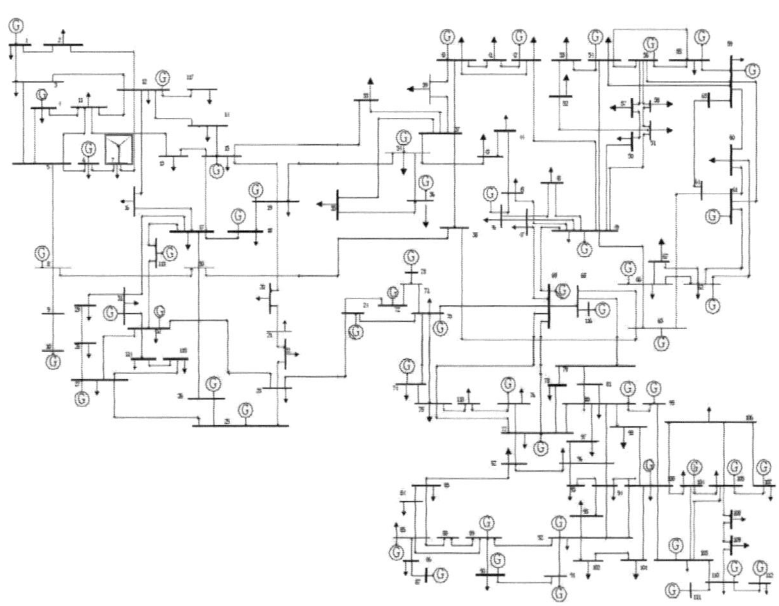

Figure1. The modified IEEE 118-bustest system.

3. Probabilistic forecast model

The probabilistic forecastresult is the summary of the point forecast result and the error sample of the point forecast error.

Firstly, the point result of the Nth wind power is predicted using the Persistence model

$$\hat{S}_1(N) = S(N-1)\,. \tag{1}$$

Secondly, predict the wind power time series $S(t)$, t=1,2,…,N-1 by the Persistence model, andobtain the error time series $e(t)$:

$$\hat{S}(t) = S(t-1) \text{ and } e(t) = S(t) - \hat{S}(t)\,. \tag{2}$$

According to probability statistics, an estimation curve of the kernel density of$e(t)$ can be obtained, from which the error samples $E(N, i)$, i=1, 2, …, M,can beobtained using linear sampling, and their probability distribution follows the distribution of $e(t)$.

Finally, the samples of wind power, i.e., the possible values of wind power,can be obtainedby adding the samples of errors to the point forecast value, which is expressed to be

$$\hat{S}(N,i) = \hat{S}_1(N) + E(N,i) = S(N-1) + E(N,i). \tag{3}$$

The probability distributionof $S(N)$ obeys the distribution of $e(t)$.

4. Dispatch model

In the conventional OPF problem, one set of objective value is obtainedon the base of the determined control values. As for the POPF,there are M sets of objective values, where M is the samplingsize. Thus, statistical properties of the M sets of objectivevalues should be considered as new objectives, such as the mean and variance, which refer to the expectation andrisk of the fuel cost, respectively. However, the risk includes the all-side risk (the whole curve in Fig. 3) and the down-side risk (only the down side in Fig. 3). With different objectives, EGC-AR and EGC-DR may have different dispatch solutions, which will be studied in the next.

4.1 Expected generation costall-siderisk (EGC-AR) model

We aim to minimize theexpectedgeneration costand all-siderisk simultaneously by adjusting the controlvariables while satisfying a set of operational constraints,which can be formulated as:

$$\begin{aligned} &\min[\mathrm{E}[F(x,u,P_{\mathrm{W}})], \mathrm{V}[F(x,u,P_{\mathrm{W}})]] \\ &\text{s.t.} \begin{array}{l} g(x,u,P_{\mathrm{W}}) = 0 \\ h(x,u) < 0 \end{array} \end{aligned} \tag{4}$$

where P_W represents the active power outputs by wind farms,P_W=$S(N)$; x denotes the state variables includingthe slack bus active power P_{G1} , the load bus voltages $V_L$andgenerator reactive powerQ_G, which can be formulated by(5); u denotesthe control variables including the active power of generatorsP_G (except P_{G1}), generator voltages V_G, and reactive poweroutputs of VAR sources Q_C, which can be formulated by(6);E[$F(x, u, P_W)$] and V(x, u, P_W) which can be expressed by(7) and (8), denote the expectation and all-side risk of the fuel cost,respectively.

$$x = [P_{\mathrm{G}_1}, V_{\mathrm{L}_1}, \ldots, V_{\mathrm{L}_{N_D}}, Q_{\mathrm{G}_1}, \ldots, Q_{\mathrm{G}_{N_G}}]^{\mathrm{T}} \tag{5}$$

$$u = [P_{\mathrm{G}_2}, \ldots, P_{\mathrm{G}_{N_G}}, V_{\mathrm{G}_1}, \ldots, V_{\mathrm{G}_{N_G}}, Q_{\mathrm{C}_1}, \ldots, Q_{\mathrm{C}_{N_C}}]^{\mathrm{T}} \tag{6}$$

$$\mathrm{E}[F(x,u,P_{\mathrm{W}})] = \frac{1}{M}\sum_{i=1}^{M} F(x_i, u_i, P_{\mathrm{W}_i}) \tag{7}$$

$$\mathrm{V}[F(x,u,P_{\mathrm{W}})] = \left\{ \frac{1}{M}\sum_{i=1}^{M} \{F(x_i,u_i,P_{\mathrm{W}_i}) - \mathrm{E}[F(x,u,P_{\mathrm{W}})]\}^2 \right\}^{\frac{1}{2}} \tag{8}$$

where $F(x,u,P_{\mathrm{W}})$ is the total fuel cost in the system which changes withthe wind power sample, and it ismodelled by the following quadratic function that takes the valve pointeffects into consideration:

$$F = \sum_{i=1}^{N_G} (a_i P_{\mathrm{G}_i}^2 + b_i P_{\mathrm{G}_i} + c_i) \tag{9}$$

whereN_G is the number of the conventionalgenerators inthe oral IEE 118-buspower system, $P_{Gi}$is the active power output of generator i, and a_i, b_i, $c_i$are fuel cost coefficients of the ith generator, andN_D is the number of loads.

4.2Expected generation costdown-side risk (EGC-DR) model

Many studies agree that all-side risk (AR) makes no distinction between gains and potential losses of investments, i.e., it regards gains that investors like are as equally undesirable as potential losses that investors dislike. Indeed, investors concern more about the potential losses. Therefore, the down-side risk (DR) is put forward to replace the all-side risk (AR), and the objective functions in (4) are changed to

$$\begin{aligned} &\min[\mathrm{E}[F(x,u,P_{\mathrm{W}})],\mathrm{V}\_[F(x,u,P_{\mathrm{W}})]] \\ &\text{s.t.}\ \begin{aligned} &g(x,u,P_{\mathrm{W}})=0 \\ &h(x,u)<0 \end{aligned} \end{aligned} \tag{10}$$

where V_ means that only the fuel cost which is lower than the average generation cost will be taken into account, i.e., (8) is changed to

$$\mathrm{V}[F(x,u,P_{\mathrm{W}})]=\left\{\frac{1}{M'}\sum_{i}\{F(x_i,u_i,P_{\mathrm{W}_i})-\mathrm{E}[F(x,u,P_{\mathrm{W}})]\}^2\right\}^{\frac{1}{2}} \tag{11}$$
$$\text{for all } F(x_i,u_i,P_{\mathrm{W}_i})<\mathrm{E}[F(x,u,P_{\mathrm{W}})]$$

where M' is the total number of the fuel costs which are lower than the average generation cost.

The flowchart of Fig. 2 shows the computing procedure of the sampling and optimization.

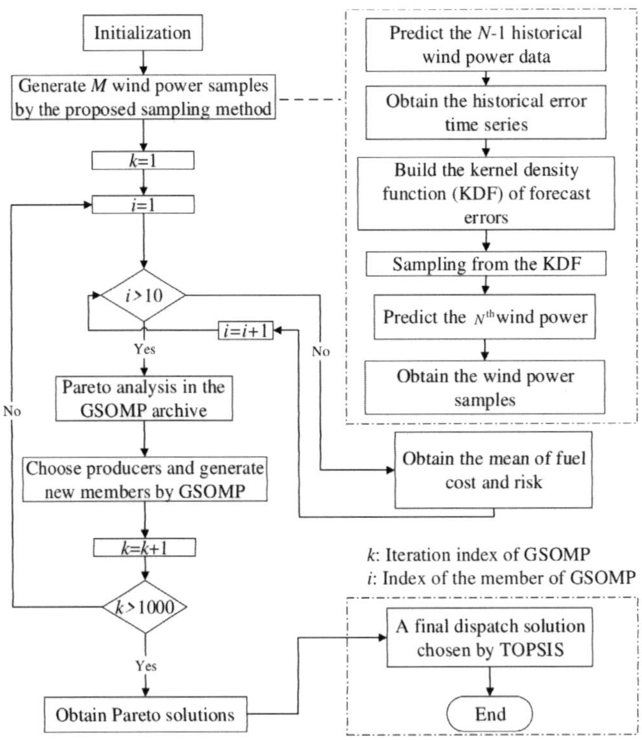

Figure 2. Detailed computing and optimizationprocedure.

5. Simulation results and discussion

The probability distribution of wind power is obtainedaccording to the kernel density estimationdescribed in Section 3 and an example is shown in Fig.3.In this paper, the length of historical wind power data is N=1000, and the sampling number is$M$=50.

The EGC-AR model concerns all the possible wind power output which occurs in the whole curve from (116, 0) to (133, 0), while the EGC-DR model pays attention to the down side of the curve from the peak to (133, 0).

As stated in Section 4, the objectives that we aim tominimize are the expectation $E[F(x,u,P_W)]$ and the all-siderisk $V[F(x,u,P_W)]$ or down-side risk $V_-[F(x,u,P_W)]$of the fuel cost, respectively.

As for the optimization algorithm of GSOMP, the population size is set to be 100. The Pareto fronts obtained by GSOMP of the EGC-AR and EGC-DR modelsare shown in Fig. 4, among which 5 typical solutionswhose EGC values spread from $6.55\times10^4\$/h$ to $6.75\times10^4\$/h$ with an interval of $0.05\times10^4\$/h$ are listed in Table 1.

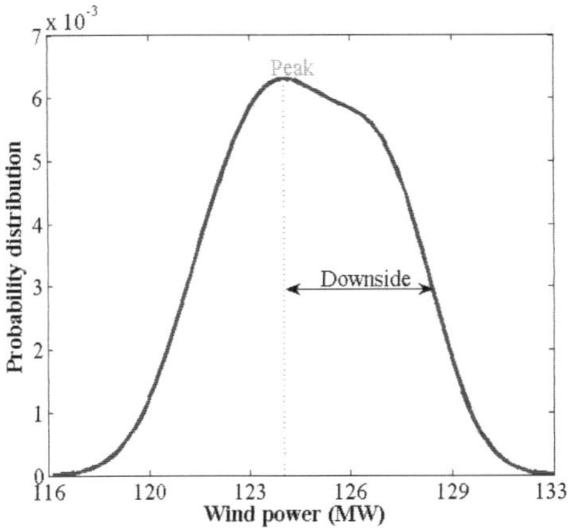

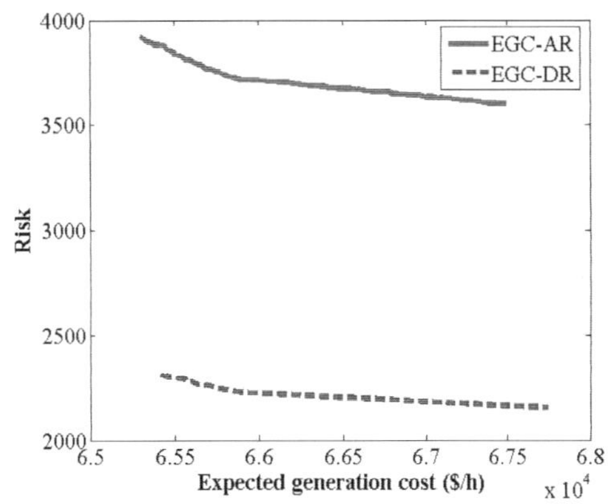

Figure 3. Probability distribution of wind power. **Figure 4.** Pareto solutions obtained by GSOMP.

Table 1. Values of EGC, AR and DR as for Pareto solutions.

Pareto solutions	PS_1	PS_2	PS_3	PS_4	PS_5
EGC($\times10^4\$/h$)	6.55	6.6	6.65	6.7	6.75
AR	3881	3704	3672	3640	3599
DR	2298	2217	2203	2178	2163

It can be seen from Fig. 4 and Table1 that with the EGC grows, the AR and DR aredeclining. The solutions to the two Pareto fronts which have the minimum EGCsand highest risks demonstrate that the dispatch solutioncannot adapt to the uncertain wind power environment well enough. Onthe other hand, the solutionswith the minimum risks indicate that the corresponding dispatch solution is adaptable, evenrobust, to the uncertain environment. However, their values ofEGC are high. Therefore, to determine the most suitable solution, the fuzzyTOPSIS is engaged. For the EGC-AR model, the best solution is (6.6404×10^4 \$/h, 3672), and the beat solution for the EGC-DR model is (6.5879×10^4 \$/h, 2216). Obviously, the final expected generation cost of the EGC-DR model is lower, which demonstrates the effectivenessof considering the down-side risk only.

6. Conclusion

This paper pays attention to the POPF problem in a power systemwith wind power integrated. The uncertainty of wind power is explainedthroughdescribing the probabilistic distribution function of forecast errorwith historical data.Besides, two POPF models with different objectives -- minimizing the expected generation cost (EGC) with theall-side risk (AR) or with the down-side risk (DR), are simulated and compared. The POPFs on a modified IEEE 118-bus power system with wind farm integrated are solvedusingthe multi-objective optimization algorithm of GSOMP. The resultshave shown that the EGC-DR model obtains a lower expected generation cost and lower risk than the EGC-AR model.

Acknowledgments

This work was supported by GuangdongInnovative Research Team Program (No. 201001N0104744201) and the StateKey Program of National Natural Science of China (No. 51437006).

References

[1] I. A. Farhat and M. E. El-Hawary. Dynamic adaptive bacterial foraging algorithmfor optimum economic dispatch with valve-point effects and wind power. IETGeneration Transmission & Distribution, 4(9):989–999, 2010.
[2] N. G. Paterakis, O. Erdin, A. G. Bakirtzis, and et al. Qualification and quantification of reserves in power systems under high wind generation penetration considering demand response. IEEE Transactions on Sustainable Energy, 6(1):88–103,Jan 2015.
[3] J. L. Wu, T. Y. Ji, M. S. Li, and et al. Multi-step wind power forecast usingmean trend detector and mathematical morphology-based local predictor. IEEETransactions on Sustainable Energy, 6(4):1216–1223, Oct 2015.
[4] D. Y. Hong, T. Y. Ji, L. L. Zhang, and et al. An indirect short-term wind powerforecast approach with multi-variable inputs. In 2016 IEEE Innovative Smart GridTechnologies - Asia (ISGT-Asia), pages 793–798, Nov 2016
[5] F. Lan, C. C. Sang, J. J. Liang, and et al. Interval prediction for wind power basedon conditional copula function. Proceedings of the Chinese Society for ElectricalEngineering, pages 79–86, 2016.
[6] C. Wan, J. Lin, J. Wang, Y. Song, and Z. Y. Dong. Direct quantile regression for nonparametric probabilistic forecasting of wind power generation. IEEETransactions on Power Systems, PP(99):1–1, 2016.
[7] G. W. Chang, H. J. Lu, Y. R. Chang, and et al. An improved neural networkbased approach for short-term wind speed and power forecast. Renewable Energy,105:301–311, 2017.
[8] D. Lee and R. Baldick. Probabilistic wind power forecasting based on the laplacedistribution and golden search. In 2016 IEEE/PES Transmission and DistributionConference and Exposition (T D), pages 1–5, May 2016.
[9] Y. M. Li, W. Y. Li, W. Yan, J. Yu, and X. Zhao. Probabilistic optimalpower flow considering correlations of wind speeds following differentdistributions. IEEE Trans. Power Syst., vol. 29, no. 4, pp. 1847–1854,Jul. 2014.
[10] M. El-Hawary and G. Mbamalu. A comparison of probabilistic perturbation and deterministic based optimal power flow solutions. IEEETrans. Power Syst., vol. 6, no. 3, pp. 1099–1105, Aug. 1991.
[11] Y Z Li, Q H Wu. Down side Risk Constrained Probabilistic Optimal Power Flow With Wind Power Integrated[J]. IEEE Transactions on Power Systems, 2016, 31(2):1649-1650.
[12] Z Q Xie, T Y Ji, M S Li, et al. Quasi-Monte Carlo Based Probabilistic Optimal Power Flow Considering the Correlation of Wind Speeds Using Copula Function[J]. IEEE Transactions on Power Systems, 2017, PP(99):1-1.
[13] A. D. Roy. Safety first and the holding of assets. Econometrica, vol.20, no. 3, pp. 431–449, 1952.
[14] C. X. Guo, J. P. Zhan, and Q. H. Wu. Dynamic economic emissiondispatch based on group search optimizer with multiple producers. Elect. Power Syst. Res., vol. 86, pp. 8–16, 2012.
[15] Elia. Wind power data [online]. Available:http://www.elia.be/en/griddata/power-generation/wind-power.

IOP Conf. Series: Materials Science and Engineering **366** (2018) 012040 doi:10.1088/1757-899X/366/1/012040

Research and Application of Three Phases Balance Technology in China Rural Power Distribution Network based on the Fast Phase Selection Method

LIU ZhanJie,CHEN Guang,WANG ZhiMin and YU Hao

State Grid Energy Research Institute Co.,Ltd.

lzhanjie@163.com

Abstract. In this paper, the objectivity and unpredictability of the three-phase load unbalance (TPLU) for the rural distribution network are proposed. Firstly, the influences of TPLU on the power quality and power loss are analyzed. And then, based on the common managements and governance measures of TPLU,whose typical problems are also introduced. Furthermore, based on the fast phase selection, the concept of the three-phase load balance is defined to describe the basic structure and complementation principles in detail. Lastly, the operating points and control strategies of the commutation equipments are proposed. The verification results show that proposed method is accurate and effective.

1. Introduction

The three-phase four-wire system was used widely in urban and rural low-voltage distribution networks, the loads are mostly single-phase or hybrid single three-phase. Because loads and system's parameters are imbalance, the unbalanced currents among the three phases exist. At the same time, the loads' irregularity and unpredictability lead to imbalance of three-phase load in a long time for the low-voltage grids. The National Standard "the maximum unbalanced degree of power quality of three-phase voltage"(GB/T 15543-2008) clearly explained: in normal circumstances, the three-phase unbalanced degree of the negative sequence voltage is less than 2% and with no more than 4% in short-term. When the common points were connected, the unbalance degree of negative sequence voltage must less than 1.3% and with no more than 2.6% in a short time. The imbalance of three-phase will lead to additional heat and vibration of the rotating motor, transformer leakage flux and local overheating, large neutral current [1], the increasing power line loss and so on.

According to statistics, more serious presence of the three-phase imbalance exceeds 85% for the low voltage distribution. Using the experience to distribute the load reasonably is the main management method at present [2-4]. In addition to the above approaches, there are few effective solutions but to take some protective measures such as increasing the neutral line section, etc. in order to reduce the three-phase unbalanced over potential problems. A classification method based on fuzzy C-means clustering is proposed [5], which divides the users based on the daily load curve so as to make the users of the same class have higher similarity and then divides each user evenly into A, B and C phase. The simulation results shows that the three-phase imbalance problem can be alleviated better.

2. Analysis on the Principle and Effect of Regulation

IOP Conf. Series: Materials Science and Engineering **366** (2018) 012040 doi:10.1088/1757-899X/366/1/012040

The three-phase unbalanced tune up device (TPUTD) is used to overcome the low-voltage three-phase unbalanced problems. The device is based on the capacitor adjustment principles, which has wide range of products[6-10], and whose working theory is similar, as shown in Figure 1.

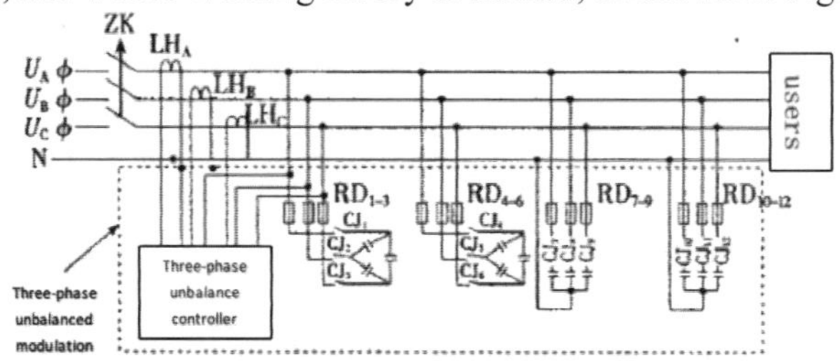

Figure 1. Schematic diagram of reactive power Compensator adjustment device

Tune up device is built with 12 single-phase 400V power capacitors. Under the terminal's control, each capacitor can connected both the phase-phase lines and phase-null lines, which can be put into operation all the angle capacitors or star capacitors. Each connection ways of angular and star capacitors have 5 types: 1) all the capacitors operate; 2) capacitors do not operate; 3) phase A and B are in the same group; 4) phase B and phase C are in a group; 5) phase A and C in a group. There are two sets of triangles and two sets of star wiring within the device, based on the permutations and combinations method, a total of 625 kinds of switching modes can be calculated. And for different capacity of the distribution transformer, capacitor banks can be increased or decreased. By properly connecting the different number of capacitors between phase-phase lines and phase-null lines, the use of the inductive characteristics of the load can not only achieve the compensation power factor but also adjust the unbalanced active current. Which the actual amount of compensation is the same with the needed reactive compensation.

However, due to the complex operating conditions of the inductive load in the low-voltage distribution network, it is difficult to achieve the desired effects as the compensation from the relatively fixed capacitors, which often results in the reverse problem of reactive power. On the other hand, in the conventional SVC TCR branch wiring use 6-pulse form, which contains the line current harmonics $6k\pm1$ (k is a positive integer), you need to install filtering while the cost increase. If the advanced STATCOM technology was used, because of the high cost, it is unreasonable in the low-pressure system. Thirdly, the TPUTD has an effect only installed on the supply side. And this product has no effect when it was installed at the outlet of the distribution transformer, which can cause enormous safety problems and low voltage line loss reduction and space. So this method has great limitations, which is not three-phase imbalance in the direction of governance research trends.

3. Adjustment Principle Based on fast phase selection technique

A new low-voltage three-phase balance monitoring system was developed for the low-voltage rural distribution network structure in China. The control terminal and commutation switch were used to realize low-voltage data acquisition and three-phase balance monitoring, while the complex management can be implemented. Three-phase load adjustment program is shown in Fig.2.

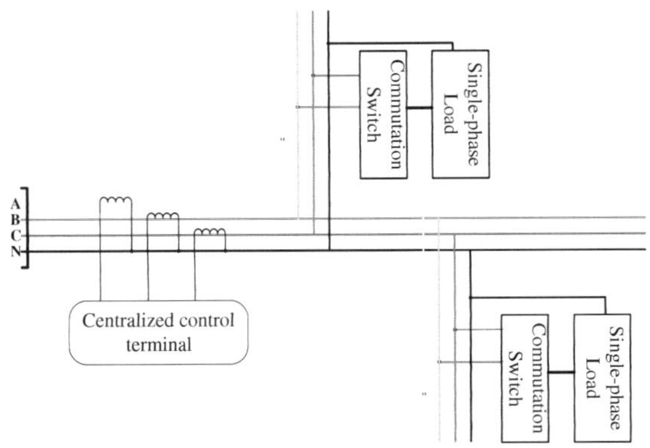

Figure 2. The sketch diagram of basic structure

It can be know from Fig,2, loads of the district of transformer and distribution of users can choose a certain number of single-phase load to install the commutation switch and these loads can be taken as the adjustable loads based on the wiring of low-voltage distribution network. Real-time acquisition monitoring this back-outlet three-phase operation in the transformer low-voltage side of each outlet installed at the control terminal. And send the collected data to the master station system for statistical calculation. When the control terminal monitor the three-phase unbalance rate exceeds the set value, the master will carry out a series of control strategies. When the control terminal detects the three-phase unbalance rate exceeds the set value, the master station will carry out a series of control strategy analysis, such as selecting the optimal action switch, check whether the time, manual setting and other operating conditions. As well as calculating the adjusted three-phase equilibrium rate. The control signal is issued when the operating condition is satisfied, which realized by the commutation control of the specified commutation switch which is carried out through the control terminal, and the load of the heavy load is transferred to the light load phase to realize the three-phase load balance.

3.1. Control terminal

The control terminal mainly is comprised by a core unit, a man-machine interface, a power supply module, a control module and a communication module. The terminal basic structures are shown in Figure 3.

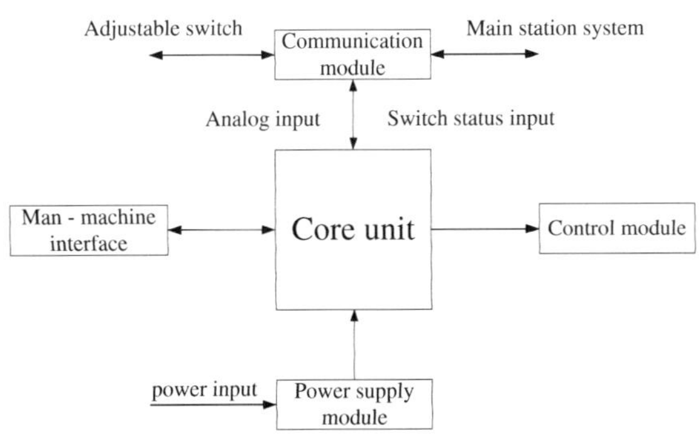

Figure 3. Structural Representation of Control Terminal

From the above Figure, firstly, the core unit usually completes the related data acquisition, the breakdown monitoring as well as the control instruction output and so on. Secondly, man-machine interface can be set the main terminal parameters. Thirdly, the power module provides DC power for the terminal. Also the control module is used to control the commutation switch to display the switch status. Lastly, the communication module can be connected with the core unit through the serial port, and the external system can be connected with the back system or the commutation switch. The control flow chart is shown in Fig.4.

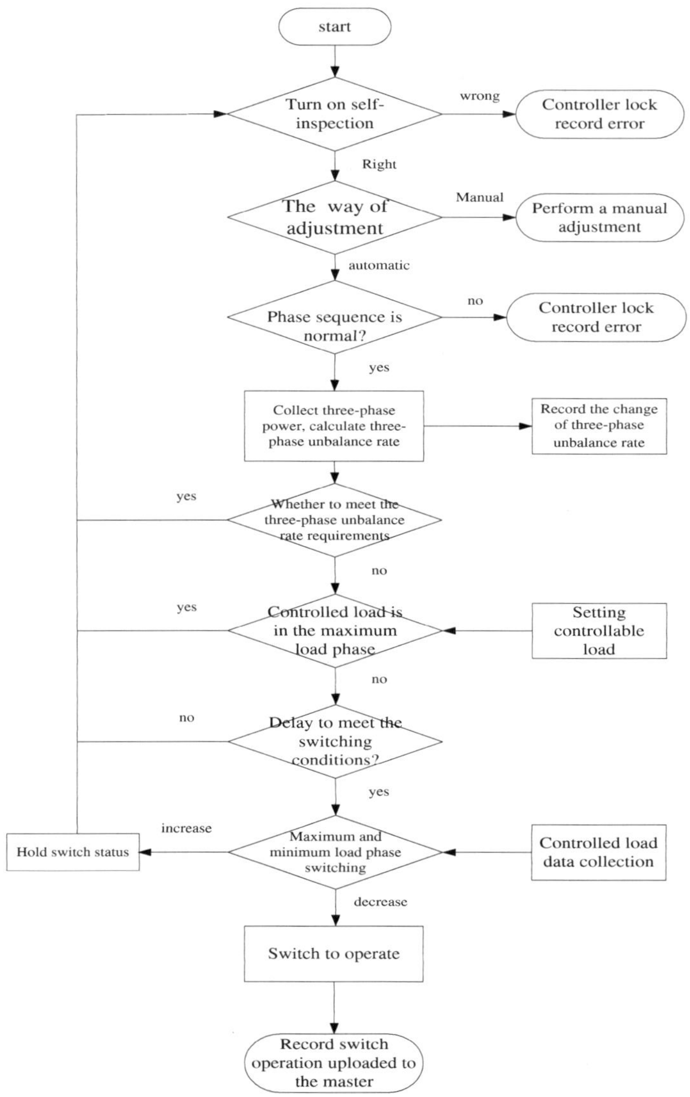

Figure 4. The Flow Chart of the Proposed Method

3.2. The communication switch

The communication switch can adjust the three-phase load quickly and safety. It is installed in the low voltage table box or the head of line on 220V. The incoming feeder is three-phase four-wire or two-phase three-wire and outgoing feeder is single-phase two-wire. Commutation switch received the commutation command, which can be completed under the load conditions. And commutation does not affect the normal operation of electrical equipment. the three-phase four-wire incoming feeder is shown in Figure 5.

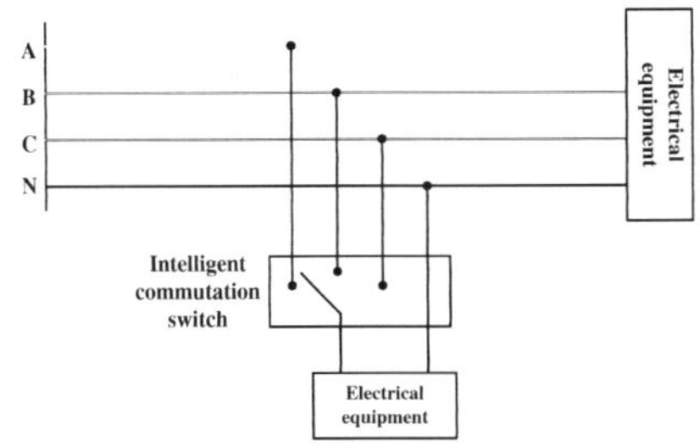

Figure 5. Structural representation of commutation switch

4. Actual Applications

In order to verify the effect of research results, combined with regional load characteristics, a typical district of transformer in China's western regions were selected. The three-phase voltages before deployment are shown in Table 1.

Table 1. Integral point three-phase voltage numerical before Adjustment

Value of the whole point	A phase voltage (V)	B phase voltage (V)	C phase voltage (V)	Three-phase unbalance degree (%)
1h	224	231	227	3.08%
2h	223	221	230	4.01%
3h	227	224	233	3.95%
4h	221	219	229	4.48%
5h	228	217	225	4.93%
6h	226	216	223	4.51%
7h	218	221	211	4.62%
8h	220	227	207	9.17%
9h	217	230	**197**	15.37%
10h	215	**237**	**196**	18.98%
11h	220	**236**	**190**	21.36%
12h	222	228	**192**	16.82%
13h	218	219	200	8.95%
14h	215	212	**196**	9.15%
15h	215	223	**191**	15.26%
16h	221	230	**193**	17.24%
17h	217	229	**189**	18.90%
18h	211	224	**187**	17.85%
19h	209	217	**192**	12.14%
20h	203	219	199	9.66%
21h	214	202	217	7.11%
22h	221	211	225	6.39%
23h	231	219	224	5.34%
24h	230	220	229	4.42%

Note: (1)Three-phase imbalance is calculated according to IEEE std 936-1987 [11]; (2)The bold part of the figure is the time when the voltage crosses the line.

The voltage curve of three-phase before Adjusting is shown in Fig.6.

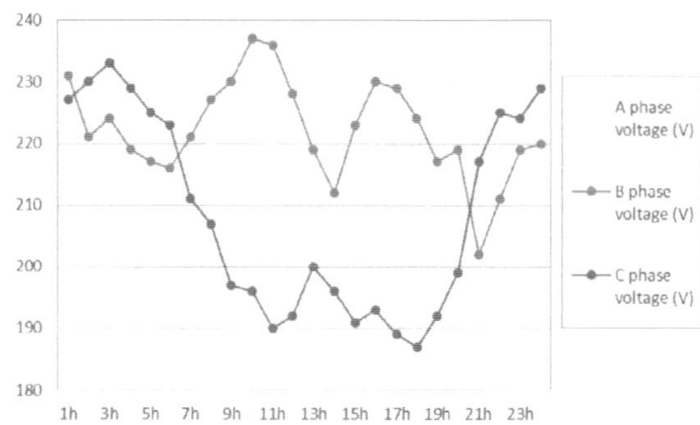

Figure 6. The three-phase Voltage Values before Adjustment

It can be known form the table 1, phase B has the serious three-phase unbalance phenomenon in this supplying district. Considering the deployment principles, such as three-phase line arrival, load size step changes, and avoid critical loads, 4-5 sets of commutation switches were installed in the each low-pressure outline. The three-phase voltage results after the adjusting are listed in Table 2.

Table 2. Integral point three-phase voltage numerical after adjust

Value of the whole point	A phase voltage (V)	B phase voltage (V)	C phase voltage (V)	Three-phase unbalance degree (%)
1h	227	225	223	1.78%
2h	230	226	230	1.75%
3h	225	224	221	1.79%
4h	226	223	227	1.78%
5h	224	225	226	0.89%
6h	221	225	222	1.80%
7h	220	216	219	1.83%
8h	221	223	218	2.27%
9h	226	220	223	2.69%
10h	226	222	225	1.78%
11h	220	218	222	1.82%
12h	231	224	228	3.07%
13h	226	224	226	0.89%
14h	226	220	219	3.16%
15h	222	217	216	2.75%
16h	221	218	218	1.37%
17h	219	213	215	2.78%
18h	219	217	213	2.77%
19h	218	221	214	3.22%
20h	224	223	216	3.62%
21h	226	220	225	2.68%
22h	233	231	224	3.92%
23h	232	229	228	1.74%
24h	231	230	228	1.31%

In order to illustrate the performance of the proposed method, three-phase voltage values are shown in Fig.7.

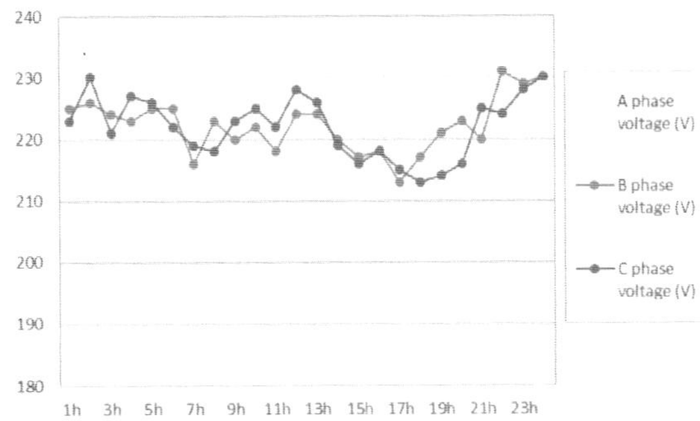

Figure 7. Three-Phase Voltage Values after the Adjustment

Next, the unbalance degrees before and after the adjustment are compared in the 24 hours. The comparison results are shown in Fig.8

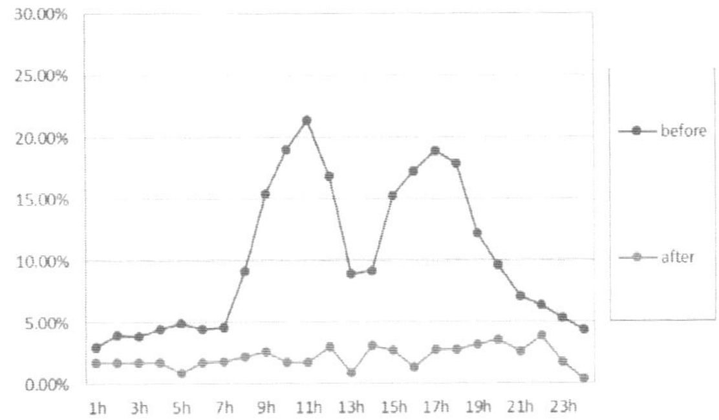

Figure 8. Three-phase voltage unbalance contrast curve chart before and after adjustment

In Fig.8, it is obviously that: (1) the unbalance degree of the selected supply district is improved greatly; (2) the suppression performance is better when the unbalance degree is serious; (3) the unbalance degree is controlled around 5%, which fits in the requirements of the guidelines.

5. Conclusion

In this paper, based on the actual situation of low-voltage distribution network in China, a new three-phase load balanced control technique is proposed. The key of the proposed method is to achieve the maximum balance of three-phase load through quick selecting phase, the uninterrupted power supply must be ensured. In the realization process, the commutation optimization strategy, the action frequencies and other factors are considered, and which can be also integrated into the unified monitoring and management.

Based on the verification, the device based on the above ideas provides a new solution to solve the problem of rural low voltage three-phase load imbalance. The device also can greatly enhance the intelligent level of low-voltage three-phase management and owns the significant values.

References

[1] PENG Chunhua, CHEN Shoukun, YU Rong. An Optimal Phase Swapping Strategy for Long-term Three-phase Balancing Using a Novel Simulated Crystallizing Algorithm [J]. Proceedings of the CSEE, 2014, 34(22): 3760–3767.

[2] Ouyang Wu,Cheng Haozhong,Zhang Xiubin,etal.Genetic algorithm based on random spanning trees indistribution network reconfiguration[J].High Voltage Engineering,2008,34(8):1726-1730(in Chinese).

[3] Yang Jianjun,Zhan Hong.Application of the improved genetic algorithm based on graph theory in distribution network reconfiguration[J].Power System Protection and

Control,2010,38(21):122-125(in Chinese).
[4] Li Dehua,Wang shao,Liu Yang,et al.Distribution network reconfiguration based on the combination of fuzzy genetic algorithm and ant colony algorithm[J].Power System Protection and Control,2009,37(17):26-30(in Chinese).
[5] Zhang Huan, based on user classification of low-voltage distribution network three-phase balanced loss reduction study [D], Changsha: Changsha University of Science and Technology, 2013.
[6] Peng Hui,Huang Yinong ,Wang Maohai.3 Phaseasymmetric load compensation in distribution system[J].Electric Power Automation Equipment,2002,22(1):32-34(in Chinese).
[7] Yuan Xufeng,Gao Lu,Wen Jingyu,et al.Unbalanced three-phase control strategy of VSC-HVDC system[J].Electric Power Automation Equipment,2010,30(9):1-5(in Chinese).
[8] Tang Jie,Wang Yueqiu,Liu Li,et al.Active disturbance rejection control for distribution static synchronous compensator[J].Power System Technology,2012,36(1):153-157(in Chinese).
[9] Zhang Maosong,Li Shangsheng,Zha Xiaoming,et al.Robust L2 performance criteria design for cascade D-STATCOM[J].Proceedings of the CSEE,2011,31(28):71-79(in Chinese).
[10] Wang Tiesong.Optimization location of power correction capacitor of distribution network based on the ANN measurements and analysis method[J].Power System Protection and Control,2011,39(2):61-64(in Chinese).
[11] Lin Xuehai. Three-phase unbalanced power system lectures (first lecture) [J] for electricity, 1998,14 (4): 51-54.

A novel model of load supply capability evaluation in active distribution network

Dayong Mu[1], Liwei Sun[2], Tianyu Li[1], Jingzhao Luan[1], Ting Sun[1] and Wei Zhou[2]

[1] State Grid Dalian Electric Power Co, Ltd, Dalian, 116001, China
[2] School of electrical engineering, Dalian University of Technology, Dalian, 116024, China

Corresponding author's e-mail address: sunliweiabc@126.com

Abstract. The output of the controllable distributed generation (DG) is determined by the economic dispatch strategy of the active distribution network (ADN). When we calculate the load supply capability (LSC) of ADN, the influence of the economic dispatch results must be taken into account. This paper proposes a bi-level optimization model for maximum power supply capacity of ADN based on chance constrained programming. The upper level model is used to calculate the LSC, and the lower level considers the economic dispatch of controllable DGs. Based on the improved IEEE-33 distribution system, the results show that the model can get the maximum power supply capacity that meets the economic operation requirements of ADN.

Key words: active distribution network; load supply capability; economic dispatch; bi-level optimization; chance-constrained programming

1. Introduction

The active distribution network (ADN) is derived from the distributed energy application and development [1]. It has the characteristics of active control and active management of distributed power compared with traditional power distribution network (DN), and has attracted wide attention and developed rapidly [2]. The output of ADN is influenced by many factors. The output of controllable distributed generation (DG) such as the diesel generator and the micro gas turbine is determined by the scheduling strategy with source-network coordination. While the output of uncontrollable DG such as the wind power and the photovoltaic power, is uncertain because it is affected by natural factors [3]. The load supply capability (LSC) is an important index to reflect the security and reliability of DN, and it is determined by the operation mode [4]. However, the factors that affect the output of DG are bound to affect the analysis of the LSC. Thus when LSC is studied, the operation characteristics of ADN must be considered.

The concept of the LSC of the DN is used for reference to that of the transmission network [5]. In view of this problem, some scholars make an economic distribution on the initial operating points according to the economic cost of power generation, without considering other operating points [6]. On this basis, some scholars coordinate the relationship between the LSC and the optimal scheduling considering the contradictions of power system security and economy [7]. For ADN, the analysis of the LSC when DGs connected to DN has achieved fruitful results. For example, the utilization of different methods to represent the uncertainty of wind power, photovoltaic power and other uncontrollable DGs can strengthen accuracy and credibility of the LSC calculation [8]. The installation

IOP Conf. Series: Materials Science and Engineering **366** (2018) 012041 doi:10.1088/1757-899X/366/1/012041

of energy storage power station can not only enhance the LSC of the network, but also increase the flexibility of DN [9]. It can adjust the output and reduce the uncertainty of the uncontrollable DGs by using complementary scheduling rules based on the prediction results [10].

In ADN, some of the DGs are controllable and they participate in the economic operation. Therefore, when establishing the LSC model of ADN, the active management of controllable DGs and the uncertainty of uncontrollable DGs should be considered.

A bi-level optimization model for the LSC of ADN is proposed in this paper, of which the upper level model is used to calculate the LSC and the lower level model make a dispatch considering the economic operation of the ADN. The opportunity constrained programming is utilized in this model to deal with the uncertainty of uncontrollable DGs.

2. Influence factors of the LSC

When studying the LSC of ADN, the main factors that affect the accuracy and rationality should be firstly considered, and the model should be built based on them.

2.1. The influence of the economic dispatch of the ADN

Compared with traditional DN, ADN has the ability to make an active management and scheduling, which will affect not only the output of DGs but also the distribution of voltage and power flow. Thus the economic dispatch is necessary for each operating point in the process of calculating the LSC. The coordination of economy and the maximization of LSC need to be considered as they have a game relationship.

2.2. The influence of uncontrollable DGs' uncertainty

The output of the uncontrollable DGs such as wind power generation and photovoltaic power generation is uncertain, which will cause the uncertainty of the power flow of ADN is uncertain. That will complicate the solution process. In this paper, random variables are utilized to express the output of uncontrollable DGs. And a model with opportunity constrained programming is built.

2.3. Logical architecture of the Double layer model

As is stated above, economic dispatch needs to be considered when the LSC is calculated. The bi-level double optimization model is suitable for this problem, which is as figure 1 shows. The upper level model is designed to calculate the LSC of ADN, while the lower level model makes the economic dispatch. Meanwhile, the chance constrained programming theory is used in the model to deal with the uncertainty of uncontrollable DGs. The LSC considering the operation of ADN can be obtained, and the contradiction between the safety and economy of operation can be solved effectively.

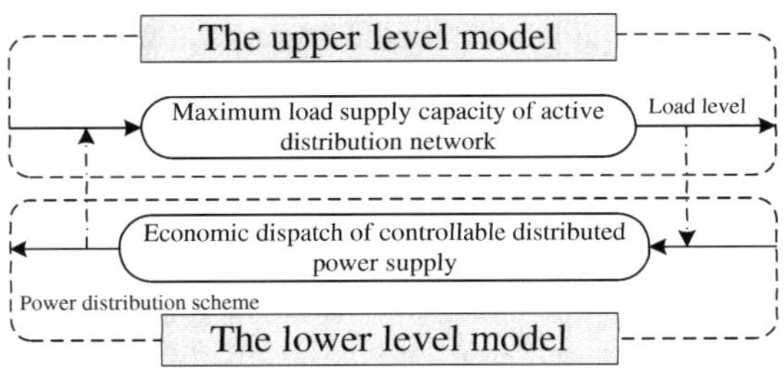

Figure 1 The frame diagram of bi-level model

3. Model establishing and solving

3.1. The upper level model

3.1.1. The objective function. According to the definition, the LSC refers to the maximum load power supply, which is determined by the mode of operation and the load growth, when all the operation constraints are satisfied. In this paper, the load is supposed to grow by equal proportions. The growth mode of the controllable DGs and the load can be expressed as follows:

$$\begin{cases} P_{\mathrm{d}i} = P_{\mathrm{d}i}^{0}\left(1+\lambda K_{\mathrm{d}i}\right) \\ Q_{\mathrm{d}i} = Q_{\mathrm{d}i}^{0}\left(1+\lambda K_{\mathrm{d}i}\right) \\ P_{\mathrm{DG}i} = P_{\mathrm{DG}i}^{0}\left(1+\lambda K_{\mathrm{DG}i}\right) \end{cases} \tag{1}$$

where $P_{\mathrm{d}i}$ and $P_{\mathrm{DG}i}$ are respectively the load demand and the active power output of the controllable DGs at node i, while $P_{\mathrm{d}i}^{0}$ and $P_{\mathrm{DG}i}^{0}$ are the initial state. λ is the load growth factor. $K_{\mathrm{d}i}$ represents the load growth mode and $K_{\mathrm{DG}i}$ is the growth mode of active power output of controllable DGs.

Thus the objective function of the upper level is to maximize the load supplying in the ADN:

$$\max P_{\max} = \sum_{i=1}^{N} P_{\mathrm{d}i}^{0}\left(1+\lambda K_{\mathrm{d}i}\right) \tag{2}$$

*3.1.2. Constraint condition.*In order to ensure the safe and stable operation of the ADN, the constraints of the upper model are as follows:

1) The constraint of power flow balance

$$P_{\mathrm{G}i} + P_{\mathrm{DG}i}^{0}\left(1+\lambda K_{\mathrm{DG}i}\right) + P_{\mathrm{RES}i} - P_{\mathrm{d}i}^{0}\left(1+\lambda K_{\mathrm{d}i}\right) - U_i \sum_{j=1}^{N} U_j \left(G_{ij}\cos\theta_{ij} + B_{ij}\sin\theta_{ij}\right) = 0 \tag{3}$$

$$Q_{\mathrm{G}i} + Q_{\mathrm{DG}i} - Q_{\mathrm{d}i}^{0}\left(1+\lambda K_{\mathrm{d}i}\right) - U_i \sum_{j=1}^{N} U_j \left(G_{ij}\sin\theta_{ij} - B_{ij}\cos\theta_{ij}\right) = 0 \tag{4}$$

2) The constraint of node voltage

$$\Pr\left\{U_{i,\min} \le U_i \le U_{i,\max}\right\} \ge \beta \tag{5}$$

3) The constraint of line capacity

$$\Pr\left\{\left|P_{ij}\right| \le P_{ij,\max}\right\} \ge \beta \tag{6}$$

4) The constraint of transformer

$$S_{\mathrm{T}} \le S_{\mathrm{T,max}} \tag{7}$$

5) The constraint of distributed generation supply

$$P_{DGi,\min} \le P_{DGi}\left(1+\lambda K_{DGi}\right) \le P_{DGi,\max} \tag{8}$$

$$Q_{\mathrm{DG}i,\min} \le Q_{\mathrm{DG}i} \le Q_{\mathrm{DG}i,\max} \tag{9}$$

where $P_{\mathrm{RES}i}$ is the active power output of uncontrollable DG at node i. U_i and θ_i are voltage amplitude and phase angle of node, respectively. P_{ij} is the transmission power for line ij. S_T is the power flow out from the transformer T. Parameters with the subscript max and min represent the maximum and minimum of the corresponding variables.

3.2. The lower level model

*3.2.1. The objective function.*The objective function of the lower level model is to minimize the operation cost of the ADN system.

$$\min F = \sum_{i=1}^{N_1} C_{\mathrm{DG}i}\left(P_{\mathrm{DG}i}\right) + C_{\mathrm{pt}} P_{\mathrm{grid}} \tag{10}$$

where N_1 is the number of the controllable DGs and C_{DG} is the operation cost of them. P_{grid} is the power purchased from the higher level grid and C_{pt} is the electricity price.

The operation cost function of a diesel generator is as formula (11) shows and that of a gas turbine is as formula (12) shows.

$$C_{\mathrm{DE}} = aP_{\mathrm{DE}}^2 + bP_{\mathrm{DE}} + c \tag{11}$$

$$C_{\mathrm{MT}} = \left(C_{\mathrm{ng}} / LHV_{\mathrm{ng}}\right) \cdot \frac{P_{\mathrm{MT}}}{\eta\left(P_{\mathrm{MT}}\right)} \tag{12}$$

3.2.2. Constraint condition

1) The constraint of power flow balance

$$P_{\mathrm{DG}} + P_{\mathrm{RES}} + P_{\mathrm{grid}} - P_{\mathrm{d}} - P_{\mathrm{Loss}} = 0 \tag{13}$$

2) The constraint of distributed generation supply

$$P_{\mathrm{DG}i,\min} \le P_{\mathrm{DG}i}\left(1 + \lambda K_{\mathrm{DG}i}\right) \le P_{\mathrm{DG}i,\max} \tag{14}$$

3) The constraint of the power interaction with the large power grid

$$P_{\mathrm{grid,min}} \le P_{\mathrm{grid}} \le P_{\mathrm{grid,max}} \tag{15}$$

3.3. The solution of the model

Alternative and iterative algorithm is utilized to solve the bi-level optimization problem in this paper. As there are random variables in the upper level model, the particle swarm optimization (PSO) algorithm based on Monte Carlo simulation is utilized. The lower level model is solved by interior point method. The solution procedure is as follow:

1) Initialize the related parameters. In this part, the parameters of the DN are obtained and the basic parameters of PSO are given.
2) Solve the upper level model. The LSC of ADN is obtained and the load growth factor is determined.
3) Information transfer. The load growth factor is transferred to the lower level optimization model.
4) Solve the lower level model. According to the transmission load growth factor, the economic dispatch and the growth mode of the controllable DGs can be achieved.
5) If the difference between the two adjacent optimization results is within the range of precision requirements, the optimal results can be obtained. Otherwise, update the growth mode of controllable DGs and return to step 2.

4. Case analysis

In this paper, the improved IEEE-33 distribution system with wind, photovoltaic and diesel generators is presented for calculation. Parameters of the DN is shown in [11]. An uncontrollable 200kw photovoltaic generator is connected to node 30. Two uncontrollable 300kw wind power generators are connected to node 7 and node 24, respectively. The light intensity and the local error of wind power prediction are both satisfies Beta distribution. Two 600kw diesel generators are respectively connected to node 14 and node 32, and a gas turbine is connected to node 25. The confidence level of the opportunity constraints is set to 0.95. The learning factors of PSO c_1 and c_2 are both 2. The population size is 40. The parameters r_1 and r_2 representing the random number within 0 to 1. The maximum iteration number is 1000, and the initial value of inertia weight is 0.9. C_{ng} is set to 2.5 yuan/m^3 and LHV_{ng} is set to 9.7 kWh/m^3.

Table 1 shows the results contrast of LSC and the cost between the paper's mode and the conventional mode.

Table 1 LSC and operation costs of ADN

model	cost/yuan	LSC /kW
the paper's mode	3098.1	4746
the conventional mode	3187.3	4769

The LSC and operation cost of ADN are reduced when the economic dispatch is considered in the model. Thus, the LSC that meets the requirements of economic operation can be obtained in this model. And the coordination of safety and economy of ADN can be achieved.

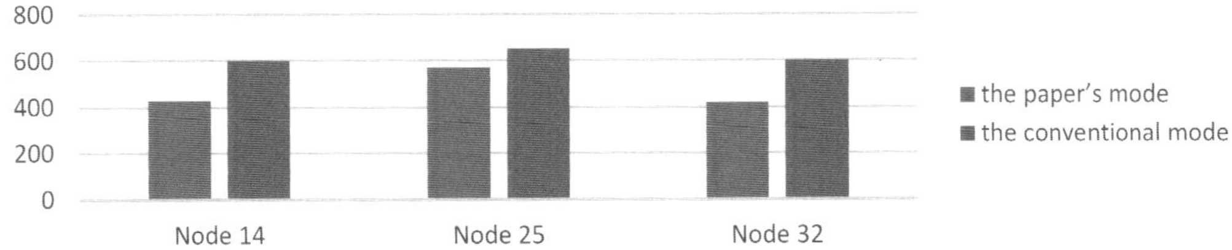

Figure 2 Active outputs of controllable DGs

Figure 2 shows the output of the controllable DGs in the two modes. The outputs of controllable DGs are allocated economically considering the influence of the economic dispatch in the model. When load reaches a certain level, more power will be purchased from the higher level grid to replace that supplied by the controllable DGs, because the operation cost of the controllable DGs become much higher. The model in this paper can enable the resources of the ADN to be economically and rationally utilized. And the operation cost is saved under the premise of ensuring safety.

The line flow distribution in two modes is as figure 3 shows. The main factor to restrict the increase of the LSC is the power flow of line 14 and 17 without considering the economic dispatch. The power flow constraints of line 14 and 17 do not reach the upper limits because of the economic operation restriction in the bi-level model.

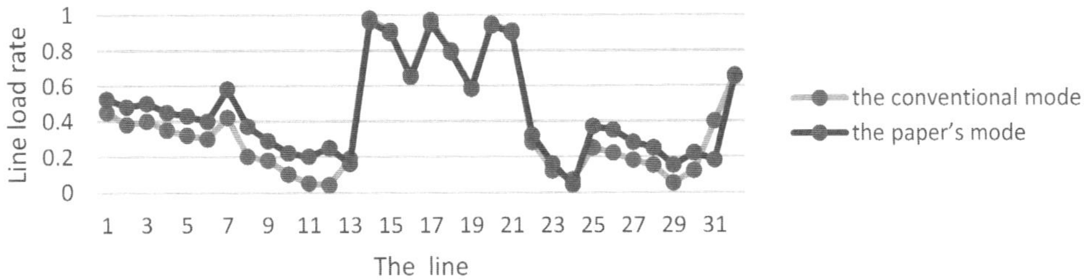

Figure 3 Comparison of line load rate under two modes

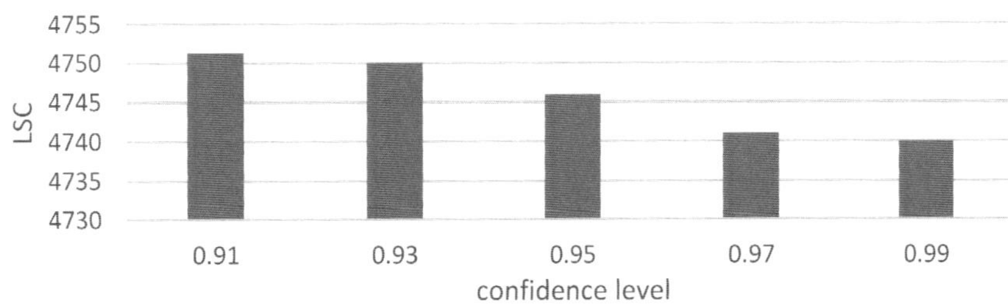

Figure 4 Comparison of LSC under different confidence level

Figure 4 lists the LSC under different confidence levels. As the figure shows, chance constrained programming model can realize the safety constraint of DN under a certain probability. But a certain expense of objective will be suffered because of the uncertainty. The higher the confidence level is, the stricter the security constraints are and the greater the expense of the objective will be.

5. Conclusion

In this paper, a model for calculating the LSC of ADN considering the influence of uncontrollable DGs with uncertainty is built. Through the analysis of the improved IEEE33 distribution system, we obtain the following conclusion.

1) A bi-level model for calculating the LSC considering the economic operation of the ADN is

established. The economic dispatch of the ADN will reduce the LSC of the ADN to a certain extent, but the economy of operation can be ensured and the rationality of the resource utilization can be improved.

2) The chance constrained programming is utilized to deal with the uncertainty of uncontrollable DGs in this model flexibly. The calculation results show that the higher the confidence level is, the more stringent the security constraint is and the smaller the LSC will be.

Acknowledgments

This work has been supported by the Science and Technology Project of State Grid (2017YF-27).

References

[1] Hidalgo R, Abbey C, Joós G 2010 *IEEE Power and Energy Society General Meeting* p 1
[2] Koutsoukis N C, Georgilakis P S, Hatziargyriou N D 2018 *IEEE Trans. on Power Systems* vol 33 p 32
[3] Ai Q, Fan S, Piao L 2016 *Protection & Control of Modern Power Systems* vol 1 p 1
[4] Miu K N, Chiang H D 2000 *IEEE Trans on Power Delivery* vol 15 p 436
[5] Wu D, Li G, Yang J, Wu J and Zhan Z 2014 *IEEE Power and Energy Engineering Conference* p 1
[6] Gnanadass R, Manivannan K, Palanivelu T G 2003 *IEEE Conference on Convergent Technologies for the Asia-Pacific Region* vol 1 p 445
[7] Li G, Tang B 2006 *Proceedings of the CSEE* vol 26 p 18
[8] Christakou K, Paolone M, Abur A 2017 *IEEE Trans. on Smart Grid* vol PP p 1
[9] Nick M, Cherkaoui R, Paolone M. 2017 *IEEE Trans. on Power Systems* vol PP p 1
[10] Martins V F, Borges C L T 2011 *IEEE Trans. on Power Systems* vol 26 p 2164
[11] BARAN M E, WU F E 1984 *IEEE Trans. on Power Delivery* vol 4 p 1401

Real time human motion recognition via spiking neural network

Yang Jing[1], Wu Qingyuan[2], Huang Maiqi[3] and Luo Ting[4]

Beijing Normal University, Zhuhai campus, Zhuhai, Guangdong, China

[1] yangjing@bnuz.edu.cn; [2] wuqingyuan@bnuz.edu.cn; [3] huangmaiqi@outlook.com;
[4] Gloria_LuoTing@163.com

Abstract. Real time human action recognition is to recognize the human motion type based on skeleton movement in real time and is always a challenging task. In this paper, a novel method is proposed to accomplish the classification by using Spiking neural network (SNN) which is biology oriented neural network dealing with precise timing spikes. First, a new temporal encoding scheme is used to encode the real time motion capture data into a series of spikes and the according type of the motion is represented by a spike time. Second, a two-layered spiking neural network is initiated and trained through a gradient descent learning algorithm. The experimental results show that this method achieves a good learning precision and generalization.

1. Introduction

With the rapid development of motion capture techniques and systems, more and more motion data are available for human action recognition in the research area of computer vision. Especially depth cameras can provide 3D depth data [1] of joint positions of the human skeleton which are showed in figure 1. That may be more helpful in recognition of the motion type than just using 2D data. However the depth map also increases the amount of the data and makes the recognition process more complex. As we know the human body is an articulated system of rigid segments connected by joints. The shape information carried by the contour can be extended to the 3D case for motion recognition [2]. The 3D depth data not only facilitates a rather powerful human motion capturing technique, but also makes it possible to efficiently model human-object interactions and intra-class variations [3].

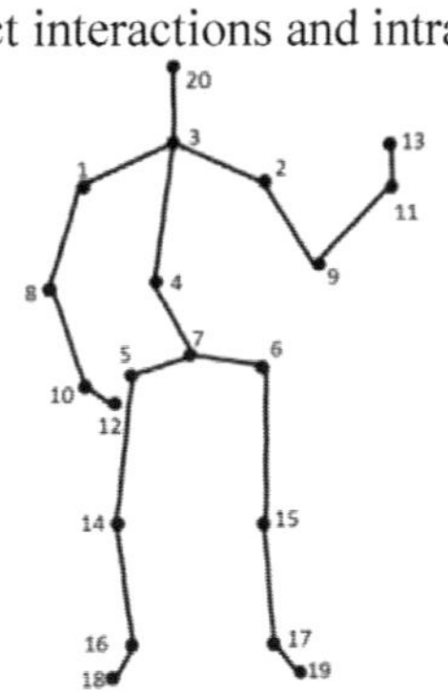

Figure 1. Skeleton of human body.

SNN is the third generation neural network model and its inputs and outputs are both multiple spikes, i.e., spike trains. It's widely accepted that artificial SNNs can stimulate any sigmoidal neural networks and even capable of exploiting time as a resource for coding and computation in a much more sophisticated manner than typical neural computational models [4]. As SNN deals with spike times the encoding of information is an important step in its applications. There are two ways of encoding, rate encoding and temporal encoding. Supervised learning algorithm is another hottest research area of SNN. For the learning of spiking neuron (SN), ReSuMe (remote supervised method)[5] is based on the rules of STDP and anti-STDP and Xu and el. [6] proposed another method based on gradient descent. For the learning of SNN, SpikeProp [7] is also a gradient-descent-based learning algorithm and there some improved algorithm based on SpikeProp. However only single spike is allowed to fire in SpikeProp and that narrows the application.

In this paper, we propose a novel real time motion data recognition approach. This method encodes the motion data into several spike trains and trains the motion samples with a two-layered SNN. The main contribution of this paper lies in the new encoding scheme of the motion data .The arrangement of the paper is as follows. The SNN model is briefly introduced in Section 2. Section 3 gives the encoding scheme of motion data and discusses the learning process of SNN. In Section 4 some experiments are conducted to demonstrate the efficiency of the classification method. Finally a conclusion is drawn in Section 5.

2. SNN model

The multilayer SNN is a fully connected feedforward network. Each neuron in the network is a spiking neuron (SN) and there are three main SN models, Leaky Integrate-and-Fire model (LIF), Hodgkin-Huxley model (HH) and Spike Response model (SRM). Because the gradient descent learning algorithms require computation of partial derivatives we take the SRM, whose internal state can be expressed intuitively, as SN model in this paper. As the figure 2 shows, the neuron receives all the weighted spikes from the presynaptic neurons via several synapses. When the neuron's internal state exceeds the threshold it fires a spike and then drops immediately to zero.

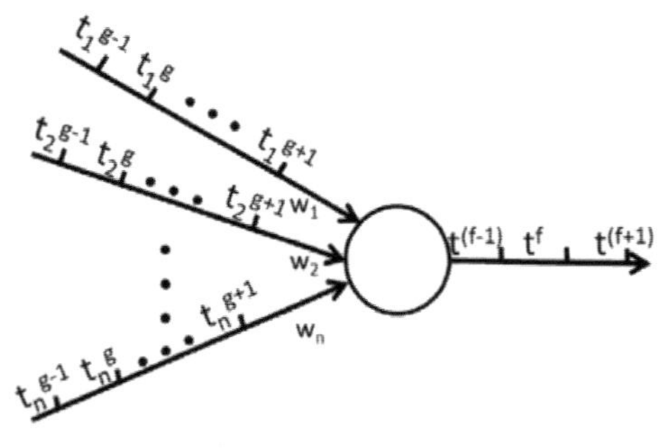

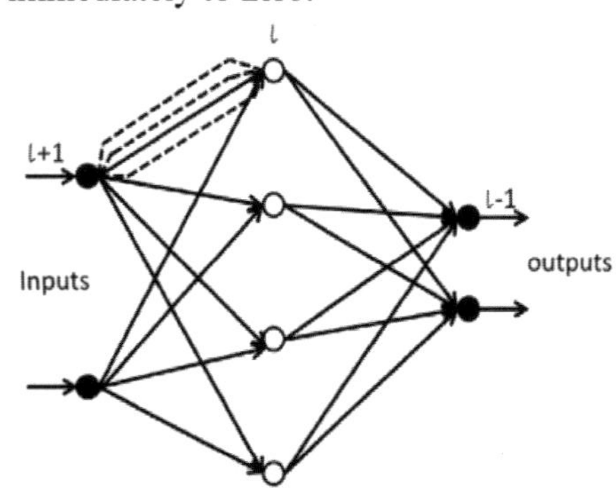

Figure 2. Spiking neuron. **Figure 3.** SNN architecture.

Figure 3 is the architecture of SNN. Between two neurons in adjacent layers there are several synapses with different transmit delays and weights, as shown in Fig 3 with dotted line. For the neuron j in layer l, its internal state at time t is expressed as,

$$u_j(t) = \sum_{i=1}^{N_{l+1}} \sum_{k=1}^{K} \sum_{\substack{t_i^{(f)} \in F_i \\ t_i^{(f)} + d^k > t_i^{(fr)} + R_a}} w_{ij}^k \varepsilon(t - t_i^{(f)} - d^k) + \eta(t - t_i^{(fr)}) \quad (1)$$

where N^{l+1} is the number of neurons in layer $l+1$, K is the number of synapses between neuron l and its presynaptic neuron, $F_i = \{t_i^{(1)}, t_i^{(2)}, ..., t_i^{(F_i)}\}$ is the spike train fired by neuron i in layer $l+1$ and $t_i^{(fr)}$ is the time of the most recent output spike for neuron j prior to the current time t , d^k is the

k th synapse's delay, w_{ij}^k is the weight of the k th synapse between presynaptic neuron i and postsynaptic neuron j, R_a is the length of the absolute refractory period which means the input spikes futile in the period of ($t^{(fr)}, t^{(fr)} + R_a$). $\varepsilon(t)$ is the spike response function expressed as

$$\varepsilon(t) = \begin{cases} \frac{t}{\tau} e^{1-\frac{t}{\tau}} & \text{if } t > 0 \\ 0 & \text{if } t \le 0 \end{cases} \tag{2}$$

where τ is the time decay constant. $\eta(t)$ expressed in formula 3 is refractoriness function and it can make internal state drop immediately after firing to prevent firing in succession. ϑ is the neuron threshold and τ^R is the time decay constant.

$$\eta(t) = \begin{cases} -2\vartheta e^{-\frac{t}{\tau^R}} & \text{if } t > 0 \\ 0 & \text{if } t \le 0 \end{cases} \tag{3}$$

3. Motion data recognition

3.1. Motion data encoding

As the inputs of the spiking neural network is a series of spikes it's an important step to encode the captured motion data into spike times. There are different ways to encode the inputs. For example, population encoding scheme is to encode each input data separately by M (>2) identically shaped overlapping Gaussian functions centered at M different locations. And the input data is finally encoded as M spike times between 0 and 10 ms. Apparently, this kind of encoding method produces artificial spike times which imply no real time sequences. As the human motion data is composed of all the joints' position information at each frame it has a close connection with time. The spiking time in SNN is a kind of relative time interval so the frame in the motion capture can just be considered as spiking time interval.

Based on these explanations we introduced a new coding method based on the timely human joint position sequences in this paper. First, we consider joint i in the human body. It has 3 coordinates information, that is, $p_i(t) = (x_i(t), y_i(t), z_i(t))$ at a frame t. For x-coordinate, the following figure shows how the position of one joint changes with time.

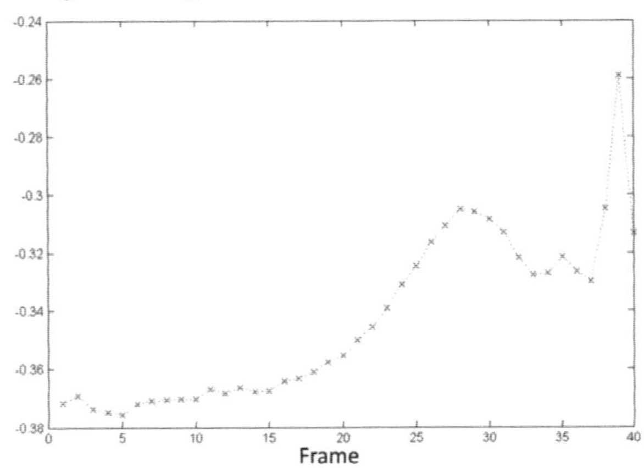

Figure 4. X-coordinate position change of on joint

From the above figure, we can see that the joint's position rises or drops from time to time. Each local lowest point means the joint's position will rise at the next frame and each local highest point means it will drop at the next frame. We name these key frames because they indicate the change in direction of joint movement. We divide the key frames into two inputs, that is, input for local highest point and input for local lowest point and set an input spike at each local extreme point. So the information in Figure.4 can be encoded into two spike trains which are showed in the following figure.

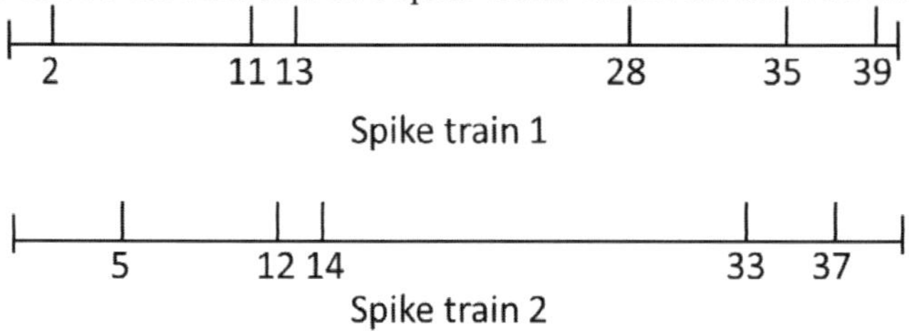

Figure 5. Parts of encoded spike train of one joint.

It can be seen in Fig 5 that the spike train 1 have six input spikes which means there are six movement shifts and each one represents the change of movement direction. And the spike train 2 means the adverse direction movement shift. In this encoding scheme the movement range and velocity information is abandoned. We think the movement range plays less important role in the motion recognition as a huge, fast wave and a small, slow wave should be put into the same category. But there is another import information should not be omitted, the initial position. This information shows where the joint begins to move so the initial position of each joint should also be encoded. Moreover, to avoid the inconsistency of position data caused by the human movement all the joints' positions are normalized based on the benchmark of joint 7 showed in Figure. 1. The initial position of each joint can be encoded by population encoding scheme as an input to SNN.

3.2. *Learning*

Supervised learning algorithm based on temporal encoding of spikes is adopted in the training of SNN in this paper. The goal of SNN learning is to make the output neuron fire spikes at desired time that is, minimize the error function defined as the following formula,

$$E=\frac{1}{2}\sum_{j=1}^{N_1}\sum_{f=1}^{F_j}(t_j^{(f)}-\hat{t}_j^{(f)})^2 \tag{4}$$

where $t_j^{(f)}$ denotes the desired output and $\hat{t}_j^{(f)}$ denotes the actual output. The weight updating rule based on gradient descent is expressed as

$$\Delta w_{ij}^k=-\alpha\nabla E_{ij}^k \tag{5}$$

The detailed weight adjustment steps have been described in reference 6. The number of synapse between two neurons should be designed carefully because few synapses results in reduced convergence rates and classification accuracy whereas more synapses reduces the learning efficiency with no improvement in convergence or classification accuracy. Actually setting synapses aims to add the number of adjustable weights to improve the convergence of learning. If a motion data has 20 joints and each one has three coordinates there would be enough inputs after encoding. So adding synapses between input layer and hidden layer can be skipped. The neuron threshold should also be considered. For this motion recognition example, there are many input spikes after encoding and the running time is relatively long so too small threshold like 1 would lead to frequently firing. In this case the threshold can be increased one by one until the frequency of firing is appropriate.

3.3. *Readout*

The readout part aims to extract information about the stimulus from responses of the learned SNN. There are different ways to represent a certain class of patterns. For example, the number of output neuron can be set equal to the number of the motion type and each output neuron represents one motion type or we can use several SNNs with each net to learn one class or we can use just one output neuron indicating the class by different output spike train. Note that the set of desired output spike train may have a significant influence on the experimental results. So a possible solution is to compute the output of the SNN before learning and observe the possible output distribution. Making the SNN's output neuron fire spikes at the exact desired time is not easy. A frequently used method is to set a tolerance t for the desired time $t_j^{(f)}$. The output spike between $[t_j^{(f)}-t, t_j^{(f)}+t]$ can be considered as a correct output.

4. Experiments

4.1. Dataset and experiment setup
The motion data used in our experiments comes from MSR-Action3D dataset available at http://www.uow.edu.au/~wanqing/#MSRAction3DDatasets. This dataset is skeleton data in real world coordinates contains twenty motion types. The 3D joint positions are extracted from the depth sequence. Every motion has twenty joints information. The SNN in this experiment has one hidden layer with 10 hidden neurons and one output layer with just one output neuron. There is no synapse between input neuron and hidden neuron but 5 synapses between hidden neuron and output neuron and the time delay of the synapse is 1, 3, 5, 9 and 13. The parameters γ and M in the population encoding are 1.5 and 4 individually. The neuron parameters used here are as follows: $\tau=11$ ms, $\tau_R=80$ ms, $\theta=4$ and $R_a=1$ ms. A SNN with one output neuron is used to learn the motions and indicate the motion type and the output spike train consists of two spikes. 60% samples are used to train the SNN and the rest samples are used to test the recognition accuracy.

4.2. Results
The following table 1 shows learning accuracy and testing accuracy to different learning epoch. As the large amount of motion data would lead to the increase of learning time we divide the dataset to three different action sets. Each action set has three types of motion data and all the data is encoded into several spike trains. The results show that after learning the SNN can recognize more than 80% training samples while the generalization is about 75%.

Table 1. Recognition accuracy to different learning epoch

Data set	epoch	Accuracy	
		Training set	Test Set
Action set 1	200	0.94	0.81
	150	0.91	0.7
Action set 2	200	0.85	0.76
	150	0.82	0.70
Action set 3	200	0.90	0.82
	150	0.86	0.75

5. Conclusion
This paper focuses on the application of SNN to do the motion recognition. The 3D coordinates motion data is encoded to several spike trains. The recognition takes advantage of the time character of the spike and the precise time of each joint's moving direction shift is kept as the key input information. Through SNN learning the network can remember and recognize different motion type. Our future work will focuses on further improving the recognition accuracy and recognizing more types of similar motions.

Acknowledgements
This research was supported by the National Science Foundation of China under grants (No.61503031), financially supported by the grant from the Soft Science Research Project of

Guangdong Province(Grant No.2014A030304013), a grant from the Guangdong province Science & Technology Plan(No.2016A040403029) and Guangdong province Philosophy Social Science "12th five" plan Interdisciplinary project（No.GD15XGL02）

References

[1] S Zhang 2010 Real-time 3D shape measurement using digital fringe projection techniques *Optics and lasers in engineering* **48(2)** 149-58
[2] W Li, Z Zhang and Z Liu 2010 Action recognition based on a bag of 3D points *Computer vision &pattern recognition workshops* pp 9-14
[3] J Wang, Z Liu, Y Wu and J Yuan 2014 Learning Actionlet Ensemble for 3D Human Action Recognition *Springer International Publishing* **36 (5)** 1290-97
[4] Maass W 1997 Noisy spiking neurons with temporal coding have more computational power than sigmoidal neurons *Advances in Neural Information Processing System* **9** 211-7
[5] Ponulak F, Kasinski A 2010 Supervised learning in spiking neural networks with ReSuMe: sequence learning, classification, and spike shifting *Neural Computation* **22 (22)** 467-510
[6] Xu Y, Zeng X, Han L and Yang J 2013 A supervised multi-spike learning algorithm based on gradient descent for spiking neural networks *Neural Networks* **43(4)** 99-113
[7] S M Bohte, J N Kok and J A La Poutre 2002 Error-backpropagation in temporally encoded networks of spiking neurons *Neurocomputing* **48(1)** 13-37

Real-time prediction of sub-item building energy consumption based on PCA-AR-BP method

Qing Qian[1], Guizhong Tang[1, 2] and Guangming Zhang [1,2]

[1] School of Electrical Engineering and Control Science, Nanjing University of Technology, Nanjing 211800,PR China
[2] Jiangsu Province Green Building Engineering Technology Research Center

Corresponding author: gztang361@126.com (Guizhong Tang)

Abstract. In this paper, a new method for real-time prediction of building energy consumption is proposed, this method solves the problem that the kinds of energy consumption are not distinguished and the prediction accuracy is low in the current energy consumption prediction algorithms. This paper divides the total energy consumption into four sections. Firstly, three main influencing factors of building energy consumption are extracted using PCA to realize real-time prediction; Secondly, the method of lighting energy consumption prediction based on time series analysis is constructed, the lighting energy consumption of the building is predicted in real time. Finally, the energy consumption prediction model based on BP network is established to predict the air conditioning, power and special energy consumption of the building. The experimental results show that the prediction model can predict energy consumption in every part of a building more accurately and effectively.

1. Introduction

With the rapidly increasing of annual energy consumption in office buildings and public buildings, building energy consumption has become the main target of building energy efficiency regulation and transformation [1]. At present, the main building energy consumption prediction methods are multiple linear regression [2], artificial neural network [3-5], time series method, genetic algorithm [6] and so on. Some scholars have also studied the prediction of building energy consumption through time series analysis [7-9]. However, the total energy consumption is used to structure the prediction model in almost all existing literatures, and most of the predictions are long-term energy consumption prediction. Therefore, the direction of the whole building energy consumption can't be predicted, and the real-time prediction of energy consumption can't be well achieved.

In this paper, a new method for real-time prediction of building energy consumption based on sub-item measurement is proposed. Firstly, 3 main influencing factor of building energy consumption was extracted by PCA. Then the real-time prediction of lighting energy consumption can be achieved. At the same time, energy consumption, energy consumption building air conditioning (lift, water supply etc.) and special energy (computer room, kitchen etc.) prediction model can also be structured, which achieve the real-time prediction of the sub term energy consumption. The real time prediction system of building energy consumption can optimize the imbalance between energy consumption systems and power supply systems (municipal power supply system, solar power system and building ontology building wind power system etc.) of large public buildings, which ensure the stability and utilization efficiency of municipal power grid renewable energy.

IOP Conf. Series: Materials Science and Engineering **366** (2018) 012043 doi:10.1088/1757-899X/366/1/012043

2. Prediction feature selection based on PCA

In order to realize the real time prediction of sub item energy consumption of construction, this paper uses PCA to extract the main prediction features.

For the given 7 prediction features $C = [C_1, C_2, \cdots, C_7]$, the normalization is made according to the formula (1).

$$T_i(k) = \frac{1}{C_{\max}(i) - C_{\min}(i)}[C_i(k) - C_{\min}(i)] \tag{1}$$

$C_i(k)$ is the *kth* sample point of *ith* influence factors, $C_{\max}(i)$ and $C_{\min}(i)$ are the maximum and minimum of all the sampling points of *ith* influence factors, $T_i(k)$ is standardized target data, The covariance of $T_i = [T_i(1), T_i(2), \cdots, T_i(n)]$ and the correlation coefficient matrix ρ_i are as follows:

$$\delta_i(s,t) = \frac{1}{n-1}\sum_{j=1}^{n}[(T_i(s) - \overline{T}_i)(T_i(t) - \overline{T}_i)] \tag{2}$$

$$\rho_i = \left[\frac{\delta_i(s,t)}{\delta_i(s,s)\delta_i(t,t)}\right] \tag{3}$$

Among them, $\overline{T}_i = \frac{\sum_{j=1}^{n} T_i(j)}{n}$ is the number of samples. By calculating the eigenvalues λ_i of the correlation coefficient matrix ρ_i, the calculation formula of the variance contribution rate and the cumulative variance contribution rate of each principal element of the PCA algorithm is as follows:

$$\alpha_i = \frac{\lambda_i}{\sum_{i=1}^{7}\lambda_i} \tag{4}$$

$$\eta_i = \sum_{k=1}^{i}\alpha_i, \quad i = 1,2,\cdots 7 \tag{5}$$

The contribution rate and cumulative contribution rate of all principal components are shown in Figure 1. In this paper, when $\eta_i \geq 85\%$, the first 3 principal components with the largest contribution rate are used as the prediction and prediction characteristics of building energy consumption. The prediction features of the selection are lighting energy consumption, indoor temperature, and 24 clocks in one day.

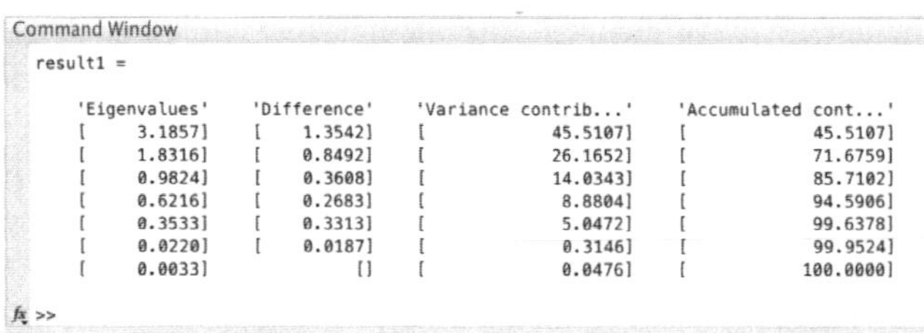
Command Window

result1 =

'Eigenvalues'	'Difference'	'Variance contrib...'	'Accumulated cont...'
[3.1857]	[1.3542]	[45.5107]	[45.5107]
[1.8316]	[0.8492]	[26.1652]	[71.6759]
[0.9824]	[0.3608]	[14.0343]	[85.7102]
[0.6216]	[0.2683]	[8.8804]	[94.5906]
[0.3533]	[0.3313]	[5.0472]	[99.6378]
[0.0220]	[0.0187]	[0.3146]	[99.9524]
[0.0033]	[]	[0.0476]	[100.0000]

fx >>

Figure 1. The main element contribution rate

3. Prediction of lighting energy consumption based on time series

3.1. The analysis of time series about lighting energy consumption

Although the lighting energy consumption often appears as a random fluctuation on the basis of past energy consumption in a short time, it belongs to the non-stationary time series, but the high-order difference of lighting energy consumption is stable. Such as, collecting the 24 hours lighting energy value of one day is $[x_1, x_2 \cdots, x_{24}] = [49.8, 51.6, \cdots, 218.2]$ unit(Kwh), which x_i represents the energy consumption value of the lighting at time i. As shown in Figure 2, the seven order difference of lighting energy consumption is shown to be stationary, so the time series analysis method can be used to predict lighting energy consumption.

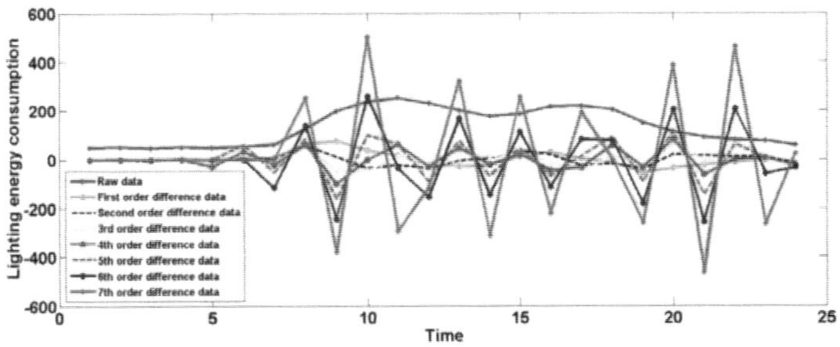

Figure 2. The high-order difference graph of lighting energy consumption

3.2. Construction of real-time prediction model for lighting energy consumption

In this section, we use the *P* order autoregressive model to build the prediction model of lighting energy consumption. The following is the definition of the auto correlation coefficient and partial auto-correlation coefficient of lighting energy consumption, so as to determine the value of order *P*.

Definition 1 Auto-correlation of lighting energy consumption

The 7-order difference value of lighting energy consumption during the 24 hours of a day is $[df_1, df_2, \cdots, df_{24}]$, and the linear dependence between df_{i+k} and df_i is defined by:

$$p(k) = \frac{r(k)}{\sqrt{Var(x_i)Var(x_{i+k})}} \tag{6}$$

where $r(k) = Cov(df_i, df_{i+k})$ is the self-covariance, $Var(df_i)$ is the variance, which represents k lag number. Because the variance of the stationary time series is equal, $Var(df_{i+k})$ is then given by:

$$Var(df_{i+k}) = Var(df_i) = \cdots Var(0) \tag{7}$$

Based on $Var(0) = r(0) = \sigma^2$, formula (6) can be written as:

$$p(k) = \frac{r(k)}{r(0)} \tag{8}$$

So, $p(k)$ can be calculated by the recursive formula (8).

Definition 2 Partial auto-correlation coefficient of lighting energy consumption

Defining the *k-1* random difference variables in the middle of all the variables as $[df_i, df_{i+1} \cdots, df_{i-k+1}]$, the correlation from df_{i+k} to df_{i-1} is defined by:

$$\phi(k) = \frac{D(k)}{D} \tag{9}$$

IOP Conf. Series: Materials Science and Engineering **366** (2018) 012043 doi:10.1088/1757-899X/366/1/012043

where, $D=\begin{bmatrix}1 & \rho_1 & \cdots & \rho_{k-1}\\ \rho_1 & 1 & \cdots & \rho_{k-1}\\ \cdots & \cdots & \cdots & \cdots\\ \rho_6 & \rho_5 & \cdots & 1\end{bmatrix}$, $D(k)=\begin{bmatrix}1 & \rho_1 & \cdots & \rho_1\\ \rho_1 & 1 & \cdots & \rho_2\\ \cdots & \cdots & \cdots & \cdots\\ \rho_{k-1} & \rho_{k-2} & \cdots & \rho_k\end{bmatrix}$

4. Experimental verification and analysis

The experimental data come from an hourly collected energy consumption from a building from July 1st to July 31st, 2016. It uses the BP neural network for prediction.

4.1. Real-time prediction results analysis of lighting energy consumption

The actual value of lighting energy consumption of 24 × 4 = 96 whole hours from July 27th to July 30th, 2016 were randomly selected as the training samples to predict the lighting energy consumption value at every 24 whole hours on July 31st.

4.1.1. Verification of lighting energy consumption prediction model. Figure 3 and Figure 4 show the auto-correlation analysis and partial auto-correlation analysis respectively. As can be seen from Figure 3, the process of the auto-correlation coefficient decays to small value fluctuations is slow and continuous, it shows a long tail dragged behind, which is in accordance with the trailing nature. As can be seen from Figure 4, the value of the partial auto-correlation coefficient is damped oscillatory, and the critical value would converge to 0 suddenly when the lag order is 7, and the partial auto-correlation coefficient values are all 0 after the *7th* order, it is consistent with truncation. Therefore, the order of auto-regressive model *AR*(*P*) is $P=7$, it further verifies that the lighting energy consumption prediction is in line with the auto-regressive model $AR(7)$.

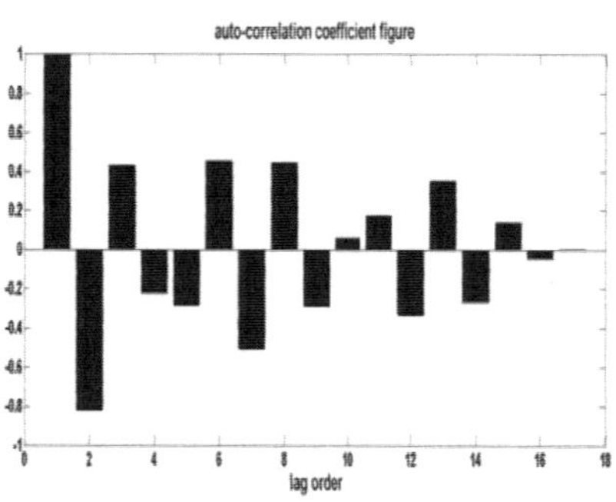

Figure 3. Auto-correlation coefficient analysis

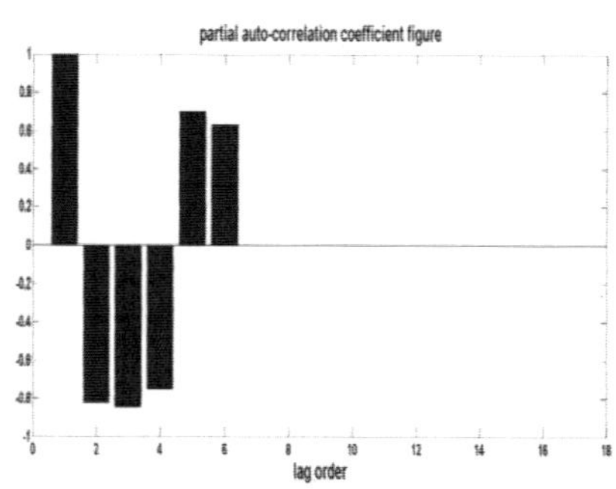

Figure 4. Partial auto-correlation coefficient analysis

4.1.2. The prediction results of lighting energy consumption. Figure 5 presents the predicted and actual lighting energy consumption profiles at 24 whole hours on July 31st. It can be concluded from Figure 5 that the predicted value is in good agreement with the actual value, and the average absolute error (MAE) is 0.0249, and the root mean square error (RMSE) is 5.6778, so it reaches the satisfactory prediction effect.

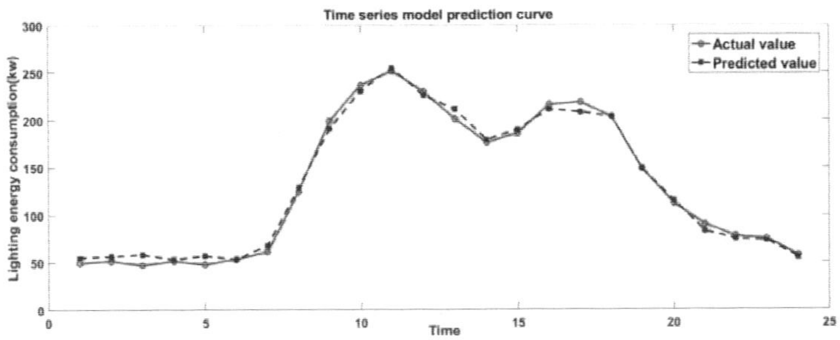

Figure 5. The results of lighting energy consumption prediction

4.2. Verification of air conditioning, power and special energy consumption prediction

The data of a week (from Monday to Sunday) are randomly selected as the training samples, and the energy consumption values of the 24 whole hours in the following week are predicted by the proposed model. The prediction results are shown in Table 1. Table 1 shows the predicted and actual values for working days (e.g. Monday and Tuesday) and non-working days (e.g. Sunday).

Table1. The results of air conditioning, power and special energy consumption prediction

		0 am		3 am		6 am		9 am		12 pm		15 pm		18 pm	
		A[1]	P[2]	A	P	A	P	A	P	A	P	A	P	A	P
	Mon.	45.50	46.81	47.17	46.23	50.12	55.14	178.11	170.09	185.71	186.03	163.34	165.09	108.50	110.12
ACEC[3]	Tue.	48.90	47.12	48.34	49.13	53.84	54.78	192.17	190.21	184.31	186.17	152.81	150.06	100.91	98.43
	Sun.	62.50	63.41	62.21	60.93	63.52	60.43	96.43	97.05	90.63	90.08	76.63	78.87	51.83	50.52
	Mon.	9.40	9.31	10.35	9.12	11.85	10.12	60.65	66.21	52.85	50.54	60.72	61.03	45.34	47.97
PEC[4]	Tue.	9.20	9.17	10.82	10.91	12.52	13.01	64.52	66.05	53.35	55.02	55.25	55.23	50.71	46.89
	Sun.	9.80	10.30	8.35	8.07	11.12	11.14	14.25	14.65	11.43	11.67	14.44	13.09	11.41	11.00
	Mon.	12.98	12.76	14.92	15.11	21.33	20.01	31.53	31.21	40.73	43.91	34.33	32.99	41.57	40.41
SEC[5]	Tue.	13.93	13.81	14.93	14.32	30.48	30.91	39.98	40.32	39.12	40.09	32.13	32.32	40.58	40.21
	Sun.	34.80	35.61	38.62	35.21	32.21	32.13	38.24	37.87	47.81	50.08	56.63	55.32	31.24	30.93

[1] A-Actual. [2] P-Predicted. [3] ACEC-Air conditioning energy consumption.
[4] PEC-Power energy consumption. [5] SEC-Special energy consumption.

The average relative error of the 7-day prediction of air conditioning energy consumption, power consumption and special energy consumption are calculated respectively. The results are shown in Figure 6, where the average relative error of the three kinds of energy consumption prediction is within 0.08, especially the prediction error of air conditioning energy consumption is generally within 0.04, it means that a better prediction effect is achieved.

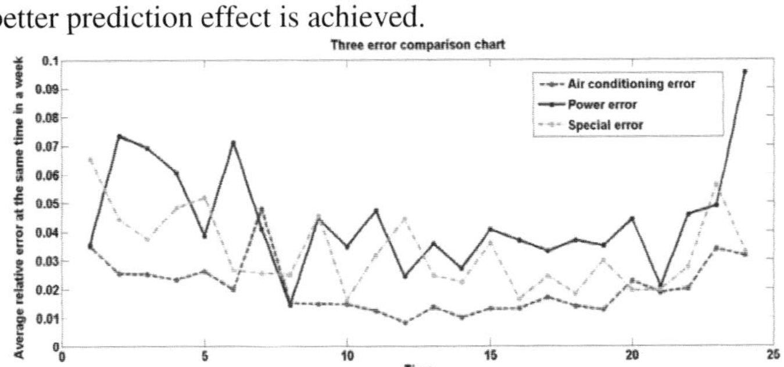

Figure 6. The average relative error of air-conditioning energy consumption, power consumption and special energy consumption prediction

5. Conclusions

This paper constructs an auto-regressive model for predicting building energy consumption, where the lighting energy consumption at 24 whole hours in a day is real-time predicted. The principal component analysis is applied to select the main factors that affect the prediction of sub-item energy consumption, and the speed and accuracy of prediction are improved. Experiments are conducted to test and verify the rationality of the auto-regressive model of lighting energy consumption and the accuracy of the sub-item energy consumption prediction.

Acknowledgements:

This work is supported by Science &Technology Development Center of Jiangsu Housing and Urban-Rural Development Department, who provided the valuable study data of this paper to us. The authors gratefully acknowledge the support of this research by the National Natural Science Foundation of China (Grant No. 51507078), the Jiangsu Provincial Department of Education University Natural Science Foundation (Grant No. 15KJB470006).

References

[1] LI Dao-yang,MENG Yan-mei, HU Ying-ning, et al. Numerical analysis of building energy consumption based on separate real-time monitoring[J]. Journal of Guangxi University: Nat Sci Ed 2012, 37(5):965-971.
[2] A Aranda, G Ferreira, M.D.Mainar-Toledo, et al. Multiple regression models to predict the annual energy consumption in the Spanish banking sector[J].Energy and Buildings, 2012, 49(49):380-387.
[3] LU Kuo, ZHONG Bocheng. Online Forecast of Building Energy Consumption Based on LM BP Neural Network[J]. Journal of shanghai university of engineering science, 2014, 28(4):347-350.
[4] Kangji Li, Chenglei Hu, Guohai Liu, et al. Building's electricity consumption prediction using optimized artificial neural networks and principal component analysis [J].Energy and Buildings, 2015, 108(10):106-113.
[5] R Mena, F.Rodri'guez, M.Castilla, et al. A prediction model based on neural networks for the energy consumption of a bioclimatic building[J].Energy and Buildings, 2014, 82(5):142-155.
[6] Zhang Zhen-ya, Cheng Hong-mei, Liu Jia, et al. Research on the Minimum Number of Mobile Node for Sweep Coverage in Wireless Sensor Network Based on Genetic Algorithm[J]. Journal of Chinese Computer Systems,2013,34(10):2388-2392.
[7] Zhou Rui-jin, Pan Yi-qun, Huang Zhi-zhong. Time Series Analysis Method and Its Application in Energy Consumption Forecast of an Actual Office Building[J]. Building energy conservation, 2012, 252(2):55-62.
[8] Zhou Rui-jin, Pan Yi-qun, Huang Zhi-zhong. Building energy consumption prediction method based on time series analysis[J].HV&AC, 2013, 43(8):71-77.
[9] OUYANG Qian-wu, LI Hui, LI Jun. Application of the ARMA Model in Energy Consumption Forecast of Commercial Buildings[J] Building Energy& Environment, 2013, 32(3):56-58.

A double closed-loop control of VIENNA rectifier based on sliding-mode of fuzzy approaching law

Li Zhanpeng [1,a], Quan Shuhai [2], Xie Changjun [2], Zhang Liyan [2] and Chen Qihong[2]

[1] School of Automation, Wuhan University of Technology, 205 Luoshi Road, Hongshan District, Wuhan 430070, China.
[2]School of Automation, Wuhan University of Technology, Wuhan 430070, China

[a] Corresponding author: 137214164@qq.com (Zhanpeng Li)

Abstract: A novel double closed-loop control of VIENNA rectifier based on sliding-mode of fuzzy approaching law is proposed in this paper. This new control strategy improves dynamic response and steady accuracy of system, chattering phenomenon of the sliding-mode is eliminated. Combined with sliding-mode and fuzzy approaching law, a new type of inner loop power control is proposed to replace traditional current inner ring. Without coordinate transformation and phase-locked-loops(PLL), control algorithm is simple. Under unbalanced three-phase power grid, output voltage still maintains small voltage ripple, network side harmonic suppression and dynamic response perform well. Sliding-mode based on fuzzy approaching law can eliminate the chattering phenomenon of the system, the overshoot produced by load transition can be reduced. The outer ring adopts voltage PI control with real time current compensation, dynamic response performs good and the control algorithm is simple.

1.Introduction[1]

The VIENNA rectifier is commonly used three-level PFC topological structure circuit at present. It has low network side current harmonic component and high power factor. No need for dead-time and voltage stress of each power device is half of output voltage of the direct current (DC)side. It is widely used in fields of communication power and new energy vehicle charging.

The study of VIENNA rectifier is mainly aimed at analysis of the topology model of VIENNA rectifier, optimization of pulse width modulation(PWM) and system control strategy improvement. Mathematical model of VIENNA rectifier was analyzed and traditional double closed-loop control strategy of VIENNA rectifier was introduce[1-4]. The disadvantages of traditional control strategy are slow dynamic response speed and larger overshoot; Double closed-loop control based on sliding-mode to improves dynamic response of VIENNA rectifier. It is difficult to eliminate chattering phenomena and overshoot is huge[5-10]. To solve problems, sliding-mode based on fuzzy approaching law was proposed. The chattering phenomena can be eliminated and advantages of traditional sliding-mode can be retained.

VIENNA rectifier is nonlinear system, traditional double closed-loop control strategy has bad control effect. Based on sliding-mode and fuzzy approaching law, a new double closed-loop control of

This work is supported by Major Projects in Technological Innovation of Hubei Province under Grant 2016AAA046.

VIENNA rectifier was proposed. The outer ring adopts voltage PI control with real-time current compensation, the power value of inner ring compensation can be obtained quick and dynamic response rate can be improved. It also simplifies selection of PI parameters; The inner ring adopts sliding-mode based on fuzzy approaching law. PLL and coordinate transformation are unnecessary. Control algorithm is simple. Under unbalanced three-phase power grid, output voltage still maintains small voltage ripple, the network side harmonic suppression and dynamic response perform well. Sliding-mode based on fuzzy approaching law can improve dynamic response and steady accuracy of system and eliminate chattering phenomena in wider load range conditions.

2. Power model of VIENNA rectifier

Figure 1 is topological structure of main circuit of three-phase VIENNA rectifier. e_a,e_b,e_c are Power grid three-phase ac power supply; i_a,i_b,i_c are three-phase power grid input side current; R_a,R_b,R_c are topological equivalent three-phase filter resistor, resistance values are all R; L_a,L_b,L_c are three phase filter inductance, inductance value is L; C_1,C_2 are DC side up and down bridge arm filter capacitance, and capacitance value is C; DC side output voltage $U_{dc}=U_{dc1}+U_{dc2}$ .In theory, because C1,C2 has same capacitance value, $U_{dc1}= U_{dc2}$;R_L are DC output side load resistance.

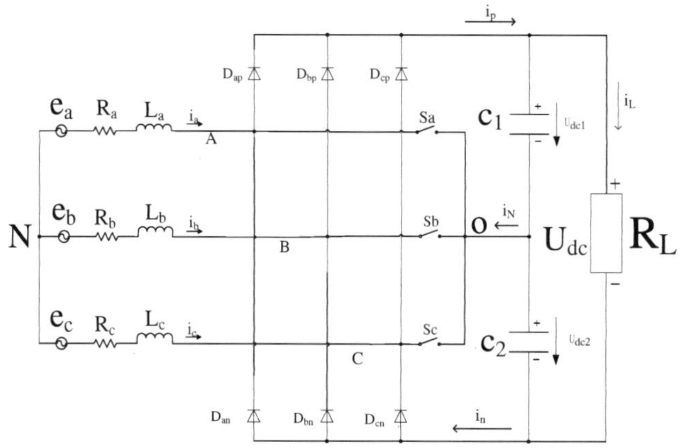

Figure 1 simplified model of the main topology of the VIENNA rectifier

When the grid is in an ideal equilibrium state and the rectifier works in continuous current mode, the following mathematical model is established in the two phase stationary alpha beta coordinate system:

$$\begin{cases} L\frac{di_\alpha}{dt} = e_\alpha - R \times i_\alpha - (V_{\alpha N} + V_{NO}) \\ L\frac{di_\beta}{dt} = e_\beta - R \times i_\beta - (V_{\beta N} + V_{NO}) \\ C\frac{dV_{dc1}}{dt} = s_{\alpha p} \times i_\alpha + s_{\beta p} \times i_\beta - \frac{V_{dc}}{R_L} \\ C\frac{dV_{dc2}}{dt} = -s_{\alpha n} \times i_\alpha - s_{\beta n} \times i_\beta - \frac{V_{dc}}{R_L} \end{cases} \tag{1}$$

When ideal voltage vector in two phase stationary α/β coordinate system rotated ωΔt(Δt approach to zero), to derive nonlinear relationship between e_α and e_β, the following equation can be listed as follows:

$$\begin{cases} \frac{de_\alpha}{dt} = -\omega e_\beta \\ \frac{de_\beta}{dt} = \omega e_\alpha \end{cases} \tag{2}$$

Taking the derivative of instantaneous active power and reactive power:

$$\begin{cases} L\frac{dP}{dt} = e_\alpha^2 + e_\beta^2 - e_\alpha V_\alpha - e_\beta V_\beta - \omega LQ - RP \\ L\frac{dQ}{dt} = (e_\alpha V_\beta - e_\beta V_\alpha) + \omega LP - RQ \end{cases} \tag{3}$$

The differential equations of instantaneous active power and instantaneous reactive power are analyzed, both sides of DC side in equation(1) multiplied by V_{dc} (Ideally, the mid-point equilibrium V_{dc1}=V_{dc2}=$V_{dc}/2$):

$$\frac{(s_{\alpha p} - s_{\alpha n})i_\alpha + (s_{\beta p} - s_{\beta n})i_\beta}{2} V_{dc} = \frac{C}{4}\frac{dV_{dc}^2}{dt} + \frac{V_{dc}^2}{R_L} \tag{4}$$

Ignore the loss in circuit(Equivalent resistance loss of circuit wire, IGBT switching loss, Inductance internal resistance loss and magnetic loss), left side of equation(4) is actually output power of dc side(P_{dc}):

$$P_{dc} = \frac{C}{4}\frac{dV_{dc}^2}{dt} + \frac{V_{dc}^2}{R_L} \qquad (P_{dc} = P = P_{ac}) \tag{5}$$

Based on the derivation of the power model of VIENNA rectifier, the following points can be analyzed:

(1) The active power is one-way flow from the ac side to the dc side, and the reactive power is only flowing in the ac side, in an ideal situation where the loss of the circuit is not considered, the active power of the ac side input is equal to the active power of the dc side output. At the same time, there is a coupling relationship between active power and reactive power, which is reflected in the ac input inductance;

(2) DC side capacitance does not consume power in a switching cycle, providing stable voltage and reducing the effect of output voltage ripple;

(3) The formula (5) can be found that the output side active power can be reflected by the square value of the voltage, Therefore, in designing the double closed loop control circuit and inner ring for power inner ring control, the input of the inner ring should be the output of the outer ring of the voltage multiplied by the load current value of the current moment, Therefore, the internal ring active power is given and the theoretical setting of reactive power is 0.

3. The design of a new closed-loop control system

Traditional current loop control needs to decouple current vector in dq coordinate system to improve power factor of the system. Therefore, additional use of the locking loop tracking technique is required. The direct control of power inner ring greatly simplifies inner ring structure and does not need to use phase-locked loop, so it also has good performances in dealing with unbalanced state of the three-phase power grid; The outer ring adopts voltage PI control with real-time current compensation, dynamic performance response is good and control algorithm is simple. Figure 2 is double closed loop control block diagram of VIENNA rectifier.

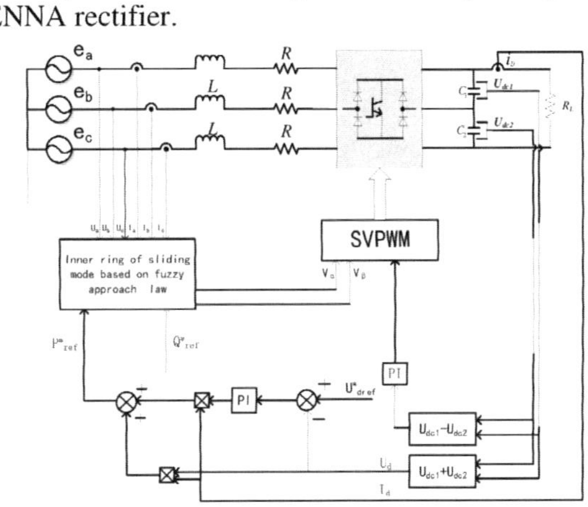

Figure 2 Double closed loop control block diagram of VIENNA rectifier

3.1 Design of sliding controller based on fuzzy approach law

3.1.1 Sliding surface selection. In the dynamic process, the sliding controller is constantly changing according to the current state of the system, the tracking error gradually converges to zero after the system reaches the sliding mode. The input of the sliding controller is the error between the current variable of the system and the ideal reference value. In VIENNA rectifier control system, the active power of inner ring should be close to ideal power P obtained through the output of the outer ring of voltage, and the reactive power value should approach ideal power Q:

$$S=\begin{bmatrix}S_1\\S_2\end{bmatrix}=\begin{bmatrix}e_P(t)\\e_Q(t)\end{bmatrix}=\begin{bmatrix}P_{ref}-P\\Q_{ref}-Q\end{bmatrix} \tag{6}$$

$P_{ref}-P$ is difference value between the active power and ideal value for the inner power ring;$Q_{ref}-Q$ is difference value between the reactive power and ideal value for the inner power ring.

3.1.2 Design of sliding mode controller. The exponential approach to the law is as follows:

$$\frac{dS}{dt}=-\varepsilon \mathrm{sgn}(S)-kS \tag{7}$$

ε>0,k>0.The rational choice of k and ε can improve the dynamic characteristics of the system and reduce the chattering phenomenon of the sliding mode controller. In equation (7), sgn(S)can be shown as:

$$\mathrm{sgn}(S)=\begin{cases}1 & S>\lambda\\ \frac{S}{\lambda} & -\lambda\leqq S\leqq \lambda\\ -1 & S<-\lambda\end{cases} \tag{8}$$

Take the derivative of equation (6) and bring the formula (7) into the equation:

$$\frac{dS}{dt}=\frac{d\begin{bmatrix}S_1\\S_2\end{bmatrix}}{dt}=\begin{bmatrix}\frac{dS_1}{dt}\\ \frac{dS_2}{dt}\end{bmatrix}=\begin{bmatrix}\frac{d(P_{ref}-P)}{dt}\\ \frac{d(Q_{ref}-Q)}{dt}\end{bmatrix}=\begin{bmatrix}-K_1\mathrm{sgn}(P_{ref}-P)-K_p(P_{ref}-P)\\-K_2\mathrm{sgn}(Q_{ref}-Q)-K_q(Q_{ref}-Q)\end{bmatrix} \tag{9}$$

In equation (1), the power mathematical model can be transformed into:

$$L\begin{bmatrix}\frac{dP}{dt}\\ \frac{dQ}{dt}\end{bmatrix}=\begin{bmatrix}-e_\alpha & -e_\beta\\-e_\beta & e_\alpha\end{bmatrix}\begin{bmatrix}V_\alpha\\V_\beta\end{bmatrix}+\begin{bmatrix}e_\alpha^2+e_\beta^2-RP-\omega LQ\\-RQ+\omega LP\end{bmatrix} \tag{10}$$

The output of power inner ring controller is V_α and V_β, therefore, equation(11) can be obtained:

$$\begin{bmatrix}V_\alpha\\V_\beta\end{bmatrix}=-\begin{bmatrix}e_\alpha^2+e_\beta^2-RP-\omega LQ\\-RQ+\omega LP\end{bmatrix}\begin{bmatrix}-e_\alpha & -e_\beta\\-e_\beta & e_\alpha\end{bmatrix}^{-1}+L\begin{bmatrix}\frac{dS_1}{dt}\\ \frac{dS_2}{dt}\end{bmatrix}\begin{bmatrix}-e_\alpha & -e_\beta\\-e_\beta & e_\alpha\end{bmatrix}^{-1}$$

$$=\begin{bmatrix}\frac{e_\alpha(F_1-DS_1)+e_\beta(F_2-DS_2)}{e_\alpha^2+e_\beta^2}\\ \frac{e_\beta(F_1-DS_1)-e_\alpha(F_2-DS_2)}{e_\alpha^2+e_\beta^2}\end{bmatrix} \tag{11}$$

which：

$$\begin{cases}F_1=e_\alpha^2+e_\beta^2-RP-\omega LQ\\F_2=-RQ+\omega LP\\DS_1=L\big(-K_1\mathrm{sgn}(P_{ref}-P)-K_p(P_{ref}-P)\big)\\DS_2=L(-K_2\mathrm{sgn}(Q_{ref}-Q)-K_q(Q_{ref}-Q))\end{cases}$$

The control parameters K_1,K_2,K_p,K_q are all positive. In terms of control effect, selecting the larger K_p,K_q can improve the response speed of the system; Under the condition of system stability, choosing the smaller K_1,K_2 can reduce the chattering phenomenon generated by the system when it enters sliding surface.

3.1.3 The design of sliding mode controller based on fuzzy approaching law. By analyzing the character of sliding mode controller and experimental data: Under the premise of stable system, the larger K_p,K_q will strengthen the robustness of the system. However, it is easy to generate large the chattering phenomenon, which is not conducive to the dynamic response of the system; On the contrary, the smaller K_p,K_q can reduce the steady state performance of the system and generate the voltage ripple, but can effectively eliminate the chattering phenomenon of the system.

In this paper, the sliding mode controller based on fuzzy approaching law was proposed. The state of sliding mode S_1 is the input of the fuzzy controller, the K_p value of response is directly estimated by the value of S_1. Based on the above ideas, a one-dimensional fuzzy controller can be established. The parameter K_p can be adjusted according to the value of S_1. The specific control block diagram is shown in figure 3.

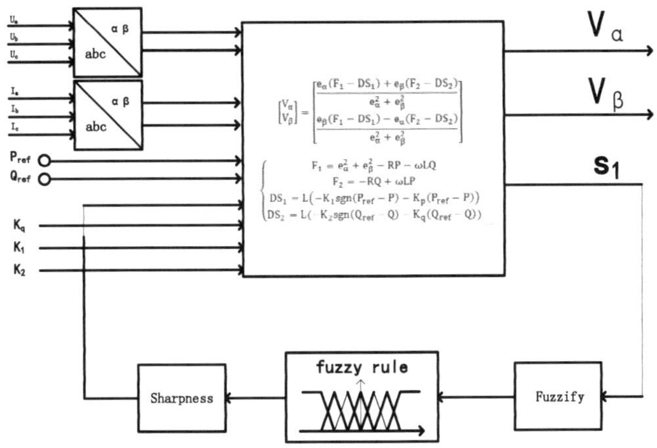

Figure 3 Fuzzy approach law control block diagram

The input variable of the fuzzy controller is S_1,the output variable of the fuzzy controller is K_p. Describe fuzzy subset of language values of input variables as {NB,NM,NS,ZE,PS,PM,PB}, describe fuzzy subset of language values of output variables as {PS,PM,PB}. Membership function curve of input variable is shown in figure 4 (a), and membership function curve of the output variable language value is shown in figure 4 (b).

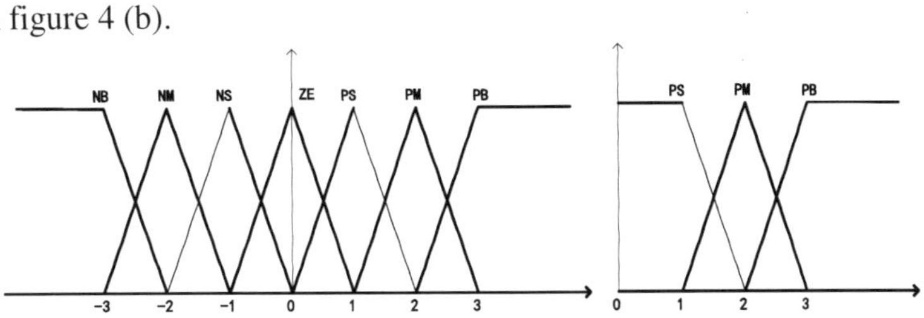

(a) (b)

Figure 4 Membership function curve

When S_1 deviates from sliding mode surface large, the sliding mode control requires larger K_p to approach sliding mode surface; When S_1 deviates from sliding mode surface small , sliding mode control requires a smaller K_p to approach the sliding mode surface. Therefore, fuzzy rules are formulated as Table 1:

Table 1 Fuzzy rules

S_1	NB	NM	NS	ZE	PS	PM	PB
K_P	PB	PM	PS	PS	PS	PM	PB

3.2 Design of the voltage outer ring

The function of voltage outer ring is mainly to maintain the stability of output voltage, and provide the given active power and reactive power input for the power inner loop control. The capacitance voltage of upper and lower bridges arm were collected, and output voltage of the DC side is obtained, which entered into PI after the difference of the given stable voltage value . The output of PI link is added to the real-time current compensation to get the difference between the real-time power and the given power. The difference of power value and real-time power value can quickly obtain active power of the inner ring of the sliding mode.

4. Simulation and result analysis

In order to verify the effectiveness of dual closed-loop control strategy of VIENNA rectifier, the power electronic simulation model of VIENNA rectifier was established by matlab/simulink simulation, and the simulation analysis was carried out. The simulation parameters are as Table 2:

Table 2 simulation parameters of VIENNA rectifier

The effective value of phase voltage E_{rms}	220V/50Hz
Output bus voltage U_{dc}	750V
filter inductance L_s	0.8mH
The upper and lower bridge arm filter capacitance	$C_{dc1} = C_{dc2} = 2000\mu F$
Max output powerP_{max}	60KW

The switching frequency is 20kHz, and simulation has added mid-point balance control strategy to ensure voltage of upper and lower bridge arm capacitor equal. In outer ring of voltage, PI parameter: $K_{pu} = 4$,K_{iu}=0.02; In power inner ring:Q_{ref}=0,K_1=0.01,K_2=0.1,K_Q=10000,λ_1=λ_2=2000. In traditional double closed loop control strategy, PI parameter of voltage outer ring:$K_{pu} = 4$、 K_{iu}=0.02; PI parameter of current inner ring:$K_{pi} = 0.5$,K_{ii}=0.005.

Under the three-phase unbalanced power: In MATLAB simulation, set A phase voltage phase voltage effective value is 175 v, the rest of the two phase voltage RMS voltage is 220 v. Compared with the simulation analysis of the traditional double closed loop PI control system and new double closed-loop control system, the specific simulation waveforms are shown in fig.5 (a) and fig.5 (b).

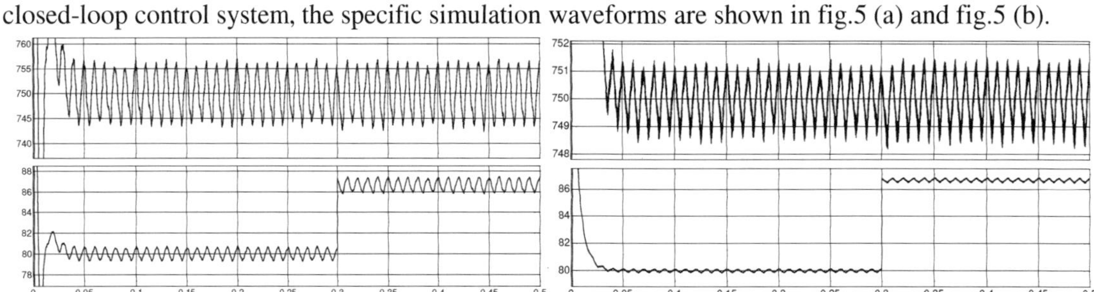

Figure 5(a) voltage and current waveform diagram of diagram traditional double closed loop PI control system

Figure 5(b) voltage and current waveform of new double closed-loop control system

In figure 5 (a), amplitude of voltage vibration is 13V under traditional double closed loop control. In figure 5 (b), voltage ripple can remain in a smaller control range, about 2.5v.

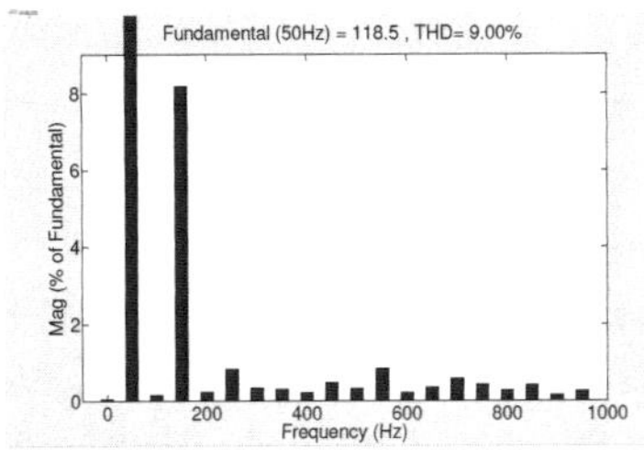

(a)

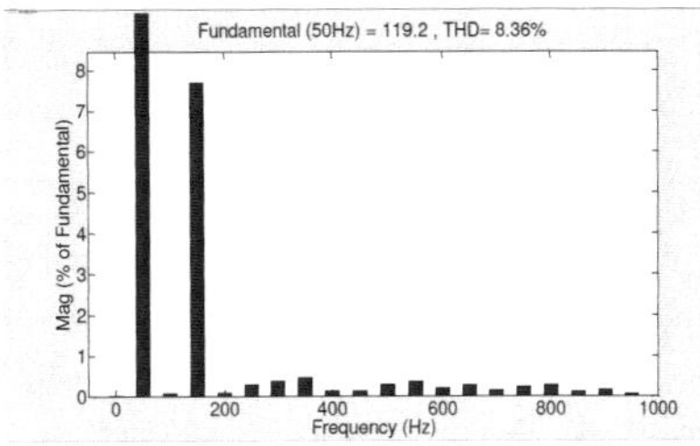

(b)

Figure 6 total harmonic distortion

Figure 6 (a) and (b) are total harmonic distortion(THD)of traditional double closed-loop control strategy and new dual-closed-loop control strategy under unbalanced three-phase power grid. The new dual closed-loop control system has a better effect on harmonic suppression when three-phase power are unbalanced.

As can be seen in Figure 7(a), without using fuzzy approaching law as power inner ring, when t=0.3s, the load increased from 5KW to 65KW. Because of fast approaching speed, the system generates chattering phenomenon, which increases overshoot ,response time, and destroys dynamic response capability of system.

In Figure 7(b), with using fuzzy approaching law as power inner ring, there are no obvious overtones, and no chattering phenomenon.

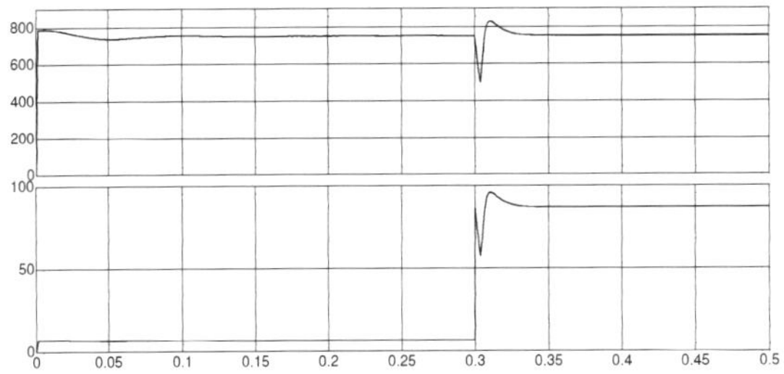

(a) The inner ring of sliding mode without fuzzy approaching law

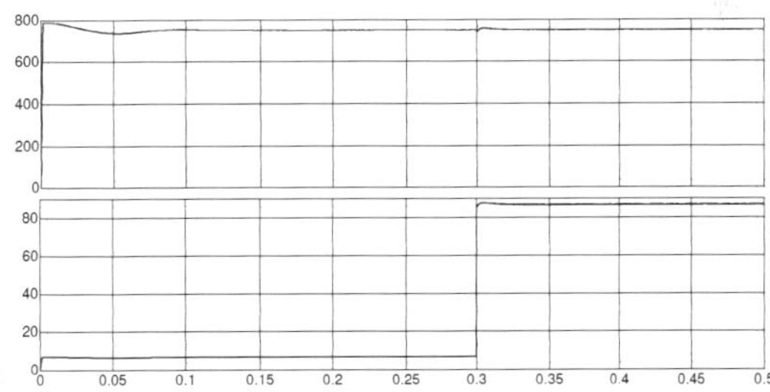

(b) The inner ring of sliding mode with fuzzy approaching law

Figure 7 the load increased from 5KW to 65KW

5. Conclusion

The power mathematical model of the VIENNA rectifier is derived. Based on sliding mode of fuzzy approaching law, the power inner ring is designed to replace the traditional inner ring. Through matlab/simulink simulation, the simulation of the control strategy was completed and the feasibility was verified. The results show that the new double-closed-loop control strategy has better dynamic performance and steady state performance. Especially in the case of unbalanced three-phase power grid, it shows better stable output voltage and the total harmonic distortion(THD). The chattering phenomenon of the sliding mode was solved, better dynamic performance was maintained when the load was fluctuating widely.

Reference

[1] Yang yubo, xie yunxiang. Research and implementation of three-phase three-phase three-level VIENNA rectifier [J]. Power electronics technology,2015,(12):21-22+38.

[2] Small-signal perturbation technique used for DSP-based identification of a three-phase three-level boost-type VIENNA rectifier. Bel Hadj-Youssef, N.,Al-Haddad, K.,Kanaan, H.Y.,Fnaiech, F. IET Electric Power Applications . 2007

[3] Wang peng, Li shan, Guo qiang, Chen min, Song lifeng, li xiao. The research on the dual closed loop control strategy and its parameters in the three-phase VIENNA rectifier [J]. Journal of power supply:1-9.
[4] Research and design of the three-phase electric VIENNA rectifier of the 7.5kW charging machine[D].University of electronic science and technology,2015.
[5] Ma hui, xie yunxiang. Research on new double closed-loop control strategy based on sliding mode structure of VIENNA rectifier[J].Journal of electrical technology,2015,(12):143-151.
[6] Li xin, Zeng qinghui, Wang junbo. Research on direct power control strategy of three-phase VIENNA rectifier [J]. Power electronics technology,2015,(11):90-92.
[7] Mo lingjun. Research and design of direct power control of the VIENNA rectifier [D]. South China university of technology,2015.
[8] Mo lingjun,Xie yunxiang. Research on the VIENNA type rectifier based on sliding mode control [J]. Electrical and energy efficiency management technology, 2014,(24):65-70.
[9] Lu xiang, Xie yunxiang, Gui cun bing, Cheng li, Yang yubo. Control strategy of VIENNA converter based on the combination of passive and sliding mode variable structure control [J]. Power automation equipment,2014,(10):110-115.
[10] Ma hui. Study on the modulation technology and non-linear control strategy of three-phase VIENNA rectifier [D]. South China university of technology,2016.

IOP Conf. Series: Materials Science and Engineering **366** (2018) 012045 doi:10.1088/1757-899X/366/1/012045

An improved method of TEF considering the saturation of inverter under VSG strategy

Wang Chenggen[1], Zhou Qian[1] and Bu Jing[2]

[1]Power Grid Technology Department State Grid Jiangsu Electric Power Company Research Institute, Nanjing, 211100, China
[2]School of Automation, Nanjing University of Science and Technology, Nanjing, 210094, China

Email: 535462103@qq.com

Abstract. In order to reduce influence of the total inertia of the system on the system stability, the virtual synchronous generator(VSG) technology is becoming a research hotspot for the penetration of distributed power supply with high permeability. Due to the existence of the current limiter in the inverter, the transient energy function which is constructed by first integral (TEF) derivative is not monotonous. It may make using TEF method to obtain the critical clearing time and system stability error. This paper proposes an improved TEF method using the tangent function to replace the saturation one. Firstly, the relationship between the virtual power angle and the power angle of the traditional system is analyzed after the improvement of the inverter current saturation link. Then, the first integral method is used to construct the TEF with improved saturation. Finally, it simulates to a single machine infinite bus system and IEEE three-machine-nine-bus system by MATLAB/Simulink, and the results show that the improved TEF is proposed, it can effectively evaluate the VSG strategy and the transient stability of AC system inverter current limiting.

1. Introduction

When power grid occurs large disturbance, the inverters using VSG strategy[1] will exhibit electromechanical dynamic characteristics similar to that of synchronous machine(SG). While the inverter has the current saturation characteristic, the transient stability analysis of the system will be more complex[2]. Therefore, it is necessary to study the transient stability analysis method considering the saturation characteristics of the inverter.

In present studies about the direct method of power system transient stability analysis which are based on the VSG strategy to improve the transient stability of the system, the references [3-5] change the rotary inertia of virtual synchronous generator. While the effect of the inverter's current saturation is ignored when using the direct method. Therefore, the reference [6] points out the problem that the inverter leads to transient instability easily under the current limiting conditions, and advances the virtual impedance, which is used to replace traditional current limiting for improving the large disturbance stability, but there is a lack of in-depth analysis on its instability rationality. As far as the electromechanical transient progress which occurs when the inverter current is saturated, the reference [7] provides the mechanism of the power angle instability from the time domain, but it only analyzes the transient power angle instability of VSG qualitatively, and the simulation verification is not provided.

As for the influence of inverter current saturation characteristics on the transient stability margin of AC system, this paper analyses the limitations of transient energy function considering the saturation of inverter, and on this basis, it proposes an improved method using the arctan function to approximate. Based on the transient stability analysis method of traditional synchronous generator, this paper deduces the relationship between the traditional power angle and the virtual power angle of the approximately improved inverter's current saturation link, and constructs the transient energy function. Finally, it uses a single machine infinite bus(SMIB) system and the system of IEEE 3 generators and 9 nodes to simulate and it verifies the validity of this method.

2. Transient model of virtual power angle of inverter using VSG

In this chapter, the virtual power angle of the inverter under the VSG strategy is modeled, and the power angle characteristics whether the inverter is saturated or not are also respectively considered.

2.1. Definition of the virtual power angle of inverter using VSG

1) The virtual power angle characteristic when the inverter is not saturated

When the inverter is not saturated, its output power expression is shown in formula (1):

$$P_e = \mathrm{Re}(|\dot{U}I^*|) = \frac{|\dot{E}||\dot{U}|}{X_\Sigma}\sin\delta' \tag{1}$$

Where, δ' is virtual power angle, it can also be considered as the angle between the inverter's output voltage and the grid voltage. From the above formula, we find that the virtual power characteristic of VSG strategy when the inverter is not saturated is similar with traditional synchronous generator, and the virtual power angle curve is shown in the Figure 1 marked by the unsaturated curve.

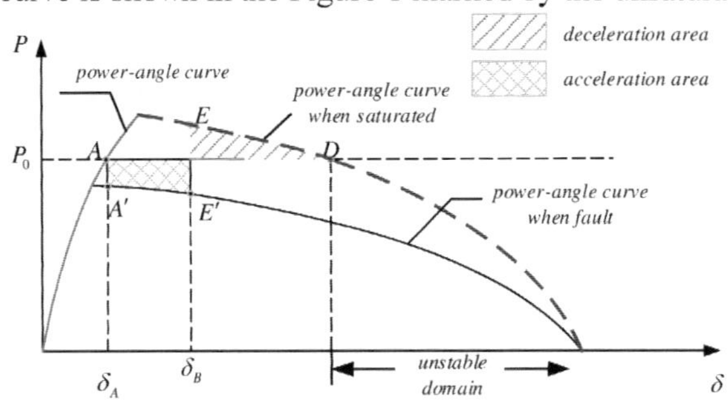

Figure 1. VSG virtual power angle curve

2) The virtual power angle characteristic when the inverter is saturated

In order to prevent the excessive current flowing in the inverter from damaging the equipment, it is necessary to set the current limiter to protect the inverter during operation [8]. When $I_{mag} \le I_{max}$, the inverter normally outputs. But when $I_{mag} > I_{max}$, its current limiting section start up, and the current is clamped at I_{max}. Therefore, it can be equivalent to the limit model shown in formula (2):

$$I = \begin{cases} I_{\min}, I < I_{\min} \\ I, I_{\min} \le I \le I_{\max} \\ I_{\max}, I > I_{\max} \end{cases} \tag{2}$$

The presentation of the power angle characteristics when the inverter current saturated is shown as formula (3):

$$\begin{aligned} P_e &= \left(\frac{I_{\max}|\dot{I}_0|}{B_\Sigma}\right)\sin(\delta' + 90^\circ) \\ &= (I_{\max}|\dot{U}|)\sin(\delta' + 90^\circ) \\ &= P_{sm}\cos\delta' \end{aligned} \tag{3}$$

According to formula (3), we can obtain the virtual power angle characteristic curve when the inverter's current is saturated shown as Figure 1. Its' cosine characteristic is no longer consistent with the sinusoidal power angle characteristic of the traditional synchronous generator.

2.2. Transient model of virtual power angle considering the inverter current saturation

We can establish the dynamic equivalent model of the inverter with VSG control strategy, and the rest of the grid can be seen as an infinite system. Thus, this SMIB system is shown in Figure 2.

For the SMIB system as above, the relationship between electrical quantities can express as follows:

$$\dot{E} = \dot{U} + j\dot{I}X_{\Sigma}$$
$$\dot{I} = \frac{\dot{E}-\dot{U}}{jX_{\Sigma}} = \frac{(E\cos\delta - U) + jE\sin\delta}{jX_{\Sigma}} = \frac{E\sin\delta}{X_{\Sigma}} - j\frac{E\cos\delta - U}{X_{\Sigma}} \tag{4}$$

Where, E is the terminal voltage of the Inverter, U is the infinity bus's reference voltage, X_{Σ} is the line's reactance, and I is the inverter's output current.

Thus, the relation between current amplitude and power angle is shown in formula (5):

$$I = \frac{\sqrt{(E\sin\delta)^2 + (E\cos\delta)^2 - 2EU\cos\delta + U^2}}{X_{\Sigma}} = \frac{\sqrt{E^2 - 2EU\cos\delta + U^2}}{X_{\Sigma}} \tag{5}$$

That is,

$$\cos\delta = \frac{E^2 + U^2 - (IX_{\Sigma})^2}{2EU}, I = \begin{cases} I_{min}, I < I_{min} \\ I, I_{min} \le I \le I_{max} \\ I_{max}, I > I_{max} \end{cases} \tag{6}$$

According to the formula (6), it can be equivalent to virtual power angle clipping to analyze the system transient stability conveniently of inverter's current saturation link under VSG strategy. For the SMIB system, the total energy after the fault stage can be expressed as (7):

$$V(\delta,\omega) = \frac{1}{2}M\omega^2 - P_M(\delta - \delta_s) - P_{em3}(\cos\delta - \cos\delta_s) \tag{7}$$

Where, ω and M are the generator's angular velocity and inertia constant; ω_0 is the synchronous electrical angular velocity; P_M is the mechanical input power of prime mover; δ_s is the stable work power angle of the post-fault system; P_{em3} is the generator's maximum electromagnetic power after fault. The system energy curve of traditional TEF can be shown in the Figure 3.

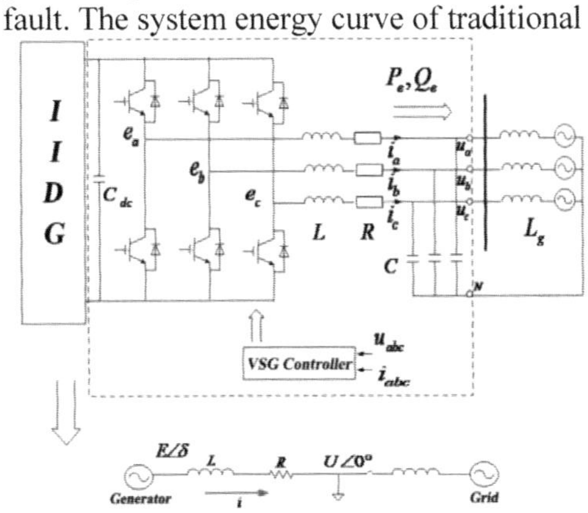

Figure 2. SMIB System Model under VSG Strategy

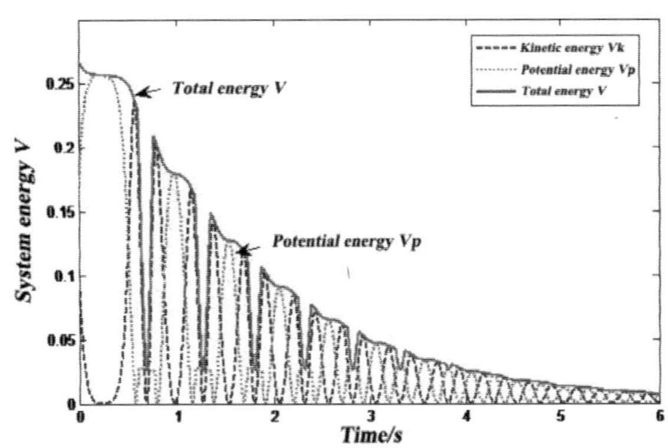

Figure 3. Transient Energy Curves with Inverter Current Limiting Link

From the figure, we can see that, the overall trend of the total system energy V is getting smaller, but its derivative is not always less than zero, which does not meet the Lyapunov stability conditions. In this case, we can't accurately determine the critical cut-off time of the fault system and the stability of the system if we use TEF method.

3. The influence of inverter current saturation on TEF

According to the previous formulas, we let $x_1 = \delta - \delta_s$, $\dot{x}_1 = \omega - \omega_s$ and $f(x_1) = P_{em3}\sin(x_1+\delta_s)\text{-}P_M$, the system equations that account for inverter saturation under the VSG strategy is as follow[9]:

$$\begin{cases} \dot{x}_1 = x_2 \\ \dot{x}_2 = -\dfrac{D}{M}x_2 - \dfrac{1}{M}f(x_1) \end{cases} \tag{8}$$

When the current is unsaturated, the derivative of the system energy V is always less than zero. But the power gets saturated along with that the current becomes saturation. We can see that when x_1 is saturated, its derivative does not exist at the point A or B (shown in Figure 4). We let $x_{10i} \in \{x_{10} \in R^n, |x_1| = x_{1\max}\}$, hence the derivative of the x_1 at the point x_{10i} is:

$$\begin{aligned} &\lim_{t\to 0^+} \frac{x_1(x_{10i} + t\dot{x}_{10}) - x_1(x_{10i})}{t} \\ &= \begin{cases} 0, |x_1(x_{10i} + t\dot{x}_{10})| > x_{1\max} \\ \dot{x}_1, |x_1(x_{10i} + t\dot{x}_{10})| < x_{1\max} \end{cases} \end{aligned} \tag{9}$$

The following formula is established when the power angle is saturated:

$$\begin{aligned} &\dot{V} = Mx_2\dot{x}_2 \\ &+D\lambda x_{1\max} + \frac{1}{2}\frac{D^2\lambda}{M}x_{1\max}^2 - P_M x_{1\max} \end{aligned} \tag{10}$$

Substitute the second equation of the formula (8) into the formula (10), we can get that:

$$h = \dot{V} = -Dx_2^2 + (P_M - P_e)x_2 + c \tag{11}$$

Where, $c = D\lambda x_{1\max} + \frac{1}{2}\frac{D^2\lambda}{M}x_{1\max}^2 - P_M x_{1\max}$.

There will be that:

$$h_{\max} = \frac{(P_M - P_e)^2}{4D} + c \tag{12}$$

From the above, we can find that the analysis method of the traditional TEF is suitable when $h_{\max} \le 0$ because the system energy V is negative, while it is not suitable when $h_{\max} > 0$. Thus ,this paper improves the TEF method for this problem.

4. TEF considering the current saturation of VSG inverter

Due to the saturation of the inverter, the system energy function is no longer negative. Therefore, it is necessary to improve the model which express the saturation for the system energy to suitable with Lyapunov conditions.

4.1. The improved virtual power angle model of the inverter considering saturation

In order to correctly study the transient stability of the system include the inverters under the VSG strategy by the traditional TEF method, this paper improves the model which describes the inverter's current saturation. We use arctangent function to approximate the characteristic of the inverter's current saturation, the curve and expression of which is shown as Figure 4 and formula (13). After that, the TEF could be derivable at any point.

$$sat(I) \approx I_M \times \tanh(\frac{I}{I_M}) = I_M \frac{e^{I/I_M} - e^{-I/I_M}}{e^{I/I_M} + e^{-I/I_M}} \tag{13}$$

After improving the saturated link, the expression can be simplified as formula (14):

$$\cos\delta_N = \frac{E^2+U^2-\left(IX_\Sigma\right)^2}{2EU} \approx \frac{E^2+U^2-\left(I_{\max}\arctan\left(\frac{I}{I_{\max}}\right)X_\Sigma\right)^2}{2EU}$$
$$= \cos\delta_O + \frac{E^2+U^2-2EU\cos\delta_O-\left(I_{\max}\arctan\left(\frac{E^2+U^2-2EU\cos\delta_O}{I_{\max}X_\Sigma}\right)X_\Sigma\right)^2}{2EU} \quad (14)$$

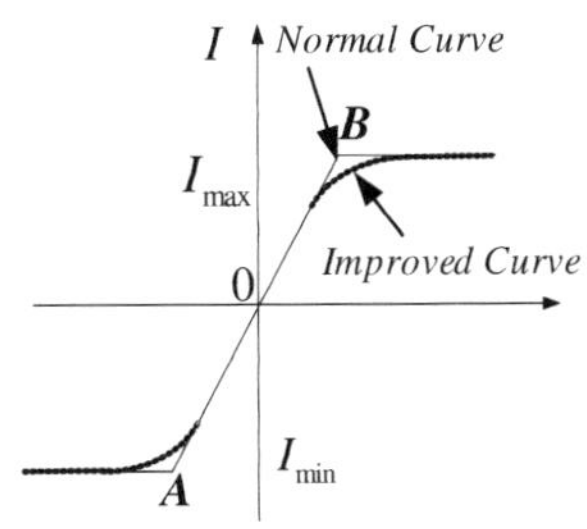

Figure 4. Current saturation

According to the formula (14), we can get the relationship between the virtual power angle δ_N of the approximately improved inverter current saturation and the power angle δ_O of the traditional power system without the saturation.

After substituting the formula (13) into the formula (14), we can get the system energy's curve of the improved limiting model shown as Figure 5.

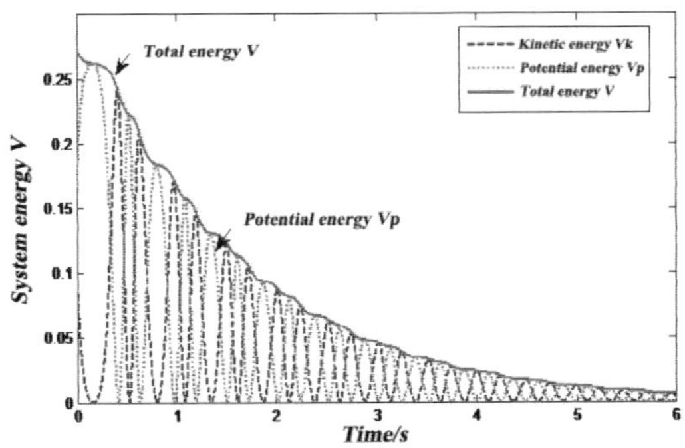

Figure 5. The transient energy curve which the improved limiting link

After approximated, we can find out that the total energy curve of the system is continuously declining from figure 5, and there is no fluctuation which satisfies the Lyapunov condition

After we improve the model expressing the inverter saturation link, the traditional TEF constructed by the first integral method can be used to analyze the transient stability of the system when the inverter's current saturation.

4.2. The improved TEF considering saturation link

Constructing the transient energy function by the first integral method[10], it's shown as formula (15):

$$V(\delta,\omega) = \frac{1}{2}\sum_i M_i(\omega_i-\omega_0)^2 + \sum_i\sum_j V_iV_jB_{ij}(\cos\delta_{ij}^s-\cos\delta_{ij})$$
$$+\sum_i[(D_i\lambda_j(\omega-\omega_0)_i(\delta-\delta_0)_i)+\frac{1}{2}\frac{D_i^2}{M_i}\lambda_i(\delta-\delta_0)_i^{\ 2}] \quad (15)$$
$$-\sum_i P_{mi}(\delta_i-\delta_0)$$

IOP Conf. Series: Materials Science and Engineering **366** (2018) 012045 doi:10.1088/1757-899X/366/1/012045

Where, M_i and ω_i are inertia constant and rotor angular velocity of the i-th generator; ω_0 is the synchronous electrical angular velocity; V_i, B_i, D_i, λ_i and P_{mi} are electric potential, susceptance, damping coefficient, weighting factor and mechanical power of the i-th generator. The value of λ_i is between 0 and 1; and δ_0 is the stable power angle of system after fault[11].

The specific procedure using this method to determine the system transient power angle stability is as follows:

1) Input data: Main parameters of electrical part, they include network parameters such as VSG, line, transformer, fault location and maximum and minimum current of limiting device.

2) Get fault trajectory: Integrate rotor equation of motion with R-K method, namely the $\delta(t)$ and $\dot{\delta}(t)$ of the VSG rotor. This procedure considers the equivalent power angle limit of the formula (14).

3) Get the transient energy V_{cl} and critical energy V_{cr} of system at the time of fault removal: they are obtained separately according to formula (12) and the critical energy formula of reference [12].

4) Determines the system stability: Compare the system transient energy V_{cl} with the system critical energy V_{cr} obtained by procedure 3) at the time of fault removal. The system will be stable if $V_{cl} \le V_{cr}$, it's unstable otherwise.

5) Calculate the critical cut-off time: Command $V_{cl} = V_{cr}$ and there will be $t_c = t_{cr}$ according to system energy trajectory.

5. Simulation verification and analysis

5.1. Simulation environment

This paper uses SMIB and IEEE 3 machines 9 nodes system to validate the effectiveness of this method. The basic data of the two systems are refer to references [11] and [13]. And D=0.1, λ=0.5, P_M=0.9, $x_{1\max}$=0.6. The topology of these two systems are shown in figure 6:

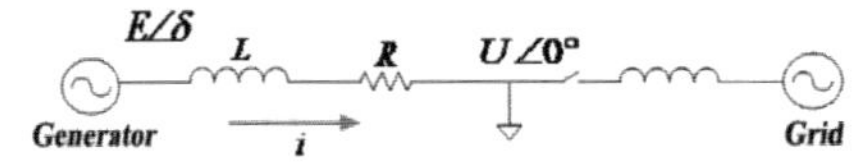

(a) single machine infinite bus system

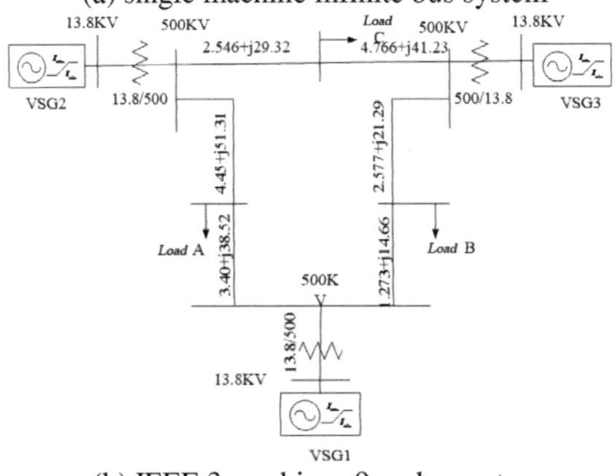

(b) IEEE 3 machines 9 nodes system

Figure 6. Schematic diagram of system structure

5.2. Simulation analysis

1) Comparison of system stability domains

Taking all the points on the boundary of the stable point's small neighborhoods as the integration initial points, have an inverse integral to the system, and the stability region of the system before and after the current saturation model's improvement is obtained, as shown in Figure 7.

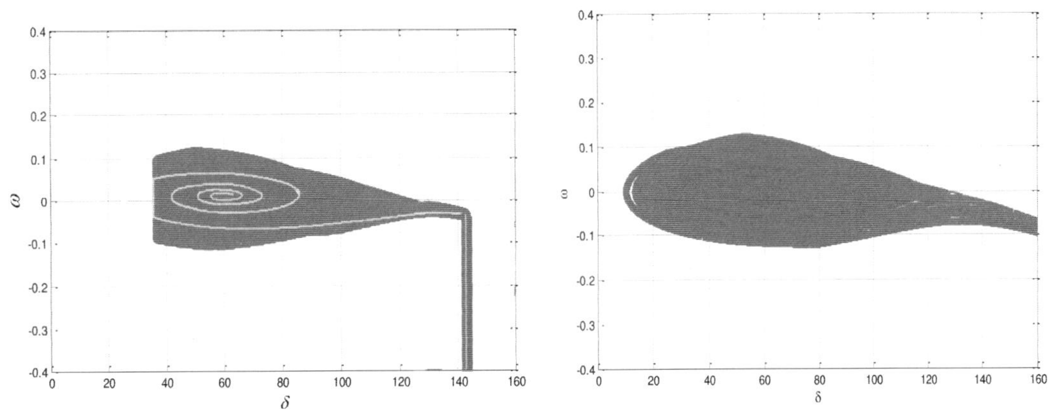

(a) Before improvement (b) After improvement

Figure 7. The stability region of the system

From figure 7, we can see that the stability region after the inverter current saturation improvement of the system become larger than the region before improvement, this is because the arctangent function in the improvement is smooth continuous, and there won't be saturation in the stability region.

2) Comparison of transient energy changes after fault removal

After calculating, we can get $h_{max} = 0.4549 > 0$ according to formula (11), and the total system energy V is not monotonically decreasing. Substituting the SMIB system's parameters into the formula (15), we can obtain the system energy curve of the unmodified saturation model shown in figure 5.

3) Comparison of system's critical cut-off time(CCT)

We use time domain method, unimproved and improved TEF method to compare CCT in both SMIB and IEEE 3 machines 9 nodes system. Simulating results are shown as Table I to IV. Where, Distance_M stands for the distance from the fault point to the head (close to the generator) of the double circuit line.

Table 1 and 2 show CCT's comparison and its error between improved and unimproved model of the saturation link in the SMIB and 3 machine 9 nodes systems. We can find that the improved method proposed increases the CCT and all the relative errors is less than 5%. Therefore, it is feasible and meets the actual requirements in the time domain simulation.

Respectively use both time domain simulation method and TEF method to analyze the system before and after the improvement of the inverter's current saturation model, the CCTs are shown in Table 3. It can be found that the CCT's value obtained by the TEF method is consistent in the fault 2, 3, 4, 5 and 6, and conservatism occurs only in fault 1. So we can get that the conservativeness of the TEF is improved when the inverter current saturation has been improved.

Table 4 shows the CCT of the three-machine system obtained respectively by time domain simulation method and TEF method. The assessment for the network faults 4 and 7 is conservative. We can find out that, for the system which inverter current saturation has been improved, the conservatism obtained by TEF has been improve.

Table 1. The CCT of the SMIB system

Fault position	CCT(s)		Error (%)
Distance_M	Unmodified	Modified	
0.65	1.78-1.79	1.86-1.87	4.49
0.6	1.62-1.63	1.69-1.70	4.32
0.55	1.48-1.49	1.54-1.55	4.05
0.5	1.32-1.33	1.37-1.38	3.78
0.4	1.04-1.05	1.07-1.08	2.76
0.3	0.74-0.75	0.75-0.76	1.35

Table 2. The CCT of the IEEE 3 machine 9 nodes system

number	Fault line	CCT(s) Unmodified	CCT(s) Modified	Error (%)
1	4*-5	0.31-0.32	0.32-0.33	3.23
2	5*-4	0.40-0.41	0.41-0.42	2.5
3	4*-6	0.30-0.31	0.31-0.32	3.33
4	6*-4	0.43-0.44	0.45-0.46	4.55
5	5*-7	0.31-0.32	0.32-0.33	3.26
6	7*-5	0.16-0.17	0.16-0.17	0
7	6*-9	0.38-0.39	0.39-0.40	2.63
8	9*-6	0.21-0.22	0.21-0.22	0
9	7*-8	0.18-0.19	0.18-0.19	0
10	8*-7	0.26-0.27	0.27-0.28	3.85
11	8*-9	0.29-0.30	0.30-0.31	3.45
12	9*-8	0.21-0.22	0.22-0.23	4.76

Table 3. The CCT's comparison of the SMIB system

	CCT(s)			
	Time domain method		TEF	
Distance_M	Before improve	After improve	Before improve	After improve
0.65	1.78-1.79	1.86-1.87	1.746	1.851
0.6	1.62-1.63	1.69-1.70	1.605	1.692
0.55	1.48-1.49	1.54-1.55	1.482	1.542
0.5	1.32-1.33	1.37-1.38	1.320	1.371
0.4	1.04-1.05	1.07-1.08	1.041	1.071
0.3	0.74-0.75	0.75-0.76	0.741	0.753

Table 4. The CCT's comparison of three-machine system

	CCT(s)			
Fault line	Time domain method		TEF	
	Before improve	After improve	Before improve	After improve
4*-5	0.31-0.32	0.32-0.33	0.322	0.313
5*-4	0.40-0.41	0.41-0.42	0.413	0.403
4*-6	0.30-0.31	0.31-0.32	0.311	0.302
6*-4	0.43-0.44	0.45-0.46	0.443	0.419
5*-7	0.31-0.32	0.32-0.33	0.324	0.313
7*-5	0.16-0.17	0.16-0.17	0.164	0.163
6*-9	0.38-0.39	0.39-0.40	0.382	0.369
9*-6	0.21-0.22	0.21-0.22	0.217	0.213
7*-8	0.18-0.19	0.18-0.19	0.183	0.182
8*-7	0.26-0.27	0.27-0.28	0.273	0.253
8*-9	0.29-0.30	0.30-0.31	0.305	0.284
9*-8	0.21-0.22	0.22-0.23	0.226	0.226

6. Conclusion

Considering the influence that the inverter's current saturation characteristic to the system's transient stability margin, this paper proposes the improved method that approximating the current saturation link model as an arctangent function, and constructs the TEF based on the first integral method. It verifies the effectiveness of the TEF constructed by the simulation in SMIB and 3-machines systems with the inverter using the VSG strategy which considering its current saturation, and gets the following conclusions:

1) Comparing the change of the stability margin the system, kinetic energy, potential energy and total energy when the fault has been removed before and after the improvement, we can get that the improved method proposed can quantitatively assess the transient stability of the system with the inverter current saturation based on the VSG strategy.

2) Comparing the CCT of system before and after the improvement, we can get that the improved method proposed can improve the conservatism of traditional TEF method.

Acknowledgments
This work is supported by the Science and Technology Project of State Grid Jiangsu Electric Power Company (J2017046).

References

[1] SIMPSON-PORCO J W, DORFLER F, BULLO F. Synchronization and power sharing for droop-controlled inverters in islanded microgrids[J]. Automatica, 2013, 49(9): 2603-2611.
[2] Williams H F, et al. Feasibility of Lyapunov function for the stability analysis of electric power system having up to generators[J]. IEEE Trans. on PAS, 1992, 91(4): 1145-1152.
[3] SHINTAL T, MURA Y, ISE T.Oscillation damping of a distributed generator using a virtual synchronous generator[J]. IEEE Trans on Power Delivery, 2014, 29(2): 668-676.
[4] ALIPOORJ, MIURA Y,ISE T. Distributed generation grid integration using virtual synchronous generator with adoptive virtual inertia[C]. Proceedings of 2013 IEEE Energy Conversion Congress and Exposition, September 15-19, 2013, 8237: 4546-4552.
[5] Cheng Chong, Yang Huan, Zeng Zheng, et al. Rotor Inertia Adaptive Control Method of VSG[J]. Automation of Electric Power Systems, 2015(19): 82-89.
[6] PAQUETTE A D, DIVAN D M. Virtual impedance current limiting for inverters in microgrids with synchronous generators[J]. IEEE Trans on Industry Application, 2014, 51(2): 1630-1638.
[7] Huang Linbin, Zhang Leiqi, Xin Huanhai, et al. Mechanism analysis of virtual power angle stability in droop-controlled inverters[J]. Automation of Electric Power Systems, 2016, 40(12): 117-123.
[8] GIROUX P, SYBILLE G. Static synchronous compensator (STATCOM) used for midpoint voltage regulation on a 500kV transmission line[R]. Natick, USA: Math Works, 2006.
[9] Athay T.A. Power System Stability Analysis by the Direct Mathod of Lyapunov[M]. New York:North Holland Publishing Company, 1981: 1828-1829.
[10] YH Moon, BK Choi, TH Roh. Estimating the domain of attraction for power systems via a group of damping-reflected energy functions[J]. Automatica, 2000, 36(3): 419-425.
[11] PRABHA KUNDUR. Power system stability and control[M]. Beijing: China electrical power press, 2002.
[12] Liu Sheng, Chen Chen. Energy function analysis of power system transient stability: Network structure preserving model[M]. Beijing: Science press, 2014.
[13] Swasti R. Khuntia. Simulation study of a SSSC-based neuro-fuzzy controller for improvement of transient stability in a three-machine power system[J]. IEEE Energy tech, 2012, 1: 1-6.

Short-term nodal load forecasting method considering redundant information

Si Yang[1], Long Zhao[1], Xueshan Han[2], Yong Wang[2], Wenbo Li[3], Donglei Sun[1] and Junxiong Wang[2]

[1]Economic & Technology Research Institute of State Grid Shandong Electric Power Company, Jinan, China
[2]Key Laboratory of Power System Intelligent Dispatch and Control of Ministry of Education, Shandong University, Jinan, China
[3]State Grid Shandong Electric Power Research Institute, Jinan, China

wajxng@163.com

Abstract. The most existing short-term nodal load forecasting methods only focused on the application of separate historical data, ignoring the abundant redundant information in power system. In this paper a new nodal load forecasting method considered physical redundancy and measurement redundancy was proposed. Firstly, the effect of redundant information on short-term nodal load forecasting was analysed and the forecasting principle was given. Secondly, in order to form additional state variable measurement equations, two kinds of existing redundant information between state variable and measured values were analysed deeply. Then based on the forecasting principle and the analysis results, the forecasting model mainly imitated the form of state estimation was established and the specific forecasting process was given. Case studies demonstrated that compared with traditional support vector machine model, the proposed method could effectively decrease forecasting errors and improve forecasting results.

1. Introduction

Load forecasting is the key and basis of power system operation and decision. In recent years, to deal with the energy and environmental crises and to meet the growing social demand for electricity, the concept of Energy Internet has been proposed and rapidly recognized [1]. As the core carrier of Energy Internet, power transmission network also shows a complex, different, multiple, flexible, associated form[2-3]. And the inherent uncertainty and flexibility of passive power supply make it more and more difficult to grasp the load law of all kinds of nodes in the power grid [4]. Therefore, it is necessary to predict and analyze the load of any node involved in grid regulation.

Nodal load forecasting techniques have evolved from initial time series extrapolation methods to non-linear regression methods and the artificial intelligence algorithm. And now nodal load forecasting technology is gradually expanding toward big data [5-6]. However, most of these researches focus on the mining of the variation of single nodal load itself from the perspective of time series. In fact, subject to the network physical laws, each load node is not isolated and there is a close physical relationship between layers and layers. At present, there are some researches on applying this kind of physical relationship to nodal load forecasting. For example, references [7-8] proposed and demonstrated that the multi-nodal load forecasting model is better than node prediction alone. In [9], according to the relationship between nodes, using the strong physical regularity that the total load is equal to the sum of all individual loads, the multi-nodal load forecasting is realized. Reference [10] refers to the idea of power system state estimation to solve the problem of uncoordinated multistage load forecasting results, which essentially utilizes the physically associated redundant information between nodes. Generally speaking, the current researches cannot sufficiently utilize the physical association of load nodes and the redundant information provided by modern measurement technology.

Aiming at the shortcomings of existing research, a short-term nodal load forecasting method considering redundant information is proposed. First of all, the principle of forecasting method is analyzed, and the physical relation of state variable is further revealed by the redundant characteristics of measured data. Then, the specific prediction models and methods are given, and the multi-group measurement equations obtained are used to realize the nodal load forecasting with the estimated correction characteristics. Finally, an actual example is used to verify the effectiveness of this forecasting method.

2. Analysis of the forecasting principle

2.1. The principle of the forecasting method

Traditional load forecasting methods usually analyze historical load data, which is actually redundant information of time series. Since nodal load forecasting uses only time-series redundant information, the prediction results will inevitably have errors. In fact, the historical information of each time section can be regarded as the measurement value [10], but these independent and redundant measurement equations are not closely related to the object to be predicted. The state estimation of the power grid is based on the assumption that all the measurements are conducted at the same time, and the state variables are estimated using the measured redundancy. The shorter the time, the higher the accuracy. Therefore, under the premise of inherited time series redundancy, nodal load forecasting can draw lessons from state estimation to deal with the correlation and redundancy information of the same time section so as to improve the prediction accuracy of nodal load.

With the idea of state estimation, nodal load is the state variable to be estimated. It is the key point of this study that how to characterize the time series and the physical correlation measurement information of each time section as the estimation function of this state variable.

Power transmission network consists of many components, is a complex network with multi-layer structure. A load node (except the terminal) exists in both the upper and lower power grids, which can be equivalent to the load of the upper grid and the power of the lower grid. Therefore, the nodal load naturally has a physical correlation and its state can be described by a number of independent measurement equations.

With the continuous deepening of power grid smart and digital transformation, many transmission and distribution components (such as transmission and distribution lines and transformers) have been able to carry out multi-terminal and high-precision measurement of transmission power, and the measurement processes are independent from each other. Therefore, it is necessary to introduce the measured values of transmission and distribution components into the nodal load forecasting system to improve the traditional nodal load forecasting methods. According to the relationship between the nodal load state and measured values, a set of measurement equations that express the state variable directly or indirectly can be extracted from independent measurements. This is the core idea of this study.

2.2. Two types of redundant information

Combined with the actual situation of power system, two kinds of redundant information between state variables and measured values are analyzed.

2.2.1. Physical correlation redundant information. The past researches on nodal load correlation prediction is based on the law of conservation of energy, that the upper load and the sum of lower nodes load are equal.

However, this process is a little rough. And because of the strong physical laws, this type of method must collect the data of all associated nodes to make a unified prediction, which cannot meet the requirements of grid flexibility in the new situation. In reality, although the network structure of the power grid is very complicated, it will spontaneously follow the constraint of node injection power balance during the operation of power system. It means that there is an additive characteristic between the nodal load state variable and the power measurement of the transmission and distribution components.

Therefore, if the transmission and distribution line power measurements are also introduced to the forecasting process, considering the physical correlation of transmission and distribution elements measurements could decouple the strong physical laws between the total load and the part. Then the nodal load forecasting can be more flexible and can sufficiently utilize the original measurement information such as line power measurements and transformer power measurement, which will be helpful to improve the load forecasting results.

2.2.2. Measurement redundancy information. In modern power system, there are multi-terminal and independent measuring devices for the same transmission and distribution components, which produces sufficient redundancy information of measurements. The distribution diagram of transmission and distribution components measuring devices is shown as figure 1.

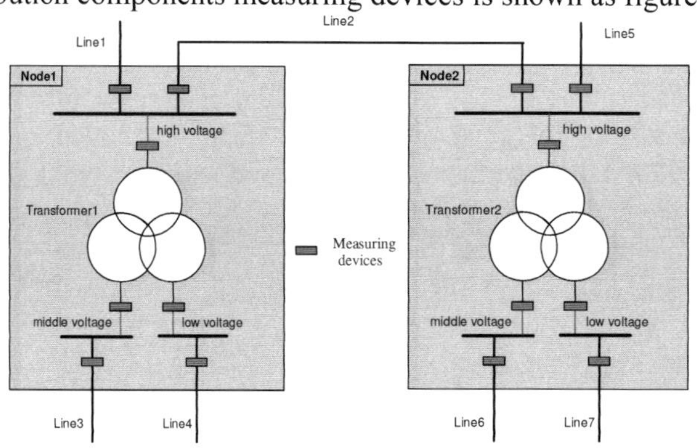

Figure 1. Distribution diagram of transmission and distribution components measuring devices

When analyzing the redundancy characteristics, the nodal load is a complete vertical structure when considering only the state variables association. There is no measurements at this time to describe the direct correlation between nodes in the same layer, which is not in accordance with the actual situation of the power grid. As the Line2 in figure 1, after the introduction of the line power measurements, the line power could be measured and predicted simultaneously on both nodes (ignoring the loss on the line). This feature can make the state variables of nodes in the same layer correlate, to make up for the deficiencies of the past researches.

In summary, this section analyzes two kinds of redundant information about nodal load state variables and power measurement values of transmission and distribution components, explains the physical rules of grid load nodes at a deeper level, and provides the following nodal load forecasting model more abundant redundant information.

3. Forecasting model and method

3.1. Nodal load forecasting model

The forecasting model mainly imitates the form of state estimation. Based on the analysis in the previous section, we choose the nodal load as the state quantity, then the previous predicted values of the transmission and distribution elements are equivalent to the measured values in the state estimation model.

Nodal load forecasting model considering redundant information:

$$\begin{aligned}\min f &= w_{\mathrm{u}}(z_{\mathrm{u}}-x)^2 + w_{\mathrm{d}}(z_{\mathrm{d}}-x)^2 \\ &+w_{\mathrm{u}}'(z_{\mathrm{u}}'-x)^2 + w_{\mathrm{d}}'(z_{\mathrm{d}}'-x)^2 \\ &+w_{\mathrm{h}}(z_{\mathrm{h}}-x)^2 + w_{\mathrm{ml}}(z_{\mathrm{m}}+z_{\mathrm{l}}-x)^2 \\ \text{s.t.}\quad z_{\mathrm{u}} &= \sum_{ku=1}^{M} z_{ku},\ z_{\mathrm{d}} = \sum_{kd=1}^{N} z_{kd}, \\ z_{\mathrm{u}}' &= \sum_{ku=1}^{M} z_{ku}',\ z_{\mathrm{d}}' = \sum_{kd=1}^{N} z_{kd}',\end{aligned} \tag{1}$$

where x is an estimate of the nodal load state variable, M is the number of lines connected to the node in the upper grid, N is the number of lines connected to the node in the lower grid, z_{ku} and z_{kd} are the upper and lower grid line measurements of local terminals, z_{ku}' and z_{kd}' are the nodal load calculated from the upper and lower grid line measurements of opposite terminals, z_{h}, z_{m} and z_{l} are the transformer high, middle and low voltage port measurements, z_{u} and z_{d} are the nodal load calculated from the upper and lower grid line measurements of local terminals, z_{u}' and z_{d}' are the nodal load calculated from the upper and lower grid line measurements of opposite terminals, w is the reliability of each measurement, which means the accuracy of the previous forecasting.

In the model shown in equation (1), there is only one state variable for nodal load of the node to be predicted, which has been described by a set of six measurement equations. The whole model can encompass the relationship between state variables and independent measurement values, and make full use of the two types of redundant information generated to correct the state variables

3.2. Nodal load forecasting method

Based on the established estimation model considering redundant information, the steps of nodal load forecasting are as follows:

(1) Retrieve historical power data and forecast reliability information of multi-terminal measurement of transmission and distribution components;

(2) According to the variation characteristics of the historical power data of transmission and distribution components, select the appropriate previous forecasting model. In this study, the previous forecasting model is support vector machine (SVM) model [11];

(3) According to SVM model, respectively forecast the transmission power of transmission and distribution components;

(4) According to the constraints of equation (1), process the pre-forecasting values of transmission and distribution components and generate multiple sets of measurement equations;

(5) Use the forecasting model to correct and estimate the nodal load state and get the final forecast value of nodal load;

(6) Collect the actual value information of the nodal load, compare the actual load with the predicted results, assess and record the reliability of each measurement;

(7) return to step (1) to carry out the next stage of nodal load forecasting.

4. Case studies

The actual operation data of some grids in Shandong Province on July 2017 are used as an example to verify the nodal load forecasting method which takes into account the redundant information. The data set covers the power measurement data of the high, middle and low voltage ports of the three-winding transformer in the substation, and the power measurement data of the transmission lines connected to the high-voltage busbar of the local and opposite nodes. The time resolution of the data set is 1h. The regional node topology is shown in figure 2.

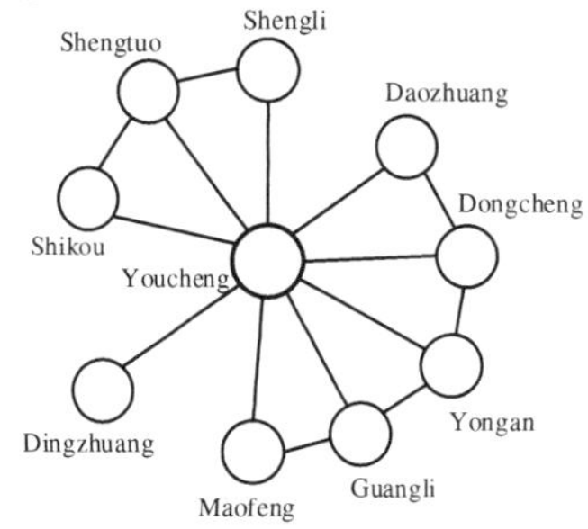

Figure 2. Regional node topology

The power measurements of the transformer high voltage port are selected as the nodal load criterion. Two models are used to predict the nodal load on July 27. On the one hand, only the nodal load criterion is selected as the historical data, and the SVM model is used to directly predict the nodal load. On the other hand, the SVM model is used to predict the power of lines and all transformer ports respectively. Then, the load forecasting method based on physical connection and redundant information proposed in this paper is used to correct and estimate the forecasting results. Taking Shikou Station as an example, the forecasting results of these two forecasting models are shown in figure 3. As can be seen from figure 3, the forecasting results of the proposed model are closer to the actual nodal load.

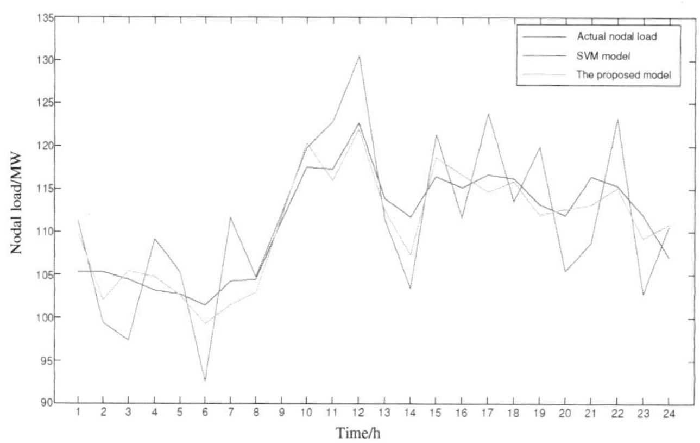

Figure 3. Comparison of nodal load forecasting results

The mean absolute percent error (MAPE) and root mean square error (RMSE) are chosen as the evaluation criteria to compare the predictive effects of the two models. The comparison of forecasting results is shown in table 1.

Table 1. Comparison of forecasting results

Node name	MAPE(%)		RMSE(MW)	
	SVM model	The proposed model	SVM model	The proposed model
Dongcheng	11.337	8.984	9.772	5.532
Yongan	9.193	6.243	9.428	6.223
Guangli	8.359	5.731	7.228	4.262
Maofeng	10.288	7.561	8.144	6.591
Dingzhuang	10.392	9.791	8.400	7.330
Shengli	6.152	5.598	10.136	5.417
Shikou	8.068	6.318	10.519	7.762

IOP Conf. Series: Materials Science and Engineering **366** (2018) 012046 doi:10.1088/1757-899X/366/1/012046

As can be seen from table 1, compared with the traditional prediction methods, the proposed method has a decline in the prediction error, and the nodal load forecasting accuracy has improved significantly.

5. Conclusions

In view of the fact that the existing nodal load forecasting methods cannot discover the physical association of the nodes deeply and utilize the redundant information provided by the modern measuring technology sufficiently, a nodal load forecasting method with the estimation and correction characteristic and considering the redundant information is proposed. The advantages of this method are the sufficient utilization of the rich original information provided by modern measuring technology, no requirement for the previous prediction model, the distributed prediction process, the flexible control of the nodal load forecasting scale and so on. The actual grid operation data are selected to validate the proposed method. The results show that the load forecasting method based on redundant information has a significant decrease compared with the SVM forecasting error. It can be seen that this method considers the actual situation of power grid in the context of Energy Internet, makes full use of the redundant information provided by modern measurement system, and based on the idea of traditional prediction methods, explores the promotion effect of redundant information on improving nodal load forecasting effect. It is helpful to improve the accuracy of nodal load forecasting results.

Acknowledgment

This paper is supported by the project of Economic & Technology research Institute, State Grid Shandong Electric Power Company, and National Natural Science Foundation of China (51407106).

References

[1] Huang A, Crow M, Heydt G, Zheng J and Dale S 2010 The future renewable electric energy delivery and management (freedm) system: the energy internet *Proceedings of the IEEE* **99**(1) 133-148.

[2] Shishebori A and Kian A 2010 Risk analysis for distribution company energy procurement with pool market, DGs and demand response *Electrical Engineering* pp 949-954.

[3] Abarghooee R and Aghaei J 2012 Stochastic Dynamic Economic Emission Dispatch considering Wind Power *Power Engineering and Automation Conference* pp 158-161.

[4] Xue Y, Lei X, Xue F, Yu C, Dong C, Wen F and Ju P 2014 A review on impacts of wind power uncertainties on power systems *Proceedings of the CSEE* **34**(29) 5029-40.

[5] Papalexopoulos A and Hesterberg T 1990 A regression-based approach to short-term system load forecasting *IEEE Transactions on Power Systems* **5**(4) 1535-47.

[6] Niu D, Gu Z, Xing M and Wang H 2006 Study on forecasting approach to short-term load of SVM based on data mining *Proceedings of the CSEE* **26**(18) 6-12.

[7] Abu-el-magd M and Sinha N 1981 Two new algorithms for on-line modelling and forecasting of the load demand of a multinode power system *IEEE Trans on Power Apparatus and Systems* **100**(7) 3246-53.

[8] Abu-el-magd M and Sinha N 1982 Univariate and multivariate time series techniques for modeling and forecasting short-term load demand *IFAC Symposium on Theory and Application of Digital Control* pp 329-334.

[9] Han L, Han X, Yun Z and Geng Y 2007 Method for ultra-short term multi-node load forecasting *Automation of Electric Power Systems* **31**(21) 30-34.

[10] Mu T, Kang C, Xia Q, John N and Liu M 2008 Power system multilevel load forecasting and coordinating: Part three correlative coordinating model *Automation of Electric Power Systems* **32**(9) 20-24.

[11] Abbas S and Arif M 2007 Electric Load Forecasting Using Support Vector Machines Optimized by Genetic Algorithm *Multitopic Conference* pp 395-399.

Reactive power optimization method of power distribution networks based on OCSFLA

Z R Wu[1,a], X Z Dong[1,b], L M Chen[1,c], Z W Liu[1,d] and X L Xu[2,e*]

[1]Electric Power Research Institute of China Southern Power Grid, Guangzhou, Guangdong , 510080, China
[2]Nanjing NARI-Relays Electric Co.,Ltd, Nanjing, Jiangsu , 211102, China

[a]wuzr@csg.cn, [b]dongxz@csg.cn, [c]chenlm2@csg.cn, [d]liuzw2@csg.cn, [e]1296685433@qq.com

Abstract: A power grid reactive power optimization algorithm was proposed in this paper to ensure that distribution network systems run efficiently, which is called the opposition-based chaos shuffled frog leaping algorithm (OCSFLA). A mathematical model of reactive power optimization was established based on the consideration of the constraint conditions. The initial swarm of the model was optimized based on the opposition strategy. The leading role changed gradually from the shuffled frog leaping algorithm to chaotic search in the iterative process by setting the appropriate chaotic variables. Thus, the global optimal solution was searched more quickly. In the simulation tests of the IEEE30 system, the results show that a better compensation effect was obtained when the reactive power compensation was performed in nodes 12 and 24. After completing the reactive power optimization based on the improved leapfrog algorithm, the active network loss value of the system was reduced by 13.85%. The algorithm was more efficient compared with other traditional optimization algorithms because OCSFLA can avoid falling into local optimization.

1. Introduction

Grid voltage control depends largely on reactive power optimization control. The reasonable distribution of reactive power has a direct impact on the safety and stability of the power system. The reactive power optimization problem is an integral part of the optimal power flow calculation in a power system.

The reactive power optimization in a power system is a mixed nonlinear programming problem with multivariable and multi-constrained characteristics. Traditional reactive power optimization algorithms such as the interior point method[1], the genetic method[2], the peak group algorithm[3], the particle swarm algorithm[4] and so on[5] were widely used. A power grid reactive power optimization algorithm was proposed in this paper to solve the problem of slow convergence in the original algorithm and to ensure that the distribution network system runs efficiently, which is called OCSFLA. OCSFLA combines the idea of shuffling complex evolution[6] and particle swarm optimization with the evolutionary properties of cultural genetic algorithms. It is a new type of post-heuristic optimization algorithm based on collective cooperative intelligence.

2. Reactive power optimization algorithm based on OCSFLA

The parameters of the reactive power constraints, according in the mathematical model of the reactive power optimization algorithm mentioned in the previous section, such as V_{GK}, T_i, C_j, V_i and Q_{Gj}, are optimized based on OCSFLA. In this section, two improved strategies were proposed.

1) *Optimizing the initial swarm.* The random initialization of the frog swarm in the shuffled frog leaping algorithm (SFLA) ensures the even distribution of the swarm to a certain extent. However, it does not guarantee the quality of the individuals. The quality of the initial swarm affects the optimization efficiency and quality of the algorithm. To this end, an improved method based on an opposite strategy was presented in this paper.

2) *Variable updating strategy.* When the worst individual in the current subgroup is iteratively updated, an appropriate chaotic variable initial value is set so the algorithm is updated mainly with the update strategy of SFLA, and the chaos search plays an auxiliary role at the beginning. As the number of local depth search iterations increases, the algorithm gradually enters the predominant chaos search to help the algorithm jump out of the local extremum.

In this paper, the parameters of the reactive power constraints in the mathematical model of reactive power optimization algorithm are simplified such that, $(V_{GK}, T_i, C_j, V_i, Q_{Gj}) = (n_1, n_2, n_3, n_4, n_5)$.

The algorithm iterative optimization steps are as follows:

1) Set the algorithm parameters, such as the total number of frog individuals in the swarm N, the frog individual dimension d, the number of individuals in each sub-swarm m, the number of sub-swarms n, the number of local iterations in sub-swarm g, mixing iteration number G, initial value of chaos variable $y(0)$, chaos factor λ, coefficient α, constants a_1 and a_2.

2) Generate the initial swarm using the opposition strategy, and set the swarm size as N (N = 5 in this algorithm).

a) Randomly initialize the frog individuals x_i, ($x_i \in (a_i, b_i)$), and find their opposite individuals $\hat{x}_i$, ($\hat{x}_i = a_i + b_i - x_i$), in order to form a frog swarm of size of $2N$.

b) Calculate the fitness values of all the individuals in the swarm and sort them from good to bad according to their values.

c) Select the individuals of N frogs randomly to form the initial swarm with a certain probability. The principle of probability selection is that individuals with better fitness values are selected into the initial swarm easily. The probability selection formula is as follows:

$$p_i = (2N - i)/2N, \quad i = 1, 2, \cdots, 2N \tag{1}$$

3) Divide N frog individuals into subgroups, and choose Q frog individuals according to the triangle selection probability of $2(m+1-i)/[m(m+1)]$.

4) Repeat the following steps g times for each subgroup.

a) Update the optimal individual position X_b and the global optimal individual position X_g of the current iteration subgroup according to the fitness values. Then, determine the worst individual position X_w of the subgroup in the current iteration.

b) Update X_w.

$$\begin{aligned} newX_{wi}(k) &= X_{wi}(k) \cdot \exp \cdot [(1 - \exp(-\alpha \cdot y(k))) \cdot \\ &(\mu \cdot X_{wi}(k) \cdot (1 - X_{wi}(k)))] + e^{-2\alpha \cdot y(k)} \cdot \Omega_i(k) \end{aligned} \tag{2}$$

where $\Omega(k)$ is the frog jump step update formula. $y(k)$ is the chaos variable.

$$\Omega(k) = a_1 \cdot rand() \cdot (X_b(k) - X_w(k)) \tag{3}$$

$$y(k) = y(k-1) \cdot (1 + \lambda) \tag{4}$$

In Eq. (3) and Eq. (4), $X_b(k)$ and $X_w(k)$ denote the best and the worst individual in the kth iteration of the current subgroup, respectively. *rand*( ) denotes a random number uniformly distributed in [0,1]. a_1 is a constant, and k is the number of iterations. λ is a chaos factor less than 1, and $X_{wi}(k)$ represents the ith dimension of the worst individual $X_w(k)$ of the current subgroup at the kth iteration. $\Omega_i(k)$ represents the ith dimension of $\Omega(k)$, and α and μ denote positive constants.

c) If the fitness value of $newX_w$ is better than X_w, replace $newX_w$ with X_w.

d) Update Ω (k) based on Eq. (5) and X_w based on Eq. (3) and Eq. (4).

$$\Omega(k)=a_2\cdot rand()\cdot\left(X_g(k)-X_w(k)\right) \tag{5}$$

where $X_g(k)$ denotes the globally optimal individual of the current sub-swarm at the kth iteration. a_2 is a constant.

e) If there is still no improvement, generate $newX_w$ according to Eq. (6).

$$newX_{wi}(k)=O_{min}+(O_{max}-O_{min})\cdot rand() \tag{6}$$

where O_{max} and O_{min} represent the maximum and minimum of the search range of the algorithm, respectively.

5) After all the subgroups have completed the local depth search, the evolution process is finished if the global mixing iteration number G is satisfied, and the global optimal value is output. Otherwise, remix all the frog individuals and return to step 3.

The diagram of the chaos shuffled frog leaping algorithm is shown in Figure 1.

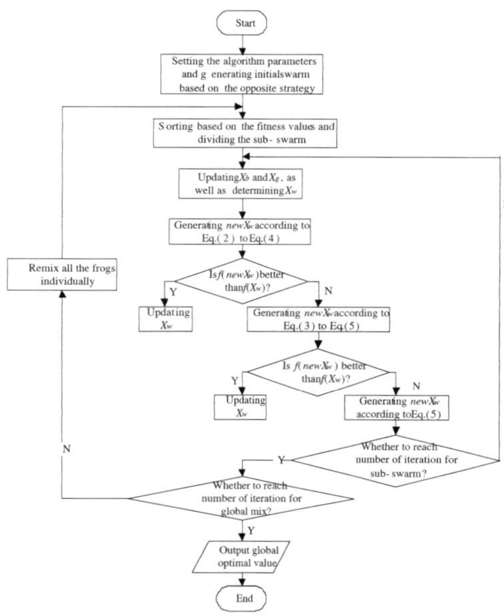

Figure 1. Diagram of the chaos shuffled frog leaping algorithm

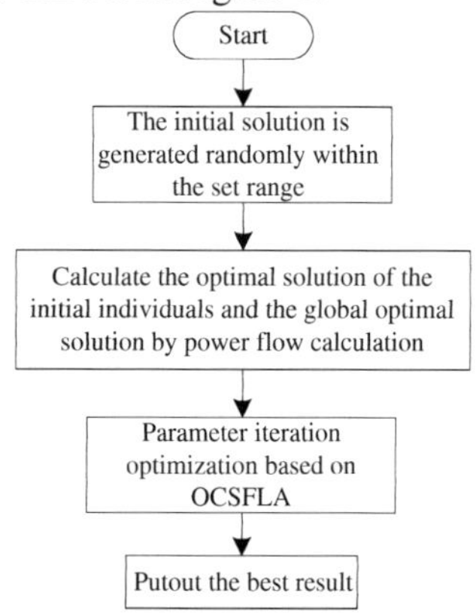

Figure 2. Diagram of the reactive power optimization algorithm

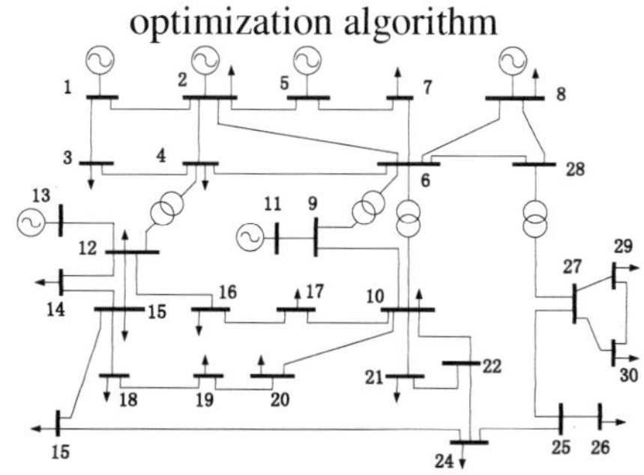

Figure 3. Structure diagram of IEEE 30 nodes system

3. Example analysis of the reactive power optimization

3.1 Example setting

The control variables used in this paper are as follows:

$$X=(V_{GK},T_i,C_j) \tag{7}$$

The state variables are as follows:

$$U=(V_i,Q_{Gj}) \tag{8}$$

where V_{GK} is the voltage amplitude of the adjustable generator node. T_i denotes the tap position of adjustable transformer. C_j is the group number of capacitor switching of reactive power compensation. V_i is the voltage of the PQ node. Q_{Gj} is the reactive power of the generator output.

The coding of all control variables can be expressed as follow, according to the above analysis:

$$X = [C, T, V_G] = [C_1, C_2, \cdots, C_i, T_1, T_2, \cdots, T_j, V_{G1}, V_{G2}, \cdots, V_{GK}] \quad (9)$$

where i, j and k represent the nodes number of reactive power compensation, the number of adjustable transformers and the nodes number of power generations, respectively.

The example initial parameter values in this paper are set as below.

Set the total number of frog individuals in the swarm as 200, the number of individuals in each sub-swarm as 20, the number of subgroups as 10, the number of local iterations in the sub-swarm as 10, the number of global mixing iterations as 30, and the initial value of the chaotic variable $y(0)$ as 10^{-7}. The chaos factor λ is 0.5, and the coefficient α is 100. $a_1 = a_2 = 3$. The diagram of the reactive power optimization algorithm is shown in Figure 2.

3.2 Example analysis of power grid optimization

The IEEE30 node system was tested as an application example in order to verify OCSFLA. The structure diagram of IEEE 30 nodes system is shown in Figure 3. Six generator nodes (PV nodes such as 1, 2, 5, 8, 11 and 13) were distributed in the system. Other nodes were PQ nodes. In addition, adjustable transformers (branches 6-9,6-10,4-12 and 27-28) were set in the IEEE30 node system. The initial transformers ratios are listed as follows: $T_{6\text{-}9}$ = 1.0155, $T_{6\text{-}10}$ = 0.9629, $T_{4\text{-}12}$ = 1.0129, $T_{27\text{-}28}$ = 0.9581.

The IEEE30 node system was simulated by the Monte Carlo method, and the six reactive compensation points were compensated. Statistical results of the reactive power compensation are shown in Table 3 after 10000 sampling iterations. It can be seen from Table 1 that the maximum compensation probability occurs at 12 nodes and 24 nodes. In other words, voltage safety problems are most likely to occur in these two nodes. Therefore, the best compensation effect will be achieved if reactive power compensation equipment is installed at these two nodes when considering the voltage safety level and economic cost.

Table 1. Statistical results of reactive power compensation in IEEE30

Node number	U_{av}/ (p.u.)	U_{max}/ (p.u.)	U_{min}/ (p.u.)	Compensation times	Compensation probability
2	1.0338	1.05	1.0309	173	17.3%
4	1.0258	1.0309	1.0058	201	20.1%
12	1.0564	1.0913	1.0883	618	61.8%
18	1.0319	1.0459	1.0307	128	12.8%
20	1.0354	1.0404	1.0307	138	13.8%
24	1.0292	1.0314	1.0298	359	35.9%

A power flow program was written based on OCSFLA to optimize the IEEE30 system. Comparison of the node voltage phase angles after 100 iterative steps of reactive power optimization are shown in Figure 4 and Figure 5. In Figure 6, changes in node network loss before and after reactive power optimization are revealed. It can be seen from Figure 6 that the voltage at each load node increases and becomes more balanced after reactive power optimization. The most voltage increase of node 7 is up to 0.02816 p.u. The loss of the IEEE30 system after reactive power optimization is reduced from 7.0901 MW to 6.1079 MW, which is a 13.85% decrease.

Figure 7 and Figure 8 show the cumulative area distributions of active power PL, reactive power QL and generator output residual power PR before and after reactive power optimization, respectively. It can be seen from the figures that the reactive power loss QL at nodes 12 and 24 is significantly reduced after reactive power optimization except for the active power loss. The total output power generator of the IEEE30 system decreased from 10.2474 MW to 8.9099 MW, a decline of 13.05%.

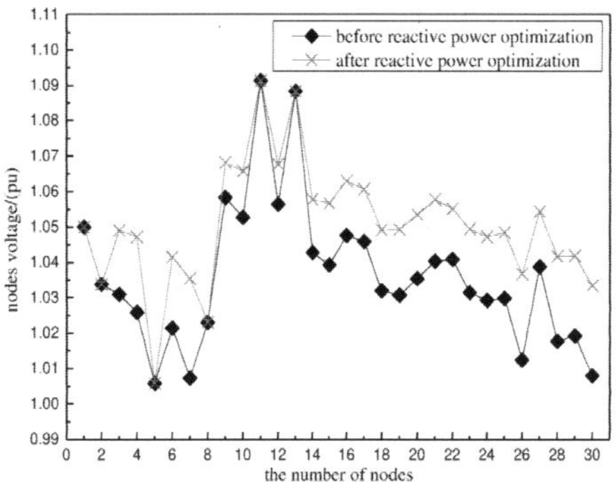

Figure 4. Comparison of node voltage values before and after reactive power optimization

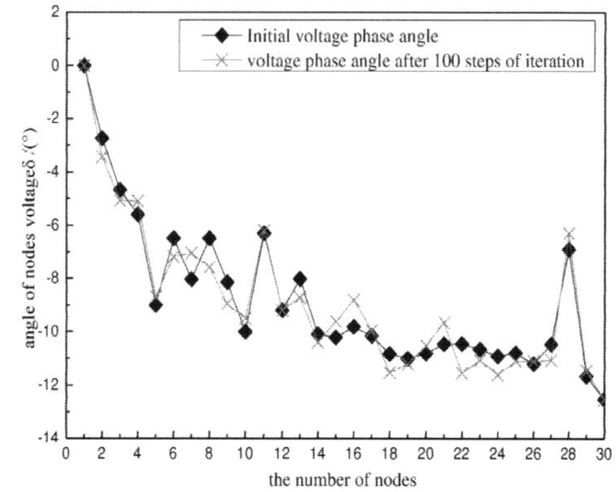

Figure 5. Comparison of nodes voltage phase angles after 100 steps iterative of reactive power optimization

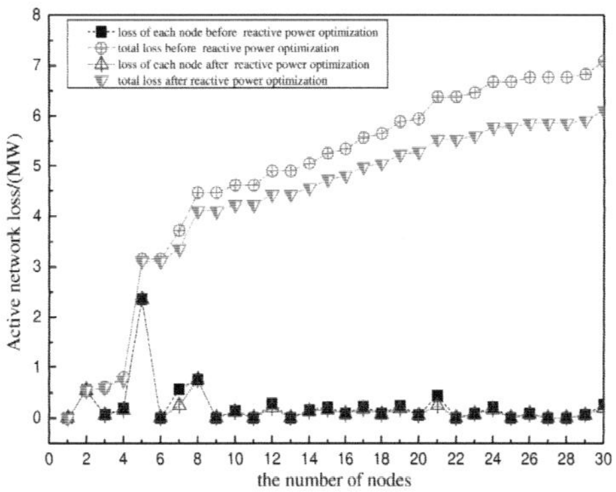

Figure 6. Changes in network loss of nodes before and after reactive power optimization

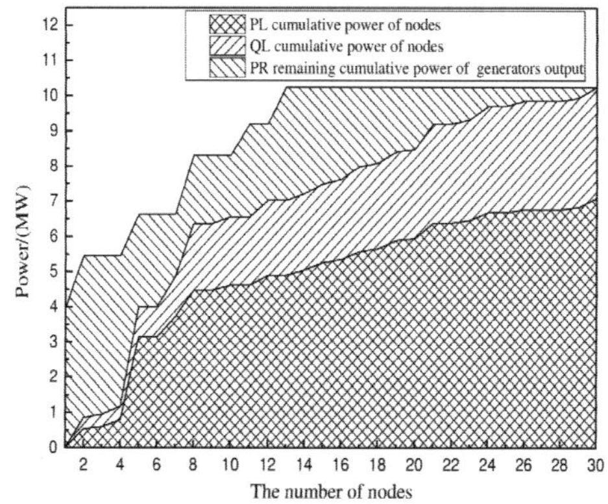

Figure 7. Cumulative power of nodes before reactive power optimization

Therefore, reactive power optimization can balance the voltage of each node within the allowable range of voltage and reduce system losses through the traditional means of reactive voltage regulator, resulting in good economic benefits.

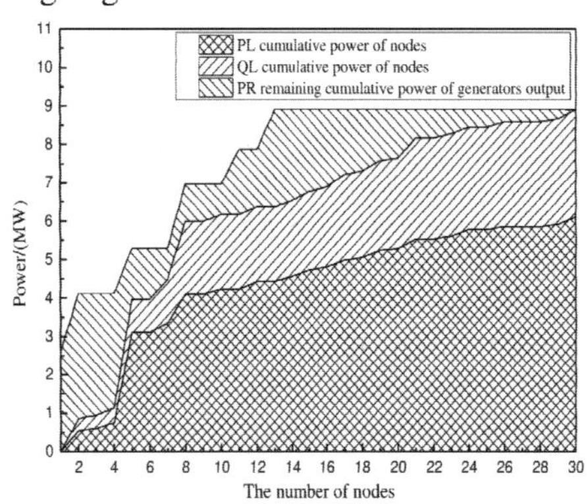

Figure 8. Cumulative power of nodes after reactive power optimization

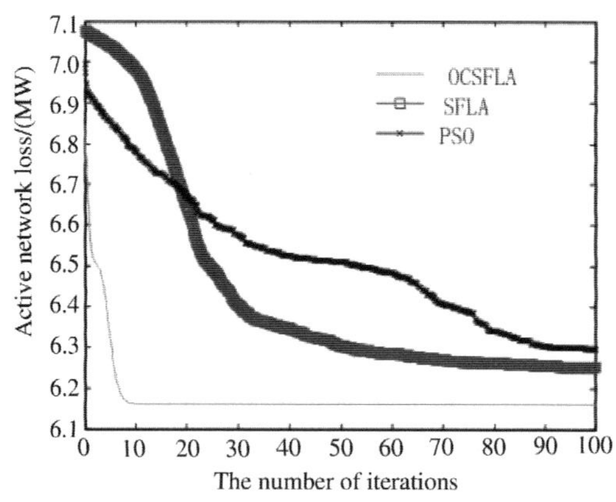

Figure 9. Contrast chart of the optimal performance for three algorithms

Due to the different performance results of the different algorithms in different stages of the evolutionary process, realizing the mutual complementarity will improve the overall optimization efficiency [7]. The traditional Particle Swarm Optimization (PSO), SFLA and OCSFLA were compared in this paper to validate the superiority of OCSFLA. These algorithms were used for simulations in the IEEE30 system. As shown in Figure 9, SFLA, which has not been improved, falls into a local optimum when it is updated several times with a minimum loss of 6.5549 MW. The convergence of PSO is very poor, having not completely converged after 100 iterations. When OCSFLA gets into the local optimum, weight variation occurs, helping the system to develop in a better direction. The minimum loss is 6.1079 MW. Thus, the OCSFLA proposed in this paper has some feasibility to improve the operational efficiency and reduce the network loss.

4. Conclusion

Reactive power optimization calculation is an important technique to improve power system analyses and operation. A power grid reactive power optimization algorithm was proposed in this paper to ensure that a distribution network system runs efficiently, which is called OCSFLA. It overcomes the shortcomings of the traditional SFLA in global search ability when solving the nonlinear optimization problem with complex constraints. The conclusions are as follows in the IEEE30 system optimization simulation:

Reactive compensation at nodes 12 and 24 will achieve better compensation in the IEEE30 system. The voltage of the IEEE 30 system increases and is more balanced after the reactive optimization of OCSFLA. The voltage increase of node 7 is up to 0.02816 p.u. The loss of the IEEE30 system after reactive power optimization is reduced from 7.0901 MW to 6.1079 MW, which is a 13.85% decrease. The load reactive power QL at nodes 12 and 24 is significantly reduced after the reactive optimization of OCSFLA. The total output power of the system generator decreased from 10.2474 MW to 8.9099 MW, a decrease of 13.85%. OCSFLA is more efficient compared with the PSO algorithm and SFLA; OCSFLA can avoid falling into local optimization, and it is feasible for use to optimize the power grid. The reactive power optimization of a power system covers a wide range of technology, and a number of areas are worth further study, including expanding the search range of OCSFLA and choosing parameters to achieve the optimal search performance that greatly influence the performance and efficiency of the system for problems of different scales and different characteristics.

References

[1] Z. Fan , X. Chang, H. Wang, T. Pu, T. Yu, G. Liu. Discrete reactive power optimization based on interior point filter algorithm and complementarity theory [C]//International Conference on Power System Technology (POWERCON).Chengdu, 2014: 210- 214.

[2] Iba, K. Reactive power optimization by genetic algorithm[J]. IEEE Transactions on Power Systems, 1994, 9(2): 685-692.

[3] Y. Li, Y. Wang, B. Li. A hybrid artificial bee colony assisted differential evolution algorithm for optimal reactive power flow[J]. International Journal of Electrical Power & Energy Systems, 2013, 52: 25-33.

[4] S. Ganguly, Multi-objective planning for reactive power compensation of radial distribution networks with unified power quality conditioner allocation using particle swarm optimization[J]. IEEE Transactions on Power Systems, 2014, 29(4): 1801-1810.

[5] W. Zheng , W. Wu, B. Zhang, H. Sun, Y. Liu. A fully distributed reactive power optimization and control method for active distribution networks[J]. IEEE Transactions on Smart Grid, 2016, 7(2): 1021-1033.

[6] L.J. Fogel, A.J.Owens, M.J.Walsh. Artificial intelligence through simulate devolution. New York: John Wiley, 1966.

[7] Santiago M, Antonio L, Jose-Maria P.A new methodology for the automatic creation of adaptive hybrid algorithms[J].Intelligent Data Analysis, 2012, 16(1): 3-23.

Visualization Research of Smart Substation Secondary Safety Measures

Y Y B[1], X M[1], H T G[1], ZH CH L[2], W K[2] and Q H CH[2]

[1]State Grid Anhui Electric Power company, Hehui 230022, China
[2]State Grid Bozhou Electric Power company, Bozhou 236000, China

1158116539@qq.com

Abstract. The smart substation adopts the network communication rather than the traditional electric cable, which makes the traditional secondary electric circuit terminal into a loop of virtual terminal and then the entire secondary system became a "black box", thus causes a lot of problems and difficulties into the operation and maintenance of smart substation. However, little attention has been paid to the visualization of secondary safety circuits for smart substations. In this paper, a portable device based on the characteristics of the secondary circuits and needs of the current secondary safety measures for smart substation was designed to realize the visualization of the device information, the control circuits and the safety circuits. Finally, the secondary security measures of smart substation were normalized and illustrated based on the first set protection of 220 kV line space.

1. Introduction

Smart substation is an important part of smart grid. In recent years, with the rapidly developed of smart substation, China has become the country with the largest number of smart substations in operation. The equipment of conventional substation is mainly connected by cable, while the smart substation based on industrial Ethernet technology mainly adopts optical fiber and network to achieve connection between substation equipment[1-2]. In this way, the open-circuit faults in the secondary circuit of current transformers and the short-circuit faults in the secondary circuit of voltage transformer can be avoided effectively. In addition, the network data stream replaces the DC secondary circuits which can eliminate the parasitic circuits and achieve the effective monitoring of DC logic [3]. Therefore, as a hub of 'Power Flow, Information Flow, Business Flow' in Smart Grid, smart substations are key components for Strong Smart Grid.

Compared with conventional substation, smart substation has no visual electrical connections. Through introducing integration units and smart terminals into smart substation, the fiber connection between secondary equipment can be achieved by the transmission of highly coupled network data. Due to the cables replaced by optical fiber, the transmission speed of smart substation is faster and the operation of secondary equipment becomes safer. However, this leads traditional secondary electric circuit terminal into a loop of virtual terminal, thus the entire secondary system becomes a 'black box' and cause difficulties to operation and maintenance of smart substations [4-5]. Therefore, the secondary security measures of conventional substation depending on physics isolation technology such as exiting hard plates of protection is no longer fully applicable to the relay protection in smart substation [6]. Moreover, the optical fiber circuits can be vulnerable in the process of frequently mechanical pull and plug. Thus, besides physics isolation technology, the secondary security measures of smart substation can mainly be realized by the logic isolation of GOOSE soft plates and maintenance mechanism.

However, the logic isolation measures based on GOOSE soft plates and maintenance mechanism of isolation measures are invisible and it requires highly reliability of secondary equipment. Moreover, the logic isolation measures were mainly realized by software which is error-prone, thus it can lead to mistake operations in the field [7, 8]. In addition, a lack of specific secondary security strategies for smart substation results in lots of problems in the process of operation and overhaul [9]. Therefore, it is essential to realize the visualization and normalization of smart substation secondary safety measures and it plays an important role in the reliable operation and large-scale construction of smart substations.

This paper firstly analyses the characteristics of secondary circuits and presents the current secondary safety isolation technologies for smart substation. Then a portable device is designed to realize the visualization of the device information, the control circuits and the safety circuits. Finally, safety isolation measures of the first set protection of 220 kV line space were put forward for the normalization of the secondary safety isolation measures in smart substation.

2. Secondary Safety Measures

2.1. Characteristics of Secondary Circuits

Compared with traditional substation, the complex secondary circuits were replaced by communication connections. The integration units, intelligent terminal and intelligent components in the process layer achieve all functions of the secondary circuits [7, 10]. The SV transmits sampling value of alternating current and the GOOSE network transmits the information of various switches. The structure of secondary circuits for smart substation was shown in Figure. 1.

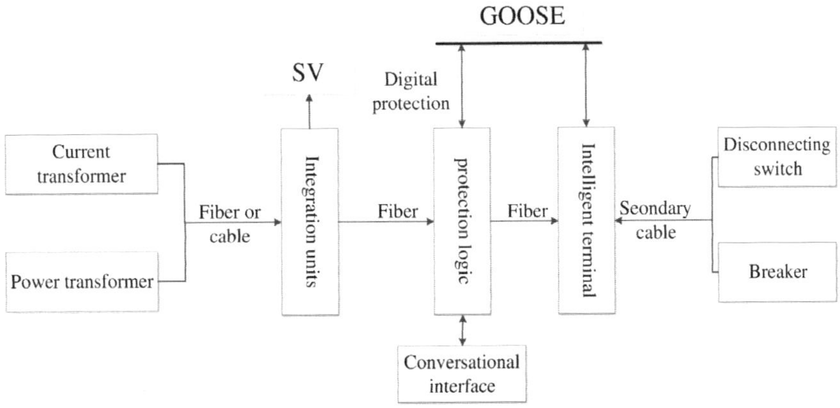

Figure 1. Structure of Intelligent substation secondary circuits

Since the point to point cable connections in the secondary circuits of conventional substations are replaced by optical fiber connections, which causes significant changes in secondary safety measures for smart substation[11,12]. Therefore, it is impossible to achieve effective hardware isolation in the security circuits. The secondary security measures of smart substation can mainly be realized by the logic isolation of GOOSE soft plates and maintenance mechanism. Figure.2 shows the specific safety measures for smart substations.

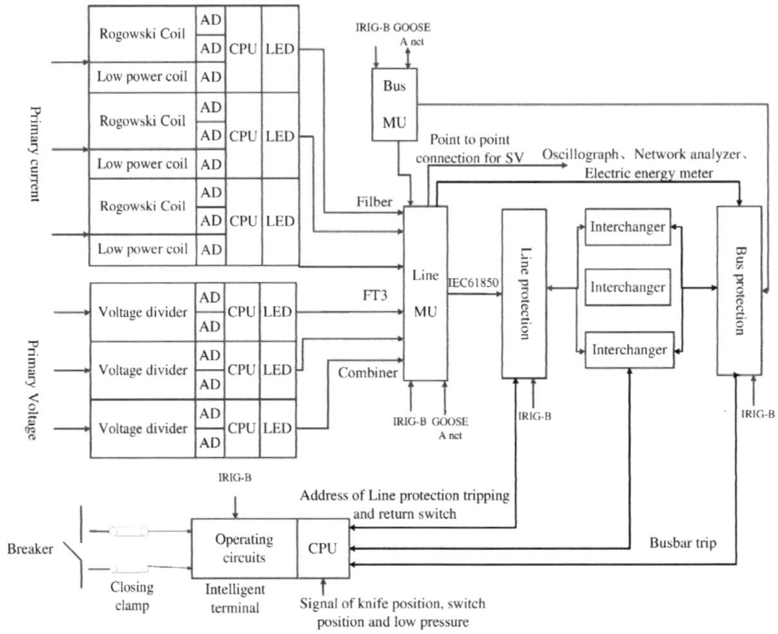

Figure 2. Security isolation measures of smart substation

3. Design of Handheld Device

To improve the work efficiency of repair personnel and improve the reliability of smart substation operation, this paper developed a handheld device to realize the visualization of the device information, the control circuits and the safety circuits, which can better meet the needs of secondary repair.

3.1. Software Design

Figure.3 shows the system structure of the visual tool. The specific work process can be shown as followings: First, import the SCD file into the system and implement the visualization of SCD file which can display the maintenance status of power equipment, the connect state of secondary virtual circuit and the flow of information between the secondary equipment. Then, the normalization of smart substation secondary safety measures can be used to confirm the implement status of relevant repair work. Finally, a variety of different situations security ticket template were formed to help repair personnel to complete the repair of smart substation. 61850 communication module mainly makes the communication with the maintenance device. Database module achieves the interactions of data with the database. Rights management module can avoid the possibility of wrong operation by grouping operation permissions.

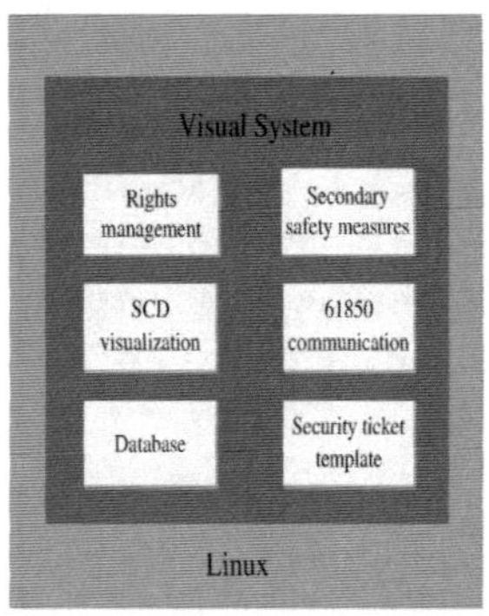

Figure 3. System structure of the visual tool

IOP Conf. Series: Materials Science and Engineering **366** (2018) 012048 doi:10.1088/1757-899X/366/1/012048

3.2. Hardware Design

The main parameters of the hardware device can be shown as followings: RJ45 interface, USB interface and touch screen. The RJ45 interface can communicate with the field maintenance equipment conveniently. The USB interface can be used to import and export the information of SCD files and database. The 10 inches touch screen can view the maintenance status of power equipment, the connect state of secondary virtual circuit and the flow of information between the secondary equipment. Figure.4 shows the physical diagram of the handheld device. The handheld device can connected with the relay protection devices automatically in the field which can improve the work efficiency of repair personnel and improve the reliability of smart substation operation.

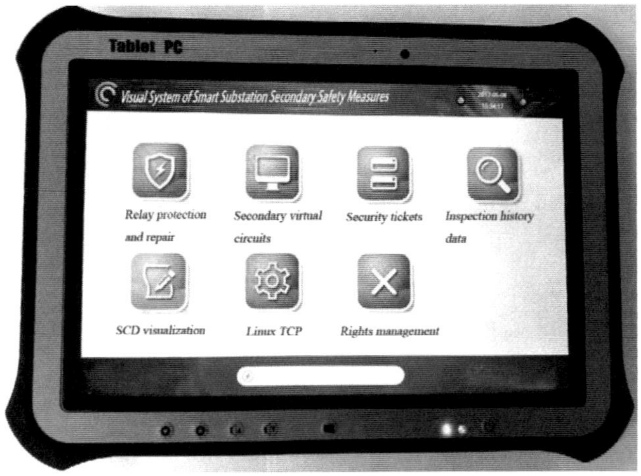

Figure 4. Figure of Handheld devices

Figure.5 shows the overview of safety circuits of smart substation. Through integration of smart substation network packet analyzer, record information, and protection logic tree, monitoring information system information and so on, it can achieve the visualization of device information, control circuits and safety circuits of smart substation. Thus, the smart protection equipment and secondary circuit security would be simpler and more clearly, which can reduce the difficulties of operation and improve the reliability and safety of substation operation and maintenance.The text of your paper should be formatted as follows:

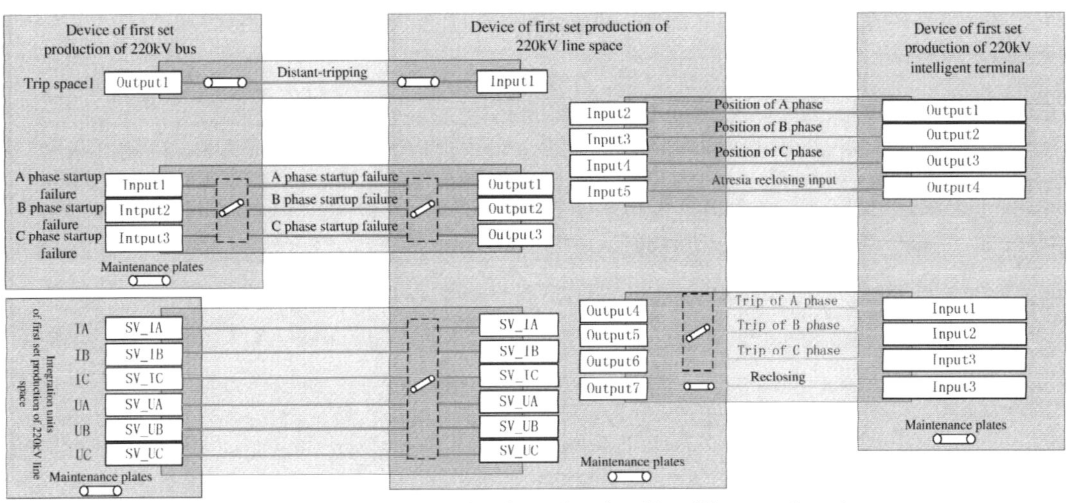

Figure 5. Visualization of safety circuit of intelligent substation

IOP Conf. Series: Materials Science and Engineering **366** (2018) 012048 doi:10.1088/1757-899X/366/1/012048

4. Secondary Safety Measures for 220 kV Smart Substation

In the first set protection of 220 kV line space, the typical configuration of SV sampling, GOOSE tripping mode and its network connections are shown in Figure 6. Meanwhile, the secondary security measures of smart substation were normalized and illustrated based on the first set protection of 220 kV line space.

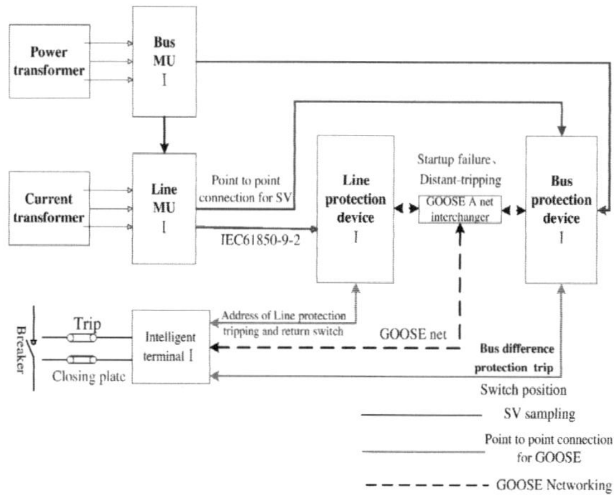

Figure 6. Figure of line space protection for 220 kV smart substation

4.1. Safety measures for outage maintenance of line space blackouts

Under the condition of a power failure of primary equipment, safety measures for the protection of 220 kV lines can be shown as followings:

a) Exit the receiving soft plates for SV and GOOSE of the first set bus protection of 220 kV line space, then input the forced divided voltage soft plates through isolating knife switches.

b) Exit the receiving soft plates for GOOSE of the first set line protection of 220 kV line space;

c) Input maintenance plates of the first set line protection, smart terminals and integration units.

Under the condition of a power failure of primary equipment, the safety measures for checking the failure circuits of the first set of 220 kV bus protection and line protection can be shown as followings:

a) Exit the soft plates for GOOSE and the failure soft plates of first set line protection of 220 kV line space, then input maintenance plates of bus protection;

b) Input maintenance plates of the first set line protection, smart terminals and integration units.

c) Shorted the secondary circuits of current transformer and made the secondary circuits of power transformer open, then separate current transformers and power transformers with the interface of integration unit.

4.2. Safety measures for charged maintenance of line space

Under the condition of charged maintenance of primary equipment, safety measures for the secondary equipment protection of 220 kV line space can be shown as followings:

When defects occurred in merging unit, it is necessary to stop the operation of related relay protection. Moreover, the primary equipment can also be out of service in some degree.

When defects occurred in line protection, safety measures can be shown as followings:

a) Exit the receiving soft plates for GOOSE of the first set bus protection of 220 kV line space;

b) Exit the failure soft plates of first set line protection of 220 kV line space, then input maintenance plate of line protection;

When defects occurred in smart terminals, safety measures can be shown as followings:

a) Exit the hard plate for smart terminals of the first set bus protection of 220 kV line space and input maintenance plate;

b) Exit the receiving soft plates for GOOSE of the first set line protection of 220 kV line space;

c) Input the forced divided voltage soft plate through isolating knife switches;

IOP Conf. Series: Materials Science and Engineering **366** (2018) 012048 doi:10.1088/1757-899X/366/1/012048

5. Conclusions

In this paper, a portable device is designed to realize the visualization of secondary safety circuits and the standardization of secondary security measures for smart substations. Therefore, the status of SV/GOOSE transmitting and receiving soft plates and hard plats of maintenance can be displayed directly. What is more, safety isolation measures of the first set protection of 220 kV line space were put forward for the normalization of the secondary safety isolation measures in smart substation.

References

[1] Sun Zhipeng, Chen Xiaogang, Ye Gangjing,etc. Security measures and visualization study of smart substation [J]. Electric technology, 2014 (4) : 19-21.
[2] Guo Shenglan Yi Shu wei, Cui Aijun. Secondary security measures of smart substation[J]. Rural electrification, 2014 (8) : 37-37.
[3] Li lei, Zhao Menglu. Security measures and visualization study of smart substation [J]. China's high-tech enterprises, 2016 (25) : 141-142.
[4] Zheng Ningmin, Huang Zhaomin, Gao Xiang, etc. Visualization operations research of secondary system for the smart substation[J]. Power supply, 2017 (1) : 55- 60.
[5]Yan Haojun, Yao Qinfeng, Xu Xin. Security measures and visualization study of smart substation [J]. Journal of Zhejiang electric power, 2015, 34 (9) : 6-9.
[6]Guo Shenglan, Yi Shu wei, Cui Aijun. Security measures and visualization study of smart substation [J]. Rural electrification, 2014 (8) : 37-37.
[7]Lan Haitao. Security measures study of smart substation [J]. Journal of smart grid, 2014, 2 (1) : 62-66.
[8] Chi Xu. Security measures and visualization study of smart substation [J]. Electric technology, 2014, 15 (12) : 147-148.
[9] Lei Young, Zhang Kan jun, Hong Meizi, Ye Pang qi, Du Zhengan. Electrical engineering society of China ,2015.
[10] Yang Luolin. Security measures study of smart substation [J]. Journal of new technology and new products in China, 2016 (16) : 187-188.
[11]Jiang Chuanfei, Gao Lingyu, Gao Guixin. Introduction to the relay protection device for smart substation safety measures [J]. Science and technology, 2015, (33) : 60.
[12] Sun Zhipeng, Chen Xiaogang, Ye Gangjing,etc. Security measures and visualization study of smart substation [J]. North China Electric Power University (Beijing), 2014.

IOP Conf. Series: Materials Science and Engineering **366** (2018) 012049 doi:10.1088/1757-899X/366/1/012049

Mobile Cyclic Accelerator Based on Ironless Pulsed Betatron. Results of Testing Powering

Yu P Kuropatkin, V I Nizhegorodtsev, K V Savchenko, V D Selemir, E V Urlin, A A Chinin and O A Shamro

Russian Federal Nuclear Center – VNIIEF, 37 Mira Ave., Sarov, 607188, Russia

mailbox@ntc.vniief.ru

Abstract. The authors describe the mobile cyclic accelerator based on the ironless pulsed betatron. The possibility of such accelerators application as a part of mobile radiographic complexes is illustrated. These complexes are intended for the radiography of dynamic objects with large optical thicknesses located in the explosion-proof chamber (EPC). Such complexes allow optimizing the radiography geometry in the hydrodynamic experiment at the expense of the possible location change of the radiator and recorder relative to the investigation object. The geometry optimization allows increasing the X-ray radiation intensity in the recorder plane at least twice if compare with the active Russian radiographic complexes. The authors provide the output parameters of the mobile cyclic accelerator at its testing powering on the test bench. The value of a capacitive storage of the pulsed powering system of the betatron electromagnet was C=900μF (C_{max}=1800 μF). The thickness of the lead test object examined with X-rays was ≈100 mm at 4 m from the tantalum target; the dimension of the target was 6x6 mm^2. The full width of the output gamma pulse at half maximum in a single frame mode was ≈100 ns. The dimension of the radiation source was ≈ 6x3 mm^2.

1. Introduction

Powerful sources of the pulsed radiation with the optimal spectral composition are required for the radiography of dynamic objects with large optical thicknesses. The generators of such radiation are either linear or cyclice electron accelerators. There are several active radiographic complexes based on the linear accelerators in the world: the American complex DARHT [1], the Chinese complex DRAGON [1], the English complex HRF AWE [2], etc.

The ironless pulsed small-sized betatrons of the BIM type [3,4] have been applied in FSUE "RFNC-VNIIEF" and FSUE "RFNC-VNIITF" for a long time in radiographic complexes to provide hydrodynamic investigations, particularly, the investigations of the substance behavior in supercritical conditions [5].

Nowadays, the active radiographic complexes are stationary ones. Their development, creation and maintenance require a lot of finance and time resources. To reduce expenses and optimize the process of preparation and hydrodynamic experimentation, FSUE "RFNC-VNIIEF" began to apply the conception of using mobile radiographic complexes (MRC) [6]. They consist of mobile cyclic accelerators (MCA) [7], an explosion-proof chamber (EPC) with an investigation object, system of X-rays collimation and shadow images recording system. One EPC can be maintained by several MCAs.

The realization of this conception allows:

1) optimizing the experiment geometry at the expense of the position change of the radiation source and recorder relatively to the test object located in the EPC; this allows increasing the x-ray radiation intensity in the recorder plane twice as compared with active Russian complexes;

2) significantly decreasing the cost of radiographic complexes if not building heavy protective casemates; thus, decreasing the expenses for their infrastructure maintenance (it is possible to use cheap rapidly erected constructions);

3) creating an efficient environment protection system at the expense of localization of dangerous explosion products and hereby, a shock wave;

4) increasing the calendar time of hydrodynamic experiments (nowadays, the experiments are prohibited during summer time because of fire risks) due to the fire safety growth.

Figure 1 presents one of the variants of the gas-dynamic experiment that allows obtaining up to three frames in one experiment.

Figure 1. The experiment sketch with the usage of a single-beam three-frame mobile radiographic complex:
1 – mobile cyclic accelerator; 2 – EPC with the test object; 3 – X-rays collimation system; 4 – shadow images recording system.

2. Mobile cyclic accelerator based on ironless pulsed betatron

Main parts of the MRC is the MCA. Figure 2 presents a sketch of a single-beam three-frame MCA based on the betatron of the BIM type that consists of two units: an accelerator unit (AU) and a unit of pulsed powering of the betatron electromagnet (UPPE).

Figure 2. The sketch of a single beam three-frame mobile cyclic accelerator:
1– accelerating unit; 2– unit of pulsed powering of the betatron electromagnet .

Symbolically, there are the oscillator elements of the radiographic facility in the van of the accelerator unit (1). In the other van (2) there is a pulsed powering system of the betatron electromagnet and technological equipment. The connection between the units and external automated control system is performed with the use of cable and fiber-optic lines (these lines are not shown in the sketch of a mobile cyclic accelerator (Fig.2)).

The base facility of the MCA BIM is a pulsed ironless betatron of a new generation. Its units have been worked out on the test bench since 2002; and the radiation parameters have been optimized. One of the main units of the facility is an injector. It is designed based on the small voltage multiplier [8] and Blumlein pulse forming line (PFL). This injector allowed significantly decreasing (2.5 times) mass-sized characteristics of the radiation source as compared with the sources of active Russian complexes [9]. The facility dimensions are 4.5×2×1.8 м^3, the total weight is about 5 tons. The major units and systems of the oscillator are similar to those of earlier developed Russian complexes.

Figure 3 illustrates the structural scheme of the MCA BIM.

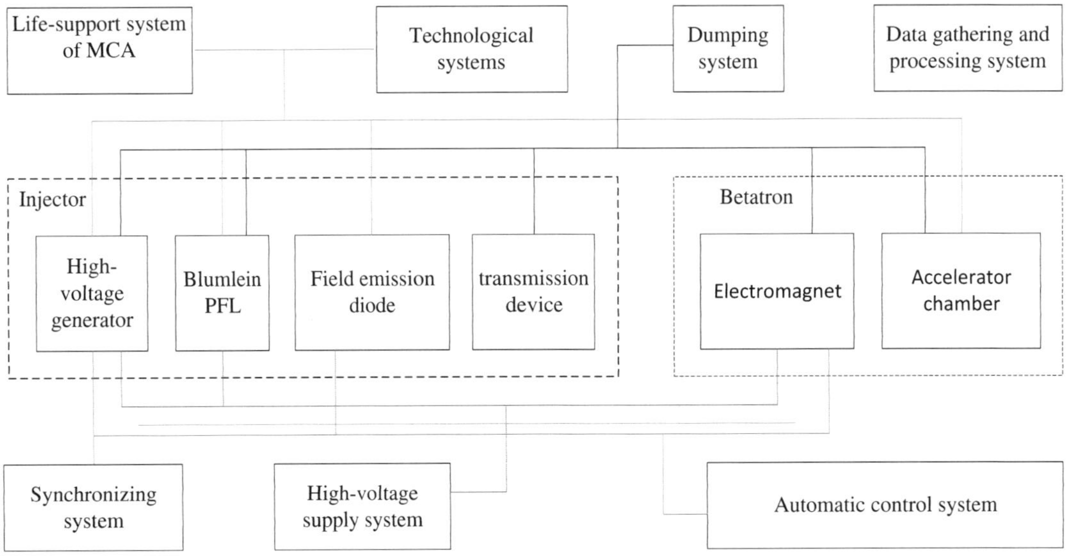

Figure 3. The structural scheme of the MCA BIM.

The MCA BIM consists of a betatron, injector, high-voltage supply system, synchronizing system (low-voltage and high-voltage), automated control system, data gathering and processing system (operate with the output parameters), dumping system ("low" and "fast" dumping), technological systems and other accessory life support systems of the MCA.

The electron acceleration in the betatron is realized by the vortex electrical field that appears because of a discharge of the capacitive storage through the electromagnet coils. Hereby, the magnetic field changes according to the sine law: $B(t) = B_0 sin(\Omega t)$, where $\Omega = 1/(LC)^{1/2}$ – circuit cyclic frequency (L and C – circuit inductance and capacity), and $B_0 = KI_{max}$ – field on the orbit at the peak current of the electromagnet I_{max} (K – correction factor). Automatically, the magnetic field inductance is changed synchronically with the acceleration by the expense of the electromagnet coils geometry. It allows holding an electron beam on the orbit. The high-voltage electromagnet power unit is switched on firstly, forming the signal "0 field". After this signal the powering time reading of some facility systems starts (Blumlein PFL, transmission device, high-voltage generator, etc.). The electron beam is generated and preliminary accelerates in the injector. In this case it is the accelerator of the direct action for E_{lim}= *1,5 MeV*. The injector operates in the following way. Blumlein PFL is charged at the high-voltage generator response. A machining voltage pulse is formed in the Blumlein PFL after a discharge arrester response. This pulse is transmitted through an intermediate discharge arrester to the field-emission diode, where an electron beam with a current amplitude of ≈ 2kA and a pulse width at half maximum of ≈10 ns is generated. A beam injection into the betatron is performed with the use of the transmission device [10] at the moment t_{inj}, when the magnetic field on the orbit achieves the value that is defined by the ratio: $B = B_0 \Omega t_{inj.} = E(300R_0)^{-1}$, here: R_0 – radius of the equilibrium orbit, E – electron beam energy. Hereby, an injection moment $t_{inj} \cong E(300B_0R_0\Omega)^{-1}$. The conduction and focusing of the electron beam is performed with the use of two magnetic lenses that are symbolically called a solenoid and lens; they are powered by their high-voltage supply units. Electrons are accelerated up to highest possible energy by the vortex electrical field. Then the accelerated beam is dumped to the target using either 'fast' or 'slow' dumping units. Both units consist of windings which are assembled inside the electromagnet and individual high-voltage generators.

The automated control system [11] according to the time algorithm provides the charging of the capacitive storage of the functional units and their parameters control during the operation time of the MCA BIM. The control of the facility output parameters is performed by the data gathering and processing system. The technological systems provide: creation of the necessary working vacuum in the betatron accelerating chamber and in the area of the field-emission diode, water-cooling of the vacuum pumps, filling of the gas-filled discharge arresters by the working composition, and filling of the high-voltage devices of the facility by the liquid dielectric. The MCA BIM is also equipped by the vision-based inspection system, light and sound alarm, firefighting, audio and climate control systems, and load lifting mechanism.

The solution to create the MCA BIM was made in 2013. The construction design of the accelerator van was being performed since 2013 up to 2015. The oscillator equipment of this van was worked out. The assembly work and commissioning were also carried out.

The oscillator testing with its electromagnet powering from the pulsed system located on the test bench was performed in 2015. All the oscillator systems located in the accelerator unit operated in the normal mode. We finished the development of the van construction of the UPPE. There is a capacitive storage for the energy of ≈ 0.5 MJ (6 capacitors per 300μF, 24kV) and a high-voltage power source 25-15 (PS25-15). The switch box adjustment was finished on the test bench; the switch box transmits the energy from the capacitive storage to the electromagnet. At the beginning of 2016 the switch box and some part of the technological equipment were assembled in the UPPE. The test powering of the pulsed powering system of the electromagnet was carried out. The load was the electromagnet located in the accelerating unit. The dimensions of the capacitive storage with the switch box were 2.6×1.4×1.8 m; the weight was ≈ 2 t. There is a photo (Fig. 4) of the capacitive storage with the switch box located in the UPPE.

1 —  — 2

Figure 4. The photo of the unit of pulsed powering of the betatron electromagnet:
1 – capacitive storage; 2 – switch box

The MCA was moved to the internal polygon of FSUE "RFNC-VNIIEF" in September 2016. The testing powering was carried out. All the systems of the MCA BIM operated in the normal mode. The value of the capacitive storage of the pulsed powering system of the electromagnet was C=300 μF.

3. Results of testing powering of MCA BIM

The MCA was moved to the internal polygon of FSUE "RFNC-VNIIEF" in June 2017. The testing powering was carried out. The value of the capacitive storage of the pulsed powering system of the electromagnet was C=900 μF. Figure 5 illustrates the MCA BIM on the test bench.

Figure 5. The photo of a single-beam three-frame mobile cyclic accelerator:
1 – accelerating unit; 2 –unit of pulsed powering of the betatron electromagnet; 3 – betatron of BIM type.

All the accelerator systems operated in the normal mode during the testing powering of the MCA BIM. The typical oscillograms of the signals from the betatron detectors are illustrated in Figure 6; the signals are recorded during the testing powering.

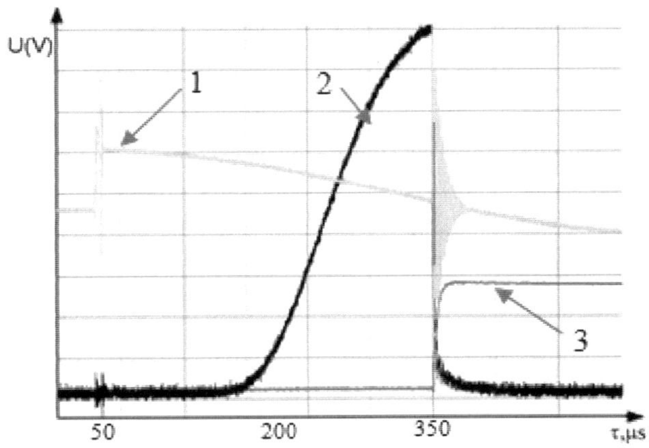

Figure 6. Oscillograms of the signals from the betatron detectors recorded during the testing powering:
line 1 – signal from the detector «0 field»; line 2 – signal from an optical detector of the synchrotron radiation; line 3 – intensity level of the bremsstrahlung.

Diagram 1 in Figure 6 is a signal from the detector «0 field» (response process of the betatron electromagnet). High-frequency oscillations appear in the diagram at 350 μs. They are caused by the operation of the electron beam dumping system. Diagram 2 is a signal from the optical detector of the electron beam synchrotron radiation in the betatron chamber. It appears after the electrons achieve some energy. Diagram 3 is a signal from the detector of the intensity level of the bremsstrahlung (detector of semi-conductor type, operates in the integration). All the provided oscillograms correspond to one powering. The charging voltage of the oscillator systems of the capacitive storage has the standard value.

The X-ray radiography of the lead test object was carried out to estimate the radiography ability of the oscillator. The test object was a lead parallelepiped with holes. The minimum thickness of the lead was 30 mm with further growing of the thickness every 10 mm. The geometry scheme of this experiment is illustrated in Figure 7.

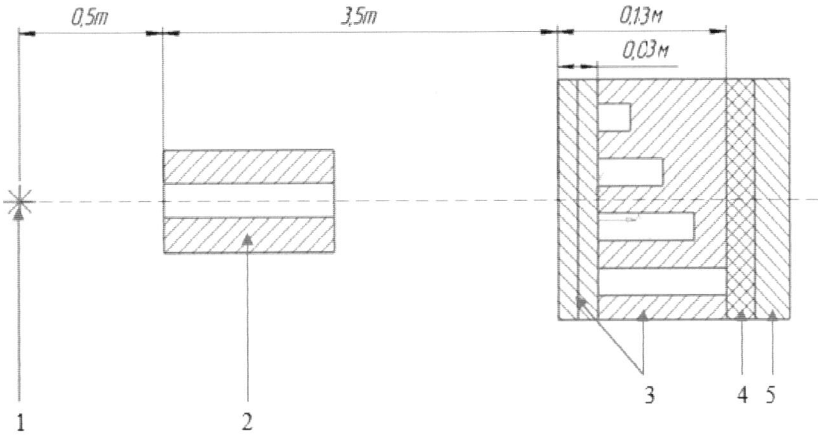

Figure 7. The experiment geometry scheme of the X-ray radiography of the lead test object: 1 – oscillator, 2 – collimation system of X-rays, 3 –lead test object, 4 - radiographic film, 5 – lead protective screen.

The recording of the lead test object shadowgraph and radiation source was carried out with the use of the system Imaging Plate. The maximum thickness of the X-rayed lead test at 4 m from the betatron tantalum target was ≈100 mm at the value of the capacitive storage of the pulsed powering system of the betatron electromagnet with C=900 μF. The X-ray picture of this test object is in Figure 8.

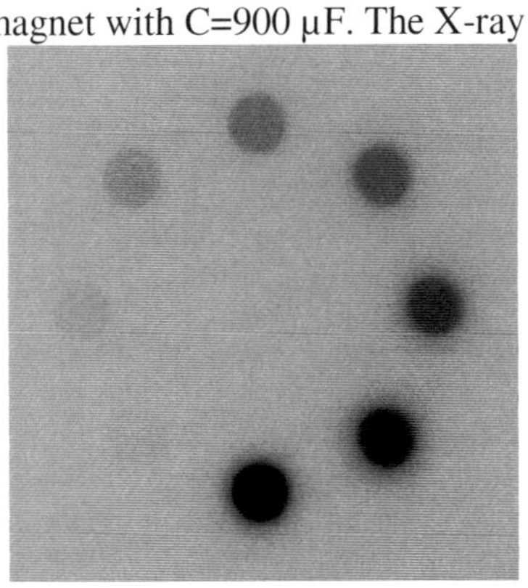

Figure 8. The X-ray picture of the lead test object.

The camera obscura was used to estimate the value of the slowing-down radiation source. Figure 9 illustrates the geometry scheme of the experiment that defines the radiation source dimension.

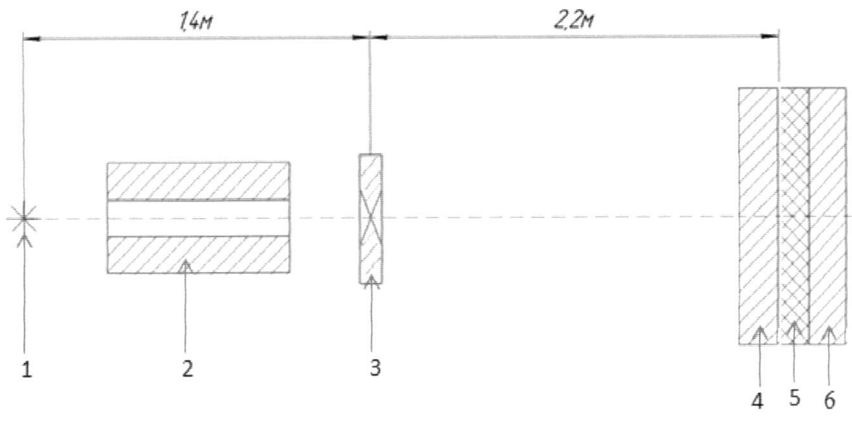

Figure 9. The geometry scheme of the experiment that defines the radiation source dimension**:** 1 – radiation source, 2 – collimation system of X-rays, 3 – camera obscura, 4 – lead protective screen, 5 – radiographic film, 6 – lead protective screen.

Figure 10 illustrates X-ray photograph of the radiation source and its densitogram.

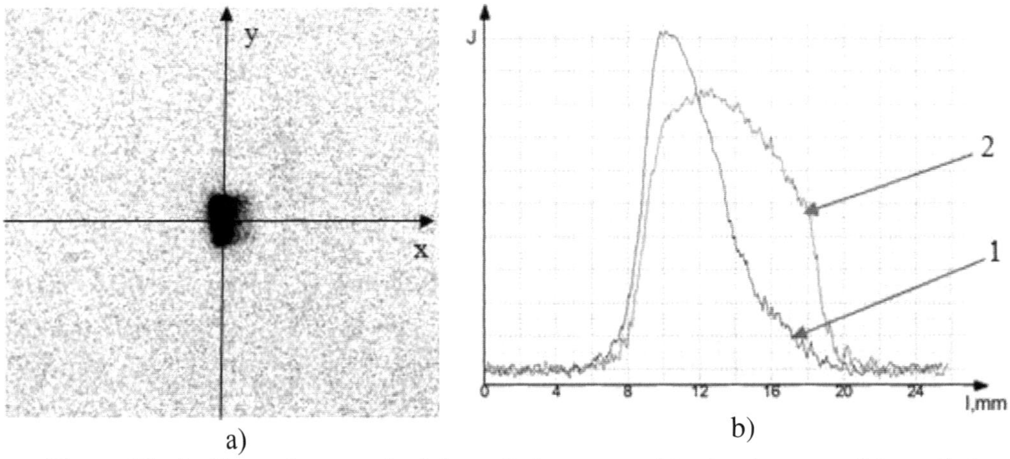

Figure 10. a) - X-ray photograph of the radiation source, b) - densitogram of the radiation source: diagram 1 – on the axis X, diagram 2 – on the axis Y.

The results analysis showed that in the geometry experiment the radiation source dimension is 6×3 mm (full width at half maximum of the densitogram diagrams) with the dimension of the tantalum target 6×6 mm. It is necessary to point out that the betatron chamber design allows changing the tantalum target dimension according to the experiment tasks. And according to the previous investigations, if the target dimension is reduced twice, the output radiation intensity is down 15-20%

A stilbene scintillator detector was used to measure the length of the slowing-down radiation pulse. Figure 11 presents a typical oscillogram of the ɣ-pulse signal from the scintillator detector recorded during the testing powering.

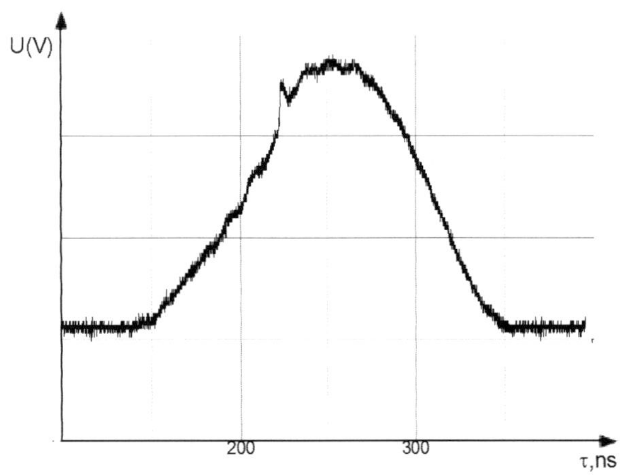

Figure 11. Signal oscillogram from the ɣ-radiation detector.

The oscillogram analysis showed that the length of the ɣ-pulse in a single frame mode at full width at half maximum was ≈100 ns. It corresponds to the normal parameters of this X-ray facility.

The MCA BIM testing is going to be continued in 2018. There are plans to achieve working parameters in a three-frame mode at the value of the battery capacity of the pulsed powering system of the electromagnet equal to C=1800 μF. Besides, it is necessary to create and assemble some equipment of the technological system in the UPPE; develop and perform assembly work and commissioning of the mobile control system. In this case we will get a fully autonomic mobile cyclic accelerator for the radiography of dynamic objects with large optical thicknesses.

4. Conclusion

The testing powering of the MCA BIM showed that the thickness of the lead test object examined with X-rays at 4 m from the tantalum target with the dimension of 6x6 mm^2 was ≈100 mm at the value of a capacitive storage of the pulsed powering system of the betatron electromagnet equal to C=900μF. The full width of the output gamma pulse at half maximum in a single frame mode was ≈100 ns. The dimension of the radiation source was ≈ 6×3 mm^2.

References

[1] A. W. Chao, Reviews of accelerators science and technology, «Accelerators for high intensity beams», W. Chou ed. Singapore: World Scientific, **6**, 126-129 (2013)
[2] J. O' Malley, J. Maenchen, G. Cooperstein, Digest of technical papers, «Status of the diode research programme at AWE», 14th IEEE International, 1 21-28 (PPC-2003)
[3] Mikhailov A.L., "Nonperturbative diagnostic techniques of fast processes", Sarov,113-114 (2015)
[4] A.R. Akhmetov, A.A. Kargin, P.A. Kolesnikov, O.A. Nikitin, S.D. Khrenkov, «BIM 234.3000M and LIA-2 installations as part of X-raying complex RGK B1» *XII Zababakhin scientific talks,*50-59 (2014)
[5] N.I. Egorov, G.V. Boriskov, A.I. Bykov, Yu.P. Kuropatkin, N.B. Lukynov, V.D. Mironenko, V.N. Pavlov, «Use of Pulsed Radiography for Investigation of Equation of State of Substances at Megabar Pressures», *Contributions to Plasma Physics*, **51**,333-338(2011)
[6] Zenkov D I, Kuropatkin Yu P, Nizhegorodtsev V I, Selemir V D, Shamro O A 2015 Patent 2548585 C1 RU MPK G03B 42/02. *Mobile radiographic complex and betatron type radiation source for radiographic complex,* **11**
[7] S G Kozlov, Yu P Kuropatkin, V I Nizhegorodtsev, K V Savchenko, V D Selemir, E V Urlin and O A Shamro, « Mobile x-ray complex based on ironless pulsed betatrons. X-ray complex conception for small-angle tomography», *IOP Conference Serials: Materials Science and Engineering,* 199-208,(2017)
[8] Zenkov D I, Kuropatkin Yu P, Nizhegorodtsev V I, Tkachuk A A, Shamro O A 2008 Patent 2317637 C1 RU MPK H03K 3/53. *Generator of pulsed voltage* **5**
[9] Kuropatkin Yu P, Mironenko V D, Suvorov V N, Zenkov D I, Tkachenko B F 1997 The 11th IEEE International Pulsed Power Conf., *Uncored betatron BIM-M a source of bremsstrahlung for flash radiography* **2**,1669-1673
[10] Kuropatkin Yu P, Zenkov D I, Tkachuk A A, Shamro O A, Nizhegorodtsev V I 2009 Patent 2356193 C1 RU MPK H05H 11/00. *The device for charged particles drive,* **14**
[11] O V Belov, R V Valekzhanin, D V Kustov, O A Shamro and T V Sharov «Control system of mobile radiographic complex to study equations of state of substances» *IOP Conference Serials: Materials Science and Engineering*, 207-218, (2017)

IOP Conf. Series: Materials Science and Engineering **366** (2018) 012050 doi:10.1088/1757-899X/366/1/012050

The Influence of New Environment on the Development of Northeast Electric Power Industry

J K Zhang[1], X Ma[1], P C Ren[1], C S Li[1], J C Liu[1], Z Y Tian[1] and W T Xiong[2,*]

[1] Northeast branch of State Grid company, Shenyang Hunnan Camp No. 1 North Street, Shenyang 110180, China
[2] School of Electric and Electronic Engineering, North China Electric Power University, Beijing 102206, China

[*] 13718899557@163.com

Abstract. Electricity is the foundation of the national economy, and the influence on the social and economic development is self-evident. At the same time, the power industry is also deeply affected by the economy and society. Firstly, this paper introduces the new environment faced by Northeast power industry from four aspects: power system reform, bilateral stochastic system, technological innovation and related policies; then It deeply analyzed the impact of the new environment on the development of the Northeast Power Industry from the above four angle; finally, the opportunities and challenges of the new environment for the future development of the Northeast power industry are summarized. To pave the way for the formulation of Northeast power grid development planning, and it is also helpful to improve and scientificity and accuracy of the Northeast Power Grid Planning.

1. Introduction

In recent years, China's economy, society has entered a new stage, the power industry is also facing a series of new environment. In order to ensure the sustainable development of the power industry and facilitate the timely adjustment of the development direction of the power enterprises, it is necessary to sort out the impact of the new environment and its impact on the power industry.

This paper introduces the new environment faced by the Northeast power industry from five aspects, and analyzed the impact of the new environment on the development of the Northeast power industry, the opportunities and challenges faced by the Northeast power industry are summed up at last, which lays the foundation for the power planning of the Northeast Power Grid.

2. The new environment of northeast electric power industry

This chapter summarized the new environment in Northeast power industry from four aspects like the reformation of electric power system Bilateral stochastic system technical innovation and relevant policies.

2.1. The new environment

During the 12th Five-Year Plan, the world economy is still in a period of profound adjustment after the crisis, showing various characteristics such as low growth and imbalance. The interweaving of this big-time background and China's "three positions" determines the "new normal" period when China's economy entered a phased decline.

In addition to the typical features of the new normal of the national economy, the economic development in Northeast China also has its own characteristics. In terms of economic growth, economic growth in Northeast China decreased by 2017, showing negative growth and overall weakness. However, in the first half of 2017, the economy in Northeast China rebounded and signs of improvement were evident.

2.2. The reformation of electric power system

In March 15, 2015, the CPC Central Committee issued a document 2015 No. 9 "on the further deepening of power system reform suggestions". The key points and paths of deepening the reform of the electric power system are clearly defined in the document, Orderly release the electricity price in competitive links outside transmission and distribution; Orderly release power distribution and saling business to social capital; orderly release power generation plan beyond public welfare and regulatory. Its core content can be summarized as follows:

- Separately check and ratify the price of electricity transmission and distribution, in order to promote price reform;
- Promote the reform of the electricity trading system, and form a fair and standard market trading platform;
- Steadily promote the reform of the sale side, and orderly release the sale of electricity to the social capital;
- Open grid fair access, establish a new mechanism for the development of distributed power.

2.3. Bilateral stochastic system

In recent years, great efforts have been made to develop clean energy in Northeast china. In 2016, the proportion of solar energy and wind energy installed capacity in Northeast China reached 1.7% and 20.5% respectively, the corresponding power generation increased by 118.4%, 12.6% compared with the same period. In the future, the proportion of renewable energy installation on large scale wind power and solar power will continue to increase in the whole power structure. But because of the existence of output volatility and intermittent, power generation side will no longer be controlled, and be more random.

In addition, Heilongjiang, Jilin and Inner Mongolia are listed as 2015~2020 electric vehicle demonstration and promotion sites. In the future, the electric vehicle industry will be developed rapidly in northeast china. At the same time, the distributed generation will continue to be popular, and the user energy storage technology will be further improved. The randomness of demand side is also increasing gradually, that is to say, the power generation side and demand side of power system have obvious randomness, forming a two-sided stochastic system [1].

2.4. Technical innovation

In recent years, China's energy technology independent innovation ability and equipment localization level improved significantly, some areas reached the international advanced level.

2.4.1. Power Technology. In terms of thermal power, China has completed the world's first million kilowatts ultra supercritical two reheat coal-fired generating units performance test. The generating efficiency of the unit is 47.82%, the power coal consumption is 256.8g/ (kW / h), and the coal consumption of power supply is 266.5g/ (kW. h). All environmental indicators are better than the national ultra-low emission limits [2].

In terms of nuclear power, China's third generation nuclear power technology "Hualong No. 1" first pile demonstration project started construction. While the power generation increases by 5%-10%, the power density in the core is reduced, and the safety of the nuclear power plant is improved [3].

In terms of wind power, the high altitude wind power generation system has also begun to develop, the latest high-altitude wind power is 600 kilowatts.

In terms of energy storage technology, researchers of Peking University have made an important breakthrough in the lithium battery material transport mechanism, which is expected to be used as electric vehicle power battery and be widely used in high temperature.

2.4.2. Grid Technology. In terms of UHV technology, China developed the world's first 1101kV controllable shunt reactor、the world's first 1500 MVA on-site assembled transformer prototype and the first UHV gas insulated transmission line, also developed the ±1100kV DC gas insulated bushing prototype [4],and has broken through the 1100kV /500 A HVDC converter valve and so on.

In terms of flexible DC transmission and DC power grid technology and equipment, our country has completed the world's first 320 thousand volts flexible DC converter transformer development, and has broken through the core technology of 500 thousand V flexible UHVDC converter valve.

In terms of micro grid and grid connection technology, China developed the micro electric network voltage control of double layer structure、Flexible switching of main power supply off grid mode、Power environment monitoring system; and developed the prototype of virtual synchronous machine [5].

2.5. Relevant policies

In recent years, China's relevant departments put forward a series of policy that are closely related to the development of the northeast electric power industry from the national level and regional level.

2.5.1. Energy conservation and environmental protection policy. In recent years, the relevant departments have issued a series of energy saving and environmental protection policy, like <Government work report of the State Council in 2016>, <Special program of action for air pollution control in Heilongjiang Province (2016 - 2018)>, <Jilin province 2017 air pollution prevention and control work plan>, <Air pollution prevention regulations of Liaoning province>and so on. In view of the electric power industry, policy requires:

- Complete the ultra low emission transformation of coal-fired generating units, and to control coal power installed.
- Increase efforts to eliminate backward production capacity of high energy consuming industries; Promote the optimization and adjustment of energy structure and replace coal with electricity and gas.
- In key areas of air pollution prevention and control, develop natural gas peak shaving power station orderly; Develop gas steam combined cycle cogeneration moderately combined with heat load demand.
- Promote the adjustment of industrial structure; Speeding up the construction of energy saving and emission reduction projects; Prohibit the direct discharge of toxic and harmful gas.
- Optimize the energy structure and encourage the development of competitive new and renewable energy sources.

2.5.2. A series of policies to revitalize the Northeast. In 2003, <Several opinions on the implementation of Northeast China and other old industrial base strategy etc > was put forward, which marks the formal implementation of the Northeast Revitalization Strategy. In 2007, the <Planning the revitalization of Northeast China>was formulated. In 2014, facing the new situation of global economy, the State Council has put forward <Some opinions on the recent major policy initiatives to support the revitalization of Northeast China>.In 2016, our country launched <The rolling implementation plan for three years to promote the revitalization of the old industrial bases in Northeast China (2016-2018 years)>. In February 2017, the General Administration of Customs has issued< Several measures to support the new round of revitalization of Northeast China >, which aim is to build the key development and opening platform in Northeast China, and to promote the optimization and upgrading of industrial structure in Northeast China.

2.5.3. "One Belt and One Road" policy. In March 28, 2015, the State Council authorized the three ministries jointly issued<Promote the construction of Silk Road Economic Belt and twenty-first Century Maritime Silk Road vision and action>, which explicitly put forward that we will Strengthen cooperation with countries along the route with the main content of Policy communication, facilities Unicom, trade unimpeded, capital financing, popular communication. In February 2017, Heilongjiang Province issued<Heilongjiang province implementation plan on "the implementation of promoting One Belt and One Road construction science and technology innovation cooperation plan">; In March 17, 2017, the general office of the State Council announced<Scheme for counterpart cooperation between Northeast China and some provinces and cities in the eastern region>, which defines the cooperation program between the northeast and the East; Focus on promoting institutional innovation, and to promote collaborative "One Belt and One Road" construction.

3. The influence of new environment on Northeast Electric Power Industry

The influence of new environment on Northeast Electric Power Industry can be seen in the figure 1 as below:

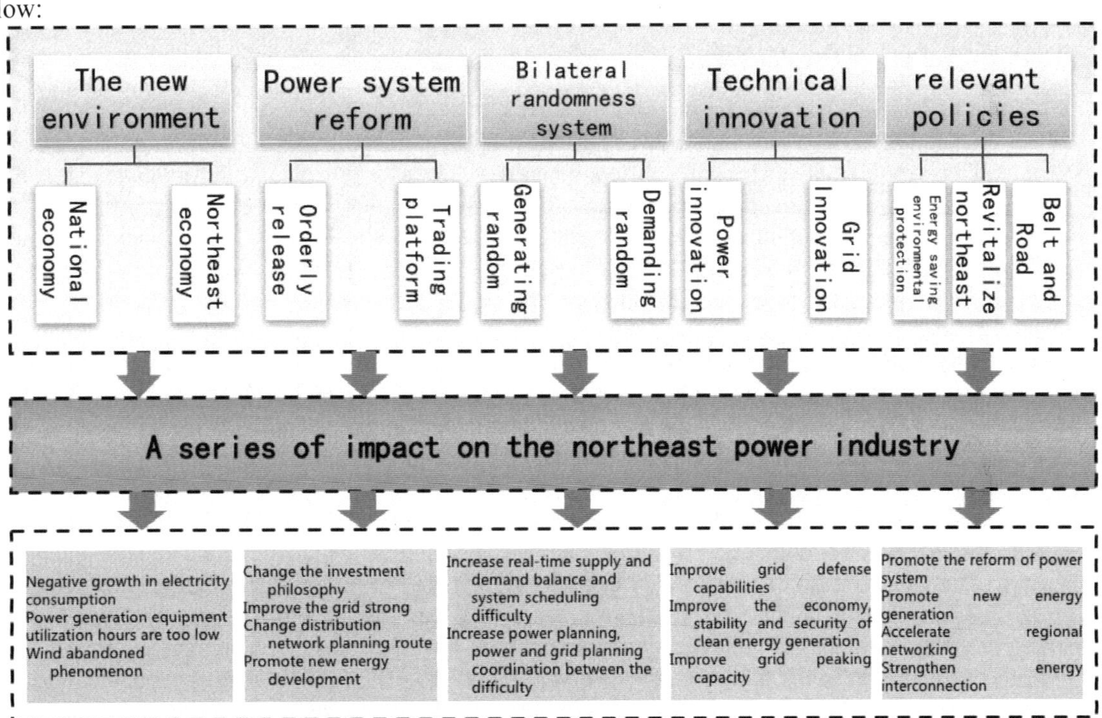

Figure 1. The influence of new environment on Northeast Electric Power Industry

3.1. The influence of electric power system reform on Northeast Electric Power Industry

The new round of power reform's core value orientation is to establish a new type of electric power governance system, which is green, low-carbon, energy-saving and emission reduction, more safe and reliable, and realize the comprehensive resource optimization allocation, promote the overall transformation of China's electricity production structure, consumption structure and technical structure. The northeast electric power industry is affected in the following aspects:

- Reform of transmission and distribution price, reshape the positioning of power grid enterprises, change the concept of power grid planning;
- Power market opening increase the competition of northeast electric power market;

- Promote multi direct transaction of electricity;
- The development of distributed power supply and micro grid will change the planning route of power grid enterprise distribution network, and change the relationship between power grid and users;
- The establishment of auxiliary service market mechanism alleviated the development of Northeast Power grid;
- Releasing the business of distributing and saling electricity gradually improved the overall management of power grid enterprises;

3.2. The influence of Bilateral randomness on Northeast Electric Power Industry

The protection of power system is faced with great challenges in the aspects of security and stability due to the randomness of two sides. Which made the real-time supply and demand balance of power system and the system scheduling more difficult [6].

Power planning can make use of the flexibility and controllability of traditional fossil power generation to compensate for the randomness, instability and other defects of renewable energy power generation, and achieve coordinated control of power generation.

At the same time, with the demand side management technology and energy storage technology, the demand side power load actively cooperates with the power generation, so as to form a relatively coordinated "bilateral controllable system".

3.3. "Mutually beneficial" Coordinated development

Power generation technology and power grid technology innovation, can not only promote the Northeast Power Grid towards a diversified, clean development, but also will further safeguard the security of the Northeast power grid operation [7].

The application and popularization of two reheat technology will improve the utilization efficiency of thermal power in Northeast Power Grid; Deep peak shaving technology for thermal storage of thermal power plant increase the flexibility of the unit operation, and lay a foundation for more renewable energy consumption [8].

The development of new energy power generation technology and grid technology, enable the better development and utilization of resources in Northeast china. The new UHV technology enables the Northeast power grid to make full use of its superior geographical position, build cross regional UHV power transmission channel, and achieve two-way power flow with the surrounding areas.

3.4. "Mutually beneficial" Coordinated development

Energy saving and environmental protection, the revitalization of the northeast, "The Belt and Road" policy promote the reform of Northeast power grid system, and promote the development of Northeast power towards clean, green and direction; accelerate the construction of regional networking, strengthen the interconnection of energy; promote renewable energy consumption, improve the industrial structure, lead into advanced science and technology, and promote industrial upgrading.

4. Conclusion

The new environment has brought a series of opportunities and challenges to the development of Northeast Power Grid. On the one hand, technological innovation has promoted the rapid development of power grid; The superior geographical position and the "One Belt and One Road" policy has promoted the wide range of energy cooperation in Northeast Asia countries. On the other hand, Clean power, distributed generation, and the popularity of electric vehicles make the grid be a Bilateral stochastic system, which further aggravate the uncertainty of the development of Northeast Power Grid under the environment of electricity market reform; energy-saving and emission-reduction has brought higher environmental protection requirements to thermal power enterprises.

Acknowledgment
2017 National Grid company Management Consulting Project: Research on coordinated development strategy of Northeast China Power Grid and national energy network under new normal conditions.

References

[1] Ming Z 2016 *J. Electr. Power Enterp. Manage.* **9** 42-5
[2] Song G, Jie Z and Ferdinand H 2017 *J. Reheat Chinese Power* **50** 6-11
[3] Donghai G and Dianyuan Z 2016 *J. China Nucl. Ind.* **5**
[4] The world's first commercial 1000kV site assembly a successful test *J. Electr. Trans. Techno.* **6** 64
[5] State Grid Energy Research Institute 2016 *M.* China power supply and demand analysis report. *China Electr. Power Press*
[6] State Grid Energy Research Institute 2016 *M.* China power generation supply and demand and power supply development report *China Electr. Power Press*
[7] State Grid Energy Research Institute 2017 *M.* China power generation supply and demand and power supply development report *China Electr. Power Press*
[8] Huijuan S, Pengcheng S, Zijie Z and et al. 2016 J. *Sci. Techno. Innov. Appli.* **16** 185

Research on Network Communication and Convergent Gateway in Terminal Communication Access Network

WANG Jinshuai[1] **,QIU Yinghui**[2] **and HUANG Hui**[3]

[1] North China Electric Power University, Beijing 102206, China
[2] North China Electric Power University, Beijing 102206, China
[3] Global energy Internet Research Institute, Nanjing 210000, China

E mail: 1053349427@qq.com

Abstract. Based on the analysis of smart grid features and business requirements and using the existing research results of heterogeneous networks for reference, this paper designs a heterogeneous network model and functional architecture that adapt to the smart grid environment. Based on this, according to different network resource status, the paper analyzes and compares the network discovery, network selection and connection management mechanism in multi-network environment which includes multi-level, heterogeneous network environment to support the transmission diversity common link layer processing technology; and business multi-network collaborative bearer technology to improve the flexibility of access to a variety of services in power communications.

1 Introduction[1]

The power terminal communication access network is an extension of the power system backbone communication network and is an important part of the power communication network. The power terminal communication access network consists of a series of delivery entities (such as line facilities and transmission facilities) between the service node interface (SNI) and the user network interface (UNI), and it provides the connection between power distribution and electricity service terminals with the power backbone communication network, and realizes the information exchange between terminals and systems equipped with electric service. It has the functions of service bearing and information transmission.

In view of the differences of various wireless / wired access technologies in the power terminal communication access network in aspects of network resources, networking modes and service provisioning, and the demand for constructing a centralized control architecture for multiservice communication networks, this paper researches on the terminal communication access network requirements analysis, network collaboration, integration gateway and other fields[1].

2 Demand Analysis of Synergetic and Converged Terminal Access Network Technology

At present, the access network faces many challenges. First of all, the energy Internet drives the new

---

National network company science and technology project:Research on Key Technologies of substation PD on-line monitoring and location based on UHF Wireless Sensor Technology business of the access network to flourish. For example, new energy access, electric vehicles, and "cloud migration" require the access network to support new services rapidly, flexibly and efficiently. Second, the access network bearer services directly to power users, with "multi-faceted" features, the network boundary of the power grid business system gradually increases with the expansion of the business, and some terminals are deployed in the street, cell and other places easy to be exposed, which puts forward higher requirements for the information security protection of the access network[2]..

Terminal communication access network construction and operation and maintenance model used is that each business department to build a single network by the business system, the formation of multiple networks, all departments to maintain their own, formed a variety of business, multiple networks, coexistence of a variety of communication methods Network status[3].. Considering the safety and real-time demand of related businesses, the distribution automation sites with control functions, intelligent power cells, charging stations for electric vehicles, marketing outlets, distributed power terminals and so on should mainly adopt optical fiber communication; The non-control function of distribution network automation station and power information collection point can choose optical fiber, carrier wave, 230MHz and public communication network according to the conditions[4].

With the development of smart grid, more and more services are carried on the power communication network, including not only the enterprise and production information business, but also the strict control of real-time and reliability requirements. The shortcomings of the previous power terminal communication access network construction mode are also gradually exposed:

(1) The power terminal communication access network has been constructed separately according to business needs. One service accesses a network and communication resources are seriously wasted.

(2) In case of network failure, traffic burst, etc., it is impossible to achieve coordinated communication and redundant backup between networks, and the network reliability is low;

(3) The access network has a large network, small traffic and is sensitive to costs. Each service has a separate network and is heavily duplicated. The high cost of network construction and operation and maintenance is not conducive to the unified operation and maintenance management of the equipment[5].

Therefore, there is an urgent need to solve the access problem of multiple communication modes of the power service terminal, to study the network coordination and convergence gateway technologies of the terminal communication access network so that the service can intelligently select among multiple communication modes or perform multiple communication modes Collaborative communication, improve network resource utilization efficiency, reduce the cost of network construction[6].

3 Power Terminal Access Network Multiple Communication Methods of Fusion Technology Research

3.1 Analysis of fusion scenarios for multiple communication modes in Smart Grid

A variety of heterogeneous network integration is the next generation of wireless communications network development direction, but also the continuous evolution of wireless communications technology, the user's communications services and the growing number of types of market demand and other factors driven by the results of multi-network convergence of communications[7]. The system can take advantage of the performance advantages of different networks to make up for their respective deficiencies.

3.2 Research on hybrid network of Optical Fiber and Wireless communication for Power Broadband Flexible Access

Due to the different network bandwidth requirements of the broadband services accessed by the power terminals in the smart grid and the external services, the bandwidth requirements of the service layer need to be analyzed according to the type of the demand of the distribution network. This paper analyzed the application of optical fiber and wireless convergence technology in power terminal access network. As shown in Figure 1 is a PON-RoF optical fiber wireless networking approach.

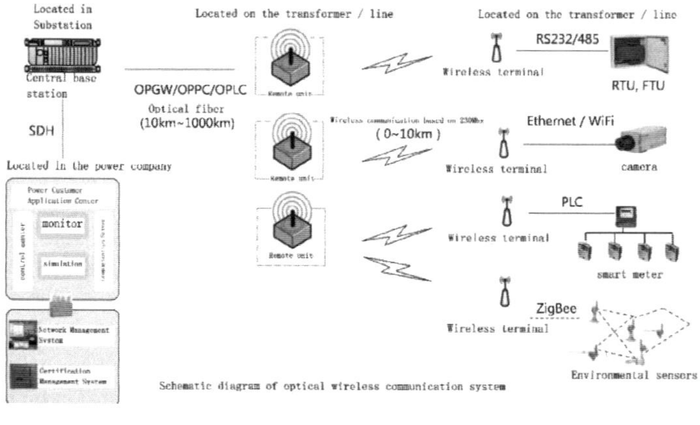

FIGURE1. Hybrid network of PON-RoF fiber and wireless communication

In PON-RoF system, the transmission of radio frequency signal is carried out directly by optical carrier. Information exchange, control and signal regeneration are done at the central station, and the base station just need to achieve the photoelectric conversion, optical amplification and signal transmission, thus complex and expensive equipment will be centralized to the central station, so that multiple remote base stations share these devices, greatly reducing the base station size and power consumption, reducing the base station construction costs and maintenance costs[8]. The radio frequency signal transmitted by the optical fiber increases the transmission bandwidth of the wireless signal.In addition, the use of optical fiber as a transmission link can reduce the loss and prevent electromagnetic interference. Therefore, PON-RoF optical fiber wireless convergence networking can realize large-capacity, low-cost RF signal cable transmission and high-speed ultra-wideband wireless signal access.

3.3 The Architecture Design of Power Line and Wireless Communication Heterogeneous Communication Network

For the smart grid backbone communication network, fiber optic communication is the main means of communication, but for terminal access and some special communication environment, both power line and wireless communication have the advantage of no need to set up a dedicated communication line. However, a single communication mode can not guarantee the reliable transmission of information in the energy Internet. The performance of low-voltage power communication is limited by the power line load, especially the new energy equipment[9]. The performance of wireless communication is affected by topography or obstacles. So power line and wireless combination application can greatly improve the communication performance. Therefore, the fusion of power line and wireless communication can effectively improve the reliability of communication, and can be an effective complement to the high-speed communication technology of optical fiber communication.

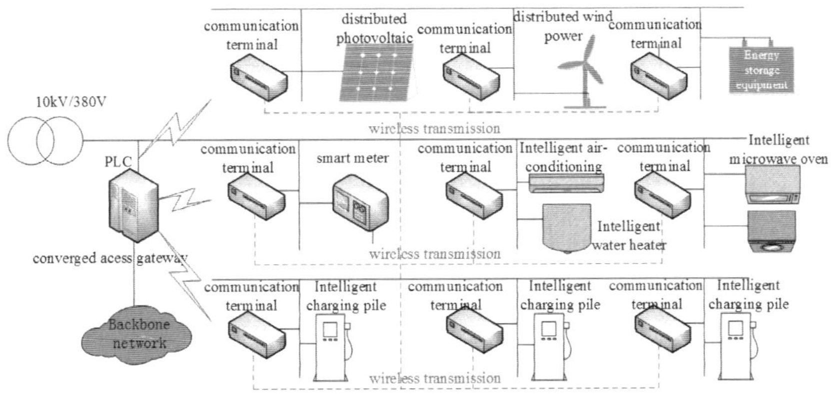

FIGURE 2. Heterogeneous Network Communication Architecture Based on Power Line Carrier and Wireless Communication

As shown in FIG. 2, for the terminal access of the smart grid, the communication architecture combining power line carrier and wireless communication can fully exert their respective advantages and effectively solve the problem of terminal access. In view of the communication demand of the terminal node of the smart grid, the research on heterogeneous fusion technology of power line and wireless communication will effectively promote the development of information communication system[10]. The technologies involved in the integration of power line carrier communication and wireless communication include QoS guarantee mechanism, load balancing, protocol conversion, hybrid path relay, and network switching. Heterogeneous network switching technology of wireless and power line carrier communication relates to how to accomplish the switching of heterogeneous networks efficiently and accurately, which is the key to accomplish the heterogeneous network convergence.

4 Research on link layer processing and network collaborative loading technology in multi level and heterogeneous network environment of smart grid

4.1 Architecture design of intelligent power communication terminal access network based on SDN Technology

This paper designs the communication network system of intelligent power architecture based on SDN technology, the architecture integrates the centralized control of network resources, the flexible configuration of power business virtual network and so on. It provides the basis for the real-time It provides the basis for the real-time bidirectional interaction between the power grid and the user's energy flow, business flow and information flow. The architecture is a power business plane that contains power business data forwarding planes, power SDN control systems and software.

The power data forwarding plane mainly includes 10KV access network such as PON, industrial Ethernet, wireless WLAN, and network equipment related to the backbone network. SDN technology is used to improve the network equipment such as OLT, ONU and switch in each access network so that it supports the unified control protocol of OpenFlow and OF-CONFIG equipment, and moves the complicated functions such as route finding and path calculation of the equipment to the power SDN Control System.

The power SDN control system is the core of the entire intelligent power communication network. It consists of the Flow Visor physical network virtualization platform, multiple access network controllers, and the master total controller. The control system mainly completes the link discovery, topology management (logical topology and physical topology), flow table generation and management, link bandwidth setting, link utilization monitoring, node traffic monitoring, global network resource control and scheduling, virtualization network creation and configuration management functions of the power communication network.

The software-based power service plane is a unified plane based on the physical network and the power SDN control system, which realizes the power communication network function and the power service function through software programming. The plane includes communication network application platform and intelligent power service platform. In the communication network application platform, different power grid operation departments can call the basic network function API of the power control system to develop the characteristic network application according to their own business needs. In the smart power business plane, users can call the network function application interface to develop various types of intelligent electricity business platform.

4.2 The heterogeneous network collaborative bearer solution based on SDN technology

Based on the idea of software-defined network, a gateway-based loosely-coupled mode is adopted without changing the network structure of various terminal access networks. A converged communication scheme based on dual-communication proxy gateway is proposed, which realizes the decoupling of terminal services and communications networks.

Heterogeneous network collaborative load bearing technology and power service visualization: The SDN intelligent power communication network can acquire the whole network resources in real time with the help of the power SDN control system. Based on the real-time status of the whole network and the detailed information of the network link, the visual monitoring of the network and the power distribution service is carried out. The main station network visualization platform can dynamically acquire and refresh the current network resources by dynamically calling the power SDN control system, and realize the network visual monitoring. Network visualization mainly includes topology information, link information and network node information visualization. The visualization of network topology information is to display the global physical topology and virtual logic networking information of power communication network in the main station network visualization platform, so that network managers can grasp the network structure integrally, and accurately determine the location of each node. The network link information visualization is to integrate and process the link resource information obtained by the master controller, and display the network link characteristic information of the link's bandwidth, packet loss rate, utilization rate, delay, connection quality and so on, and real-time feedback the network link quality. Network node information visualization is a detailed display of each node information in the network topology, which can display the details of the working state, MAC address, port situation and real-time traffic curve of the node equipment.

Centralized control of power business based on software programming and unified protocol: The intelligent power device realized by software programming replaces the proprietary power distribution device, flexibly completes the deployment of power distribution service, and unified data transmission and communication protocol for centralized control, which is the key to the intelligent power distribution business. Intelligent power device support OpenFlow unified communication protocol and OF-CNOF device configuration protocol,it can realize the data interaction with the power communication network, the unified configuration of the equipment, and has the standard electrical interface and drive control interface, which can complete the unified access and control of the new power distribution business such as distributed energy and controllable load. The deployment of new business using the software program to complete the new business function model, and then through the main station network resources and business platform to determine the location of the smart appliances of the new business, after that the master controller determines the corresponding domain controller, and the domain controller sends the software function model to the smart power device by the OF-CNOF device configuration protocol to achieve rapid deployment of new services. At the same time, when the power business in a certain area changes, it is only necessary to quickly positioning the exact position of the smart power device on the network resource and business visualization platform of the main station, rewriting the business function model in the power business functional software model development platform and re-issuing the model to overwrite the original strategy.

5 Conclusion

This article mainly studies and analyzes the realization technology of architecture of multi-service bearer access network. Firstly, based on user strategy and operation strategy of different services, combined with different network resource status,this artical analyzes and compares the fusion technologies of EPON, industrial Ethernet, wireless private network and wireless public network;and network discovery technology in dynamic network environment; and network intelligence selection technology. This paper analyzes multi-level and heterogeneous network environment to support the sending of common link layer processing technology to study the business of multi-network collaborative bearer technology to further refine the granularity of network resource allocation and improve business access flexibility.At last, the article also discusses the design of heterogeneous network collaborative bearer scheme based on SDN technology to support multi-service integrated access, and realizes the decoupling between the terminal service and communication network. Through the design of control strategy, virtualization strategy and programmable strategy, Multi-level, heterogeneous network environment, multi-network collaborative bearer.

References

[1] Supplement ITU-T Y.2000-series(2010), supplement on Multi-Connection Scenario.
[2] Ford A, Raiciu C, Handley M, et al. Architectural guidelines for multipath TCP development[S]. RFC6182, Internet Engineering Task Force, 2011.
[3] Iyengar J R, Amer P D, Stewart R. Concun-ent multipath transfer using SCTP multihoming over independent end-to-end paths[J]. Networking, IEEE/ACM Transactions on, 2006, 14(5): 951-964.
[4] Blanchet M, Seite P. and Provisioning Domains Problem Statement[S]. RFC 6418, Internet Engineering Task Force, 2011.
[5] Wasserman M, Seite P. Current practices for multiple-interface hosts[S].RFC 6419, Internet Engineering Task Force, 2011.
[6] Raiciu C, Handley M, Wischik D. Coupled congestion control for multipath transport protocols[S]. RFC6356, Internet Engineering Task Force, 2011.
[7] Recommendation ITU-T Y.2027, Functional architecture of Multi-connection Requirements[S]. International Telecommunication Union, 2012.
[8] Recommendation ITU-T Y2251, Multi-connection Requirements[S]. International Telecommunication Union, 2011.
[9] Recommendation ITU-T Y2091 Terms and definitions for next generation networks[S]. International Telecommunication Union, 2011.
[10] Ford A, Raiciu C, Handley M, et al. TCP extensions for multipath operation with multiple addresses[S]. RFC 6824, Internet Engineering Task Force, 2011.

IOP Conf. Series: Materials Science and Engineering **366** (2018) 012052 doi:10.1088/1757-899X/366/1/012052

Intelligent power service and improvement analysis of communication network based on SDN

ZHANG Shanhui[1] ,SONG Yanbin [2] ,ZHEN Chong[3] and CHEN Yang[4]

[1]North China Electric Power University, Beijing 102206, China
[2]Beijing wisdom Microelectronics Technology Co., Ltd.Beijing 100192,China
[3]North China Electric Power University, Beijing 102206, China
[4]North China Electric Power University, Beijing 102206, China

E-mail: 1758494561@qq.com

Abstract. As the carrying network of electric energy, the power grid is the foundation of the national industry. For the implementation of the "Internet plus" strategy, there is an urgent need for a new network communication technology to improve the traditional power communication network, and to make up for the original defects. This paper mainly analyzes the new demands of new intelligent power business of communication network. Then, the limitations of existing power communication network and power distribution equipment are explored. At last, It put forward the basic scheme of using SDN technology to improve the existing power communication network, which provides a train of thought for the design of intelligent power distribution system based on SDN Technology.

1. Introduction[1]

Power communication network is the basis for the development of strong smart grid. In recent years, the new power business has many characteristics, such as a variety of services, high flexibility and strong interactivity. It requires that the power communication network not only has a powerful core network, but also has a high flexibility, high scalability edge access network. But the current power communication network is mainly SDH/MSTP backbone network, which network flexibility and business scheduling ability is insufficient. It is difficult to meet the demand for intelligent distribution and electricity business[1].Therefore, this paper uses the SDN technology to improve the existing power communication network, and establish an open, active intelligent power communication network, which will create favorable conditions for the future development of intelligent power business.

2. Research on new intelligent power service

Smart power service is a new generation of power business with high intelligence, real-time interaction and informatization, relying on advanced measurement technology, mobile Internet, artificial

---

Wisdom Microelectronics Technology Co.:Deepened development and application of substation terminal in smart platform based on impedance line aging and low voltage topology

intelligence, big data, cloud computing and other emerging technologies. In recent years, with the rapid development of new energy technology and Internet technology, the new network communication technology is rapidly integrated into the grid, resulting in a variety of new power distribution business [2], as shown in the Figure 1.

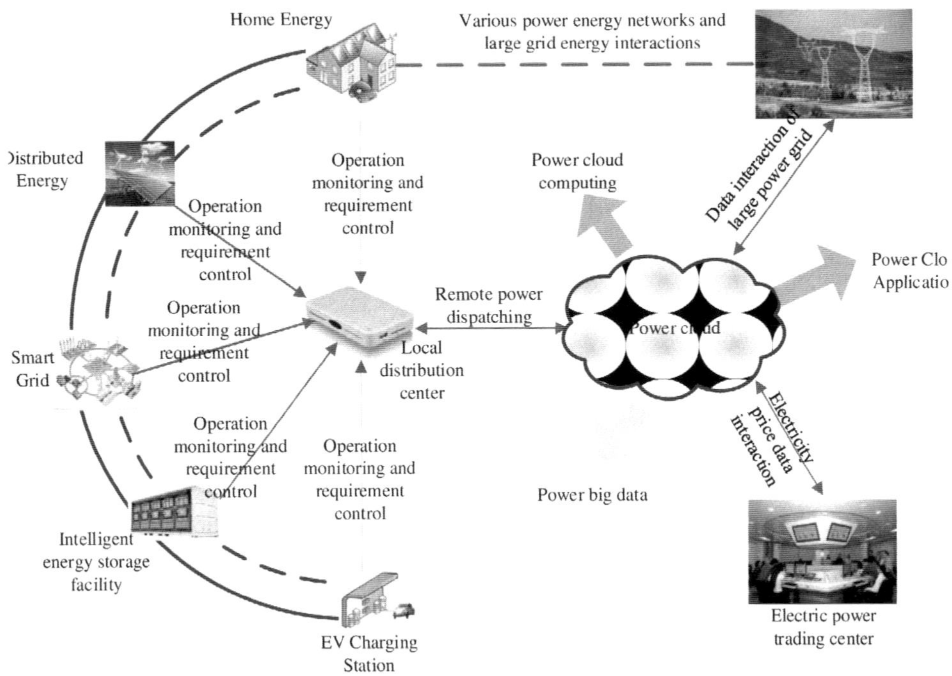

Figure 1. New intelligent power business.

In the future, the communication network is essential for efficient, centralized management of a large number of intelligent power business and unified coordination of multiple energy sources.The flexible and open features of intelligent power services have brought severe challenges to the existing power communication networks.

3. Demand analysis of communication technology for smart power services

3.1. Network real time requirement

In the expanding and increasingly complex smart grid and energy Internet, to ensure the transmission delay of the power communication network is crucial for the reliability of the distribution automation and the experience of intelligent power distribution service. Intelligent transmission delay requirements for different power services are shown in Table 1 and table 2.

Table 1. Real time index of intelligent power power businesses.

Businesses	Index
Feeder automation service	Time delay ≤60s
Monitoring of intelligent energy storage station	<1s
Distributed new energy station control	<1s
Load forecasting of new energy station	<60s
Smart electricity meter	<60s
Operation state measurement of power equipment	<3s

Table 2. Ductility index of power monitoring system.

Communication protocols	Packet loss rate	Network delay（ms）	Delay jitter（ms）
TCP	≤1/100	≤200	≤50
UDP	≤1/1000	≤500	≤100

Different power distribution services have different requirements on the delay size. Therefore, the intelligent power distribution service needs the communication network to configure the delay of the network link according to the real-time requirements of different business[3], so as to ensure the service quality of the intelligent power business.

3.2. Network reliability requirements

Most of the outdoor communication facilities need to ensure stable operation in severe weather, so as to provide high reliability information transmission channels. Especially, the development of smart power distribution business depends entirely on the communication network, Once the communication network fails, all services will be paralyzed. Therefore, the intelligent power distribution service requires that the communication network has the ability of flexible migration and fault isolation. Once a part of the physical network fails, the network can intelligently isolate the fault area and quickly restore the network service.

3.3. Network scalability requirements

Communication network is the foundation of intelligent power business, and the existing communication network of distribution automation and power service is constructed separately. Once the business is increasing or changing, it is necessary to re-establish the physical network and configure the network devices one by one. With the explosive growth of the types of smart power services, the original mode can not meet the demand, which requires the communication network architecture and function can be flexibly extended to shorten the new power business deployment cycle [4].

3.4. Network compatibility requirements

Intelligent power distribution including power grid operation, power marketing, network maintenance and other services, including data communications, business monitoring, customer interaction and other diverse functions. At present, all kinds of power business and application come from independent and scattered platforms. The communication protocols and information interaction methods of each system are different, which leads to the difficulty of joint operation between various services and functions, and increases the difficulty of the unified implementation of intelligent power business [5-7].Therefore, it is very important to establish a new type of power distribution communication network which is compatible with protocol and bidirectional interaction.

4. Analysis of the limitations of power communication network for intelligent business

The current electric power communication system in the construction of backbone communication network management and maintenance system has been perfect, and all the 10kV communication access network has EPON, carrier, wireless and many other ways. There are many equipment manufacturers and different communication systems[8]. In supporting the future of the energy Internet in the smart power distribution business development, the limitations of the existing power communication network include the following aspects:

In architecture, the existing communication network contains a huge number of dedicated network nodes and equipment functions, many network vendors have their own hardware and software gateway platform unique, can only realize data exchange, but difficult to achieve unified control network and equipment, resulting in complex network, poor flexibility[9].

In terms of technology, the current power communication network formed a block of hardware and software integration tightly coupled structure, resulting in new business development for a long time, new technology development is complex, network scalability is insufficient.

In the network construction and operation maintenance, network managers face a large number of equipment, their time, manufacturers, models are different, leading to subsequent deployment, configuration, operation, maintenance and transformation and upgrading can not be centralized and unified, resulting in network construction and operation and maintenance is very complex

In terms of power distribution business support, there is a big difference in the business service quality requirements of various power services, and the existing power communication network is difficult to configure flexibly, which can not meet the personalized requirements of various services [6].

The traditional communication network is difficult to effectively solve the intelligent electric business deployment requirements. Therefore, an innovative power communication network is needed to overcome the shortcomings of the existing communication network that can not be centralized control, lack of openness and flexibility, and create good network conditions for the development of intelligent power services.

5. The characteristics of SDN technology and its application advantages in power communication

SDN is an innovative network model based on software. The most obvious feature is that it separates the management and control part of the network and the data forwarding layer, and supports the centralized control of the network. It can be transparent to the application of the underlying physical facilities relative to the upper layer network. The software defined network architecture based on OpenFlow consists of three layers, as shown in Figure 2.

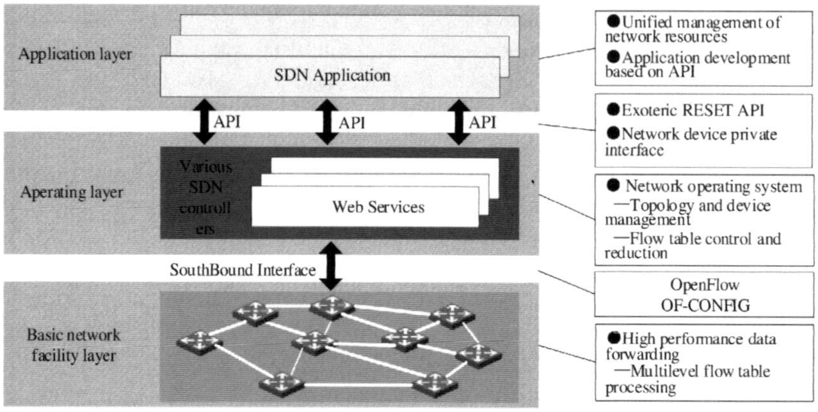

Figure . 2 Software defined network three layer model.

One of the key technologies of SDN is the OpenFlow protocol. It is a protocol of device specification, defines the components and functional requirements of the OpenFlow standard switch in the basic network facilities layer, and the OpenFlow protocol used to accomplish the remote controller's control and management of the switch.

6. Improved scheme of intelligent communication network for electric power service based on SDN

According to the needs of the new electric business, through the analysis of the previous sections, you can use the SDN network architecture model to replace the existing electric power communication network, So that the network operator can control the network resources from the overall situation, can dynamically configure the network delay, software configuration network equipment parameters, real-time expansion of network scale and so on. Therefore, the use of SDN technology to improve the current power distribution communication network, can achieve centralized control of the network, unified deployment of resources. At the same time, its programmable features can also enhance the openness and flexibility of the network, accelerate business innovation.. The improvement idea of power communication network is shown in Figure 3.

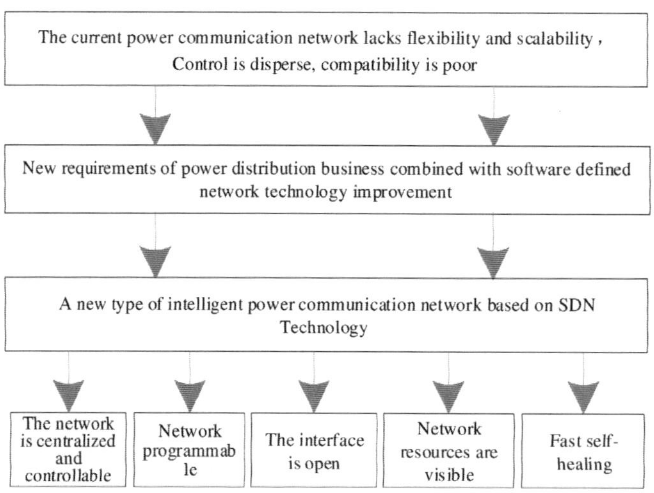

Figure 3. Basic idea of improving power communication network by SDN Technology.

7. conclusion

This paper first explored the intelligent electric business in new energy Internet demand for electric power communication network. Then, analyzed the limitations of the existing power communication systems on the development of intelligent power services. Combined with the characteristics and advantages of SDN technology, this paper put forward the basic scheme to improve the current communication network to adapt to the development of intelligent power business by using SDN technology, and created favorable conditions for the future development of intelligent power business.

References

[1] Guo Yunfei, Liu Shidong, Wang Yao. Application of SDN technology in power optical communication network [J]. *Microcomputer and application*, 2014, 23: 68-71.

[2] Li Yang, Wang Beibei, Li Fangxing. Flexible interactive smart power outlook and thinking [J]. *Automation of electric power systems,* 2015, 17: 2-9.

[3] Huang Sheng. Requirement analysis and technical scheme of intelligent distribution network communication service [J]. *Power system communication*, 2010, 31 (212): 10-12.

[4] Faycal Bouhafs, Michael Mackay, Madjid Merabti. Links to the future: communication requirements and challenges in the smart grid[J]. *IEEE Power and Energy Magazine*, 2012, 10(1): 24-32.

[5] Liu Yu, Chen Xingying, Yang Yongbiao, Zhu Hong, Yu Kun. Typical models of information exchange for intelligent power distribution [J]. *Electric power construction*, 2015, 36 (8): 1-6.

[6] Hao Hanyong, Hu Ziwei, Li Jianqi. Typical applications of SDN in the next generation power communication network [J]. *Power information and communication technology*, 2016 (5): 65-68.

[7] Marvin Moser, Fabio Jaramillo. Extending Software Defined Networking to End User Devices[C]. *2015 24th International Conference on Computer Communication and Networks (ICCCN)*, 2015: 1-8.

[8] Wu J, Ni H, Zeng X, et al. A storage approach for OpenFlow switch based on Protocol Oblivious Forwarding[C]. *2016 IEEE Symposium on Computers and Communication (ISCC)*, 2016: 1145-1150.

[9] Yukihiro Nakagawa, Kazuki Hyoudou, Chunghan Lee, Shinji Kobayashi, Osamu Shiraki, Takeshi Shimizu. Domainflow: Practical flow management method using multiple flow tables in commodity switches[C]. *Proceedings of the ninth ACM conference on Emerging networking experiments and technologies. ACM*, 2013: 399-404.

IOP Conf. Series: Materials Science and Engineering **366** (2018) 012053 doi:10.1088/1757-899X/366/1/012053

Energy Internet Oriented Services Awareness in Elastic Optical Networks

ZHAO Biyou[1] **,LIN Shengjin**[2] **and WANG Jinshuai**[3]

[1] Maoming Power Supply Bureau of Guangdong Electric Power Co., Ltd., Guangdong 525000, China
[2] North China Electric Power University, Beijing 102206, China
[3] North China Electric Power University, Beijing 102206, China

E-mail: 1076212844@qq.com

Abstract. Aimed to improve the ability of EON (Elastic Optical Networks) to support multiple services brought by Energy Internet, this paper proposes a Echo-State-Network Based Services Awareness schem of EON system for Energy Internet. The proposed scheme is able to be aware of characteristics of multiple services and to match their requirements with better quality. Thus, resources allocation optimization is enabled by the proposed scheme in each edge node. Simulation results show that this proposed approach can greatly improve the supporting ability for multiple services of Energy Internet with better performance from aspects of both packet loss rate and time delay performances.

1. Introduction

With the rapid development of the energy Internet, energy interconnection and communication access business types are increasingly diversified, Elastic Optical Networks (EON), with its high bandwidth and high flexibility, is expected to become an important technology for energy interconnection service communication network.[1]

The optical communication technology of energy Internet is not only to meet the requirements of bandwidth, flexibility and reliability of energy interconnection, but also to support the diversification of energy and new business. Therefore, the operational awareness of power elastic optical networks is particularly important. Business awareness is usually defined as identifying the service type of service flow in a perceptual communication network, the QoS requirements of the service, the traffic priority, and its bandwidth and real-time[2].[1]

At present, most of the service sensing technology of optical communication network adopts the method of detecting protocol word, port number and IP address included in the data frame. However, with the diversification and complexity of the new smart grid business, the traditional method is becoming more and more difficult to adapt to the development trend of the business. Therefore, the use of business flow transmission characteristics of business awareness has become a new research direction[3-8]. The literature [3-4] shows that there is a certain correspondence between the statistical properties of the network traffic and the application protocol, but these mechanisms have insufficient

China Southern Power Grid:Research on advanced operation and maintenance of resilient optical network and optical network.

accuracy for the perception of the business type. In [5], the feature set extraction method of service flow is proposed to realize the awareness of service flow. In the literature [6], the IP packet length, packet arrival interval and order are extracted from the service flow, and the traffic flow recognition is realized by the neural network algorithm. However, the complexity of the method is higher and the computation is larger.

Therefore, a highly accurate and complex business awareness approach is particularly important. In this paper, we propose a flexible optical network service flow sensing technology based on Echo State Network (ESN), which can ensure the awareness and support of diversified services while playing the flexible bandwidth of flexible optical network ability.

2. Echo state network classification algorithm

2.1 Echo state network principle

The structure of the echo state network is shown in Fig. 1, which consists of input layer, core layer and output layer[9-13]. Assuming that the echo state network consists of K input units, N reserve pool processing units, and L output units. The basic equations of the echo state network are equations (1) and (2):

$$x(n+1) = f\left(W^{in}u(n+1) + Wx(n) + W^{back}y(n)\right) \tag{1}$$

$$y(n+1) = f^{out}\left(W^{out}u(n+1) + Wx(n+1) + W^{back}y(n)\right) \tag{2}$$

Where u (n) is the input variable of ESN, x (n) is the state variable, y (n) is the output variable; In addition, f and foutt can be regarded as the excitation function of the processing unit and the output unit in the ESN model. The input variable is connected by Win and ESN processing unit, W is the weight between ESN internal processing unit, Wback represents the connection weight of output layer and core layer, Wout is the connection weight of core layer and output layer. In addition, Win, W and Wback are usually set to constant, and Wout can be obtained through a certain training.

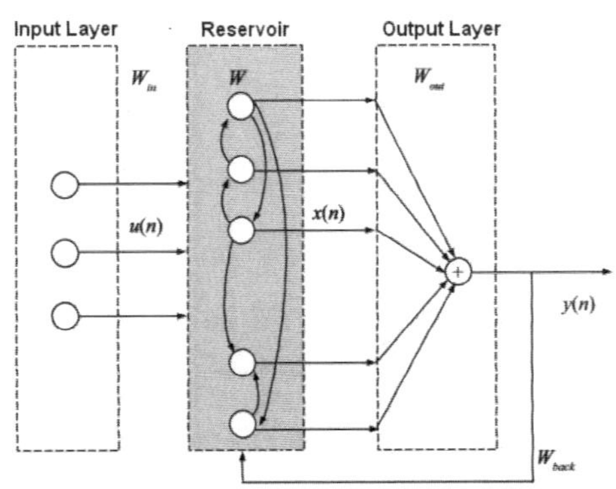

Figure 1. Echo state network diagram

In the training of ESN, the sample data can be excited by the linear regression to correct the internal parameters of ESN after each round of training, and the mean square error can be minimized by using the randomly generated weight matrix Win and Wback to excite the processing unit of the core layer.

2.2 Echo state network classification algorithm

The basic principle of the echo state network classification method is shown in equation (3), where n represents only different samples, not time. In the process of classification training, the input samples must always be kept constant until the pool state variables become stable, making the difference between the two iterations before and after the minimum[14].

$$\begin{cases} x(n+1)^{(i)} = W^{in}u(n+1) + Wx(n+1)^{(i-1)} \\ x(n+1)^{(0)} = 0 \end{cases} \tag{3}$$

The advantage of this algorithm is that it uses the reserve pool processing unit to activate the function to stabilize the state variables before processing it, and still maintain the network training process of echo state simple, with excellent performance.

3. Business - aware Mechanism Based on Echo State Network Algorithm

3.1 ESN - based business classification awareness principle

In order to further improve the operational awareness of the elastic optical network, this paper proposes a service perception based on the echo state network algorithm. Business-based ESN algorithm is essentially a mapping of business characteristics to business type. Its essence is to determine the classification process of decision attributes (business types) based on condition attributes (business characteristics). In this paper, the business is divided into packet loss-sensitive business and the delay-sensitive business business sense, each business according to the priority is divided into high, medium and low three grades.

Service perception based on echo state network algorithm is divided into three parts: feature extraction, ESN training and ESN decision. The edge node of the elastic optical network extracts the main characteristic parameters of the bidirectional service stream Pi, mainly including frame length, service flow duration and service packet flow time interval, which forms a feature set of service flow u={u1,u2,…,uk,…}. The feature set u of the service flow is input to the trained echo state network model, and the classification type of the service belongs is obtained by calculating $\boldsymbol{W}^{out}$.

3.2 Implementation of business-aware mechanisms

According to the characteristics of the network system composed of the edge node and the intermediate node, the service awareness scheme is proposed according to the realization mode of the service perception mechanism. The process of realizing the business-aware mechanism of the power elastic optical network is shown in Fig2.

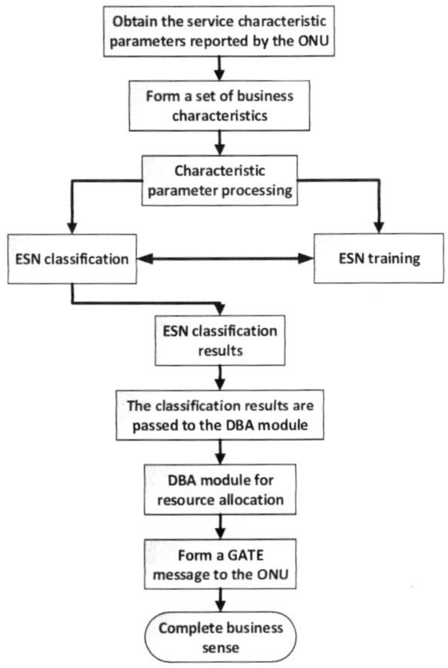

Figure 2. Power Elastic Optical Network Business Perception Process

In the flexible optical network, the service-aware mechanism proposed in this paper is mainly composed of the service-aware module embedded in the edge node. The service awareness module is responsible for initializing the ESN configuration and complex ESN training. The edge node can independently assume the local service awareness function after the ESN classifier is trained. The edge node extracts the characteristic parameters of the service flow, and allocates the corresponding bandwidth and queue priority according to the QoS requirements of the service flow to optimize the multi-service awareness and efficient matching capability of the flexible optical network.

3.3 Analysis of complexity

Complexity is one of the key performance indicators of ESN-based multi-service aware algorithms. Only a high-accuracy and moderate complex core algorithm model can make business-aware technology achievable. Using the ESN algorithm model as the core, to obtain the result of business flow recognition calculation, only one addition and one subtraction and two multiplication are needed. Therefore, the computational complexity can be regarded as O (1).

Complexity is usually proportional to the number of members of the input parameter set. Assuming that the members of the input parameter set have m classes, the computational complexity of ESN is O (m) × O (1). Therefore, the algorithm model proposed in this paper has low complexity, while ensuring the perceived accuracy of the traffic flow.

4. Simulation results and analysis

In order to verify the proposed mechanism, this paper builds a simulation system of elastic optical network based on NS2 simulation software platform. The simulation system is simulated by an elastic optical network with 32 nodes. The local communication access with energy interconnection is taken as a typical business scenario, and the flexible optical network with business perception mechanism and the flexible optical network without service perception are compared by simulation. In the simulation, the service type is divided into two categories: packet loss rate sensitive service and delay sensitive service. Each type of service internal priority is divided into two levels: high priority service and low priority service.

Figure 3 shows the relationship between the accuracy of business perception and the number of training. It can be seen that with the increase of the number of training, the accuracy of business perception is improved. The well-trained echo state network algorithm can guarantee the accuracy of classification of business classification.

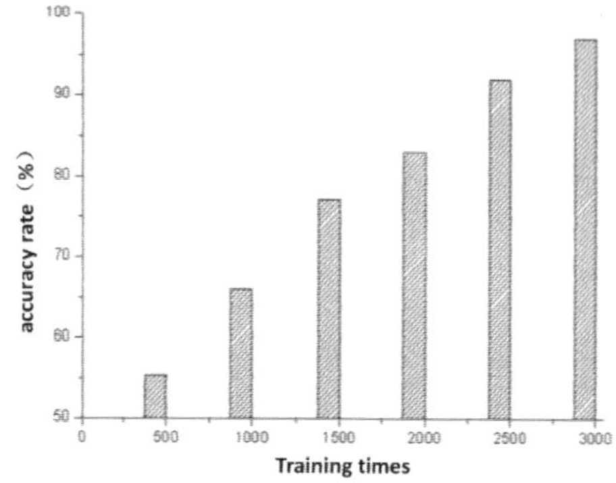

Figure 3. Relationship between the accuracy of business perception and the number of training

In the flexible optical network system, this paper uses the mechanism (business perception) and the traditional non-business perception mechanism to simulate. Figure 4 and Figure 5 is the comparison of the results.

Compared with the simulation results of Fig.4 and Fig.5, the packet loss rate and real-time performance of the elastic optical network tend to be degraded as the network traffic load increases. Under the condition of heavy load of elastic negative network, this paper is superior to the traditional elastic optical network mechanism in the two important indexes of packet loss rate and transmission delay of high priority service. On the other hand, since the low priority class service requires lower transmission delay and packet loss rate, the mechanism of this paper is in a way to reduce the cost of such business performance in exchange for the overall business service quality, especially high priority business.

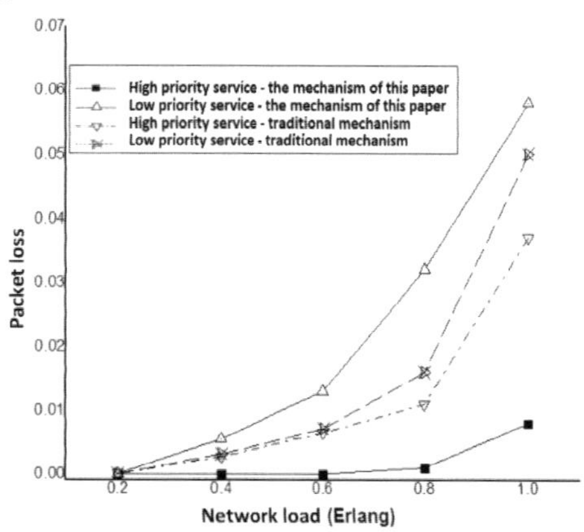

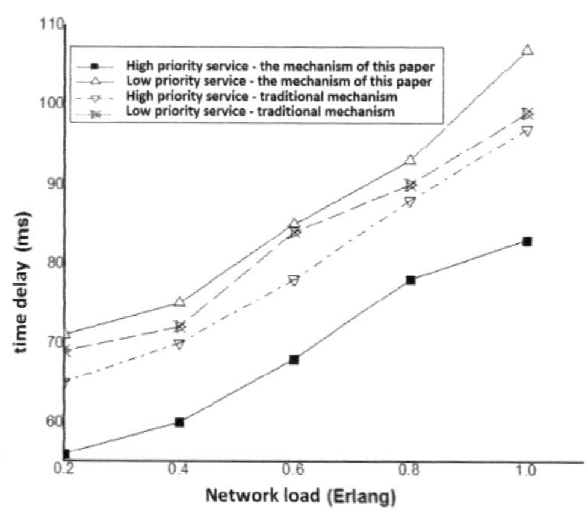

Figure 4 Comparison of packet loss rate **Figure 5** Delay comparison

To sum up, the service-aware approach mentioned in this paper enables different types and priorities of services to be matched with support, and to ensure the overall quality of service to the business. Therefore, compared with non-service-aware flexible optical network, the mechanism mentioned in this paper can effectively enhance the ability of power elastic optical network to support multi-service differentiation and optimize its service-oriented resource allocation effect.

5. Conclusion

In order to improve the active support capability of energy interconnection multi-service and the typical algorithm of echo state network of artificial intelligence, this paper proposes a service-aware method of power elastic optical network based on echo state network algorithm. The simulation results show that the service-aware technology proposed in this paper can improve the bandwidth and real-time performance of the flexible optical network, and can effectively enhance the resilience of the elastic optical network to the active capability of the energy interconnection multi-service, and finally realize the new business needs of the optical network and energy Internet The efficient match.

References

[1] Tian S, Luan W, Zhang D, et al. Technical forms and key technologies on energy internet[J]. Zhongguo Dianji Gongcheng Xuebao/proceedings of the Chinese Society of Electrical Engineering, 2015, 35(14):3482-3494.
[2] Tsoukalas L H, Gao R. From smart grids to an energy internet: Assumptions, architectures and requirements[J]. Smart Grid & Renewable Energy, 2008, 01:94-98.
[3] Talebi S, Alam F, Katib I, et al. Spectrum management techniques for elastic optical networks: A survey ☆[J]. Optical Switching & Networking, 2014, 13(9):34-48.
[4] Yin Y, Wen K, Geisler D J, et al. Dynamic on-demand defragmentation in flexible bandwidth elastic optical networks.[J]. Optics Express, 2012, 20(2):1798-804.
[5] Gong L, Zhu Z. Virtual Optical Network Embedding (VONE) Over Elastic Optical Networks[J]. Journal of Lightwave Technology, 2013, 32(3):450-460.
[6] CROTTI M, GRINGOLI F, PELOSATO P, et al. A statistical approach to IP-level classification of network traffic. In: Proceedings of IEEE International Conference on Communications, Istanbul, Turkey, 2006. 170-176.
[7] Moore A W, Zuev D. Internet traffic classification using bayesian analysis techniques[J]. Acm Sigmetrics Performance Evaluation Review, 2005, 33(1):50-60.
[8] BAI Hui-Feng, LI Ming-Wei, WANG Dong-Shan. Bayesian Classifier Based Service-aware Mechanism in 10G-EPON for Smart Power Grid[J]. Acta Photonica Sinica, 2013, 42(6): 668-673.
[9] Jaeger H, Lukosevicius M, Popovici D, et al. Optimization and applications of echo state networks with leaky-integrator neurons[J]. Neural Networks the Official Journal of the International Neural Network Society, 2007, 20(3):335-352.
[10] Jaeger H. Echo state network[J]. Scholarpedia, 2007, 2(9):1479 - 1482.
[11] Jaeger H, Maass W, Principe J. Special issue on echo state networks and liquid state machines. Neural Networks, 2007,20(3): 287-289.
[12] Rodan A, Tino P. Minimum complexity echo state network.[J]. Neural Networks IEEE Transactions on, 2011, 22(1):131-144.
[13] Jaeger H. Reservoir riddles: suggestions for echo state network research. In: Proceedings of the IEEE International Joint Conference on Neural Networks. Montreal, Canada: IEEE, 2005. 1460-1462.
[14] Alexandre L A, Embrechts M J, Linton J. Benchmarking reservoir computing on time independent classification tasks. In: Proceedings of International Joint Conference on Neural Networks, Atlanta, Georgia, United States of America, 2009:89-93.

IOP Conf. Series: Materials Science and Engineering **366** (2018) 012054 doi:10.1088/1757-899X/366/1/012054

Research on Network Resource Management Technology in Multi Access Network Environment of Terminal Communication Access Network

YU Zhipeng[1],GAO Zhiyuan[2],WANG Jinshuai[3],WANG Qiang[4] and QIU Yinghui[5]

[1]North China Electric Power University, Beijing 102206, China
[2]China Electric Power Research Institute, Nanjing 210000, China
[3]North China Electric Power University, Beijing 102206, China
[4]State Grid Shanxi electric power company Yangquan power supply company, Shanxi 045000, China
[5]North China Electric Power University, Beijing 102206, China

E-mail: 1053349427@qq.com

Abstract : In view of the traditional electric power communication network terminal access to the lack of unified management and control of the situation, based on the analysis of heterogeneous network management technology and channel protection technology of current mainstream access network communication methods in resource dynamic changes of frequency, time slot and cyber source; smart grid multi services in heterogeneous network integration under the bearing design scheme and choose to create favorable conditions for the optimization of communication terminal access.

1. Introduction

The problem that traditional power terminal access network of communication lacked of unified management and control of the concept of networking, and often set up network according to their business needs which are different departments, maintain the network by themselves, and formed a number of networks, a variety of communication modes coexist power access network status, have caused a serious waste of communication resources and reduced the reliability of communication network and improved the costs of communication system network construction, operation and maintenance[1]. This paper studies the cyber source management technology communication terminal access network under network environment of the terminal access network of communication.

2. Network resource management architecture and requirement analysis in smart grid communication system

According to the characteristics of the distribution network, distribution network communication mainly includes the communication between the distribution master station and the distribution station, the electronic distribution station and the distribution terminal, which is defined as a wide range[2].Communication system is a key of distribution network automation, and it is also a difficult point. The communication mode which chosen by the different region, different geographical environment and different conditions is various[3]. Communication can be divided into two categories: wired communication and wireless communication. In practice, they can be used alone or in combination. First of all, the cable communication includes: optical fiber communication, distribution carrier communication, communication cable, special line, local telephone network, paging, etc.. Second, Wireless communication includes GPRS/CDMA, wireless spread spectrum, microwave communication, satellite communication, etc.[4]. Because of the interference of clutter and the blocking of high-rise buildings in the city, wireless communication will cause many blind spots in communication, so it is not suitable to use wireless communication in urban power distribution communication, but mainly used in rural power distribution network, ground dispatching and county dispatching. At present, the application of the power system is mainly as a supplement to the wired communication.

3. Research on Network Discovery, Network Selection and Connection Management Mechanism for Smart Grid Communication System

The smart grid connects the client and different units through the access network to achieve intelligent measurement and control. When different cells in the access network are located in different heterogeneous networks, how to find and choose to access the network to achieve efficient data communication is extremely important.[5] As shown in the figure 1,this chapter focuses on the network discovery strategy and selection algorithm to carry out research to achieve the optimization of data communications, at the same time, with the current intelligent communication requirements for heterogeneous networks, the connection management mechanism of multi-media heterogeneous communication networks for smart grid environments and production needs is studied.

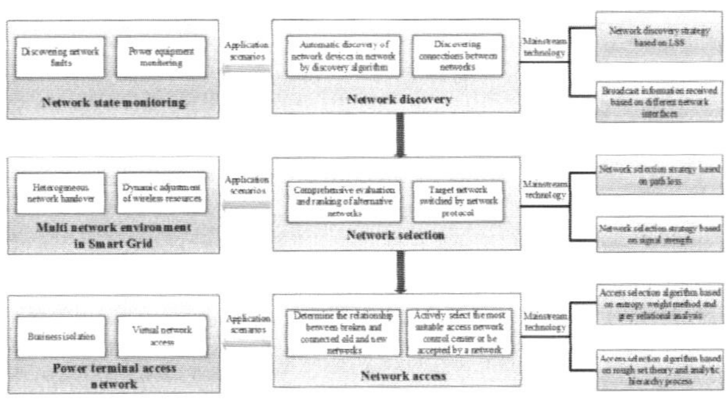

Figure 1:Discovery, Selection and Connection Method of Communication Networks in Data Sharing System

3.1 Analysis of Network Selection and Access Strategy in Public Communication

In the network discovery phase of the smart grid communication system, the power mobile terminal searches and finds the currently available wireless network, existing network discovery strategies can be divided into two categories:

(1) Network Discovery Policy Based on Location Service Server (LSS). The LSS stores various wireless network information in different areas and provides the surrounding network system and its parameters, including bandwidth, delay and jitter, according to the geographical location of the mobile terminal (available from GPS)[6]. The mobile terminal opens the corresponding network interface according to the received information. Although this method can reduce the energy consumption well, it needs to save the precise data of each network in the LSS, and the cost of establishing and maintaining the database is relatively large, it's difficult to realize.

(2) Broadcast Messages Received based on Different Network Interfaces. Different wireless networks regularly send broadcast messages, and the mobile terminals of multi network interfaces know the current network environment according to the broadcast messages received on different interfaces.

This topic is intended to adopt network discovery strategy based on broadcast messages received from different network interfaces. Different wireless networks regularly send broadcast messages, and the mobile terminals of multi network interfaces know the current network environment according to the broadcast messages received on different interfaces.

3.2 Design of Multi - media Heterogeneous Network Selection for Smart Grid Business Requirements and Resource Characteristics

Traditional network selection methods are aimed at wireless heterogeneous networks. At present, in the case of wireless and optical communication network discovery technology is very mature, when the wireless network discovery technology is applied to the heterogeneous network environment of power system, on the one hand, it needs to meet the needs of multi service communication, bandwidth demand, and communication reliability of the power grid[7]. On the other hand, it also needs to solve the technical problem that the time-varying PLC channel is strong.

In view of this, this paper analyzes the time-frequency characteristics of PLC based on Wigner distribution and utilizes the characteristics of fast convergence of family eugenics based evolution (FEBE) to combine real-time network communication parameters of various communication modes and select the real- Excellent communication method.

4. Performance Improvement of Smart Grid Mainstream Communication Technology

The mainstream communication access technologies of smart grid mainly include optical fiber communication technology, power line carrier technology and wireless communication technology. Optical fiber communication technology uses light as an information carrier. Currently, EPON technology and Industrial Ethernet technology are the most widely used on the access side. Different power services have different requirements on transmission bandwidth, rate and delay. This requires that the communication technology applicable to the automation of intelligent power distribution should possess the features of real-time and reliability, advanced technology, and independent intellectual property rights.

4.1 Improvement of Mainstream Communication Technology in Smart Grid

In the current mainstream access network communication technology, in the optical fiber communication, due to the large difference in demand of each node in the power grid, the unreasonable allocation of bandwidth, and the difficulty in construction, EPON technology is adopted to dynamically allocate the uplink bandwidth so as to reasonably enable the bandwidth allocation is rationalized and the network utilization rate is improved. At present, the transmission distance and reliability of PLC communication are shortcomings. This paper proposes to adaptively aggregate the OFDM subcarriers based on the channel conditions to solve the problem of access to multiple communication modes of the power service terminal , Enabling the service to intelligently select among multiple communication modes or perform coordinated communication among multiple communication modes to improve the utilization efficiency of network resources and reduce the cost of network construction.

4.2 EPON network dynamic bandwidth optimization

With the continuous development of smart grid, communication access is the key to supporting smart grid communication. EPON technology supports single-fiber bidirectional high-bandwidth service bearer and site-to-terminal 1 + 1 protection and 50ms protection switching, therefore, EPON technology networking is an ideal communication method for distribution links. In the distribution network automation system, digitalization substation and a series of applications will play an important role. In the current fiber-optic communications, due to the large differences in service requirements among nodes in the power grid, the unreasonable bandwidth allocation and the difficult construction, EPON technology is adopted to allocate the dynamic bandwidth of the network bandwidth[8].

By adopting a flexible period restriction policy and then using the prediction method to perform bandwidth authorization calculation for the high-priority services of different ONUs, the situation that high-priority services occupy the medium-priority services when the ONU data is sent is reduced occurs[9]. Finally, by using the early sending strategy, Further improve network performance under heavy load. This article's bandwidth allocation scheme is as follows:

1)Authorization scheduling and periodicity determination: This article uses centralized authorization to complete the authorization of all ONUs to achieve the fairness and integrity of the bandwidth allocation. Under the centralized authorization mode, the OLT can allocate bandwidth to different levels according to the current overall status of the network. At the same time, the bandwidth allocation among the ONUs can obtain higher fairness.

2)Power Business Distribution: According to the different power services carried by the power grid communication network, the data can be divided into three priority services, corresponding to three types of services in the service differentiation. The first type of service is highly demanding in real time. It has the highest priority for expedited forwarding (EF) services. The second type is medium priority services called assured forwarding (AF) services. The lowest priority of the three types of business, known as best effort (BE, besteffort) business.

3)Early sending policy: The OLT needs to collect the bandwidth request information of all the ONUs before performing unified bandwidth allocation calculation, and then sends the authorization frame to all the ONUs.

4.3 PLC sub-carrier frequency allocation scheme based on quantum genetic algorithm

Quantum genetic algorithm (GGA) borrows some basic ideas of quantum computing and applies the probability amplitude representation of quantum bits to chromosome coding so that a chromosome can express the superposition of multiple states and use quantum logic gates to realize the chromosome updating operation, so as to achieve the goal Optimization.

(1) Quantum bit coding: In genetic algorithm, chromosomes are expressed as deterministic values, whereas in quantum genetic algorithms, chromosomes are represented by quantum bits.

(2) Quantum state measurement: Quantum state measurement measures a quantum superposition state from a random variable of 0 to 1, causing it to collapse from a probabilistic state to a specific state.

(3) Individual Evaluation: Set appropriate fitness function to evaluate the current individual's merits.

(4) Individual Evolution: By adopting quantum door evolutionary strategy, quantum gates act on the ground state of quantum superposition state or entangled state to change the probability magnitude of each ground state in order to maintain the diversity of population and objectively simplify the algorithm itself.

5. Protection Scheme of Heterogeneous Network Channel in Power Grid Emergency State

5.1 Optical Fiber 1+1 Protection Mechanism for Power Grid Business Requirements

In order to solve the problem of false alarm of active fault alarm technology, this paper proposes an active fault alarm mechanism based on Bayesian model algorithm: firstly, an initial graph G0 is generated randomly, and the network Net0 is generated according to the initial graph, and the initialization parameters are set randomly; then the total iteration number of the external loop is set; the structure of Net0 is studied by EM algorithm, and the maximum likelihood function measure value V0 is calculated; then, the structure graph Gall, which is adjacent to the initial topological graph and is different from each other, is generated; then, the internal loop parameters are set to enter the inner loop, and the number of internal loops is the number of neighboring networks; the maximum likelihood function measure Vi of adjacent graphs is computed by traversing, and the graph Ghao with maximum value is searched; if Vi is greater than V0, the corresponding Gi and node order are recorded, and the Bayesian network Net0 and its parameters are retained, and then go to step 8; if Vi is less than V0, let Gi = G0, Vi = V0, go to step 10; then let Net0 = Neti, as the initial network when the next outer loop, go to step 3; at the end of the outer loop, the Gi is generated; finally, based on the network obtained by Gi, EM algorithm is used to study the parameters.

After the structure learning and parameter learning of the Bayesian model, the real-time network state parameter feature set can be input into a sufficiently trained Bayesian model. Once it is determined that an impending failure occurs, a fault alarm is triggered, and the protection and restoration of the business connection will be carried out ahead of time, and the active fault alarm is implemented.

5.2 Spectrum resource defragmentation algorithm based on OFDM

In order to solve the existing spectrum fragmentation of power line communication OFDM technology and proposes a sorting method of OFDM power line communication based on spectrum fragmentation: first statistics of all subcarriers "continuous idle frequency band set", and are arranged according to size; then all the continuous idle frequency gap into the set S (Csub); set the threshold calculation of S (Csub) "spectrum fusion set"; in order to meet the request of new bandwidth, spectrum space given "continuous idle frequency gap" needs to be emptied as $[FSP_{MAX}, FSP_{MAX}-(B-CIFSS)]$ (where B is the service request, bandwidth) to check the use of the need to clear each sub carrier wave frequency according to the gap; status of frequency gap, select the new business location of maximum spectrum utilization rate of subcarrier to reach; business bandwidth is allocated in the adjacent spectrum position occupied by the "maximum frequency gap" or The "minimum frequency gap" is used as the starting or ending position; finally, the steps are carried out in accordance with the size of the continuous idle frequency gap until the newly arrived service is received, or all successive idle frequency gaps have been checked".

The advantages of this spectrum defragmentation method are as follows: 1) The combination of spectrum defragmentation and new arrival resource allocation is dynamically combined to achieve the dual functions of defragmentation and dynamic resource allocation. 2) The computational complexity of this method is low, easy to implement, enabling fast defragmentation and providing as much idle resources as possible for new arrivals.

6. Conclusion

This paper focuses on the research of heterogeneous network convergence in multi network environment. The characteristics of heterogeneous converged networks and their interconnection schemes are discussed. A heterogeneous converged network model and a resource management framework are established. Research on network discovery technology and network selection technology in multi network environment. On the basis of the existing programs, put forward a strategy of received signal strength based on the network, and put forward a network access algorithm based on family eugenics to realize the selection of access in heterogeneous converged networks.

References

[1] Damnjanovic A,Montojo J,Wei Y B,et al. A survey on 3GPP heterogeneous network[J]. IEEE Wireless Communications,2011(6): 10-21

[2] Jeounglak H,JiYeon K,Jin-Up K,et al. Dynamic Load Balancing Architecture in Heterogeneous Wireless Network Environment[C]. 9th International Symposium on Communications and Information Technology. Icheon Korea: IEEE Computer Society,2009: 248-253

[3] Jie X,Yuming J,Perkis A.Multi-service Load Balancing in a Heterogeneous Network[C]. Proceedings of the Wireless Telecommunications Sysposium. America New York: IEEE Communication Society,2011: 1-6

[4] Sorr H,LEE S,Kurt S,et al. Soft load balancing over heterogeneous networks[J]. IEEE Transactions on Vehicular Technology,2008,57(4): 2632-2638

[5] Selvakumar A,Vaidehi V,Palanivelu T G. Adaptive load balancing handoff scheme for heterogeneous wireless network[C]. Proceedings of the Third International Conference on Advanced Computing. India Chennai: IEEE Computer Society,2011: 327-332

[6] M. Ismail,Weihua Zhuang,"Network cooperation for energy saving in green radio communications,"IEEE Wireless Communications,vol. 18,no. 5,2011,pp. 76-81

[7] Boucher J. Traffic system design handbook: Timesaving telecommunication traffic tables and programs[M]. American: Wiley-IEEE Press,1993,33-78

[8] Shakkottai S, Stolyar A L. Scheduling algorithms for a mixture of real-time and non-real-time

data in HDR[R]. Bell Laboratories Lucent Technology,2000

[9] Yu F, Krishnamurthy V. Optimal joint session admission control in integrated WLAN and CDMA cellular networks with vertical hand off[J]. IEEE Transactions on Mobile Computing,2007,6(1):126-139

IOP Conf. Series: Materials Science and Engineering **366** (2018) 012055 doi:10.1088/1757-899X/366/1/012055

A Review of Distributed PV Operation Analysis and Demand Response Research

Jiancheng Yu[1], Lei Wu[2], Cheng Yao[1], Jing Chen[3,4], Siyuan Li[3] and Xiufan Ma[3]

[1] State Grid Tianjin Electric Power Company, Tianjin 300010, China.
[2] State Grid Tianjin Electric Power Research Institute, Tianjin 300384, China.
[3] School of Electric and Electronic Engineering, North China Electric Power University, Beijing 102206, China.

[4]charris@126.com

Abstract. The large-scale access of the distributed photovoltaic power supply to the grid will bring great challenge to the safe operation of power grid, thus adversely affecting the further development of distributed photovoltaic. First of all, from the aspects of structure, power quality, network loss and relay protection, this paper summarizes the influence of distributed photovoltaic system on distribution network. Secondly, the distributed photovoltaic output is summarized and summarized from two aspects of distributed photovoltaic output performance analysis and output forecast. In addition, from the perspective of photovoltaic user demand response can improve the photovoltaic extinction rate, this paper also combs the analysis and research of PV user demand response modelling and electricity price mechanism. Finally, some suggestions on the key areas for future research are put up.

1. Introduction

The development prospect of distributed photovoltaic power generation is broad [1], it advocates the spontaneous use of electricity, the rest of the internet, there is no step-up and long-distance transportation problem, directly facing the terminal power consumers[2], in the power market reform and energy-saving emission reduction situation, has got more and more attention and development. With the increase of the distributed PV network capacity, the safe and stable operation of the power system will be influenced by the non-negligible [3]. The accurate characteristic analysis and output prediction of distributed PV are the important foundation to solve these problems and ensure the safe operation of the power grid [4]. At the same time, a lot of researches have been done on demand side participation in distribution network, the purpose of which is to adjust shifting and increase PV extinction rate by using demand response of PV users [5-9].

Based on the influence of photovoltaic power grid on distribution network, this paper collated the research of PV power generation characteristics analysis and output prediction, and then it has further combed the research on PV user demand response, which has provided some reference for readers.

2. The Influence of Distributed Photovoltaic on Distributed Network

Due to the influence of solar radiation intensity, the output of PV power has a strong randomness and intermittence, which has important influence on power quality, voltage fluctuation, economy and relay protection [10-14]. First of all, the grid-connected distributed photovoltaic power system changes the distribution network structure, and when not coordinated with the load distribution grid, it will increase the difficulty of the voltage regulation of the distribution grid [15-16]. Secondly, the large-capacity photovoltaic grid-connected under different operating modes will greatly influence the power flow size and direction of distribution network, thus causing the change of voltage distribution [17-19] and network loss [20-21], and changing the economy. In addition, the distribution of distributed photovoltaic power supply network, will inevitably provide a certain short-circuit current to short-circuit point, which may have an impact on the distribution automation fault location [22-24].

The above research shows that large-scale grid-connected photovoltaic power supply will bring power quality problems, affect the stability of power grid, and cause economic loss to power users of distribution network, and endanger the safe and reliable operation of PV system. Therefore, it is necessary to accurately analyse and forecast the output characteristics of distributed photovoltaic system, and put forward a reliable countermeasure of eliminating the solar energy to reduce the impact of distributed photovoltaic on grid.

3. Analysis and Prediction of Distributed PV Output Characteristics

Due to the randomness and volatility of photovoltaic power generation, a large number of photovoltaic power grids will not only increase the system net load fluctuation range, but also reduce the accuracy of load forecasting, and make the fluctuation of net load of system more uncertain. Accurate analysis of PV output characteristics and generation forecasting will have an important impact on daily power generation scheduling and power system operation planning.

3.1. Analysis of Distributed PV Output Characteristics

The analysis of the distributed PV output characteristics mainly aims at the fluctuation of PV power in different time periods and different conditions, and the weather factors such as sunshine and temperature, and the structure and installation mode of PV system are also the important factors that influence the output.

For the study of volatility, literature [25-26] studied the effects of cloud movement and different configuration combinations of photovoltaic arrays on photovoltaic output power. Literature [27] presented a method to extract the random components of photovoltaic output, and accurately described the continuity of PV output fluctuation. Literature [28] presented a method for analyzing the stochastic characteristics of the output of photovoltaic power station. The deterministic factors of PV power generation and the uncertain factors were separated. The stochastic characteristics of PV output occlusion factors were studied from the aspects of probability distribution and fluctuation characteristics, and the spatial correlation of photovoltaic output was analysed by means of empirical analysis. Literature [29] defined the photovoltaic output state according to the PV output operating point and its fluctuation limit value, then it quantitatively studied the time length of PV power maintenance in a certain state, and put forward the concept of the duration distribution of PV power.

For PV System Research, literature [30] took the distributed maximum power point tracking (DMPPT) photovoltaic system as the research object, aiming at the optimization problem of multiple distributed maximum power tracking module networks, analysed the structure of multiple PV modules cascade based on the static output characteristics of the system, and presented a new distributed maximum power point tracking PV system structure, besides, the output characteristics of photovoltaic under this structure were analysed.

3.2. Distributed Photovoltaic Output Forecast

At present, the main principle of photovoltaic power forecasting is based on the future weather conditions combined with the solar radiation and photovoltaic panel layout of the solar panels to receive radiation prediction, and then estimate the power of photovoltaic panels [30].

From the point of view of weather condition and solar radiation law, literature [31] used the autoregressive sliding average model to construct the stochastic time series of the PV output sequence, and gave a consideration to the correlation of the output sequence of the multi power station by using the characteristics of the autoregressive sliding average mathematical model. Literature [32] put forward a systematic method to optimize the operation of the feeder and to ensure the voltage quality, to determine the photovoltaic active and reactive power output of residential roofs. Literature [33] used the framework of stochastic differential equations to predict the probability of solar irradiation. Literature [34] first predicted the probability distribution of the hourly clear sky index through Bayesian autoregressive time series model, and then calculated the probability distribution of PV output by random sampling of the clear sky index by Monte Carlo simulation. Literature [35] used the copula function to model the joint probability distribution of PV actual output and point prediction, and realized the estimation of conditional probability distribution of the corresponding photovoltaic actual output by any point prediction. Literature [36] in view of the timing and randomness of PV power generation, the output model of PV power system was improved by stratified sampling of weather type sampling and equipment outage probability.

From the point of view of other distributed power sources research, literature [37] used the copula theory of stochastic mathematics to set up the conditional prediction error model of PV power generation based on the wind power forecasting error model. From the point of view of power network planning. Literature [38-40] made a deep study on the distributed Photovoltaic capacity prediction and evaluation in the planning area, and put forward a method of distributed photovoltaic generation and spatial forecasting for active distribution network planning.

4. PV User Demand Response

From the above analysis, distributed photovoltaic power grid will bring a variety of adverse effects on PV, on the basis of accurate analysis and prediction of photovoltaic output, the demand response in PV user group can improve the self-elimination rate of PV power and effectively adjust the influence of PV output power on power grid [41].

In the research of demand response modeling of PV users, based on PV generation curve and demand response characteristics, literature [42] established the real-time pricing model and the user's price response model. Combined with real-time pricing and user response model, the realization method and process of load transfer were constructed, which encouraged the user to transfer the transfer load to the photovoltaic power generation and increase the net load value of PV output period. Under the peak-Valley time-sharing price, literature [43] constructed a multiple-time tariff response model based on the energy price elasticity matrix, and put forward the strategy of energy charging and discharging and the optimal operation of optical storage micro-network. Literature [44] took the load curve minus the equivalent load curve obtained from the PV curve as the optimal object of the time-sharing price, and established and solved the optimization model of the peak-Valley time-sharing electricity price. Based on the characteristics of users' power, literature [45-46] constructed a multi buyer-seller pattern within the PV user group, combined with China's distributed PV Internet policy, and put forward a demand-side response model based on game theory, and proved that the game problem has a unique Nash equilibrium strategy. Literature [47] presented an internal price model based on photovoltaic power supply and demand ratio (SDR). Based on the consideration of economy and comfort, the utility cost model of user participating in demand response was put forward.

In the research of PV user price mechanism, literature [48] studied the user's response to the price, and the marginal cost fixed price or the fixed price of the given rule was to develop the optimal pricing strategy by collecting the power generation and load of the whole cluster. Literature [49] studied the optimal operation of photovoltaic micro-power grids containing energy storage. Through the relationship between supply and demand in the micro-grid, this paper put forward the two pricing strategies based on the pricing rules, and gave two kinds of game models. Literature [50] for the Peak-Valley time-sharing electricity price and the ladder electricity price 2 kinds of mainstream electricity purchase policy, taking account of the factors such as the profit of electricity sales, the cost of buying electricity and government subsidy, the optimal dispatching model of grid-connected PV power system was established, which was based on the distribution ratio of surplus to the optimal variable and the user benefit. Literature [51-52] calculated the economical reasonable PV power price by establishing the net present value model, then according to the difference of the solar resource endowment and the system construction condition in the domestic area, changed the electricity generation and the installment cost in the model to set up the PV network guiding price.

The above research from the net present value, the game, the optimization and so on launches, has carried on the comprehensive and the detailed analysis to the PV user demand response modelling and the electricity price mechanism, has the important reference significance to the future research. However, the analysis of the response behavior characteristics of PV users is not deep enough to reflect the dynamic relationship between the electricity price mechanism and the user's power behaviour.

5. Conclusion

This paper sums up the current research situation from the aspects of the impact of large-scale grid-connected photovoltaic power supply on power grid, analysis and prediction of PV output characteristics and demand response of PV users. At present, most of the literatures on PV users' demand response are based on the research of ordinary resident users, and the characteristics of PV users are not detailed and thorough enough, some models fail to consider the actual installation and operation of the distributed PV system. Then the paper offers the suggestions on the research point, to provide reference for the further study.

1) In the study of distributed photovoltaic grid-connected, we should consider the safe and stable operation of power grid, and further research on grid-connected detection and grid-connected capacity control.
2) In the study of PV user demand response, we should deeply analyze the user configuration of PV demand response behavior, to achieve the dynamic correlation with electricity price.
3) We should discuss the influence of PV users' consumption ability, economic level, and time and seasons on the usage of photovoltaic users, to make the research more realistic.
4) We can further discuss the selection behavior of the user under the demand of response mechanism when combining PV with energy storage.

Acknowledgments

2017 State Grid Corporation technology project funding: research and application of multi-load active response and predictive control technology for stochastic power supply.Contract number: SGTJDK00DWJS1700034.

References

[1] Sicheng. Current situation and development trend of distributed photovoltaic power generation policy [J]. Solar energy, 2013 (8): 8-19.
[2] Bo, Liaoqiangqiang, Liu, etc. economic analysis of the hybrid system of distributed photovoltaic energy storage battery [J]. Power construction, 2016, 37 (8): 102-107.
[3] Chang Jiale. The influence of distributed photovoltaic power supply on distribution network and the study of grid-connected planning [D]. Power, 2016.
[4] Wang Hongkun, Greid, Li Hongwei, and so on. A review of the characteristic analysis and prediction methods of distributed photovoltaic power generation [J]. Power construction, 2017, 38 (7): 1-9.
[5] Kim T, Kim S, Shin D, et al. Design and Implementation of an Open, Interoperable Automated Demand Response Infrastructure[J]. Lawrence Berkeley National Laboratory, 2008, 12(4):537-550.
[6] Gao Wei, Liang Sweet, Li. A review of the theory and practice of automatic demand response [J]. Power Grid technology, 2014, 38 (2): 352-359.
[7] Shenwancheng, Zhang, di Hongyu, etc. application of immune optimization algorithm based on dynamic antibody memory in automatic demand response [J]. Journal of China Electrical Engineering, 2014, 34 (25): 4199-4206.
[8] Wang Dongjong. The theory and empirical Study of DSM response [D]. Power (Beijing), 2011.

[9] Zeng Ming, Han Xu, Rae. Support mechanism of demand side response for active distribution network operation [J]. Power construction, 2015, 36 (1): 110-114.

[10] Huang, Dong, Requin, and so on. Comprehensive impact analysis of large capacity distributed PV grid on distribution network [J]. Journal of power system and automation, 2016, 28 (11): 44-49.

[11] Requin, Li Hawk, Luzehan, etc. distributed generation technology and its influence on power system [J]. Southern Power grid Technology, 2011, 05 (4): 46-50.

[12] Yu Lei, Xu Lili, Zhou Xin. The influence of distributed photovoltaic power generation on distribution network and its countermeasures [J]. North China Power Technology, 2013 (10): 13-17.

[13] Tong, Lancenlin. The influence of distributed generation on distribution network [J]. East China Power, 2010, 38 (7): 1038-1039.

[14] Zhu Zhongli, Ying, Cao Wei, and so on. The impact of large-scale distributed access on power grid [J]. East China Power, 2012 (12): 2224-2227.

[15] Miuliheng. Study on the effect of distributed PV access on the steady state operation of distribution network [D]. Power (Beijing), power, 2015.

[16] Liqingfeng. Effect of distributed photovoltaic system on power quality of distribution network [D]. Power, 2016.

[17] Li Yaling, Weile, Zhao Jingtao, and so on. The effect of distributed PV system on distribution network voltage [J]. Power technology, 2016, 40 (6): 1257-1259.

[18] Wang Zhiqun, Zhu Shoujien, double happiness, etc. effect of distributed generation on voltage distribution of distribution network [J]. Power system Automation, 2004, 28 (16): 56-60.

[19] Xu Xiaoyan, Sui Fai, Liu Chun, etc. the influence of distributed photovoltaic on distribution network voltage and the solution of voltage limit [J]. Power grid Technology, 2010 (10): 140-146.

[20] Ho, He Huachen, Li. Influence of distributed PV access on network loss of distribution network [J]. Modern electronic technology, 2014 (6): 158-162.

[21] Li, Peng, clear, and so on. The power flow calculation of distributed grid connection [J]. Power system Protection and control, 2009, 37 (17): 78-81.

[22] Lin Tao, Liu Jian, Tong Xiangqian and so on. Simulation analysis of the effect of distributed photovoltaic power supply on short-circuit current of distribution network [J]. Power grid Technology, 2013, 37 (8): 2080-2085.

[23] Kong Xianping, Jangchul, Yin Xiangen, etc. research on fault current characteristics and fault analysis method of power system with inverse variant distributed power supply [J]. Journal of China Electrical Engineering, 2013 (34): 65-74.

[24] Mrs. Liu, Zhang Xiaoqing, Tong Xiangqian, and so on. Fault location of distribution network with distributed power supply [J]. Power system Automation, 2013, 37 (2): 36-42.

[25] Lappalainen K, Valkealahti S. Output power variation of different PV array configurations during irradiance transitions caused by moving clouds[J]. Applied Energy, 2017, 190:902-910.

[26] Belhaouas N, Cheikh M S A, Agathoklis P, et al. PV array power output maximization under partial shading using new shifted PV array arrangements[J]. Applied Energy, 2017, 187:326-337.

[27] Zhao Liang, Li Jiaming, Aixiao, et. Analysis of the stochastic components extraction and statistical characteristics of PV output [J]. Power system automation, 2017, 41 (1): 48-56.

[28] Xi, Kang Chongqing, Ning, etc. analysis of the medium and long term stochastic characteristics of solar photovoltaic generation [J]. Power system Automation, 2014, 38 (6): 6-13.

[29] Li Jiaming, Aixiao, Wen Jingyu, et. Quantitative analysis of the probability distribution of PV power duration characteristics [J]. Power system automation, 2017, 41 (6): 30-36.

[30] Wang, Kong Pengju, Fred, etc. analysis of output characteristics of PV system based on distributed maximum power tracking [J]. Journal of Electrotechnical Technology, 2015, 30 (24): 127-134.

[31] Klöckl B, Papaefthymiou G. Multivariate time series models for studies on stochastic generators in power systems[J]. Electric Power Systems Research, 2010, 80(3):265-276.
[32] Dall'Anese E, Dhople S V, Giannakis G B. Optimal dispatch of photovoltaic inverters in residential distribution systems[C]// Pes General Meeting | Conference & Exposition. IEEE, 2014:1-1.
[33] Iversen E B, Morales J M, Møller J K, et al. Probabilistic forecasts of solar irradiance using stochastic differential equations[J]. Environmetrics, 2014, 25(3):152–164.
[34] Bracale A, Caramia P, Carpinelli G, et al. A Bayesian Method for Short-Term Probabilistic Forecasting of Photovoltaic Generation in Smart Grid Operation and Control[J]. Energies, 2013, 6(2):733-747.
[35] Chang, Ning, Kang Chongqing, and so on. The conditional prediction error probability distribution estimation method for PV generation output [J]. Power system automation, 2015 (16): 8-15.
[36] Wang Beautiful, Wu Zechen, Qu Chong. Reliability analysis and confidence capacity calculation of photovoltaic power system [J]. Journal of China Electrical Engineering, 2014, 34 (1): 15-21.
[37] Zhang N, Kang C, Xia Q, et al. Modeling Conditional Forecast Error for Wind Power in Generation Scheduling[J]. IEEE Transactions on Power Systems, 2014, 29(3):1316-1324.
[38] Fu Yang, Tianzan, Li Zhenkun, Diliang, Jin Shanhong, Zhang Jingwei. Distributed photovoltaic spatial-temporal distribution prediction considering spatial capacity saturation [J/ol]. Power grid technology: 1-8. (2017-07-04).
[39] Villavicencio J, Melo J D, Feltrin A P. Estimation of photovoltaic potential on residential rooftops using empirical Bayesian estimator[C]// Innovative Smart Grid Technologies Latin America. IEEE, 2016:242-247.
[40] Rowley P, Leicester P, Palmer D, et al. Multi-domain analysis of photovoltaic impacts via integrated spatial and probabilistic modelling[J]. Renewable Power Generation Iet, 2015, 9(5):424-431.
[41] Ding Z, Lee W J, Wang J. Stochastic Resource Planning Strategy to Improve the Efficiency of Microgrid Operation[J]. IEEE Transactions on Industry Applications, 2015, 51(3):1978-1986.
[42] Shi Lei, Hu Jianzhao, Han Shengfeng, et. Study on the strategy of increasing the power price incentive load transfer for PV grid-connected penetration [J]. Power system protection and control, 2016, 44 (23): 153-157.
[43] Zhou, Fan Wei, Liu Nian, and so on. Multi-Objective capacity optimization configuration of PV micro-network energy storage system based on demand response [J]. Power grid technology, 2016, 40 (6): 1709-1716.
[44] Fan Wenfei, Feng Fei, Wang Delin. Study on optimization strategy of time-sharing electricity price after PV power access network [J]. Electrical switches, 2016, 54 (4): 46-50.
[45] Wang, Liu Nian, Cheng Minhao, etc. the optimal pricing model of PV user group based on Stackelberg game [J]. Power system automation, 2017, 41 (12): 146-153.
[46] Mary, Liu Nian, Jianhua, etc. optimal operation model of PV user Group under Automatic demand response mode [J]. Journal of China Electrical Engineering, 2016, 36 (13): 3422-3432.
[47] Liu Nian, Wang, Requin. Power sharing and demand response model of PV user Group in market mode [J]. Power system automation, 2016, 40 (16): 49-55.
[48] Ding Z, Lee W J, Wang J. Stochastic Resource Planning Strategy to Improve the Efficiency of Microgrid Operation[J]. IEEE Transactions on Industry Applications, 2015, 51(3):1978-1986.
[49] Maity I, Rao S. Simulation and Pricing Mechanism Analysis of a Solar-Powered Electrical Microgrid[J]. IEEE Systems Journal, 2010, 4(3):275-284.
[50] Ma Lin, Zhang Shirong. Optimal operation of home-connected PV power system under time-sharing price/step electricity price [J]. Power grid technology, 2016, 40 (3): 819-825.
[51] Rigter J, Vidican G. Cost and optimal feed-in tariff for small scale photovoltaic systems in China[J]. Energy Policy, 2010, 38(11):6989-7000.

IOP Conf. Series: Materials Science and Engineering **366** (2018) 012055 doi:10.1088/1757-899X/366/1/012055

[52] Chan Wei. Research on PV power Price and its policy [D]. East China University of Technology, 2013.

IOP Conf. Series: Materials Science and Engineering **366** (2018) 012056 doi:10.1088/1757-899X/366/1/012056

A Virtual Array Extension Method for Multiple Partial Discharge Sources Localization Based on Fourth-order Cumulant and Signal Sparse Decomposition

Qiang Gao[1], Qi Liu[1], Zhannan Guo[2], Nan Zhou[3], Zhen Li[3] and Lingen Luo[3]

[1] Electric Power Research Institute of State Grid Liaoning Power Supply Co. Ltd., Shenyang, China
[2] Shenyang Kekai Electric Power Technology Co. Ltd., Shenyang, China
[3] Shanghai Jiao Tong University, Shanghai, China

Corresponding author's e-mail: llg523@sjtu.edu.cn

Abstract. The localization of multiple partial discharges based on UHF signal is difficult to accomplish due to the excessive localization error. A virtual array extension method based on four-order cumulant and signal sparse decomposition is proposed in this paper. The virtually extended array has higher directional resolution and it could detect partial discharge sources with higher accuracy. The 2×2 UHF array that commonly used in existed partial discharge localization methods is virtually extended to 3×3 and 5×5 array. The virtual array extension is accomplished by extending the corresponding steering vector matrix of PD signal spatial spectrum to a overcomplete dictionary, which is then used to realize the sparse decomposition of the PD signal. Eventually the DOA estimation of multiple partial discharge sources is achieved. Simulation and field test results show that the virtually extended UHF array can detect and locate two PD sources simultaneously. The directional angel estimation error is about 6° and the pitch angle error is lower than 4°, which has proved the feasibility of the proposed method.

1. Introduction

The Partial Discharge (PD) localization based on Ultra-High Frequency (UHF) signals has been widely studied in recent years, due to its high detection accuracy and strong ability of anti-interference. Many innovative PD detection methods have been proposed by researchers [1-4] and several successful applications have been reported. However, most studies focus on the detection or accurate localization of single PD source, which has limited their practical application.

The detection and localization of multiple PD sources is an important research area while few studies are addressed. In [5], each PD source is represented as a feature vector and a clustering algorithm is applied to separate the PD sources. Reference [6] proposed a self-adaptive PD separation algorithm based on optimized feature extraction of cumulative energy function. A probability-based algorithm is presented in [7], which could locate multiple PD sources in electrical equipment. These researches generally follow a similar roadmap of "signal processing, feature extraction, cluster analysis and multiple sources separation." Those algorithms based on feature extraction and clustering analysis can successfully separate multiple PD sources while suffer to its algorithm complexity and huge computation burden.

IOP Conf. Series: Materials Science and Engineering **366** (2018) 012056 doi:10.1088/1757-899X/366/1/012056

In this paper, an effective multiple UHF PD DOA estimation method based on virtual array extension is proposed. The fourth-order cumulant matrix [8-9] is applied to the PD signals received by the sensor array to accomplish the sparse decomposition. The weighted l1-norm method is used to solve the DOA estimation equations. Eventually, the spatial spectrum of multiple PD sources is established and the DOA estimation is achieved. As most of the multi-channel synchronous GHz sampling rate devices that having an acceptable price are usually four-channel, the UHF sensor array is limited to a 2×2 array. Therefore, the 2×2 actual array is virtually extended by the proposed method, and simulation and fields tests are performed. Test results show that the extended array could estimate the PD source more accurately than the actual array. The directional angel estimation error about 6° and the pitch angel error lower than 4°, which has verified the feasibility of the proposed method in practical application.

2. Signal Sparse Decomposition Based on Fourth-order Cumulant Matrix

2.1. Application of Fourth-order Cumulant Matrix

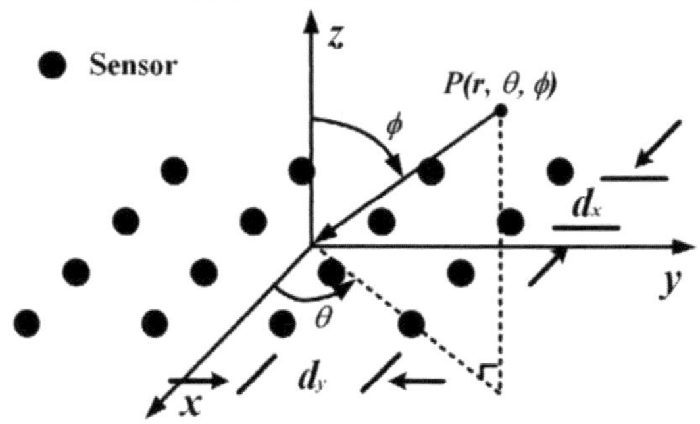

Figure 1. Partial Discharge DOA Estimation by Sensor Array.

Suppose the plane UHF sensor array is composed of M×M UHF sensors, as shown in Fig. 1. The sensors are distributed evenly with the interval as dx along x-axis and the interval as dy along y-axis. Suppose P stands for the PD source, the distance from P to axis origin is r, the directional angel is □ and the pitch angel is □.

Suppose K independent narrowband signals are received by the M×M sensor array, the PD signal received by the mth sensor in the array could be expressed as:

$$y_m(t) = \sum_{k=1}^{K} s_k(t) \cdot a_m(\theta_k, \phi_k) + v_m(t) \tag{1}$$

And the vector form could be expressed as:

$$\boldsymbol{y(t)} = \boldsymbol{A(\theta, \phi)} \cdot \boldsymbol{s(t)} + \boldsymbol{v(t)} \tag{2}$$

where $\boldsymbol{y(t)}$ stands for the signal received by the array. $\boldsymbol{A(\theta,\phi)} = (a(\theta_1, \phi_1), a(\theta_2, \phi_2), \ldots, a(\theta_N, \phi_N))$ is the overcomplete steering vector and N is the number of divided angles. $a(\theta_i, \phi_i) = (1, e^{j\pi \sin\theta_i \cos\phi_i}, \ldots, e^{j\pi(M-1)\sin\theta_i \cos\phi_i})^{\mathrm{T}}$. $\boldsymbol{v(t)}$ stands for the environment noise. $\boldsymbol{s(t)}$ stands for the signal source vector. The diagram of DOA estimation with overcomplete steering vector is shown in Fig. 2.

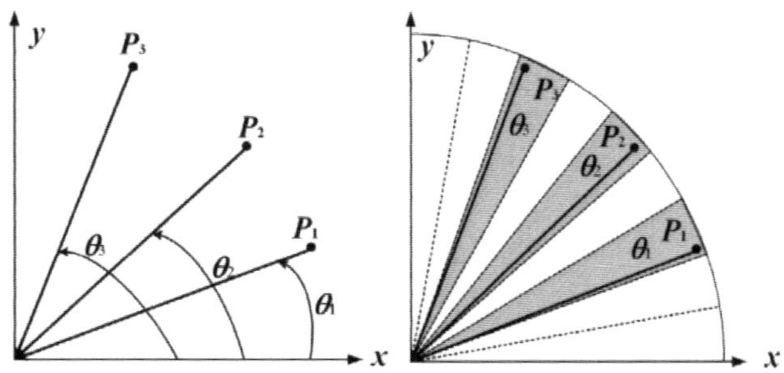

Figure 2. DOA Estimation with Overcomplete Steering vector.

The $M^2 \times M^2$ fourth-order cumulant of the received signal $\boldsymbol{y(t)}$ is calculated as follows [10]:

$$\begin{aligned}\boldsymbol{C}_Y &= cum\{\boldsymbol{Y}(t), \boldsymbol{Y}^{\mathrm{T}}(t), \boldsymbol{Y}^*(t), \boldsymbol{Y}^{\mathrm{H}}(t)\} \\ &= (\boldsymbol{A}(\theta,\phi) \otimes \boldsymbol{A}^*(\theta,\phi))\boldsymbol{C}_s(\boldsymbol{A}(\theta,\phi) \otimes \boldsymbol{A}^*(\theta,\phi))^{\mathrm{H}}\end{aligned} \tag{3}$$

where $\boldsymbol{C}_s$ is expressed as follows:

$$\begin{aligned}\boldsymbol{C}_s &= cum\{\boldsymbol{S}(t), \boldsymbol{S}^{\mathrm{T}}(t), \boldsymbol{S}^*(t), \boldsymbol{S}^{\mathrm{H}}(t)\} \\ &= E\{(\boldsymbol{S}(t) \otimes \boldsymbol{S}^*(t))(\boldsymbol{S}(t) \otimes \boldsymbol{S}^*(t))^{\mathrm{H}}\} - \\ &\quad E\{(\boldsymbol{S}(t) \otimes \boldsymbol{S}^*(t))\}E\{(\boldsymbol{S}(t) \otimes \boldsymbol{S}^*(t))^{\mathrm{H}}\} - \\ &\quad E\{(\boldsymbol{S}(t)\boldsymbol{S}^{\mathrm{H}}(t)\} \otimes E\{(\boldsymbol{S}(t)\boldsymbol{S}^{\mathrm{H}}(t))^*\}\end{aligned} \tag{4}$$

$\boldsymbol{C}_s$ is the fourth-order cumulant matrix of the signal source vector $\boldsymbol{s(t)}$. The value of the matrix element at the $((i_1\text{-}1)M+i_3)^{th}$ row and $((i_2\text{-}1)M+i_4)^{th}$ column of $\boldsymbol{C}_s$ is $cum(s_{i_1}(t), s_{i_2}(t), s_{i_3}^*(t), s_{i_4}^*(t))$, $i_1, i_2, i_3, i_4 \in \{1, 2, \ldots, N\}$.

The equations could be expressed as:

$$\begin{aligned}&\boldsymbol{C}_s((i_1-1)M+i_3, (i_2-1)M+i_4) = \\ &cum(s_{i_1}(t), s_{i_2}(t), s_{i_3}^*(t), s_{i_4}^*(t))\end{aligned} \tag{5}$$

2.2. *Signal Sparse Decomposition and Array Extension*

Suppose the signal source vector is non-correlation signal. According to the principle of fourth-order cumulant, (3) could be simplified as:

$$\begin{aligned}\boldsymbol{C}_Y = &\sum_{i=1}^{N} \boldsymbol{C}_s((i-1)M+i, (i-1)M+i)(a(\theta_i,\phi_i) \otimes \\ &a^*(\theta_i,\phi_i))(a(\theta_i,\phi_i) \otimes a^*(\theta_i,\phi_i))^{\mathrm{H}}\end{aligned} \tag{6}$$

In (6), the Kronecker product of the steering vector, $a(\theta_i,\phi_i) \otimes a^*(\theta_i,\phi_i)$, would form the virtual array sensors, extending the sensor array from M×M to (2M-1)×(2M-1). However, the size of vector $a(\theta_i,\phi_i) \otimes a^*(\theta_i,\phi_i)$ is $M^2 \times 1$. It is because that many elements in the vector carries the same signal information received by the same sensor [11]. The location of sensor is expressed as *P*. The relationship between *P* and the number of elements related to it, N_{num}, is shown as follows:

$$N_{num} = M - \frac{|P|}{d} \qquad -(M-1)d \le P \le (M-1)d \tag{7}$$

where *d* is the interval between sensors in the array.

As there are many elements in $a(\theta_i,\phi_i) \otimes a^*(\theta_i,\phi_i)$ that are representing the same received signal, add these elements together. Similarly, add the same elements in $\boldsymbol{C}_Y$ together. Thus the $M^2 \times M^2$ is turned into a (2M-1)×(2M-1) matrix. Express it in the form of $b(\theta_i,\phi_i)b^{\mathrm{H}}(\theta_i,\phi_i)$, where $b(\theta_i,\phi_i)$ is the newly designed steering vector:

$$b(\theta_i,\phi_i)=(e^{-j\pi(M-1)\sin\theta_i\cos\phi_i},\dots,e^{-j\pi\sin\theta_i\cos\phi_i},\\ 1,e^{j\pi\sin\theta_i\cos\phi_i},\dots,e^{j\pi(M-1)\sin\theta_i\cos\phi_i})^{\mathrm{T}} \tag{8}$$

Thus (6) can be simplified as:

$$\begin{aligned}\boldsymbol{C}_{\mathrm{Y}}' &= \sum_{i=1}^{N}\boldsymbol{C}_s((i-1)\mathrm{M}+i,(i-1)\mathrm{M}+i)b(\theta_i,\phi_i)b^{\mathrm{H}}(\theta_i,\phi_i)\\ &= \boldsymbol{B(\theta,\phi)}\cdot diag\{\boldsymbol{C}_s(1,1),\dots,\\ &\boldsymbol{C}_s((N-1)M+N,(N-1)M+N)\}\cdot\boldsymbol{B}^{\mathrm{H}}\boldsymbol{(\theta,\phi)}\end{aligned} \tag{9}$$

Suppose $\boldsymbol{X}$ = diag{$\boldsymbol{C}_s$(1, 1) , ... , $\boldsymbol{C}_s$((*N*-1)M+*N*, (*N*-1)M+*N*)}$\boldsymbol{B}^{\mathrm{H}}(\theta,\phi)$, the signal sparse decomposition could be written as:

$$\boldsymbol{C}_{\mathrm{Y}}' = \boldsymbol{B(\theta,\phi)}\cdot\boldsymbol{X} \tag{10}$$

Compared to (2), (10) is the signal equations received by the extended to a (2M-1)×(2M-1) sensor array.

3. The Solution to the Extended Array Model and the Spatial Spectrum PD Estimation

3.1. Model Solution

The extended signal source vector **X** in (10) is the solution to the partial discharge DOA estimation in this paper. To every row in matrix **X**, calculate the l_2-norm vector $x_{l_2}=(\|\boldsymbol{X}(1,:)\|_2,\dots,\|\boldsymbol{X}(N,:)\|_2)^{\mathrm{T}}$. The l_1-norm of vector x_{l_2} is the objective function of the minimization problem. Thus, the solution to (10) is transferred to the convex optimization problem as shown below [12]:

$$\begin{aligned}&\min\sum_{i=1}^{N}\omega_i\|\boldsymbol{X}(i,:)\|_2\\ &\text{s.t. } \boldsymbol{C}_{\mathrm{Y}}'=\boldsymbol{B(\theta,\phi)}\cdot\boldsymbol{X}\end{aligned} \tag{11}$$

where ω_i is the weight, which could be calculated by the signal subspace and the noise subspace [13]. In this paper, it is calculated as below:

$$\omega_i=b^H(\theta_i,\phi_i)\boldsymbol{U}(:,(K+1):(2M-1))\\ \boldsymbol{U}^{\mathrm{H}}(:,(K+1):(2M-1))b(\theta_i,\phi_i) \tag{12}$$

where $\boldsymbol{U}$ is the noise subspace of the $\boldsymbol{C}_{\mathrm{Y}}'$.

3.2. PD DOA Estimation by Spatial Spectrum Estimation

The spatial spectrum $P(\theta,\phi)$ is of PD sources are established to estimate the DOA. It is calculated as below [14]:

$$P(\theta,\phi)=\frac{1}{b^{\mathrm{H}}(\theta,\phi)\cdot\boldsymbol{U}\cdot\boldsymbol{U}^{\mathrm{H}}\cdot b(\theta,\phi)} \tag{13}$$

where $\boldsymbol{U}$ is the noise subspace in (12) and $b(\theta,\phi)$ is the steering vector designed in (8).

The corresponding (θ,ϕ) to the maximum point of $P(\theta,\phi)$ is the DOA estimation results of the partial discharge. According to the principle of spatial spectrum DOA estimation, the number of sources K should be smaller than the number of sensors in the array M [15]. Thus, the extended (2M-1)×(2M-1) array could only detect (2M-2) sources at a time.

3.3. Simulation Tests of PD DOA Estimation

Simulation tests based on MATLAB are conducted to verify the feasibility of the proposed method. The double exponential model, added with Gauss noise, is used to simulate the partial discharge signal. Two PD sources are simulated simultaneously, with their DOA coordinates as $(\theta_1,\phi_1)=(-35°,20°)$ and $(\theta_2,\phi_2)=(-40°,55°)$. The propagation speed of the PD signal is set as the speed of light, *c*. The frequency range is 300~3000MHz, with the center frequency as 1000MHz. The wavelength is 30mm and the interval between sensors is 15mm. The sampling frequency is 5GHz and the sampling point is 1024. The SNR of the simulated signal is 10dB.

In the simulation tests, the simulated sensor array is set to be a 3×3 UHF array, which could locate up to two PD sources simultaneously. By applying the array extension method proposed in this paper, the array is extended to a 5×5 virtual array, which could locate up to four PD sources. Fig. 3 shows the DOA estimation with 3×3 UHF array. Detailed localization results are shown in Table 1.

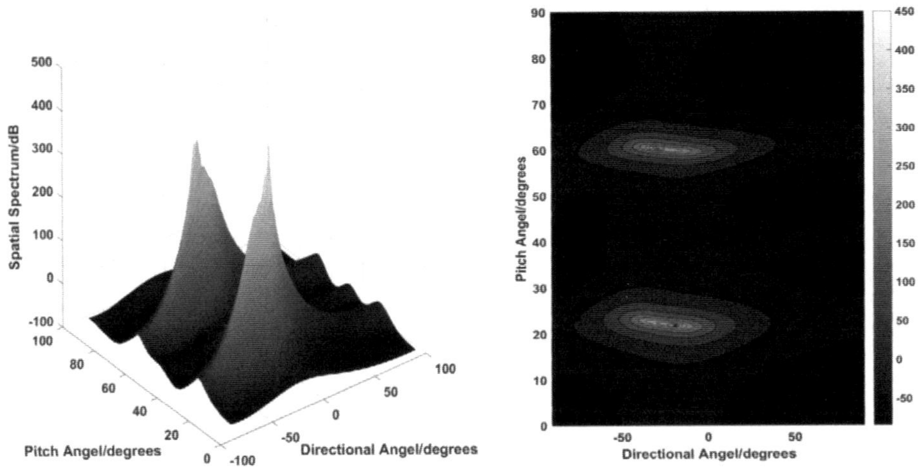

Figure 3. PD DOA Estimation Results with Extended Array.

Table 1. PD DOA Estimation Results with Both Array

Array Model	Simulated PD DOA Estimation		Estimation Error /(°)
	Simulated PD Coordinates (θ, ϕ) /(°)	DOA Estimation Results (θ', ϕ') /(°)	
Simulated Array (3×3)	(-35.0, 20.0) (-40.0, 55.0)	(-25.39, 27.93) (-31.59, 61.20)	9.61, 7.93 8.41, 6.20
Extended Array (5×5)	(-35.0, 20.0) (-40.0, 55.0) (35.0, 70.0)	(-31.57, 22.14) (-36.27, 58.43) (-33.59, 72.93)	3.43, 2.14 3.71, 3.43 -1.41, 2.93

It can be seen from Table. 1 that the DOA estimation error of the 3×3 array is about 8 degrees. Besides, it could only locate up to two PD sources at the same time. By comparison, the DOA estimation error of the extended array is less than 4 degrees. Furthermore, the extended array could locate more PD sources than the actual array. The simulation results have verified the feasibility of the proposed method.

4. Field Tests and Result Analysis

4.1. Experiment Setup

Field tests are conducted to verify the performance of the proposed array extension method in practical application. The PD DOA estimation system is established under actual electromagnetic environment, whose schematic diagram is shown in Fig. 4. The system is composed of the sensor array, the signal pre-processing unit, the UHF signal collector unit and the computer.

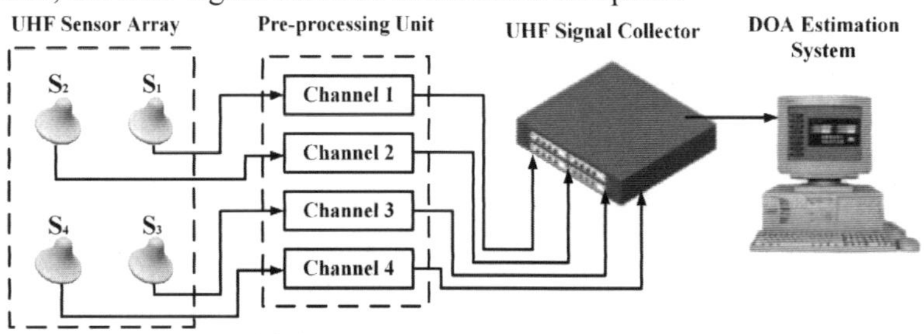

Figure 4. The Multiple UHF Partial Discharge DOA Estimation System.

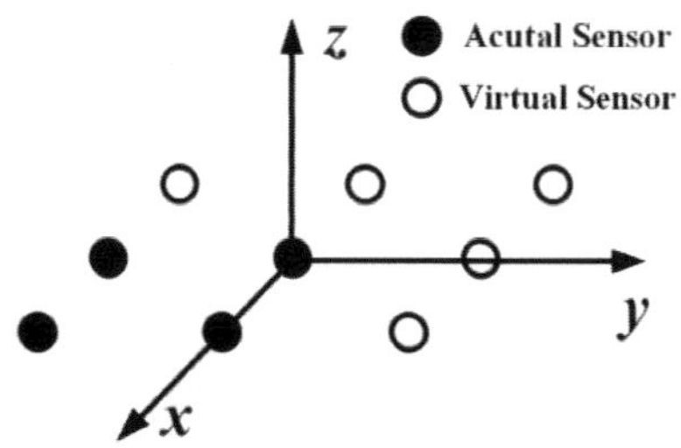

Figure 5. The Schematic Diagram of UHF Array Extension.

As mentioned before, most of the synchronous GHz sampling rate devices have only four channels. Therefore, the UHF sensor array is a 2×2 array and their relative coordinates are shown in Fig. 5. The 2×2 actual UHF array will be extended to 3×3 virtual array by the proposed method.

An EM TEST DITO ESD gun is used to simulate partial discharge. The frequency detection range of the UHF sensor is 300M-1.5GHz. The frequency detection range of the signal amplifier is 300M-1.5GHz. We discharge at different directional angles and pitch angles, while the distance is fixed to 3 meters without changes. The typical two PD sources signal received by the sensor array is shown in Fig. 6 and the localization result is shown in Fig. 7.

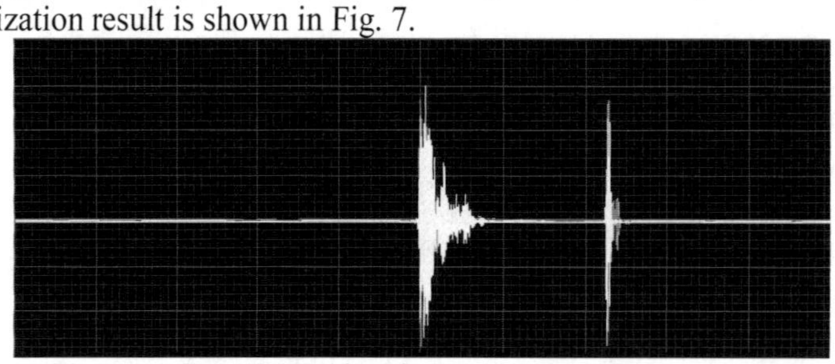

Figure 6. Two UHF Signals Received by the Sensor Array.

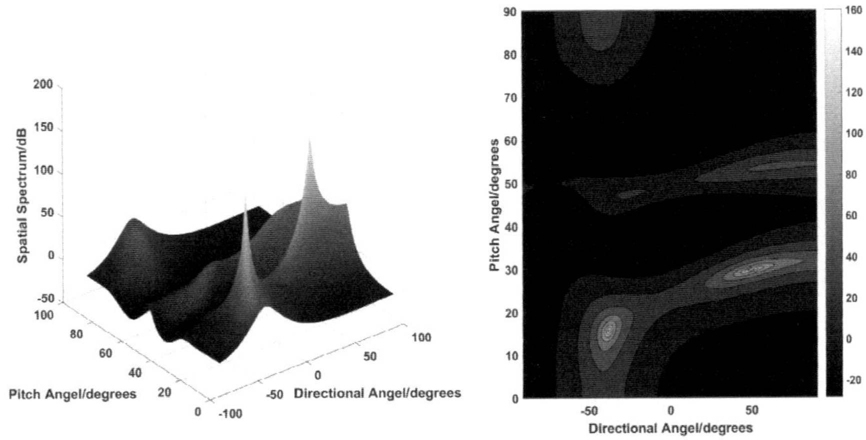

Figure 7. DOA Estimation of Actual PD Sources.

It can be seen from Fig. 7 that the extended UHF array could locate the two PD sources, and the localization results are (-43.19°, 13.30°) and (54.63°, 17.48°) respectively.

To further evaluate the performance of proposed array extension method, several field tests are done and the averaged estimation results are shown in Table 2. For comparison, the DOA estimation of a single PD source with the actual 2×2 array is also shown in Table 2.

Table 2. PD DOA Estimation Results with Both Array

Array Model	No.	Simulated PD DOA Estimation		Estimation Error /(°)
		Simulated PD Coordinates (θ, ϕ) /(°)	DOA Estimation Results (θ', ϕ') /(°)	
Actual Array (2×2)	1	(30.0, 20.0)	(-15.91, 11.38)	14.09, 8.62
	2	(40.0, 25.0)	(-24.46, 13.23)	15.54, 11.77
Extended Array (3×3)	1	(-30.0, 20.0)	(-23.16, 15.78)	-6.84, 4.22
		(30.0, 20.0)	(-24.66, 17.49)	5.34, 2.51
	2	(-40.0, 20.0)	(-33.28, 23.41)	-6.72, -3.41
		(40.0, 20.0)	(-45.28, 17.24)	-5.28, 2.76

As shown in Table. 2, to estimate the DOA of a single PD source with the actual 2×2 array, the estimation error is about 15 degrees. Besides, the 2×2 array could only locate one PD source at a time, which has limited it utilization in practical applications.

By applying the proposed array extension method, the actual 2×2 array is extended to a 3×3 array, which could locate up to two PD sources at a time. More importantly, the extended array could estimate the PD source more accurately, with the directional angel estimation error about 6 degrees and the pitch angel error lower than 4 degrees. The extended array has a better performance in PD localization than the actual array.

5. Conclusion

(1) Aiming at the seldom addressed multiple PD sources localization problem, a multiple UHF PD DOA estimation method is proposed in this paper. The actual UHF sensor array is extended to a virtual array based on the fourth-order cumulant matrix and signal sparse decomposition. Results of

simulation tests and field tests have shown that the proposed method could increase the PD estimation number and simultaneously increase the DOA estimation accuracy.
(2) The steering vector of the PD signal is extended to the overcomplete steering vector based on the fourth-order cumulant matrix. The signal sparse decomposition method is applied to simplify and extend the signal DOA equation. Lastly, the spatial spectrum of the PD signal is established and the PD sources are located.
(3) The proposed method could be applied in other devices such as ultrasound sensor array, which should also increase the accuracy of PD localization.

References

[1] P. J. Moore, I. E. Portugues, and I.A. Glover, "Radiometric location of partial discharge sources on energized high-Voltage plant", *IEEE Trans. Power Del*, Vol. 20, No.3, pp. 2264-2272, 2005.
[2] S. Biswas, D. Dey, B. Chatterjee and S. Chakravorti, "Cross-spectrum analysis based methodology for discrimination and localization of partial discharge sources using acoustic sensors", *IEEE Trans. Dielectr. Electr. Insul*, Vol. 23, No.6, pp. 3556-3565, 2016.
[3] M. Mondal and G.B. Kumbhar, “A technique based on an archimedean copula for the localization of partial discharge in a transformer winding”, *IEEE Trans. Dielectr. Electr. Insul*, Vol. 23, No.5, pp. 2908-2916, 2016.
[4] H. Hou, G. Sheng and X. Jiang, "Robust time delay estimation method for locating uhf signals of partial discharge in substation", *IEEE Trans. Power Del*, Vol. 28, No.3, pp. 1960-1968, 2013.
[5] R. D. Nimmo, G. Callender, and P. L. Lewin, “Methods for Wavelet-Based Autonomous Discrimination of Multiple Partial Discharge Sources”, *IEEE Trans. Dielectr. Electr. Insul*, Vol. 24, No.2, pp. 1131-1140, 2017.
[6] M. X. Zhu, Q. Liu, J. Y. Xue, J. B. Deng, and G. J. Zhang, “Self-adaptive separation of multiple partial discharge sources based on optimized feature extraction of cumulative energy function” *IEEE Trans. Dielectr. Electr. Insul*, Vol. 24, No.1, pp. 246-258, 2017.
[7] M. X. Zhu, Y. B. Wang, Q. Liu, J. N. Zhang, and J. B. Deng, “Localization of multiple partial discharge sources in air-insulated substation using probability-based algorithm” *IEEE Trans. Dielectr. Electr. Insul*, Vol. 24, No.1, pp. 157-166, 2017.
[8] E. Masry, “The Estimation of the Fourth-Order Cumulant for Dependent Data: Consistency and Asymptotic Normality” *IEEE Press*, Vol. 58, No.4, pp. 1991-1998, 2010.
[9] A. Ahmed, Y. D. Zhang, and B. Himed, “Effective nested array design for fourth-order cumulant-based DOA estimation” *2017 IEEE Radar Conference (RadarConf)*, pp.0998-1002.
[10] P. Boaz, and F. Benjamin, “Direction findings algorithms based on high-order statistics”, *IEEE Trans. on Signal Process.* Vol. 39, No.9, pp. 2016-2024, 1991.
[11] B. Jiang, and L. J. Zheng, “Fourth-order music algorithm based on minimum redundant virtual extended array”, *IEEE Trans. on Signal Process.*, Vol. 42, No.6, pp. 517-600, 2006.
[12] B. Stephen and V. Lieven, *Convex Optimization*, Translated by S. N. Wang and X. L. Huang, vol. I. Beijing: *Tsinghua University Press*, 2013, p. 149-153.
[13] C. D. Zheng, G. Li, and Y. M. Liu, “Subspace weighted $l_{2,1}$ minimization for sparse signal recovery”, *EURASIP Journal on Advances in Signal Processing,* Vol. 98, No.1, pp. 1-11, 2012.
[14] S. Cai, “A normalized spatial spectrum for DOA estimation with uniform linear arrays in the presence of unknown mutual coupling”, *2016 IEEE International Conference on Acoustics*, pp. 3086-3090.
[15] O Cholnam, D Kang, and J Cholyong, “Spatial spectrum for direction of arrival estimation of multiple non-stationary wideband sources in the time–frequency plane”, *IET Signal Processing*, Vol. 10, No.9, pp. 1105-1111, 2017

Analysis on Control Stability of STATCOM Applied on Weak Power System

Fang Taixun[1,2], Wang Yu[1] and Xie Yeyuan[1]

[1]NARI-RELAYS Electric Co., Ltd, No.69,SuyuanRoad,Jiangning District, Nanjing City, Jiangsu Province, China
[2]Shanghai Jiaotong University, No. 800, Dongchuan Road, Minhang District , Shanghai, China

Corresponding author's e-mail: fangtx@nrec.com

Abstract—Control stability when Multi-level STATCOM is applied on weak power system is discussed in this paper. In order to research the stability of weak system quantitatively, the current inner open-loop transfer function with system impedance is presented. Nyquist curve is used to demonstrate the relationship between the system impedance and the system control performance. Then a new control strategy - adding the current's proportion-differential (PD) to the feed forward voltage, is proposed. This control strategy can eliminate the positive feedback action caused by system impedance. Finally, test result proves that the above theoretical analysis is correct.

1. Introduction

STATCOM is now used more widely with the development of power electronics and microcomputer technologies. In China it is mainly used for voltage control in different industries, such as wind farm power station, Photovoltaic power station, substation, mill, subway traction [1-3].

In most cases, the power system connected with STATCOM has a large capacity. But some wind farm power stations or Photovoltaic power stations are far from the strong power system. Those stations are connected to weak power system. In practical applications, weak power grids have great influence on STATCOM operating stability, and the following two cases may happen:

i) STATCOM can't start up because of oscillations

ii) STATCOM can start up but may oscillate due to minor disturbance of system voltage.

When STATCOM starts up, the current injected into the grid may cause the voltage collapse. High amplitude oscillation may cause protection trip and converter block. So it is necessary to research the control stability of weak system quantitatively.

2. System modeling

In order to simplify the study, Circuit model includes only the PWM rectifier with output filter, and power grid.

2.1Infinite Grid Model

First, the case of infinite grid without system impedance is researched [4]. Simplified circuit model without considering system impedance is as shown in Fig. 1

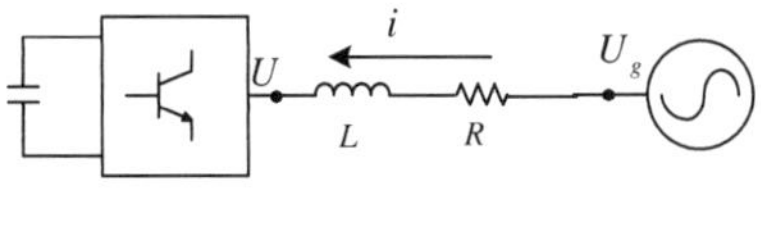

Figure 1. Circuit model without system impedance

Where U_g is Voltage of infinite grid, U is STATCOM output voltage ,R is Resistance of STATCOM interface, L is Inductance of STATCOM interface.

If the system impedance is not considered,a current loop transfer function control diagram using PI regulator is shown as Fig. 2 [5]:

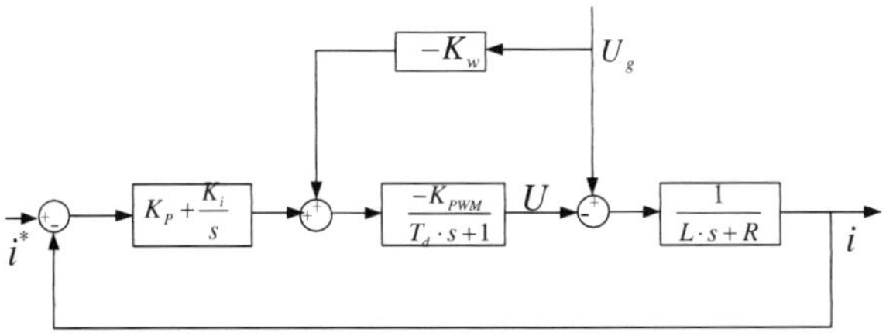

Figure 2. Control diagram without system impedance

Where K_W is Feed forward coefficient, K_P is Proportional coefficient, K_i is Integral coefficient, K_{PWM} is Modulation coefficient, T_d is Control delay time.

Fig. 2 shows the control strategy with feed forward grid voltage. That control strategy is composite control, witch can eliminate the grid voltage's disturbance. Using composite control method, one can improve control system's robust performance [6,7].

2.2Model with System Impedance

Compared with the model without system impedance, the simplified circuit model with system impedance is as Fig. 3.

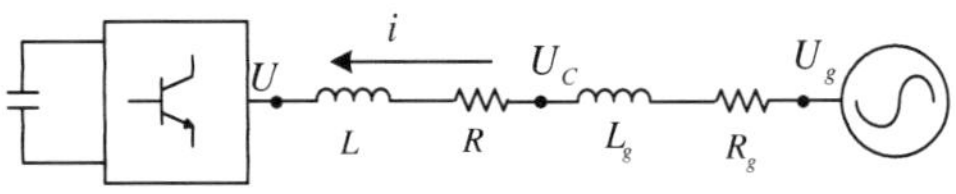

Figure 3. Simplified circuit model with system impedance

Where U_C is Common node voltage, L_g is Inductance of equaling power system, R_g is Resistance of equaling power system. one can get the control diagram with system impedance, as Fig. 4 shows.

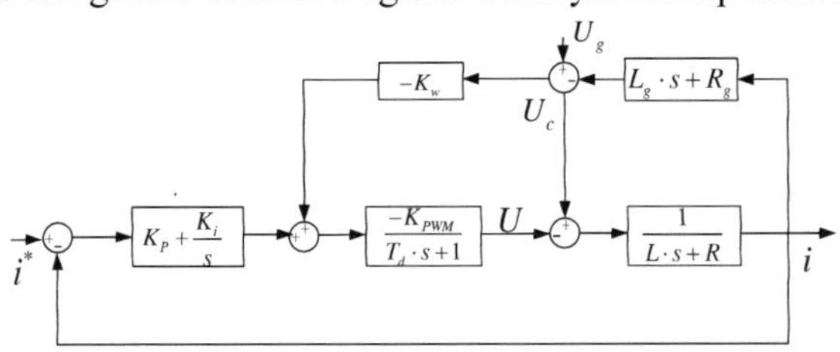

Figure 4. Simplified control diagram with system impedance

Fig. 4 shows that the feed forward part contains a current transfer function, and it is a positive feedback part. Positive feedback part can make the control system less stable or even unstable. There is a delay part within the control system, which is caused by sampling, modulating and communicating. So feed forward part can't eliminate the influence caused by system impedance, instead it introduces a positive feedback part for control system.

IOP Conf. Series: Materials Science and Engineering **366** (2018) 012057 doi:10.1088/1757-899X/366/1/012057

2.3 Proportion-differential (PD) Compensation Model

From the above analysis, one can see that if the system impedance cannot be ignored, feed forward part can produce a positive feedback loop. To solve the above problem, a method with new feed forward part is proposed, which contains a current's proportion-differential (PD) compensation. The aim of adding the PD compensation is to eliminate the effect of the system impedance. So the new feed forward part's transfer function can be described as equation (1).

$$U_c^{'} = U_c + (L^{'} \cdot s + R^{'})i \tag{1}$$

Where $U_C^{'}$ is New feed forward voltage, $L^{'}$ is Estimated resistance of equivalent power system, $R^{'}$ is Estimated inductance of equivalent power system.

Adding the current's proportion-differential compensation to the feed forward voltage, the control diagram with system impedance (as Fig. 4 shows) is changed to Fig. 5.

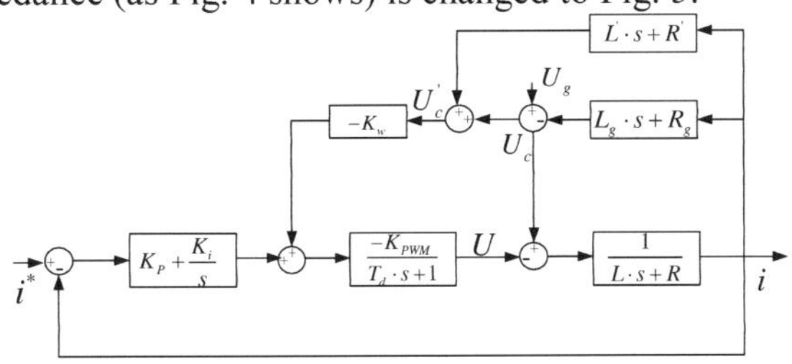

Figure 5. Control diagram with PD compensation feed forward

Under ideal conditions, if $L^{'} = L_g, R^{'} = R_g$, the control diagram can be simplified as Fig. 6.

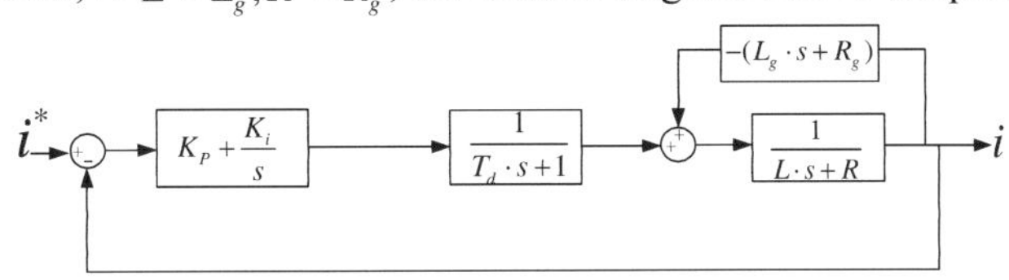

Figure 6. Simplified control diagram with PD compensation

Compared with Fig. 4, PD compensation can eliminate the positive feedback part caused by system impedance. One can get system's open loop transfer function from Fig. 6,as equation (2) shows.

$$G(s) = \frac{K_P \cdot s + K_i}{(L+L_g)\cdot T_d \cdot s^3 + [L + L_g + (R+R_g)T_d]\cdot s^2 + (R+R_g)\cdot s} \tag{2}$$

3. Further analysis and experiment

3.1 Further analysis

In order to verify that the new feed forward part with PD compensation can improve control system stability, nyquist curves of two control strategies are compared. The two control strategies adopt same parameters under same conditions. Matlab tool is used to draw the nyquist curve: Where G1 is the nyquist curve of traditional feed forward control strategy, G5 is that of the prposed new feed forward with PD compensation control strategy.

IOP Conf. Series: Materials Science and Engineering **366** (2018) 012057 doi:10.1088/1757-899X/366/1/012057

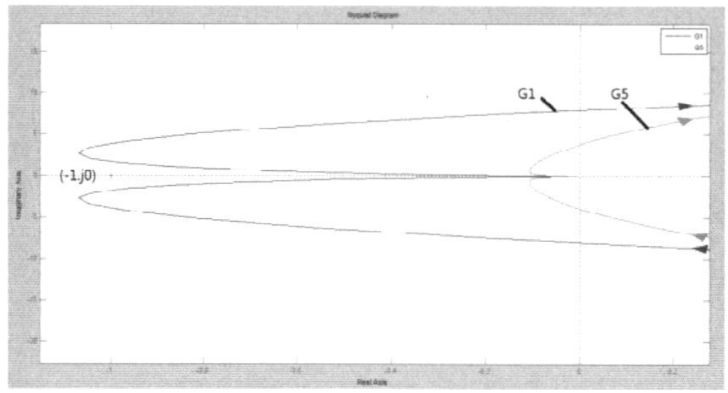

Figure 7. Nyquist curves with different control strategies

Fig. 7 shows that adding the PD compensation, the distance from the curve to the point (-1, j0) increases. It proves that the new feed forward with PD compensation control strategy leads to a larger stability margin than traditional feed forward control strategy.

3.2 Experiment

Take Korla Photovoltaic power station as an example. The previously mentioned when STATCOM starts up, the current injected into the grid causes the voltage collapse. The waveforms have been recorded by Waveform Recorder. Fig. 8 shows the STATCOM output voltage waveform and the STATCOM output current waveform. It causes protection trip and converter block.

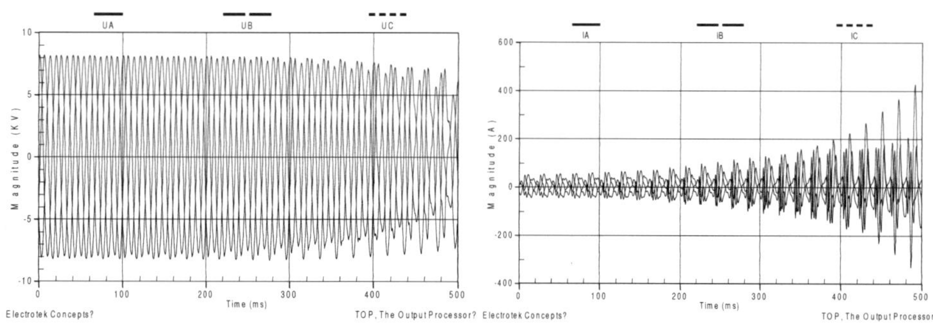

Figure 8. Phase voltage and current waveform of STATCOM (Trip)

To solve above problem, PD compensation is added on the basis of the original feed forward voltage. Fig. 9 shows how PD compensation is added based on the original control strategy.

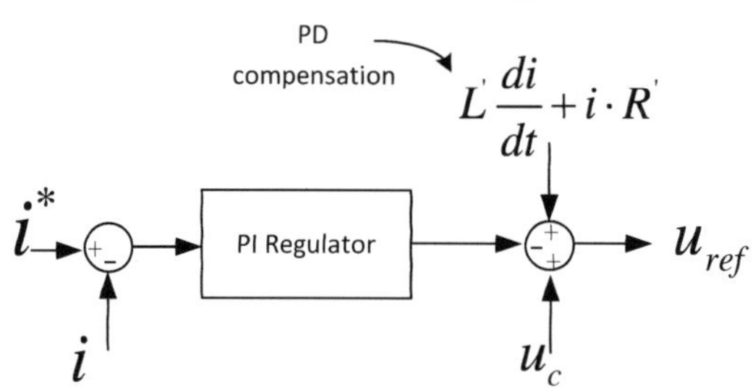

Figure 9. Control Strategy with PD compensation

In practical applications, PD coefficient need be regulated. In this case, differential coefficient K'_L = 0.022, proportion K'_R =0.01. Fig. 10 shows the waveforms after PD compensation is added. From voltage and current waveform of STATCOM, One can see the system can work normally. It proves that PD compensation can improve system stability margin.

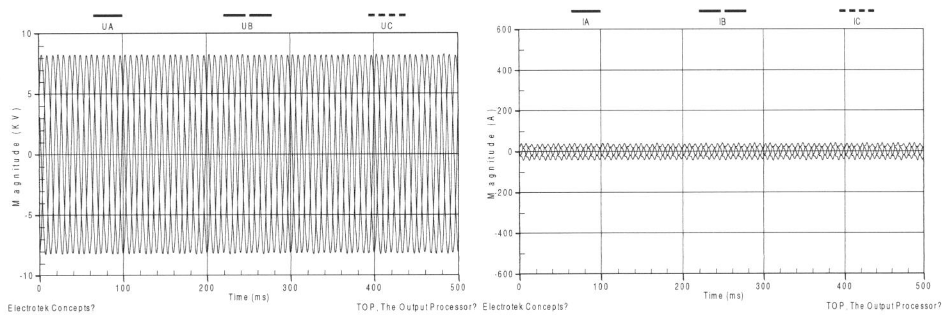

Figure 10. Voltage and current waveform of STATCOM with PD compensation

4. Conclusion

Control stability when Multi-level STATCOM is applied on weak power system is discussed in this paper. In practical applications, weak power grids can cause STATCOM trip. In order to research the stability of weak system quantitatively, Korla Photovoltaic power station is taken as an example, and the current inner open-loop transfer function with system impedance is presented. Nyquist curve are used to demonstrate the relationship between the system impedance and the system control performance. Then a new control strategy is proposed that the current's proportion-differential (PD) is added to the feed forward voltage to eliminate the positive feedback action caused by system impedance. The theoretical analysis and experiment results show that PD compensation can improve stability margin.

References

[1] C.Schauder, H.Mehta. Vector analysis and control of advanced static VAR compensator, IEEE Proceedings, Vol 140, July 1993, pp 299-306

[2] J. Espinoza, and G. Joos, "State Variable Decoupling and Power Flow Control in PWM Current-Source Rectifiers", IEEE Transactions on Industrial Electronics, vol. 45, no. 1, February 1998, pp. 78-87.

[3] J. R. Espinoza, G. Joos, and L. Moran, "Decoupled Control of the Active and Reactive Power in Three-phase PWM Rectifiers based on Nonlinear Control Strategies", 30th Annual IEEE Power Electronics Specialists Conference, 1999, pp. 131-136.

[4] P. Rao, M. Crow, and Z. Yang, "STATCOM control for power system voltage control applications," IEEE Trans. Power Delivery, vol. 15, pp.1311–1317, Oct. 2000.

[5] Zhang Xing, Zhang Chongwei, PWM Rectifier and its Control, ISBN 978-7-111-36822-9

[6] Soto D, Green T C. A Comparison of high-power converter topologies for the implementation of FACTS controllers[J]. IEEE Trans. On Industrial Electronics, 2002, 49(5):1072-1080

[7] XIE Hua-an, WANG Yong-yuan, WAN Si-wei, et al. Research on the control and protection system of Dongguan Substation cascade static synchronous compensator based on RTDS[J]. Power System Protection and Control, 2013,41(4)：117-122

IOP Conf. Series: Materials Science and Engineering **366** (2018) 012058 doi:10.1088/1757-899X/366/1/012058

MVCR Performance Improvement Research Based on ANSYS

LIU Hai-peng and DUAN Shao-mi

Kunming University of Science and Technology, Kunming 650500, China

42227324@qq.com

Abstract. Magnetic valve controllable reactor (MVCR) is a kind of reactive power compensation equipment with good control flexibility and high reliability, which as a result making MVCR being applied more and more widely. Therefore it is crucial to improve the performance of MVCR. In this paper, a method is presented for reducing the fringing effect by dividing one single magnetic valve of MVCR into several series-connected parts, which will reduce the corresponding inductance of AC winding. Consequently the equipment will generate more reactive power with the same working voltage and the control voltage as well as improving its performance. By theoretical analysis and simulation with software ANSYS, an optimal separation of MVCR's iron-core is presented in this paper which calculates the number of magnetic valves and the space length between every two of them. The validity of this design has been fully confirmed by specific experiments that under the same working voltage and control voltage, the current of the AC winding increases by 7.4% when a single magnetic valve is divided into two parts.

1.Introduction

Reactive power compensation device and technology play important roles in the current power system[1] -[2]. Magnetic valve controllable reactor (MVCR) is an important component in the field of static reactive power compensation since it compensates large reactive current flexibly with high reliability[3].

In this paper, an optimum design of MVCR's iron-core is presented based on the conventional structure by dividing the single magnetic valve of MVCR into several series-connected parts, which effectively reduce the fringing effect of MVCR. Consequently the equipments adopted the optimum design will generate more reactive power than before if given the same working voltage and control voltage, and the performance of MVCR will improve as well. By theoretical analysis and ANSYS simulation, an optimal design is presented in this paper which calculates the effective number of magnetic valves and the space length between every two of them and the validity of this design has been fully confirmed by specific experiments.

2.The structure and working principle of MVCR

Figure.1 shows the profile structure of MVCR according to conventional design which is widely used currently. Ⅰ and Ⅱ are the main iron-cores, Ⅲ is the side yoke [4]. It is known that there are several parts in MVCR's iron-core possessing smaller cross section than that of the other parts, and these narrow sections are called magnetic valves. It should be noted that within the adjustment scope of the MVCR's capacity, only these sections with smaller cross section will get saturated, while the other

IOP Conf. Series: Materials Science and Engineering **366** (2018) 012058 doi:10.1088/1757-899X/366/1/012058

sections are in the unsaturated linear state. Each main iron-core is equipped with an AC winding and a DC winding. Where L1, L2 and L3, L4 present the profile of AC winding 1 and AC winding 2 respectively, L6, L7 and L8, L9 present the profile of DC winding 1 and DC winding 2 respectively. The direction of the current at a certain moment in each winding is shown in Figure.1. During every frequency cycle, these two iron-cores become saturated in turn with the alternation of AC flux's direction.

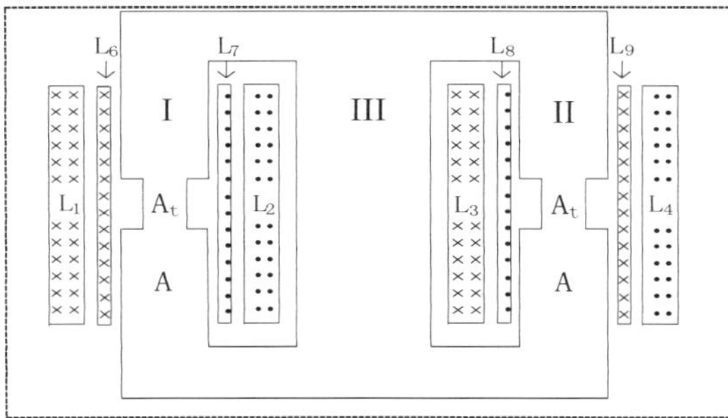

Figure 1. The structure of MVCR .

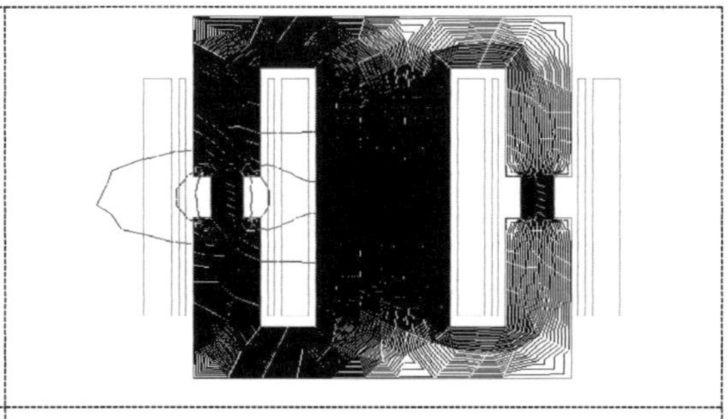

Figure 2. Distribution of magnetic flux lines at the moment of 0.005s.

3.The analysis of MVCR's ac winding output reactive power

When the MVCR's magnetic valve is not saturated, almost all of the magnetic flux lines are within the magnetic valve. In comparison, when the MVCR's magnetic valve is saturated, some magnetic flux lines will overflow into the air around the magnetic valve instead of staying inside. And with the valve getting more and more saturated, the number of flux lines passing through the air will increase correspondingly. Among these flux lines going through the air, some of them are in a straight path and enter the iron-core vertical to its cross section, while the other bend their path in the air thus cannot enter the valve vertically, which makes the collection of them form the fringing flux.

Assuming that the main core Ⅰ of MVCR is in a certain state of saturation and the working voltage of AC winding and control voltage of DC winding are kept constant. According to the principle of magnetic circuit, the total permeance of MVCR's magnetic valve can be expressed as:

$$\Lambda_t = \Lambda_i + \Lambda_a = \frac{\mu A_t}{l_g} + \frac{\mu_0 (A - A_t)}{l_g} \tag{1}$$

Where Λ_t, Λ_i, Λ_a, μ, μ_0, A_t, A, $A - A_t$ and l_g represent the total permeance of the magnetic valve, the permeance of magnetic valve's iron-core and the permeance of the air around the magnetic valve respectively, the permeability of magnetic valve after it becomes saturated, the permeability of the air around the magnetic valve, the cross section of magnetic valve, the cross section of the iron-core, the cross section of the air around the magnetic valve and the length of the magnetic valve.

The inductance of AC winding surrounding the main core Ⅰ can be expressed as:

$$L = N^2 \Lambda_t \tag{2}$$

Where N represents the number of AC winding turns.

4.The analysis of MVCR's ac winding output reactive power

Firstly we build MVCR model in ANSYS according to its conventional single-valve structure which has been illustrated in Figure. 1. Considering the MVCR's rated frequency is 50 Hz. The simulation starts from the moment of 0s and the AC working voltage reaches the first peak of sine wave after 0.005s. At this point, if DC control voltage exists, one of the magnetic valves of MVCR will be saturated. The distribution of magnetic flux lines of saturated magnetic valve is shown in Figure.2 considering there is only one magnetic valve in each AC winding.

As is shown in Figure.2, there are a large number of magnetic flux lines overflowing from the saturated magnetic valve and forming the fringing flux. If there is only one magnetic valve in each AC winding, its magnetic resistance can be expressed as:

$$R_m = \frac{l_g}{\mu A_t} \tag{3}$$

According to (3), if the magnetic valve is divided into several parts, the length of each will become smaller, thus the magnetic resistance of each also become smaller. And this change of resistance will let more magnetic flux lines pass inside the iron-core of magnetic valve instead of the air around, which makes the fringing flux decrease. As a result, the corresponding magnetic flux density inside magnetic valve, as well as the saturation degree of magnetic valve increases.

Assuming that one magnetic valve is divided into n parts of equal length, the magnetic resistance of each part can be expressed as:

$$R_{mn} = \frac{l_g/n}{\mu_n A_t} \tag{4}$$

Where μ_n represents the permeability of each magnetic valve parts after being divided.

Assuming that each of them will still be regarded as an independent magnetic valve and they are connected in series. So the total reluctance of these n magnetic valves can be expressed as:

$$R_{mt} = nR_{mn} = \frac{l_g}{\mu_n A_t} \tag{5}$$

When taking the permeance of air around each magnetic valve into consideration, the total permeance can be expressed as:

$$\Lambda_{tt} = \Lambda_{it} + \Lambda_{at} = \frac{\mu_n A_t}{l_g} + \frac{\mu_0 (A - A_t)}{l_g} \tag{6}$$

Where Λ_{it} and Λ_{at} represent the total permeance of magnetic valve and the total permeance of the air around magnetic valve.

According to the working principle of MVCR and the B-H curve of iron-core, we know that when the magnetic valve is divided, the saturation degree of each magnetic valve will increase, consequently μ_n will be less than μ mentioned in (1) where there is only one magnetic valve in each AC winding. As a result the total permeance Λ_{tt} calculated above is less than the total permeance Λ_t calculated in (1).

Therefore, in order to reduce the fringing magnetic flux of MVCR, one single magnetic valve can be divided into several equilong magnetic valves in series. In addition, the length of the several magnetic valves mentioned above summing up together should equal to the length of one single magnetic valve in order to keep the MVCR's capacity unchanged.

5.The different division of MVCR magnetic valve and the optimal settings

A three-dimensional model of MVCR was built by ANSYS in order to simulate and calculate the AC winding inductance of MVCR accurately. The length of the iron-core and the side yoke is 0.25m. The length of the magnetic valve is 0.04m. The width of the iron-core is 0.05m. The width of the magnetic valve is 0.025m. The width of the side yoke is 0.1m. The length of the upper yoke and the lower yoke is 0.28m. The width of the upper yoke and the lower yoke is 0.05m. AC and DC windings are racetrack coils.

In this model, the voltage of AC winding, the frequency and the voltage of DC winding is set as 220V, 50Hz and 10V respectively. The method of transient analysis is applied in the simulation since it includes AC voltage and DC voltage at the same time. And in order to simulate actual working condition of MVCR, the field circuit coupled method is used to simulate the external AC and DC voltage.

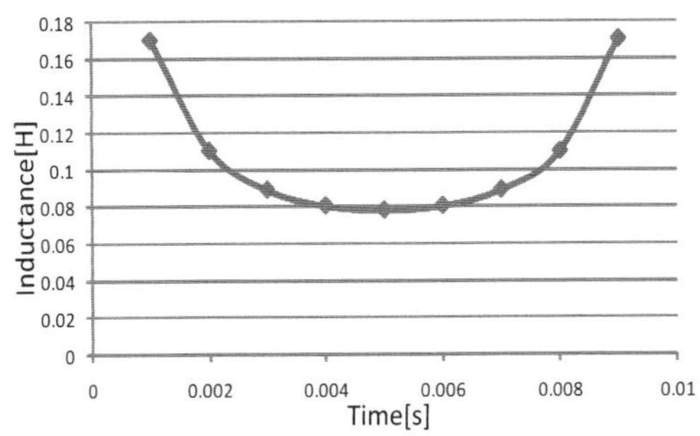

Figure 3. The inductance curve of winding 1 with single magnetic valve.

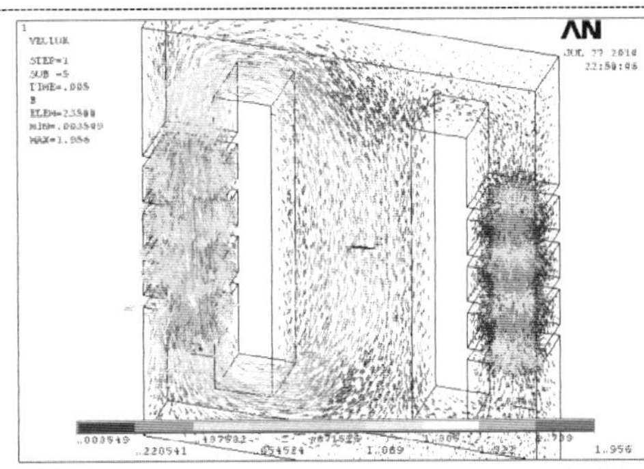

Figure 4. The flux density when every single magnetic valve is divided into 4 parts at the moment of 0.005s, the space length between two adjacent magnetic valve is 0.03m.

Firstly, set the structure of MVCR to be single magnetic valve and the length of the magnetic valve to be 0.04m. The corresponding inductance of AC winding 1 at different moment is shown in Figure.3. According to Figure.3, we know that the minimum inductance of winding 1 is gained at 0.005s. And at this moment, the current of winding 1 is maximal, the value of magnetic flux inside iron-core which is surrounded by winding 1 is the largest and MVCR's saturation degree arrives at the peak point. It can be observed that the inductance of winding 1 decreases during the period of 0.001s to 0.005s, and then increase symmetrically during the period of 0.005s to 0.009s, making the moment of 0.005s a symmetrical point. During the period of 0.011s-0.019s, the inductance of AC winding 2 will be the same with what is shown in Figure. 3.

Then the model of MVCR with every single magnetic valve divided into four equilong parts is built by ANSYS. Figure. 4 shows the flux density at 0.005s with the space length between two adjacent magnetic valves being 0.03m.

When one magnetic valve is divided into n parts, the fringing effect of the flux going across two adjacent parts can be reflected by the permeance of the space around the edges of magnetic valves, which is approximately expressed as follows[5]:

$$\Lambda_{fe2} = \mu_0 D(1+\ln\left(\frac{\pi(2l')}{2l_g{}'}\right)) \tag{7}$$

Where D represents the diameter of iron-core of which the cross section is A, l' represents the length of iron-core. It should be noted that l' is expressed as (11), which is shown in figure 5.

$$l' = \left(H_f - nl_C - l_g\right)/2 \tag{8}$$

Where H_f, l_C and l_g represents the space length between the upper and the lower iron yokes, the space length between two adjacent magnetic valves and the total length of magnetic valves.

$l_g{}'$ represents the space length of two adjacent magnetic valves parts after divided, which can be expressed as:

$$l_g{}' = l_C + \frac{2}{n} l_g \tag{9}$$

According to (7), when the space length between two adjacent magnetic valves gradually increases, the permeance Λ_{fe2} will gradually reduce and finally comes to 0, which means the magnetic flux lines crossing two or more magnetic valves will gradually disappear. Therefore the corresponding space length acquired at that zero point is regarded as the minimum optimal value. Only when the length is larger than this minimum value will the fringing flux be minimal. Assuming the minimal optimal space length mentioned above is l_{Co}, it can be expressed as:

$$l_{Co} = \frac{(\pi e/2)H_f - (\pi e/2 + 2/n)l_g}{1+(\pi e/2)n} \tag{10}$$

In addition, it can be concluded according to the simulation that when the magnetic valve is divided into 5 parts or more, the change of the inductance is so insignificant that it can be ignored. Taking the manufacturing cost of iron-core into consideration, it would be more appropriate to divide magnetic valve into 4 sections or less. The corresponding inductance of AC winding 1 is shown in Figure.6 when one magnetic valve is divided into n parts with the space length being the minimum optimal value respectively. where n ranges from 2 to 5.

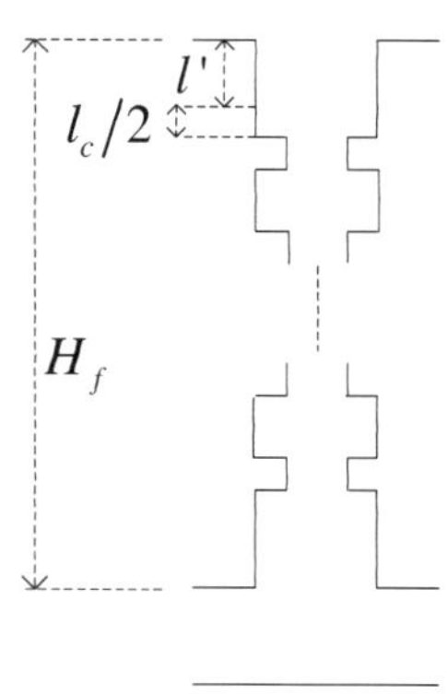

Figure 5. Schematic diagram of iron-core

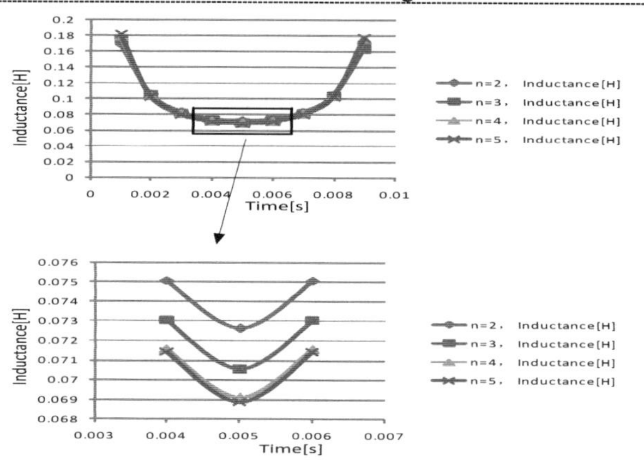

Figure 6. Curve of AC winding inductance with every single magnetic valve divided into n parts.

By means of detailed calculation, we came to the following conclusion: after dividing the magnetic valve into 2 to 5 parts, the minimal inductance decreases by 7%, 9.7%, 11.5%, 11.8% respectively comparing to that before dividing.

6.Experimental verification

In order to confirm the theory mentioned above, two MVCRs of the same size are manufactured to carry on specific experiments. One MVCR has only one magnetic valve in each AC winding and the length of them is 0.04m, with the rated voltage and current being 220V and 10A respectively. The other MVCR has two magnetic valves in each AC winding and the length of them is 0.02m. Additionally, the space length between the two valves is 0.09m, which equals to the minimum optimal length. The current in AC winding 1 was measured, which is shown as the waveform in Figure.7. It can be observed from that during the positive half-cycle, the current reached its peak point and had larger amplitude since the iron-core is saturated. In comparison, the amplitude of current is smaller during the negative half-cycle since the core is unsaturated. And the saturation degree of single-valve MVCR is $\pi/2$ under this condition.

IOP Conf. Series: Materials Science and Engineering **366** (2018) 012058 doi:10.1088/1757-899X/366/1/012058

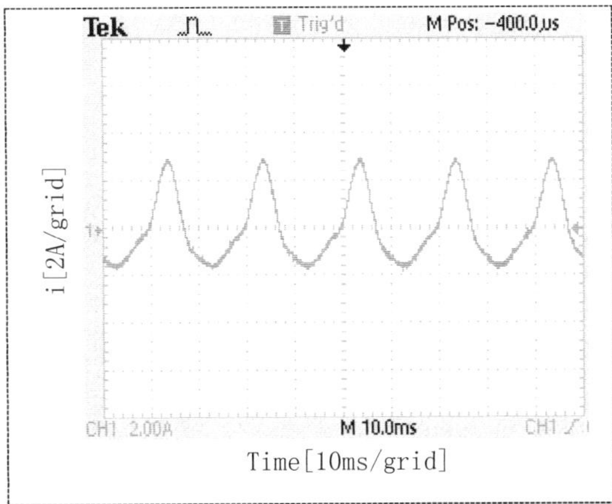

Figure 7. The waveform of current flowing inside one single AC winding.

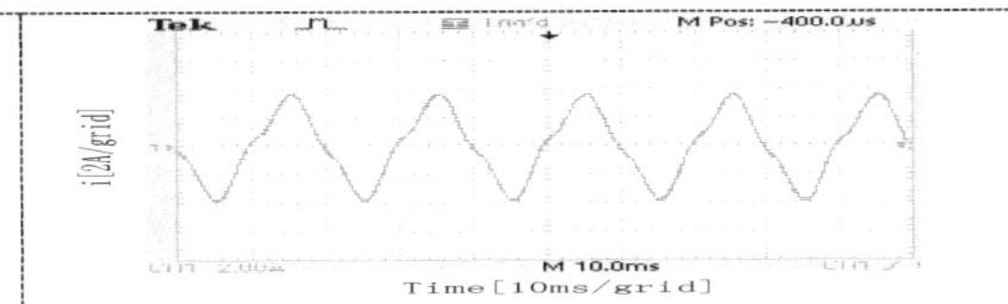

(1) Waveform of current generated by single-valve MVCR

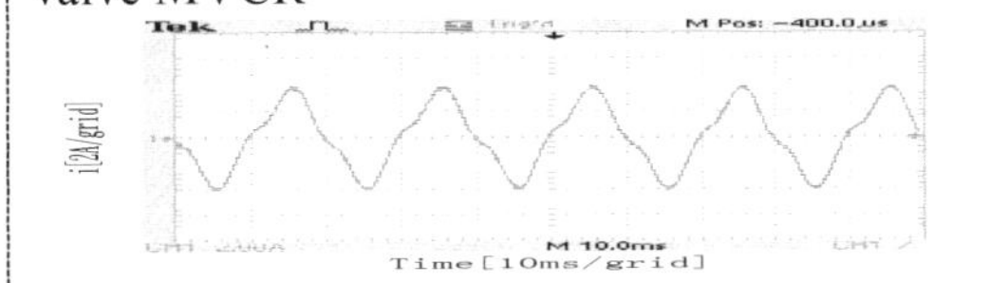

(2) Waveform of current generated by two-valve MVCR

Figure 8. Waveform of current generated by individual MVCR.

When the two MVCR are saturated given the same AC voltage and DC voltage, the reactive current generated by two-valve MVCR was 7.4% more than that generated by single-valve MVCR. The waveforms of current are shown respectively in Figure.8, from which can be observed that the amplitude of current generated by two-valve MVCR is slightly greater than that by single-valve MVCR under the same working condition. And the saturation degree of MVCR with single magnetic valve is $\pi / 2$.

7.Conclusion

In this paper, an optimum design of iron-core of MVCR is presented including the method of dividing magnetic valves and the optimal space length between two adjacent valves. Initially keeping the total length of magnetic valves of MVCR unchanged, the experiment results showed that by dividing the single magnetic valve into several series-connected parts and adopting the optimum space length, it can effectively reduce the fringing magnetic flux and cut down the inductance of the AC winding of MVCR. Therefore the MVCR will generate more reactive current under the same working and control voltage, which improving the performance of MVCR. In this paper, the conclusions mentioned above are well elaborated by mathematical analysis and ANSYS simulation, and optimal design equations are given as well. At the same time, the reducing of the MVCR's fringing magnetic flux can also reduce the stray loss. It has a positive significance for decreasing MVCR's loss.

References

[1] B. Leonardo, A. Sérgio, and G. Alessandro, "Application of shunt active power filter for harmonic reduction and reactive power compensation in three phase four-wire systems," IET Power Electronics, VOL. 7, ISS. 11, pp. 2825-2836, MAY 2014.

[2] B. Leonardo, A. Sérgio, and G. Alessandro, "Application of shunt active power filter for harmonic reduction and reactive power compensation in three phase four-wire systems," IET Power Electronics, VOL. 7, ISS. 11, pp. 2825-2836, MAY 2014.

[3] H. Qiu, and M. Yu, "Supervisory Control System of a Novel 110 kV Magnetically Controlled Reactor," High Voltage Apparatus, Vol. 46, No. 6, pp. 69-73, Jun. 2010.

[4] B. Wang, L. Jin, Y. Luo, et al., "Dynamic reactive power compensation equipment based on magnetically controlled reactor," Electric Power Automation Equipment, Vol. 30, No. 4, pp. 97-100, Jun. 2010.

[5] A. Balakrishnan, W. Joines, T. Wilson. "Air-gap reluctance and inductance calculations for magnetic circuits using a Schwarz-Christoffel transformation," IEEE Transactions on Power Electronics, Vol. 12, No. 4, pp. 654-663, Jul. 1997.

IOP Conf. Series: Materials Science and Engineering **366** (2018) 012059 doi:10.1088/1757-899X/366/1/012059

Study on Unified Reference Source Phase-Check Technology Based on Synchronized Phasor Measurement

Pei Li[1], Hongbo Li[1], Ning Kong[2], Nannan Gao[1], Lin Niu[1], Ying Pei[1] and Liangfeng Guo[1]

[1] State Grid of China Technology College, 500 Erhuan Nan Road, Jinan, Shandong;

[2] Heze Electric Power Supply Company of National Power Grid, Heze, Shandong

287322937@qq.com

Abstract. In the power system, the phase-check is a simple but very important work. The newly built, transformed and expanded power plant, transformer substation electrical equipment and power transmission lines before put into operation must be carried out phase-check test, in order to ensure voltage phase sequence and phase correct. At present, there are some limitations of traditional phase-check technology in China. In this paper, we study the method of unified reference source based on synchronized phasor measurement technology. Through the GPS time synchronization signal to realize synchronous sampling of source voltage signal in different regions, by comparing to the phase of a fixed reference source signal in the power network to determine the phase relationship between any two different power points, and thus to complete the work of phase-check. The unified reference source phase-check method based on the synchronized phasor measurement technique is simple, easy to operate, and has high efficiency, which has unique advantages. And wide area power grid phase-check can be realized.

1. Introduction

With the concept of the global energy Internet,As The development of power grid planning in 13th Five-Year,We should speed up the problems of UHV and distribution network and enhance the energy optimization and allocation capability. For this reason, The State Grid is expected to build the Eastern and Western synchronous networking projects by 2025. During the "13th Five-Year" period, it will accelerate the upgrading of synchronous power grid integration. The "Four AC Four DC" UHVDC projects that is integrated into the national air pollution control action plan will be completed in 2017. By 2020, the Eastern and Western UHV main network frame will be formed. The new transformation and upgrading of the agricultural power grid and the investment in the urban distribution network have a total of 128 billion 100 million yuan. The task of building and reforming the power grid is very heavy. High strength and large scale power grid construction will be through the "13th Five-Year" period[1].

The phase-check test must be done before power plants, substations and transmission lines are put into operation when they are built, reconstructed and expanded, in order to ensure that the three-phase phase sequence of the power equipment and the transmission line is the same and the phase is correct. The phase of the two three-phase power supply should be ensured for the three-phase AC power system which needs parallel or closed loop operation[2]. If the phase sequence is not consistent, the generator will not be connected to the power grid. The forced integration of the power grid will cause serious damage to the power generation equipment. The phase-check test is a basic and important work in power system. Because of the new-build and expansion of substations every year, Power supply enterprises invest a lot of manpower and material resources in nuclear power. Therefore, it is a very meaningful thing to study how to improve the efficiency of the phase-check test, reduce the cost, and reduce the risk. The main limitations of traditional phase-check test mainly include the limitation of space distance for phase-check equipment, and simultaneity of signal sampling must be satisfied, and it also takes time to operate[3].

The aim of this paper is to improve the efficiency of phase-check test in substations, In view of the problems of low efficiency, complex operation and large limitation in traditional phase-check method, the technology and method of unified reference source phase-check based on synchronous phasor measurement technology are proposed. In order to improve the efficiency of phase-check and to simplify the operation process. At the same time, when the GPS unified clock is used, the phase-check method is not limited by the space distance of three-phase power, and it can solve the single power phase-check problem in substation well, so it has high popularization and application value.

2. Current methods, principles and limitations of phase-check

2.1 The Phase-check Method and Principle from Primary System

A phase-check from primary system is usually composed of two parts: a phase-check instrument hand-held terminal and a high voltage phase-check probe. As shown in Figure 1, the principle diagram of the phase-check from the high voltage primary system is shown. Wireless communication is usually used to transmit data between high voltage probe and phase-check instrument handheld terminal. Because the handheld terminal of the phase-check instrument requires high real-time sampling data, and the distance of data transmission is not the same in different wireless communication modes. Therefore, it is required that the distance between the high voltage probe 1 and the probe 2 of the phase-check meter should not be too far away, and should be within the range of the technical requirements of the phase-check instrument. Otherwise, the real time data sampling and the distance of data transmission will not be met.

The handheld terminal makes a comparison of the phase and amplitude of the data after transmitting the sampled data detected by the high voltage probe 1 and 2 to the handheld terminal, then the phase-check of 1 and 2 of the high voltage power supply is completed.

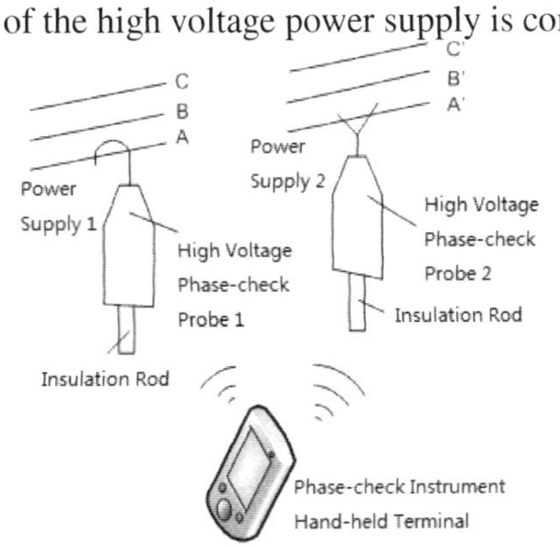

Figure 1. The Principle Diagram of Phase-check from Primary System

2.2 The Phase-check Method and Principle from Secondary System

The phase-check from secondary system generally uses the phase-check meter in the secondary side of the voltage transformer in substations. The principle diagram of the phase-check from secondary system is shown in Figure 2. The phase-check from secondary system is a kind of indirect phase-check method, and the correctness of the results of the phase-check is based on the correctness of the primary and secondary side connection of the voltage transformer TV1 and the TV2. It is necessary to guarantee phase congruency about High and low voltage side voltage U_A and U_a、$U_{A'}$与$U_{a'}$ of two voltage transformer. In this way, the accurate phase-check of the two power supply can be carried out indirectly.

The phase-check from secondary system usually uses two sets of test lines to simultaneously connect voltage transformers TV1 and TV2, then pass its secondary voltage to a phase-check instrument G. A phase-check instrument is used to do phase-check tests from secondary system on two power sources directly. Therefore, the phase-check method requires that the secondary lead terminal distance between TV1 and TV2 which is connected with the primary system cannot be too far away. Otherwise, the test towing line will be too long, the phase-check is very inconvenient, or the phase-check cannot be completed at all.

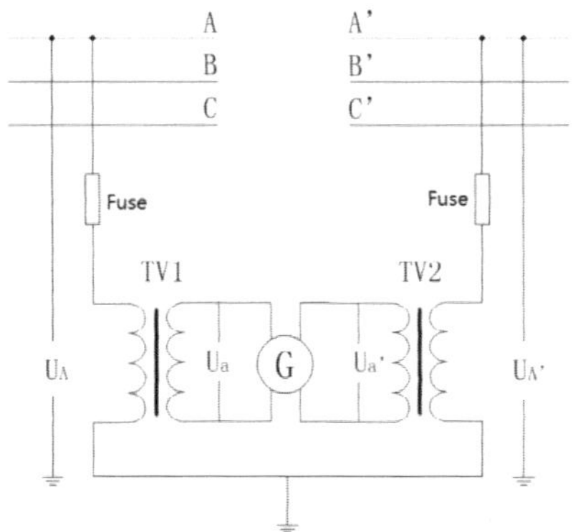

Figure 2. The Principle Diagram of Phase-check from Secondary System

2.3 The Advantages and Limitations of Traditional Phase-check Method

There are some limitations in the analysis of the principle of traditional phase-check method and whether it is a method from primary system or that from secondary system.

(1) The signal of two power source that needed phase-check must be collected at the same time to keep the synchronization of the AC voltage signal sampling.

(2) The space distance between the two power- sources needed phase-check is not too far away, and the phase-check of the wide area power grid cannot be realized.

(3) The phase-check tests need to be involved in more people, and the steps of the phase-check are complicated, the cost of human and material resources is high and the time consuming is long.

(4) There is a certain danger in the phase-check from primary system test. It is necessary to consider the coordination and cooperation between the experimental personnel and take corresponding safety precautions.

To sum up, it is known that the traditional phase-check method has many limitations, and it needs to be improved to improve the efficiency of phase-check.

3. Phase-check Method Based on Synchronous Phasor Measurement

In order to realize the wide area power grid phase-check, a phase-check scheme based on GPS synchronous phasor measurement technology is proposed. The principle diagram of this design idea is shown in figure 3. The power supply 1 and the power 2 are those which will be done phase-check. In order to ensure the phase sequence and phase relation between power source 1 and power source 2. A power point in the synchronous grid is introduced as a unified phase-check reference source for the whole power grid. At the power point, the sampling host computer with reference source voltage signal is installed. The sampling host computer has the function of the network communication and the synchronization of the synchronization pulse of the satellite GPS. The synchronous sampling of the reference source voltage signal can be carried out according to the GPS pulse signal. And the data information is stored on the local hard disk and network server. At the same time, the voltage signal is sampled by the portable handheld phase-check instrument 1 and 2 for the two power supply 1 and 2, respectively. The handheld phase-check instrument 1 and 2 have the same functions as the reference source sampling host computer. At the same time, it can be used to sample the sampling data of the host sampling host and perform the function of comparison and analysis. Through the analysis of the sampling signal of the power point voltage, the voltage phasor relationship between 1 and 2 of the power points based on the unified reference source is obtained.

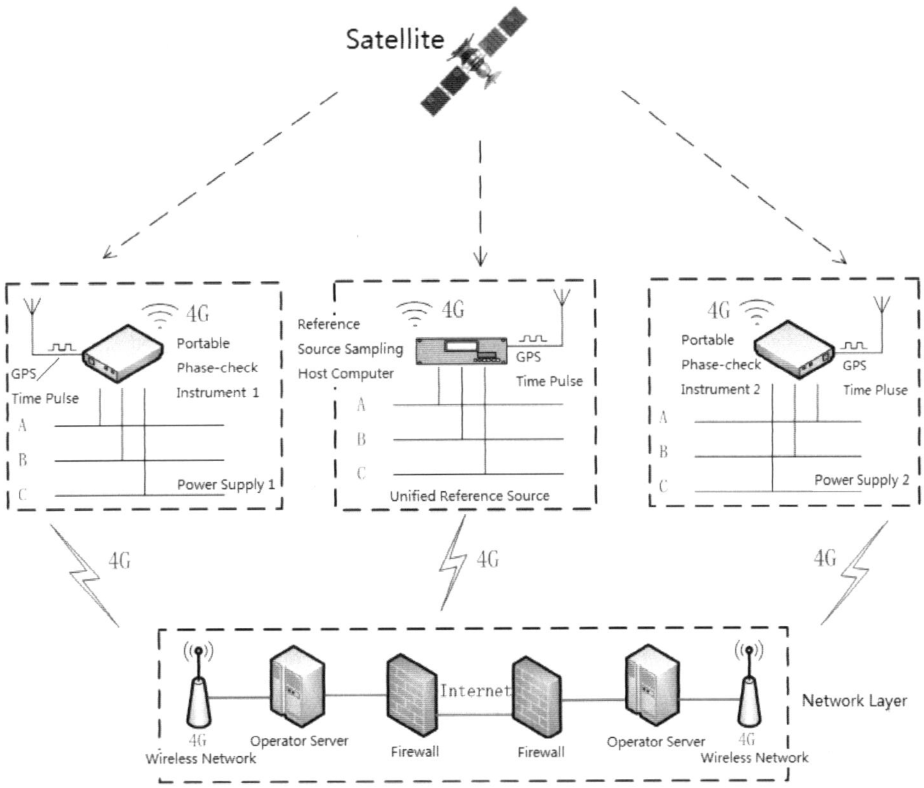

Figure 3. The Principle Diagram of Phase-check Instrument Based on Unified Reference Source

As a result of high precision real-time signal with global positioning system (GPS), the synchronous sampling of the wide area power point of the power system can be realized, and there is a unified time standard for the voltage measured by each node, the corresponding time scales are added to the sampling data of voltage signals at different locations. The comparison and analysis of the voltage phase of different power supply points in the synchronous grid is realized. And finally the phase-check work of power point 1, 2 is achieved[4-7].

At present, GPS is used to build the whole network in the same time, 99.87% of the time synchronization error can be accurate to 1s. For the 50Hz power system, the phase error will be less than 0.018 degrees, providing an accuracy guarantee for phase-check work[8]. At the same time, high precision, high speed A/D converter and DSP digital signal processor can be used to solve the problem of synchronous sampling in wide area power grid.

4. The structure and hardware of Phase-check Instrument Based on Unified Reference Source

4.1 The Structure of the Hardware System of the Phase-check Instrument

The core equipment in the unified reference source phase-check instrument is the base source sampling host computer and the portable phase-check terminal. They basically have the same function, and the hardware structure is basically the same. The schematic diagram of its hardware structure is shown in Figure 4. The core panel is composed mainly of a high-speed digital signal processor DSP+FPGA. It also has USB, RS485 data interface module, TCP/IP network, 4G wireless communication interface module, GPS timing and B code timing module, remote switch on and voltage sampling A/D conversion module, and display module, keyboard and man-machine interface module and other expansion modules. The GPS time module realizes the output of the time pulse signal, and then produces a high precision analog synchronous sampling clock after the processor's own crystal oscillator completes the frequency division. A/D analog conversion module completes synchronous sampling of voltage signals based on synchronous sampling clock. Finally, the DSP digital signal processor is used to realize the accurate calculation and processing of the voltage quantity digital signal, and then output the nuclear phase result to the user through the human-machine interface device.

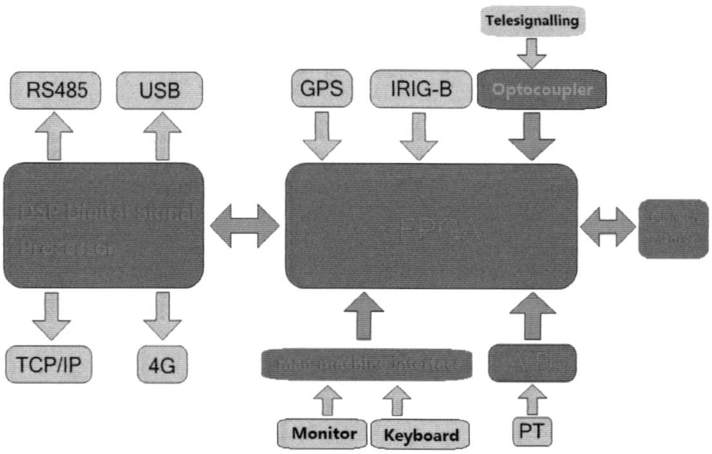

Figure 4. the Schematic Diagram of the Hardware Structure of a Unified Reference Source Phase-check Instrument

4.2 The Design of the Hardware Module of the Phase-check Instrument

The core device hardware of phase-check detector is mainly composed of DC power module, digital signal processor module, analog acquisition module, GPS timing module, display and keyboard man-machine interface module, as shown in figure 5.The DC power supply module provides power supply for signal conditioning circuits, analog to digital conversion circuits, and DSP digital signal processors. There are 4 power output, 5V, 24V, +12V, -12V, respectively. Voltage transformer, operational amplifier IC and 14 bit MAX1320 analog to digital converter and other components constitute a voltage signal sampling circuit, which completes the synchronous sampling of the power voltage signal to the digital quantity. The TMS320F206 DSP digital signal processor and FPGA integrated circuit can measure the voltage amplitude and phase angle based on the DFT algorithm, and calculate the voltage phase amount.

Figure 5. Hardware Module of Unified Reference Source Phase-check Instrument

5. Unified Reference Source Phase-check Instrument System Experiment

In order to verify the performance of the whole system and its core equipment, a phase-check test project is designed. Feasibility and superiority of phase-check scheme based on GPS synchronous phasor measurement is achieved.

A phase-check test was carried out about high and low voltage side power supply for a main transformer in an 110kV substation. The connection of main transformer winding in this substation adopts Yd-11 connection mode. Primary side voltage phasor delay secondary voltage side voltage phasor of about 30 degrees. The reference source sampling host computer sampling the voltage of high voltage side bus is installed at the high voltage bus bar of the substation. And the voltage is used as the reference source of the phase-check. The portable phase-check terminals 1 and 2 synchronously sample the high and low voltage side voltage of the main transformer, respectively. The voltage phase of the high and low voltage side of the main transformer based on the voltage of the high voltage side bus is obtained. The results are shown in Table 1.

Table 1. Phase-check Result of Power Supply Voltage

Supply Voltage	Phase	Amplitude(V)	Phase Angle(°)
110kV Bus Voltage	Phase A	57.84	0.0
	Phase B	57.75	-120.0
	Phase C	57.79	120.0
Main Transformer 110kV Side Voltage	Phase A	57.83	0.0
	Phase B	57.79	-120.0
	Phase C	57.69	120.0
Main Transformer 10kV Side Voltage	Phase A	57.80	29.85
	Phase B	57.72	-120.0
	Phase C	57.76	120.0

The voltage recording of portable phase-check terminals 1 and 2 is shown as shown in Figure 6, The T1 and T2 in the diagram are absolute and relative time, respectively. The time difference between the left and right two scales from the graph (1.655ms) and it can be calculated that the phase angle difference between the primary voltage phase and the secondary voltage of the main transformer. It is about:

$$1.655\text{ms}\times 18\ {}^{\circ}/\text{ms}=29.79\ {}^{\circ} \quad (1)$$

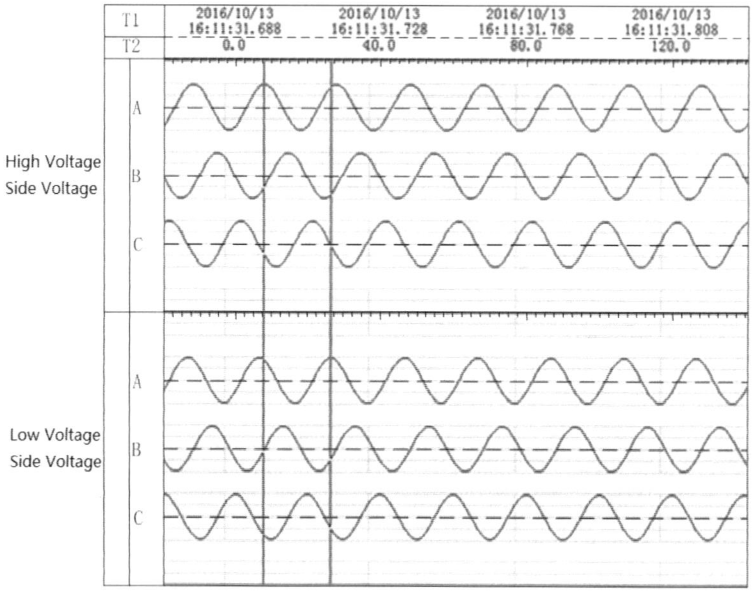

Figure 6. Recorded Diagram from Unified Reference Source Phase-check Instrument

6. Conclusions

This paper applies the synchronous phasor measurement technology based on GPS to the power system. The power phase-check of the wide area power grid is realized. It can make the work of the power system phase-check easier and simpler, and meet the needs of a single power supply substation high-voltage power phase-check work. It has made up many defects and shortcomings of the traditional phase-check, and greatly improved the efficiency of the phase-check of the professional and technical personnel of the power supply enterprises.

Reference

[1] Liu zhenya. Global Energy Internet. China Electric Power Press, 2003.
[2] Wang Genshui,Pan Lei,Zhan Linsong. Technical evolution and analysis and comparison of power network phase-check. High technology and industrialization,2009,5(2):80-82.
[3] Du Wenwei, the study of generalized nuclear phase technology and its application in the operation of Pudong power grid. Shanghai: Shanghai Jiao Tong University, 2013.
[4] Zhang Chun. Key technology research and engineering implementation of remote wireless phase-check system. Hangzhou, Zhejiang: Zhejiang University, 2006.
[5] Hu Shaoqian. Research on synchronous phasor measurement device of power grid based on GPS. Measurement and Control Technology and Equipment. Xi'an, Shanxi: Xi'an Jiao Tong University.
[6] Hu ShaoQian, Wang Xiaoru. A synchronous phasor measurement device for power system based on GPS Technology. Measurement and Control Technology and Equipment, 2003,29(11):21-23.
[7] Kong Wuyang. Research on synchronous phasor measurement device of power grid based on GPS. Nanjing, Jiangsu: Hohai University,2007.
[8] Hofmann-Wellenhof B,Lichtengegger H,and Collins J . GPS,Theory and Practice[C]. Springer-Verlag Wien New York,2003.

A Top-oil Temperature Prediction Method for the Split Transformer Based on PSOGSA

XinZi[1], Wei Bengang[1], Liang Yongliang[2], Xu Yanshun[2], Guo Ruochen[2] and Lou Jie[2]

[1]Electric Power Research Institute, SMEPC, 171# Handan Rd., Shanghai, China
[2]School of Electrical Engineering, Shandong University,27# South Shanda Rd., Jinan, China

391818095@qq.com

Abstract. Because of the complicated structure of cooling system, the heat radiation principle and top-oil temperature prediction of the split transformer is attracting more and more attention. This paper proposes a top-oil temperature prediction method for split transformer based on PSOGSA. Based on the thermal circuit of ordinary transformers, a thermal circuit for split transformers is proposed taking the heat radiation principle into account. Besides, the model is modified with the consideration of the non-linear heat resistance, load ratio and oil viscosity. Further, the parameters of the model are optimized by PSOGSA. Finally, the method proposed is tested by on-site temperature-rise experiment. The test results prove that the method could provide a high-precision prediction result of the top-oil temperature for split transformer.

1. Introduction

With the development of economy and urban construction, more and more split transformers have been used in the cities because of excellent structure. Now most underground substations which have large-scale application of split transformers concentrate in Beijing and Shanghai and other major cities [1-2]. However, as the load increases, the issue of transformer heat radiation has become increasingly prominent. Because of the complicated structure of split transformer and its cooling system, the key components of the heat transfer mode and heat radiation principle remains to be further studied. So it is of great significance to research the thermal circuit of split transformers for improving the operational reliability of split transformer [3-4].

For ordinary oil-immersed transformer, the most valuable reference temperature are winding hot-spot temperature, top oil temperature and bottom oil temperature[5-6], this is still suitable for split transformer. In order to ensure the heat radiation effect, split transformers are usually installed by upper and lower or horizontal layout, and often use water, oil, air and other means to cooling. Transformer heat radiation calculation of conventional cooling system has been carried out a large number of research works at home and abroad. In the past 90 years of last century, the classical calculation model of transformer thermal state parameters has been proposed in GB/T15164-1994[7-9].There is a way of predicting the transformer oil temperature by using the thermal circuit model method which is obtained by the thermoelectric simulation method. The heat source is equivalent to the current source, and the voltage source represents the temperature difference that causes the heat transfer. This method has fewer parameters and the calculation process is simple [10-11]. There is also a way that using intelligent algorithms to predict. T-S fuzzy model is applied to the top of the power transformer oil temperature prediction, and the nonlinear system is linearized. The prediction accuracy has been improved compared with the IEEE empirical model [12].Kalman filter algorithm is also applied to the top oil temperature prediction. On the basis of the thermal circuit model, the post-Eulerian formula is used to discretize the differential equation of the top oil temperature of the transformer linearly, which also improves the prediction precision [13-14].The latter two methods are highly targeted but lack a clear physical meaning and hide the mapping relationship between influencing factors and internal temperature in the network. However, for the split cooling system, due to technical barriers and little in-depth relevant research, scientific basis is in sufficient to calculate the top oil temperature and winding hot temperature.

This article is based on the thermal circuit models which has actual physical meaning, proposed a split transformer top oil temperature thermal circuit model by analogizing with the thermal circuit of ordinary transformers. The model considers the influence of nonlinear thermal resistance, actual load, oil viscosity variation and other parameters on the thermal resistance and heat capacity of the model. After the model parameters are reasonably simplified, the corresponding parameters are optimized based on the PSOGSA algorithm to obtain the optimal Model parameters.

2. The top oil temperature thermal circuit model for split transformer

2.1. The top oil temperature prediction thermal circuit model for split transformer

Split transformer applied in underground substation generally take the body and radiator vertical distribution of the structure. The structure and flow of oil are shown in figure 1.

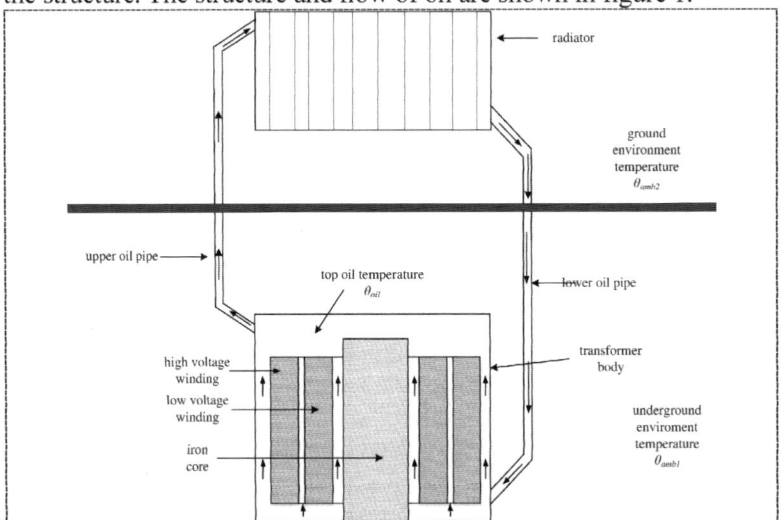

Figure 1. Vertical structure and oil flow of split transformer.

The key thermal parameters of the split transformer also select the hot spot temperature and the top oil temperature, namely "ambient temperature - top oil temperature - hot spot temperature" double-layer thermal circuit model. Among them, the part of "top oil temperature - hot spot temperature" is located in transformer body, it can be assumed with the ordinary cooling transformer, so the paper mainly research in the establishment of "ambient temperature - top oil temperature" thermal circuit and calculation of parameters. The difference of cooling process between split and ordinary transformer mainly comes from the structure, where there is a large vertical height difference between the radiator and the body of split transformers, distance between the two parts is far and their working environments are different. About the thermal calculation method, we can refer to ordinary transformers, but also to fully distinguish the difference. As for the selection of transformers body structure model, since there is no significant difference between the two structures, the ordinary transformer thermal circuit model can be directly used. The heat radiation process of radiator and transformer body were considered separately, parallel analyzing through the path of heat radiation, and the radiator part taking a simplified treatment similar to the traditional ordinary transformer radiator. We may think that winding outlet temperature and the radiator inlet temperature of split transformer are the same, the transformer winding inlet temperature and radiator outlet temperature are the same. The temperature of the transformer oil in the body heated by the heating element is equal to the temperature at which it is cooled in the radiator so as to achieve the balance of heat generation and heat dissipation of the split-type transformer in steady-state operation. That is to say, the transformer oil flowing through the closed heating and cooling oil system and the temperature profile along the geometric height of the oil circulating system of the transformer are closed curves when the split transformer operates in steady state.

Transformer and radiator parts are used in the form of RC circuit, and the two branches are in parallel. Top oil temperature thermal circuit of split transformer established are shown in figure 2.

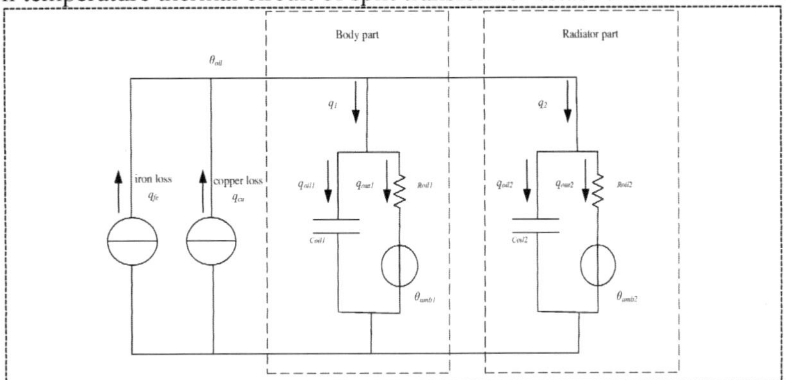

Figure 2. Top oil temperature thermal circuit of split transformer

Where qFe is the transformer iron loss, qCuis the transformer copper loss, θamb1 and θamb2 represent ambient temperature underground and above ground, the heat dissipated through the body and heat sink is q1 and q2. Among them, qoil1 and qoil2 represent heat dissipated to body transformer oil and radiator transformer oil, qout1 and qout2 represent heat dissipated to the environment underground and above ground. Coil1、Coil2 and Roil1、Roil2 are the capacitance and resistance of the body transformer and heat sink separately.

By simplifying the capacity branch, the simplified top oil temperature thermal circuit of split transformer is shown in figure 3.

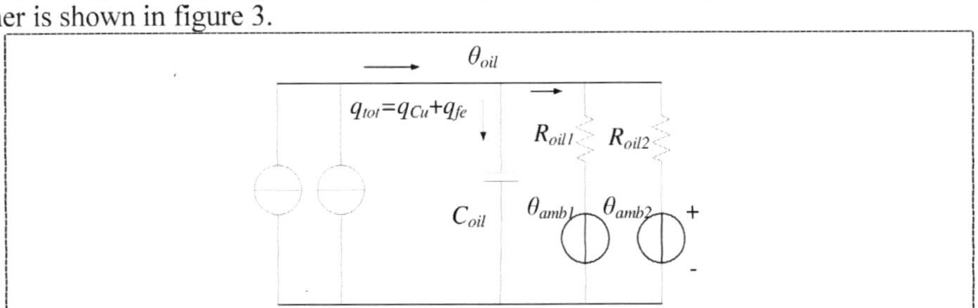

Figure 3. Simplified top oil temperature thermal circuit of split transformer.

Based on the simplified top oil temperature thermal circuit of split transformer, According to Kirchhoff's law:

$$q_{Fe} + q_{Cu} = C\frac{d\theta_{oil}}{dt} + \frac{(\theta_{oil} - \theta_{amb1})}{R_1} + \frac{(\theta_{oil} - \theta_{amb2})}{R_2} \tag{1}$$

2.2. Analysis and correction of influencing factors of the thermal circuit

As the oil temperature changes during the operation of the transformer, the oil density and viscosity change with heat transfer. Therefore, it is necessary to consider the influence of the non-linear heat resistance, load ratio and oil viscosity [15-16].

(1)Non-linear heat resistance correction. The factor 1/n is introduced to describe the non-linear variation of heat resistance, so:

$$q = \frac{\Delta\theta_{oil}^{\frac{1}{n}}}{R} \tag{2}$$

Where n is a constant, often take n = 0.8 ~ 1, 1 / n is used to characterize the heat resistance and the non-linear relationship added at both ends of the heat resistance.

Substituting (2) into (1):

$$q_{Fe} + q_{Cu} = C\frac{d\theta_{oil}}{dt} + \frac{(\theta_{oil} - \theta_{amb1})^{\frac{1}{n}}}{R_1} + \frac{(\theta_{oil} - \theta_{amb2})^{\frac{1}{m}}}{R_2} \tag{3}$$

Among them, m and n have the same value range, but the values may not be the same.

(2)Correction of the actual load: In non-rated load conditions, it is necessary to correct the equation of transformer model based on real-time load, the relationship between load current and rated current as follows:

$$\dot{K} = \frac{I}{I_{rated}} \tag{4}$$

$$\beta = \frac{q_{Cu}}{q_{Fe}} \tag{5}$$

In practice, the iron loss basically does not vary with the load; the copper loss is proportional to the square of the current, so the total loss after correction is:

$$q_{Fe} + q_{Cu} = (q_{Fe} + q_{Cu})_{rated} \times \frac{\beta K^2 + 1}{\beta + 1} \tag{6}$$

(3)Correction ofoil viscosity: among the parameters which determine the heat resistance, the oil viscosity,which changes the most with temperature, has the greatest impact on the changes of heat resistance.As the temperature changes, the variety of oil viscosity is reflected through the heat resistance.

$$R = \frac{1}{\gamma_1 \cdot \left(\frac{\Delta\theta_{oil}}{\mu}\right)^n \cdot A} \tag{7}$$

where$\gamma 1$ is a constant, determined by the oil density, oil heat expansion coefficient, oil heat conductivity, oil specific heat capacity and so on; μ is the oil viscosity which changes with temperature.

In actual operation, oil viscosity is usually difficult to obtain, and its variation is reflected by heat resistance. Therefore, in actual treatment, heat resistance is directly taken as model identification parameter. And the formula (3) becomes as follows:

$$(q_{Fe} + q_{Cu})_{rated} \times \frac{\beta K^2 + 1}{\beta + 1} = C^* \frac{d\theta_{oil}}{dt} + \frac{(\theta_{oil} - \theta_{amb1})^{\frac{1}{n}}}{R_1^*} + \frac{(\theta_{oil} - \theta_{amb2})^{\frac{1}{m}}}{R_2^*} \tag{8}$$

3. Parameter optimization of a top-oil temperature prediction method for split transformer based on PSOGSA

In equation (8), once the parameters of the model can be determined, the value of the top oil temperature can be solved. The parameters to be optimized are [C *, R1 *, R2 *, n, m]. PSOGSA is used to optimize the parameters.

3.1. Parameter optimization of a top-oil temperature prediction method for split transformer based on PSOGSA

PSOGSA composed by particle swarm optimization algorithm and gravitation search algorithm is aimed to combine the optimization ability of PSO and the search ability of GSA[17-18]. There are unique advantages in solving the optimization problem. The particle acceleration a(t) and the global extreme mumber gbest are introduced into the calculation formula of the particle velocity. The final gbest is the optimization result we need. The convergence speed is better than PSO and GSA.

In a D-dimensional search space, supposing there are N particles, so the position of the ith particle is:

$$X_i = (x_i^1, \cdots, x_i^d, x_i^D) \qquad i = 1, 2, \cdots, N \tag{9}$$

At the tth iteration, the inertial mass Mi(t) of the ith particle is defined as:

$$m(t) = \frac{f_i(t) - f_{worst}(t)}{f_{best}(t) - f_{worst}(t)} \tag{10}$$

$$M_i(t) = \frac{m_i(t)}{\sum_{j=1}^{N} m_j(t)} \tag{11}$$

where fi(t)is the fitness value of particle i at the tth iteration, and fbest(t) and fworst(t) are the optimal value and the worst value of fitness in the t-th iteration.

The gravitational force of particle j on particle i is defined as:

$$F_{ij}(t) = G(t)\frac{M_{pi}(t) \times M_{aj}(t)}{\left\| p_i(t), p_j(t) \right\|_2}[X_j(t) - X_i(t)] \tag{12}$$

Where G(t) is the gravitational time constant, Mpi(t) and Maj(t) are the passive inertial mass of particle i and the active inertial mass of particle j, respectively. It is generally assumed that the passive inertial mass is equal to the active inertial mass.

For particle i, its attraction by the rest of the particles is represented by a random weighted sum:

$$F_i(t) = \sum_{j \neq i} rand_j F_{ij}(t) \tag{13}$$

Based on Newton's second law, the acceleration generated by particle i is:

$$a_i(t) = \frac{F_i(t)}{M_i(t)} \tag{14}$$

Then the particle's velocity and position can be updated as:

$$v_i(t+1) = rand_i \times v_i(t) + a_i(t) \tag{15}$$

$$p_i(t+1) = p_i(t) + v_i(t+1) \tag{16}$$

Where pi is the position of the ith particle, vi is the velocity of the ith particle, ai is the acceleration of the ith particle, and randi is a random number between [0,1].

In the traditional GSA, each update is based on the result of one iteration calculation, which leads GSA to have strong search ability but not global optimization ability. Therefore, considering the global optimal concept in PSO, the optimal location gbest is obtained and (15) is changed to:

$$v_i(t+1) = \omega v_i(t) + c_1 \times r \times a_i(t) + c_2 \times r \times (gbest - p_i(t)) \tag{16}$$

Where ωis the inertia factor; the learning factors c1 and c2 are non-negative constants, r is a random number between [0,1], gbest represents the global optimum.

3.2. Optimization of parameters based on PSOGSA

The parameters to be optimized are [C*, R1*, R2*, n, m], which form a 5D search space, and the number of particles is set 20. Before the parameters are optimized, the range of parameters should be set rationally.

The variation range of C*, R1*, R2* should be near their rated value, and the calculation method of rated R value is:

$$R_{rated} = \left(\frac{\Delta\theta_{oil}}{q_{Fe} + q_{Cu}} \right)_{rated} \tag{17}$$

Therefore, the optimization range of R1*and R2*is [0.5Rrated, 2Rrated].The calculation formula of rated Crated is:

$$C_{rated} = 0.48 \cdot M_{FLUID} \tag{18}$$

The MFLUID is the quality of the oil. Therefore, the optimization range of C*is [0.5Crated, 2Crated].

The range of n and m values is generally [0.8, 1.2].

The optimal fitness function is the average absolute percentage (MAPE) error of the calculated and tested values, which is:

$$MAPE = \frac{1}{k}\sum_{i=1}^{k}\left|\frac{y_i - y_{it}}{y_i}\right| \times 100\% \tag{19}$$

4. Verification based on an example

In this paper, a split transformer which has not been put into operation is chosen as an example to validate the model proposed in this paper.

The transformer parameters are shown in table 1.The transformer rated capacity is 80000kVA, refrigeration mode of which is ONAN. The weight of its oil and winding respectively are 28400kg and 14818kg.Specific volume heat of the transformer oil and wingding (copper) respectively are 2569 (20℃) [J / kg·℃] and 390 (20℃) [J / kg·℃].

Table 1. The parameters of SZ11-80000/110 split transformers.

Item	Content
Refrigeration mode	ONAN
Type	SZ11-80000/110
Rated Capacity	80000kVA
Rated Voltage	110kV/10.5kV
Rated Current	419.9A/2539.7A
Oil Weight	28400kg
Winding Weight	14818kg
Specific volume heat of oil	2569(20°C)[J/kg.°C]
Specific volume heat of winding(copper)	390(20°C)[J/kg.°C]
Item	Content

According to the test results in transformer factory, because the split transformer has not yet been put into operation, still in the pilot phase, the body and the radiator are at the same ambient temperature. And oil viscosity and other factors cannot be measured .According to the definition of the thermal resistance concept, the rated thermal resistance is the ratio of the steady temperature change (under rated loss) and the total loss. The rated loss is 370360W (P=370360W), the rated thermal resistance can be obtained according to the existing formula:

$$R_{rated}=\left(\frac{\Delta\theta_{oil}}{q_{Fe}+q_{Cu}}\right)_{rated}=\frac{59.8\text{-}13.5}{370.36\times10^{3}}=1.25\times10^{-4}{}^{\circ}C/W \tag{20}$$

According to the transformer parameters, the rated heat capacity can be obtained:

$$C_{th-oil}=0.48\cdot M_{FLUID}=0.48\times28400=13632\text{Wh}/{}^{\circ}\text{C} \tag{21}$$

so the optimal range of model parameters is

$$\begin{gathered}0.625\times10^{-4}{}^{\circ}C/W\leq R_1*,R_2*\leq2.5\times10^{-4}{}^{\circ}C/W\\ 6816\text{ Wh}/{}^{\circ}\text{C}\leq C*\leq27264\text{ Wh}/{}^{\circ}\text{C}\\ 0.8\leq n,m\leq1.2\end{gathered} \tag{22}$$

Based onsite temperature-rise experimentdatasampled every 15 minutes from 9:00p.m. to 7:00 a.m.the next day, under the constant power state of the transformer, the optimization of the model parameters is finished by PSOGSA. The results are showed in table 2.

Table 2. Optimized parameters of model.

R1*	R2*	C*	n	m
1.7×10-4	1.28×10-4	19268	1.01	1.09

The results of the calculation of the top-oil temperature are shown in figure 4, which is using the optimized parameter model (n=m=1).

The orange indicates the measured value and the blue indicates the calculated value of the hot-circuit model. It's obvious that the final steady state temperature is both consistent in figure 4. But the temperature rise fitting of the optimization parameter is more accurate than that of the rated values model. It illustrates that the optimization of the parameter model is effective.

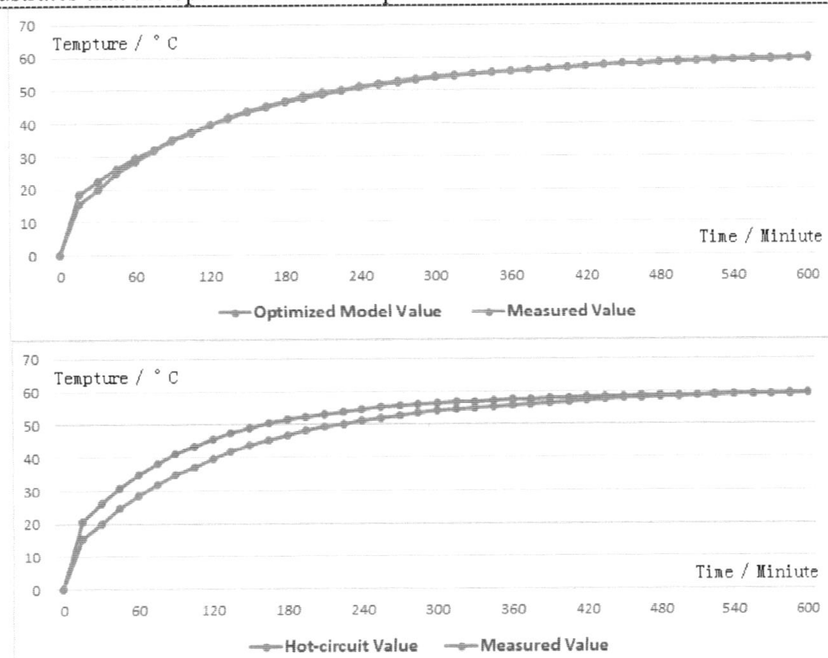

Figure 4. Prediction results of optimized model and rated model.

5. Conclusion

Aiming at the prediction of top-oil temperature of split transformer, a method based on PSOGSA is proposed. Firstly, according to the split transformer structure, heat radiation principle and the ordinary transformer thermal circuit model, the thermal circuit model of the split transformer top-oil temperature is proposed. Besides,model parameters are revised bynon-linear heat resistance, load ratio and oil viscosity. Then, on the premise of not affecting the calculation results, the model is further simplified and the five model parameters to be optimized are extracted. Further, PSOGSA hybrid optimization algorithm is used to optimize the model parameters. The model proposed in this paper is verified by on-site temperature-rise experiment data. The results show that the proposed method can effectively improve the calculation accuracy of the thermalcircuit model of the top-oil temperature. The proposed method provides a feasible solution for the calculation of the top-oil temperature and the hot spot temperature of the split transformer, which will help to improve the operational reliability of split-type transformers.

Reference

[1] Ye Jun, Hu Wangrong,Xu Ping. Arrangement and Application of Split Transformer in 110 kV and 35 kV Substation[J]. Electrical World, 2007, 48(7):14-15.

[2] Yang Zhengyan. Improvement the Operation Condition of 35kV Substation with Split Transformer. Shanghai Electric Power, 2001, 5:55-56.

[3] Shao Zhiwei. Investigation on Loss and Heat Dissipation Characteristics for Power Transformer Systems[J]. Electric Power, 2016, 49(6):132-136.

[4] Fu Chenzhao,JiShengchang, Wang Shishan,et a1. 2D Numerical Calculation of Temperature Field of Winding in Transformer[J]. High Voltage Engineering, 2002, 28(5):10-12.

[5] Pan Chong. Study on insulation fault diagnosis and thermal state parameters prediction model of power transformers[D]. Chongqing University, 2009.

[6] Zhao Yinhan. Application of Optical Fiber Temperature Measurement Technology to Transformer[J]. Electric Power, 1990(11):48-49.

[7] GB/T 15164-1994 loading guide for oil-immersed power transformers[S],d94.

[8] Huang Xiaofeng, ZhouLijun, Wu Guangning. Comparative Analysis on Classical Thermal Models for Calculating Transformer Hot-spot Temperature and Its Improvement[J]. Modern Electric Power, 2013, 30(2):69-73.

[9] Su Xiaoping. Investigation into the calculation model and prediction method of hot-spot temperature for oil-immersed transformers[D]. Chongqing University, 2012.

[10] Jiang Taosha, Li Jian, Chen Weigen, Sun Caixin . Thermal model for hot spot temperature calculation in oil-immersed transformers[J]. High Voltage Engineering, 2009, 35(7): 1635-1640.

[11] Tang Yan, Liu Chengyuan,HaoZhongyan,et a1. Calculation and test of hot—point temperature-rise in transformer windings[J]. Tansformer, 2001, 38(2): 1-5.

[12] XiongHao, Chen Weigen, Du Lin, Sun Caixin, Liao Ruijin. Study on prediction of top-oil temperature for power transformer based on T-S model[J]. Proceedings of the CSEE, 2007, 27(30): 15-19(in Chinese).

[13] Wang Yongqiang,YueGuoliang, He Jie, Liu Hongliang, Bi Jiangang, Chen Sufen. Study on prediction of top oil temperature for power transfbrnler based on Kalman filter algorithm[J]. High Voltage Apparatus, 2014, 50(8) :74-79+86.

[14] Su xiaoping, Chen Weigen, Xi Hongjuan. Winding Hot-spot Temperature Prediction for Oil-immersed Power Transformer Based on Kalman Filter Algorithm[J]. High Voltage Engineering, 2012, 38(8):1909-1916.

[15] Wei Bengang, Huang Hua, Fu Chenzhao, Li Honglei, Liu Jiayu. Algorithm for transformer top-oil temperature and winding hot-point temperature based on modified thermal circuit model[J]. East China Electric Power, 2012, 40(3):444-447

[16] Wang Fenghua, Zhou xiang,Gao Pei, et al. Improved Thermal Circuit Model of Hot Spot Temperature in Oil-immersed Transformers Based on Heat Distribution of Winding[J]. High Voltage Engineering, 2015, 41(3):895-901.

[17] MaLi, Liu Litao. Analysis and Improvement of Gravitational Search Algorithm[J]. Microelectronics & Computer, 2015(9):76-80.

[18] Mirjalili S,Hashim S Z M. A new hybrid PSOGSA algorithm for function optimization[C]// International Conference on Computer and Information Application. IEEE, 2012:374-377.

IOP Conf. Series: Materials Science and Engineering **366** (2018) 012061 doi:10.1088/1757-899X/366/1/012061

Factors Affecting the Discharge Characteristics of Rod-Rod Long Air Gaps at Different Altitudes

Zhang Hao, Duan Yubing, Liu Hui, Liu Rong, Zhang Yang, Zhou Chao and Jia Ran

State Grid Shandong Electric Power Research Institute , Jinan 250002, China;

sdqzzh@163.com

Abstract. The impulsive discharge characteristics of the air gap is one of the main bases for the design of the external insulation of high-voltage electrical equipment; meanwhile, air gap discharge characteristics at different altitudes have been of concern for a long time. In this paper, the impact discharge tests of rod-rod gaps at different altitudes were carried out to study the influences of elements such as electrode shape, voltage polarity, altitude and temperature. It can be found that, with the increase of the gap distance, the 50% impulse voltage of the switching impulse tends to be saturated, and the discharge characteristic curve can be fitted by exponential function. The 50% impulse voltage of the lightning impact and the gap distance shows a good linear relationship which can be fitted by the linear function. This is no direct relationship with temperature and altitude. The saturation tendency of switching impulse's 50% discharge voltage decreased with the decrease of temperature. With the increase of the altitude and the decrease of the 50% impulse discharge voltage, the polar effect of the air gap between the rod-rod and the rod-plate decreases. The results are of great reference value for Insulation configuration of electrical equipment in high altitude.

1. Introduction

Reasonable insulation coordination is the basic guarantee for the safe and reliable operation of the power system[1]. China is a large country with a large altitude difference especially in the east and west regions where the altitude is between 3000 and 5300 meters in Qinghai-Tibet region; therefore, in the construction of high-voltage transmission projects in high altitude areas, we must solve the problems of outer insulation, corona and electromagnetic as required[2~6]. In engineering practice, most of the electric fields encountered are inhomogeneous electric fields. Since the rod-rod air gap electric field gets perfect symmetry in all kinds of extremely inhomogeneous electric field air gaps, its breakdown characteristics can be used as reference for other types of inhomogeneous electric fields[7,8].

In this paper, the impact discharge tests were conducted in two regions: a place in the coast area of China (altitude of 50m) with the temperature range of 5~5°C and 15~25°C in the discharge tests fields while the second located somewhere in Qinghai-Tibet region (altitude of 4300m) with the temperature in the range of 15~25°C. The impulsive discharge tests include switching impulse and lightning impact tests. Those tests are conducted to analyze air gap impulse discharge characteristics based on test results, such as the polarity effect of impulse voltage, saturation phenomenon, and dispersion.

2. Impact test principle

Impulse voltage generator use multi-stage capacitor to charge in parallel and discharge in series to generate the required voltage, while its waveform can be adjusted by changing the resistance of R_f and R_t.

Air gap impact test principle of wiring is shown in Figure 1. The test of wiring mainly consists of the impact of voltage generators, capacitive voltage divider, digital recorder, and other discharge gap components.

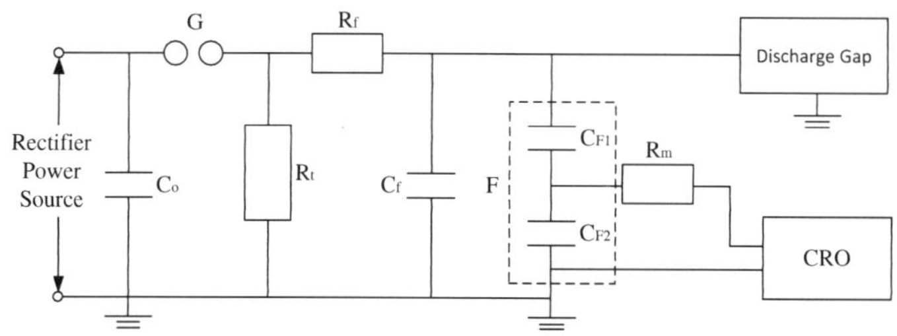

Co:Main Capacitor of Impulse Voltage Generator; G:Isolation Gap;
Rf:Wavefront Resistance; Rt:Stroke Resistance; Cf:Wavefront Capacitor;
F:Shock Capacitor Divider; Rm:Protective Resistance; CRO:Oscilloscope

Figure 1. Air Gap Impact Test Principle of Wiring

Using the lifting method, the voltage's rise and drop magnitude is estimated to be about 3~5% in the impact flashover voltage test where the interval between each impact is about 3~5min. Atmospheric pressure, humidity, temperature and other meteorological parameters were measured every time before and after each test, and the average is taken as the experimental data of the atmospheric parameters[9,10].

50% discharge voltage U_{50} is calculated as:

$$U_{50} = \frac{\sum (n_i \times U_i)}{n} \tag{1}$$

In the equation above: U_i stands for Impact voltage value, kV; n_i is the number of valid tests at the same surge voltage (U_i); n equals the number of valid tests.

Generally speaking in each data point, the number of operational impact pressure is 40 times and the effective number of lightning impulse is 30 times. If the impulse voltage series is less than four (i<4), the impact test operation can be carried out 30 times, while the lightning impulse test can be carried out 20 times. Standard deviation of operational impact does not exceed 5%, and the standard deviation of lightning impulse does not exceed 3%. For the test whose standard deviation does not meet the standard data points, test is re-runned.

The minimum of 50% impulse discharge voltage in each test is taken as the reference value, and the 50% impulse discharge voltage in the test is taken as the per unit value. Per unit value for the 50% of the percussive discharge voltage (U_{50*}) is as below:

$$U_{50*} = \frac{U_{50}}{U_{50B}} \tag{2}$$

In the equation above: U_{50} is the actual value of the 50%discharge voltage, kV; U_{50B} is the reference value of 50% discharge voltage, taking the minimum of 50% impact discharge voltage in each set of tests, kV.

The dispersion of the data of the test data can be showed by the coefficient of variation. The coefficient of variation is calculated as follows:

$$z = \frac{\sigma}{U_{50}} \times 100\% \tag{3}$$

In the equation above: U_{50} is the 50% discharge voltage, kV; σ is the standard deviation.

The diameter of the rod electrode used in the experiment was 60 mm. There are three shapes at the ends, hemispherical, flat, and conical, respectively. The cone angle of the conical rod electrode was 30°. The shape of each end is shown in Figure 2.

(a) Hemispherical rod electrode (b) Flat rod electrode (c) Conical rod electrode

Figure 2. Rod electrodes

Positive polarity impact discharge tests were performed on the above three end-shape rod-rod electrodes' air gap.

Fitting the test results to get the relationship between the U_{50*} of the rod-rod electrode and the gap distanced of the three end shapes, as is shown in the curve in Figure 3.

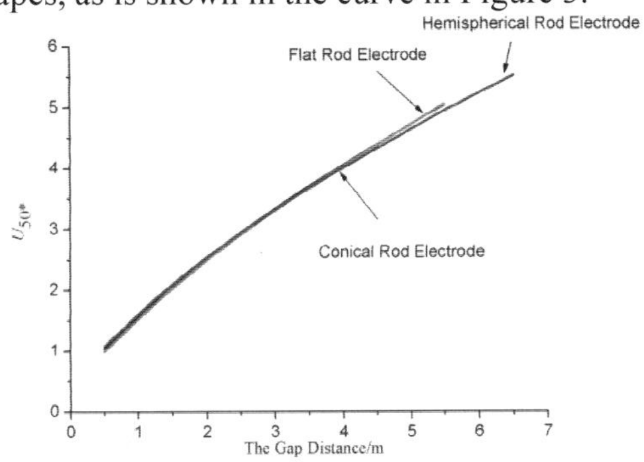

Figure 3. The relationship between the switching impulses of the electrode of different shapes U_{50*} and the gap distance d

As can be seen from Fig.3, the impact discharge curves of three different end-shape rod-rod electrodes are basically coincident, with little difference between the three test results. It can be seen that when the long air gap is more than 1 meter (d>1m), the shape of the end of the rod electrode is not the main factor affecting the discharge voltage. As the hemispherical rod electrode is the most typical form used in the power system, the semi-spherical rod electrode is chosen as the follow-up test rod electrode.

3. Factors affecting the discharge characteristics of the air gap

Air gap discharge test was carried out in different environments to analyze the influencing factors of rod-rod air gap. The environmental parameters of the test are shown in table 1.

Table 1. The Environmental Parameters of the Test

	Place A in the coast area	Place B in the coast area	somewhere in Qinghai-Tibet region
Altitude	50m	50m	4300m
Impact type	Switching impulse/ Lightning impact	Switching impulse/ Lightning impact	Switching impulse/ Lightning impact
Voltage Polarity	+ /-	+ /-	+ /-
Temperature Range	-5~5℃	15~25℃	15~25℃
Average Temperature	-1℃	21℃	18℃

Absolute Humidity range	2~5 g/m3	4~11 g/m3	3~7 g/m^3
Average Absolute Humidity	4.1 g/m3	6.0 g/m3	5.3 g/m^3

3.1. Effect of Temperature on Air Gap Discharge Characteristics

It is generally believed that when the temperature raises, the average kinetic energy of air molecules increases, and the molecular activity increases, so the discharge voltage of the air gap decreases. In order to investigate the effect of temperature on the discharge characteristics of air gap, we compared the experimental data of low temperature (-5~5℃) and high temperature (15~25℃) both conducted in coastal area to analyze the influence of temperature on the discharge characteristics of air gap. The curves of rod-rod electrode impact discharge U_{50*} changing with the gap distance are as following.

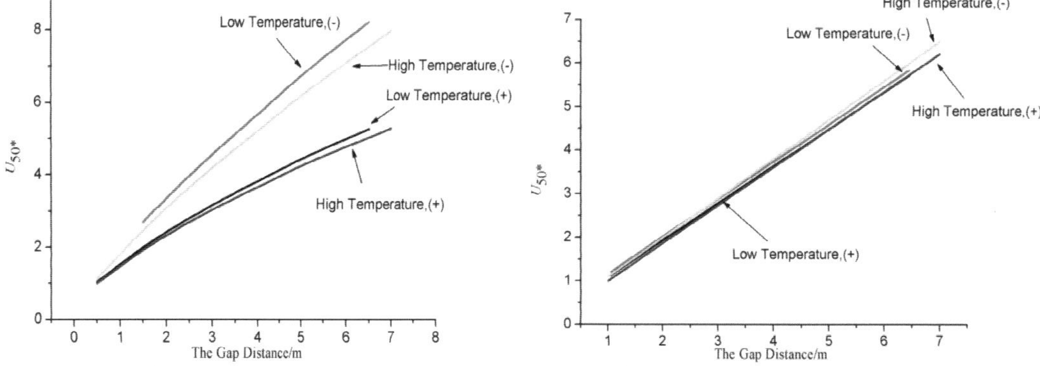

Figure 4. The relationship between switching impulse U_{50*} and the gap distance d as well as the relationship between lightning impact U_{50*} and the gap distance d in different temperature.

Table 2. Coefficient of Variation of Rod-rod Air Gap Impact U_{50}

Impact Type	Coefficient of Variation (z)
switching impulse (low temperature)	3.6%~4.8%
switching impulse (high temperature)	2.9%~4.8%
lightning impact (low temperature)	1.8%~2.9%
lightning impact (high temperature)	1.8%~3.0%

As can be drawn from Figure 4 and Table 2:

(1) As the temperature decreases, the breakdown voltage of the rod-rod air gap increases. The effect of temperature on the impact polarity of the rod-rod air gap is great: when the air gap distance is in the range of 2m-6m, the switching impulse of rod-air gap at low temperature$U_{50\,(-)}$ increases by about 6% ~ 9% comparing to it at high temperature $U_{50(-)}$. The switching impulse $U_{50\,(+)}$ of the rod-rod air gap at low temperature increased about 1%~4% compared with the $U_{50\,(+)}$ at high temperature. The greater the gap distance, the greater the difference between them. The effect of temperature on the polarity effects of the lightning impulses of the rod-rod air gap is less significant: at both temperature levels, the positive and negative U_{50} can be considered equal.

(2) When the temperature rises, the saturation tendency of the switching impulse voltage of the rod-rod air gap increases, and the effect of temperature on the saturation trend of the switching impulse $U_{50\,(-)}$ is greater than that of the $U_{50\,(+)}$.

(3) As the temperature decreases, the rod-rod air gap's switching impulse dispersibility of the U_{50} increases; while the temperature has less effect on the dispersibility of the lightning impact U_{50}.

3.2. Influence of Altitude on Air Gap Discharge Characteristics

With the increase of altitude, the density of air decreases, and the mean free path of free electrons in air increases, hence the ionization process of gas increases, while the discharge voltage of air gap decreases. In this part, we compare the experimental data at low altitudes (50m, 15~25℃) and the one at high altitude (4300m, 15~25℃) somewhere in the Qinghai-Tibet region to analyze the influence of altitude on the air gap discharge characteristics. The curves of rod - rod electrode impact discharge U_{50*} changing with the gap distance are as following.

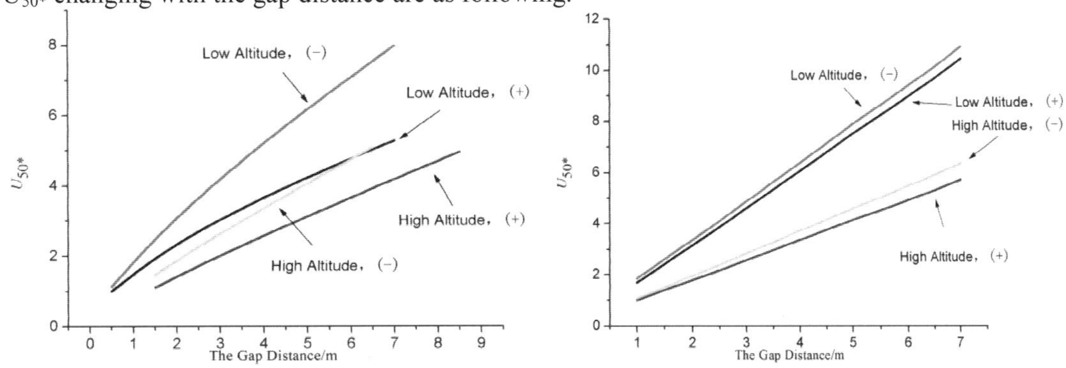

Figure 5. the relationship between switching impulseU_{50*} and the gap distance d as well as the relationship between lightning impact U_{50*} and the gap distance d in different altitude

Table 3. Coefficient of variation of rod- rod air gap impact U_{50}

Impact Type	Coefficient of Variation (z)
Switching Impulse (50m)	2.9%~4.8%
Switching Impulse (4300m)	2.6%~4.8%
Lightning Impact (50m)	1.8%~3.0%
Lightning Impact (4300m)	1.4%~3.0%

As can be drawn from Figure 5 and Table 3:

(1) As altitude increases, the U_{50} of the rod-rod air gap decreases; the saturation trend of switching impulse U_{50} drops. Altitude has a large effect on the impact discharge characteristics of the rod-rod air gap: U_{50} at an altitude of 4300 m is about 60% to 80% of it at an altitude of 50 m for a positive-polarity switching impulse; for a negative-polarity switching impulse at an altitude of 4300 m U_{50} is about 60%-70% of it at the sea level of 50m. For a lightning impact, the U_{50} at an altitude of 4300m is about 55% of the U_{50} at an altitude of 50m; therefore, the effect of altitude on positive and negative polarity lightning impulse voltage is basically the same.

(2) As the altitude increases, the polar effect of the rod-rod air gap decreases.

(3) According to the test data dispersion analysis, it is found that the dispersion of the impact discharge U_{50} of the rod- rod air gap in the high altitude area is generally lower than that in the low altitude area. The difference between the two is so tiny that it can be taken as there is no difference at all.

4. Conclusions

In this paper, the impact-discharge tests of air gap between rod and rod at a temperature range of -5 ~5℃ and 15~25℃ in a certain place along the coast and those at the altitude of 4300min the Qinghai-Tibet region are conducted. The characteristics of impingement discharge of air gap are analyzed from several aspects, such as polarity effect, saturation phenomenon, and dispersibility. When the other environmental conditions are close to each other, temperature and altitude are chosen as two parameters for studying the influence of single environmental factor on the air-gap discharge voltage. The main conclusions of this thesis are as follows:

(1) For a long air gap, the end shapes of the electrode have no effect on the air gap impulse discharge characteristics when the rod electrode diameter is 60 mm.

(2) When the gap distance is longer, there is a polar effect on the impact discharge of the rod-rod air gap under different experimental conditions. Usually the 50% impulse discharge voltage of the negative polarity is greater than it of the positive polarity. With the increase of the gap distance, the 50% impulse voltage of the switching impulse tends to be saturated, and the discharge characteristic curve can be fitted by exponential function. The 50% impulse voltage of the lightning impact and the gap distance shows a good linear relationship which can be fitted by the linear function. The breakdown voltage data dispersion of the switching impulse is greater than the lightning impact's; while the voltage polarity has less effect on the data dispersion of the 50% impulse discharge voltage.

(3) The saturation tendency of switching impulse's 50% discharge voltage decreased with the decrease of temperature. The increase of discharge voltage is large in the rod-rod air gap's switching impulse. The temperature influence on rod-rod air gap of the lightning impact is relatively small. When the temperature decreases, the depressiveness of 50% discharge voltage of the rod-rod air gap's switching impulse increases; the temperature has little effect on the dispersion of the 50% discharge voltage of the lightning impact.

(4) With the increase of the altitude and the decrease of the 50% impulse discharge voltage, the polar effect of the air gap between the rod-rod and the rod-plate decreases. The saturation tendency of the 50% discharge voltage decreases with the switching impulse, and the increase range of 50% discharge voltage decreases. The dispersion of 50% impulse discharge voltage at high altitude is slightly less than it at low altitude.

References

[1] Zhao Wanjun. High Voltage DC Transmission Engineering Technology[M]. BeiJing:China power press, 2004: 233-266.

[2] Zhang wenliang, liao weiming, ding yujian, etc. Selection of the Minimum Air Clearance for ± 660 kV Double-circuit DC Transmission Line Tower in Different Altitude Areas [J]. Proceedings of the CSEE, 2008, 28(34): 1-6.

[3] Du Shuchun.Research on Air Clearance of Tower for AC 750kV Single Circuit Transmission Line[J]. Power System Technology, 2008, 32(19): 1-4.

[4] Jiang Xing-liang, Wang Jun,Fan Jihe, etc. Correction Method of Positive Switching Impulse Discharge Voltage for 0.5~1.0 m Air Gap Under Artificial and Natural Conditions[J]. Proceedings of the CSEE, 2008, 28(28): 13-17.

[5] Sun Cai-xin, Jiang Xing-liang, Sima Wen-xia, etc. Ac Discharge Performance and Voltage Correction of Short Air Gaps in 4000m and Above Altitude DAistricts[J]. Proceedings of the CSEE, 2002, 22(10): 116-120.

[6] Shu Yinbiao, Zhang Wenliang. Research of key technologies for UHV transmission[J]. Proceedings of the CSEE, 2007, 27(31): 3-8(in Chinese).

[7] Rizk A M.Critical switching impulse strength of phase to phase air insulation[J]. IEEE Trans. on Power Delivery, 1993, 18(8): 1492-l506.

[8] Calva P A, del Moral V, M.G.Marquez, et al. New proposal of correction factors for DC Voltages[C]. Annual Report conference on electrical insulation and dielectric Phenomena, Albuquerque, New Mexico, USA, 2003: 455-458.

[9] GB/T16927.1-1997 High-voltage test techniques. Part 1: General definitions and test requirement [S]. The People's Republic of China Standard. Put in practice in 1998.

[10] IEC, Insulation coordination Part 2 Application Guide, IEC600071-2-1996, 1996.

Study on Influencing Factors of Technical-economy Effect of VSC-HVDC System Based on Improved ISM Method

Hongzhi Li[1], Weihua Xu[1], Huicong Xia[2], Dongxiao Niu[2] and Yijia Yuan[1]

[1] Global Energy Interconnection Research Institute, Beijing, China
[2] Economics and Management School, North China Electric Power University, Beijing, China

lihongzhi@geiri.sgcc.com.cn, xuweihua@geiri.sgcc.com.cn, 1656184583@qq.com, niudx@126.com, yuanyijia1@126.com

Abstract. With the higher controllability and power supply reliability, VSC-HVDC transmission technology can support the grid connection of offshore wind power and other distributed energy resources, which conforms to the trend of energy revolution. For investors, the technical-economy effect of VSC-HVDC transmission project is the most attractive issue, and furtherly, the influencing factors are necessary when analyzing the technical-economy effect level of a certain project. To help investors identify the main influencing factors, arrange the internal relationship of influencing factors, and make investment decisions, this paper firstly builds an influencing factor set of technical-economy effect of VSC-HVDC transmission system using brainstorming method, and then proposes an improved ISM method to divide the hierarchical structure of influencing factors. The result shows that influencing factors can be divided into 9 categories and four levels, of which design requirements, equipment requirements and project features are superficial factors, while natural environment and policy environment are fundamental factors. The case study proves that improved ISM method is applicable and practicable for influencing factor analysis.

1. Introduction

Currently, most countries regard renewable energy as a key solution to energy and environmental problems. However, renewable energy has a general characteristic of intermittence and randomness, thus affecting the power quality, which highlights the importance of transmission technology [1]. VSC-HVDC (voltage source converter based high-voltage direct current) is a reliable guarantee to make full use of renewable energy because of its unique advantages: flexible operation, high transmission efficiency, high power quality, less occupied area, fast recovery, and enhanced capacity for distributed renewable energy [2]. VSC-HVDC is now widely used in the following fields: offshore wind power integration, distributed energy resources integration, power supply to passive network, and urban power grid renovation [3]. Generally speaking, the technical-economy effect needs to be considered before investing VSC-HVDC transmission projects, and furtherly, the influencing factor analysis is the basis of technical-economy effect calculation, Therefore, this paper firstly summarizes factors affecting the technical-economy effect of VSC-HVDC transmission system, and then uses the improved ISM method (interpretative structure mode) to identify the internal structure and hierarchical relationship of influencing factors.

There are many methods can be used to classify, sort, and analyze influencing factors from different point of view, such as AHP (analytic hierarchy process), factor analysis, ISM, multiple linear regression model, sensitivity analysis and correlation analysis. AHP is a subjective analysis method, while factor analysis method is based on objective data, and both of them divide influencing factors into different categories [4]. ISM method resolves complex factors into a set of elements with logic relation, and arranges the hierarchical relationship by the aid of practical experience and knowledge [5]. The result of multiple linear regression, sensitivity analysis and correlation analysis method is the mathematical relationship between influencing factor set and analysis objective, which reflecting the influence degree of each factor. Due to the lack of comparable VSC-HVDC projects and the ubiquity of qualitative factors, it is difficult to achieve quantitative analysis, thus the ISM method is selected in this paper for its comprehensiveness.

ISM method is widely used in the process of investment decision, physical distribution management, teaching management, risk management, quality control and so on [5-7]. ISM can be combined with DEMATEL method (decision-making and trial evaluation laboratory) to improve the reliability, the key factor can be obtained using DEMATEL method, and the multilevel hierarchical structure of the complex factor set can be built based on ISM method [8]. On the other hand, it is difficult to judge the accurate causality between factors just by definition, but it is possible to convert the simple 0-1 function into a fuzzy binary relation, so we can express the existence of influencing relationship between factors and determine the influencing degree [9]. Different experts have their own subjective judgments on the issue of whether there is any influencing relationship between factors. Based on the thought of fuzzy analysis, this paper uses questionnaire survey to synthesize the opinion of experts and draw the fuzzy relation between factors, and calculates the binary relation according to membership function. The case analysis result shows that the improved ISM method has good practicability and feasibility for analyzing the influencing factors of technical-economy effect of VSC-HVDC system.

2. Influencing factors of technical-economy effect of VSC-HVDC system

Technical level and price level are direct factors affecting the technical-economy effect of VSC-HVDC projects. In addition, the design, construction, operation and maintenance process of projects are inevitably restricted by the external factors consisting of economic environment, natural environment, social environment and policy environment. Moreover, unique design requirements, equipment requirements and project features of a certain VSC-HVDC project are also important influencing factors [10,11].

2.1. Economic environment

Almost all the technical-economy effect of power grid construction projects will be affected by regional economic development level, inflation rate, interest rate, exchange rate and so on. The regional economic development level determines local power load and transmission capacity; the inflation rate, interest rate and exchange rate affect the cost of projects.

2.2. Natural environment

The nature environment factors include resource condition, climate condition, topographic condition, geological condition, hydrologic condition and so on. The resource condition determines the capacity and distribution of power supply, and affects the power fluctuation of grid; the climate condition influences local electricity demand to a certain degree; topographic condition, geological condition, and hydrologic condition are the key issues to be considered during the design of transmission line and converter station, and they also put forward certain requirements for the equipment performance.

2.3. Social environment

The development of VSC-HVDC is greatly influenced by the social environment. If VSC-HVDC transmission line for offshore wind farm involves fishing areas or shipping routes, it will be needed to consider the attitude of fishermen, maritime companies and other related personnel. Government can also participate in coordination activity to help projects accepted by popular. The design, construction and operation of VSC-HVDC transmission system need to take the stability system and acceptable intermittent power capacity into account. The frequency, outlet voltage, phase sequence and voltage phase of generators must be in accord with that of power system.The technical-economy effect of VSC-HVDC transmission system is affected by the quality, performance and parameters of material and equipment. All the system equipment and material, construction machinery and tools, and spare parts need to be obtained in time and meet technical requirements. The construction and operation of VSC-HVDC transmission of offshore wind farms still exist a certain transportation difficulty. Equipment and parts can only be supplied by air and sea transport, resulting in higher cost.

2.4. Policy environment
The development and construction of power system must obey the energy policy and environmental protection policy, whose target is to improve China's resource environment and energy structure, and reduce pollution. VSC-HVDC can be used for distributed renewable energy integration, which is consistent with the requirements of renewable energy development policy. On the other hand, to ensure the orderly construction and stable power supply of power grid, the construction of transmission system must also accord with the overall power grid planning.

2.5. Price level
Price level is the economic factor that directly affects the cost of VSC-HVDC transmission system, including equipment price, material price, labor price, construction machinery price and so on. The main profit source of VSC-HVDC transmission project is the transmission and distribution income, that is, the transmission and distribution price also affect the economic benefit.

2.6. Technical level
Personnel involved in VSC-HVDC transmission system mainly include designer, construction crew and manager. Their technical level directly affects the engineering quality of VSC-HVDC transmission system. Personnel quality of operation and maintenance staffs also affects the reliability and operation cost of power grid.

2.7. Design requirements
VSC-HVDC transmission system should be designed to meet some technical-economy indicators. At present, the localization of VSC-HVDC technology is relatively low, some demonstration projects have higher requirements for technological innovation, so the technology cost is slightly higher. VSC-HVDC transmission system for distributed renewable energy resources integration has strict requirements for the intelligent level. For VSC-HVDC transmission system with modular structure, the standardization degree and extensibility should be taken into consideration, so as to facilitate the future improvement. In addition, the insulation and lightning protection requirements should also be taken into account.

2.8. Equipment requirements
At present, VSC-HVDC is mostly applied to the offshore wind power integration. The special and rugged marine environment determines that the moisture proofing, salt spray proofing, fungus proofing, water proofing, anti-corrosion, typhoon and crazy wave resistance and mechanical strength of equipment needs to reach a higher level. Equipment with longer life period is also required as a result of restrictions of traffic and construction condition on replacing damaged equipment and ageing facilities.

2.9. Project features

IOP Conf. Series: Materials Science and Engineering **366** (2018) 012062 doi:10.1088/1757-899X/366/1/012062

Projects are mainly divided into new projects, rebuilt projects and expansion projects, different projects have different requirements for design, construction and purchase. The voltage and transmission capacity of VSC-HVDC transmission system determine the required specification and parameter of material and equipment; construction season directly affects the schedule and plan; transmission distance directly determines the length of transmission line.

The above factors are summarized as shown in Table 1 and Table 2.

Table 1. Influencing factors of technical-economy effect of VSC-HVDC system (a).

Influencing factors		Influencing factor content
Economic environment	S_1	Regional economic development level Inflation rate Interest rate Exchange rate
Natural environment	S_2	Resource condition Climate condition Topographic condition Geological condition Hydrologic condition
Social environment	S_3	Popular acceptance Condition of grid connection Transportation condition Material supply condition
Policy environment	S_4	Environmental protection requirements Power network planning Energy development planning
Price level	S_5	Equipment price Material price Labor price Construction machinery price Transmission and distribution price
Technical level	S_6	Personnel quality of designers Personnel quality of construction units Personnel quality of supervisors Personnel quality of operation and maintenance staffs
Design requirements	S_7	Technological innovation requirements Intelligent level Standardization degree Extensibility

IOP Conf. Series: Materials Science and Engineering **366** (2018) 012062 doi:10.1088/1757-899X/366/1/012062

		Insulation requirements
		Lightning protection requirements
Equipment requirements	S_8	Moisture proofing, salt spray proofing, fungus proofing, water proofing, anti-corrosion, and typhoon and crazy wave resistance
		Mechanical strength
		Life period

Table 2. Influencing factors of technical-economy effect of VSC-HVDC system (b).

Influencing factors		Influencing factor content
Project features	S_9	Project classification
		Voltage level
		Transmission capacity
		Transmission distance
		Construction season

3. ISM methodology

In 1973, Professor J. N. Warfield developed an ISM method in order to analyze the structure of complex social and economic system. The basic idea of this method is that the system is made up of elements, and there are a large number of interactions between these elements. Therefore, to establish the structural model, we need to comprehend the relationship between elements and divide the hierarchical structure. Because of the existence of qualitative influencing factors, and considering the different determination of experts on relationship between factors, the concept of fuzzy matrix is introduced to improve the usability and accuracy of ISM method applied in this paper.

Implementation steps of the improved ISM method are as follows:

Step 1: Identify the influencing factors of technical-economy effect of VSC-HVDC system, classify these factors and label them as S_i. In this step, there is no need to analyze the internal relationship for the time being.

Step 2: Organize experts to judge the relationship between pairs of S_i and S_j. The score can be 0 or 1: 1 indicates that row factor affects column factor, 0 indicates that row factor has no influence over column factor.

Step 3: Arrange the result of expert survey and establish a fuzzy adjacency relation matrix $A=(a_{ij})_{n\times n}$.

$$a_{ij} = \frac{m}{M} \tag{1}$$

a_{ij} indicates the influence of S_i on S_j: m means the number of experts believing the influence of S_i on S_j; M means the number of experts; n means the number of influencing factors.

Step 4: Determine the threshold λ and calculate the final adjacency relation matrix $B=(b_{ij})_{n\times n}$.

$$b_{ij} = \begin{cases} 1, a_{ij} \geq \lambda \\ 0, a_{ij} < \lambda \end{cases} \tag{2}$$

Step 5: Calculate the reachability matrix $C=(c_{ij})_{n\times n}$ by MATLAB.

Adjacency relation matrix belongs to boolean matrix, the reachability matrix satisfies the following condition.

$$(B+I)^1 \neq (B+I)^2 \neq \cdots (B+I)^{r-1} \neq (B+I)^r = (B+I)^{r+1} = C \tag{3}$$

I is a unit matrix. Reachability matrix indicates the channel from one factor to another.

Step 6: Divide the reachability matrix by MATLAB, and draw the multilevel hierarchical directed graph.

The structure division is to obtain a clear understanding of the internal hierarchical relation among factors. Factors belonging to the first layer represent ultimate goal, while underlying factors are the reason for upper factors. Before the division of reachability matrix, the reachability set $R(S_i)$ and antecedent set $Q(S_i)$ are needed to be analyzed. Reachability set $R(S_i)$ represents the column factor with row elements in reachability matrix of factor S_i being 1, while antecedent set $Q(S_i)$ represents the row factor with column elements in reachability matrix of factor S_i being 1. Hierarchical structure division is the gradual extraction of factors according to $R(S_i) \cap Q(S_i) = R(S_i)$. The earliest extracted factors belong to the first layer, and the latest ones are basic factors.

Technology roadmap the process is shown in Figure 1.

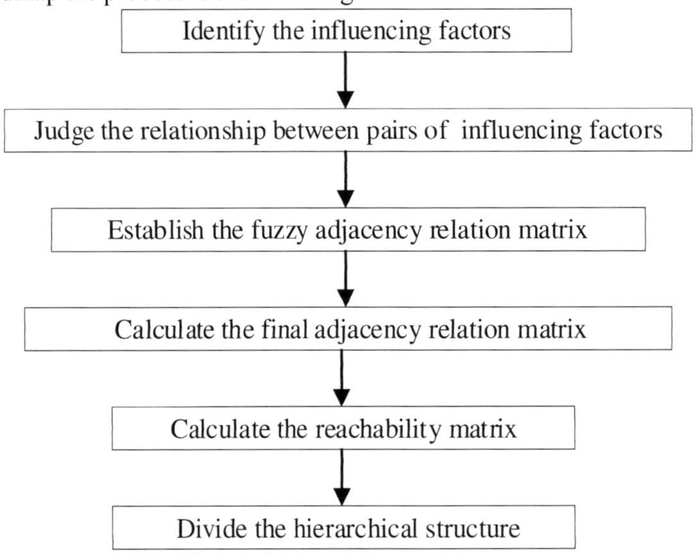

Figure 1. Technology roadmap.

4. Case study

VSC-HVDC plays a prominent role in optimizing the energy structure and improving the environment condition. To provide investors, government and other related institutions with decision support, this paper summarizes the influencing factors of technical-economy effect of VSC-HVDC system, and analyzes the hierarchical structure based on improved ISM method.

Step 1: Establish the influencing factor set.

This paper divides the influencing factors of technical-economy effect of VSC-HVDC system into 9 categories, each category has its own corresponding emphasis. The influencing factor set is shown in Table 1 and Table 2.

Step 2: Questionnaire survey.

To improve the reliability of analysis result, this paper sends 30 questionnaires to 30 experts, and all of them are valid. Because of the limited space, the details of these questionnaires will not be reflected.

Step 3: Summarize the result of questionnaire survey according to equation (1).

$$A=\begin{bmatrix}0.00 & 0.23 & 0.73 & 0.67 & 0.83 & 0.77 & 0.47 & 0.47 & 0.57\\ 0.77 & 0.00 & 0.53 & 0.53 & 0.33 & 0.40 & 0.70 & 0.67 & 0.70\\ 0.73 & 0.30 & 0.00 & 0.67 & 0.60 & 0.70 & 0.57 & 0.57 & 0.50\\ 0.73 & 0.33 & 0.67 & 0.00 & 0.63 & 0.83 & 0.63 & 0.57 & 0.63\\ 0.67 & 0.30 & 0.53 & 0.47 & 0.00 & 0.43 & 0.60 & 0.70 & 0.53\\ 0.67 & 0.50 & 0.60 & 0.60 & 0.57 & 0.00 & 0.80 & 0.83 & 0.80\\ 0.13 & 0.40 & 0.27 & 0.07 & 0.30 & 0.50 & 0.00 & 0.80 & 0.73\\ 0.07 & 0.47 & 0.27 & 0.23 & 0.37 & 0.53 & 0.73 & 0.00 & 0.77\\ 0.17 & 0.40 & 0.33 & 0.23 & 0.43 & 0.43 & 0.77 & 0.77 & 0.00\end{bmatrix}$$

Step 4: Let the threshold λ be 0.7, the final adjacency relation matrix is:

$$B=\begin{bmatrix}0&0&1&0&1&1&0&0&0\\1&0&0&0&0&0&1&0&1\\1&0&0&0&0&1&0&0&0\\1&0&0&0&0&1&0&0&0\\0&0&0&0&0&0&0&1&0\\0&0&0&0&0&0&1&1&1\\0&0&0&0&0&0&0&1&1\\0&0&0&0&0&0&1&0&1\\0&0&0&0&0&0&1&1&0\end{bmatrix}$$

Step 5: The reachability matrix is:

$$C=\begin{bmatrix}1&0&1&0&1&1&1&1&1\\1&1&1&0&1&1&1&1&1\\1&0&1&0&1&1&1&1&1\\1&0&1&1&1&1&1&1&1\\0&0&0&0&1&0&1&1&1\\0&0&0&0&0&1&1&1&1\\0&0&0&0&0&0&1&1&1\\0&0&0&0&0&0&1&1&1\\0&0&0&0&0&0&1&1&1\end{bmatrix}$$

Step 6: Reachability sets and antecedent sets are shown in Table 3.

Table 3. Reachability sets and antecedent sets.

S_i	$R(S_i)$	$Q(S_i)$	$R(S_i) \cap Q(S_i)$
S_1	1,3,5,6,7,8,9	1,2,3,4	1,3
S_2	1,2,3,5,6,7,8,9	2	2
S_3	1,3,5,6,7,8,9	1,2,3,4	1,3
S_4	1,3,4,5,6,7,8,9	4	4

S_5	5,7,8,9	1,2,3,4,5	5
S_6	6,7,8,9	1,2,3,4,6	6
S_7	7,8,9	1,2,3,4,5,6,7,8,9	7,8,9
S_8	7,8,9	1,2,3,4,5,6,7,8,9	7,8,9
S_9	7,8,9	1,2,3,4,5,6,7,8,9	7,8,9

According to the rule $R(S_i) \cap Q(S_i) = R(S_i)$, the final level division result is described in Table 4 and Figure 2.

As can be seen from Table 4 and Figure 2, the 9 categories of influencing factors can be divided into 4 levels. Natural environment and policy environment are fundamental factors, which will not only affect the technical-economy effect of VSC-HVDC system, but also affect the economic environment and social environment within the influencing factor set. Price level, technical level, economic environment, and social environment belong to the intermediate level, and the price level and technology level can be affected by the latter two. Design requirements, equipment requirements and project features are superficial factors, which can be affected by all the other factors in the influencing factor set.

Table 4. Hierarchical structure of influencing factors.

Level	Influencing factors
I	S_7, S_8, S_9
II	S_5, S_6
III	S_1, S_3
IV	S_2, S_4

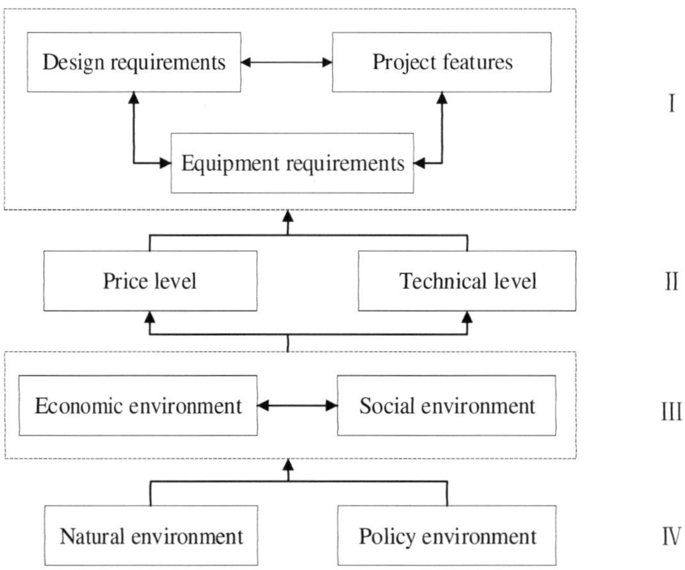

Figure 2. Multilevel hierarchical directed graph.

Based on the above study, we can summarize the technical-economy analysis procedure of VSC-HVDC system as follows:

Firstly, consider the consistency between VSC-HVDC project, natural condition and policy direction; secondly, discuss whether the economic and social environment can provide a smooth channel for VSC-HVDC project; thirdly, consider whether the technical condition can provide sufficient support, and whether the price level is within acceptable range; finally, formulate reasonable design scheme.

5. Conclusion

To provide decision support for investors, government and relevant institutions, this paper establishes a influencing factor system for technical-economy effect of VSC-HVDC system, and divides the hierarchical structure within the system based on improved fuzzy ISM method. The main conclusions are as follows:

(1) There are 39 influencing factors of technical-economy effect of VSC-HVDC system, which can be divided into 9 categories, including economic environment, natural environment, social environment, policy environment, price level, technical level, design requirements, equipment requirements and project features.

(2) Influencing factors of technical-economy effect of VSC-HVDC system can be divided into 4 levels. Natural environment and policy environment are fundamental factors; economic environment and social environment belong to the second level; price level and technical level belong to the third level; design requirements, equipment requirements and project features are superficial factors.

This paper gives a general summary for influencing factors of technical-economy effect of VSC-HVDC system. With the development of VSC-HVDC, the influencing factor system in this paper can be further complemented and perfected, the hierarchical structure division can be more thorough and comprehensive, and the research area can be further broadened.

Acknowledgments

This paper is based on the technology project：key equipment developing and engineering technology of VSC-HVDC for large-scale further offshore wind farm. The support of Global Energy Interconnection Research Institute of State Grid is greatly appreciated.

References

[1] Wang LF, Singh C and Kusiak A 2012 Special issue on integration of intermittent renewable energy resources into power grid *J. IEEE Systems Journal* **6** 2-3

[2] An J, Chai XZ and Wang H 2016 Prospective study of Henan power grid optimized operation with VSC-HVDC *J. Power System Technology* **40** 86-91

[3] Pei ZY, Liu P, Zhang AJ and Zhou YP 2016 An overview on VSC-HVDC power transmission systems *J. IJCA* **9** 33-44

[4] He YX 2011 *Comprehensive Evaluation Method of Electric Power Its Application* (Beijing: China Electric Power Press) p 131

[5] Wu H and Niu DX 2017 Study on influence factors of electric vehicles charging station location based on ISM and FMICMAC *J. Sustainability* JS9(2017)484

[6] Sun QH and Yang Y 2015 On the risk management of the project portfolio of construction enterprises based on ISM *J. Value Engineering* **16** 227-30

[7] Jin WL 2015 Analysis of influence factors of logistics management specialty experimental teaching practice of independent colleges based on ISM *J. Logistics Technology* **17** 295-299

[8] Zhang PK and Luo F 2017 Influencing factors of runway incursion risk and their interaction mechanism based on DEMATEL-ISM *J. Tehnicki Vjesnik-Technical Gazette* **24** 1853-61

[9] Zhang XQ, Yuan YL and Sun ZC 2017 Research of improved interpretive structural model on influencing factors of resident travel cost *J. Journal of Chongqing Jiaotong University:Natural Science* **7** 101-5

[10] Yang XP 2012 *HVDC transmission and VSC-HVDC transmission* (Beijing: China Electric Power Press) p 13

[11] Zhao WJ 2011 *High voltage direct current transmission engineering technology* (Beijing: China Electric Power Press) p 20

IOP Conf. Series: Materials Science and Engineering **366** (2018) 012063 doi:10.1088/1757-899X/366/1/012063

Analysis and Countermeasure of Burnout of Electromagnetic Voltage Transformer of High Altitude Area in Western China

Hongbo Li[1], Yongbin Bao[2], Pei Li [1], Wenqi Xu[2], Yanyan Zhang[1],Tailong Si[1] and Jiadong Guo[1]

[1]State Grid of China Technology College, Jinan 250002, China
[2]Yushu Electric Power Supply Company of National Power Grid,Yushu 815000, China.

13305318311@163.com

Abstract. 3 burning accidents of electromagnetic voltage transformer of Qinghai power grid in the southern high altitude area was introduced in this paper, and the cause of the accident was analyzed, which is due to a phase grounding line, then the other two phases voltage rising caused voltage transformer damaged; occurrence of Ferro resonance over-voltage, which resulting in serious core saturation, and the over-current caused voltage transformer damaged; due to open delta is short circuit, when the single-phase grounding fault occurred, because the voltage output of 100 V from open delta winding, a very large current is formatted, and a lot of heat is made, also led to the current in primary side very large, resulting in burning TV. In view of these three reasons, the corresponding prevention measures are put forward to ensure the normal operation of the power system.

1. Introduction

In the 3~35kV neutral point ungrounded grid, a large number of electromagnetic voltage transformers are installed on the bus line to monitor the Three-Phase phase voltage of power system. Because the power distribution units mostly adoTV indoor cabinet structure, the installation of the equipment in the cabinet is compact, and the small volume voltage transformer is required. After many years of practical operation, the pouring type voltage transformer is small in size and reliable in operation, which is suitable for switch cabinet operation, and also meets the development requirements of switch cabinet without oil. In this paper, several 10kV electromagnetic voltage transformer burnout accidents in high altitude area of Qinghai power grid are analyzed and summarized; at last, corresponding preventive measures are put forward[1].

2. Accident Process

2.1 35kV Substation A

One day, 22:00, * line B phase line in 110kV substation * is grounding, At the same time, the C phase voltage is abnormal, and the #3521 circuit breaker is opened in the distance, and the equipment is required to be checked in the field. Through on-the-spot checking, Operation and maintenance personnel found that the cabinet closed door 35kV I bus open, the whole cabinet body deformation, side opening. After examination, the appearance of TV in A phase was normal, The B phase TV has a burn mark on the side of the C phase, and the C phase TV burst, harmonic elimination device crack.

And Figure 1 is the burning accident diagram of this TV.

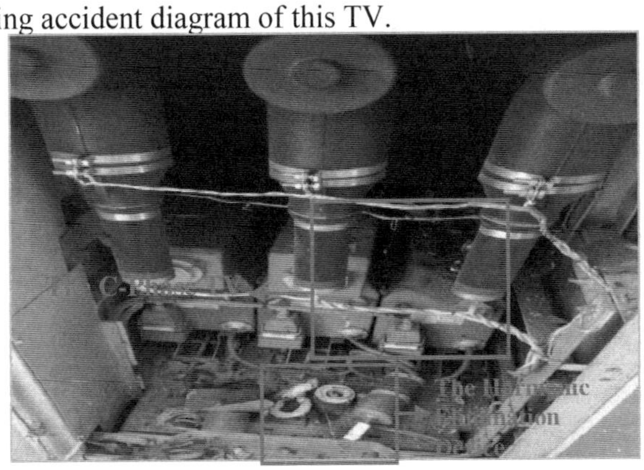

Figure.1 the 35kV Voltage Transformer Burning Accident Diagram

2.2 110kV Substation B

The substation is 4200 meters above sea level. A dispatch notification is received at 15:16 a day, and the voltage of the B10kV bus voltage of the substation B phase voltage disappears. At 19:20, Operation and maintenance personnel arrive at the station to check 10kV and 35kV high voltage rooms, after entering, there was a strong smell of smoke. The protection action information of overcurrent I section on the protective device of * Road is found at 10:08. The operation and maintenance personnel apply for the 10kV bus TV to be checked carefully after the operation of the reacting scene. At 20:15, the operation and maintenance personnel opened the 10kV cabinet door, and found B phase TV burst. And figure 2 is the burning accident diagram of this TV.

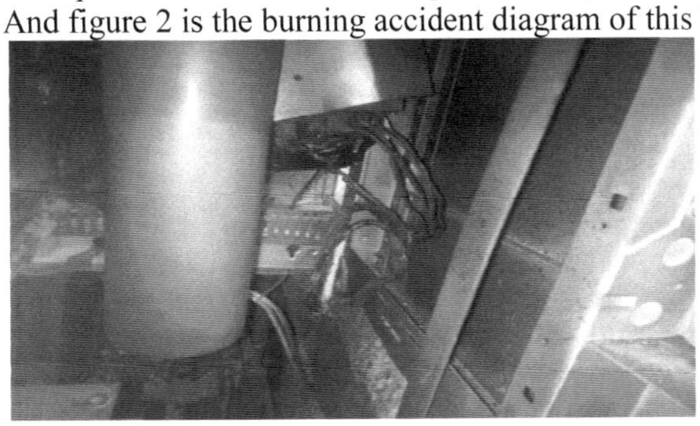

Figure.2 the 10kV Voltage Transformer Burning Accident Diagram

2.3 35kV Substation C

The substation is 3800 meters above sea level. At 22:19 one day, the dispatch department noticed that low backup protection of #1 main transformer in this substation have actioned. Measurement and control device display the bus-bar TV broken line and alarm. At second days 09:00, the dispatching department notifies the voltage anomaly of the 10kV bus in the substation. At 13:16, operation and maintenance personnel arrive at the scene, and found the 10kV TV door opened by burst. At 13:37, apply for the dispatch department to transfer the 10kV bus TV to the overhaul. At the same time, open the door to check, found TV explosion, at 14:40 the 10kV bus TV isolation after the restoration of power supply. And figure 3 is the explosion scene diagram of this TV.

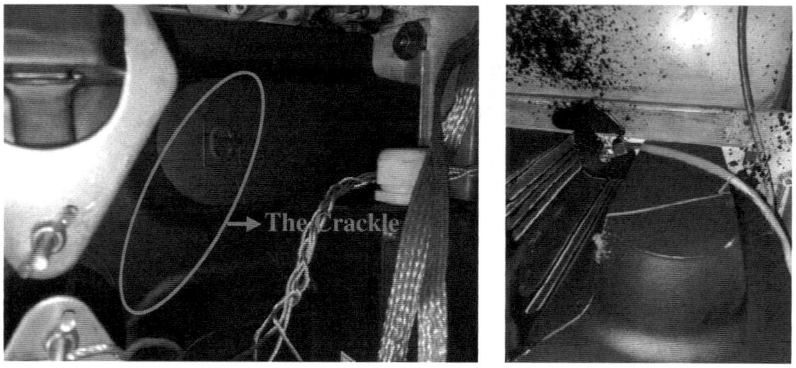

Figure.3 the 10kV Bus TV Explosion Scene Diagram

3. Analysis of the Cause of the Accident

3.1 35kV Substation A-caused by Single-phase grounding fault.

After investigation, we found that B phase of the single-phase ground fault occurred at 22:00, the phase voltage of A and C phase changes into line voltage, resulting in C phase 35kV voltage transformer explosion. When B is connected to the ground, the equivalent circuit diagram is shown in Figure 4.

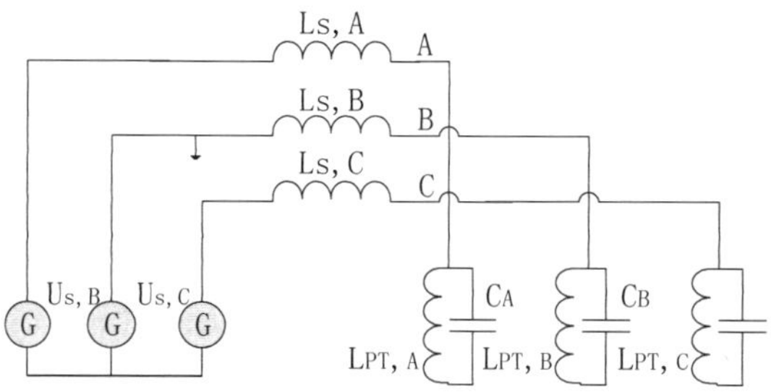

Figure.4 Equivalent Circuit Diagram of B-phase Grounding

The instantaneous flux of the C phase ΦU is as follows:

$$\Phi_U = \sqrt{3}\Phi_m \sin(wt+\alpha-\pi/3)+[\Phi_r-\sqrt{3}\Phi_m \sin(\alpha-\pi/3)]e^{-(R_1/L_1)t} \quad (1)$$

From the formula (1) we can obtained that: The main parameters affecting the instantaneous magnetic flux in the core are the initial phase angle α and the remanence Φr when the failure occurs. It can be seen that when single phase grounding occurs, the voltage of the non-fault phase is abrupt to the line voltage; the core enters the saturation area and produces inrush current in the winding. When the core enters the depth saturation, its inrush can reach dozens of times of the steady current, and its heat is enough to cause the damage of the voltage transformer [2].

3.2 110kV Substation B-caused by ferromagnetic resonance

From analysis of the information collected from the protection device and its fault recorder, at 14:55 we found that Ua,Ub,Uc all change suddenly;Ub,Uc transcends the upper limit, and Ua goes beyond the lower limit. A-phase voltage disappears, indicating the occurrence of grounding fault in A-phase.

Because the power neutral point of the system is insulated to the ground, the neutral point of the primary winding of the voltage transformer is directly grounded, and the open triangle zero sequence winding open circuit. The 35kV line is unloaded line, the line to ground capacitance is larger than the line impedance, and line voltage will increase. And when the bus-bar is unloaded or less line, closing the charge or eliminating the ground fault during operation, the ground capacitance of the power grid is matched to the excitation inductance of the B phase voltage transformer. When a grounding fault occurs in A-phase, the condition of the excitation of ferromagnetic resonance is produced, which makes the core of the voltage transformer saturate and produces the neutral point displacement. These conditions are meted, so the judge is ferromagnetic coil voltage transformer Ferro-resonance, excitation inductance decreases, the saturation voltage transformer excitation current depth will increase sharply, and finally burned B-phase TV. Figure 5 is the voltage waveform diagram when a metal grounding fault occurs in A-phase.

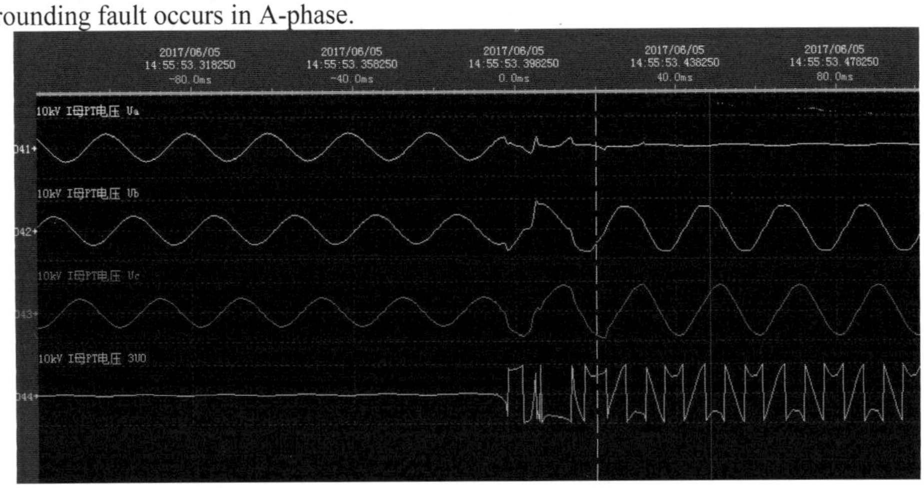

Figure.5 Voltage Waveform Diagram When a Metal Grounding Fault Occurs in A-phase

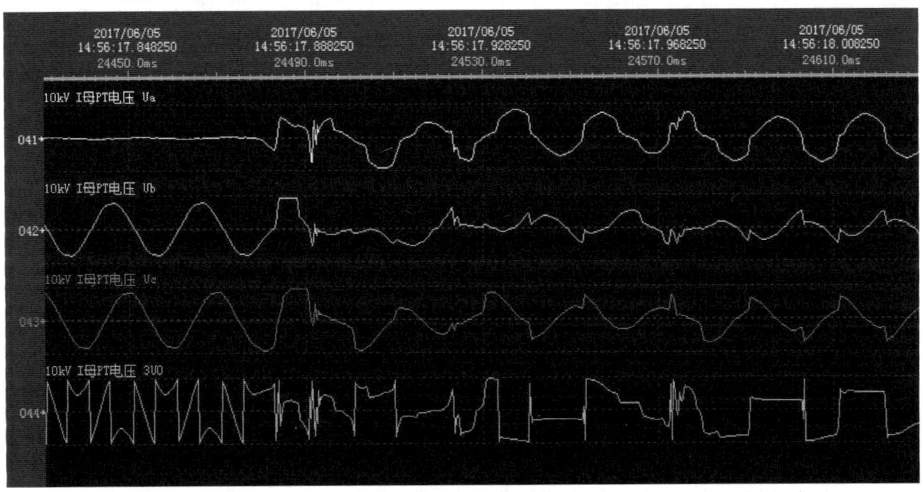

Figure.6 Voltage Waveform Diagram with Ferromagnetic Resonance

In the medium voltage power grid, the voltage transformer is directly installed on the bus. Because of the capacitance of the circuit, the TV will form a zero sequence circuit with the inductance of the transformer, and the system equivalence diagram is shown in Figure 6.

The operation and experiment of the power system show that when the inductance of the TV and the line are matched to the ground capacitance, it produces Ferro-resonance at different frequencies under certain conditions. With the increase of line length, 3 harmonic harmonic resonance, fundamental frequency resonance and frequency division resonance will occur in turn [3]. An equivalent circuit diagram of the system is shown in Figure 7.

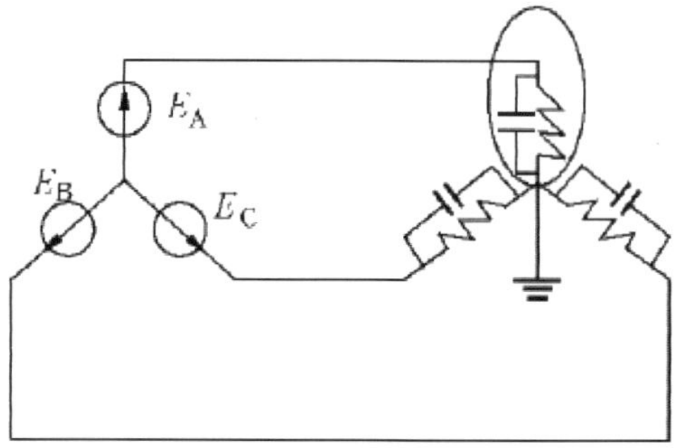

Figure.7 Equivalent Circuit Diagram of the System

3.3 35kV Substation C-caused by open-delta short circuit

After the field investigation, it was found that the two lines of C601 in the TV cabinet were bonded together with the N600 insulated wire skin as shown in figure 8. Unlock the secondary line of the TV side and check it. It is found that the secondary line N600 in the TV cabinet is connected with the L650, and the discharge coils of the N600 and the capacitor in the 10kV capacitor cabinet are found together. The open triangle is short circuited. Under normal conditions, the output voltage of the open triangle winding is basically 0, so the 3Uo is 0. The short circuit of the open triangle winding does not affect the normal operation of the voltage transformer. When a single-phase grounding fault occurs, the voltage of the open triangle winding output is 100 V. Once the connection is short circuited, that is, the secondary side of the voltage transformer is short circuited, which will form a very large current and heat, and will lead to a large current in primary side. It eventually led to the burning of TV.

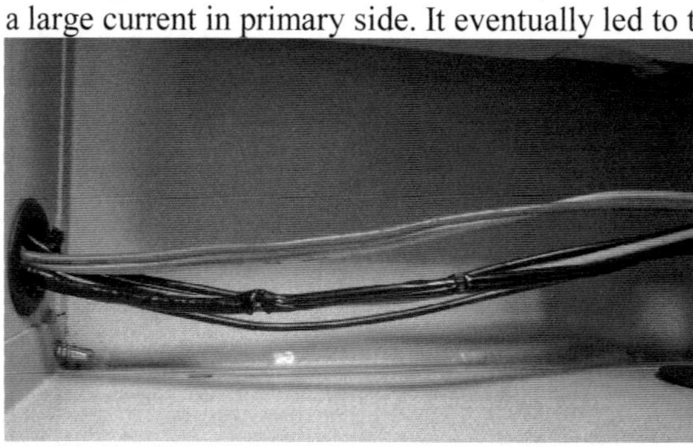

Figure.8 Secondary Connection Short Connection Schematic

4. Measures to Prevent the Burning of Voltage Transformers

4.1 Prevention and control measures of transformer damage caused by single-phase grounding fault

In this case, the measures taken are:

1) To improve the insulation level of the distribution network and reduce the condition of single-phase grounding;

2) To add appropriate arc suppression coil to restrain the occurrence of intermittent arc grounding;

3) To improve the excitation characteristics of the voltage transformer so that the voltage transformer is not easy to enter the saturation area.

4) To adopt a new type of heat resistant material to increase the heat resistance of voltage transformer.

4.2 Prevention and control measures of transformer damage caused by ferromagnetic resonance

In this case, the measures taken are:

1) Installing a micro-computer harmonic elimination device at the open triangle terminal of the remaining voltage winding.

This device can identify various forms of ferromagnetic resonance overvoltage, including frequency division, fundamental frequency, and three frequency doubling. When a certain resonance is determined, a specific harmonic elimination program is introduced. The open delta winding is short circuited, and the damping current is output to the grid, so that the resonance overvoltage is eliminated quickly. At the same time, it is noted that the voltage transformer is a full insulation design.

2) Parallel one resistance (incandescent lamp) in the remaining voltage winding open triangle terminal

Linear damping resistors or incandescent lamps are installed at the opening end of the TV open triangular winding. This method is simple and is especially easy to be realized. When the excitation characteristic of TV is good, it can play zero sequence damping effect, but because the resonance excited by non-metallic grounding cannot be restrained, the high voltage fuse of TV is easy to fuse.

3) Grounding the neutral point of the high voltage side of the transformer through the high impedance zero sequence transformer

The zero sequence TV is connected to a neutral point at TV, which is called the 4TV connection mode. The zero sequence measurement loops in Figure 9 is a measurement coil of the open triangle and zero sequence voltage transformer of a three-phase voltage transformer in series with positive polarity. It contains a few partial zero sequence voltages of the three-phase voltage transformer. At the same time, the zero sequence circuit is not too short, and the zero sequence current is not too large to avoid the burn out of the zero sequence loop winding. This method is more effective for eliminating resonance and restraining ultra-low frequency oscillation.

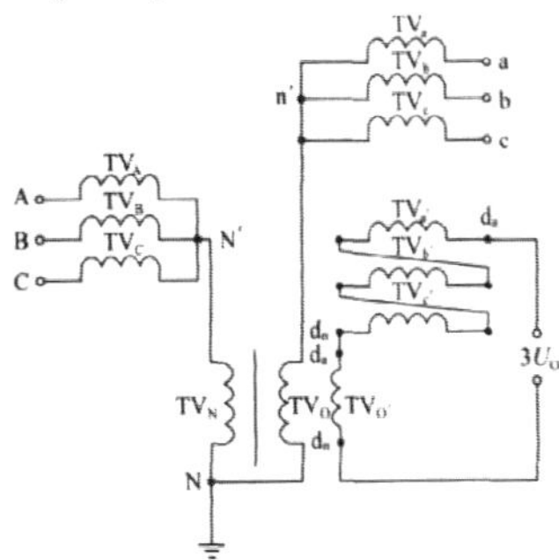

Figure.9 4TV Open Triangle and Zero Sequence Voltage Transformer Series Principle Diagram

4) To connected a LXQ type harmonic eliminator between the neutral point and the ground of the primary side of TV

The most important feature of the LXQ harmonic eliminator is that it can effectively limit the primary side inrush of TV, prevent the fusion of TV high voltage fuse and eliminate the resonance phenomena of various harmonics. The resonance overvoltage excited by non-metallic grounding can also be suppressed. In comparison, the principle of primary side harmonic elimination is to suppress the inrush, so that the resonance is not easy to expand at the beginning, and it is eliminated in the bud. The secondary side harmonic elimination is eliminated after the resonance, and there is a response time problem, so the effect of primary side harmonic elimination is better than that of the two harmonic elimination effects [4].

4.3 Prevention and control measures of mutual inductor caused by open triangle short circuit

The direct cause of the 10 kV voltage transformer fault is the short circuit of the two connection of the open triangle winding of the voltage transformer. If the secondary wiring of the secondary opening trigonometric winding of the voltage transformer is short circuited by the construction unit, it is difficult to find the conventional inspection and acceptance. When the equipment is running normally, the external check will not find any exception. Or the short connection of the secondary connection leads to the open triangle short circuit. In this case, there is no problem when the system is running normally. Only when the single phase grounding fault occurs will cause serious damage to the voltage transformer equipment. Therefore, it is suggested that in the process of substation completion and acceptance and operation inspection, we should strengthen the wiring inspection of the secondary opening triangular circuit of voltage transformer, so as to avoid the occurrence of open triangular short circuit [5].

5. Conclusions

About the phenomenon of frequent burnout of 10~35kV electromagnetic voltage transformer in high altitude area of Qinghai power grid, It should be combined with the characteristics of high altitude, The reasons such as grounding fault of line, ferromagnetic resonance and open triangle short circuit are taken into consideration. The corresponding prevention and control measures are taken to avoid similar accidents and ensure the normal operation of the system.

Reference

[1] Zhu Tao,Zhang Hua. Practical Operation Technology of Substation Equipment [M]. China Electric Power Press,2014:105.

[2] Che Weihong,Chen Runying etc. Analysis of the cause of voltage transformer damage caused by single-phase grounding fault in 10kV power grid [J]. Guangdong Electric Power, 2016,29（1）: 113-117.

[3] Lin Quan. The cause and solution of the explosion of 10kV electromagnetic voltage transformer [J]. Power Equipment, 2005,6(12)：71-73.

[4] Zhang Li. Anti-resonance measures of 10kV voltage transformer [J]. Jiangsu electrical, 2007,5（4）：40-42.

[5] Chen Zhiyong. Fault analysis of 10kV voltage transformer caused by single-phase grounding [J]. Power safety technology, 2016,18（6）：23-24.

IOP Conf. Series: Materials Science and Engineering **366** (2018) 012064 doi:10.1088/1757-899X/366/1/012064

Research on Distribution Network Reconfiguration Method Considering Switching Operations

YangFan[1], LiuJun[1] and LuBingbing[1]

[1]Electric Power Research Institute, SMEPC, 171# Handan Rd., Shanghai, China

1156070293@qq.com

Abstract. The traditional distribution network reconfiguration method does not consider the impact of switching operations, resulting network equipments voltage instability and user blackouts in actual implementations. In this paper, the loop operation security constraints have been fully taken into account as establishing distribution network reconfiguration model by introducing constraints into the traditional model. The greedy random adaptive search procedures (GRASP) is adopted to obtain optimal solution and detailed steps of reconfiguration scheme and switching operation for distribution network, which ensures the solution is realized without network equipments voltage instability and user blackouts. After IEEE 14-bus system verification, this method shows its better adaptability and practical value, which fully combines actual work statues to improve distribution network reliability.

1. Introduction

Distribution network reconfiguration plays a crucial role in grid optimization, which refers to changing segment switches and contact switches status to adjust distribution network structure, aiming network loss decrease, balance and other purposes[1]. Link and unlink operations are commonly used in grid dispatching department to achieve network structure adjustment, as closing a contact switch to form a ring and then open the ring by switching off a certain segment breaker to restore its radial wiring, without user power loss during load transfer process. With the increasing complexity of distribution grids and demanding power supply reliability, the link and unlink operations have become a routine for network dispatchers to adjust grid structure to realize power flow redistribution[2].

However, in items of security constraints, the current distribution network reconfiguration methods basically only verify the final network structure after the model is solved, which lacks detailed analysis for each switch operation process and its relevant network security constraint. In particular, the node voltage and line power flow constraints required to meet switching operation during reconfiguration process are generally not taken into consideration.

Therefore, although the final network structure can meet security constraints, problems like network equipments voltage instability and user blackouts may still occur during the reconfiguration implementation causing infeasibility in areas where the demand for power supply reliability is high.

In summary, this paper researches the distribution network reconfiguration method considering switching operations in order to ensure its rationality and feasibility.First, it establishes distribution network reconfiguration model by introducing constraints into the traditional model. Then, the greedy random adaptive search procedures (GRASP) are used to solve the model based on the Thevenin equivalence theorem. Finally, an example of a 14-node system is given to verify the correctness and validity of the proposed method.

2. Distribution network reconfiguration method considering switching operations

After link and unlink operations, the system security constraints of distribution network reconfiguration model include both radiation and ring connection.To ensure system security during the whole reconfiguration procedures, corresponding constraints have to be included after each switch operation in the model. According to the daily actual operation of distribution network, only single link or unlink operation is commenced at the same time.Based on the considerations above, the distribution network reconfiguration method considering switching operations is shown as follows:

$$\min P_{loss} = \sum_{ij=1}^{N_l} r_{ij} \left(\frac{p_{ij(2g)}^2 + q_{ij(2g)}^2}{v_{k(2g)}^2} \right) \tag{1}$$

s.t.

$$\begin{cases} \sum_{i=1}^{k} (p_{ij(2t-1)} + jq_{ij(2t-1)}) = p_{dj} + jq_{dj} \\ v_{i(2t-1)} - v_{j(2t-1)} = z_{ij} i_{ij(2t-1)} \\ \sqrt{\left| p_{ij(2t-1)}^2 + q_{ij(2t-1)}^2 \right|} \le s_{ij\max} \\ v_{i,\min} \le v_{i(2t-1)} \le v_{i,\max} \\ N_b = N_l \end{cases} \tag{2}$$

$$\begin{cases} \sum_{i=1}^{k} (p_{ij(2t)} + jq_{ij(2t)}) = p_{dj} + jq_{dj} \\ v_{i(2t)} - v_{j(2t)} = z_{ij} i_{ij(2t)} \\ \sqrt{\left| p_{ij(2t)}^2 + q_{ij(2t)}^2 \right|} \le s_{ij\max} \\ v_{i,\min} \le v_{i(2t)} \le v_{i,\max} \\ N_b = N_l - 1 \end{cases} \tag{3}$$

$$t = 1, 2, \cdots g \tag{4}$$

Among them, P_{loss} for system actual loss, N_l for number of lines, N_b for number of nodes, r_{ij} , p_{ij} , q_{ij} , i_{ij} , z_{ij} , $s_{ij\max}$ for resistance, active power, reactive power, current, impedance, maximum apparent power of line ij respectively, v_i , v_j for voltage value of node i and node j. $v_{i,\max}$, $v_{i,\min}$ for upper and lower voltage limits of node i. p_{dj} , q_{dj} for active load and reactive load value of node j. g for total number of link and unlink operations, one operation represents one step, t for t_{th} step.

In the model, equation (1) is the objective function which aims to minimize line loss in this chapter.The variable with a subscript of 2g in the objective function represents all the operational variables of the final reconstructed radial grid after all link and unlink operations have been commenced.

Equation (2) is the set of network constraints for the single-loop network after link operations, where all the variables with the subscript 2t-1 represent the operational variables after link operations. This constraint set includes the power balance constraint, Kirchhoff's voltage law constraint, line current limit constraint, voltage limit constraint and ring topology constraint for the single-loop network after link operations.

Equation (3) is the set of network constraints for the single-loop network after unlink operations, where all the variables with the subscript 2t represent the operational variables after unlink operations.

This constraint set includes the power balance constraint, Kirchhoff's voltage law constraint, line current limit constraint, voltage limit constraint and ring topology constraint for the single-loop network after unlink operations.

Equation (4) for the step of link and unlink operations, t for t_{th} step.

Comparing equation (2) and (3), we can see that except variable subscript, the only difference between two equations is the relationship between the number of nodes N_b and the number of branches N_l. As a single ring network in the distribution grid is generated after link operation, the number of nodes N_b and the number of branches N_l are exactlythe same.

The model designed in this paper includes more variables and constraints comparing traditional distribution network reconfiguration model,which satisfies voltage, power and load balance limit constraints in each link and unlink operation during reconfiguration process.Therefore, it is practicable to solve the safety constraints in link and unlink operation by adopting the distribution network reconfiguration methodpresented in this chapter.

3. Closed-loop distribution network power flow calculation

Closed-loop distribution network power flow calculation is an important way to verify the operation safety of distribution network when the loop is closed.

This chapter uses the Thevenin equivalence theorem for the distribution network closed-loop power flow calculation considering its advantage of fast calculation speed and no need to modify the original forward and backward substitution power flow algorithm. The illustration is shown in figure 1:

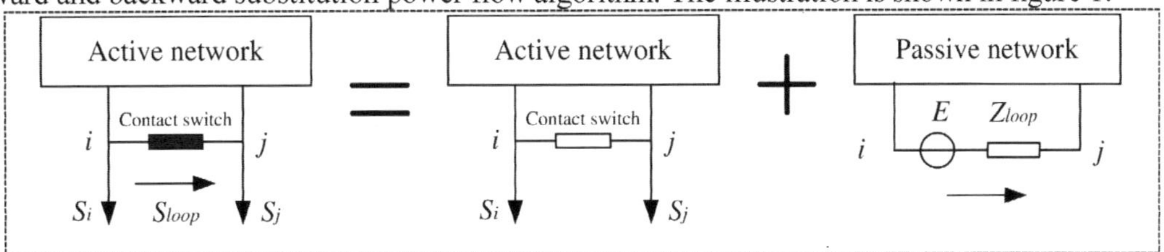

Figure 1. Closed-loop distribution network power flow calculation based on Thevenin equivalence theorem.

The calculation steps of Closed-loop distribution network power flow calculation based on Thevenin equivalence theorem are shown as follow:

(1) Obtaining voltage values v_i , v_j of the nodes on both sides of loop by using forward and backward substitution method to calculate power flow of initial distribution grid;

(2) Calculating Thevenin equivalent impedance $Z_{loop} = \sum_{i=1}^{loopb} Z_i$, as *loopb* for the number of branches and Z_i for the impedance of the i_{th} branch in the loop;

(3) Calculating closed-loop current: $I_{loop} = \frac{v_i - v_j}{Z_{loop}}$. Obtaining equivalent injection power of node i,j in the closed-loop branch according to closed-loop current value. Adding equivalent injection power values to node i,j, generating radiation equivalent circuit as shown in figure 2:

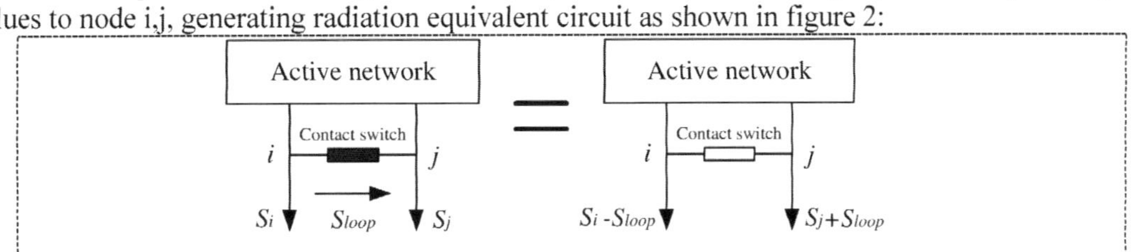

IOP Conf. Series: Materials Science and Engineering **366** (2018) 012064 doi:10.1088/1757-899X/366/1/012064

Figure 2. Equivalent closed-loop circuit.

(4) Calculating power flow of radial network shown in figure 2byforward and backward substitution method;

(5) Verifyingthe network voltage and line power constraints according to power flow calculation results after equivalent process.

As can be seen from the closed-loop power flow calculation method mentioned above, the closed-loop branches finally become equivalent to the nodes on both sides the branch in actual calculation, which can be directly calculated by using forward and backward substitution method. The safety of link operation can be verified by closed-loop power flow calculation and it will be not feasible if the safety constraints are not satisfied.

4. Model solving

At present, there are three kinds of algorithms to solve distribution network reconfiguration model. The first one is the heuristic algorithm, which is easy to achieve but difficult to get optimal solution. The second one is the classical mathematical algorithm, which can get the global optimal solution but the solution speed will become slow when system gets huge. The third one is the modern optimization algorithm, which is more suitable for solving large-scale combinatorial optimization problems and it has been widely used in power system optimization problems[3]. The greedy random adaptive search procedures (GRASP) presented in this paper is such an algorithm.

Compared with other modern optimization algorithms such as genetic algorithm and simulated annealing algorithm, the GRASP has advantages with less parameter to be set and easy to make configurations[4]. In addition, since the GRASP is based on line-by-line construction to generate initial feasible solution, the specific operation process of reconfiguration switch can be reflected and achieve security verification at construction stage[5]. The other types of modern optimization algorithms generally provide final reconfiguration solution directly without detailed reconfiguration process.The distribution network reconfiguration flow chart considering closed-loop constraints is shown in figure3:

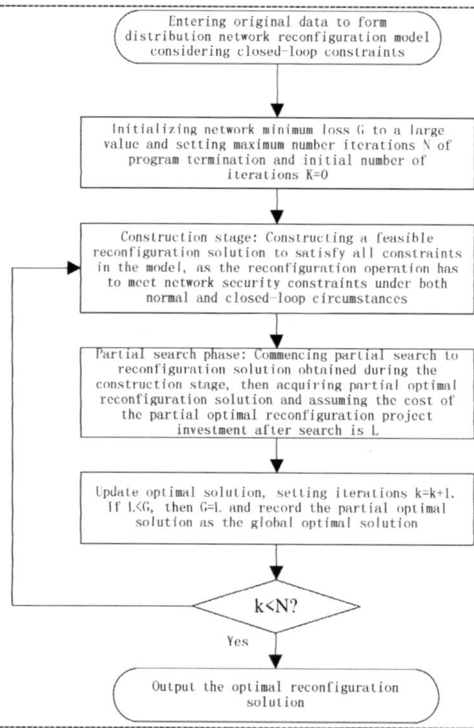

Figure 3. The distribution network reconfiguration flow chart considering closed-loop

constraints.

The main steps of using GRASP to solve distribution network reconfiguration model considering closed-loop constraints are shown as follows:

(1) Entering original data to form distribution network reconfiguration model considering closed-loop constraints as equation (1)-(4);

(2) Initializing network minimum loss G to a large value and setting maximum number iterations N of program termination and initial number of iterations K=0;

(3) Construction stage: constructing a feasible reconfiguration solution to satisfy all constraints in the model, in other words, the reconfiguration operation has to meet network security constraints under both normal and closed-loop circumstances;

(4) Partial search phase: Commencing partial search to reconfiguration solution obtained during the construction stage, then acquiring partial optimal reconfiguration solution and assuming the cost of the partial optimal reconfiguration project investment after search is L;

(5) Update optimal solution, setting iterations k=k+1. If L<G, then G=L and record the partial optimal solution as the global optimal solution;

(6) If the current number of iterations k is less than the maximum number of iterations N set before, then goes to step (3).Otherwise, the program will terminate and output the optimal reconfiguration solution with its investment cost G.

As can be seen from the above calculation steps of the GRASP algorithm, the construction phase and the partial search phase are two crucial components of the algorithm. The detailed steps of two stages are explained in the next part.

4.1. Construction phase

(1) Using forward and backward substitution method to calculate current radial connection distribution network power flow;

(2) Calculating the voltage difference between two sides of all contact switches and sort them in descending order, selectingpre-sequence CL switches to form an alternative ring switch set;

(3) Randomly selecting a contact switch from the set of alternative closed-loop switches and closing it;

(4) Calculatingpower flow in closed-loop distribution network and check safety constraints. If the constraints are met, turn to (5). Otherwise, turn to (3);

(5) Selecting the segment switch that minimizes network loss in the ring formed by contact switch, and then turning it on. Among them, the degree of loss reduce enabled by different sections of switches can be immediately obtained by equation (5);

$$\Delta P = \mathrm{Re}\left\{2(\sum_{i\in D} I_i)(E_m - E_n)^*\right\} + R_{loop}\sum_{i\in D}|I_i|^2 \tag{5}$$

(6) Calculating the power flow in open-loop distribution network by forward and backWard substitution method, if the safety constraint is satisfied, keep the operation and turn to (7).Otherwise, mark the segment switch not for open in this loop and turn to (5);

(7) If the loss changing indicators $\Delta P > 0$ turn to (1). Otherwise, terminating the program and output the candidate reconfiguration solution.

An initially feasible distribution network reconfiguration solution will be generated through construction phase.

4.2. Partial phase

(1) Selecting the reconfiguration solution generated in construction phase as the initial solution in partial search phase;

(2) Select one of the link and unlink operations in the reconfiguration solution and temporarily cancelling it to restore grid connection before any operation;

(3) Calculating the power flow of distribution network and solving the change of network loss;

(4) If $\Delta P > 0$, the network loss will increases, then the link or unlink operation is retained in the reconfiguration solution, marking it as a non-cancellable operation and turning to (2). Otherwise, if the network loss is unchanged or reduced, the link or unlink operation will be permanently cancelled in the reconfiguration solution and turning to (5);
(5) If all the link and unlink operationshave completed temporary cancellation and verification, the final reconfiguration solution will be presented.Otherwise, turn to (2).
Based on the N times iterations, the global optimal solution with the least loss is obtained and the model solving process is completed.

5. Example analysis

This chapter uses 14 nodes system as an example to verify the method in this paper. The example assumes that the lower voltage limit for all nodes is 0.95 pu. The power limits for lines 1-2, 2-4, 4-5, 5-14, 2-3, 3-9 are 10 mW. The revised load data is shown in table 1:

Table 1. 14 nodes system load data.

Node number	Active power (mW)	Reactive power(mVar)	Node number	Active power (mW)	Reactive power(mVar)
1	0.21	0.10	8	0.95	0.87
2	2.13	1.41	9	0.46	-0.55
3	2.94	0.16	10	4.73	-1.76
4	2.01	-0.33	11	1.16	0.82
5	1.67	0.99	12	0.80	-1.15
6	3.77	2.92	13	0.97	0.78
7	4.98	2.00	14	1.06	-2.73

The initial grid topology of a 14nodes power distribution system is shown in figure 4-a, where the solid lines represent segmented lines and dashed lines for contact lines. The initial grid satisfies security constraints and the total active loss is479.3kW.

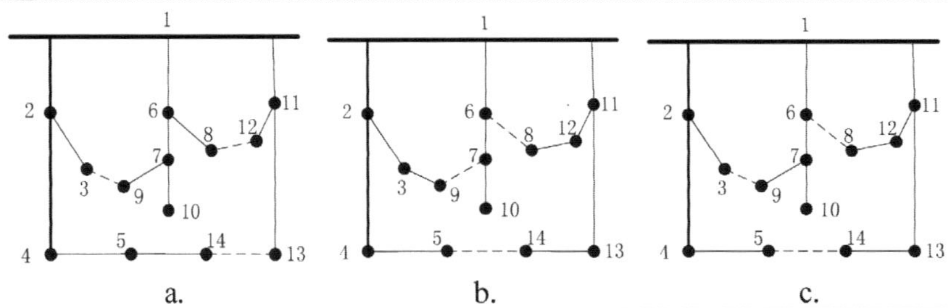

Figure 4. The initial grid / The reconfiguration without considering the closed-loop constraints / The reconfiguration considering the closed-loop constraints.

First of all, the reconstruction results of the initial grid are obtained by using the traditional distribution network reconfiguration method to get the optimal reconfiguration grid connection without considering the constraints of closed-loop is shown in figure 4-b.The final grid without considering the constraints of closed-loop can satisfy the safety requirements under regular radial grid connection. The total active loss is 431.0 kW.

It can be seen from figure 4-b that regardless of the closed-loop constraints, the states of six switches in reconfiguration solution have changedin total.However, according to the conventional calculation method, it is impossible to obtain the detailed operation steps of six switches.In order to avoid power cut, according to the location of the ring where the switch is located, the final reconfiguration solution can be achieved by the following operations: close the contact switch of lines 3-9 and open the segment switch of lines 7-9;close the contact switch of lines 8-12 and open the segment switch of lines 6-8;close the contact switch of lines 13-14 and open the segment switch of lines 5-14.

However, the flow current of line 1-2 is 11.7 mW, when the link operation is commenced based on closed-loop power flow calculation method, which is higher than the line limit of 10mW.The optimal network frame shown in figure 4-bcannot be achieved by link or unlink operation. To avoid line overrun, the segment switch of line 7-9must open and the contact switch of lines 3-9 must close, but it will lead to power cut in node 9.Therefore, it is not possible to realize the optimal reconfiguration solutionshown infigure 4-b through link and unlink operations without power cut.

The optimal reconfiguration solution is obtained as shown in figure 4-c using the method proposed in this paperconsideringclosed-loop constraints.Four switching states have changed based on the optimal reconfiguration solution in this chapter.The specific steps use of switch operations during the reconfiguration process can be obtained by improved GRASP algorithm. The optimal reconfiguration solution is shown as following process:close the contact switch of lines 8-12 and open the segment switch of lines 6-8;close the contact switch of lines 13-14 and open the segment switch of lines 5-14.

The link and unlink operations mentioned above have been verified by closed-loop power flow which meets the requirements of the closed-loop safety constraints.The network loss of the optimal reconfiguration solution is 458.2 kW. The improved GRASP algorithm proposed in this chapter has faster calculation speed when solving the model.When the population number N is set to 10, the calculation can be completed in only 0.53 second to obtain optimal reconfiguration solution. The calculation results of 14nodes system in all cases are summarized in table 2:

Table 2. 14 nodes system calculation results comparison.

Grid Type	Active power loss (kW)	Network contact switch	Closed-loop constrains status
Initial network	479.3	3-9,8-12,13-14	/
Without considering closed-loop constrains	431.0	6-8,7-9,5-14	Line 1-2 will overrun as the contact switch of line 3-9 is closed
Considering closed-loop constrains	458.2	3-9, 6-8,5-14,	No line will overrun

6. Conclusion

From the reconfiguration solution given in table 2, it can be seen that the optimal reconfiguration solution obtained by the method in this paper can meet the requirements of the closed-loop safety constraints. Compared with the solution without considering the constraints, it has higher loss, but it can achieve uninterrupted power supply during the whole reconfiguration process, which has good operation feasibility in areas with high requirements on power supply reliability.

Reference

[1] Yang Fan, Liu Jun, Si Wenrong, Xu Li, Research on theoretical line loss calculation of low-voltage power distribution area in Shanghai, 2014 China International Conference on Electricity Distribution (CICED), pp 119 - 122

[2] Yang Fan, Liu Jun, Lu Bingbing, Design and application of integrated distribution network line loss analysis system, 2016 China International Conference on Electricity Distribution (CICED), pp 1-4

[3] Edgar Manuel Carreno,Rubén Romero. An efficient codification to solve distribution network reconfiguration for loss reduction problem, IEEE Trans. Power Syst., vol. 23, no. 4, 2008, pp 1542-1551

[4] Benoit Bernay, Samuel Deleplanque, Alain Quilliot. Routing on dynamic networks: GRASP versus genetic, 2014 Federated Conference on Computer Science and Information Systems (FedCSIS), pp 487 - 492

[5] S. Binato, G.C. de Oliveira, J.L. de Araujo. A greedy randomized adaptive search procedure for transmission expansion planning, IEEE Transactions on Power Systems, vol. 16, no.2, May 2001, pp 247 - 253

Research on estimating method for the smart electric energy meter's error based on parameter degradation model

Z Zhang[1], H Gong[2,a], C Li[1], ZWang[1] and X Xiang[2]

[1]State Grid Shandong Electric Power Research Institute, Jinan 250001, China
[2]School of Electrical and Electronic Engineering of Huazhong University of Science and Technology, Wuhan 430074, China

[a] Email: 2636178594@qq.com

Abstract. The performance of the smart electric energy meter deteriorates during the operation, which will affects the accuracy of energy metering. In order to estimate the smart meter's error during the operation, a method based on parameter degradation model is proposed. The meter's degradation parameters and degradation acting parameters are determined, aiming at building parameter degradation model and putting forward error estimation constraint. Big data analysis methods are adopted in the process of solving degradation network. As the data are of multiple categories and data changing rates are variable, pre-processing method of differential normalized data is employed. Additionally, feed-forward neural network is adopted to approximate degradation characteristics, because elementary function is incapable of describing degradation network. Therefore, the smart meter's error can be estimated according to the error estimation constraint, assuming degradation acting parameters pre-determined. Case analysis results show that estimation results using the proposed method is in accordance with operating state in short time, and absolute error is less than 0.1%, demonstrating that the error of smart meters can be estimated with this method effectively and dynamically.

1. Introduction

The smart meter is the key equipment for electricity measurement in the power system and the basis for settlement of electricity trade. The accuracy of smart meter directly relates to the economic benefits of the power grid and the trade fairness of the users. While, regular tests are still taken on most of the meters or off-line maintenance only when the measurement error exceeds the error limit, which affects the reliability of power supply and increases the cost of operation and maintenance[1-4].

In this paper, an estimating model for the smart meter's error is proposed according to the metrological characteristics, basing on temperature, humidity, load and historical data. The neural network algorithm is adopted to estimate the error, then the historical calibration data is applied to modifying the estimating results, and the validity has been verified in the end.

2. Parameter degradation model

2.1. The principle of the model

The parameter degradation of a system or equipment generally means that the performance deteriorates. At present, the state estimating models and methods for most power equipments already have a complete system[5]. For complex equipments or systems, there are usually several parameters describing the state. For example, the state parameters of a transformer include dissolved gas in the oil, insulation resistance, dielectric loss, moisture content in oil, etc[6-8].While for a circuit breaker, the state parameters usually include contact abrasion, opening/closing coil current, breaking stroke and so on[9]. Each parameter needs to be considered when a state estimating model is established.

Suppose the set of a certain device's state parameters is defined as $\{P_1,P_2,...,P_n\}$, then a simple and effective estimating model can be described as: if $\max|P_i|(i=1,2,...,n$) exceeds the threshold, some performance of the device will deteriorate and the normal operation of the device will be influenced. The above estimating model is not suitable for the smart meter's error, mainly for three reasons.

1) Generally, the smart meter's parameters degradation under the influence of some degradation acting parameter, will not lead to malfunction of the meter;
2) The degradation of smart meter's parameters has directionality. That is, the effect of each parameter degradation on the meter's measurement error isn't the same, which may be complementary thus not affect the overall accuracy;
3) Under the above premise, the performance parameters can't be used as the basis for estimation.

The theoretical research suggests that many aspects affect the performance of the smart meter, including the temperature and humidity of environment; electrical parameters such as load, frequencies and harmonics; electromagnetic fields, vibration and communication anomalies; power grid events such as pressure loss and interruption[10-13]. According to whether or not the smart meter's normal work is affected, they can divide into reliability and stability factors; according to the influence time, they divides into the persistent and the occasional influencing factors.

Among them, three factors which influence the smart meter's error largely are selected as the degradation acting parameters to build the model, which are the temperature, humidity of the environment and the load, respectively. The three factors influence the smart meter at the same time, which affect the measurement error.

To simplify the model, only single-phase energy metering is considered. Single-phase energy metering can be equivalent to a multi-input single-output system, of which the inputs are actual energy W_0, temperature, humidity and load, and the output is the energy shown in the smart meter. Its equivalent diagram is shown in figure1.

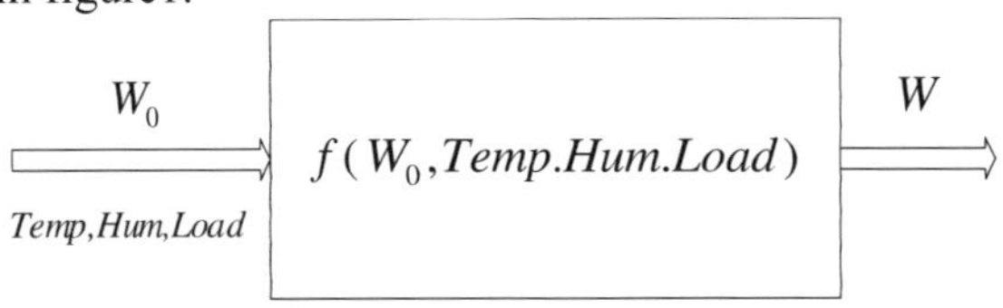

Figure1. Equivalent schematic for input and output of single-phase energy metering system.

The degradation acting parameters including temperature, humidity and load will lead to the smart meter's error. Therefore, the degradation parameters of the smart meter include the following three parts: the smart meter's ratio error under the influence of temperature, the meter's ratio error under the influence of humidity and the ratio error under the influence of load. For the system shown in figure 1, the measurement error of the energy ratio error depends on the three parts, which is shown in figure 2.

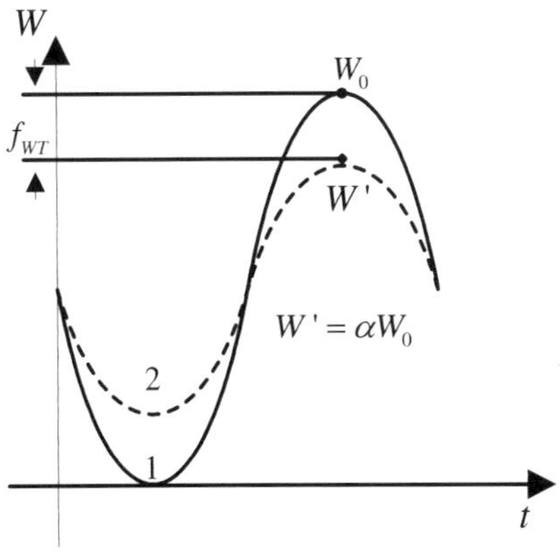

Figure 2(a). Ratio error of energy under the temperature's influence.

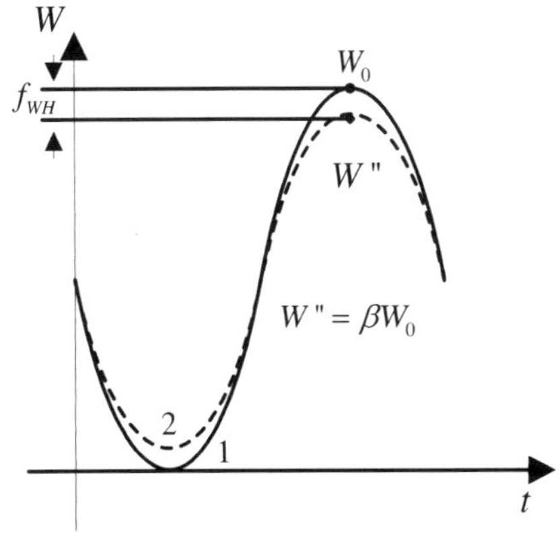

Figure 2(b). Ratio error of energy under the humidity's influence.

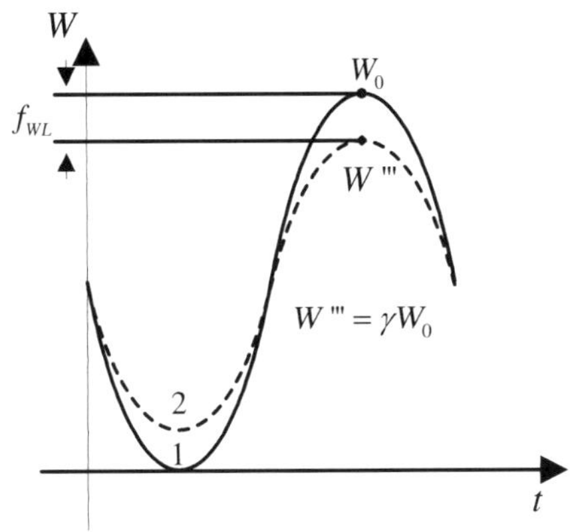

Figure 2(c). Ratio error of energy under the load's influence.

Curve 1 shows an actual input signal, and curve 2 shows the signal with ratio error. Assuming that the degradation parameter of the smart meter is zero at the initial time. During the operation, they are degrading in their respective dimensions and the direction is uncertain. When three parameters are degraded to f_{WT}, f_{WH} and f_{WL} at some point, the discrete electric energy is:

$$W = \sum_{i=0}^{N}(\alpha_i + \beta_i + \gamma_i)W_0 \tag{1}$$

In the equation (1), α_i is the energy measurement error coefficient caused by f_{WT}, β_i is the measurement error coefficient caused by f_{WH}, and γ_i caused by f_{WL}. The range of α_i's, β_i's and γ_i's value are: $-1\leq\alpha_i\leq1$, $-1\leq\beta_i\leq1$, $-1\leq\gamma_i\leq1$. When $\alpha_i + \beta_i + \gamma_i = 1$, the measurement accuracy of the smart meter is unchanged, indicating that the influence of each degradation acting parameter on the smart meter's error is complementary. Based on the above analysis, the following conclusion can be obtained: compared with the existing state estimating model for the equipments or systems, the parameter degradation of the smart meter has directivity.

2.2. Model establishment

In order to simplify the model, the three degradation parameters f_{WT}, f_{WH} and f_{WL} are merged into one degradation parameter, namely the energy's measurement ratio error f_W. The three factors influence the smart meter at the same time, which affect the degradation parameter f_W. The degradation process can be described as follows:

$$f_W(t) = g\left(Temp(t), Hum(t), Load(t)\right) \tag{2}$$

In the upper equation, *Temp(t)* is a function of temperature over time, *Hum(t)* is a function of humidity and *Load(t)* of load. Let:

$$\boldsymbol{\varepsilon} = f_W(t) \tag{3}$$

$$\boldsymbol{\alpha} = (Temp(t), Hum(t), Load(t)) \tag{4}$$

The solution of the equation(2) is A, and then equation(2) can be written as:

$$\varepsilon = A\alpha \tag{5}$$

Equation(5) is defined as the parameter degradation equation of the smart meter, where ε is the degradation parameter, α is the degradation acting parameter, and the solution matrix A is the degradation network under the degradation acting parameters. The meter's measurement ratio error is used as the estimated constraint:

$$\eta_W = \frac{W - W_0}{W_0} \tag{6}$$

In the upper equation, η_W is the average energy's measurement error. Equation (5) and (6) constitute a model for smart meter's error estimating.

3. The solution method of the parameter degradation network

Using reasonable data processing method to obtain the parameter degradation network is the key to estimating the smart meter's error according to the historical operation data.

3.1. Data pre-processing

The degradation acting parameters of the degradation model include temperature, humidity and load. When the data being obtained, data volume varies according to different sampling frequencies. For example, the temperature and the humidity are sampled every three minutes, while the ratio error of the measured energy is sampled every minute. The cardinality and dimensions of all kinds of data are not consistent. For example, the base of temperature is 25, and the base of load is 0.5. In order to synchronize the data, the sampling rate of them should be unified. In order to unify the weights and dimensions of variables, all kinds of data should be dealt with normalization processing. Because the temperature and humidity will not mutate, a first-order linear interpolation method can be used to expand the data of temperature and humidity. Taking the temperature data processing as an example, assuming that the sampling temperature of i and j is $T(i)$ and $T(j)$, the amount of data is increased to N times. When the data is expanded, the interpolation sequence between the two temperature sampling values is:

$$T(i+k) = \begin{cases} T(i) + \dfrac{T(j) - T(i)}{N} \bullet k & k = 1, 2, \mathrm{L}\ , N-1 \\ T(j) & k = N \end{cases} \tag{7}$$

Humidity data is also dealt in the same way, and the sampling rate of temperature and humidity after treatment is as same as that of load.

Suppose that the variables sequence is $\{W_1, W_2, \ldots, W_n, W_{n+1}\}$, then its maximum change gradient with the method of differential normalization is defined as：

$$\Delta W = \max\{\Delta W_i\}(i = 2, 3, \ldots, n, n+1) \tag{8}$$

The variable sequence is dealt with differential normalized as follows:

$$W_i = \frac{W_{i+1} - W_i}{\Delta W} \tag{9}$$

The equation (5) can be rewritten as the normalized form:

$$\Delta\varepsilon = B \bullet \Delta\alpha \tag{10}$$

In the above equation, $\Delta\varepsilon$ is the differential normalized degradation parameter, $\Delta\alpha$ is the differential normalized degradation acting parameter, and ***B*** is the corresponding differential normalized degradation network.

3.2 The solution method of degradation network

The essence of solving the parameter degradation network is to solve the coefficient matrix B of equation (10) based on the known degradation acting parameters and the degradation parameter. The elements of matrix B are not analytical, and can not be described by constants or elementary functions. An artificial intelligence learning method is adopted to approximate the input and output characteristics of the actual coefficient matrix B. This paper uses BP neural network to complete, as shown in figure. 3 [14].

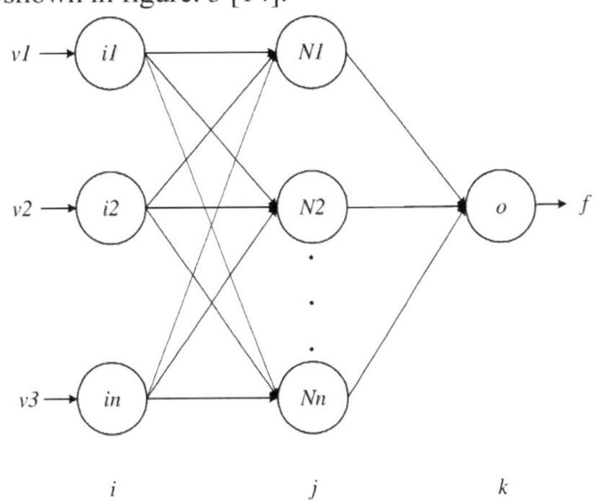

Figure 3. Structure diagram of network.

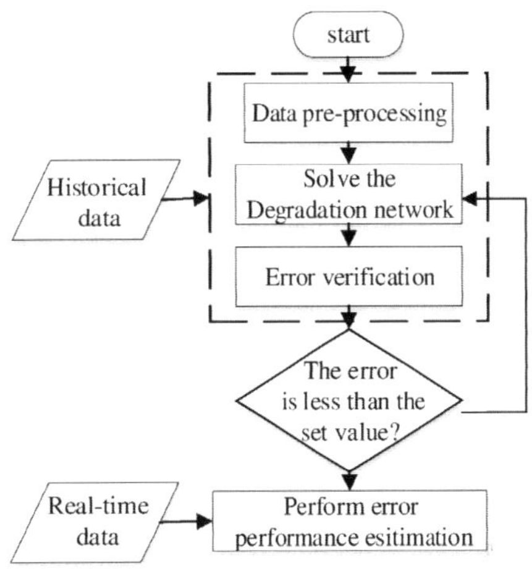

Figure 4. Estimation procedure based on parameter degradation model.

In figure 3,*i*, *j* and *k* correspond to input layer, hidden layer and output layer, respectively. In this paper, the input is the differential normalized degradation acting parameter $\Delta\alpha$, and the output is the differential normalized degradation parameter $\Delta\varepsilon$ .The transformation function of the degradation network to the input approximates the coefficient matrix B in equation (10).

4. Estimation steps based on a parameter degradation model

In this paper, the steps of smart meter's error estimation based on the parameter degradation model is shown in figure.4.

(1)Data pre-processing. Data samples are treated with sampling rate unification and differential normalization, getting $\{\Delta\text{Temp}(1), \Delta\text{Temp}(2),..., \Delta\text{Temp}(n)\}$, $\{\Delta\text{Hum}(1), \Delta\text{Hum}(2),..., \Delta\text{Hum}(n)\}$, $\{\Delta\text{Load}(1), \Delta\text{Load}(2),..., \Delta\text{Load}(n)\}$, and $\{\Delta f_W(1), \Delta f_W(2),..., \Delta f_W(n)\}$ respectively.

$\{\Delta\text{Temp}(i), \Delta\text{Hum}(i), \Delta\text{Load}(i)\}$ is the input of figure 3, and $\{\Delta f_W(i)\}$ is the output of figure 3.

(2) Solve the degradation network. According to the input and output sample obtained by (1), the BP neural network is adopted to iteratively approximate the coefficient matrix B and verify the mean square error until it is less than the set value.

(3)Estimate the meter's error. The degradation acting parameters $\{\Delta\text{Temp}, \Delta\text{Hum}, \Delta\text{Load}\}$ of the real time measurement is the input to the degradation network, and the change of the degraded parameter is obtained. Combined with the initial value of the degraded parameter, the state parameters of the smart meter are obtained, and the smart meter's error is estimated according to equation (7).

5. Case analysis

In order to verify the effectiveness of the estimating model and method, an on-line monitoring system of smart meter's measurement error is established, the structure diagram of which is shown in figure 5. The monitoring method can be described as follows.

The signal passes through two channels to calculate the smart meter's error. One of the channels is a standard channel, which uses the standard voltage transformer and the standard current transformer to measure the voltage and current signals respectively. The sampling unit collects the signal and calculates the standard electrical energy. The other one is a testing channel. In this channel, the tested smart meter measures the voltage and current signals and outputs the pulses proportional to the electrical energy. The PPS (Pulse per second) acquisition unit combines the pulse number with the pulse constant of the smart meter to calculate the measured electrical energy. The standard electric energy and the measured electric energy are input to the error calculation unit for calculation, and finally the measurement electric energy's ratio error is obtained.

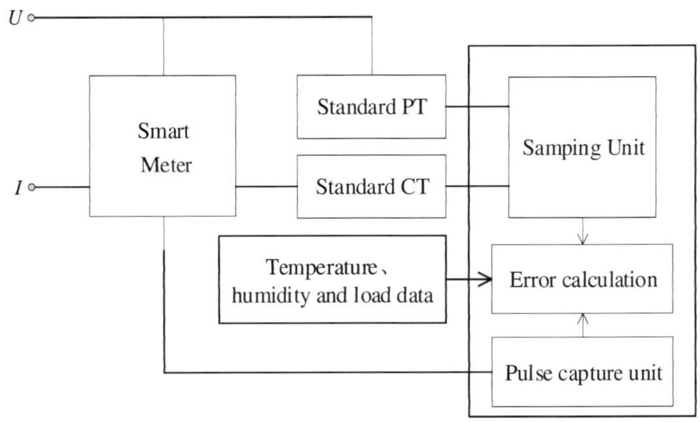

Figure 5. Online monitoring system structure diagram.

(1) Intercept the temperature, humidity, load and smart meter's ratio error data from 10 a.m. to 5 p.m. on April 7, 2015, as samples. The temperature and humidity data were collected every 3 minutes using the Rotronic Instruments HP22 temperature and humidity meter, and the load was recorded every minute by the electrical energy shown in the smart meter.

The data is shown in figure 6.

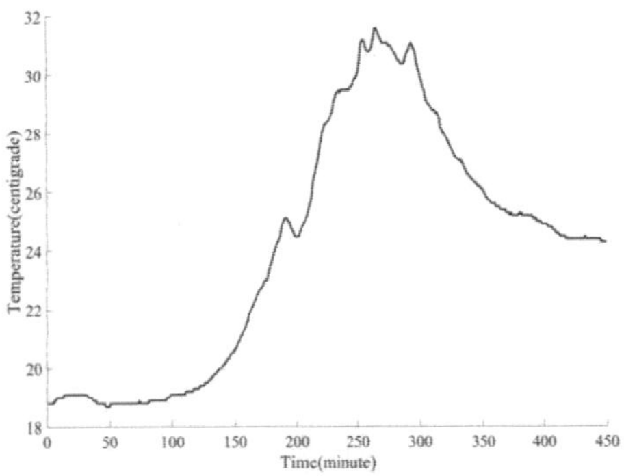

Figure 6(a). The temperature data.

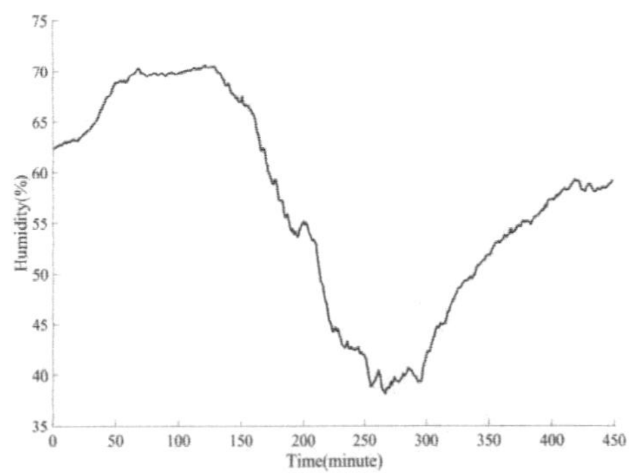

Figure 6(b). The original humidity data.

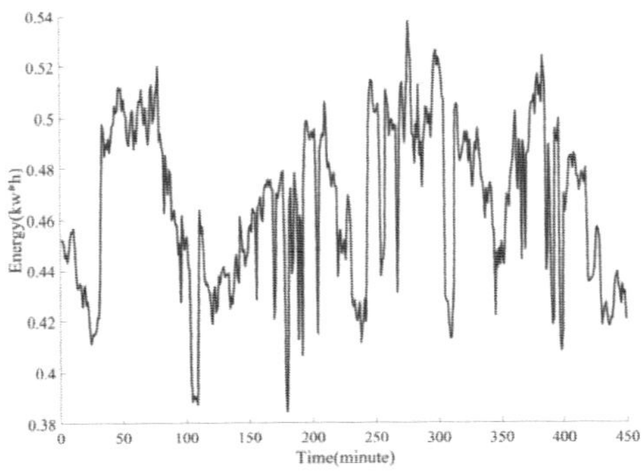

Figure 6(c). The original load data.

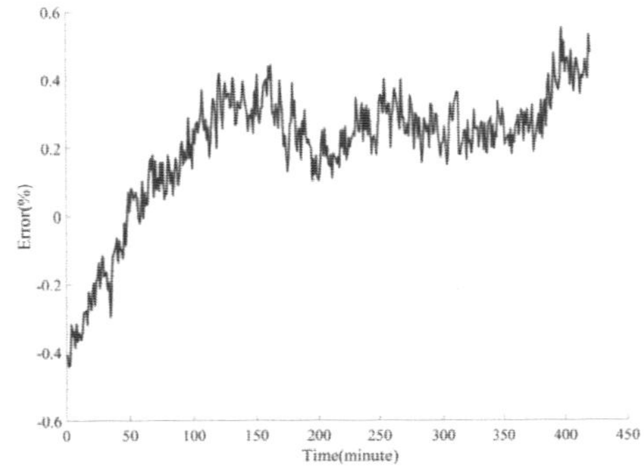

Figure 6(d). The original ratio error data.

As shown in figure 6(a)-(c), the original temperature, humidity and load change in the day smoothly. Figure 6(d) shows the original data of smart meter's error fluctuating in a day.

(2) Use data samples to solve the degradation network, and set the approximation error to 0.01%. The result of iterative solution is shown in figure 7.

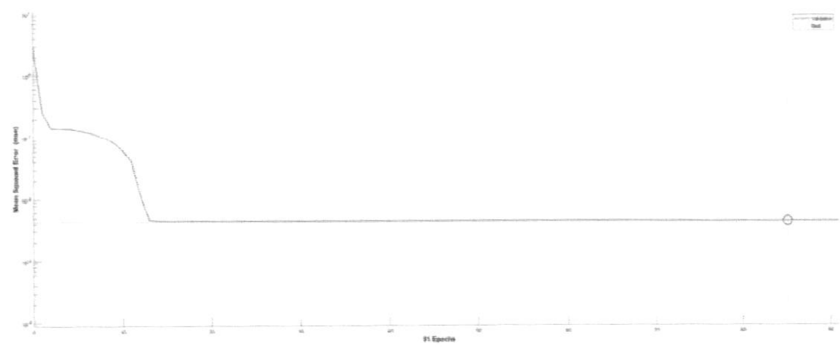

Figure 7. The iterations and error.

It can be seen from figure7 that the mean square error is less than 0.01% after 85 iterations.

(3) The real-time measurement of the degradation acting parameter is adopted to estimate the short-time error of the smart meter after 5:00 pm on April 7, the result is shown in figure8.

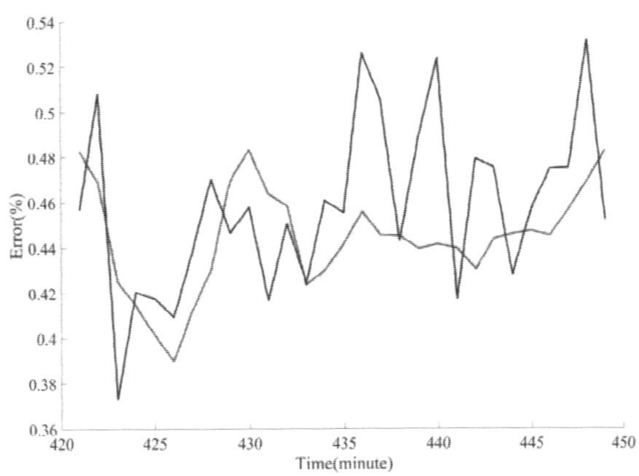

Figure8. Comparison between the estimation errors and the actual errors.

In figure 8, the blue line is the estimation error, and the red line is the error actually obtained. It can be seen that the estimating results based on the parameter degradation model are basically in line with the trend of actual results, and the absolute error of the short-term estimating error is no more than 0.1%

compared with the actual error.

6. Conclusion

In this paper, we propose a smart meter's error estimating method based on parameter degradation model to estimate the smart meter's error during the operation, considering the combined effect of various error factors such as temperature, humidity and load under the actual operating conditions, and an on-line monitoring system for smart meter's error was established. Through the case analysis, it is found that the estimating error is basically the same as the actual error, the absolute error is less than 0.1%, which shows that this error estimation method based on parameter degradation model for the smart meter can effectively estimate the smart meter's error during the operation.

References

[1] Chen S, Song S, Li L and Shen J 2009 Survey on smart grid technology *Power System Technology***33**1-6
[2] Yu Y and Luan W2009 Smart grid and its implementations. *Proceedings of the CSEE***29**1-7
[3] Lu G, Liu Ji and Zhang C2006 The technology development of substation digitization *Power System Technology***30**501-503
[4] Wang s, Zuo J, Chu Y, Wang Y and Cui X 2013 Research on the smart meter's accelerated life test *Electrical Measurement & Instrumentation***50**30-33
[5] Cheng Y, Wu H, Yang H,Hou X, Xiao J, Li C and Yao C2012 Fuzzy Comprehensive Status Evaluation on Electric Energy Metering Device and Test Strategy Research *Electrical Measurement & Instrumentation***49**1-6
[6] Liao R, Zhang Y, Huang F, Zheng H and Yang L 2012 Power Transformer Condition Assessment Strategy Using Matter Element Analysis *High Voltage Engineering* **38**521-526
[7] Hao J, Liao R, Yang Land Ma Z 2011 Aging Status Assessment of Power Transformer Oil-Paper Insulation by Frequency-Domain Dielectric Spectroscopy *Power System Technology***35**187-193
[8] Li L, Zhang D, Xie LYu B and Lin F 2013 A Condition Assessment Method of Power Transformers Based on Association Rules and Variable Weight Coefficients *Proceedings of the CSEE***33**152-159
[9] Guo L, Li K, Liang Y and Sun L 2014 HV circuit breaker state assessment based on gray-fuzzy comprehensive evaluation *Electric Power Automation Equipment***34** 161-167
[10] Tang Y, Jiang B, Li H2014 Handling Methods for Sampled Value Frame Missing in Digital Power Metering *Transaction of China Electro technical Society***29**444-448
[11] Zhou l, Liu K005 Error Analyzing for Electric Metering and Discussion about Pricing of Electric Energy *Transaction of China Electrotechnical Society***20**63-68
[12] Liu B, Ye G, Guo KTong Y, Hu B and Wan G2012 Quality test and problem analysis of electronic transformer *High Voltage Engineering***38**2972-2973
[13] Wang J, Guo Z, Zhang G, Shen Y and Yu W 2012 Experimental investigation on optical current transducer's long-term operation stability *Power System Technology***36**37-41
[14] Huang L2008 The Study of Algorithm of BP Neural Network Chong Qing *Chong Qing Normal University*

IOP Conf. Series: Materials Science and Engineering **366** (2018) 012066 doi:10.1088/1757-899X/366/1/012066

Fault Detection Algorithm for Multi-terminal LCC HVDC Interconnected Grid

M M Eissa, S Hasan and R Mamdouh

Department of Electrical Machine and Power Engineering, Faculty of Engineering - Helwan University at Helwan – Egypt

mmmeissa@yahoo.com, ahsaady500@gmail.com, ramy.ali@h-eng.helwan.edu.eg

Abstract--The protection of Multi-terminal DC (MTDC) system is not straightforward. This paper proposes a novel protection technique for the proposed configuration topology of interconnection project between Egypt and Saudi Arabia. The proposed scheme depends on the second derivative of dc line current with the time which represents the acceleration of dc current with the time. A model of three-terminal Line Commutated Converter (LCC) MTDC with polarity reversing availability is used to study fault behavior and apply the proposed algorithm to detect and classify the fault. The proposed scheme is evaluated via an extensive simulation study for 500kV dc transmission network using Matlab©/Simulink©. The suggested algorithm detects a dc fault without communication link between the relays and can respond to different fault resistances and different fault distances in few milliseconds.

1. Introduction

Multi-terminal DC (MTDC) refers to an HVDC system that consists of more than two converter stations, and it is more complex compared to that of a two-terminal point-to-point system.

HVDC transmission can be based on two alternative technologies: Line Commutated Converter (LCC) using thyristors, and Voltage Source Converter (VSC) using IGBT. The most of the HVDC transmission systems used today in the world are based on the LCC, which is a quite mature technology and it has been in operation for more than thirty years with little maintenance. Many papers proposed different techniques for fault detection and fault location in LCC and VSC HVDC systems for two terminals HVDC and especially for multi-terminal HVDC (MTDC). These techniques can be categorized into magnitude processing techniques and signal processing techniques. In the magnitude processing techniques, the protection technique is set up based on direct measurement of voltage (V) and current (I) signals and their derivatives (dV/dt) and (dI/dt) or signals derived from both voltage and current as power (P) and energy (E). In the signal processing techniques, the protection technique is set up based on the changes that occur in the wave shape of the electrical quantities immediately after fault occurrence, these techniques aim to extracting a unique characteristic which exists in the wave shape due to a certain fault type. These techniques as ABC to dq0 transformation, wavelet transform techniques, traveling waves, and artificial intelligence techniques as Neural Network. A fault detection technique is developed for HVDC transmission lines based on transient energy (ΔE) [1] it depends on a communication channel between relays at both sides. In [2], a protection technique is proposed based on the pilot principle, it also requires a communication channel to be built. A backup protection scheme based on the integral of reactive power for HVDC systems is proposed in [3]. In [4], a proposed method of limiting fault current in a meshed HVDC grids is presented, this method depends on the superconducting fault current limiters (SFCLs) in the protection of HVDC system. Another magnitude processing techniques are presented in [5], [6].

Although the magnitude processing techniques are straight forward and don't require any complicated processing, but they are subjected to noise and incorrect data samples and require a communication channel for accurate detection of fault which leads to increase the cost of the protection system.

Wavelet analysis for fault detection and fault location is discussed in different papers as in [6] and [7]. In [8], a fault location on multi-terminal DC systems is presented based on traveling wave. Although the signal processing techniques are advanced and accurate methods for fault detection and fault location, but these techniques require more sophisticated mathematical equations, long processing time, and some artificial intelligence techniques need training. In this paper, a stand-alone protection algorithm is proposed based on the second derivative of dc line current which represents the acceleration of dc current according to the time to detect the fault based on local measurements without communication channel.

2. MTDC system description and modeling

Matlab®/Simulink® has been successfully applied for modeling and simulation of the three terminals HVDC system for studying the protection of multi-terminal of HVDC system. The simulated system represents one operating case of the network with two rectifiers and one inverter as shown in figure 1.

2.1 AC supply

The AC supply of the two rectifiers and inverter is modeled as a thevenin equivalent voltage with equivalent source impedance. The voltages of AC side of three terminals are 500kV, 50Hz, and 380kV, 60Hz, and 380kV, 60Hz.

2.2 Converter

The 12-pulse converter station for rectifiers and inverter are constructed by two universal bridge blocks connected in series.

2.3 Filters and DC Lines

Filters are provided for 11th, 13th, and 24th harmonics at each converter station. The DC side of the rectifiers and inverter consist of smoothing reactors, DC filter and DC circuit breakers (DC CB).

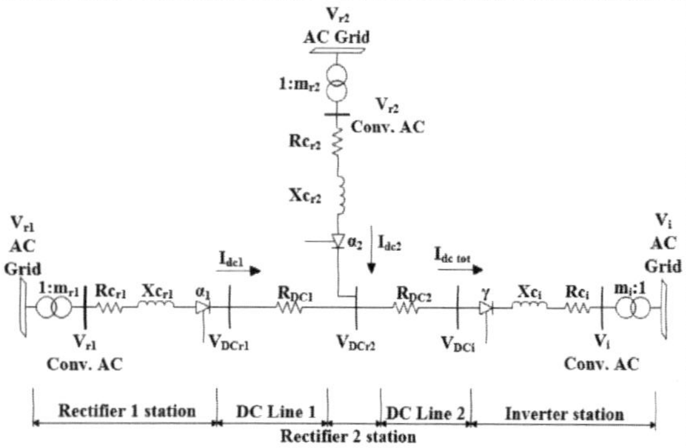

Figure 1. Representation of MTDC link model

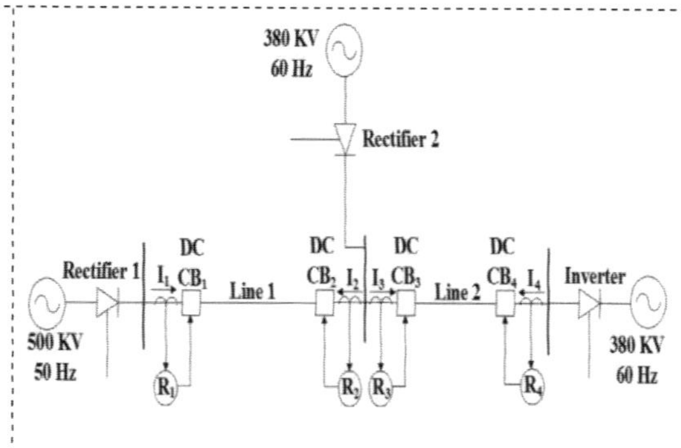

Figure 2. Schematic diagram of the studied MTDC system

IOP Conf. Series: Materials Science and Engineering **366** (2018) 012066 doi:10.1088/1757-899X/366/1/012066

3. The proposed protection algorithm

The concept of the relay which is applied in the protection system is based on obtaining the second derivative of the DC current which represents the acceleration of DC current.

The acceleration of DC current with time can be calculated digitally as follow:

$$\frac{\Delta^2 I_d}{\Delta t^2} = \frac{\Delta}{\Delta t}\left(\frac{\Delta I_d}{\Delta t}\right) \quad (1)$$

$$\frac{\Delta I_d}{\Delta t} = \frac{I_d(k) - I_d(k-1)}{t(k)-t(k-1)} \quad (2)$$

Where;

$\frac{\Delta I_d}{\Delta t}$: the first derivative of the DC link current.

I_d : the DC link current.

K: the most recent sample.

The acceleration of DC current can differentiate between fault case and transient case. As the DC current is constant value during normal operation and has zero rate of change with the time, so any disturbance that may occur in the DC system will lead to producing a change of current with the time so it is so difficult to make a protection decision in a DC system based on the rate of change of the DC current (First derivative) but the second derivative of the DC current which represents the acceleration of DC current with the time can be applied to discriminate between fault case and any other transient case.

The protection system consists of four relays located at two terminals of transmission lines as shown in figure 2, each relay applies the second derivative on its terminal current to determine the acceleration of current with time to detect the fault based on the polarity of the first spike of $\left(\frac{\Delta^2 I_d}{\Delta t^2}\right)$ and its value.

For relay (1), faults in line (1) (internal faults) and faults in line (2) (external faults) will lead to a huge increase of current detected by relay (1), and the first spike of the acceleration will be positive. But the magnitude for external faults is lower than the internal one because the equivalent impedance which seen by relay (1) for faults in line (2) will be greater than the equivalent impedance for faults in line (1) due to existinig of rectifier (2).

For relay (2), faults in line (1) (internal faults) will lead to a huge increase of current in the same direction of measurement of relay (2) the polarity of the first spike of the acceleration is positive. But faults in line (2) (external faults) will lead to a huge increase of current but in the reverse direction of measurement of relay (2), so the polarity of the first spike of the acceleration is negative.

For relay(3), faults in line (2) (internal faults) will lead to a huge increase of current in the same direction of measurement of relay (3), so the polarity of the first spike of the acceleration is positive. But faults in line (1) (external faults) will lead to a huge decrease of current flowing through relay (3), so the polarity of the first spike of the acceleration is negative.

For relay (4), faults in line (2) (internal faults) and faults in line (1) (external faults) will lead to a huge increase of current detected by relay (4), and the first spike of the acceleration will be positive. But the magnitude for external faults is lower than the internal one because the equivalent impedance which seen by relay (4) for faults in line (1) will be greater than the equivalent impedance for faults in line (2) due to existinig of rectifier (2).

The principle of the protection technique which is set up in the digital relay can be summarized in the following flow chart shown in figure 3.

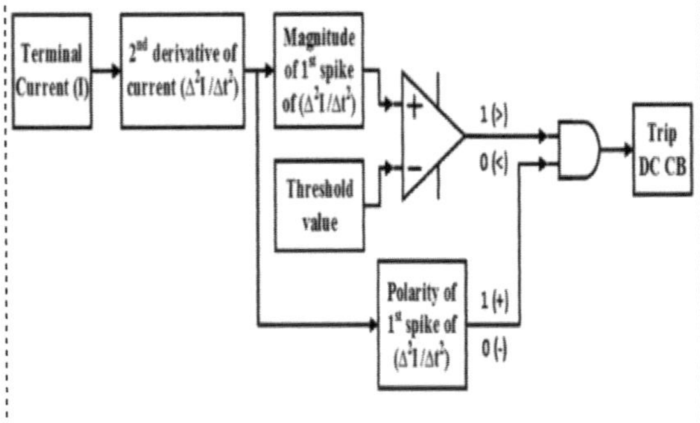

Figure 3. Block diagram of operation of the relay.

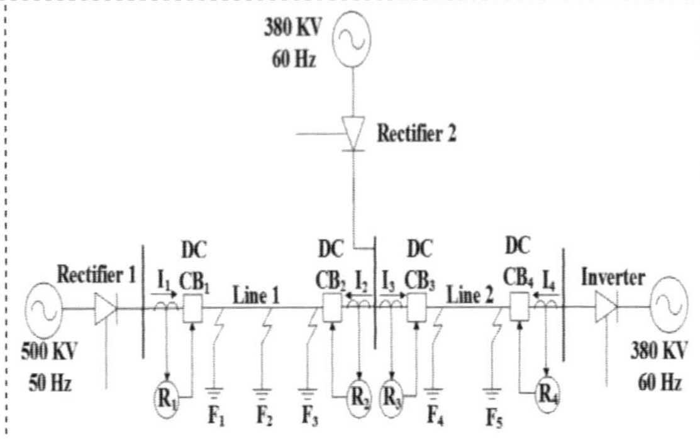

Figure 4. Overall scheme of MTDC system with different events.

4. Threshold selection

Comprehensive test studies for different cases such as normal case, switching case, effect of source impedance, AC fault case and DC faults at different distances with different fault resistances are simulated. There are many of case studies are taken and the threshold boundary is selected as follows; 4.4 MA/ms^2 for relay (1), 10.5 MA/ms^2 for relay (2), 9.5 MA/ms^2 for relay (3) and 5.2 MA/ms^2 for relay (4).

5. Simulation results

The simulation set-up shown in figure 4 has been used to implement the transient protection technique for multi-terminal HVDC lines with various disturbances, such as DC faults (internal and external) with different fault resistances, different fault distances are applied.

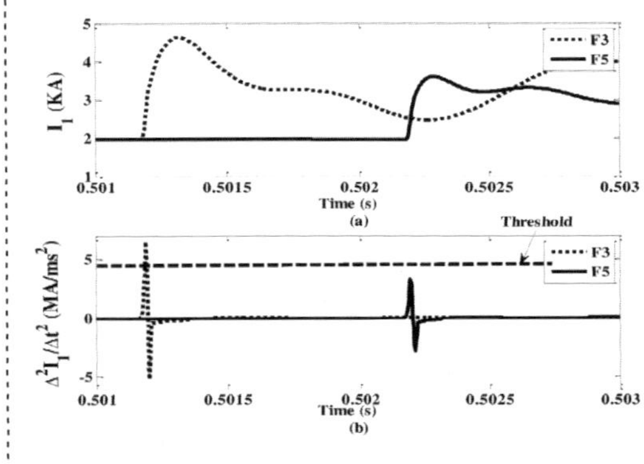

Figure 5. Response of relay (1) for DC faults (F3) and (F5) with R_f=0Ω. (a) DC current of relay (1)(b) Acceleration of DC current processed by relay (1).

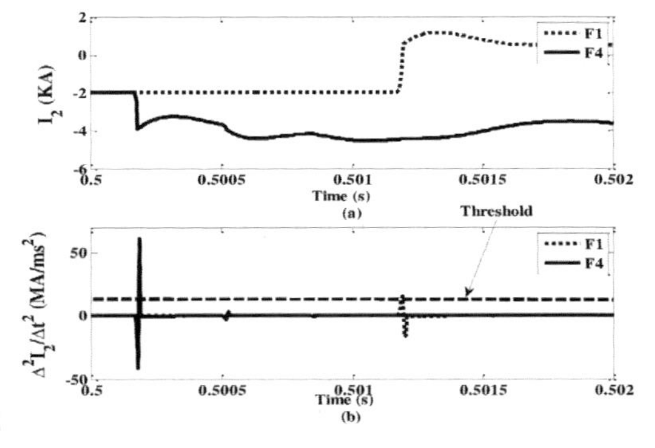

Figure 6. Response of relay (2) for DC faults (F1) and (F4) with R_f=0Ω. (a) DC current of relay (2). (b) Acceleration of DC current processed by relay (2).

5.1 Effect of fault position

Response of relay (1) for DC fault (F3) in line (1) and DC fault (F5) in line (2) is shown in figure 5. A fault resistance (R_f) of zero Ω is included. The polarity of the first spike of acceleration of the both faults is positive, but its value for (F5) is lower than the threshold.

Another case study is performed to test the response of relay (2) for DC fault (F1) in line (1) and DC fault (F4) in line (2) is shown in figure 6. A fault resistance (R_f) of zero Ω is included. The polarity of the first spike of acceleration of the DC current the acceleration for (F1) is positive and its value is greater than the threshold, but for (F4) the polarity is negative.

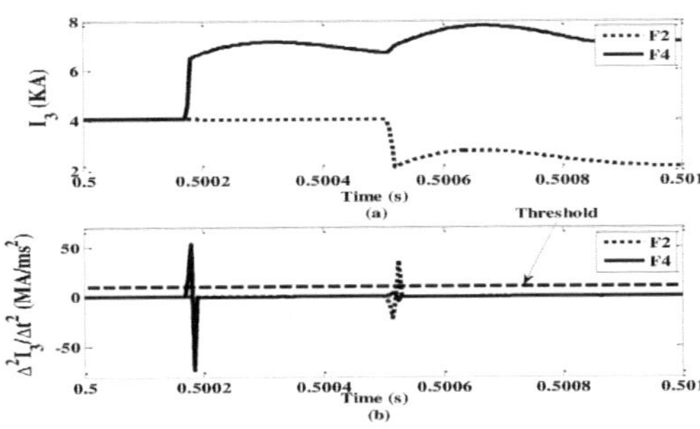

Figure 7. Response of relay (3) for DC faults (F2) and (F4) with R_f=0Ω. (a) DC current of relay (3) (b) Acceleration of DC current processed by relay (3).

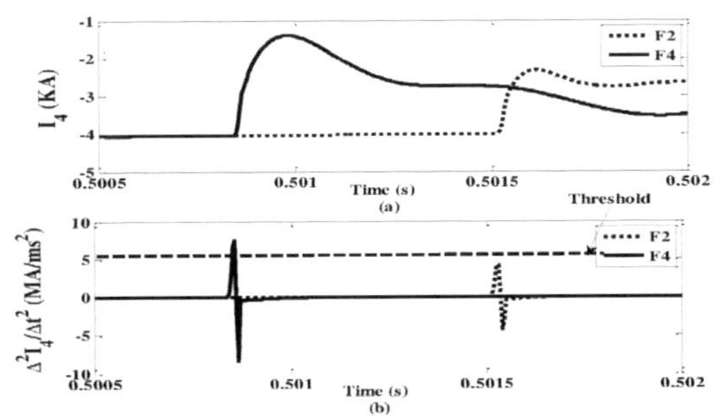

Figure 8. Response of relay (4) for DC fault (F2) and (F4) with R_f=0Ω. (a) DC current of relay (4). (b) Acceleration of DC current processed by relay (4).

Another case studies are performed to test the responses of relay (3) and relay (3) for internal and external faults with fault resistance (R_f) of zero Ω as shown in figure 7 and figure 8 respectively.

5.2 Effect of fault resistance
Another case studies are performed to test the responses of relay (1), relay (2), relay (3) and relay (4) for internal and external faults with fault resistance (R_f) of 50 Ω as shown in figure 9, figure 10 figure 11and figure 12 respectively.

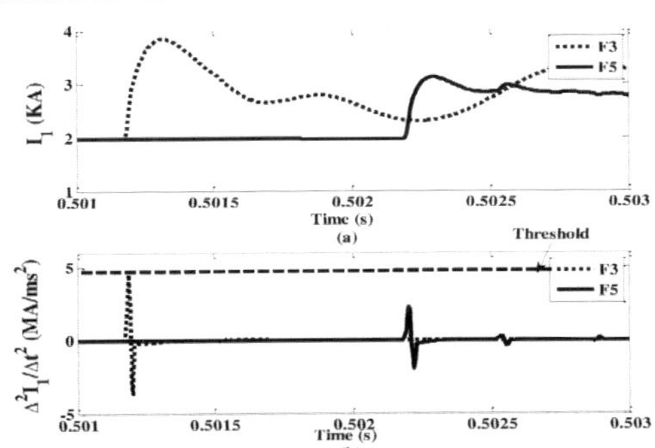

Figure 9. Response of relay (1) for DC faults (F3) and (F5) with R_f=50Ω. (a) DC current of relay (1). (b) Acceleration of DC current processed by relay (1).

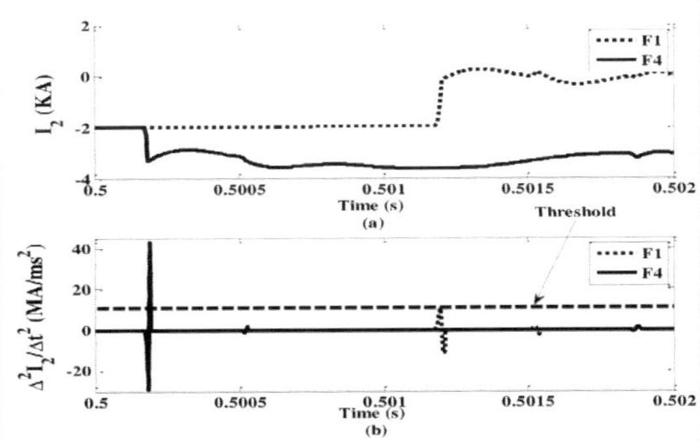

Figure 10. Response of relay (2) for DC faults (F1) and (F4) with R_f=50Ω. (a) DC current of relay (2). (b) Acceleration of DC current processed by relay (2).

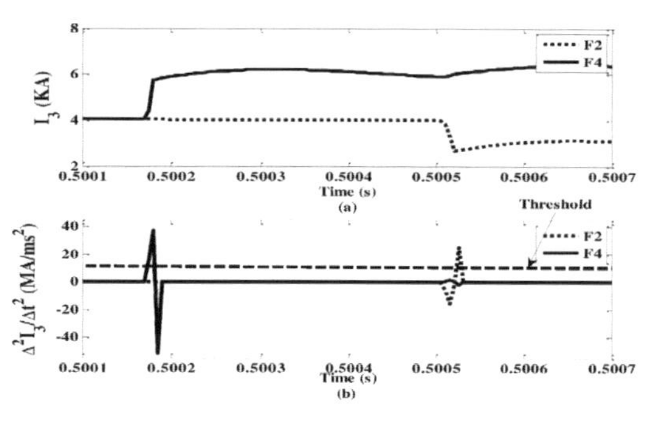

Figure 11. Response of relay (3) for DC faults (F2) and (F4) with R_f=50Ω. (a) DC current of relay (3). (b) Acceleration of DC current processed by relay (3).

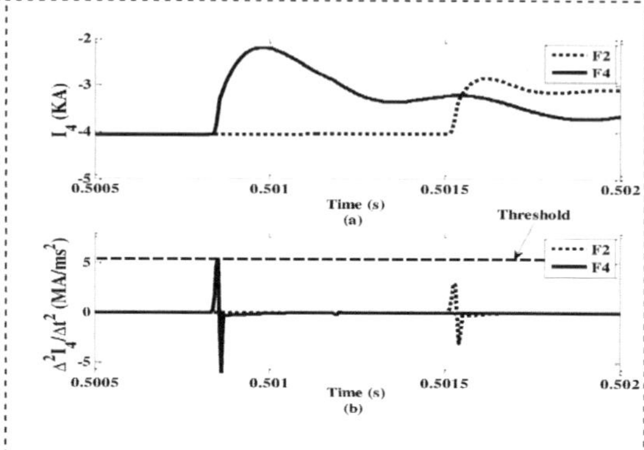

Figure 12. Response of relay (4) for DC faults (F2) and (F4) with R_f=50Ω. (a) DC current of relay (4). (b) Acceleration of DC current processed by relay (4).

6. Conclusion

A novel protection algorithm based on the acceleration of fault current is proposed for the multi-terminal HVDC system, this technique is straight forward, doesn't require any complicated processing, it requires one electrical quantity to detect the fault in MTDC system and doesn't require a communication channel. The performance of the proposed in-dependably algorithm of protection is satisfied and can discriminate between internal and external faults in a few milliseconds without communication link between the relays.

References

[1] Xiao-dong Z, Member S, Neng-ling T, Thorp J and Fellow L 2012 Scheme for HVDC Transmission Line *IEEE Trans. Power Deliv.* vol **27** pp 2278–2285.

[2] Song G, Chu X, Cai X and Gao S 2013 A novel pilot protection principle for VSC-HVDC cable lines based on fault component current *Int. J. Electr. Power Energy Syst.* vol **53** pp 426–433.

[3] Luo S, Dong X, Shi S and Wang B 2016 A Directional Protection Scheme for HVDC Transmission Lines Based on Reactive Energy *IEEE Trans. Power Deliv.* vol **31** pp 559–567.

[4] Mokhberdoran A, Carvalho A, Silva N, Leite H and Carrapatoso A 2017 Application study of superconducting fault current limiters in meshed HVDC grids protected by fast protection relays *Electr. Power Syst. Res.* vol **143** pp 292–302.

[5] Kontos E, Pinto R, Rodrigues S and Bauer P 2015 Impact of HVDC Transmission System Topology on Multiterminal DC Network Faults *IEEE Trans. Power Deliv.* vol **30** pp 844–852.

[6] Sneath J and Rajapakse A 2016 Fault Detection and Interruption in an Earthed HVDC Grid Using ROCOV and Hybrid DC Breakers *IEEE Trans. Power Deliv.* vol **31** pp 973 – 981.

[7] Abu-Elanien A, Elserougi A, Abdel-Khalik A, Massoud A and Ahmed S 2016 A differential protection technique for multi-terminal HVDC *Electr. Power Syst. Res.* vol **130** pp 78–88.

[8] Abu-Elanien A, Elserougi A, Abdel-Khalik A, Massoud A and Ahmed S 2017 A non-communication based protection algorithm for multi-terminal HVDC grids *Electr. Power Syst. Res.*, vol **144** pp 41–51.

Fault Identification for Transformer Axial Winding Displacement Using Nanosecond IFRA and SFRA Experiments

J. J. Huang[1], W. H. Tang[1,a], Y. L. Xin[1], J. J. Zhou[1] and Q. H. Wu[1]

[1] School of Electric Power Engineering, South China University of Technology, Guangzhou, 510641, China.

[a]Corresponding author: wenhutang@scut.edu.cn.

Abstract. Nanosecond IFRA has the potential to realize online detection of power transformer winding deformation and displacement. But this method is not mature even in offline condition, and the reliability, accuracy and repeat ability of which are in doubt. To verify the reliability and accuracy of nanosecond IFRA a comparison experiment is conducted between it and a mature method which is applied widely all over the world, SFRA. In this experiment, three levels of axial displacements are experimentally simulated on a dry-type distribution transformer in the lab and both a nanosecond IFRA system and a commercial SFRA analyser are applied to obtain the frequency responses at each fault level separately. To verify the repeatability of Nanosecond IFRA, two Nanosecond IFRA measurements conducted in a time interval of 30 days with all the test condition remaining the same are compared. The results are analysed by visual inspection and quantitative comparison.

1. Introduction

Power transformers are of essential significance, a failure of which may affect the stability and reliability of a power system and its maintenance is costly. According to statistics, a mechanical defect is one of the major causes of power transformer failure. Mechanical defects such as winding deformation and displacement can be caused by short circuit events, transportation and earthquake [1]. These defects may be slight in the beginning, but they are cumulative and may lead to serious incidents in the end. Therefore, it is very important to detect them in its early age and take corresponding measures to diminish the loss.

Frequency response analysis (FRA) is increasingly being used to detect winding mechanical defects [2, 3]. FRA is established upon the fact that the frequency response of a transformer winding depends on its internal distances and profiles [4]. Mechanical defects change the internal distances, thus, cause frequency response variation.

According to the way to obtain winding frequency responses, two kinds of FRA exist, i.e., sweep frequency-response analysis (SFRA) and impulse frequency-response analysis (IFRA). For SFRA, input and output signals are both sinusoidal signals, and a frequency response is obtained dot by dot. For IFRA, input and output are impulse signals which have rich high frequency harmonic components. Usually, Fast Fourier transform (FFT) is employed to transform those impulses into the frequency domain, after that a frequency response is calculated as a whole at one time [5]. SFRA has been applied in industry since 1975, and related standards have been proposed and commercial products have been manufactured [6]. While IFRA has seldom been practiced in comparison with SFRA, it has worse repeatability, lower signal-to-noise ratio and more requirement of measurement equipment [7, 8].

With the development of the digital measurement techniques, the interest in online FRA has increased in recent years. Many researches propose that the transient overvoltage in the electric power system can be used to obtain the frequency response [5]. While other researches state that controlled test signal is more suitable for condition monitoring of transformers. Reference [9] proposed that the test signal can be injected into the transformer in service through the bushing tap. In reference [10], a ferrite transformer was used to isolates the injection circuit from the transformer under test. Reference [11] designed a non-invasive capacitive sensor (NICS) to inject signals into a transformer in service. As recent international standards recommend and many literature suggest the frequency range of FRA test shall be from tens of Hz to 2 MHz [12]. However, reference [13] suggests that the range shall be extended to 10 MHz where the frequency response is more sensible to small changes of winding mechanical defects. In reference [14] by employing NICS for signal injection and acquisition, a Nanosecond IFRA test using an impulse with the width of 400 ns was conducted on a power transformer online and the frequency range was extended to 10 MHz, while the reliability of the results needs verification.

To verify the reliability, accuracy and repeatability of the nanosecond IFRA and to extract the frequency response feature of the axial displacement, this paper conducts an experiment on axial winding displacement within a power transformer by using SFRA and the nanosecond IFRA methods. The results are discussed both by visual inspection and quantitative analysis.

2. Basic principle of FRA

In practice, FRA is usually conducted using the end-to-end measurement connection (EE) [15, 16], as shown in Figure 1 and in this paper, EE is adopted. For SFRA, $V_{in}(t)$ and $V_{out}(t)$ are sweep sinusoidal signals and for IFRA $V_{in}(t)$ and $V_{out}(t)$ are impulse signals, which are transformed into the frequency domain, namely, $V_{in}(f)$ and $V_{out}(f)$ by FFT. An FRA system is connected using coaxial cables, the characteristic impedance of which is 50 Ω and those 50 Ω resistors in Figure 1 are used for impedance matching.

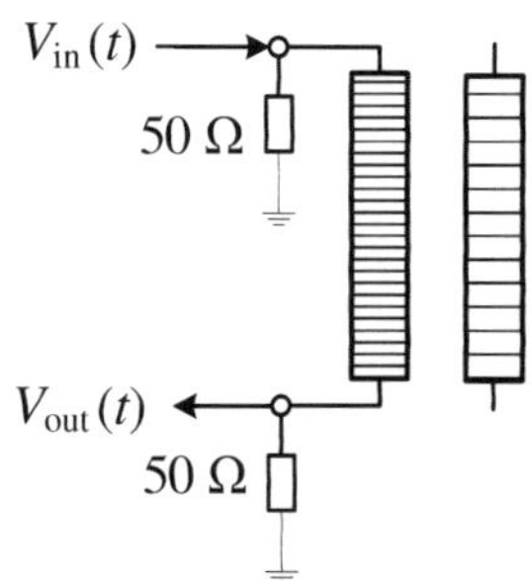

Figure1. End-to-end measurement connection for FRA

The frequency response is determined by the following transfer function:

$$H(f) = 20\ \lg \frac{|V_{out}(f)|}{|V_{in}(f)|} \tag{1}$$

FRA compares a frequency response measured at present with one measured at an early stage when the winding is confirmed normal and the difference between them may reveal internal damages of this winding. Above mentioned is so-called time-based (reference) comparison and in many cases due to lack of fingerprints, the construction-based (phase) comparison and type-based (sister unit) comparison are practiced [17].

3. Experiment setup

3.1. Test object
The test object in this research is a dry-type three-phase power transformer (50 kVA, 10/0.4 kV) shown in Figure 2(a). Its high-voltage (HV) winding has 60 discs, and each disc consists of 25 turns of coils. The low-voltage (LV) winding consists of 60 turns of coils. The top 15 discs of the HV winding of each phase is movable and replaceable for experimental purposes, and the rest is fixed. Gaskets can be added between the removable and fixed parts of the winding, as shown in Figure 2(b). A gasket is 10 mm of thickness, and more numbers of gaskets reflects higher severity of axial displacement fault. Three gaskets can be added at most, and the case without any gasket is defined as normal, so, in total, 3 degrees of axial displacement can be experimentally added to the test transformer.

Both the SFRA and IFRA tests are conducted on the same HV winding of the test object in the EE measurement connection with all the other terminal float, and the measurement environment is the same for every test.

(a) (b)

Figure 2. Test object: (a) Transformer: dry type three-phase power transformer (50 kVA, 10/0.4 kV), (b) Experimental setup for axial displacement

3.2. Test systems

3.2.1. SFRA system The configuration of SFRA test is illustrated in Figure 3. FRAnalyzer has three terminals, that is, a source terminal, a reference terminal and an output terminal. The first two terminals are connected to the input terminal of the test winding. The source terminal injects the sweep sinusoidal signal and the reference records the signal that been injected into the test winding. The output terminal obtains the output signal at the other end of the test winding.

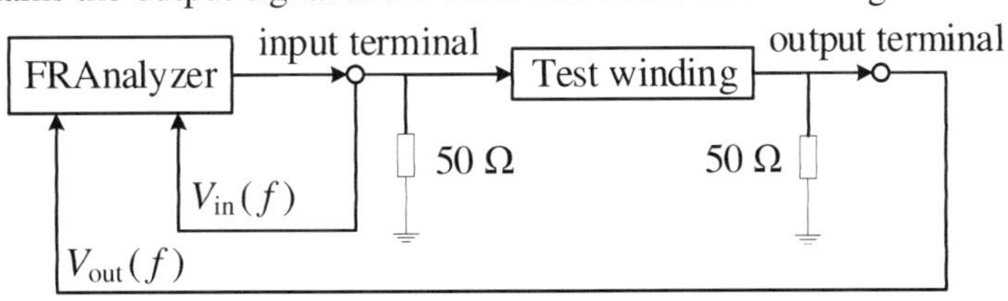

Figure 3. SFRA measurement connection

3.2.2. IFRA system The IFRA system, as shown in Figure 4, is more complicated than the SFRA system. An impulse generator provides a nanosecond double exponential voltage impulse with the rise time of 30 ns and the width of 300 ns and the open circuit peak value is 600 V. This impulse is injected to one terminal of the test winding as the source signal and the reference signal is measured at the same spot by an oscilloscope. The output signal is also measured by the oscilloscope whose sampling frequency is set to 1 GSa/s and the memory length is set to 1 MSa. Since the frequency resolution of frequency response equals the sampling frequency divided by the memory length, the frequency resolution is 1 kHz. The data obtained by the oscilloscope is processed in a computer.

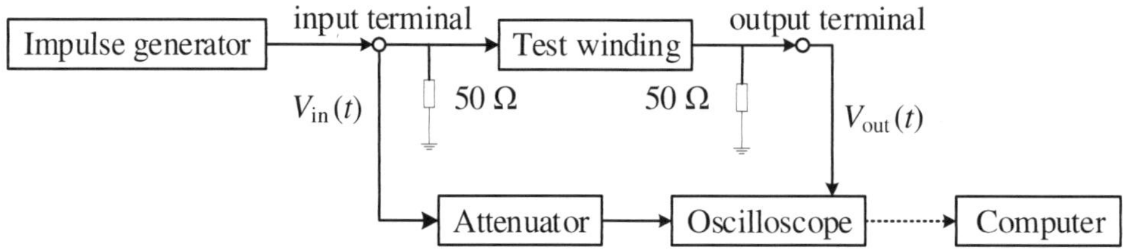

Figure 4. IFRA measurement connection

4. Results and discussion

Figure 5(a) shows a comparison of frequency responses measured by SFRA and IFRA at the same fault level in the same day. The two curves are almost the same to each other, especially in the frequency range under 4 MHz, while there is an obvious difference around 8 MHz. Figure 5(b), (c) show the frequency responses obtained by SFRA and IFRA measured at different fault levels, respectively. They illustrate that the results of SFRA and IFRA have the same trend when the fault degree changes. Two frequency responses obtained by the IFRA system in a time interval of 30 days with other conditions as the same is presented in Figure 5(d). Except the difference around 9 MHz, those two curves are almost the same to each other.

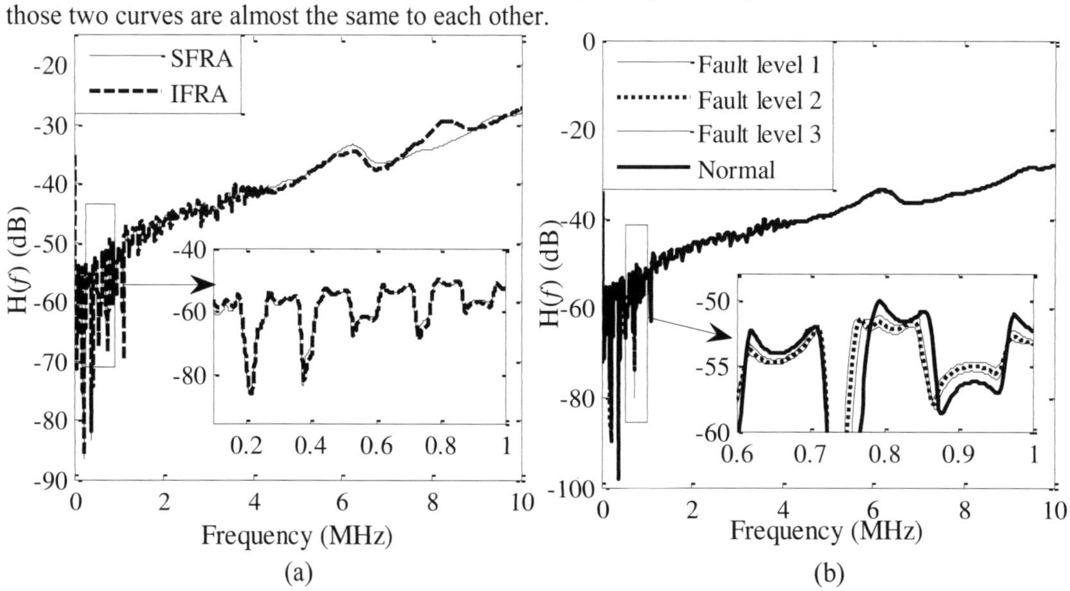

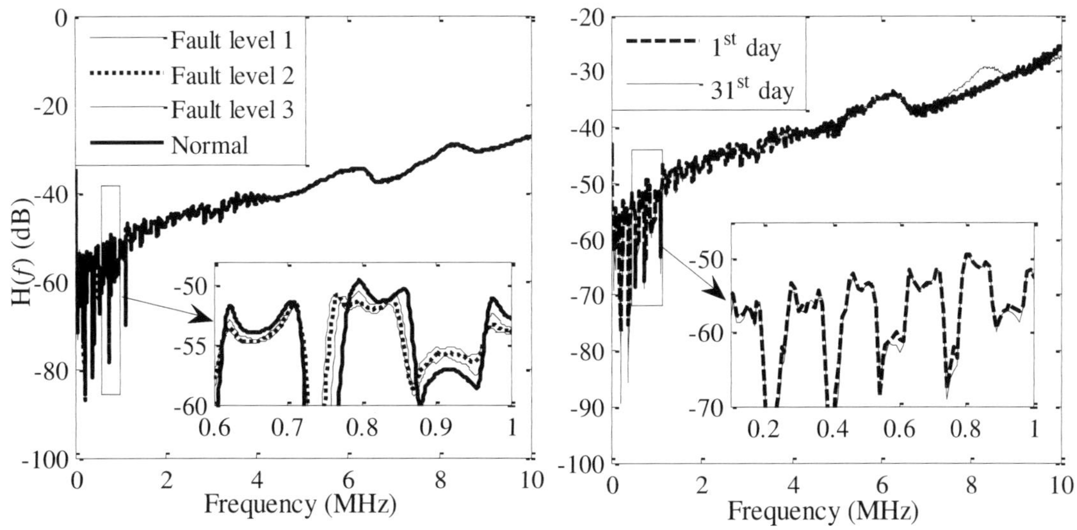

(c) (d)

Figure 5. Frequency responses: (a) Frequency responses obtained by SFRA and IFRA with the same defect degree, (b) Frequency responses obtained by SFRA at different defect degrees, (c) Frequency responses obtained by IFRA at different defect degrees, (d) Frequency responses obtained by IFRA at the same defect degree in a time interval of 30 days.

Apart from the visual inspection, a quantitative analysis can also help assess the severity of mechanical defects. Various statistic indexes can be applied for such purpose [18]. In this paper, the Spearman correlation coefficient, ρ and the Euclidean distance, *ed* are employed. Equations (2) and (3) show how to calculate ρ and *ed* on two frequency responses. x and y are two frequency responses, and each of them contains N points, where based i is the serial number of the points, $\bar{x}$ and $\bar{y}$ are the average of x_i and y_i, respectively. The closer to 1 ρ is or the closer to 0 *ed* is, the more similar to each other x and $y$are. If x is the fingerprint and y is the measured result, higher similarity between them indicates less severity of a fault.

$$\rho = \frac{\sum_{i=1}^{N}(x_i - \bar{x})(y_i - \bar{y})}{\sqrt{\sum_{i=1}^{N}(x_i - \bar{x})^2 \sum_{i=1}^{N}(y_i - \bar{y})^2}} \tag{2}$$

$$ed = \sqrt{\frac{1}{N}\sum_{i=1}^{N}(x_i - y_i)^2} \tag{3}$$

As Figure 5 (a) and (b) shows, in different frequency ranges, the resonance point densities are different, i.e., in the frequency range below 1 MHz, resonance points are closer to each other and the oscillation amplitude of the frequency response is greater, as the frequency becomes higher, the resonance points gradually moves away from each other and the oscillation amplitude decreases. Based on the distribution of resonance point and oscillation amplitude, in this paper, the concerned frequency range is divided into four parts, that is, 1 kHz~1 MHz, 1MHz~2 MHz, 2 MHz~5MHz, 5MHz~10MHz.ρ and *ed* of IFRA results are calculated in each frequency range separately, as presented in Table 1, the minimum value of ρ and the maximum value of *ed* in each row are in bold. The bold numbers only occur in the first column, which shows the axial displacement simulated in this research influences the 1 kHz~1 MHz range the most. And as the severity of the defect increases, ρ decreases and *ed* increases, which is sensible and thus verifies that the results obtained by IFRA are reliable. Table 2 presents ρ of two IFRA frequency response curves measured sequently in a time interval of 30 days, which verified that the nanosecond IFRA has reliable repeatability.

Table 1. Correlation coefficient and Euclidean distance of IFRA frequency responses at different fault levels

index	fault level	1 kHz~1 MHz	1 MHz~2 MHz	2 MHz~5 MHz	5 MHz~10 MHz
ρ	1	**0.9375**	0.9408	0.9925	0.9987
	2	**0.8569**	0.8815	0.9863	0.9985
	3	**0.7959**	0.8351	0.9817	0.9982
ed	1	**2.0063**	0.6518	0.2492	0.1175
	2	**3.0158**	0.9519	0.3383	0.1338
	3	**3.5053**	1.1324	0.3964	0.1361

Table 2. Correlation coefficient of two IFRA frequency response curves in a time interval of 30 days

index	1 kHz~1 MHz	1 MHz~2 MHz	2 MHz~5 MHz	5 MHz~10 MHz
ρ	0.9719	0.9900	0.9661	0.9479

5. Conclusion

This paper has investigated axial winding displacement for a power transformer using SFRA and the nanosecond IFRA experimentally. The visual inspection and quantitative analysis shows that SFRA and the nanosecond IFRA results have great similarity with respect to each other, which proves that the nanosecond IFRA has the same reliability and accuracy as SFRA and its result is repeatable. The experimental also reveals that an axial winding displacement fault mainly effects the 1 kHz~1 MHz frequency range. Moreover, the variation of amplitude of the spectrum is more detectable than the FRA shift of resonance points.

Acknowledgments

This work was supported by National Natural Science of China (No. 51477054).

References

[1] Zhao Z, Yao C, Zhao X, et al. Impact of capacitive coupling circuit on online impulse frequency response of a power transformer. IEEE Transactions on Dielectrics & Electrical Insulation, vol. 23, no. 3, pp. 1285-1293, 2016.

[2] Bengtsson C. Status and trends in transformer monitoring. IEEE Transactions on Power Delivery, vol. 11, no. 3, pp. 1379-1384, 1996.

[3] Leibfried T, Feser K. Monitoring of power transformers using the transfer function method. IEEE Transactions on Power Delivery, vol. 14, no. 4, pp.1333-1341, 1999.

[4] Ryder S A. Diagnosing transformer faults using frequency response analysis. IEEE Electrical Insulation Magazine, vol. 19, no. 2, pp. 16-22, 2003.

[5] Gomez-Luna E, Mayor G A, Gonzalez-Garcia C, et al. Current Status and Future Trends in Frequency-Response Analysis with a Transformer in Service. IEEE Transactions on Power Delivery, vol. 28, no. 2, pp. 1024-1031, 2013.

[6] Dick E P, Erven C C. Transformer Diagnostic Testing by Frequuency Response Analysis. IEEE Transactions on Power Apparatus & Systems, PAS, vol. 97, no. 6, pp. 2144-2153, 2007.

[7] Wang M, Vandermaar A J, Srivastava K D. Review of condition assessment of power transformers in service. IEEE Electrical Insulation Magazine, vol. 18, no. 6, pp. 12-25, 2003.

[8] Liu X, Qiang S. Test research on power transformer winding deformation by FRA method. International Symposium on Electrical Insulating Materials. IEEE, pp. 837-840, 2001.

[9] Rybel T D, Singh A, Vandermaar J A, et al. Apparatus for Online Power Transformer Winding Monitoring Using Bushing Tap Injection. IEEE Transactions on Power Delivery, vol. 24, no. 3, pp.996-1003, 2009.
[10] Gomez-Luna E, Mayor G A, Guerra J P. Application of Wavelet Transform to Obtain the Frequency Response of a Transformer From Transient Signals—Part II: Practical Assessment and Validation. IEEE Transactions on Power Delivery, vol. 29, no. 5, pp. 2231-2238, 2014.
[11] Setayeshmehr A, Akbari A, Borsi H, et al. On-line monitoring and diagnoses of power transformer bushings. Dielectrics & Electrical Insulation IEEE Transactions on, vol. 13, no. 3, pp.608-615, 2006.
[12] BS EN 60076-18:2012, "Power transformers: Part 18 measurement of frequency response," BSI Standards Limited, 2013.
[13] Wang M, Vandermaar A J, Srivastava K D. Improved detection of power transformer winding movement by extending the FRA high frequency range. IEEE Transactions on Power Delivery, vol. 20, no. 3, pp.1930-1938, 2005.
[14] Yao C, Zhao Z, Chen Y, et al. Transformer winding deformation diagnostic system using online high frequency signal injection by capacitive coupling. IEEE Transactions on Dielectrics & Electrical Insulation, vol. 21, no. 4, pp.1486-1492, 2014.
[15] IEEE Standard, "IEEE guide for the application on interpretation of frequency response analysis for oil-immersed transformers," IEEE STd C57.149-2012, Mar.2013
[16] Std. DL/T911-2004, "Frequency response analysis on winding deformation of power transformers," The Electric Power Industry Standard of People's Republic of China, ICS27.100, F24, Document N0.15182-2005, Jun.2005.
[17] Ji T Y, Tang W H, Wu Q H. Detection of power transformer winding deformation and variation of measurement connections using a hybrid winding model. Electric Power Systems Research, vol. 87, no. 87, pp. 39-46, 2012.
[18] Behjat V, Mahvi M, Rahimpour E. New statistical approach to interpret power transformer frequency response analysis: non-parametric statistical methods. Iet Science Measurement & Technology, vol. 10, no. 4, pp.364-369, 2015.

Research on Safety Assessment for Power Line Design

Jun Ding[1], Yong Wang[1], Rui Guo[1], Jia Shi[1], Yan Shen[1], Wen Tao Tan[1] and Jun Da Qu[1]

[1] State Grid Suzhou Power Supply Company, No.555, Laodong Road, Suzhou, Jiangsu, China

568483238@qq.com

Abstract. As the source of power line project, line design is a key part of life cycle management. The safety and reliability of line design can provide necessary support for construction, operation and maintenance. This paper proposes a risk-based approach for safety management in line design, decomposing risks into two dimensions, the probability and the consequence. Based on factors, such as risk assessment, number of historical incidents, consequence of historical incidents, qualification of proofreading and auditing, this approach in line design can predict risk status in real time, provide scientific basis for risk control, and minimize safety hazards.

1. Introduction

With the transformation of economic structure, China has vigorously promoted the construction of smart cities in recent years, putting forward higher requirements for grid construction in cities [1]. As the source part of grid project, line design is related to the whole process quality control, including construction, operation and maintenance [2-3]. Therefore, improving the safety management of line design can reduce safety hazards and ensure the stable operation of the grid.

2. Key point of power line design [4-8]

Route selection is the primary element considered in power line design [9]. Besides municipal planning and field environment, it also needs to take the following factors into account, safety distance, tower form, foundation form, cable channel form, overall investment, transportation convenience, and construction influence on surrounding facilities.

Based on the long-term grid planning, type design of the wire and cable must comprehensively consider the current density and transmission capacity in the line, as well as mechanical strength calculated, natural conditions and geographical environment along the route [10-11].

The tower and corresponding foundation are the core parts of overhead line projects, both cost accounts for about 40% of the total project investment. The design of the tower, the foundation and the insulator must take the followings into account, voltage grade, load size, mechanical strength, filth intensity and geological conditions [10-11].

Cable channel is the core of cable line project, which accounts for about 45% of total investment. At present in south area of China, the laying of cables is mainly the combination of cable trench and cable duct [12]. Channel design is affected by voltage grade, route path, cable quantity, mechanical strength, and geological condition.

3. Quantitative evaluation of risks for power line design

During the line design cycle, we need to take risks into account, such as tower risk, wire risk, foundation risk, insulator risk, cable risk, channel risk, investment risk and environment risk [13]. These factors are not only related to the whole life cycle of the line, but also closely related to the safe operation of the grid and the comprehensive development of the city. Therefore, during design process, relevant designers should consider the possible consequence of these risks and strive to minimize the occurrence probability and potential consequence of these risks.

According to the two-dimensional evaluation theory, risks can be decomposed into two dimensions, the occurrence probability and the potential consequence [14]. Referring to classification standards for incidents in "Investigation procedures for safety incidents of State Grid", we carry out risk's quantitative evaluation based on three probability rankings and eight consequence rankings [15]. The evaluation matrix is shown as table 1. Larger number means higher probability and worse consequence.

In this paper, all incidents mentioned are caused by corresponding design risks.

Table 1. Evaluation matrix for probability and consequence.

Probability	Consequence							
	1	2	3	4	5	6	7	8
3								
2								
1								

4. Quantitative evaluation of risks for power line design

We must correctly deal with risks in line design, trying to find out it's regularity and rationality, to explore the balance among investment control, environment protection and safety control, and finally to realize the optimization of line design. Hence, we need to carry out quantitative control at different levels, based on the probability and consequence of risks [16-21].

According to the existing risks and previous incidents of line design in an area, based on incidents' frequency, risks' potential consequence and weight factors, the real-time probability and consequence of risks is calculated. Thus, risk control can be targeted. This paper takes cable risk as an example to show the detailed calculation procedure.

4.1 Average calculation of the probability value and consequence value for existing risks by assessments.

We evaluate each cable risk to get its residual probability and residual consequence. In other words, the probability and consequence of current cable risk is calculated by numeric values, and the result is rounded, calculation process is shown below.

- $L_{av}=(\Sigma\ L_{pi})/n$, L_{av} is the average probability, L_{pi} is the probability of the single assessment, and n is the quantity of all assessments.
- $C_{av}=(\Sigma\ C_{pi})/n$, C_{av} is the average consequence, C_{pi} is the consequence the single assessment, and n is the quantity of all assessments.

Table 2. Average calculation for cable risk by assessment.

Assessment	Assessment Probability	Assessment Consequence
Ass. 1	3	2
Ass. 2	1	1

Ass. 3	1	3
Ass. 4	1	1
Average	1.5(round to 2)	1.75(round to 2)

In table 2, the calculation shows that the average probability of the cable risk is 2, and the average consequence is 2.

4.2 Adjustment of this average value to make sure it reflects actual incidents' consequences over the last 12 months.

We evaluate all cable incidents (in last 12 months) and find out the actual consequence of each incident based on the evaluation matrix. If the highest actual consequence over the last 12 months is higher than the average consequence calculated in step 1, the average consequence automatically jumps up to reflect the highest one.

As shown in table 3, 6 cable incidents occurred in last 12 months and the highest consequence value is 3. Compared with the highest value, average consequence jumps up to 3.

Table 3. Consequence of actual cable incidents.

Cable Incident	Actual Consequence
Incident 1	1
Incident 2	3
Incident 3	1
Incident 4	1
Incident 5	2
Incident 6	1

Hence, probability of the cable risk remains 2, and the consequence is 3 in step 2.

4.3 Calculation of severity index based on statistical analysis on all incidents

The actual incidents over the last 12 months are classified in table 4 (the horizontal axis is the incident type, and the vertical axis is the potential consequence level).

Table 4. Statistics of all actual incidents (in last 12 months)

Realistic Consequence	Incident Type (corresponding to risks)								
	Cable	Cable channel	Tower	Foundation	Insulator	Wire	Investment	Environment	Total
8	0	0	0	0	0	0	0	0	0
7	0	0	0	0	0	0	0	0	0
6	0	0	0	0	0	0	0	0	0
5	0	1	0	0	0	0	1	0	2
4	0	1	0	0	0	0	1	0	2
3	1	2	0	1	0	0	2	0	6
2	2	1	1	2	2	1	3	0	12
1	3	2	2	5	2	1	4	1	20

In the same consequence ranking, incident frequency is calculated while the quantity of cable incidents is divided by all incidents quantity. The severity index in each ranking is calculated by the multiplication of incident frequency and weight factor, as shown in table 5 [18].

Table 5. Calculation of severity index for cable incidents

Realistic Consequence	Cable Incident Quantity	Total Quantity	Incident frequency		Weight Factor	Calculation	Severity Index
8	0	0			64		
7	0	0			49		
6	0	0			36		
5	0	2	0/2	0	25	0 * 25	0
4	0	2	0/2	0	16	0 * 16	0
3	1	6	1/6	0.1667	9	0.1667 * 9	1.5003
2	2	12	2/12	0.1667	4	0.1667* 4	0.6668
1	3	20	3/20	0.1500	1	0.1500 * 1	0.1500
						Total	2.3171

Severity threshold is defined as 50% of the top incident index among all incidents. If severity index of one risk is equal or larger than the threshold, the probability and consequence of this risk will have an increase of one ranking, as shown below.

- Severity index of the risk≥ Severity threshold

Table 6. Severity index of all risks.

Risk Type	Severity Index
Cable	2.3171
Channel	23.9333
Tower	0.4333
Foundation	2.4167
Insulator	0.7667
Wire	0.3833
Investment	24.7000
Environment	0.0500

We get the severity threshold, 12.35, from table 6. Thus, the probability and consequence of cable risk remains unchanged.

4.4 Calculation of nonconformity index for proofreading and auditing

The nonconformity index is the quantity of nonconformity for proofreading and auditing in all line design projects over the last 12 months, classified by risk types. The nonconformity threshold is defined as 50% of the top nonconformity index.

The probability of the risk will have an increase of one ranking by the following two conditions.

- Nonconformity index of the risk≥ Nonconformity threshold
- Nonconformity index of the risk≥ 40

As shown in table 7, nonconformity index of cable risk is 69, meeting the above two conditions. Thus, the probability of cable risk increases from ranking 2 to ranking 3.

IOP Conf. Series: Materials Science and Engineering **366** (2018) 012068 doi:10.1088/1757-899X/366/1/012068

Table 7. Severity index of all risks.

Risk Type	Nonconformity Index for Proofreading and Auditing
Cable	69
Channel	122
Tower	57
Foundation	82
Insulator	45
Wire	29
Investment	96
Environment	15

Finally, the probability and consequence of cable risk is revised from (2,2) to (3,3).

5. Conclusion

Safety management of line design is the source part of life cycle management. The quantification of risks based on the probability and consequence can effectively improve the source prevention of safety management, timely evaluate possible design flaws and improve designers' ability. In addition, it can also help the construction, operation and maintenance departments to carry out hazard management to ensure the safe and stable operation of the grid.

References

[1] Fu Lijun. Analysis of Smart Grid Development Direction Based on Smart City[J]. China Instrumentation, 2017, 7(1):21-25.
[2] Shi Junqiang. Research and Application of Life Cycle Management of Transmission Line Project[J]. Power & Energy, 2017, 5(1): 533-537.
[3] Liu Jian, Zhang Yong, Du Zhiye, Ruan Jiangjun, Liu Kaipei. Analysis of full life cycle cost sensitivity in ac transmission line design[J]. High Voltage Engineering, 2010, 6(1): 1554-1559.
[4] DL/T 5221-2015. Technical rule for design of ubran power cables[S], in Chinaese.
[5] DL/T 5220-2015. Coad for designing over-head distribution transmission line up to 10kV[S], in Chinsese.
[6] GB 50545-2010. Coad for designing of 110kV-750kV overhead transmission line[S], in Chinese.
[7] Chen Xiaodong. The key point of transmission line design and the analysis of safety accident prevention countermeasures[J]. Telecom World, 2013, 10(1): 16-17.
[8] Chai Changqing, Sun Shouwei. Key points analysis in transimission line design[J]. Proceedings of the CSEE, 2009, 1: 613-614.
[9] Li Tiejun. Choice and orientation of path in design of electric line[J]. Ningxia Electric Power, 2006, 3(1): 20-23.
[10] Liu Qingfeng. Analysis and calculation of unbalanced tension on transmission line[J].Electric Power Automatic Equipment. 2006, 26(1): 93-95.
[11] Jiang Xingliang, Ma Jun, Wang Shaohua, Sun Caixin, Shu Lichun. Analysis of the ice damage on power line[J]. Electric Power, 2005, 11(38): 27-30.
[12] Chen Jianguang. Analysis of 10kV power cable line design technology[J]. Electronic Test, 2016, 10(1): 130-131.
[13] Sun Fei, Wang Ning. Transmission network planning design and the impact on power grid safety[J]. Electronic Test, 2017, 12(1): 129-130.
[14] Zhang Yuhong. Quantitative analysis and reserarch on security risk of newclear power unit[J].

Guangdong Electric Power, 2017, 30(6): 30-35.
[15] No. 599 Decree of the State Council. Regulations on emergency disposal and investigation of power safety accidents[S], in Chinese.
[16] What is risk management[EB/OL]. http://www.whatisriskmanagement.net/.
[17] Maurizio Bevilacqua, Filippo Emanuele Ciarapica. Human factor risk management in the process industry: A case study[J]. Reliability Engineering & System Safety, 2018, 169(1): 149-159.
[18] Pual Adams. How to use risk tool--Application of quantitative safety evaluation[Z]. Austrilia: Jemena Company, 2017.
[19] Alexander J. McNeil, Rüdiger Frey, Paul Embrechts. Quantitative Risk Management: Concepts, Techniques and Tools[M]. Princeton and Oxford: Press of Princeton University, 2015.
[20] Akintola S. Akintoye, Malcolm J. MacLeod. Risk analysis and management in construction[J]. International Journal of Project Management, 1997, 15(1): 31-38.
[21] Sun Sha. Quantitative analysis of risk in project process[D]. Shanghai:Fudan University, 2013.

IOP Conf. Series: Materials Science and Engineering **366** (2018) 012069 doi:10.1088/1757-899X/366/1/012069

Research of reclosing failure of transmission lines caused by multiple lightning strikes

SI Wenrong[1], JIN Heng[1], TAN Fali[2], PENG Yang[2], LU Bingbing[1] and ZHAO Chun[2]

[1]State Grid Shanghai Electric Power Research Institute, Shanghai, 200437, China
[2]Wuhan sunshine power Science &Technology co., LTD, Wuhan, 430074, China

625778539@qq.com

Abstract: Lightning faults of transmission lines are usually instantaneous faults, and most of them can automatically reclose successfully, but reclosing failure will also occur sometimes. The present paper analyses a reclosing failure of a 500kV transmission line after lightning strike. In this paper, the lightning monitoring system is used to monitor power frequency fault current and lightning transient traveling wave current, and the fault location, the cause identification and the reasons for the reclosing failure are obtained based on the monitoring data. The results show that there are several successive lightning strikes near the fault point of transmission line during the switch to closing period, which makes the fault frequency continuous current unable to return to zero, so the insulation cannot be recovered, which ultimately leads to reclosing failure.

1 Introduction

Thundercloud discharge to the earth is generally accompanied by multiple lightning strikes[1-2]. During the initial stage of thundercloud discharge, downward leader and upward leader between thundercloud and the earth meet together and a discharge channel forms, and then a strong discharge occurs, this process is called main discharge[3-4]. In the vicinity of the main discharge area there are often a number of dense charge centers, when the main discharge occurs, multiple charge centers will meet along the initial discharge channel and then multiple lightning strikes occur[5]. Most of the lightning strike faults are instantaneous faults. When the fault occurs, most of the operating mechanisms in the substation can automatically reclose the fault line and resume power supply within a short time, but in some extreme cases, for example, when a line is struck continuously for a short period of time, it is easy to happen reclosing failure and power supply cannot be restored[6-9].

Lightning fault reclosing failure is relatively rare, and related research are seldom reported at home and abroad. Based on a reclosing failure of a 500kV transmission line after lightning strike, the paper analyzed the precise location of the lightning strike point, the reason identification function and then found out the reason of the failure of the reclosing, which may provide some reference for the reason analysis of similar complex faults.

2. Working principle of lightning monitoring system

As shown in Fig 1, lightning monitoring system is composed of workstation, data center and monitoring terminals. Monitoring terminals are distributed on the conductor of transmission lines, monitoring power frequency current and high frequency transient current in real time, sending such

IOP Conf. Series: Materials Science and Engineering **366** (2018) 012069 doi:10.1088/1757-899X/366/1/012069

data to data center by the way of GPRS. Data center is used to collate and analyze the data and output the diagnostic results.

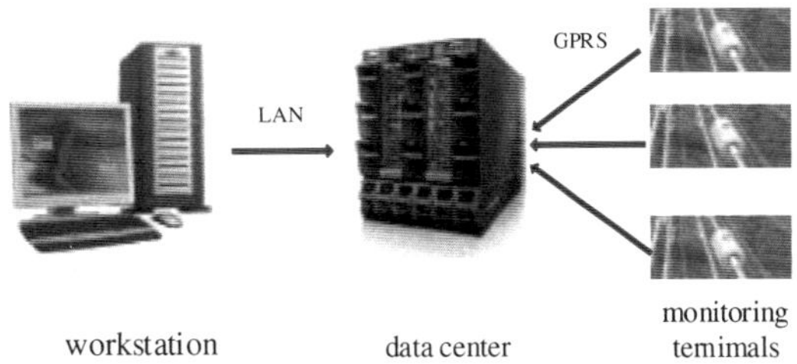

Figure 1. Structure diagram of the system

The main components of monitoring terminal include high-frequency current sensor, conditioning circuit, acquisition and high-speed processing circuit, battery, GPS clock module and wireless transmitting module. The monitoring terminal shell is a cylindrical aluminum alloy metal container with all the electronic components installed inside it. In actual operation, the terminal shell and the wire are in equipotential connection, which can effectively protect the internal functional modules from the interference of high pressure and electromagnetic environment. The sensor adopts the self integral Roche coil, which has the advantages of good high frequency response, good linearity, wide measurement band and strong anti-interference ability[10-12]. The principle of measuring current by Roche coil is shown in Fig 2.

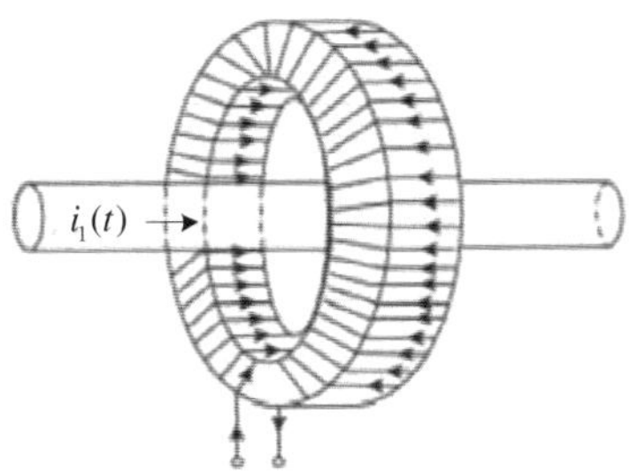

Figure 2. Principle of rogowski coil current measurement

In Fig 2 $i_1(t)$ is the current value on the wire. The output of the sensor is a voltage signal, and also the acquisition of the monitoring terminal. The value of the output of the sensor can be obtained as formula (1):

$$e(t) = -M\frac{di_1(t)}{dt} = L_0\frac{di_2(t)}{dt} + (R_0 + r)i_2(t) \tag{1}$$

In formula (1), M represents the mutual inductance between the coil and the tested conductor. L_0 and R_0 are the self inductance and internal resistance of the coil respectively. C_0 and R are the distribution capacitance and the sampling resistance of the coil respectively, and $i_2(t)$ is the current of the sampling resistor[13-14]. When frequency of measured current is very high, $wL_0 \rangle\rangle R_0 + r$, it can be deduced that

$$e(t) = i_2(t)r \approx -\frac{Mr}{L_0}i_1(t) \tag{2}$$

By formula (2), the output of the sensor is approximately proportional to the current of the conductor. Acquisite output of the sensor, and lightning current waveform can be obtained.

3. Lightning strike Fault monitoring and diagnosis

3.1. Monitoring terminals installation

August 19, 2016 22:59:41, a 500kV line of China Southern Power Grid lightning trip occurred, and reclosed unsuccessful. As shown in Fig. 3, the total length of the line is 141km. Lightning monitoring terminals are installed at # 1 tower, # 73 tower, # 136 tower, # 209 tower, # 253 tower and # 302 tower of the line respectively, and # 1 tower ,as well as # 302 tower are located at both ends of the substation exit.

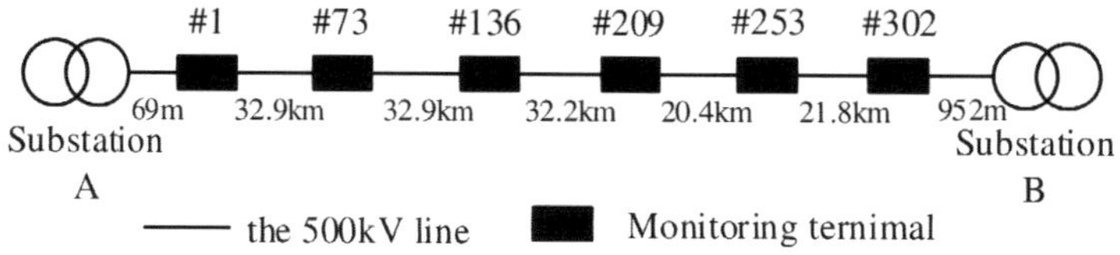

Figure 3. Schematic diagram of the installation of monitoring terminals

3.2. Trip process record

When flashover occurs, power frequency current will change suddenly at a point of time, which is the time of fault. At 2016-8-19 22:59:41, the monitoring terminal of A phase of the #209 tower monitors the mutation of line frequency current, and the waveform of frequency current is shown in Fig 4.

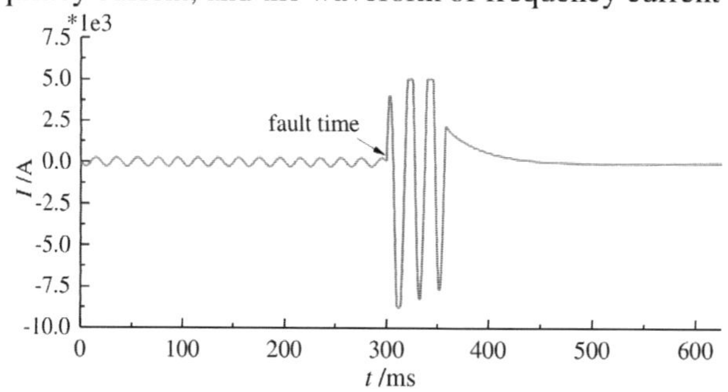

Figure 4. Power frequency split current waveform

Fig 4 shows that in 22:59:41 sec. 471 ms, power frequency current of the monitored line suddenly increases, with the amplitude over 5000A, far exceeding the normal operation current of the line, and it changes to 0 after maintaining two cycles, concluding that the line has a fault and trip.

Fig 5 is a frequency current waveform recorded by the A phase monitoring terminal of #1 tower at 22:59:42 time. The amplitude of the front 290ms is 0, while it increases suddenly at 292ms, and then goes back to zero after two cycles. Besides, the time interval between the waveform and the fault reclosing waveform of Fig 4 is 811ms, which is consistent with the characteristics of the reclosing process of the transmission line, so it can be concluded that Fig 5 is frequency current waveform of reclosing, and reclosing is unsuccessful.

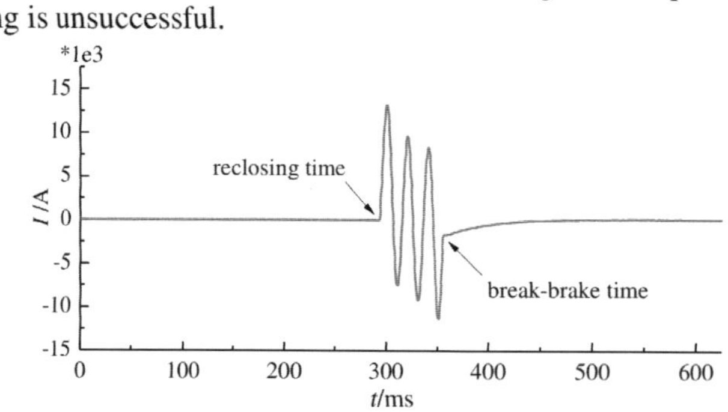

Figure 5. Power frequency waveform of the circuit breaker Reclosing

3.3. Fault location

IOP Conf. Series: Materials Science and Engineering **366** (2018) 012069 doi:10.1088/1757-899X/366/1/012069

The fault interval positioning is realized by the power frequency current, what's more, fault point can be accurately located by the traveling wave current of the fault interval.

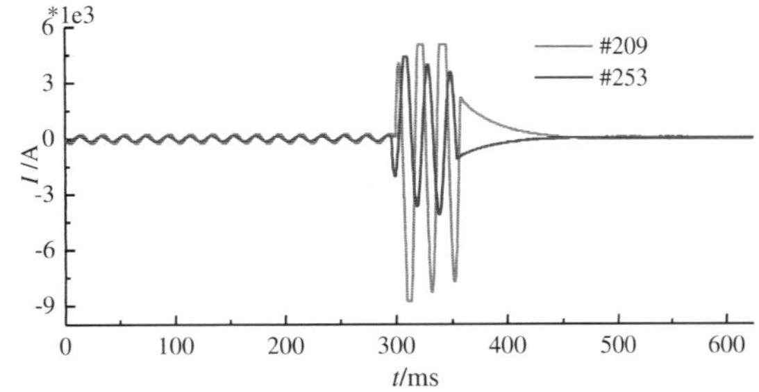

Figure 6. Power frequency switching current waveform of different point

From Fig. 6, when the fault occurs, the polarity of short circuit power frequency current of #209 tower and #253 tower is the opposite, which indicates that the short-circuit fault point is between #209 tower and # 253 tower.

Two fault points are found by traveling wave positioning, as is described as follows.

(1) The first fault point

Traveling waves current monitored of #209 tower and # 253 tower at the time of failure are as shown in Fig.7.

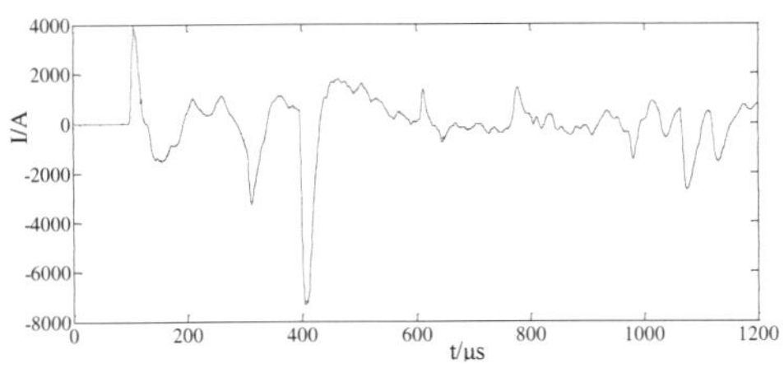

(a)fault current traveling wave of #209 tower

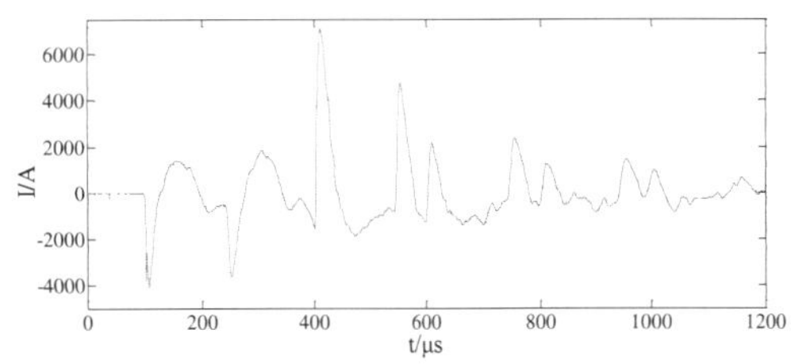

(b)fault current traveling wave of #253 tower

Figure 7. Fault traveling wave

In Fig 7, the time difference between the fault point current traveling wave from the fault point to the #209 tower and the #253 tower is 8 microseconds, and the distance between the two tower is 20.3 kilometers. Through the double end traveling wave positioning method, the fault point distance can be calculated that it is near the # 233 tower, which is about 9 km away from # 253 tower.

(2) The second fault point

The reclosing of the circuit breaker also produces high frequency traveling wave, which can be applied to locate the second fault point. Reclosing traveling wave is as shown in Fig.8.

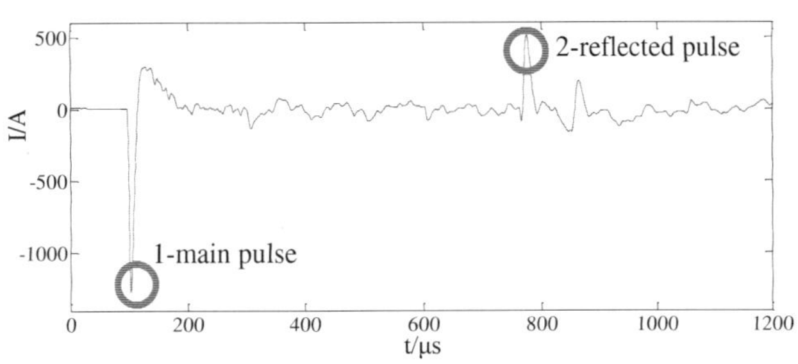

Figure 8. Reclosing traveling wave

Fig.9 shows a schematic diagram of the reflection process of the reclosing traveling wave. The reclosing traveling wave is produced when the circuit breaker of substation B recloses, and is recorded by the monitoring terminal after the fault point, such as the first main pulse 1 in Fig 8. On the other hand, reflection occurs when traveling wave arrives at the fault point or substation, when traveling wave is reflected at the fault point, it will undergo second reflection after passing through the substation B. After two times reflection, the wave will be refracted through the fault point and monitored by the monitoring terminal, which is corresponding to the reflected pulse 2 in Fig.8. From Fig.8 and Fig.9, we can calculate the second fault point is near the #236 tower by single end traveling wave location method.

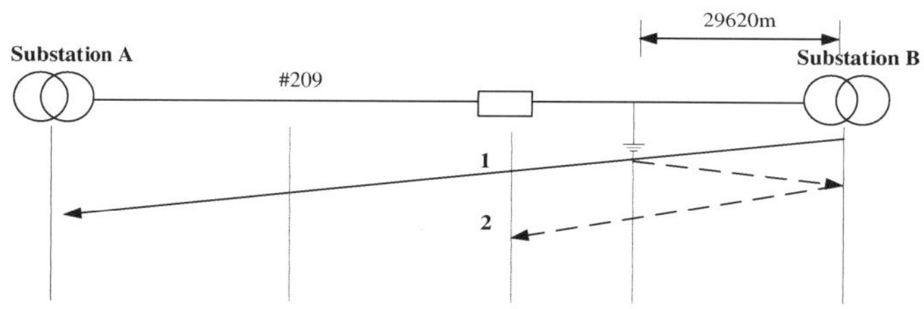

Figure 9. Location sketch map by reclosing traveling wave

3.4. Analysis of the cause of failure

According to Fig.7, the amplitude of traveling wave reaches thousands of amperes, and the wave tail time is within 40 microseconds, which is consistent with the characteristics of lightning fault traveling wave, and there is no reverse polarity pulse in the front part of the main wave, which is initially judged as shielding failure.

3.5. Patrolling result

The power supply department inspected #233 and the nearby area. Results show that there were lightning flashover traces of #233 tower A phase and #236 pole tower A phase insulator, which are shown in Fig 10. Combined with the weather conditions at the time of failure, it is basically determined that lightning flashover occured in #233 and #236 towers.

(a)Flashover trace of A phase insulators in #233 tower (b) Flashover trace of A phase insulators in #236tower

Figure 10. Location sketch map by reclosing traveling wave

4. Failure analysis of reclosing

A lightning trip fault occurred in the A phase of this line has at 22:59:41 471 ms, and then circuit breaker disconnected at 22:59:41 516ms, reclosed unsuccessful at 22:59:42 327ms. The break from the circuit breaker disconnecting and reclosing is about 811 milliseconds, during which the insulation can be recovered in normal conditions.

According to the monitoring data of lightning monitoring system and lightning location system, there are 8 lightning strikes near the #233 tower of the transmission line from closing to reclosing of the circuit breaker, all of which are around the A phase conductor, as shown in Tab 1. I_L is the amplitude of lightning current.

Table 1. Lightning stroke process record

T	22:59:41 471	22:59:41 553	22:59:41 593	22:59:41 665	22:59:41 686	22:59:41 722	22:59:41 887	22:59:41 990
I_L/kA	-19	-23.6	-11.9	-7.9	-8.8	-14.7	-9.5	-14.4
Lightning strike position	#234	#235	#236	#232	#234	#234	#233	#236

Fig.11 is the power frequency current waveform recorded 300ms before the reclosing failure recorded by the lightning monitoring system. In Fig.11 (a), t=300ms is the reclosing time, and power frequency current increases sharply, while the spike near 130ms is subjected to the lightning time. Fig.11 (b) is the part of Fig.11 for t=0-120ms. From Fig.11 (b), it can be seen that the current on the line before reclosing is not 0, and presents a certain periodicity.

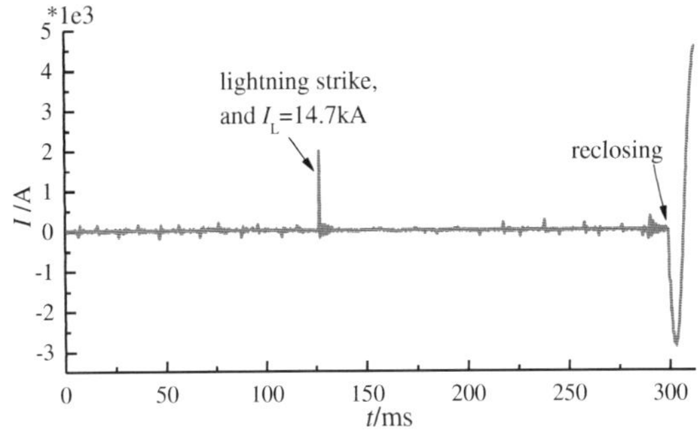

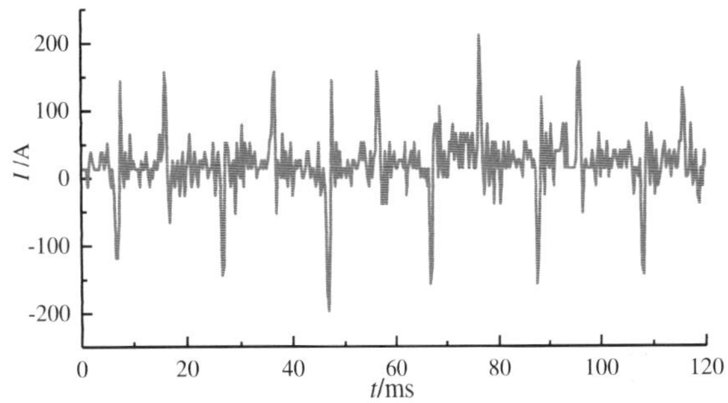

(a)A phase line current of #209 tower before reclosing (b) waveform of (a) at t=0-120ms

Figure 11. Current waveform before reclosing

In order to eliminate the high frequency component in the waveform in fig.11(b), the filter cut-off frequency 200Hz is set up, and the waveform after filtering is shown in Fig.12.

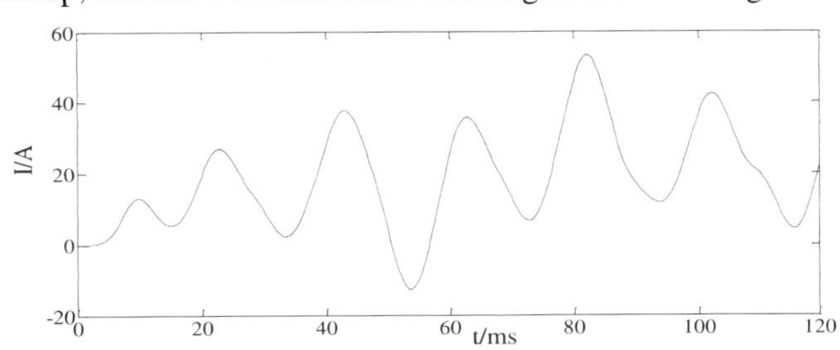

Figure 12. Current waveform after filtering

It can be seen from Fig. 12 that the waveform has obvious periodicity, and its period is about 20ms. It is deduced that the secondary supply current has not disappeared before reclosing, and its effective value is about 10~20A. Due to the coupling capacitance and coupling inductance between line to line and line to earth, the capacitive coupling(inductive) current will not be interrupted immediately after the fault line tripping. It will form a continuous flow at the fault point arc passage, which is called the secondary supply current. When secondary supply current doesn't completely disappear, the arc channel will be in a free state with good conductivity. Under normal circumstances, the secondary supply current will soon disappear, enabling the line to restore good insulation.

From the monitoring results, during the trip to reclosure, 8 lightning strikes near the fault point make it difficult to extinguish the supply arc of the fault point, and the line can not restore the insulation, so the reclosing fails.

According to the above analysis, it is inferred that the reasons of the failure of the lightning reclosing of this line are described as follows:

1) when lightning fault occurs, the circuit breaker is sluice, and secondary supply current exists in the

fault point;

2) during the closing to reclosing of the breaker, 8 lightning strikes stroke near the fault point, which leads to the failure of the short-circuit point secondary supply current to be completely extinguished and the insulation can not be recovered.

3) after the automatic reclosing, the fault still exists because of the continuous arc burning in the short circuit point, so reclosing failed.

5. Conclusion

1) The fault between #209 and #253 tower is judged by power frequency current. Then two fault points are calculated by traveling wave location method, which correspond to #233 and #236 respectively, which is consistent with the actual inspection results, indicating that the fault diagnosis results are correct.

2) The amplitude of traveling wave reaches thousands of amperes, and the wave tail time is within 40 microseconds, which is consistent with the characteristics of lightning fault traveling wave, and there is no reverse polarity pulse in the front part of the main wave, so the failure is shielding failure;

3) During the closing to reclosing of the breaker, 8 lightning strikes stroke near the fault point, which leads to the failure of the short-circuit point secondary supply current to be completely extinguished and the insulation can not be recovered, resulting in reclosing failure.

References

[1] A. T. Johns and S. K. Salman, *Digital Protection for Power Systems*,1st ed. London, U.K.: Peregrinus, 1997.

[2] H. Suad, G. Ferdinand, and F. G. Andrej, "Prediction of power system security levels," *IEEE Trans Power Syst*, vol. 24, no. 1, pp. 368–377, Feb. 200

[3] *IEEE Standard Dictionary of Electrical and Electronics Terms*, ANSI/ IEEE Standard 100, 1996.

[4] A. A. M. Zin and S. P. A. Karim, "Protection system analysis using fault signatures in Malaysia," *Int J Elect. Power Energy Syst.*, vol. 45,pp. 194–205, 2013.

[5] G. Diendorfer and W. Shulz, "Ground flash density and lightning exposure of power transmission lines," presented at the IEEE Power Tech Conf., Bologna, Italy, 2003.

[6] G. Andersson, P. Donalek, R. Farmer, N. Hatziargyriou, I. Kamwa, P. Kundur, N. Martins, J. Paserba, P. Pourbeik, J. Sanchez-Gasca, R.Schulz, and A. Stankovic, "Causes of the 2003 major grid blackouts in North America and Europe and recommended means to improve system dynamic performance," *IEEE Trans Power Syst*, vol. 20, no. 4, pp. 1922–1928, Nov. 2005.

[7] EventsAnalysis: System Disturbance Reports NorthAmerican Electric Reliability Corporation (NERC), USA, 1992–2010.

[8] M. Evan, "Extreme grid disruptions and extreme weather," presented at the U.S. Disaster Reanalysis Workshop, Asheville, NC, USA, May 3–4, 2012.

[9] J. O. Adepitan and E. O. Oladiran, "Analysis of the dependence of power outages on lightning events within the ijebu province, Nigeria,"*J. Environ. Earth Sci.*, pp. 850–856, 2012.

[10] Magnago F H, Abur A. Fault location using wavelets[J]. IEEE Transactions on Power Delivery, 1998, 13(4): 1475-1480.

[11] Y. F. Li and P. Chen, "The Principle of the Modem Travelling Wave Based on Fault Location and the Applications to Actual Fault Analysis," in Proc. 2009 Asia-Pacific Power and Energy Engineering Conference.

[12] F. H. Magnago and A. Abur, " Fault location using wavelets," IEEE Trans. Power Delivery, vo13, pp. 1475-1480, Oct. 1998.

[13] de T ourrei l C , Roberge R, Bourdon P .Long-Term Mechanical Properties of High Voltage Composite Insulators [J] .IEEE Transactions on Power Apparatus and Systems , 1985 , 104 (10):2918-2921 .

[14] M. Vitins, "A Correlation Method for Transmission Line Protection," IEEE Trans. Power Apparatus and Systems, vol. PAS-97, pp. 1607-1617, Sep. 1978.

IOP Conf. Series: Materials Science and Engineering **366** (2018) 012070 doi:10.1088/1757-899X/366/1/012070

Operational Analysis of Dynamic Line Ratings

Saifal Talpur[1], T.T. Lie[1], R. Zamora[1] and N.K.C. Nair[2]

[1] Dept. of Electrical & Electronic Engineering, Auckland University of Technology, New Zealand
[2] Dept. of Electrical and Computer Engineering, University of Auckland, New Zealand

saifal.talpur@aut.ac.nz

Abstract. Overhead line is an important asset for power industries. Rising demand in electricity, environmental concerns and cheaper electricity prices are some important factors behind operating the lines at their maximum potential. Replacing overhead lines and/or constructing new lines are not only expensive but also at the same time raise environmental concerns. However, rating overhead lines based on their natural capacity besides being cost effective is a technically feasible solution. In this project, the operational analysis of dynamic line ratings (DLR) is conducted to analyze potential of DLR when practically implemented across overhead line in a sub transmission network. It further guides system operators in fully utilizing the line capacity to its maximum potential. The project also investigates the impact of Flexible AC Transmission System (FACTS) device in the presence of DLR technique to control power flow from wind generators for stabilized and controlled transfer of electricity.

1. Introduction

In times of facing economic and environmental challenges, power industries are looking towards effective solutions in making power components more effective and fully usable to save money and protect the environment. Rating overhead lines based on their natural capacity makes them an affordable solution for power industries to adopt. Dynamic line rating (DLR) is not only based on real-time weather and line loading but also is dependent on line sag and its tension. Previous research on DLR has mainly focused on either basing the line rating on real-time weather scenarios or line sag and/or tension [1]. In this paper both approaches are implemented together.

High wind speed and low ambient temperature are two main weather factors in line capacity enhancement. In New Zealand, due to windy summers and cold winters, the DLR technique is quite beneficial for transmission system operators in enhancing the line capacity. Additionally, the role of FACTS devices in presence of DLR improves system's stability besides making active and reactive power flows regulated and controlled. In this project, thyristor controlled series capacitor (TCSC) was employed to control and regulate the active and reactive power flows based on bus terminal voltages.

The advantages of employing FACTS device with DLR involve the followings: 1) Regulating the AC power through voltage control; 2) Transmitting additional power at controlled voltage; 3) Maintaining the small-signal stability by damping power oscillations; 4) Transmitting excess power with no violation in line-sag limit. The advantage of DLR technique also involves distribution and transmission system operators to transfer additional power across existing network without investing on upgrading or building new lines [1]-[2].

Dynamic rating of overhead lines varies significantly with weather parameters and conductor temperature. The change in conductor temperature further affects the line sag and therefore it needs to be monitored continuously. Besides, due to excess wind energy, wind farms are mostly installed in rural areas with no direct connection with strong grid [3]. Therefore, the only options available are to either build a new line or to up rate the existing overhead lines based on dynamic ratings. However, the environmental concerns, high costs and time required in building new lines (as compared to installing wind farms) and more importantly the association of conductor cooling with the wind power restricts from building the new lines.

Electrical distribution network operators face number of challenges, like increment in distributed generation, load growth and ageing infrastructure. Due to these reasons, the need for developing and implementing those techniques which will allow more efficient asset utilisation has arisen. DLR is mainly used to increase the capacity of overhead conductor in terms of transmitting the maximum electric current besides transferring the electricity during peak load and emergency states [4].

The heat-balance equation working under heat absorption and heat loss decides the conductor rating. For conductor to transmit higher electricity, it is required to dissipate more and absorb less heat energy. The more the heat absorbed by it, the lesser it becomes efficient in transferring the power and vice versa. Similarly, from the reliability perspective, the absorbed heat energy should instantly be radiated to not affecting sag of the line (mid span point). Dynamic line rating calculation is based on type of conductor (mainly its diameter), its real-time loading (i.e. the temperature rise), real-time ambient temperature, real-time wind speed and its direction [5].

Adopting the dynamic thermal line rating technique is considered a challenge for control room operators in case of limited knowledge about monitoring the steady state and post-contingent line loadings [6]. Besides, before implementing the DLR technique in the electricity network, the comparison of results from normal and emergency states is important in terms of helping the control room operators to allow exact wind energy from the wind farms to avoid sag violations and maintaining the system reliability.

Conductor sag however affects operational line ampacity and is mainly dependent on terrains and span length. In New Zealand, due to extensive terrains, the span length of conductor besides ground clearance vary significantly and therefore it is a decisive factor in determining the operational capacity of the line (relatively different from its maximum capacity) [7]. This project develops a technique that determines the operational line ampacity for the power system operator to direct the power flow and loading the line accordingly. However, the challenges associated with operational implementation of dynamic ampacity across overhead conductors are power system security, availability of spinning reserves and dynamic ampacity variations [7]-[9].

DLR helps system operators in knowing the true thermal limit of transmission lines before loading them. By doing so, the delivery of excess power through overhead lines during high demand helps in unnecessary load shedding during contingencies in the network. Consequently, additional power can be transferred economically due to minimal expenditure associated with implementing the DLR in the network. This practice is considered important when considering the demand of transmission open-access rules of operation as it encourages wheeling of power with economic energy transfers [10].

From both the installation cost and energy transfer capacity viewpoints, the DLR approach may be able to offer the greatest potential benefits at minimal cost. In comparison with line re-tensioning technique, the implementation of DLR system can provide 67% gain in energy transfer capacity at 62% of the cost associated with line re-tensioning technique [11]. Moreover, by calculating the hours of potential thermal overload, the DLR approach may allow the overhead line to transfer power up to twice the value of its static rating [3]. The analytical and experimental results show that the DLR technique has the potential to accommodate excess energy in the existing infrastructure, with maintaining the operational security [11].

2. Mathematical Modelling

Dynamic ampacity of the overhead conductor depends upon two main factors, i.e. its physical characteristics and the environmental parameters [12], with sub categories such as conductor diameter, conductor temperature, ambient temperature, wind speed, angle between wind speed and conductor, and the solar radiation. Dynamic Line Rating is calculated by taking into account the mathematics behind heat energy absorption and consumption as represented in Equations (1)-(9) [13]. According to heat balance Equation (1),

$$Heat_{GAIN} = Heat_{LOSS} \tag{1}$$

Heat gain is related to heat absorbed by conductor and the solar radiation, i.e.

$$P_{loss} + Q_{solar} = Q_{convection} + Q_{radiation} \tag{2}$$

Furthermore, the absorbed heat depends on the amount of current and the conductor resistivity, i.e.

$$P_{loss} = I_i^{\,2} * R_{T_C} \tag{3}$$

Conductor resistivity as shown in Equation (4) is based on difference between low and high conductor resistance as well as on temperature differences, i.e.

$$R_{T_C} = \left[\frac{R_{T_{high}} - R_{T_{low}}}{T_{high} - T_{low}}\right] * \left(T_{C,i} - T_{low}\right) + R_{T_{low}} \tag{4}$$

Solar absorptivity depends on many factors like conductor's projected area (A), absorptivity constant (β_i), global solar radiation (φ) and the effective incidence angle (θ) between conductor and sun's rays, time and day, i.e.

$$Q_{solar} = \beta_i \varphi \sin(\theta) A_{p,i} \tag{5}$$

Heat radiation from the conductor is divided into two types, i.e. high and low. The low heat radiation is related to lower wind speed, mainly lesser than 0.5 m/sec [7], i.e.

$$Q_{FC\_low} = \left[1.01 + 0.0372\left(\frac{D_i * \sigma * V_W}{\varepsilon}\right)^{0.52}\right]\alpha * K_{angle} * \left(T_{C,i} - T_A\right) \tag{6}$$

Whereas, the higher heat radiation from conductor is dependent on wind speed above 0.5 m/sec [7], i.e.

$$Q_{FC\_high} = \left[0.0119\left(\frac{D_i * \sigma * V_W}{\varepsilon}\right)^{0.6} * \alpha * K_{angle}\left(T_{C,i} - T_A\right)\right] \tag{7}$$

The higher the difference between conductor and ambient temperatures, the higher it can be loaded.

$$Q_{radiation} = 0.0178 * D_i * \epsilon_i * \left[\left(\frac{T_{C,i} + 273}{100}\right)^4 - \left(\frac{T_A + 273}{100}\right)^4\right] \tag{8}$$

Overall, the dynamic ampacity of overhead conductor as shown in Equation (9) is increased when difference between heat absorption and heat radiation is reduced with decreased conductor resistance and vice versa.

$$I_i = \sqrt{\frac{\left(Q_{convection} + Q_{radiation}\right) - Q_{solar}}{R_{T_C}}} \tag{9}$$

Both span and line length have direct relation with changes in line temperature ($T_f - T_i$), its tension (H_f -H_i), cross-sectional area (A) and the thermal expansion coefficient (σ) as represented in Equations (10) and (11). Considering that, the line sag varies proportionally with line length as represented in

Equation (12).

$$L_2 = L_1 * [1 + \sigma * (T_f - T_i)] * (1 + \frac{H_f - H_i}{E * A}) \quad (10)$$

$$new_{sag} = (new_{length})^2 * \frac{per\ unit\ weight}{8 * (new_{tension})} \quad (11)$$

$$\Delta sag = \sqrt{\frac{3 * Span\ length * (Change\ in\ condcutor\ length - Span\ length)}{8}} \quad (12)$$

Equation (13) represents the TCSC response in controlling the active power through the voltage phasor by compensating reactance of the line connected between both wind farms. Moreover, the difference between static and dynamic ampacities as given in Equation (14) results in additional power flow (P_Δ) that otherwise is limited by adopting the worst weather based line ampacity. The network was modelled in power factory with wind generators connected at both end so the overhead line as shown in Fig.1.

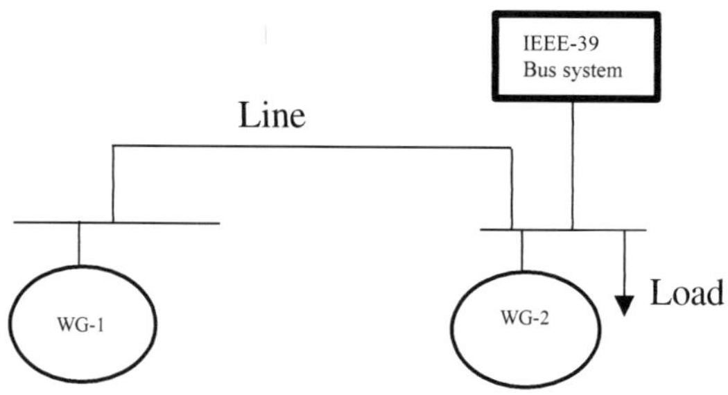

Fig. 1. Single line diagram of the modeled network

Moreover, as represented in Equation (15), the active power from wind generator 1 in Fig.1 is wind dependent so is the line ampacity hence resulting in additional power to be reliably dispatched through the overhead line. Similarly, the active power from wind generator 2 in Equation (16) is transferred through the same overhead line that based on line sag as shown in Equation (11) does not result in line elongation and helps system operators load the line safely.

$$P_{im} = \frac{U_i U_m \sin(\phi_{im})}{j(X_{TCSC} + X_{line})} \quad (13)$$

$$P_\Delta = ((I_{DLR} - I_{SLR}) * V_{L-L}) \cos(\phi) \quad (14)$$

$$\begin{cases} u_i < u < u_r \rightarrow P_{w,1} = \frac{P_{w,r,1}(u - u_i)}{(u_r - u_i)} \\ P_{w1} = 1.5 * (V_{ds,1} I_{ds,1} + V_{qs,1} I_{qs,1}) \end{cases} \quad (15)$$

$$\begin{cases} u_i < u < u_r \rightarrow P_{w,2} = \frac{P_{w,r,2}(u - u_i)}{(u_r - u_i)} \\ P_{w2} = 1.5 * (V_{ds,2} I_{ds,2} + V_{qs,2} I_{qs,2}) \end{cases} \quad (16)$$

The active power produced from wind generator 1 results in 'I_1' and is directed to pass through the overhead line towards the load center. Similarly, the active power from wind generator 2 resulting in I_2 passes through the same overhead line resulting in the added power flow.

Now,

$$I_1 + I_2 = \frac{(P_{w1} + P_{w2})}{V_{L-L}} \quad (17)$$

$$P_{added} = \sqrt{3} * [(I_{DLR}) - (I_1 + I_2)] * V_{L-L} \quad (18)$$

Similarly, P_{added} in Equation (18) determines the added power flow through conductor as shown in Fig.2.

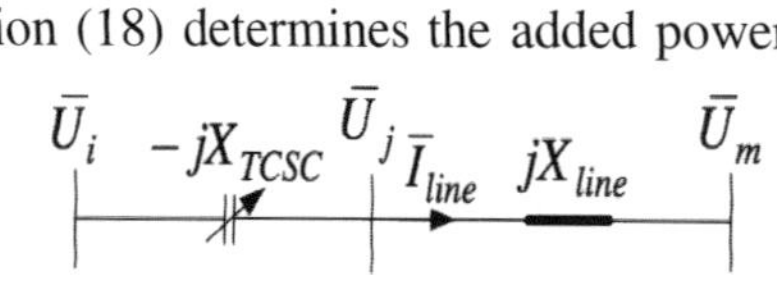

Fig. 2. TCSC steady-state circuit representation

As shown in Fig.2, TCSC controls the power flow through X_{TCSC} that provides controlled I_{line}. Both wind generators in Fig.1 are connected at both terminals of the line, where the line is operated under DLR technique. The modelled grid is considered weak where inconsistent wind speed causes voltage fluctuations at both connected wind generators. TCSC therefore helps in improving the grid stability and avoiding the voltage fluctuations.

3. Simulation and Results

In this project, the operational analysis of conductor ampacity is investigated under sag and voltage limits, where, its voltage limit is set by TCSC as shown in Fig. 1. The modeled network is equipped with two wind power plants and one FACTS device 'TCSC.' The projected overhead conductor moreover connects both wind generators to the load-side and is loaded until it reaches its maximum capacity to find the actual line potential in transferring the maximum generated electricity to the loads as shown in Fig.3.

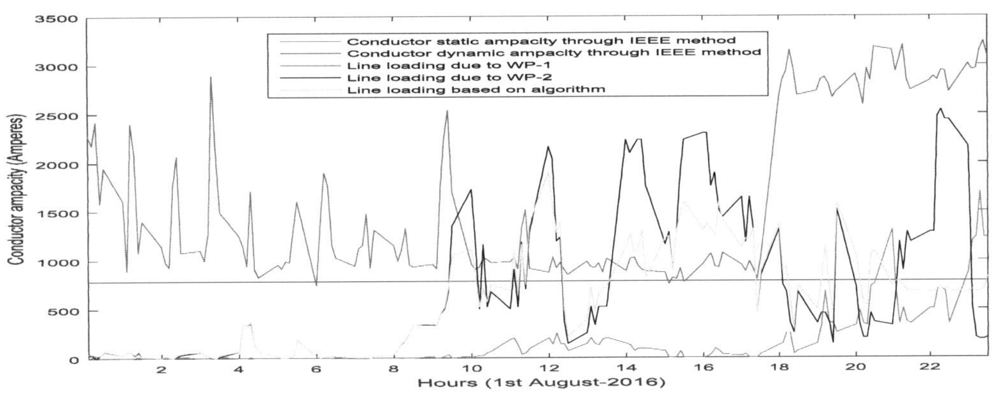

Fig. 3. Conductor ampacity versus line loading

Table I shows line loading and its dynamic ampacity versus wind power generation. Based on the line loading, conductor temperature and line sag are monitored as shown in Table I. The weather parameters were collected for one particular hour in 1st August-2016 at the overhead line location near the area of Nelson in New Zealand. Moreover, based on the sag limit, the overhead conductor is allowed to pass the adjusted power flow from both wind generators simultaneously.

TABLE I. 1-Hour Line data (at every 10-minute interval) based on Conductor Loading

Power from WG-1 (MW)	Power from WG-2 (MW)	Loading by algorithm (due to power from both wind generators) (A)	DLR (A)	Conductor temperature (^{0}C)	Conductor sag (m)
15.05	10	169.5	1111	31.4	0.941
1	5	40.6	962	23.3	0.694
1.5	4.5	40.6	821	23.7	0.7
4	0	27.1	800	22.5	0.675
0	4.5	30.45	583	23.9	0.72

The proposed algorithm selects dynamic ampacity if found lower than actual line loading to allow maximum power from wind generators through the line.

4. Conclusion

In this project, the overhead conductor is considered to transfer active power from two wind generators in the presence of FCATS device. The resulting current based on the line power flow is compared with line's static and dynamic ampacities under conductor sag as a constraint. The resulting sag moreover suggests the maximum line loading and the maximum power, the overhead line can transfer in the designed electricity network. In order to achieve the maximum controlled power flow through the overhead conductor, a FACTS device is then connected in series with the line that moreover is observed in reducing the power oscillations when overhead line experienced different power flows due to dynamic ampacity.

REFERENCES

[1] Wallnerström, C. J., Huang, Y., & Söder, L. (2015). Impact from dynamic line rating on wind power integration. IEEE Transactions on Smart Grid, 6(1), 343-350.

[2] Shu, Z., & Jirutitijaroen, P. (2011). Latin hypercube sampling techniques for power systems reliability analysis with renewable energy sources. *IEEE Transactions on Power Systems*, *26*(4), 2066-2073.

[3] Talpur, S., Wallnerström, C. J., Flood, C., & Hilber, P. (2016). Dynamic line rating for wind power. *Journal of Renewable and Sustainable Energy*, *8*(1), 013114.

[4] Talpur, S., Wallnerström, C. J., Hilber, P., & Saqib, S. N. (2014, December). Implementation of dynamic line rating technique in a 130 kv regional network. In *Multi-Topic Conference (INMIC), 2014 IEEE 17th International* (pp. 477-482). IEEE.

[5] Zhang, X. P., Yao, L., Chong, B., Sasse, C., & Godfrey, K. R. (2005, November). FACTS and HVDC technologies for the development of future power systems. In *2005 International Conference on Future Power Systems*(pp. 6-pp), IEEE.

[6] Raniga, J. K., & Rayudu, R. K. (2000). Dynamic rating of transmission lines-a New Zealand experience. In *Power Engineering Society Winter Meeting, 2000. IEEE* (Vol. 4, pp. 2403-2409). IEEE.

[7] Raniga, J. K., & Rayudu, R. K. (2000). Dynamic rating of transmission lines-a New Zealand experience. In *Power Engineering Society Winter Meeting, 2000. IEEE* (Vol. 4, pp. 2403-2409), IEEE.

[8] Stephen, R, "Real Time Monitoring - revised document for final comment 18 December 1998", Convenor Cigre Panel SC22, WG12. December 1998

[9] Engelhardt J S and Basu S P, "Design Installation, and Field Experience with an Overhead Transmission Dynamic Line Rating System, IEEE paper 0-7803-3522-8/96, 1996

[10] Roberts, D., Taylor, P., & Michiorri, A. (2008). Dynamic thermal rating for increasing network capacity and delaying network reinforcements.

[11] Roberts, D., Taylor, P., & Michiorri, A. (2008). Dynamic thermal rating for increasing network capacity and delaying network reinforcements.

[12] Douglass, D. A., Lawry, D. C., Edris, A. A., & Bascom, E. C. (2000). Dynamic thermal ratings realize circuit load limits. *IEEE Computer Applications in power*, *13*(1), 38-44.

[13] Standard for Calculating the Current-Temperature Relationship of Bare Overhead Conductors, IEEE 738-2012, Jun. 2012, Power Energy Soc.

Numerical Simulation Studies on the Influence of Metallurgical Contamination on the Electric Field of RTV Coating

SHEN Yang[1], MU Lin[2a], YUAN Xiaoqing[2], WANG Xiao[1], ZHOU Kai[1], YAN Jing[2], HAO Lu[2] and Li Ji[1]

[1]State Grid Anhui Electric Power Maintenance Branch, Hefei, 230061, China
[2]Wuhan Fenjin Electric Power Technology co., LTD, Wuhan, 430205, China

[a]mu_lin@whu.edu.cn

Abstract. The metallurgical industry in China has developed rapidly, which has caused an increasing amount of exhausted gas emission. After a long period of operation, a layer of metallurgical contamination is deposited on the surface of insulators near the metal work, which may have an effect on the distribution of electric field on the surface of insulator. The chemical test indicates that the metallurgical contamination layer contains particles of Fe_3O_4 with strong conductivity, which may become suspended potential conductor. In this paper, the influence of thickness of contamination layer and diameter, arrangement of conductive particles on the electric field are discussed by the simulation model.

1. Introduction

The contamination deposition on the insulator surface is more serious. The contamination on the insulator surface can reduce the insulation performance and increase the risk of contamination flashover with high conductivity and high dielectric constant. Currently, contamination flashover has become the most dangerous fault in the power system. The study on the distribution of electric field under the influence of contamination is essential for the research of contamination flashover [1-3]. The calculation of electric field distribution on composite insulators is helpful to the study of partial discharge occurs on the surface of composite insulators and relating surface aging problems and is also helpful for the insulation design, operation and maintenance of composite insulators [4]. The distribution of electric field of the insulator has become one of the research focuses for researchers around the world. Kontargyri V calculated the distribution of electric field of glass insulator string using finite element method and verified the result with the experiment [5]. In literature [6], the electric field and potential distribution under the effect of partial contamination and dry band were studied. However, current research mainly focuses on the influence of insulator structure and pollution distribution on the electric field distribution. The electric field distribution of insulator with water drop and dry band were simulated and analysed in literature [7]. The study of numerical and experimental results regarding the effects of pollution on the flashover performance of glass suspension-type insulators was presented in literature [8]. However, the study on the effect of metallurgical pollution layer and the suspended potential conductor on the distribution of electric field are relatively rare.

In this paper, the model of small ball array in the contamination layer is proposed to simplify the metallurgical contamination simulation. The geometric model is built on a finite element analysis

IOP Conf. Series: Materials Science and Engineering **366** (2018) 012071 doi:10.1088/1757-899X/366/1/012071

software. The distortion of electric field under the influence of metallurgical contamination is calculated and analysed. Electric field and potential distribution of dry contamination layer are compared with that of wet contamination layer. The influence of thickness of contamination layer, diameter and arrangement of conductive particles on the electric field are discussed.

2. Modelling procedure

2.1. Geometric model

The geometric model was built utilizing finite element method. The simulation model was an RTV plate attached a layer of metallurgical contamination. The conductive particles were inlaid in the contamination layer in a uniform arrangement with 0.003m longitudinal gap and 0.0013m lateral gap. The outside of the contaminated RTV plate was air layer. The integral geometric model is shown as Figure 1. The geometric parameters of the model are shown in Table 1.

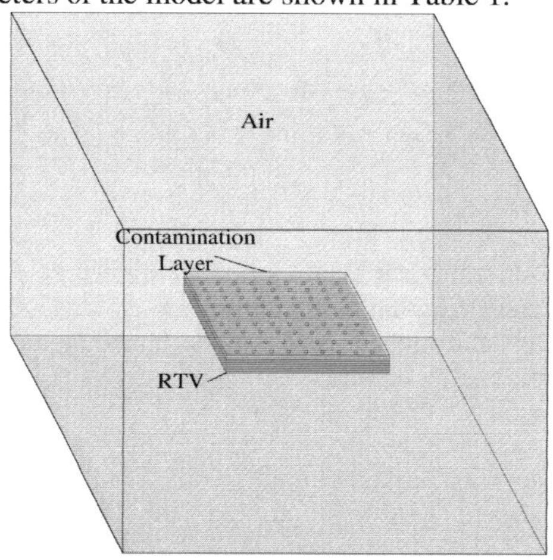

Figure 1 The integral geometric model

Table 1 The geometric parameters of the model

Geometric entity	Length (m)	Width (m)	Height (m)	Radius (m)
RTV	0.02	0.01	0.002	\
Contamination	0.02	0.01	0.0005	\
Air	0.2	0.2	0.2	\
Conductive Particles	\	\	\	0.0002

2.2. Calculation setting

The surface of longitudinal end of RTV plate was applied with 1kV AC voltage. The other end was grounded.

The wavelength of 50Hz AC voltage is far greater than the length of RTV plate, the electric field can be regarded as stable at any one moment. Therefore, only steady state need to be considered. Frequency domain was used to calculate. The boundary conditions of the model are shown in Table 2.

Table 2 The boundary conditions of the model

boundary conditions	Surface of RTV longitudinal end 1	Surface of RTV longitudinal end 2	Surface of air layer	Surface of conductive particles
setting	1kV (50Hz)	grounded	grounded	Suspended-potential

The electric field distribution of RTV specimen with dry and wet contaminated layer were comparatively studied. Supposing that the contamination layer was in ideal dry state, the conductivity of the contamination layer was low. The conduction current was so small that the influence on the electric field can be neglected. The electrostatic field was used to calculate the electric field of dry contamination layer. The material parameters of dry contamination layer model were set according to relative literatures as shown in Table 3.

Table 3 The material parameters of dry contamination layer model

Geometric entity	Dielectric constant
RTV plate	3
Contamination layer	7
Air layer	1
Conductive particles	1e8

In actual circumstances, the air around the insulator has a certain humidity. The contamination layer can't be in absolute dry state absorbing moisture from the air. In order to make the difference more obvious and more convenient for comparative analysis, the wet contamination layer was defined to have fully absorbed water and in the saturated humid state. The wet contamination layer was highly conductive and had characteristics of resistance and capacitance. The current field was used to calculate the electric field of wet contamination layer considering the influence of conduction current and dielectric properties in contamination layer on electric field distribution. The material parameters of wet contamination layer model was set as Table 4.

Table 4 The material parameters of wet contamination layer model

Geometric entity	Conductivity (S/m)	Dielectric constant
RTV plate	6e-15	7
Contamination layer	0.1	20
Air layer	1e-14	1
Conductive particles	1e7	1e8

Finally, the number of longitudinal particles, the number of lateral particles, the thickness of contamination layer and the diameter of conductive particles were modified respectively on the basis of the wet layer model. The influence of the model parameters on the electric field distribution was studied.

3. Results and discussion

3.1. Difference between dry contamination layer and wet contamination layer

The influence of the conductivity of contamination layer on the electric field distribution was neglected in the model of dry contamination layer. The electric field distribution on the surface of contamination layer was determined by the capacitance distribution. The influence of dielectric characteristics and resistance characteristics of contamination layer on electric field distribution were both considered in the wet contamination model. The horizontal section of the contamination layer was taken as the subject of research. Cloud picture of electric potential distribution is shown as Figure 2. Cloud picture of electric field distribution is shown as Figure 3.

IOP Conf. Series: Materials Science and Engineering **366** (2018) 012071 doi:10.1088/1757-899X/366/1/012071

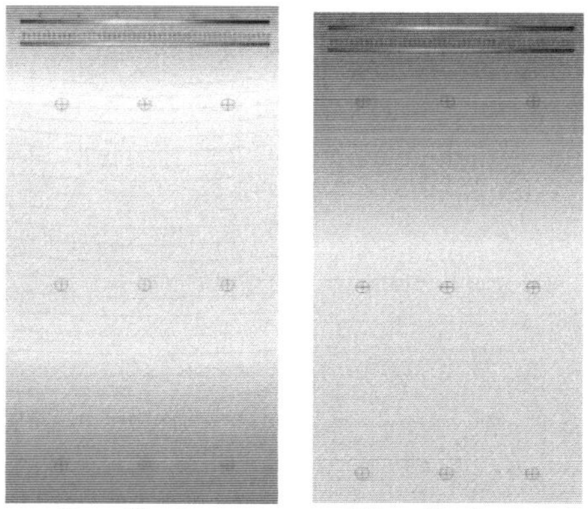

Dry dirt layer Wet dirt layer

Figure 2 Cloud picture of electric potential distribution

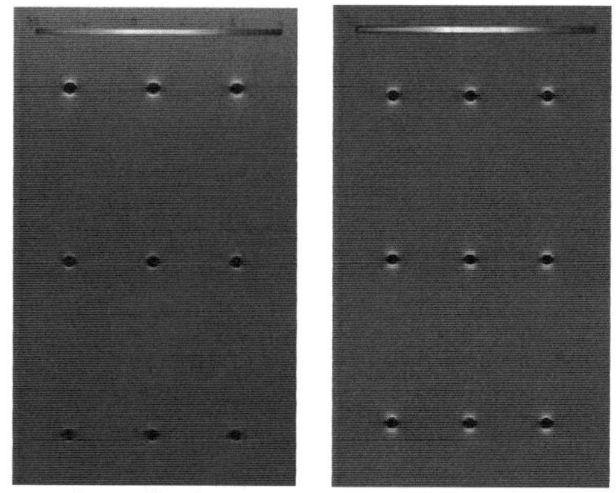

Dry dirt layer Wet dirt layer

Figure 3 Cloud picture of electric field distribution

When the contamination layer is dry, the potential and electric field distribution is affected by the capacitance distribution of contamination layer. The equipotential line on the whole specimen is bent to the side of the grounding electrode. The electric field near the high voltage electrode side of the specimen is larger than that of the ground electrode. The electric field on the surface of conductive particles is distorted in the direction of applied electric field.

When the contamination layer is wet, the potential and electric field distribution is affected mainly by the conductivity distribution of contamination layer. Leakage current is uniformly distributed in contamination layer. Therefore, the equipotential lines are uniformly distributed and parallel to the electrode. The electric field on the surface of conductive particles is distorted in the direction of applied electric field.

The 3D line passed through the center of contamination layer along the direction of electric field The electric field distribution on the path of dry and wet contamination layers with and without suspended potential particle were studied respectively. Variation of electric field along the longitudinal study path is shown in Figure 4. Change of electric field along the horizontal study path is shown in Figure 5.

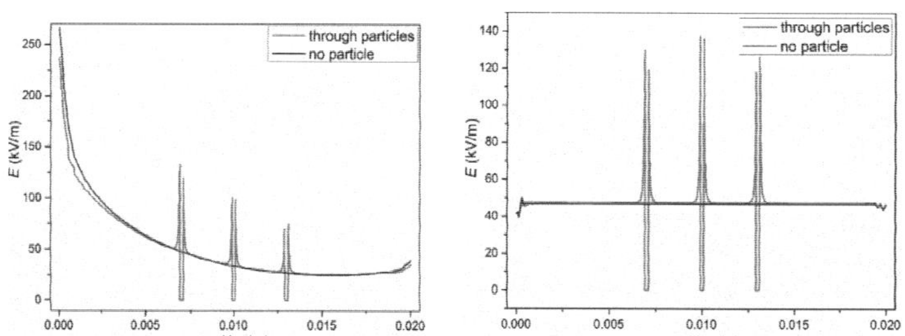

(a) Electric field distribution curve of dry layer (b) Electric field distribution curve of wet layer

Figure 4 Distribution of electric field along the longitudinal study path

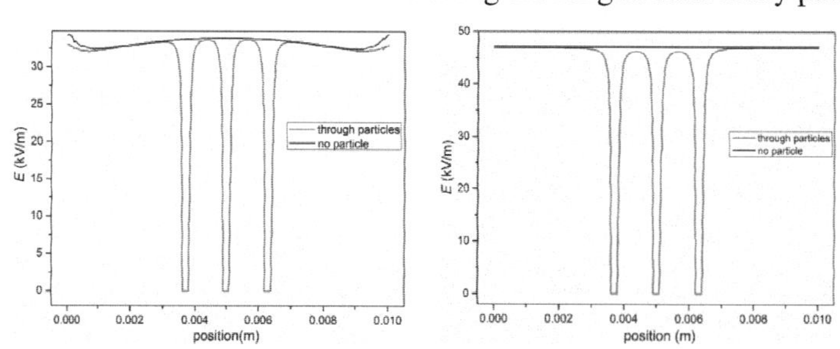

(a) Electric field distribution curve of dry layer (b) Electric field distribution curve of wet layer

Figure 5 Variation of electric field along the longitudinal study path

The blue line indicates the variation of electric field along the research path with the absence of conductive particles in contamination layer. The red line indicates the variation of electric field along the research path with the presence of conductive particles in contamination layer.

3.2. Influence of thickness of contamination layer

The distribution of electric field in contaminated layer of different thickness was calculated respectively. The influence of thickness of contamination layer on the distribution of electric field was studied. The geometric parameters and the arrangement of the particles are shown in Table 5. The variation curve of the electric field on the longitudinal path is shown in Figure 6.

Table 5 The geometric parameters and the arrangement of the particles

	Model a	Model b	Model c	Model d
Thickness of layer(mm)	0.4	0.5	0.6	0.7

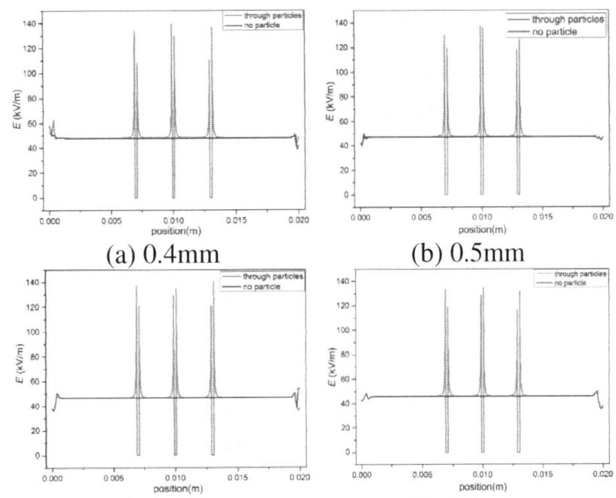

(a) 0.4mm (b) 0.5mm

(c) 0.6mm (d) 0.7mm

Figure 6 The distribution of electric field with different thickness of contamination layer

The larger the thickness of contamination layer, the more uniform the distribution of conduction current is, the further the distance from the center of the layer to the specimen is. The result shows that the thickness of contamination layer has little influence on the overall electric field distribution. The maximum electric field distortion decreases with the increase of the thickness of layer.

3.3. Influence of diameter of conductive particle

The distribution of electric field in contamination layer with particles of different diameter was calculated respectively. The influence of diameter of conductive particles on the distribution of electric field was studied. The geometric parameters and the arrangement of the particles are shown in Table 6. The variation curve of the electric field on the longitudinal path is shown in Figure 7. The variation curve of the maximum distorted electric field is shown in Figure 8.

Table 6 The geometric parameters and the arrangement of the particles

	Model a	Model b	Model c	Model d
Particle diameter (mm)	0.1	0.15	0.2	0.3

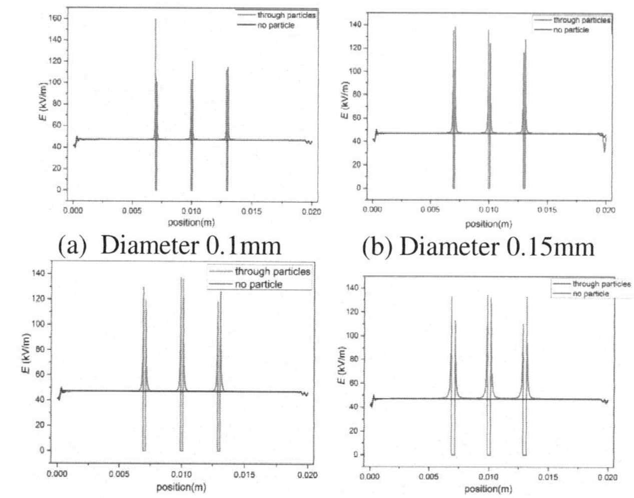

(a) Diameter 0.1mm (b) Diameter 0.15mm

(c)Diameter 0.2mm (d) Diameter 0.3mm

Figure 7 The distribution of electric field with different diameter of particles

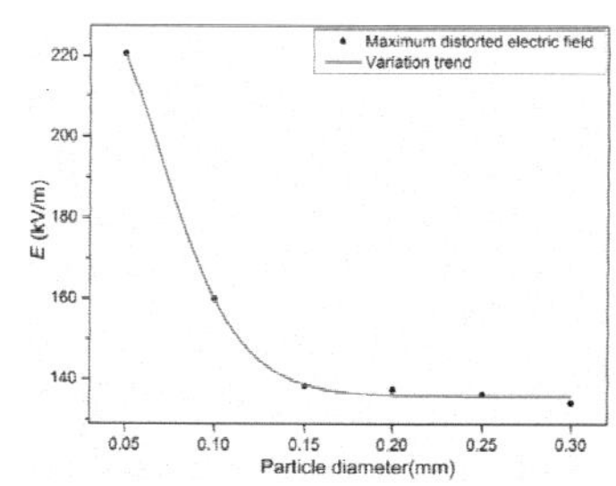

Figure 8 Variation of maximum distortion with diameter of particles

The result shows that when the diameter of the particle is 0.1mm, the electric field distortion of particle near the high voltage end is more serious, and the electric field distortion of other particles are relatively small. The maximum electric field distortion decreases with the increase of the diameter of conductive particle.

4. Conclusion

(1)The electric field distributes in exponential form along the longitudinal path in the dry contamination layer. The closer the particle is to the high voltage end, the larger the distorted electric field is. The electric field distortion ratio of different positions on longitudinal path is constant. The electric field distributes uniformly along the longitudinal path in the wet contamination layer. The electric field distortion ratio is 2.7 to 2.92. The electric field distortion ratio of the intermediate particle is larger. Distorted electric field and distortion ratio is larger than that of dry contamination layer.

(2)The larger the thickness of contamination layer, the more uniform the distribution of conduction current is. The thickness of contamination layer has little influence on the overall electric field distribution. The maximum electric field distortion decreases with the increase of the thickness of layer.

(3)The electric field distortion of particle near the high voltage end is more serious. The maximum electric field distortion decreases with the increase of the diameter of conductive particle.

References

[1] Benguesmia H, M'Ziou N, Boubakeur A. AC Flashover: An Analysis with Influence of the Pollution, Potential and Electric Field Distribution on High Voltage Insulator[M]// Multiphysics

Modelling and Simulation for Systems Design and Monitoring. Springer International Publishing, 2015.

[2] Kumar S M, Kalaivani L. Electric field distribution analysis of 110 kV composite insulator using Finite Element Modeling[C]// International Conference on Circuit, Power and Computing Technologies. IEEE, 2015:136-141.

[3] Liu P, Zhang Z, Cheng L, et al. Electric field distribution characteristics of composite insulator with internal defect[C]// Properties and Applications of Dielectric Materials. IEEE, 2015:1011-1014.

[4] Liang M, Wong K L. Study of electric field distribution on 22kV insulator under three phase energisation[C]// International Symposium on Electrical Insulating Materials. IEEE, 2014:140-143.

[5] Kontargyri V T, Plati L N, Gonos I F, et al. Measurement and simulation of the voltage distribution and the electric field on a glass insulator string[J]. Measurement, 2008, 41(5):471-480.

[6] Arshad, Nekahi A, Mcmeekin S G, et al. Numerical Computation of Electric Field and Potential Along Silicone Rubber Insulators Under Contaminated and Dry Band Conditions[J]. 3d Research, 2016, 7(3):25.

[7] Vosloo W L, Holtzhausen J P. The electric field of polluted insulators[C]// Africon Conference in Africa, 2002. IEEE Africon. IEEE, 2004:599-602 vol.2.

[8] Ilhan S, Ozdemir A, Jayaram SH, Cherney EA. Numerical and Experimental Investigation of the Effects of Pollution on Glass Suspension-type Insulators[J]. Ieee Transactions on Dielectrics and Electrical Insulation. 2015;22:2987-94.

Design and development of a portable partial discharge generator

Feng XUE[2], Hongyi HUANG[1], Yue HU[1*], Dongliang WEI[2] and Zhi WANG[2]

[1]State Energy Smart Grid R&D Centre (Shanghai), School of Electronic Information and Electrical Engineering, Shanghai Jiao Tong University

[2]Guangdong Power Grid Corporation Limited, Dongguan Power Supply Bureau, Dongguan, China

*yuehu@sjtu.edu.cn

Abstract. Intensive study of Partial Discharge (PD) phenomenon is of great importance for detecting and distinguishing different types of partial discharge. A portable device for partial discharge generation is designed and developed in this paper, which possesses advantage of miniaturization, dependability and flexibility. Such device is composed of power supply module, PD generating module and PD output interface. An implementing example of the partial discharge generator is given here in detail. Simulated high voltage signal is generated by a waveform generator and applied on physical PD generating model after being amplified by power amplifier and being boosted by a high voltage transformer with high transformation ratio. The generated PD pulse sequence are acquired and further analysed by MATLAB. Point to plane discharge is used as an example and verifies the performances of this device.

1. Introduction

Partial Discharge (PD) is a main cause of insulation fault in electrical equipment. Intensive study of PD phenomenon is of great significance for PD detection in order to ensure safety, reliability and stability of electrical equipment [1]. Due to the randomness and uncontrollability of PD in practical operation of electrical equipment [2], PD generator becomes a vital tool for PD research, especially for experimental study [3].

At present, the simulation of PD aiming at obtaining discharge pattern in lab mainly take advantage of high voltage experimental system combined with various real PD test model [4]. However, such method has many limitations: the hugeness results in immovability which restrict simulation in specific environment such as high-voltage testing hall, the long period for constructing experimental equipment and single form of experimental power source such as power frequency voltage or DC voltage.

A portable device for partial discharge generating is proposed to avoid these shortages mentioned above, which possesses advantages of miniaturization, dependability and flexibility. Miniaturization means that the experimental device is portable and movable, the test is not limited in specific environment any more. Dependability possesses great importance in experiments, proposed method can veritably simulate partial discharge. What's more, flexibility not only refers to various kinds of partial discharge, but also the diversity of power source and physical models.

Supported by NSFC (51777122) and Natural Science Foundation of Shanghai, China (17ZR1414400)

In this paper, the fundamental of this portal PD generator is described in detail in second part, constructing method and details are illustrated as well. Result obtained from generator, is also analyzed. Finally, a brief conclusion is drawn in the last part.

2. Design of a partial discharge generator

The device comprises of three parts: power supply module, PD generating module and PD output interface. The structure of it is shown in Fig.1.

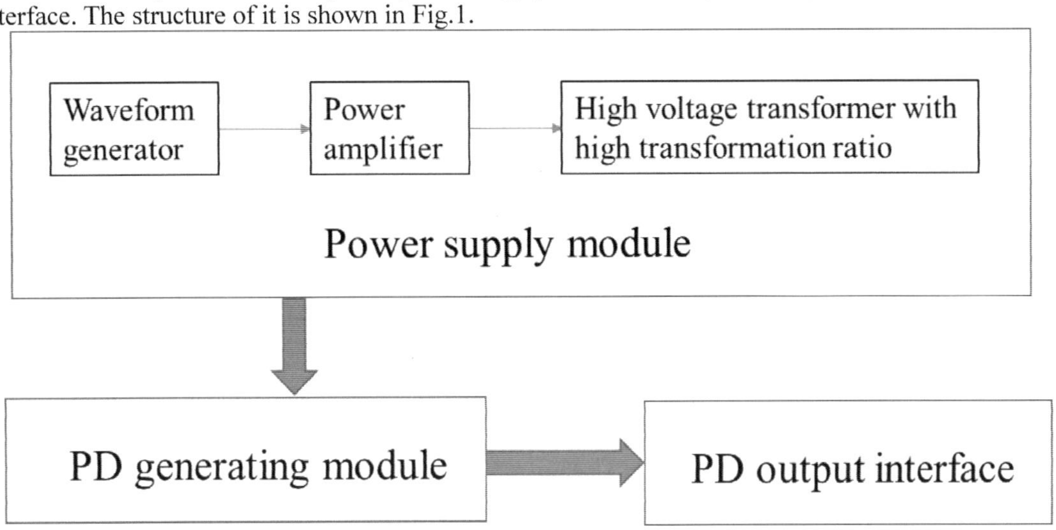

Figure 1. Functional structure of a portable partial discharge generator

2.1 Power supply module

Power supply module plays a role as generating various forms of high voltage signal on partial discharge physical models, which can be divided into three components.

The first component is waveform generator, which generates programmable and amplitude adjustable voltage waveform, such as power frequency signals, shockwave signal, and other programmable signals. The signal is then sent to power amplifier to ensure that the generated partial discharge current by PD generating as supposed. After being amplified, such signal is handled by the last component named high voltage transformer. The amplitude of voltage is raised to high enough to reach boundary of partial discharge generation.

2.2 PD generating module

PD generating module includes three major typical kinds of physical models for surface discharge, air-gap discharge and point to plane discharge, respectively. Specific types can be determined according to different discharge types in different electrical equipment. Physical model is a reduced version of the practical model to decrease required PD voltage on physical models, so that the experiments can be carried under relatively low voltage. This module includes protective resistance in series, which can limit maximum current generated during PD to protect power module and guarantee safety of users.

2.3 PD output interface

PD output interface including serial interface for common detection resistance, current detection port for high frequency pulse and detection for electromagnetic wave signal and ultrasonic signal, mainly deals with detection problem.

3. Implementation of the partial discharge generator

An example of the implementation of the partial discharge generator is given in this section.

IOP Conf. Series: Materials Science and Engineering **366** (2018) 012072 doi:10.1088/1757-899X/366/1/012072

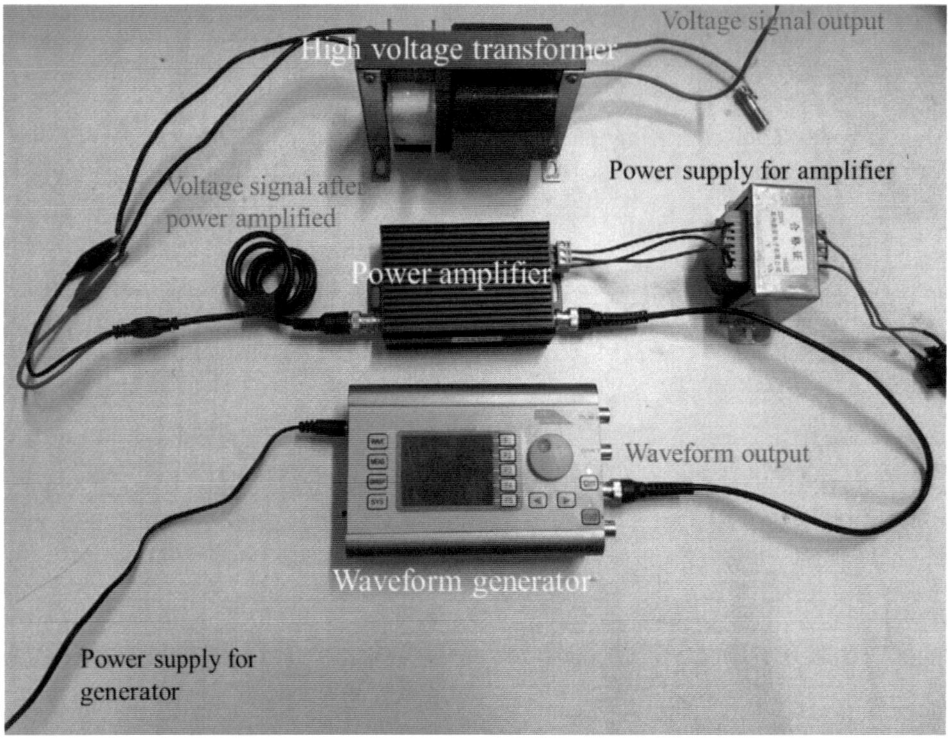

Figure 2. An example of power supply module

Fig.2 shows power supply module of the device, where FPA performs as power amplifier and self-defined 1:1000 ratio transformer works as high voltage transformer. Circuit in this example can generate square wave, triangle wave and sinusoidal wave with adjustable frequency. In this way, the frequency, amplitude and waveform can be decided according to requirement. FPA owns output overcurrent protection and internal temperature abnormal protection, which provides maximum 2A output current, 50V voltage and 30W power. High voltage transformer with high transformation ratio is used to boost voltage from 10V to 10kV to meet requirement of generating partial discharge on test model. The performance parameters of it are listed as follows. Transformation ratio is 1:1000 as described before, working frequency ranges in low band, which is 50Hz here. The size is 125mm*96mm*75mm, which take advantage of epoxy potting syrups technology to guarantee dampproof property.

After constructing reasonable power supply module, the generating type of partial discharge is closely related with spatial electric field distribution and electrode shape. Therefore, reasonable partial discharge generating model is of great importance for generating desired partial discharge phenomenon.

In this example, we design physical PD generating model for point to plane discharge, surface discharge and air-gap discharge, respectively. The structure of point to plane discharge is shown in Fig.3 filled with air, where left side is iron probe with 7kV voltage of 50Hz, right side is copper rivets with ground connection, insulation shell is made of acrylic.

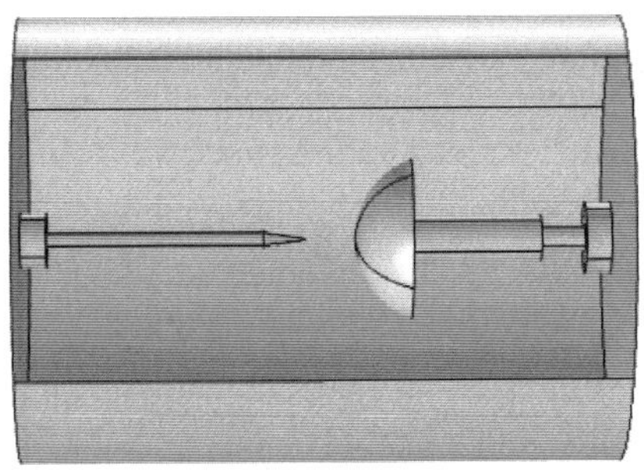

Figure 3. Model for point to plane discharge

Fig.4 shows the model for surface discharge, left side of the geometric model is an iron screw model with a diameter of 3mm, with an acrylic plate with a thickness of 5mm in the middle. Right side is an aluminium plate electrode and the insulating shell around is made of acrylic as well. The AC potential on left side is 7.5kV with frequency of 50Hz, right side is grounded.

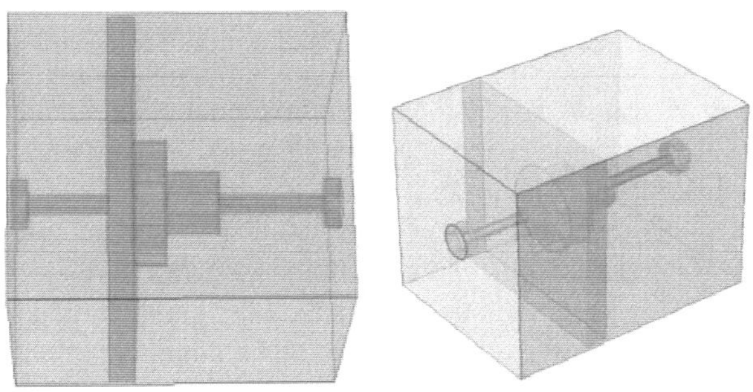

Figure 4. Model for surface discharge

The air-gap model is shown as Fig.5 that left side is a smaller iron plate electrode while right side is a larger plate electrode, middle of them is an organic glass plate with the exist of an air-filled circular hole with a diameter of 2mm. The AC potential on left side is 5kV with frequency of 50Hz, right side is grounded.

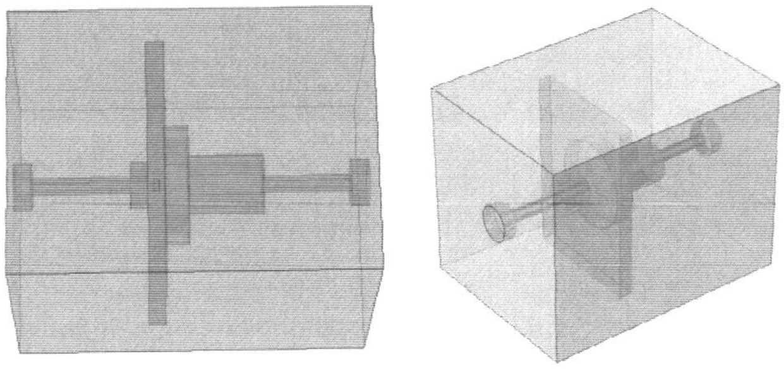

Figure 5. Model for air-gap discharge

All parts described above are integrated into Pelican1500 safety box, whose outer size is 47cm*35.7cm*17.6cm and inner size is 42.5cm*28.4cm*15.5cm. The width of box cover is 4.6cm and depth of box is 10.9cm. There are three bending strips with width of 2cm and thickness of 3mm aluminium that build a frame shaped working as bearing and fixed fulcrum of PD generator equipment platform. At the bottom of aluminium frame, epoxy resin plate, which works as insulating substrate and supporting system for components, is fixed on the punch with aluminium screws.

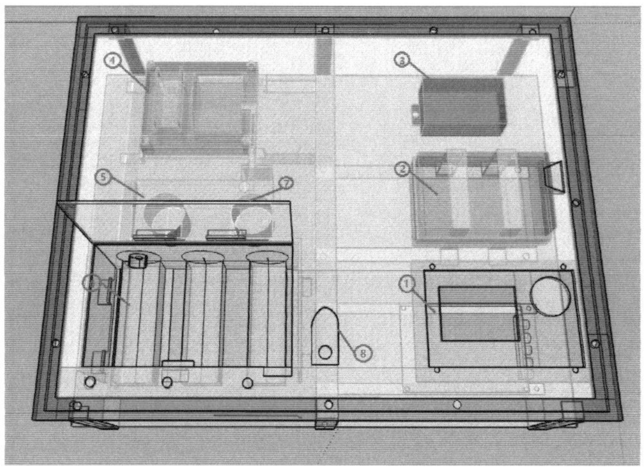

Figure 6. An example of PD generator platform

The integrated platform is shown in Fig.6, where the components are introduced as follow: (1) Waveform generator; (2) Rechargeable battery: power supply for waveform generator and power amplifier; (3) Power amplifier: input port receives waveform signal from waveform generator and output port is connected to high voltage transformer; (4) High voltage transformer: boost voltage signal from power amplifier; (5) High voltage bus: insulation supported by insulators; (6) Exchangeable PD model groove; (7) Glass retainer; (8) Slot for glass retainer

In this way, the PD simulation platform can integrate all required devices in a portable safety box.

4. Performance analysis of the partial discharge generator

Here oscilloscope is used to acquire output of PD simulation (Fig. 7), point to plane discharge is considered here as an example of assessing the performance of this device.

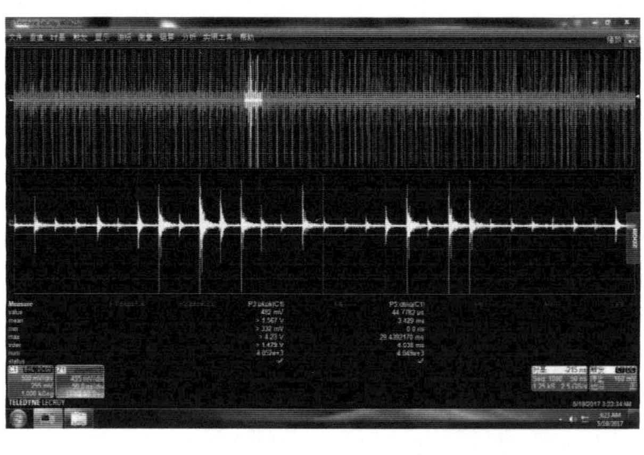

Figure 7. Waveform of point to plane discharge recorded by oscilloscope

IOP Conf. Series: Materials Science and Engineering **366** (2018) 012072 doi:10.1088/1757-899X/366/1/012072

The voltage on point to plane discharge model is adjusted to 7kV with frequency of 50Hz, simulation is repeated until obtaining current pulse data of 1000 times. The waveform acquired by MATLAB is shown as Fig.8, we can further get waveform after filtering and generate Phase Resolved Partial Discharge (PRPD) pattern, Phase Resolved Pulse Sequence (PRPS) pattern by using dlmread function of MATLAB.

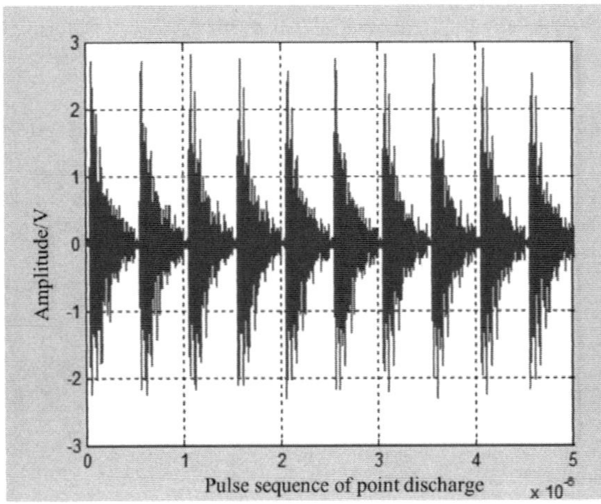

Figure 8. Waveform sequence of point to plane discharge

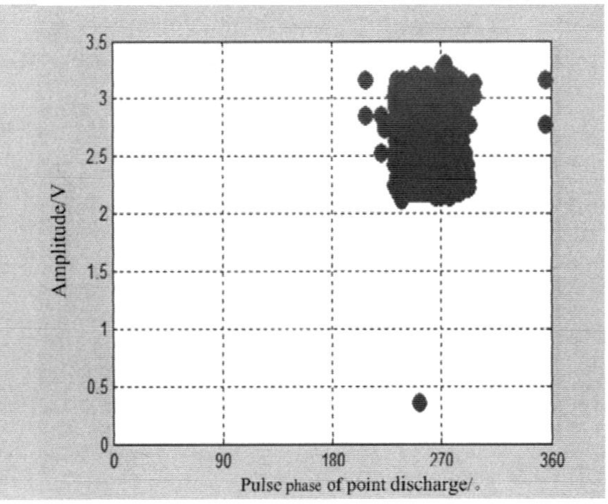

Figure 9. Flame grey pattern

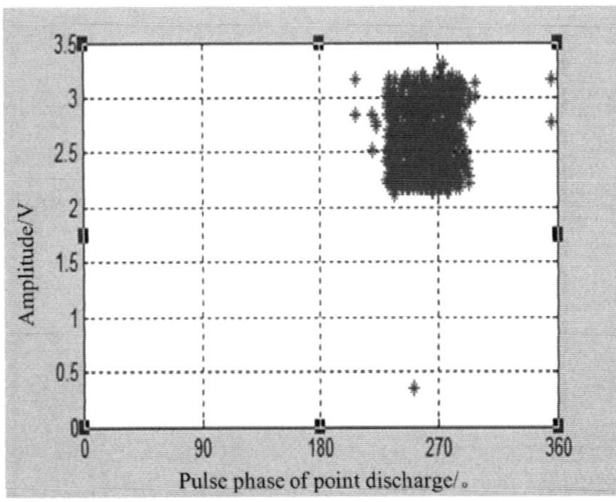

Figure 10. PRPD pattern

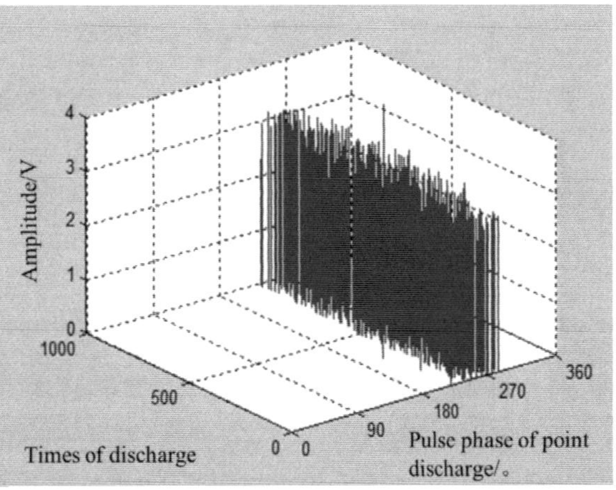

Figure 11. PRPS pattern

By referring to related papers about partial discharge [5][6], large amount of experimental data indicates that point to plane discharge first occurs only around negative half peak of test voltage, and the discharge time points symmetrically distribute on both side of peak point [7]. Only when test voltage is too high can second group discharge with similar distribution appear near the positive peak of test voltage waveform.

Compared with what's described above, it's obvious that the distribution characteristics obey characteristics of point to plane discharge well. In flame grey pattern (Fig.9), PRPD pattern (Fig.10) and PRPS pattern (Fig.11), most occurrences of discharge are distributed symmetrically around phase of 270 degree, and their amplitude is close.

It is reasonable to reach the conclusion that the generator platform of partial discharge can accurately simulate the occurrence of point to plane discharge with little interference and obvious discharge characteristics. These characteristics obtained from generator mean a lot for detecting and distinguishing different types of partial discharge.

5. Conclusion

In this paper, a method to completely integrate modules into a portable partial discharge generator is proposed, which can generate multiple types of partial discharge under different types of power sources.

It's noteworthy that the greatest advantage is the simplification, mobility and reliability of such platform. Its environmental dependence to external equipment is low while the operability is quite strong, which means that the test can be carried out anytime and anywhere by any method. The reliability is also verified through comparing simulation results and known partial discharge characteristics, the point to plane discharge is taken as an example here.

References

[1] Li J, Si W, Yao X, et al. Measurement and simulation of partial discharge in oil impregnated pressboard with an electrical aging process[J]. Measurement Science & Technology, 2009, 20(10):105701.

[2] Noskov M D, Sack M, Malinovski A S, et al. Measurement and simulation of electrical tree growth and partial discharge activity in epoxy resin[J]. Journal of Physics D Applied Physics, 2001,34(9):1389-1398(10).

[3] Champion J V, Dodd S J. Simulation of partial discharges in conducting and non-conducting electrical tree structures[J]. Journal of Physics D Applied Physics, 2001, 34(8):1235-1242.

[4] Pu L, Ni H, Zhao X F, Ju Z L, Study on partial discharge characteristic in XLPE accessory based on pulse sequence analysis[J]. Insulation Materials, 2016,49(12) :45-51.

[5] Li H J, Si W R, Yao X, Guo Z F, Li Y M. Measurement and simulation of partial discharge in disc-void of oil-pressboard insulation with electrical degradation process[J]. Proceedings of the CSEE, 2009,29(31):128-134.

[6] Yu H, Research on GIS defects recognition based on partial discharge pattern[A]. CSEE. 2013 Proceedings of the annual meeting of CSEE[C]. 2013:8

[7] Zhou Y J, Zhang J X, Di S G. Study on Insulation Defects in GIS Partial Discharge[J]. Inner mongolia electric power, 2013,31(5):106-110.

Performance evaluation system for partial discharge detector based on Tesla coil structure

Yu LIU[1], Hongyi HUANG [1], Yue HU*[1], Tong LI [2] and Jiangtian CHEN [2]

[1]State Energy Smart Grid R&D Centre (Shanghai), School of Electronic Information and Electrical Engineering, Shanghai Jiao Tong University
[2]Guangdong Power Grid Corporation Limited, Dongguan Power Supply Bureau, Dongguan, China

*yuehu@sjtu.edu.cn

Abstract. Partial Discharge (PD) detection aiming at finding the insulation defects make great effort for ensuring the reliability of the electrical equipment. Thus the performances, such as sensitivity and classification and recognition of PD pattern of PD detector need verified. This paper offers an effective tool to evaluate these performances of PD detectors. Design of the system consists of three modules, including PD pattern simulation module, PD pulse sequence generating module, and high voltage simulation module. Example of realization is also given in detail, which takes advantage of Tesla coil structure to generate programmable pulse sequence under the control of embedded system according to PD pattern. After that, the performance analysis of this proposed evaluation system is discussed, the performance evaluation system can simulate PD pulse sequence with controllability of pulse position distribution and randomness of amplitude in a range referring to PRPD pattern from database.

1. Introduction

Partial discharge detection is a non-destructive method to effective detect insulation defects of high voltage electrical equipment [1], which take advantage of the characteristics of insulation defects that its exist enhances electric field around and further results in local insulation breakdown with quick charge transfer [2]. Partial discharge detection is based on the evaluation of electrical or non-electrical signals in the discharge process, which is essential part of insulation assessment for high voltage electrical equipment [3].

At present, PD detectors have been developed from analogy type that purely measures apparent charge to digital equipment that can realize partial discharge pattern recognition. The magnitude of apparent charge only indicates the existence and severity, while by analysing Phase Resolved Partial Discharge (PRPD) pattern can we obtain much more detailed information [4]. Therefore, detectors adopt Digital Signal Processing (DSP) method to implement statistical or time-frequency analysis to achieve classification and recognition of PD, also interference elimination and noise suppression [5].

However, most traditional partial discharge calibrator only output magnitude-aware charge pulse with fixed frequency to verify the linearity and influence of polarity, while programmable calibrator can carry out additional test, such as pulse response sequence and pulse resolution time, which mainly focus on calibration and evaluation of apparent charge [6]. This paper aims at providing a reasonable

Supported by NSFC (51777122) and Natural Science Foundation of Shanghai, China (17ZR1414400)

IOP Conf. Series: Materials Science and Engineering **366** (2018) 012073 doi:10.1088/1757-899X/366/1/012073

method to evaluate PD detector, which makes use of Tesla coil to generate PD pulse by referring to high voltage signal and PRPD pattern from simulation module

2. Design of the performance evaluation system

According to what's mentioned above, the performance evaluation, which is on the basis of programmable PD calibrator, proposed in this paper have several outstanding advantages. It can miniaturise experimental equipment, also generate various types of PD pulse sequence, including programmable calibration pulse sequence, which therefore can simulate PD by generating pulse sequence according to patterns of different insulation defects. The performance evaluation system is shown in Fig.1, which includes three modules, PD pattern simulation module, PD pulse sequence generating module and high voltage simulation module, as follow.

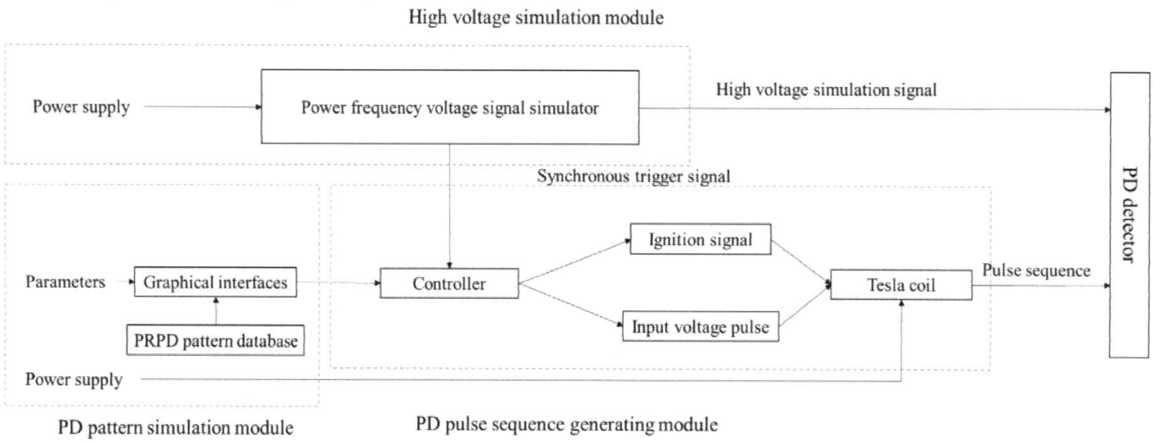

Figure 1. Structure of an evaluation system for PD detector

2.1 PD pattern simulation module

PD pattern simulation module, which is used to input power supply and characteristic information of each pulse of PD pulse sequence to be simulated, including amplitude, polarity, pulse phase angle, time-frequency parameters and so on. These information can be referred from PRPD pattern and receive parameters from graphical interfaces by users.

2.2 PD pulse sequence generating module

PD pulse sequence generating module, which is composed of control part, Tesla coil circuit and synchronous trigger output circuit. Controller plays a role as receiving information from PD pattern simulation module to provide Tesla coil input voltage and ignition signal to match desired pulse sequence [7]. Tesla coil is triggered by ignition signal and amplitude of pulse is controlled by input voltage. In this way, this module can generate and output pulse sequence to simulate PD signal.

2.3 High voltage simulation module

High voltage simulation module, which takes advantage of isolation transformer to couple power supply. The coupled signal generates synchronous sinusoidal signal and then becomes square wave signal through circuit, detailed process is illustrated in Section 3. The rising edge of square wave signal works as synchronous signal that triggers discharge of each cycle of the output.

PD detector serves as a terminal part that receives PD pulse sequence and synchronous trigger signal, after data acquisition, interference recognition, PD classification and recognition are carried out.

3. Realization of performance evaluation system

3.1 Simulation of PD pattern

Different types of defects possess different statistical characteristics of pulse time distribution and amplitude distribution, which is resulted from the mechanism of partial discharge. Such specificity can be used to classify and identify insulation, therefore, the gather of discharge pattern is of great importance, which can be achieved through following three methods: partial discharge on line detection, digital simulation based on mathematical model and experiments based on typical defects. The PD pattern database can be constructed through gathering information to restore waveform of partial discharge, information of PD test, and the characteristics of each pulse in PD pulse sequence, such as amplitude and phase.

When the system generates PRPD pattern, synchronous output is generated by trigger circuit, which is mainly used for simulating high voltage signal. It takes advantage of isolation transformer to couple power supply. The coupled signal generates synchronous sinusoidal signal after low-pass filter, which further generates square wave signal by comparing with referred DC voltage signal in comparator. The rising edge of square wave signal works as synchronous signal that triggers discharge of each cycle of the output. By changing referred DC source, the phase difference between trigger signal and zero point of rising edge ranges from 0 degree to 90 degree. And the high voltage simulating signal comes from amplify of synchronous signal.

Tesla coil circuit is controlled by embedded system, input trigger signal connects circuit and input controlling voltage to define amplitude of pulse. PRPD pattern pulse sequence is read from database by programming, and this sequence is divided into pulse fragments according to power frequency cycle, these fragments perform as reference source to control trigger circuit. Fig.2 shows an example of pulse sequence, each of pulse contains information of phase position and amplitude.

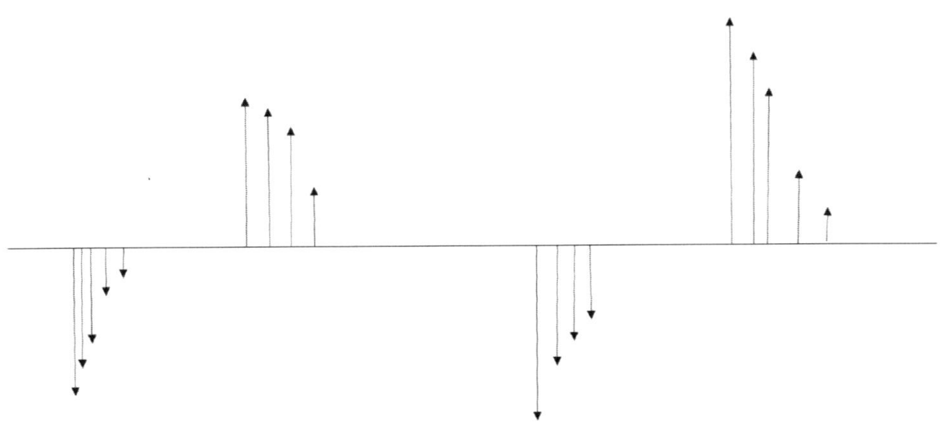

Figure 2. Pulse sequence example

3.2 Programmable pulse sequence

Partial discharge pulse sequence can be reproduced by PD pulse sequence generator, the Tesla coil circuit is shown in Fig.3.

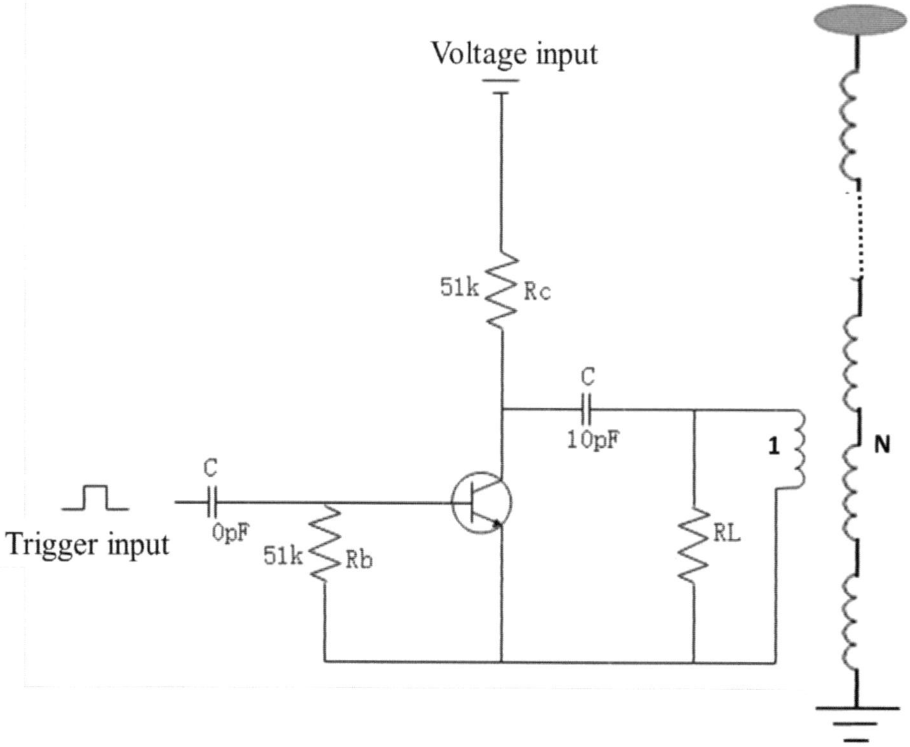

Figure 3. Tesla coil circuit

Inside drive circuit, the avalanche tube remains cut off condition until trigger pulse from control circuit arrives, the capacitor C is charged by input voltage V. When a trigger pulse that is large enough arrives, the operating point of the transistor is moved to an unstable avalanche negative resistance region. Then avalanche breakdown of transistor produces a rapidly increasing current, which further results in a rapid discharge of capacitor through transistor, then a narrow pulse occurs on load RL.

The narrow pulse possesses high amplitude because of the high avalanche current. And due to the result that limited charge is stored in capacitor, which is usually not larger than hundreds of picofarad, thus the pulse width is also limited. With the influence of circuit distribution parameters, the avalanche current gradually increases till a certain peak value, then the decrease of charge in capacitor leads to the decrease of discharge current. Former process generates the rising edge of pulse while latter process causes the falling edge. In this way, a sub nanosecond UWB narrow pulse with a very short rising time is obtained on the load. The trigger signal and voltage controlling signal to Tesla coil circuit are controlled by embedded controller, and the generated narrow pulse is output after being boosted through Tesla coil transformer.

4. Performance analysis of the device for partial discharge simulation

This performance evaluation system can simulate all kinds of PD pattern theoretically, point discharge is taken as an example here.

Point discharge first occurs only around negative half peak of high voltage signal, and the discharge time points symmetrically distribute on both side of peak point, Fig.4 is a typical example of point to plane discharge pulse sequence.

IOP Conf. Series: Materials Science and Engineering **366** (2018) 012073 doi:10.1088/1757-899X/366/1/012073

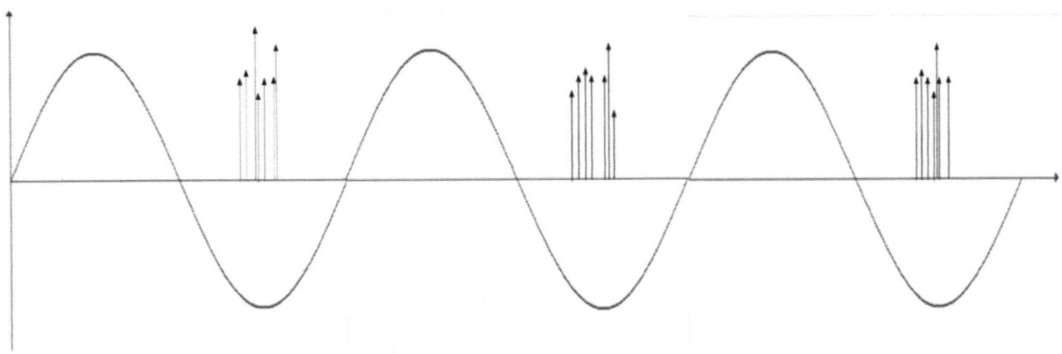

Figure 4. Pulse sequence of point discharge

The simulation of performance evaluation system is based on the PRPD pattern stored in database, where program read pulse sequence data including pulse amplitude (A_i) and phase angle (T_i). Then these sequences are divided into groups according to primitive periods, each pulse is triggered by trigger signal and the amplitude is controlled by input voltage on Tesla coil circuit. Fig.5 is the output PRPD pattern observed in oscilloscope, it can be clearly seen generated PD pulse sequence obeys the distribution characteristics of point discharge. The high comparability verifies the simulation function of evaluation system.

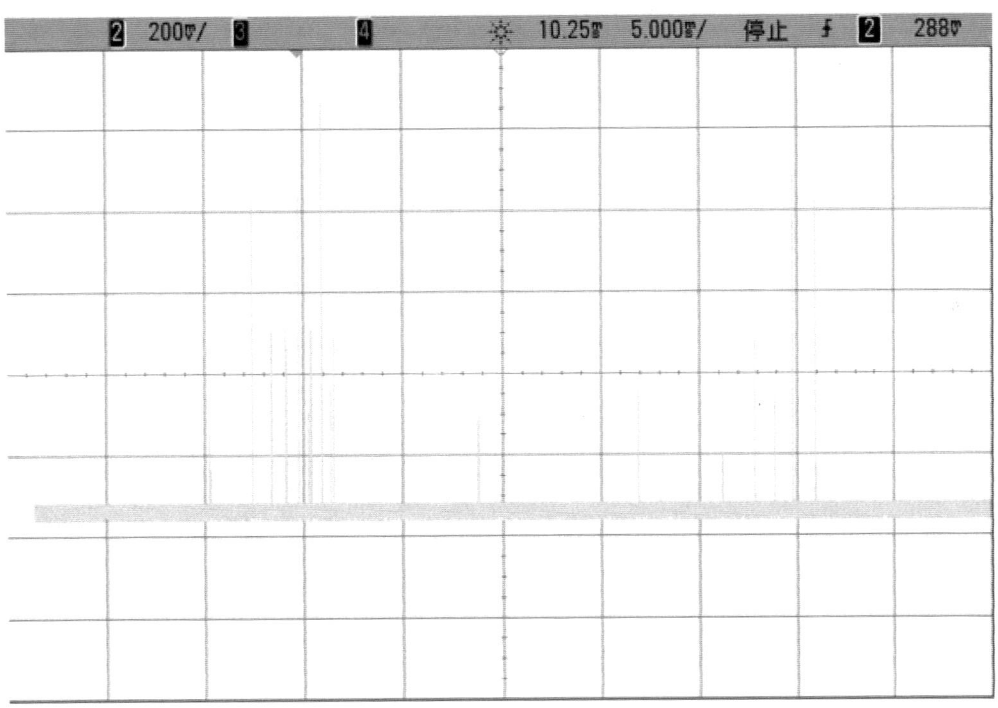

Figure 5. PRPD pattern observed in oscilloscope

5. Conclusion

In this paper, a method to construct performance evaluation system of partial discharge detector based on Tesla coil is proposed, which can reproduce pulse sequence of various types of partial discharge. The system is a miniaturization and integration of experimental equipment that can satisfy requirement

IOP Conf. Series: Materials Science and Engineering **366** (2018) 012073 doi:10.1088/1757-899X/366/1/012073

for verifying the feasibility and accuracy of PD detector. The design and realization details are given, performance analysis of this system is also discussed above.

The most outstanding advantage of this method is that simulation realized by this performance evaluation system possesses high reliability, which can guarantee not only the controllability of discharge process but also the randomness of reality simulation process.

References

[1] Hikita M, Okabe S, Murase H, et al. Cross-equipment evaluation of partial discharge measurement and diagnosis techniques in electric power apparatus for transmission and distribution[J]. IEEE Transactions on Dielectrics and Electrical Insulation, 2008, 15(2): 505-518. Nattrass D A. Partial discharge measurement and interpretation[J]. IEEE Electrical Insulation Magazine, 1988, 4(3): 10-23.

[2] Champion J V, Dodd S J. Simulation of partial discharges in conducting and non-conducting electrical tree structures[J]. Journal of Physics D Applied Physics, 2001, 34(8):1235-1242.

[3] Judd M D, Cleary G P, Bennoch C J J. Applying UHF partial discharge detection to power transformers[J]. IEEE Power Engineering Review, 2002, 22(8): 57-59.

[4] Lim Y S, Koo J Y. Comparative analysis of partial discharge patterns from different artificial defects by means of conventional phase-resolved partial discharge analysis and a novel chaotic analysis of partial discharge[J]. JOURNAL-KOREAN PHYSICAL SOCIETY, 2003, 42: 755-764.

[5] Hu Y, Chiampi M, Crotti G. Characterisation system for the evaluation of digital partial discharge measuring instruments[J]. IET Science, Measurement & Technology, 2015, 9(7): 817-825.

[6] Gunnarsson O, Bergman A, Rydler K E. A method for calibration of partial discharge calibrators[J]. IEEE Transactions on Instrumentation and Measurement, 1999, 48(2): 453-456.

[7] Azure L. Complex frequency pulsed electromagnetic generator and method of use: U.S. Patent 5,908,444[P]. 1999-6-1.

IOP Conf. Series: Materials Science and Engineering **366** (2018) 012074 doi:10.1088/1757-899X/366/1/012074

Design of Driving and protection circuit for Submodule based on cascaded H-bridge STATCOM

Xuan Wang[1, 2], Qingguang Li[1, 2], Shuai Wang[3, a)], Xiaoxing Wang[3] and Yongsheng Fu[1, 2]

[1]NARI Group Corporation/State Grid Electric Power Research Institute, Nanjing, Jiangsu 211106, China.
[2]China EPRI Science & Technology Co., Ltd., Changping District, Beijng 102200, China.
[3]New Energy Research Institute, School of Electrical Engineering, Beijing Jiaotong University, Haidian, Beijing 100044, China.

[a]Corresponding author: 15126037@bjtu.edu.cn

Abstract. Driving and protection circuit is an important part of power electronics, which is related to safe and stable operation issues in the power electronics. The driving and protection circuit is designed for the cascaded H-bridge STATCOM in this paper. The circuit can realize flexible dead-time setting, operation status self-detection, fault priority protection and detailed fault status uploading. It can help to improve the reliability of STATCOM's operation. Finally, the proposed circuit is tested and analysed by power electronic simulation software PSPICE (Simulation Program with IC Emphasis) and a series of experiments. Further studies showed that the proposed circuit can realize drive and control the H-bridge circuit, meanwhile it also can realize fast processing faults and have advantage of high reliability.

1. Introduction

At present, Statics Synchronous Compensator (STATCOM) is an ideal reactive compensation device, meanwhile it is an important part of the high-voltage DC (HVDC) transmission systems. [1-3] It is a circuit composed of Insulated Gate Bipolar transistor (IGBT), MOSFET or other full-controllable power electronic devices, which can control the amplitude and phase of AC grid current. So, it has the ability to compensate reactive power to the grid.

For full-controllable power electronic devices, some driving and protection methods have been proposed and investigated in recent years. Many companies have designed a dedicated driving chip, but designers always focus only on their own devices, resulting in weakness general-ability. The driving and protection function of chip is not perfect due to many chips' driving power is too small to drive switching devices. Meanwhile, high-power drive protection circuit is too expensive. In addition, the driving and protection circuit should have ability of multiple signal isolation, dead-time setting freedom, external protection interface, over-voltage/over-current/over-load or other faults detection. [4-6]

This paper focuses on the above issues for driving and protection circuit of large-scale cascaded H-bridge STATCOM. The driving and protection circuit is used to 4 IGBTs of an H-bridge, which is easy to modularized design. In addition, the circuit is powered by two ways: source power supply and battery power supply. And it uses optical fibers as the input interfaces. In the output side of signal, the optocoupler and power isolation can realize twice driving signals isolation, which greatly improves the operation reliability of driving circuit. In the aspect of protection, it can locate, code and classify the faults quickly, which can achieve protection of switching devices effectively.

2. The Cascaded H-bridge STATCOM

The cascaded H-bridge converter is different from other chain converters, which consists of the same H-bridges in series, so it is easy to expand. The figure 1 shows the circuit configuration of cascaded H-bridge STATCOM. Three phase clusters are Y-connected, then connected to AC grid through inductances. STATCOM injects the reactive current opposite to the load current, in order to compensate the reactive power. Then there are only active power will be supplied by grid, the power transmission quality can be improved. Each phase cluster consists of N H-bridge Sub-Modules (SM), which composes cascaded multilevel converter. Each H-bridge SM consists of 4 IGBTs and a capacitor. By using the H-bridge construction for the SMs, the capacitor voltage (U_{sm}) can be inserted or bypassed, thus, the SM voltage can be zero or $\pm U_{sm}$, which is added or subtracted from the phase voltage. [7-9]

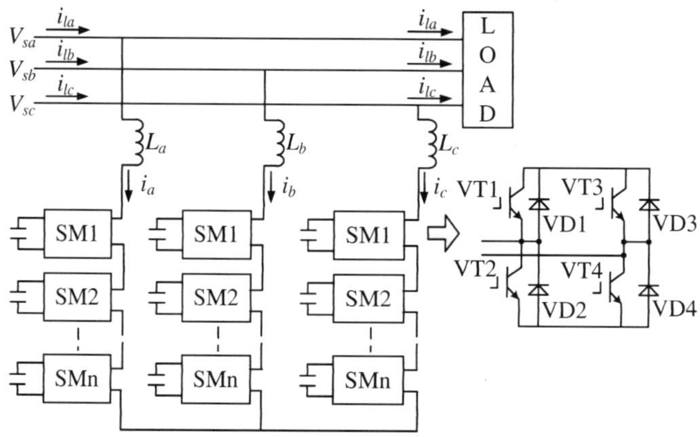

Figure 1. Scheme of the cascaded H-bridge STATCOM.

In order to keep cascaded H-bridge STATCOM stable operation, it is critical for designing the driving and protection circuit. In general, the drive and protection circuit should satisfy the followings:

- Fault detection. The drive and protection circuit need to locate, code and classify IGBT faults quickly, do different processing, in order to achieve the rapid protection of IGBT.
- Isolation. The driving and protection circuits are usually low voltage signal circuits, which consists of analogic devices and digital chips. Hoverer, the IGBTs operates on high-voltage bus, and IGBTs potential may be different, necessary electrical isolation must be provided, such as control signal isolation and drive power isolation. The control signal isolation need to ensure that isolation delay time as low as possible or even without delay time. The electrical isolation need to ensure that isolation voltage as high as possible in order to avoid leakage of insulation breakdown.
- Dead-time setting freedom. The parasitic capacitor on IGBTs will lead to the delay of IGBT-on or IGBT-off. Therefore, in order to keep stable operation, avoid the upper and lower IGBTs from conducting at the same time, the dead-time must be set.

3. Design of Driving and Protection Circuit

3.1. Basis circuit structure

The driving and protection circuit of H-bridge was shown in figure 2. The circuit is mainly composed of 6 parts, which are power supply circuit, optical fiber interface circuit, fault feedback circuit, pulse output and detection circuit, display circuit and CPLD control circuit. CPLD is the core of the circuit, mainly used to achieve the control and protection of the entire drive circuit.

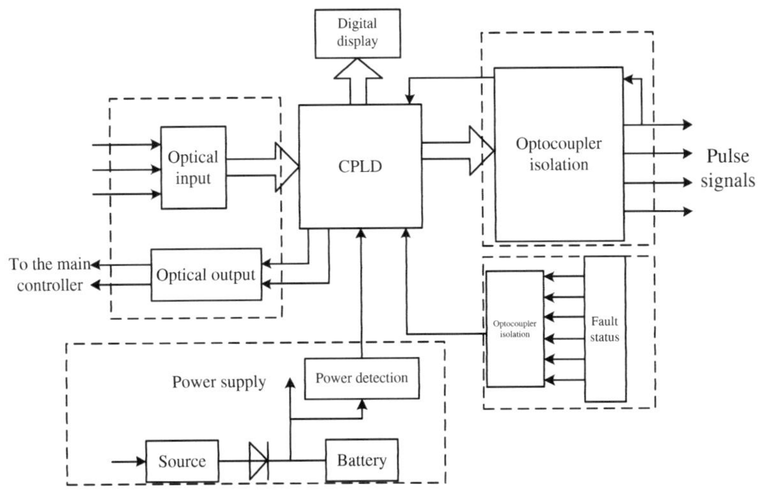

Figure 2. Structure of driving and protection circuit.

For the driving and protection of H-bridge, the reliability of power supply is very important. The circuit is designed by dual power supply (source power supply and battery power supply). In the case of normal power supply, the source supply power for the entire circuit. And when the battery is not full, the source power also charges the battery. When the source power fail, the battery will replace the source supply power to the driving circuit. Meanwhile the power detection circuits will output the fault signal to CPLD.

The driving signals are delivered to CPLD by the main controller. Considering the symmetry of the driving signals of H-bridge (the IGBT driving signals of the same bridge arm are complementary), the main controller will send three signals to CPLD, two of them are driving signals. CPLD divides two driving signals into four paths, and sets dead-time in the driving signals. Then the driving signals were send to IGBTs through the optocoupler isolation circuit. The other signal is a locked pulse signal which the main controller sends to the CPLD. The signal is low when STATCOM is in standby or fault condition, and pulse will be blocked. On the contrary, when the signal is high, the drive circuit will operate normally and STATCOM is running. The parasitic capacitor will be different because of different types of switching devices, resulting in different delay-time. Therefore, dead-time can be set according to the relevant parameters of switching devices. Thereby, these disadvantages can be avoided, which dead-time of the conventional driving circuit is not adjustable. In order to ensure the driving signals accuracy, the pulse signals will also be fed back to CPLD through optpcoupler, which is send to IGBT. Then the driving signals send to IGBT and the driving signals fed back will be compared by CPLD, in order to verify that signal send to IGBT is normal.

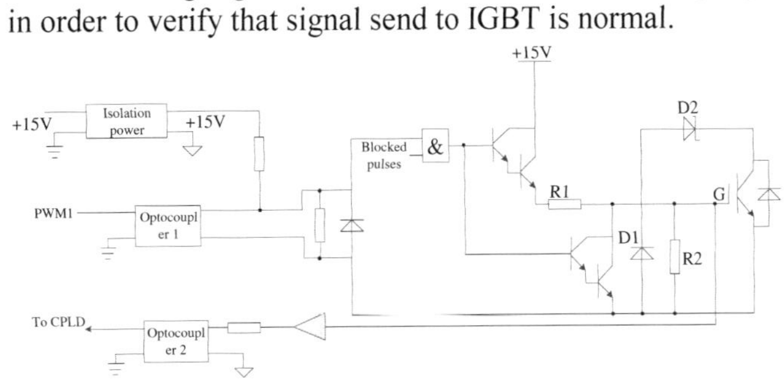

Figure 3. Structure of IGBT driving and pulse detection circuit.

IGBT driving and pulse detection circuit is shown in figure 3. The circuit achieve a complete isolation by using isolated power and aptocouplers. The output impedance is low because of the amplifying circuit using transistors composition, which can effectively reduce impact of load changing on the voltage gain. Meanwhile, it will have a larger current gain, which can basically meet IGBT

IOP Conf. Series: Materials Science and Engineering **366** (2018) 012074 doi:10.1088/1757-899X/366/1/012074

gates driving power requirements. The forward voltage of the drive circuit is determined by the gate voltage of IGBT, typically about 15V. Diode D1 is used to clamp the voltage between gate and emitter in order to ensure safe and reliable operation of IGBT. Diode D2 consists of 6 regulators, which is used to achieve over-voltage protection. R2 is a 5.1 kΩ resistor which is used to drain the voltage of IGBT junction capacitor. The drive circuit is simple, with strong anti-interference and high reliability advantages.

The fault status feedback circuit is used to detect whether the drive circuit fault. If there is a fault detected, fault signals will be uploaded to CPLD; then the fault information will be displayed on the digital tube. The fault signals that need to be uploaded to the main controller through the optical fiber.

Figure 4 shows the circuit of over-current/over-voltage fault detection, which is composed of a hysteresis comparator and some peripheral resistors and capacitors. U_1 is the voltage signal of corresponding to the current flowing through IGBT or the voltage of between gate and emitter of IGBT. The low-pass filter circuit of R_1 and C_1 can filter out the high-frequency components in the voltage signal. R_2, R_3 and R_4 are used to set a threshold voltage, the different resistance can achieve different threshold voltage. For this circuit, the circuit outputs a high level when no faults occurs, and outputs a low level if there is a fault. After a fault is detected by CPLD, the fault will be processed according to the priority. The specific handling methods will be introduced in Section 3.2.

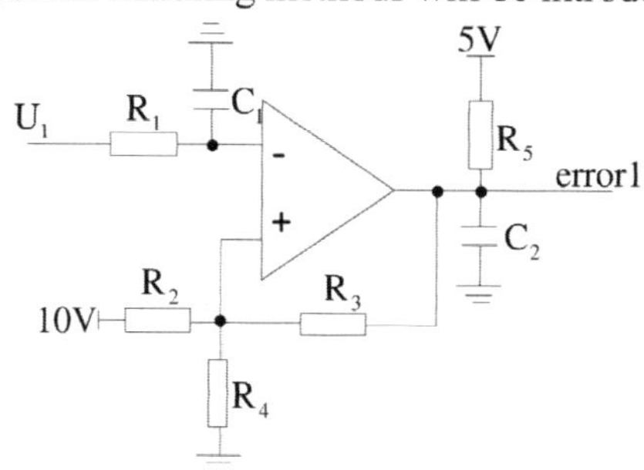

Figure 4. Structure of the fault detection circuit.

A common cathode seven-segment digital display circuit is designed according to costs and utilization rate of CPLD. The displays units shows two digits, the first is the number of the faulty IGBT and another is the fault type. The numbering rules will be described in the next section.

3.2. The protection and procession of fault

In order to achieve the protection of IGBT, some serious faults need to be located quickly by the drive and protection circuit. The fault signals will be processed by CPLD, and the fault signals will also be uploaded to the main controller.

IGBT faults can be roughly divided into five categories in general, namely, under-voltage, over-temperature, over-current, over-voltage, drive power failure. In addition, the status of power supply should also be monitored. The faults are prioritized according to the degree of IGBT faults, because of the faults can't be responded by CPLD at the same time. The sort results as shown in Table 1. If multiple faults occurs at the same time, CPLD will respond according to the priority of the faults.

When a fault signal is detected by hardware circuit, a pulse signal will be generated. After CPLD detects the pulse signal, it will determine and deal with the fault according to the input information. Driving power supply fault, over-current, over-voltage and other faults, will cause damage to IGBT easily. Therefore, when these three types of faults detected by CPLD, the switching pulses will be blocked immediately in order to protect IGBT. Meanwhile, CPLD will send the fault signal to the main controller in the form of code. The main controller will take some fault measurements, such as bypassing the fault SMs, enabling redundant SMs or the fault-tolerant state. For the source supply fault, because of the existence of batteries, the fault information will only display. Similarly, the fault information of under-voltage and over-temperature will be uploaded to the main controller by CPLD;

IOP Conf. Series: Materials Science and Engineering **366** (2018) 012074 doi:10.1088/1757-899X/366/1/012074

the fault will display on the digital display, while the main controller will take the appropriate treatment.

Table 1. The fault code.

IGBT	IGBT Number	Fault Type	Fault Code	Priority
VT1	00	Over-current	000	1
VT2	01	Over-voltage	001	2
VT3	10	Driving power fault	010	3
VT4	11	Over-temperate	011	4
		Under-voltage	100	5
		Source supply fault	1	6

In order to locate the fault quickly, the types of fault are used binary encoding in this paper in order to deal with faults easily. The fault code consists of 6 bits of data, and the data representation is shown in figure 5. The data bit encoding is shown in Table 1.

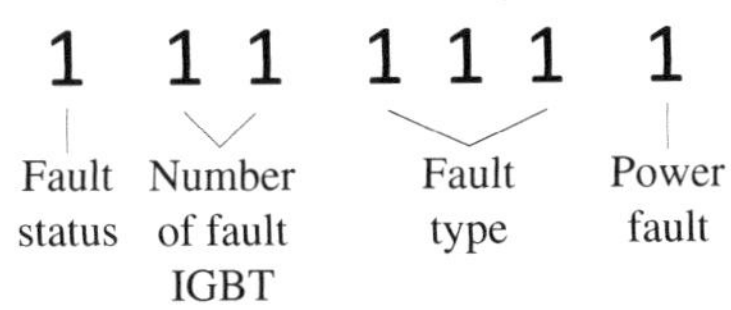

Figure 5. The fault code.

The CPLD returns the fault signals to the main controller through two optical fibers. When a fault occurs, one of the fiber optic signals will jump to a continuous low signal. At the same time, another way will start simultaneously send a 115200 baud rate signal to the main controller. The main controller will judge the type of faults according to the signal fed back from the protection circuit and process the faults.

4. Simulation and Experiment Results

4.1. Simulation results

In order to verify whether the proposed circuit in this paper meets the requirements of the design, the circuit is simulated with the power electronics simulation software (PSPICE) firstly. Figure 6 shows the simulation results. Figure 6 (a) shows the waveform of dead-time $t_{dead\text{-}time}$=4μs and $t_{dead\text{-}time}$=6μs at the switching frequency of 10 kHz. Figure 6 (b), (c) and (d) shows the waveforms of fault signals and pulse signals in the fault simulations. From figure 6 (b), we can know that over-current/over-voltage and other serious problems has been detected at t_1. Then the pulse is blocked, all pulse signals are set low, so that IGBT is locked. Figure 6 (c) shows the waveform of source supply fault. At t_2, there is a source fault detected, then CPLD send the fault signal to the main controller, but the drive pulse is not blocked. The waveform of drive power fault is shown in figure 6 (d). At t_3, there is a drive power supply fault occurs, the drive pulse will be blocked by CPLD.

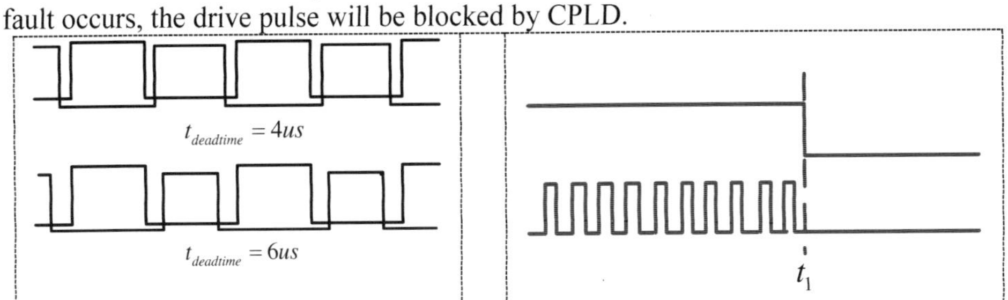

(a) Dead-time test **(b)** Serious fault test

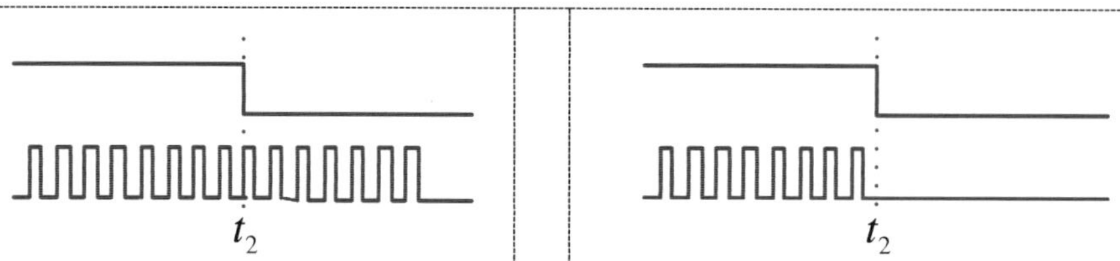

(c) Source power supply fault test **(d)** Drive power supply fault test

Figure 6. The simulation results of fault test.

In order to further verify the reliability of the design, the Drive and protection circuit that proposed in this paper have been tested in experiments. Figure 7 shows the experiment results of fault test, VT1 and VT4 share the same pulse signal, and the VT2 and VT3 share another pulse signal. At t_1, VT1 and VT4 are off; at t_3, VT2 and VT3 are on. Due to the existence of the parasitic capacitor, the switching devices will not turn on immediately. Therefore, after VT1 and VT4 are off, VT2 and VT3 will be triggered to conduct again. The time between t_2 and t_3 is dead-time. In order to verify the dead-time can be adjusted freedom, the dead-time $t_{\text{dead-time}}$=4μs and $t_{\text{dead-time}}$=6μs were tested in experiment, the waveform as shown in figure 7 (a). In order to verify handling the faults in time, the driving and protection circuit have been tested for faults, as shown in figure 7 (b). At t_5, a serious fault occurs (over-voltage/over-current, driving power fault), the pulse signals will be blocked by CPLD.

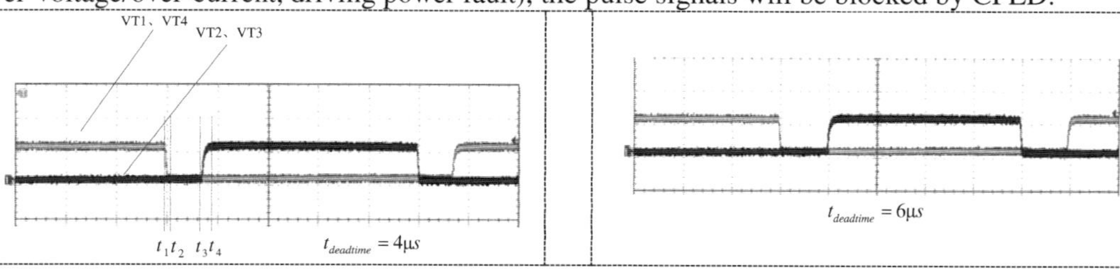

(a) Dead-time test

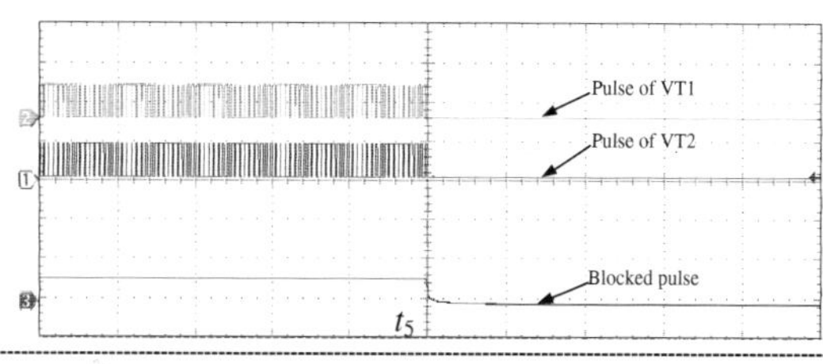

(b) Serious fault test

Figure 7. The experiment results of fault test.

Figure 8 shows the voltage and current waveforms when 1200V/600A IGBT is turned on and off. The DC voltage is 850V, and load current is 380A in the experiment. Where V_{ce} is the voltage between collector and emitter, V_{ge} is the voltage between gate and emitter, and the current flows through the emitter. And from figure 8, we can get that the driving and protection circuit proposed in this paper can effectively drive IGBT.

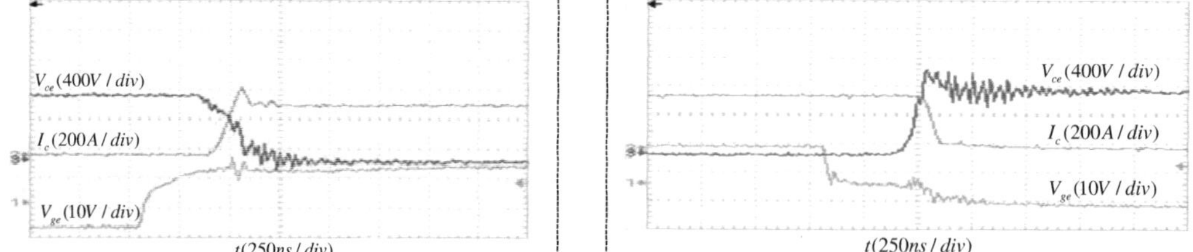

(a) IGBT turn on	**(b)** IGBT turn off

Figure 8. The test experiment results of IGBT driving capability.

Acknowledgments
This work was supported by State Grid Corporation of China (No. 52466F150008).

References

[1] Zheng, Q., Wang, X., Fu, Y., Yan, H., Ou, Z., & Wang, G, A STATCOM compensation scheme for suppressing commutation failure in HVDC, C. IECON 2016 Conference of the IEEE. (2016) 1081-1086.
[2] Bayliss, T., Elgarrista, F., Thompson, S., Yedla, R., Perrier, J., & Arani, S., Development of a modular cooling solution for STATCOM and HVDC transmission schemes, C. AC and DC Power Transmission. (2017).
[3] Chivite-Zabalza J, Perrier J, Boden M, et al, Development of a Full-Bridge Sub-Module for HVDC and STATCOM Markets, C. International Exhibition and Conference for Power Electronics, Intelligent Motion, Renewable Energy and Energy Management. (2017).
[4] Qi F, Xu L, Development of a High-Temperature Gate Drive and Protection Circuit Using Discrete Components, J. IEEE Transactions on Power Electronics, **32** (2016) 2957-2963.
[5] Weichert H, Benz P, Liberto S, Application of (Motor Protection) Circuit Breakers in Combination with Variable Frequency Drives, C. Electrical Contacts. (2012).
[6] Zhou D, Liu Z, Kong P, et al, An improved driving and protection circuit for reverse blocking IGBT, C. Power Electronics Specialists Conference. (2004).
[7] Mohamed A A S, Allen D, Youssef T, et al, Optimal design of high frequency H-bridge inverter for wireless power transfer systems in EV applications, C. Environment and Electrical Engineering. (2016).
[8] Ehsan Behrouzian, Massimo Bongiorno, Impact of capacitor balancing strategies on converter ratings for star-connected cascaded H-bridge STATCOM, C. Power Electronics and Applications. (2017).
[9] Jung J J, Lee J H, Sul S K, et al, DC capacitor voltage balancing control for delta-connected cascaded h-bridge STATCOM considering the unbalanced grid and load conditions, C. Energy Conversion Congress & Exposition (2017) 1-8.

Research on the Choice Strategy of Electric Vehicle Charging Users

Y Zhu[1,a], D Zhou[2], Z Den[2], S J Chen[2] and X You[2]

[1]Hubei Key Laboratory of Advanced Technology for Automotive Components, Automobile Engineering Institute, Wuhan University of Technology, 205 Luoshi Road, Hongshan District, Wuhan 430070, China.
[2]School of Automation, Wuhan University of Technology, Wuhan 430070, China

[a] Corresponding author e-mail address: 861089246@qq.com (Y Zhu).

Abstract. Starting from the perspective of electric vehicle users, this paper analyzes the service needs of different users for charging electric vehicles and considers various objective conditions of electric vehicles: the remaining electric vehicles, the driving distance from the charging pile / station, the charging of electric vehicles Charging time, the location of the charging equipment, etc., gives several user selection strategies that minimize the energy loss of electric vehicles, minimize the time spent by electric vehicle users, and optimize the comprehensive cost of users. Through the theoretical analysis and experimental simulation results, the advantages and disadvantages of several user selection strategies are compared to provide users with a good charging service for electric vehicles, and at the same time, the optimal choice of charging options for users is realized, The service quality of the car is conducive to the promotion and sustainable development of electric vehicles and charging equipment.

1. Introduction

With the continuous advancement of science and technology in our country, the urbanization and the improvement of people's material life quality, the automobile industry has developed rapidly. Energy crisis, environmental pollution is worsening, people's health and depleting fossil fuels are threatened[1,2]. In recent years, a number of studies have shown that automobile exhaust pollution has risen to the major air pollution, threatening the natural environment upon which human beings depend[3]. In order to reduce pollution and reduce dependence on fossil fuels to the maximum extent under the intensification of contradictions between economic development, energy crisis and environmental protection, China will igorously develop new energy vehicles and relying on new energy technologies to realize the automobile industry The transition to new energy vehicles[4].

In this paper, aiming at the various needs of a single EV charging, it recommends the best charging pile charging for the user, reduces the user charging cost and improves the user satisfaction. And research on the best state of charge for a single charging pile, improve the utilization rate of charging pile, and realize the best choice scheme of charging pile for electric vehicle users[5,6]. Considering the remaining capacity of electric vehicles, mileage, charging time, charging pile facilities, design a variety of decision-making tools for different optimization objectives for the appropriate strategy to meet the corresponding needs of different roles to provide users with convenient and efficient charging service , To minimize social costs and improve higher social benefits[7,8].

IOP Conf. Series: Materials Science and Engineering **366** (2018) 012075 doi:10.1088/1757-899X/366/1/012075

2. electric car charging model analysis

2.1 electric car battery model

In this paper, in order to facilitate the study, electric car battery power in 0-100% of the range considered. The total battery capacity is the sum of the remaining capacity of the battery and the amount of power to be replenished:

$$E\_all = e\_y + e\_w \tag{1}$$

Where *E_all* is the total battery capacity, *e_y* is the remaining battery capacity, *e_w* is the battery to be fully charged.

2.2 electric car charging model

The characteristic matrix of electric vehicle charging behavior is characterized by electric vehicle number, coordinates, distance from charging pile, time spent on the road, energy consumption on the road, charging time of electric energy, ending time of charging, time cost, cost cost and comprehensive cost [9] The characteristic matrix is as follows:

$$Che = \begin{Bmatrix} y & Che\_d & Dis_{yx} \\ Td_{yx} & Tchar_y & Tend_{yx} \\ CT_{yx} & CP_{yx} & W_{yx} \end{Bmatrix} \tag{2}$$

$$x \in [1,i],\ y \in [1,\ j]$$

Where, x, i, respectively, said the number and number of charging pile, y, j, respectively, that the number and number of electric vehicles. *Che_d* is the coordinates of the car, Dis_{yx} is y the distance from the electric car to x charging pile, Td_{yx} is the time it takes for the car to charge the pile, *T_char* charging time of the electric car, $Tend_{yx}$ charging end time, *CT* electric car charging time cost ,*CP* for the cost of charging electric vehicles. W_{yx} is the total cost of charging electric vehicles.

Initial position of electric vehicle user; initialized place is randomly generated by the system, with the initial coordinates as the starting point of the user:

$$Che\_d = (X_y, Y_y),\ j \in [1, n(t)] \tag{3}$$

X_y and Y_y represent the abscissa and ordinate of the electric vehicle, respectively, and $n(t)$ represents the electric vehicle time function.

Electric car to charging pile distance:

$$Dis_{yx} = \left| Che\_d_y,\quad CS\_d_x \right| \qquad x \in [1,i],\ y \in [1,\ j] \tag{4}$$

Dis_{yx} means that the electric car y to the charging pile x distance, CS_d_x said charging pile x position coordinates.

Car to charging pile time; Assuming the normal speed of each car is constant and the same as v, the distance and time proportional relationship as follows:

$$Td_{yx} = Dis_{yx} / v \qquad x \in [1,i],\ y \in [1,\ j] \tag{5}$$

The energy consumption of the car to the charging pile; electric car energy consumption per hundred kilometers to 19.5kw.h, the energy consumption of electric vehicles on the road:

$$Ed_{yx} = Dis_{yx} * 19.5/100 \qquad x \in [1,i],\ y \in [1,\ j] \tag{6}$$

Ed_{yx} represents the energy used by the electric car y to the charging pile x road.

Electric vehicle charging time; the value of the remaining charge at the moment of charging request, to the charging pile road energy consumption and electric car charging power decision :

$$Tchar_{yx} = \frac{Ed_{yx} + e\_w_y}{P_x(t)} \quad (7)$$
$$t \in [0,24], x \in [1,i], y \in [1,j]$$

$P_x(t)$ is the charging power of charging pile x at t moment.

User end of charging time; This calculation method to reach the electric vehicle charging pile and the charging pile last car at the end of charging the maximum, and after the electric energy and charging time as the electric vehicle charging end time:

$$Tend_{yx} = \max(t + Td_{yx}, Tend_{kx}) + Tchar_{yx} \quad (8)$$

t is the current moment, $Tend_{yx}$ is the charging end time of the last electric vehicle k charged to the charging pile x.

The waiting time [10] is determined by the charging end time of the last arriving electric vehicle of the charging post and the moment when the current vehicle reaches the charging post:

$$Twait_{yx}(t) = \begin{cases} endt_x - (t + td_{yx}), t + td_{yx} \le endt_x \\ 0, t + td_{yx} \ge endt_x \end{cases} \quad (9)$$
$$t \in [0,24], x \in [1,i], y \in [1,n(t)]$$

Time cost CT mainly includes: driving time on the road Td_{yx} charging waiting time $Twait_{yx}$ power supplement time $Tchar_{yx}$.

$$CT_{yx}(t) = Td_{yx} + Tchar_{yx} + Twait_{yx} \quad (10)$$
$$t \in [0,24], x \in [1,i], y \in [1,j]$$

Charging cost CP by the cost of electricity on the road and the user sends a charging request at the moment to be replenished charge charging composition:

$$CP_{yx} = J(t) * (ed_{yx} + e\_w) + J_0 * ed_{yx} \quad (11)$$
$$t \in [0,24], x \in [1,i], y \in [1,j]$$

Where CP is the cost of charging electric vehicle y to charging pile x, $J(t)$ is the price of electricity at time t, and J_0 is the price of last charging of electric vehicle.

Electric vehicle users choose to charge a comprehensive cost by the cost of costs and time costs constitute:

$$W_{yx}(t) = a * \frac{CP_{yx}(t)}{\min CP_{yx}(t)} + b * \frac{CT_{yx}(t)}{\min CT_{yx}(t)} \quad (12)$$
$$t \in [0,24], x \in [1,i], y \in [1,n(t)], a + b = 1$$

W_{yx} represents the total cost of charging users, a, b, respectively, the cost of costs and time costs of the weight coefficient. User charge the lower the overall cost of higher customer satisfaction.

2.3 charging pile model established

The charging characteristics of electric vehicle charging pile are characterized by charging pile number, charging pile coordinate, charging power and charging pile charging end time as follows:

$$CS_x = (x, CSd_x, P_x, Tend_x), x \in [1,i] \quad (13)$$

Where, x, i, respectively, represent the number and number of charging pile; CS, said charging pile coordinates, P_x charging pile charging power, $Tend$ said charging end time.

Electric vehicle charging pile position coordinates and its corresponding one-to-one fixed value, this paper in order to facilitate the study of the layout of the manual settings:

$$CSd_x = (X_x, Y_y), x \in [1,i] \quad (14)$$

This article assumes that the charging pile to the electric car charging power $Px=50kw/h$ constant and convenient research, the market by the mainstream of the three kinds of charging research as follows:

$$P_x = \begin{cases} 27.3 - 41kw/h, slow \\ 41 - 131.2kw/h, normal \\ 131.2 - 262.4kw/h, fast \end{cases} \quad (15)$$

Charging pile charging end time charging terminal used to charge the last electric car charging end time:

$$Tend_x = Tend_{yx} \quad (16)$$

$Tend_x$ indicates the charging end charging time of the charging post x, and real-time dynamic updating is performed from the charging ending time of the last electric vehicle to the charging post.

3. electric car user strategy analysis

3.1 User Cost Analysis

For EV users, the satisfaction of charging behavior of users is measured by time cost and cost cost. The lower the user charging time cost and the cost cost are, the higher the user satisfaction is. The composition of user costs is shown in Figure 1 below:

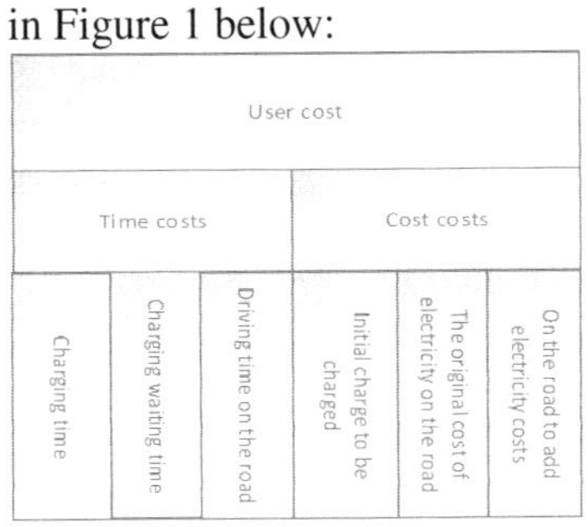

Figure 1 user cost structure

3.2 The user selects the evaluation index

The method of calculating the time cost and the cost cost for electric vehicle users is based on the analysis and calculation of the total cost cost and the time cost objective function.

The cost cost objective function is as follows:

$$CP(t) = \sum_{t=0}^{24} \sum_{y=1}^{n(t)} CP_{yx}(t) \quad (17)$$

$$t \in [0,24], x \in [1,i], y \in [1,n(t)]$$

$n(t)$ represents the number of electric vehicles in different time periods, and $CP_{yx}(t)$ is the cost of choosing charging pile x for charging when the electric vehicle y is $J(t)$ at a certain time under a certain scheduling strategy cost. The $CP(t)$ can be transformed as follows:

$$\begin{aligned} & CP_{yx}(t) = f(Dis_{yx}) + f(e\_w_y) \\ & f(Dis_{yx}) = (J_0 + J(t)) * (Dis_{yx} * 19.5/100) \\ & f(e\_w_y) = J(t) * e\_w_y \\ & t \in [0,24], x \in [1,i], y \in [1,n(t)] \end{aligned} \quad (18)$$

The user cost includes the on-street cost $f(Dis_{yx})$ and the initial time power supplement $f(e\_w)$. The former is related to the distance between the electric vehicle and the selected charging pile, the latter has nothing to do with the selection of the charging pile. Therefore, the farther from the electric vehicle to the charging pile $Dist_{yx}$, the higher the charging cost.

The objective function of time cost is:

IOP Conf. Series: Materials Science and Engineering **366** (2018) 012075 doi:10.1088/1757-899X/366/1/012075

$$CT(t) = \sum_{t=0}^{24} \sum_{y=1}^{n(t)} CT_{yx}(t) \tag{19}$$
$$t \in [0,24], x \in [1,i], y \in [1,n(t)]$$

CT is the time cost of charging the charging pile *x* in the electric car under a certain choice scheme of the user. *CT* can do the following changes:

$$\begin{gathered} CT_{yx}(t) = g(Dis_{yx}) + g(Twait_{yx}) + g(e\_w_y) \\ g(Dis_{yx}) = \frac{Dis_{yx}}{v} + \frac{19.5 Dis_{yx}}{100 P_x} \\ g(Twait_{yx}) = Twait_{yx} \\ g(e\_w_y) = \frac{e\_w_y}{P_x} \\ t \in [0,24], x \in [1,i], y \in [1,n(t)] \end{gathered} \tag{20}$$

$g(Dis_{yx})$ represents a function related to the distance *Dis* from the electric vehicle *y* to the charging pile *x*, $g(Twait_{yx})$ represents a function related to the charging latency of the electric vehicle, and *g (e_w)* represents a function related to the initial charge of the electric vehicle. Since each charging pile has the same charging power at the same time, *g(e_w)* is independent of charging pile selection. The user's time cost *Dis* is determined by *Twait*. The user time cost is proportional to the distance from the electric car to the charging pile and the charging waiting time.

4. program simulation analysis

4.1MATLAB simulation process

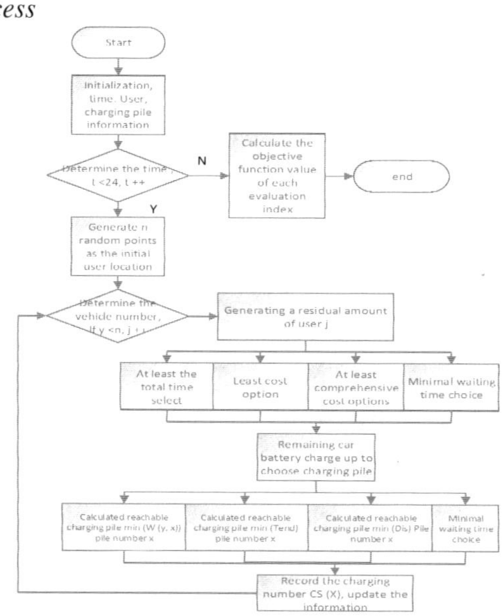

Figure 2 user selection strategy simulation process

In Figure 2 above, the selection and comparison process of the entire user selection strategy is described, and the simulation results are compared with the results of four different decision-making methods.

4.2 Simulation Results

In this paper, 16 charging piles are selected as the research object to simulate the user's travel patterns at different periods of time, leading to 30 users for 0-4h, 10 users for 4h-6h, 20 users for 6-9h, 40 users

IOP Conf. Series: Materials Science and Engineering **366** (2018) 012075 doi:10.1088/1757-899X/366/1/012075

for 9-16h, 20 users for 16h-22h ; 22h-24h cited 20 users, only consider the single constant power charge of the case, the four kinds of user selection strategy simulation.

According to the four kinds of user selection schemes described in this article, the number of EVs selected to charge each charging pile within 24 hours is calculated respectively. After carrying on the emulation of four kinds of different schemes, the overall distribution of 4 kinds of vehicles of different schemes is shown as in Figure 3 below:

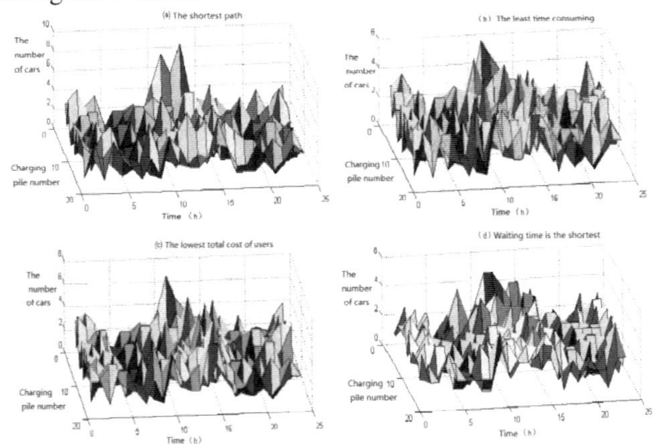

Figure 3 (a) the shortest path (b) the total time consumption of at least (c) the user the least cost (d) shortest waiting time

Since the number of vehicles put in each time period is not the same within 24 hours, the number of vehicles is not evenly distributed on the time axis, that is, the number of vehicles in charging piles varies from one distribution to another.The variance of the number of vehicles for each scenario is shown in Figure 4 below:

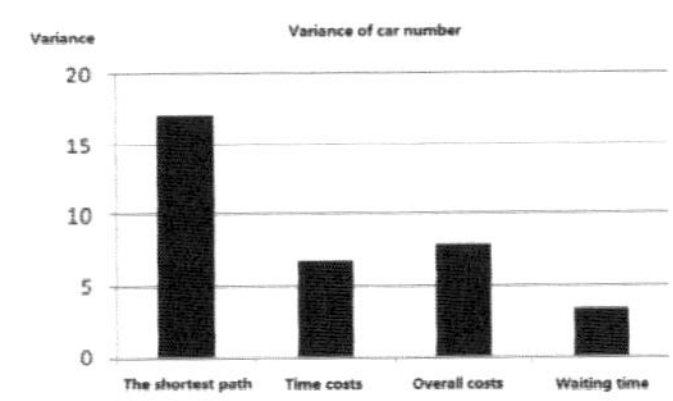

Figure 4 Variance distribution of the four strategies

Since the integrated strategy is to combine the variance between the two, in order to facilitate the study, this paper selected a = 1 for simulation analysis.

For the day when the peak use of electric vehicles, the following two selected periods, t = 11 and t = 12, combined with the overall distribution of the law for analysis:

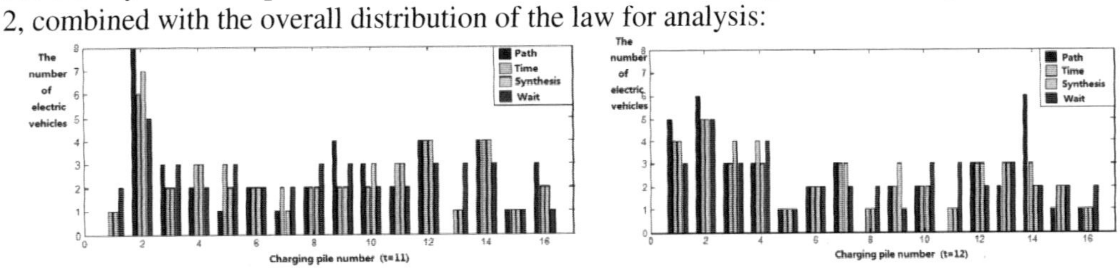

(a) (b)

Figure 5 (a) User selection distribution at t=11(b) User selection distribution at t=12

From the simulation results in Figure 5 we can see:The first (a) scheduling (shortest path), the distribution of electric vehicles is fragmented, because this scheduling method only considers the driving distance factor of the charging pile scheduling, while ignoring the user time costs, as well as the charging pile operating conditions , Did not show a good time effect, do not have the regularity. At t=11, the number of electric vehicles that choose charging post 2 for charging reaches 8, and when t=12, the number of charging vehicles that choose charging post 2 and 14 reaches 6, which greatly increases the waiting time for user charging Cost, and lead to overcharged operation of the charging pile in areas where the distribution of traffic volume is relatively intensive, increasing the charging pressure of the charging pile.

The third (c) scheduling (the smallest user integrated cost), although the number of electric vehicles at various times the distribution of scattered, but its effect is better than the first scheduling. At t=11, the number of vehicles supplemented by electric charging pile with number 2 is 7, and the distribution is even when t=12. According to the overall distribution, the number of vehicles served by a single charging pile is less than that of the first scheduling method under the scheduling cost with the lowest total cost.

The second (b) scheduling method (the total time consuming), Show a more even distribution at t=11 and t=12. At t=11, compared with the first and the third scheduling schemes, the number of vehicles in charging station No.2 is reduced. At t=12, the number of vehicles charged by each charging station does not exceed 5 vehicles. Compared with the shortest path scheduling scheme, The status of the charging posts 1,2 and 14 has been improved. Charging pile is a reasonable distribution of use.

User costs include cost and time costs. According to these two evaluation indexes, the correlation values of the four scheduling methods are respectively calculated, the result is shown in Figure 6 below:

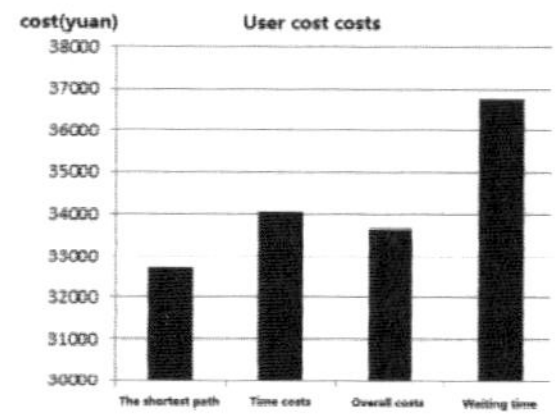

Figure 6 total cost of cost

Four kinds of scheduling mode user cost costs were: 32731.52 yuan, 34075.74 yuan, 33639.73 yuan, 36753.78 yuan. Because the user charging cost is proportional to the distance from the electric vehicle to the charging pile, the shortest path cost of the first scheduling method has the lowest user cost.

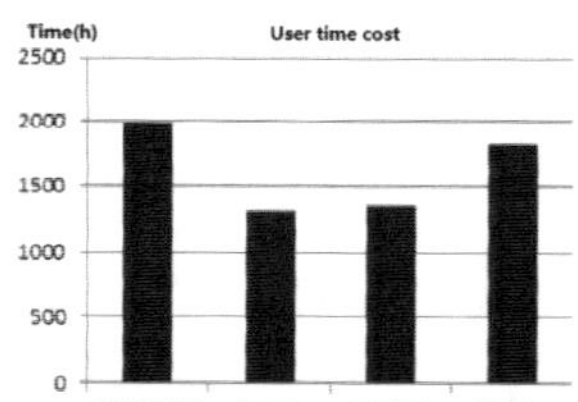

Figure 7 Total time cost chart

According to the results shown in Figure 7 above.Four kinds of scheduling mode user time costs were: 1979.85h, 1313.91h, 1362.58h, 1830.82h. In the first scheduling mode, the user charges the nearest charging pile, which saves the user time-consuming on the road. However, due to the dense distribution of vehicles in a certain area, a certain charging pile is charged with too many vehicles and increases the waiting time for the user to charge. Therefore, User time costs higher. Therefore, the total time is the longest, while the time cost and comprehensive cost selection strategy save more time, there will be better results when charging.

5. Conclusion

With the continuous development of electric vehicle technology, the user's demand for DC charging pile is also growing. After a large number of electric vehicles are put into the market, the charging needs of users will certainly be more complicated and diverse. Through the comparison and analysis of the four user strategies of the electric vehicle users with the above design, the advantages and disadvantages of the three schemes of lowest cost cost, lowest time cost and lowest comprehensive cost are obtained, The optimal choice of the key needs of users, for the general case, comprehensive cost options for the user has a good guiding significance.

References

[1] Liu Jinpeng. Study on the development of China's energy supply and demand pattern based on resource and environment constraints [D]. North China Electric Power University, 2013.

[2] Zhang Tao. Research and Development Trend of Motor Vehicle Exhaust Emissions in China [J] .Resources Development and Market. 2013 (02): 159-161.

[3] Geng Yongsheng. Automobile exhaust pollution and control technology [J]. Environmental Science Guide. 2010,29 (6): 62-69.

[4] Wang Wenchao.Analysis and Prediction of Drivers of Energy Consumption and Carbon Dioxide Emissions in China's Provinces [D]. Dalian University of Technology, 2013.

[5] Boulanger AG, Chu A C, Maxx S, et al. Veeicle electrification: status and issues [J]. Proceedings of the IEEE 2011, 99 (6): 1116-1138.

[6] Wang Jiacheng. China's new energy automotive industry development analysis [D]. Shanghai Normal University master's degree thesis, 2012.

[7] Quinn C, Zimmerle D, Bradley T H. The effect of communication architecture on the availability,reliability, and economics of plug-in hybrid electric vehicle-to-grid ancillary services [J]. Journal of Power Sources. 2010, 195 (5): 1500-1509.

[8] Li Chunhui. Development status and trends of electric vehicles [J]. Automotive Industry Research .2005 (5): 44-46.

[9] ang YNOU, GUO Chuang-xin, WANG Li-cheng, et al.Cluster scheduling strategy for electric vehicles considering customer satisfaction [J] .Automation of Electric Power Systems.2015.3.9 (17): 183-191.

[10] Luo Yugong, YI Yi-yao, ZHU Tao, et al.Optimal charging path planning method for smart electric vehicles [J]. Engineering Research: In the field of interdisciplinary Engineering. 2014, 6 (1): 92-98.

Intelligent elevator control and safety monitoring system

Shi Dan[1] and Xu Bixi[1]

[1]College of Electrical Engineering and Control Science, Nanjing Tech University, Nanjing, China

[a]Corresponding Email:2276596470@qq.com

Abstract: Nowadays, China, as the largest market in the global elevator industry, has become a closely related vertical transportation tool. The intelligence, safety and reliability of elevators are becoming a common concern among the state, elevator manufacturers and users[1-3]. In recent years, due to the excessive reliance on traditional technology, the user has caused casualties in the actual use due to improper operation. However, the use of elevators causes many casualties, which can be avoided from a technical point of view. Therefore, in this paper, on the basis of the traditional Internet technology, increase the microprocessor technology, automatic sensing technology, remote control technology, speech recognition technology and the establishment of the elevator safety operation regulation system, put forward the concept of "intelligent elevator" system, this system in practice, will greatly facilitate People's Daily travel and improve the safety and reliability of the elevator, the elevator more intelligent, better service to the society.

1. Introduction
Today, the elevator has been one of the important tool of People's Daily travel, but how to make the elevator is more convenient, smart, has plagued the people. Therefore, in this paper, we put forward the concept of "intelligent elevator".

2. Fundamental theories

2.1. Design objectives
The elevator control system can achieve the following three objectives.

2.1.1. Automatic induction of elevator doors. Using infrared induction technology, users can only make a stop at the automatic sensing area in front of the elevator, and the doors of the elevator can be opened automatically without manual manipulation.

2.1.2. Speech control. After the user enters the elevator, the elevator door shut down automatically triggers the elevator speech control system, voice guide users to select floor, according to user's voice response decisions and perform the corresponding instruction, it is also without any difficulty[4].

2.1.3. Remote control technology. On the base of ordinary elevator, the elevator call function and elevator arrival feedback function are added. Before the user goes out to take the elevator, through the common mobile terminal (e.g. app). The device can remotely call the elevator, and when the elevator

IOP Conf. Series: Materials Science and Engineering **366** (2018) 012076 doi:10.1088/1757-899X/366/1/012076

reaches the designated floor, the terminal will receive feedback (for example: vibration or ring tones), alert the user in real time, and avoid a long wait[5].

2.1.4. Elevator safety operation supervision system. Inside and outside the system software in e-government construct multiple application subsystem, network and Internet to the competent department of industry, the lift owner, pipe unit, the elevator manufacturing unit, elevator maintenance unit and the social public to provide information about the elevator operation and management of all kinds of services[6].

2.2. Design scheme

In the elevator control system based on the original add a set of infrared sensor systems, a voice control system, a remote control system and a set of remote monitoring system for the above four target can be realized, as shown in fig.1[7].

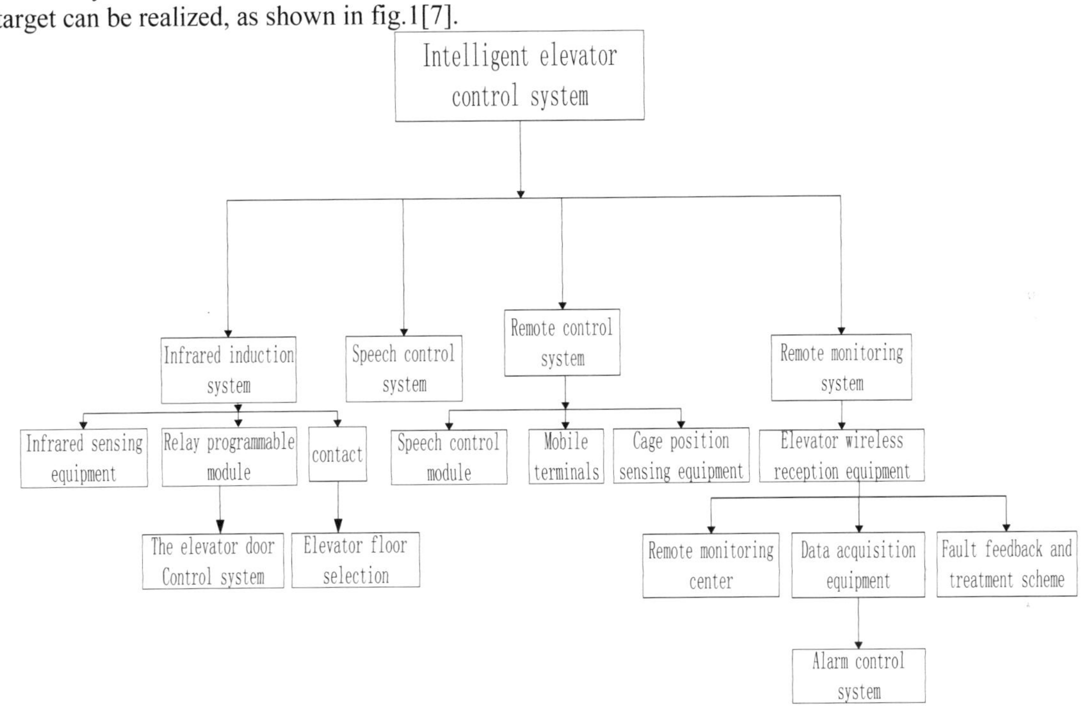

Figure 1. block diagram of intelligent elevator control system

2.2.1. Infrared sensor system. Programmable infrared sensing system by relay module and original elevator of elevator door machine system to achieve seamless docking, voice control system via voice control module and original floor choice of elevator system achieve seamless docking, remote control system through the relay and original elevator elevator system achieve seamless docking[8].

2.2.2. Speech control system. At the moment when the elevator door is closed, can trigger the elevator floor voice control system began to work, until you receive the two effective signal or close the door after more than 20 seconds (mainly first come), elevator floor voice control system to stop working. The workflow of the speech control system, as shown in Fig.2[9].

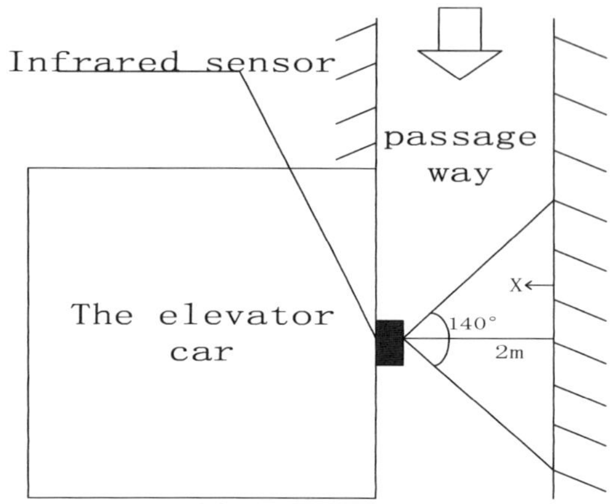

Figure 2. workflow of speech control system

2.2.3. Remote control technology. The working flow of elevator remote control system is shown in fig.3.

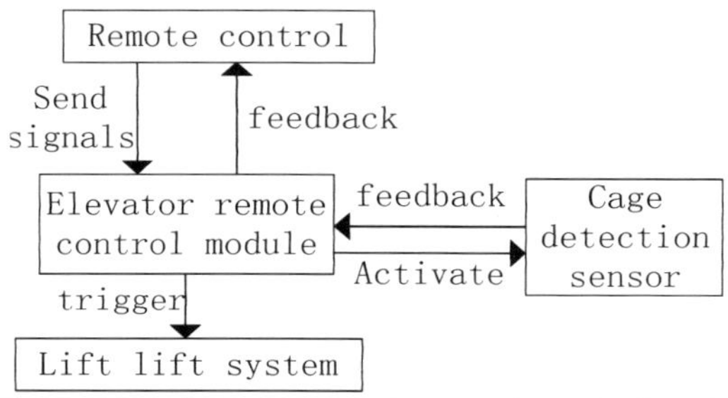

Figure 3. workflow of elevator remote control system

2.2.4. Remote monitoring system. During the normal operation of the elevator, the situation of elevator car can be learned at any time. When the elevator failure, installed in the elevator control cabinet linkage of data collector, signal alarm, report to the server at the same time, the field staff and remote control room can timely know which lifts malfunctioned, when failure happens, in order to take timely measures.

3. Implementation of the scheme

3.1. Circuit diagram of elevator control system

3.1.1. Elevator door open and closed system. The circuit diagram of elevator door open and closed system is shown in fig.4 Its functions are realized mainly through 51 MCU programming.

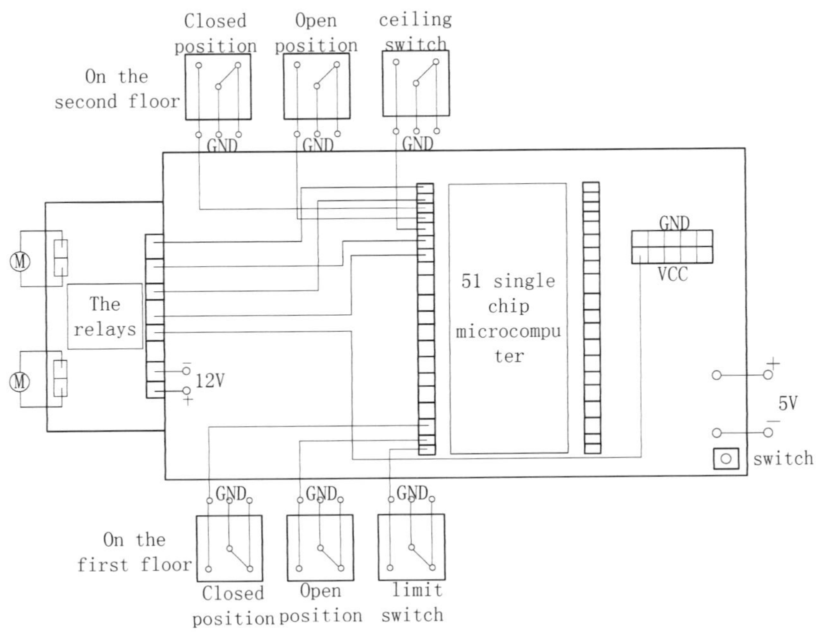

Figure 4. elevator door open and closed system

3.1.2. Speech control system. The circuit diagram of the speech control system is shown in fig.5. The voice module can be triggered when the door of the elevator is closed (that is, the elevator door closed on the first or second floor of the picture).

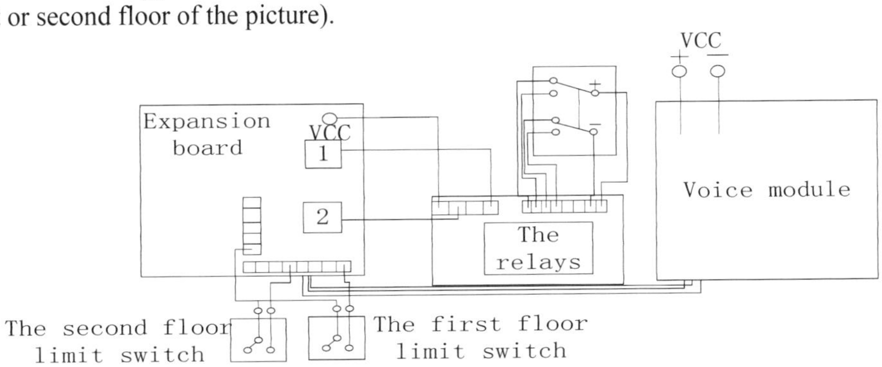

Figure 5. speech control system

3.1.3. Elevator remote control system. The circuit of the remote control system is shown in fig.6. Using the commonly used mobile terminal, the receiver's KD terminal can guide the motor, which drives the elevator cage up. When capsules up close to the second floor, close to the switch is closed, the feedback loop current launch, feedback to remote transmitters, make as shown in fig.7.

IOP Conf. Series: Materials Science and Engineering **366** (2018) 012076 doi:10.1088/1757-899X/366/1/012076

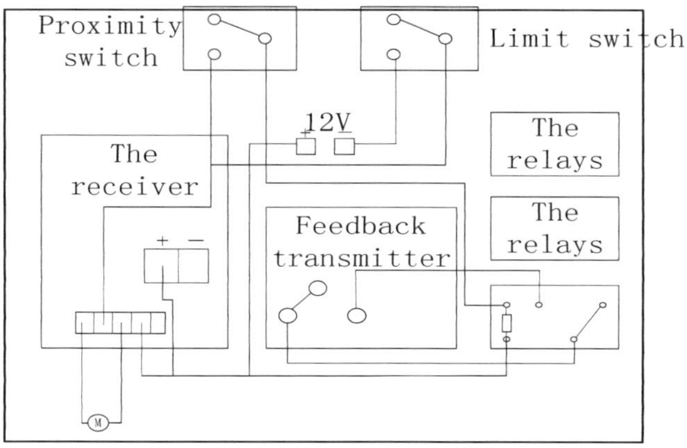

Figure 6. remote control system

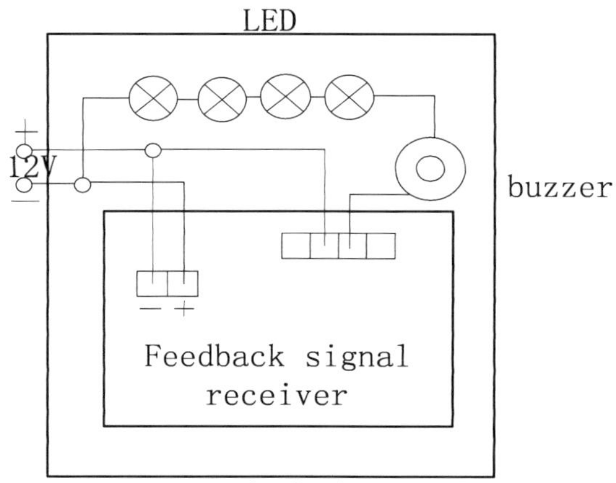

Figure 7. feedback system

3.1.4. Auxiliary manual control system. For the convenience of manual control model of the rise and fall of the elevator, manual control elevator lifting system is designed, and concatenated in the loop on the first floor limit switch and the limit switch on the second floor, circuit diagram is shown in fig.8.

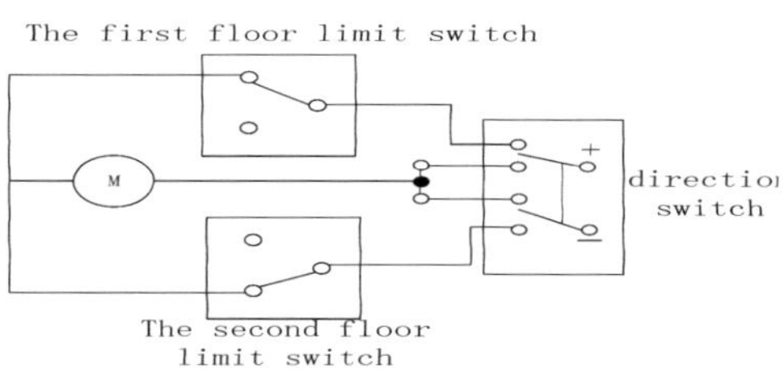

Figure 8. manually control elevator elevator circuit

3.1.5. Remote monitoring system. Realize the fault alarm, tiring rescue, daily management, quality assessment, hazard prevention functions such as integrated management platform, elevator is a kind of safety system based on Internet of things technology, as shown in fig.9.

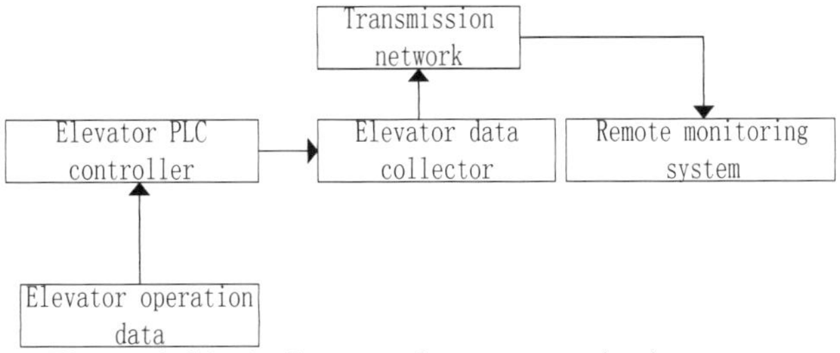

Figure 9. block diagram of remote monitoring system

4. Conclusion

The advantages of the system include: 1. The elevator door infrared sensor system and the floor selection speech control system can realize the full automatic control of the elevator. 2. The elevator infrared induction system can distinguish whether the user needs to take the elevator and avoid the invalid opening of the elevator door to realize the purpose of energy conservation. 3, voice control system of the trigger mode and other elevator speech control system, which can realize the doors close automatically trigger the voice control system, even if the first time to ride the elevator, the user can also easy to use, versatility is stronger; 4. By remote control, we can save valuable waiting time for users, so that users can wait at home to make it easy for users to travel. 5. Using the remote monitoring system, the safety and reliability of the elevator are increased, so that the user can be ensured in the elevator.

References

[1] STMicroelectronics.STM32F10xxx Reference manual,REV 9.June,2009:21-23.

[2] NXP Semiconductors.LPC17xx User manual.Rev00.06.June ,2009:68-70.

[3] Embedded Systems Conference Papers, San Francisco, 2001:65-68.

[4] N. Fragopanagos,and J. G.Taylor, " 2005 Special Issue: Emotion recognition in human-computer interaction," Elsevier Science Ltd. vol. 18, May 2005:389-405.

[5] R. Cowie, E. Douglascowie, N. Tsapatsoulis, G Votsis, and Votsis, "Emotion recognition in human-computer interaction," IEEE Signal Processing Magazine, vol. 18, 2001:32-80.

[6] L. Malta, C. Miyajima, and K. Takeda, "Analysis of real-world driver's frustration," IEEE Transactions on Intelligent Transportation Systems, vol. 12, Sept 2011:109-118.

[7] D. J. France, R. G. Shiavi, S. Silverman, M. Silverman, and D. M. Wilkes, "Acoustical properties of speech as indicators of depression and suicidal risk," IEEE Transactions on Bio-Medical Engineering, vol. 47, 2000:829-837.

[8] Wei Kongping, roni. Elevator technology [M]. Beijing: chemical industry press, 2006.6:56-58.

[9] zhu devan. Intelligent control and application of elevator traffic system [M]. Changchun:Jilin university press, 2002:17-19.

IOP Conf. Series: Materials Science and Engineering **366** (2018) 012077 doi:10.1088/1757-899X/366/1/012077

Optimized Economic Dispatch of Active Distribution Network with Electric Vehicle Group

Jianhong Kong[1], Zhonghao Shen[3], Mingzhu Niu[1], Tao Zhang[2], Bing Liu[1], Shubo Hu[4] and Hui Sun[4]

[1] State Grid Dalian Electric Power Co, Ltd, Dalian, 116001, China
[2] State Grid Liaoning Electric Power Research Institute Co, Ltd, Shenyang, 110006, China
[3] State Grid Zhengzhou Electric Power Co, Ltd, Zhengzhou, 450005, China
[4] School of electrical engineering, Dalian University of Technology, Dalian, 116024, China

Corresponding author's e-mail address:shubo_hu@hotmail.com

Abstract. With increasing penetration of distributed generation system and electric vehicles in the distribution network, new active power flow management and economic dispatch are required to cope with uncertainty and intermittence. An electric vehicle (EV) group management agency is introduced in this paper in the context of Chinese electricity market, considering the flexibility of EVs and distributed generations (DGs). An active distribution network (ADN) coordination optimal dispatch model is set up to minimize the generation cost of the distribution network and the charging cost of the electric vehicles in Matlab environment. An IEEE33-bus test system model with electric vehicle group is built up to improve electric vehicle charging and discharging management. The results verify the effectiveness of the proposed method on improving the satiability and safety of the system.

1. Introduction

In order to establish a low-carbon economy, global effort has been made to promote EVs and EVs will be a new important type of load in power grids [1]. Large scale of EVs will be connected to power grids and greatly impact in the operation and planning of distribution networks [2]. As energy storage systems, EVs can also support the network and actively participate in economic dispatch technology of ADN.. Distribution system operator (DSO) can then make use of new economic dispatch technologies and coordinate power flow between DGs, loads and energy storage system (ESS) [3].

In ADN, EVs can connect to grid as distributed energy storage and can interact with power grids as "load" and "supply". With large scale EVs connected to the power grids, centralized power supply and peak load will appear in the distribution network or even cause the network congestion if there are no constraint and management of the behaviors of EVs. However, if the charging/discharging behavior of each EV was under the control of DSO, the dimensionality of variables in the optimization model will increase sharply. And the threshold power level of users participate in electricity market is MW, which cannot be reached by a small number of EVs [4, 5]. Therefore, a charging-discharging control strategy in the distribution network is needed to coordinate optimal economic dispatch, so as to make advantage of the flexible charging/discharging characteristics of EVs and generation characteristics of DGs.

IOP Conf. Series: Materials Science and Engineering **366** (2018) 012077 doi:10.1088/1757-899X/366/1/012077

In AND, there are many kinds of DGs. Wind power and photovoltaic power are intermittent renewable energies. Micro turbine (MT) and diesel engine (DE) are controllable distributed generations (CDGs), whose power outputs are flexible and adjustable. The uncorrelated EVs and CDGs are connected through the DSO to become correlative scheduling objects. A management agency and the DSO are needed to coordination dispatch the generation of CDGs and the charging/discharging of EVs. That is to prevent the congestion caused by the large scale EVs unordered charging/discharging and ensure the safety and stability of the system operation. Controlled EV charging and discharging in ADN can improve the valley to peak of load [6], economy of power systems [7] and reduce the charging cost of EVs [8]. In practice, it is impossible for the DSO to control charging-discharging at a vehicle level, and this needs to be done at a group level (hundreds of EVs).

In this paper, an optimal dispatch model aiming to minimize the charging cost of EV group and the generation cost of the system is proposed. The chance-constrained programming theory is used to conduct the stochastic optimization. An IEEE33-bus test system model with electric vehicle group is built up to improve electric vehicle charging and discharging management and reduce the related costs.

2. Supply and demand coordination economic dispatch

2.1. ADN market structure with DGs and EV group

A bidirectional interact control system is built up in this section. As a link between DSO and the users, a charging/discharging operator can integrate the EVs charging/discharging power and provide dispatchable power electricity to power systems. EVs in aggregator is a group of EVs with near spatial distances like a housing estate or a street. The aggregator can promote the interaction between DSO and users and to realize management of the charging/discharging time of EVs by charging/discharging operators.

2.2. Operation model of DGs

Traditional objective model of DE generators is a quadratic polynomial equation related to fuel cost and active power outputs. The MT operation fuel cost functions are as equation (2) shows.

$$F_{\mathrm{DE}}(P_{\mathrm{DE}}(t)) = c_2 P_{\mathrm{DE}}(t)^2 + c_1 P_{\mathrm{DE}}(t) + c_0 \tag{1}$$

$$F_{\mathrm{MT}}(P_{\mathrm{MT}}(t)) = C_{\mathrm{MT}} \frac{P_{\mathrm{MT}}(t)}{\eta_{\mathrm{MT}}(P_{\mathrm{MT}}(t))} \tag{2}$$

$$C_{\mathrm{MT}} = \frac{C_{ng}}{L} \tag{3}$$

$$\eta_{\mathrm{MT}}(t) = m_3\left(P_{\mathrm{MT}*}(t)\right)^3 + m_2\left(P_{\mathrm{MT}*}(t)\right)^2 + m_1 P_{\mathrm{MT}*}(t) + m_0 \tag{4}$$

where $P_{\mathrm{DE}}(t)$ and $P_{\mathrm{MT}}(t)$ are the DE andMTpower outputs at time t; c_2、c_1、$c_0$are DE consumption characteristic parameters;C_{MT} and $P_{\mathrm{MT}*}$ are the fuel unit cost and the per-unit value in MT, respectively; $\eta_{\mathrm{MT}}(P_{\mathrm{MT}}(t))$ is the efficiency of MT operation. C_{ng} is the price of natural gas, and L is the low heating value. m_3、m_2、m_1、m_0 are the relation characteristic parameters of MT efficiency and power outputs.

2.3. Logical architecture of the Double layer model

As a mobile energy storage system, EVs can optimize the power system operation by storing energy and outputting power. Refer to the driving cycles, the numbers of dispatchable EV $N_{avi}(t)$ by the DSO and numbers of EV in the state of driving $N_{dri}(t)$ in time t are provided. The per kilometre energy consumption is E_{C} ,and energy consumption is $P_{dch}(t)$. The state of charge (SOC) at the start of time period $t$is $S(t)$, and the SOC at the start of time period t+1is $S(t+1)$. The related equations are given as follows.

$$N_{avi}(t)=\lambda_{acc}N_{ev}P_{avi}(t) \tag{5}$$

$$N_{dri}(t)=\lambda_{acc}N_{ev}-N_{avi}(t) \tag{6}$$

$$P_{dch}(t)=N_{dri}(t)E_{C}(t)v_{e}(t)\Delta t \tag{7}$$

$$S(t+1)=S(t)+\frac{\eta_{ch}P_{ch}(t)\Delta t-P_{dch}(t)\Delta t}{N_{acc}B_{C}} \tag{8}$$

$$P_{ch}(t)=\sum_{i=1}^{\lambda_{acc}N_{avi}}p_{ch}(i,t) \tag{9}$$

$$N_{acc}=\lambda_{acc}N_{ev} \tag{10}$$

where, N_{ev} is the number of EV; λ_{acc} is the factor of acceptable dispatch; $P_{avi}(t)$ is the driving probability; $v_{e}(t)$ is the driving speed; η_{ch} is the efficiency of charging and discharging; $P_{ch}(t)$ is the total charging power of lay-off vehicles. $p_{\mathrm{ch}}(i,t)$ is the charging power; N_{acc} is numbers of dispatchable EV; B_{C} is the average capacity of EV batteries.

3. Supply and demand coordination optimal dispatch model

3.1. Objective function

The minimum generation cost objective functions contains the electricity purchasing cost, producing cost of direct dispatched DGs, purchasing cost and compensation of the flexible load. The objective of charging/discharging operators is the minimum EV group charging cost. Equations are as follows:

$$\min F_1=\sum_{t=1}^{N_t}(f_{grid}(t)+f_{agent}(t))+\Delta f_{agent}+\sum_{i=1}^{N_{\mathrm{DG}}}\sum_{t=1}^{N_t}f_{\mathrm{DG},i}(t)+M_{\mathrm{DG},i}(t)+D_{\mathrm{DG},i}(t) \tag{11}$$

$$f_{grid}(t)=C_{grid}(t)P_{grid}(t)\Delta t \tag{12}$$

$$f_{agent}(t)=C_{agent}(t)P_{agent}(t)\Delta t \tag{13}$$

$$M_{\mathrm{DG},i}(t)=K_{\mathrm{MT}}P_{\mathrm{MT}}(t)\Delta t \tag{14}$$

$$\Delta f_{agent}=\lambda_{com}\max\left\{\sum_{t=1}^{N_t}(P_{agent}^{0}(t)-P_{agent}(t)),0\right\} \tag{15}$$

$$f_{\mathrm{DG},i}(t)=\begin{cases}F_{\mathrm{MT}}(P_{\mathrm{MT}}(t))\\ F_{\mathrm{DE}}(P_{\mathrm{DE}}(t))\Delta t\end{cases} \tag{16}$$

$$D_{\mathrm{DG},i}(t)=\frac{C_{\mathrm{az,MT}}}{8760k_{\mathrm{MT}}}\frac{r(1+r)^{n_{\mathrm{MT}}}}{(1+r)^{n_{\mathrm{MT}}}-1}P_{\mathrm{MT}}(t)\Delta t \tag{17}$$

$$\min F_2=\sum_{i=1}^{N_d}\sum_{t=1}^{N_t}C_{\mathrm{P}}(t)P_{\mathrm{ch},i}(t) \tag{18}$$

$$C_{\mathrm{P}}(t)=\alpha(t)+C_{\mathrm{S}}(t) \tag{19}$$

where, N_{DG} and N_d are the number of DG and Aggregator, respectively; N_t is the dispatchable time period; $f_{grid}(t)$ and $f_{agent}(t)$ are electricity purchasing cost from power system and controllable DG agents, respectively; Δf_{agent} means the compensation to controllable DG agents; $C_{grid}(t)$、$P_{grid}(t)$ are the ADN electricity purchasing price and power from the power system, respectively; $C_{agent}(t)$、$P_{agent}(t)$ are the ADN electricity purchasing price and power from the controllable DG agents, respectively; λ_{com} is the factor of compensation; $P_{agent}^{0}(t)$ is the purchasing power from controllable DG agents before the adjustment of charging/discharging service fee; $f_{\mathrm{DG},i}(t)$, $M_{\mathrm{DG},i}(t)$ and $D_{\mathrm{DG},i}(t)$ are the fuel cost, operating maintenance cost and depreciable cost of DGs, respectively; K_{MT} is the per-unit operating maintenance cost coefficient; $C_{\text{az,MT}}$ is per-unit capacity installation cost; k_{MT} is the factor of MT capacity; r is annual interest rate; n_{MT} is the investment payback period of MT; $C_{\text{P}}(t)$ is the charging price of EV; α is the time-of-use (TOU) power price; C_{S} is the charging/discharging service fee; $P_{\text{ch},i}(t)$ is the charging power.

3.2. Constraint function

The constraint functions of power balance, EV group charging and discharging power, ESS and SOC and line power flow constraints are as follows:

$$P_{Gi,t} - P_{Di,t} - P_{\text{ch}i,t} - V_{i,t}\sum_{j=1}^{N} V_{j,t}(G_{ij}\cos\theta_{ij,t} + B_{ij}\sin\theta_{ij,t}) = 0 \tag{20}$$

$$Q_{Gi,t} - Q_{Di,t} - V_{i,t}\sum_{j=1}^{N} V_{j,t}(G_{ij}\sin\theta_{ij,t} - B_{ij}\cos\theta_{ij,t}) = 0 \tag{21}$$

$$P_{Gi}^{\min} \le P_{Gi,t} \le P_{Gi}^{\max} \tag{22}$$

$$Q_{Gi}^{\min} \le Q_{Gi,t} \le Q_{Gi}^{\max} \tag{23}$$

$$V_{i}^{\min} \le V_{i,t} \le V_{i}^{\max} \tag{24}$$

$$P_{ch}^{\min} \le P_{\text{ch},t} \le P_{ch}^{\max} \tag{25}$$

$$P_{dch}^{\min} \le P_{\text{dch},t} \le P_{dch}^{\max} \tag{26}$$

$$S_{\min} \le S(t) \le S_{\max} \tag{27}$$

$$S_{need} \ge S_{need,\min} \tag{28}$$

$$S(t_0) = S(t_{end}) \tag{29}$$

$$P_{l,t} = V_{i,t}V_{j,t}(G_{ij}\cos\theta_{ij,t} + B_{ij}\sin\theta_{ij,t}) - V_{i,t}G_{ij} \tag{30}$$

$$-P_{l}^{\max} \le P_{l,t} \le P_{l}^{\max} \tag{31}$$

where, $P_{Gi,t}$ and $Q_{Gi,t}$ are the active and reactive power at node i and time t, respectively; $V_{i,t}$, G_{ij}, B_{ij} and $\theta_{ij,t}$ are the voltage amplitude of node i, conduction of branch ij, susceptance of branch ij, and phase angle difference of node i and node j at time t, respectively. For EV group, the $P_{ch}^{\max} = \lambda_{acc}N_{\text{avi}}(t)p_{ch}^{\max}$, where $p_{ch}^{\max}$ is the maximum charging power of a single EV. And $P_{ch}^{\min} = \lambda_{acc}N_{avi}(t)p_{ch}^{\min}$, $p_{ch}^{\min} = -p_{dch}^{\min}$, where $p_{ch}^{\min}$ is the maximum discharging power of a single EV. S_{need} is the SOC which satisfies the EV group driving demands, and $S_{need,\min}$ is the minimum limit of S_{need}. $S(t_0)$ and $S(t_{end})$ are the SOC at the dispatching beginning time and dispatching ending time, respectively; $P_{l,t}$ is the active power of the l th electric line at time t.

A chance constrained programming is used to conduct the stochastic simulation of the intermittent renewable energy power outputs. And a MCS-PSO (Particle Swarm Optimization based on Monte Carlo Simulation) method is used to calculate the optimal solutions [9].

4. Test results and analysis

4.1. Data information

In the modified IEEE33-bus test system with EV group, each operator administers an EV Aggregator, and the number of EV in each Aggregator is 360, 270 and 270, respectively. A battery energy storage with capacity of 800 kW•h is connected at node 7. The maximum power of battery energy storage is 0.4MW, and the initial SOC is 50%. A 300 kW diesel engine is connected at node 32 and a 300kW micro turbine is connected at node 5. Two wind turbines with 500kW capacity are connected at node 25 and node 28. A local error beta-distributed model is used to simulate the wind power outputs. Two photovoltaic generators with 500kW capacity are connected at node 15 and node 22 respectively and the illumination intensity distribution fits the Beta distribution [10].

4.2. Results analysis

There are 4 scenarios in the optimization process: Scenario I is free charging/discharging model; Scenario II is the optimal power dispatch with EV group aiming to minimize only the generation cost of the system; Scenario III is the optimal power dispatch with EV group aiming to minimize only the EV charging cost; Scenario IV is the coordination optimal dispatch aiming to minimize both generation cost and EV charging cost.In the free charging/discharging model, the EV group charging operations will be conducted randomly upon the EV owners' habits. The charging power of each Aggregator and original electricity load are shown in figure 1.In Scenario I, the peak charging power is added on the peak load, which makes the valley-to-peak change from 1580kW to 3185.33kW. The utilization efficiency of electricity resource is also decreased. However, this condition is improved after the EV group accepting the dispatch like Scenario IV shows. There is an obvious concave region in the optimal load curve in Scenario IV at peak load times. This concave region comes from the discharging operations of EV group at peak load times. The owners prefer to perform EV charging at valley load times and discharging at peak load times to reduce the charging cost, which can realize the peak and valley complementation with the original load.

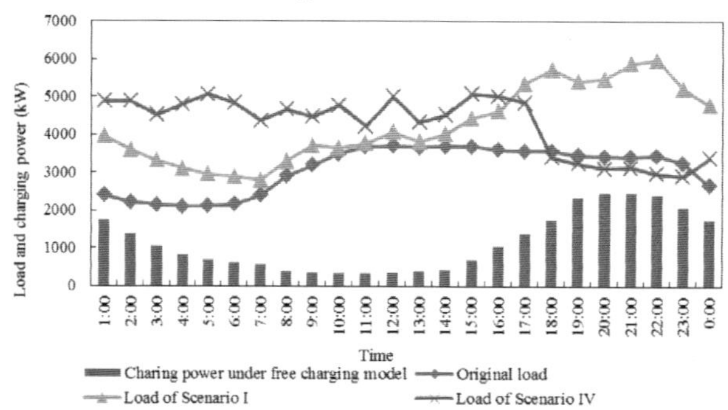

Figure 1. The charging/discharging power distribution of each Aggregator under free charging/discharging mode

The results of load rate of transmission line are shown in figure 2. In Scenario I, the branch 1 is overloaded and congestion occurs in the power system from 18:00 to 23:00. The maximum overload probability is 125%. On the contrary, the load rate from 6:00 to 10:00 is low because the EV group charging power and electricity consumption are low at this time period.

In Scenario IV, the overload of Branch 1 is reduced to under the datum line, and the system congestion problem is solved. The load rate from 18:00 to 23:00 is decreased due to the dispatchable EV group discharged to supply energy to the system to relieve the stress of transmission lines at this original peak load time period.

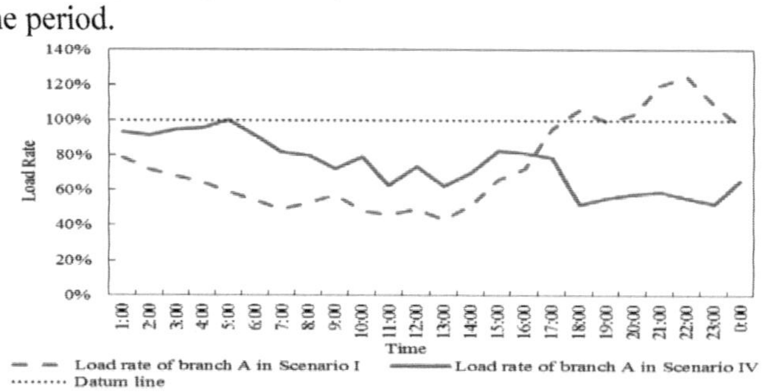

Figure 2. The load rate of Branch A

The generation cost and charging cost in different Scenarios are shown in table1. Both of the costs in Scenario IV are lower than those in Scenario I. Compared with the single-objective optimization results in Scenario II and III, the generation cost and EV charging cost in Scenario IV are higher. In order to minimize this generation cost, the EV charging/discharging strategy is restricted at some time periods to satisfy the constraints of the generation. The EV charging/discharging strategy focuses on its own benefits and the scheduling of generating units is operated passively. The results show the local conflicts of the two objective functions. Therefore, the proposed coordination optimal dispatch model that considers both generation cost and EV charging cost in the optimization problem gives the best results.

Table1. Theoptimization results

Scenario	Generation cost/yuan	Charging cost/yuan
I	6.31×10^4	4.13×10^4
II	5.72×10^4	3.96×10^4
III	6.58×10^4	3.48×10^4
IV	6.25×10^4	3.83×10^4

5. Conclusion

A supply and demand coordination ADN optimal dispatch model with EV group flexible charging/discharging response is proposed in this paper. The experiments are conducted on a modified IEEE 33-bus test system. The results show that the ADN market structure can realize the coordination dispatch more conveniently and quickly. The optimization model with multiple objectives, minimum generation cost and minimum EV charging cost, can make full use of the response capability of EV group and can also offer dispatching accepted chances to the dispatch operators, charging/discharging operators and EV owners. Four scenarios of the optimization model are given in the experiments and the test results show that the proposed coordination optimal dispatch strategy is more effective which can reduce both the generation cost and charging cost of the EV group, which can improve the positivity of EV owners participating in the power dispatch. In general, the proposed coordination multi-objective optimal dispatch model provides an optimal operating mode to the power system.

Acknowledgments

This work has been supported by the Science and Technology Project of State Grid (2017YF-27).

References

[1] W. Yao, J. Zhao, F. Wen, Y. Xue and G. Ledwich, 2013,*IEEE Trans. Power Syst.*, **28**p2768.
[2] J. Rolink and C. Rehtanz, 2013, *IEEE Trans. Power Delivery*, **28**894
[3] C. D'Adamo, S. Jupe and C. Abbey,2009,*CIRED 2009 - 20th International Conference and Exhibition on Electricity Distribution - Part 1, Prague, Czech Republic*, pp 1-4
[4] X. F. Wang, C. C. Shao, X. L. Wang, C. Du, 2013, *Proceedings of the CSEE*,**33**1
[5] P. H. Divshali, B. J. Choi. 2016, *Energies*, **9**
[6] M. Alonso, H. Amaris, J. G. Germain, J. M. Galan, 2014, *Energies*, **7**2449
[7] S. Shao, M. Pipattanasomporn and S. Rahman, 2009, *2009 IEEE Power & Energy Society General Meeting, Calgary*, pp 1-8
[8] N. Rotering and M. Ilic, 2011,*IEEE Trans. Power Syst.*, **26**1021
[9] J. Bai and Y. Guo, 2009, *Tianjin,2009 Fifth International Conference on Natural Computation*, pp. 565-568
[10] E. Kaplani, S. Kaplanis, 2012, *Applied Energy*, **97**970

Typical Power Load Profiles Shape Clustering Analysis Based on Adaptive Piecewise Aggregate Approximation

Quan TANG[1], Quanming ZHANG[1], Yingying DENG[1], Mi ZHU[1], Yunche SU[1], Junyong LIU[2] and Xiaodi WANG[2]

[1]Sichuan Electric Power Corporation Power Economic Research Institute, Chengdu 610000, China
[2]College of Electrical Engineering and Information Technology of SCU, Chengdu 610065, China

Corresponding author: 414363359@qq.com

Abstract. With the trend of increasing dimensions of collected power load data, the data dimension reduction and classification become essential pre-processing steps for data mining and further application. On the basis of adaptive piecewise aggregate approximation(APAA) and k-Shape algorithm, a novel method is proposed. In light of the fluctuation degree and shape characteristics of the original load profiles, the new load dataset with variable temple resolution replace the old one by APAA, and further k-Shape algorithm is adopted for lower dimension load profiles clustering. K-Shape algorithm cluster curves with distance SBD as similarity measurement and also a novel method to extract the representative centroids is mentioned. The experiment testifies that APAA-kShape algorithm has shorter calculating time, higher accuracy and represent better load patterns than other cluster algorithms.

1. Introduction

With the popularity of smart meters, huge amount of detailed end-user energy consumption data at different granularity are generated. Clustering as a useful tool to simplify the data, groups similar data sets into aggregated clusters by weighing the the similarity of load profiles[1]. Valuable information can be extracted from databases and pattern discovery for further application can be developed by clustering[2].

The shape of a curve is of great important, the most common clustering methodologies such as k-means, fuzzy C-means, self-organizing map, etc and their combination algorithm[3,4], most of which measure the similarity of profiles by traditional euclidean distance. [5] points out that all above methods have difficulty in distinguishing curves with close distance but huge shape discrepancy.

High temporal resolution is required to elaborately analyze daily load data, which is also accompanied by time-consuming computation. Dimension reduction methodologies are applied such as symbolic aggregate approximation, piecewise aggregate approximation, principal component analysis, sammon map and etc[6,7]. Most of them re-express the original data with the constant temporal resolution, which is unreasonable because constant temporal resolution ignores the fluctuation degree and will lead to loss of useful information.

In this paper, we propose a novel algorithm APAA-kShape. First, adaptive piecewise aggregate approximation(APAA) is applied to obtain the re-expressed data sets with variable temporal resolution, which retain the fluctuation features and the morphological characters of load profiles. Then, k-Shape algorithm is used to cluster the low-dimension data sets. k-Shape algorithm belongs to the partitional clustering algorithm but has the capability of keeping the shape feature of time series data set. It puts forward a novel distance measurement SBD and a way to extract centroids on the basis of Steiner's sequence. Further more, the experiment shows that our APAA-kShape algorithm can reconstruct original data effectively and our distance measurement is scale-, shift- and noise invariant. APAA-kShape algorithm can cluster load profiles with less computation time, higher accuracy and shows an outperforming than traditional time-series clustering method. The study proves the algorithm practical for high dimension power load profiles.

2. Adaptive Piecewise Aggregate Approximation

2.1. load profiles characteristics

2.1.1 Climbing accident. Within the certain period of time, a sharp change of a curve is called climbing accident, which can measure the fluctuation characteristic of that curve[8]. At the time window *T*, if the difference between the maximum and minimum value of a curve x_i excesses a certain threshold, we define the variable ϕ_i :

$$\phi_i = \begin{cases} 1, & x_i \in \text{Climbingaccident} \\ 0, & x_i \notin \text{Climbingaccident} \end{cases} \tag{1}$$

For *N* load profiles, at the time window *T*, we define the total number of climbing accidents as ψ .

2.1.2 Edge point extraction based on slope. Piecewise aggregation approximation(PAA) based on the change rate of slope is put forwards to detect the edge points of time series, which can retain the main shape characteristic of the curves and remove detail interference.

The slope rate $tg1$, $tg2$ of one point x_i of the time series, can be calculated by comparing the point x_i with its adjacent points x_{i-1} , x_{i+1} . If the change rate of slope exceeds the threshold d, we define the point x_i as the edge point.

$$\mu_i = \begin{cases} \left\| tg1 \right| - \left| tg2 \right\| > d = 1 \ (x_{i-1}、 x_{i+1} \text{ on thesameside of } x_i) \\ \left\| tg1 \right| + \left| tg2 \right\| > d = 1 \ (x_{i-1}、 x_{i+1} \text{ on thedifferentsides of } x_i) \\ 0 \end{cases} \tag{2}$$

For *N* load curves, at the time window *T*, we define the total number of the edge points as ε .

2.2. APAA detailed steps

Adaptive piecewise aggregate approximation(APAA) divides the time series into several segments of different length based on the curves' characteristics without supervision. Take the normalized data set $X = (x_1, \ldots, x_N)$ of length n and the threshold τ and d as the input, here's the detailed steps:

1) Divide *n* dimensions of data set X into *n/4* segments with four data in each segment, calculating the number of ψ and ε in each .

2) For all the segments, if $\psi < \sigma 1$ and $\varepsilon < \sigma 2$,replace the the four data in the segment with their average data; if $\psi < \sigma 1$ and $\varepsilon > \sigma 2$, replace the four data with their maximum and minimum value.

3) Calculate the data dimensions

4) If not satisfied with the dimensions, enlarge the threshold and repeat step 2.

5) if satisfied with the dimensions, return data set $\overline{X}$ with *m* dimensions and output the results.

The value of $\tau, d, \sigma_1, \sigma_2$ are chosen according to the practical situation.

3. k-shape algorithm

3.1. Time-series theoretical background

Time series data are sometimes distorted. Hence, a number of invariances(scaling invariance, shift invariance, noise invariance) have to be taken into account when it comes to compare time series data sets. An ideal distant measurement can achieve scaling-, shift- and noise invariance.

The two state-of-the-art distance measures for time-series comparison to determine their similarity and possibly capture more invariances are ED and DTW. ED is most widely used one, but is sensitive to the noise, scaling and shift distortions.

DTW offers a local optimal alignment by constructing an *n*-by-*m* matrix *M* to achieve distance measuring. A warping path is a set of matrix elements that defines a mapping between $X=(x_1,\ldots,x_n)$ of length *n* and $Y=(y_1,\ldots,y_m)$ of length of *m*. Dynamic programming detect the best path to obtain the minimum distance DTW, which realizes the similarity comparison of two time series of different length.

$$DTW(x,y)=\tau(m,n) \tag{3}$$

$$\tau(i,j)=ED(i,j)+\min\{\tau(i-1,j-1),\tau(i-1,j),\tau(i,j-1)\} \tag{4}$$

However, the same weight is vested to the observed values, which may lead to misclassification. Also DTW is too time-consuming to cluster power load profiles with high dimensions.

3.2. Time series similarity

Each time series must be pre-processed before comparing the similarity.

Cross-correlation measure: Cross-correlation is a statistical measurement with which we keep $\tilde{y}=(y_1,\ldots,y_n)$ static and slide $\tilde{x}=(x_1,\ldots,x_n)$ to align the two sequences to achieve shift-invariance and calculate the similarity. Considering all possible shift s, the cross-correlation sequence $CC_w(\tilde{x},\tilde{y})$ with length *2m-1* is produced.

$$CC_w(\tilde{x},\tilde{y})=R_{w-m}(\tilde{x},\tilde{y})=\begin{cases}\sum_{l=1}^{m-k} x_l+k\cdot y_l, & k\geq 0\\ R_{-k}(\tilde{y},\tilde{x}), & k<0\end{cases} \quad w\in\{1,2,\ldots,2m-1\} \tag{5}$$

Where $CC_w(\tilde{x},\tilde{y})$ is maximized is the place w that the optimal shift s can be calculated: $s=w-m$. The cross-correlation coefficient is normalized to limit its range from -1 to 1 and 1 means highly correlated. And we derive the following distance measure SBD:

$$SBD(\vec{x},\vec{y})=1-\max_{w}\frac{CC_C(\vec{x},\vec{y})}{\sqrt{R_0(\vec{x},\vec{x})\cdot R_0(\vec{y},\vec{y})}} \tag{6}$$

Which takes value between 0 to 2, and 0 indicates perfect similarity for time-series sequences.

3.3. Time-Series Shape Extraction

Extracting meaningful centroids is a challenging task which critically depends on the choice of distance measurement and each cluster centroid represents the typical shape characteristics. Traditional k-means centroid is to compute the arithmetic mean of the corresponding coordinates of the sequences, but these centroids do not effectively represent the cluster shape characteristics.

The centroid computation can be an optimization problem where the objective is to find the minimum value of the sum of squared distances to all other sequences, which is Steiner's sequence:

$$c_j=\underset{\varpi}{\operatorname{argmin}}\sum_{x_i\in p_j} dist(\mu,x_i)^2,\ \vec{w}\in R^{\mathrm{m}} \tag{7}$$

As cross-correlation captures the similarity, we express the computed sequence as the maximum value and use the previously computed centroid as reference to align all sequences.

$$\begin{aligned}\vec{c}_k^{\,*}&=\underset{\vec{c}_k}{\operatorname{argmax}}\sum_{\vec{x}_i\in p_k} NCC_C(\vec{x},\vec{c}_k)^2\\ &=\underset{\vec{c}_k}{\operatorname{argmax}}\,\vec{c}_k^{\,T}\cdot\sum_{\vec{x}_i\in p_k}\left(\vec{x}_i\cdot\vec{x}_{\mathrm{i}}^{\mathrm{T}}\right)^2\vec{c}_k\end{aligned} \tag{8}$$

$$\vec{c}_k^{\,*} = \underset{\vec{c}_k}{\arg\max} \frac{\vec{c}_k^{\,T} \cdot Q^T \cdot S \cdot Q \cdot \vec{c}_k}{\vec{c}_k^{\,T} \cdot \vec{c}_k} \tag{9}$$

Where $\vec{c}_k = \vec{c}_k \cdot Q$, $Q = I - O/m$, I is the identity matrix and O is a matrix with all ones. The maximizer $\vec{c}_k^{\,*}$, as the eigenvector that corresponds to the largest eigenvalue of the real symmetric matrix M, is also our extracted centroids.

3.4. k-Shape algorithm detailed steps

Taking the load curve matrix X as input, the centroid of each cluster is extracted by cross-correlation measure and the data of each cluster is updated in each iteration.

1) Initial each cluster centroid $\vec{c}_k$ and $\vec{c}_k$ is 0 vector.
2) The SBD distance from each sample $\vec{x}_i$ to all the centroids c_k is calculated by equation 8. Assign $\vec{x}_i$ into the cluster c_k with smallest distance.
3) Calculate the centroid of each cluster and extract the shape characteristics.
4) Repeat the step 2 and the step 3. Stop iterating when the maximum number of iterations t is reached or the samples of each cluster do not change anymore;
5) Output cluster results.

4. Experience and analysis

4.1. Simulation four types of typical daily load profiles

Four types of typical daily load curves are simulated, a certain ratio of random noise $\alpha$is added and the curves are partially shifted. We simulate 200 load curves which are grouped into four typical clusters with 50 curves in each cluster.

200 simulating load profiles with 10% noise in four clusters are shown in figure 1 and re-expressed low-dimension load curves are shown in figure 2, where the time window T is 1h, τ is 5 and d is 2.

The original 96-dimension data is reduced to 63-dimension by APAA, which can clearly represent the morphological characteristics of each cluster(bi-peak, tri-peak, peak-period, peak avoidance). We compare APAA with the constant temporal resolution PAA and measure the cluster results by CP value. The average CP value of each re-expressed cluster by our method is 1.891 which is significantly smaller than the CP value 2.548 of PAA , slightly larger than the CP value 1.797 of the original data, which proves the effectiveness of our method, which validated our method.

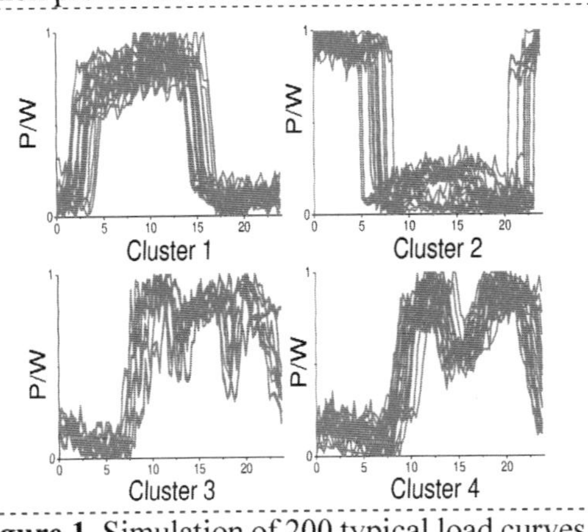

Figure 1. Simulation of 200 typical load curves with 10% noise

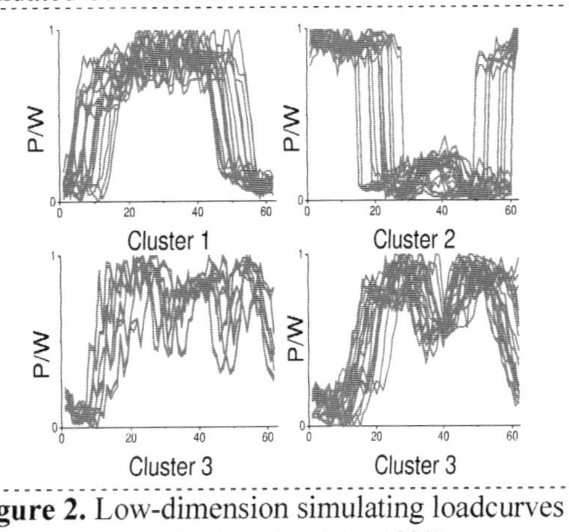

Figure 2. Low-dimension simulating loadcurves withAPAA method (α=10%)

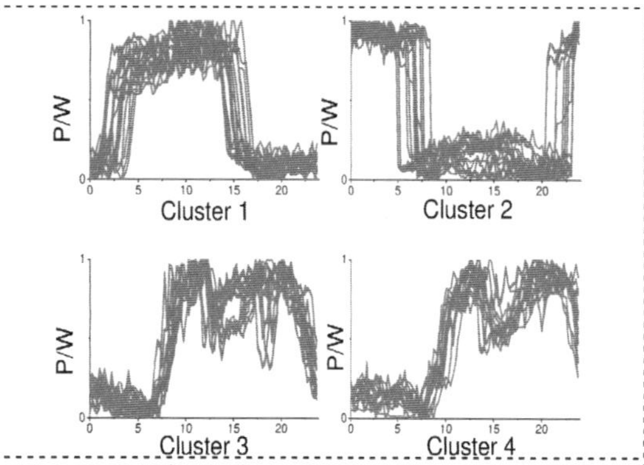

Figure 3. Clustering results of k-means method

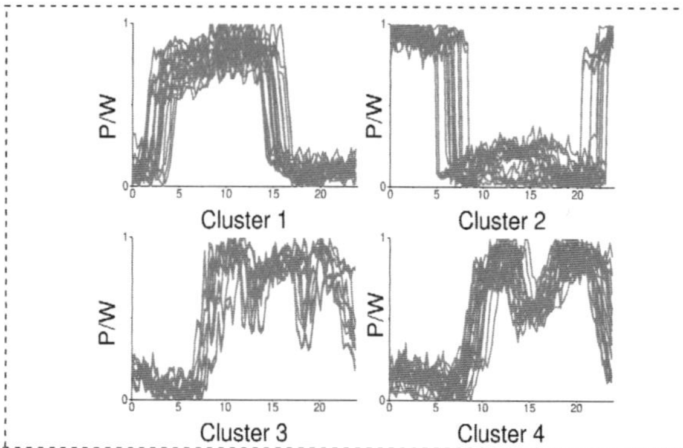

Figure 4. Cluster results of APAA-kShape method

The APAA-kShape detailed steps are shown in figure 5, and the clustering results of the k-means and the APAA-kShape algorithms are shown in figure 3 and figure 4 respectively. k-means fails to distinguish the curves in the cluster 3 and cluster 4 and shows some deviation with its accuracy 86%, while the cluster accuracy rate of APPA-kShape reaches 100%.

With the certain noise ratio, each algorithm has been run 10 times, and the accuracy results are shown in figure 6. k-means clustering accuracy rate is inconsistent in 10 times clustering, which is noise sensitive and poorly stable., while APAA-kshape average cluster accuracy rate is much higher. The experiment confirms APAA-kshape with noise invariance and good robustness.

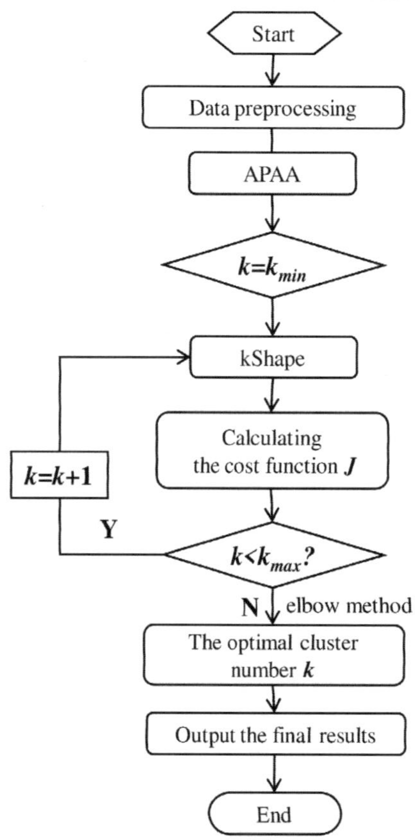

Figure 5. APAA-kshape processing

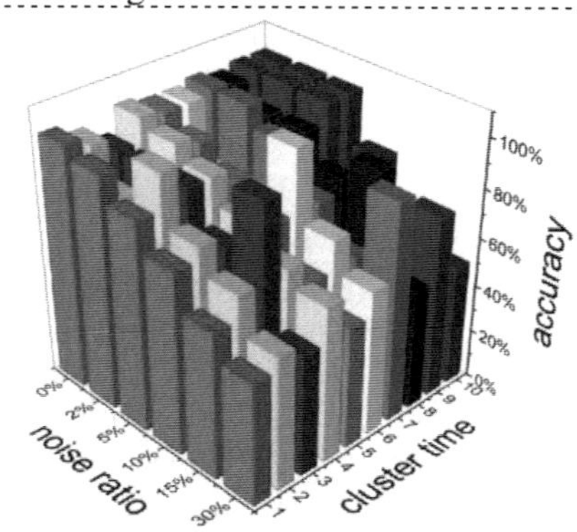

(a)k-means

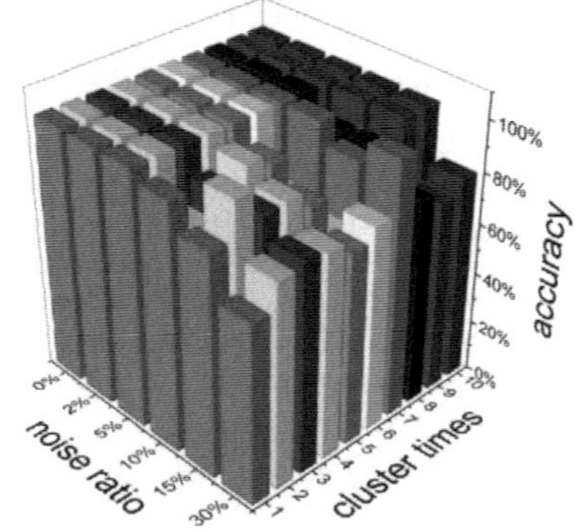

(b) APAA-kShape

Figure 6. Comparison of cluster accuracy with different noise ratio

4.2. Measured data

Taking the daily load curves of 1600 users on a working day in April, 2015 in a Chinese city as the research object, the energy consumption data is collected at every 15 min interval so that 96 data points are obtained from one user in that day. All the data has gone through data cleaning process to identify the outlines and impute the missing data. After data cleaning. 1589 practical daily load curves are collected. At the stage of APAA, we set time window T as $1h$, τ as 20, d as 5.

Ding, Wang[9] have made a uniform comparison experiment on the existing methods for measuring the similarity of time series, and pointed out that the DTW distance measurement may be the best method at present. The APAA-kShape and the DTW are adopted to cluster 1589 daily load curves respectively, the optimal cluster number of APAA-kShape is 8 based on elbow method[10] , as well as DTW. The clustering results are shown in figure 7 and figure 8.

These 8 clusters of APAA-kShape can be classified as the following 4 type: single-peak, double-peak, multi-peak and avoidance peak. The typical daily load curves extracted by our algorithm is reasonable and each one has obvious shape feature. Each cluster are different in peak value, peak valley time and basic load of all the clusters. The extracted typical load curves can represent the typical user behavior patterns and load characteristics. The optimal cluster number of DTW is also 8, but the shape characteristic of the curves in each cluster is vague. Though the avoidance-peak type users(the cluster 2 and the cluster 3) in figure 8 can be distinguished accurately, for the single- peak, double-peak, multi-peak type users, DTW performs worse than APAA-kShape. Choosing 100 curves from 1589 daily load curves, the performance of DTW, k-Shape, APAA-kShape are comprehensive compared by cluster validity index and running time, the results are shown in the Table 1. Among these three algorithms, the APAA-kShape shows great advantage.

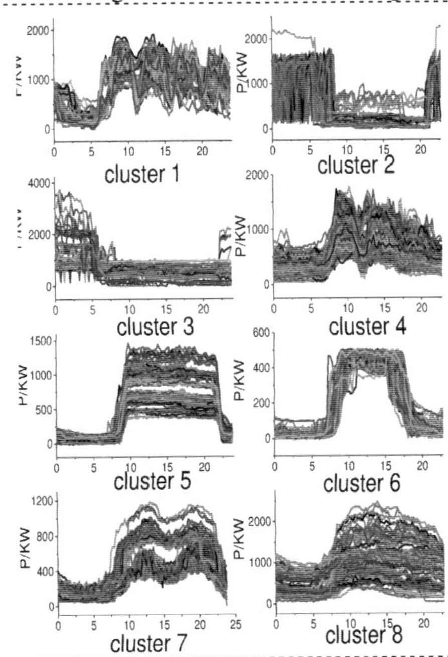

Figure 7. Load curves clustering results of APAA-kShape method

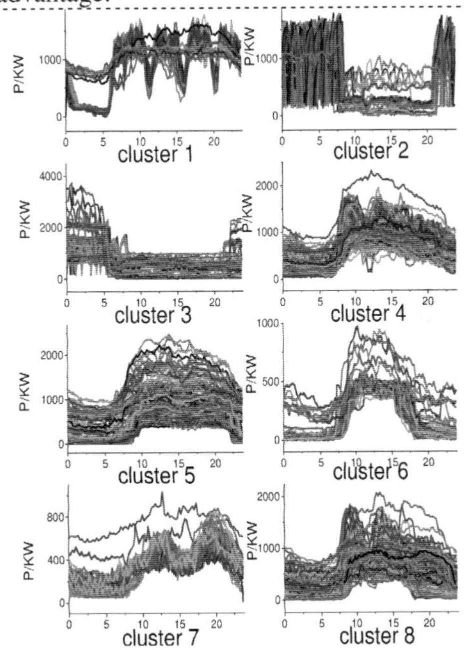

Figure 8. Load curves clustering results of DTW method

Table 1. Clustering property comparison among APAA-kShape, k-Shape and DTW method

cluster	APAA-kShape				k-Shape				DTW			
numbe	Time	SIL	DBI	CH	Time	SIL	DBI	CH	Time	SIL	DBI	CH
3	18.41	0.312	2.050	154.2	24.47	0.455	1.501	116.1	36.70	0.194	2.518	170.8
4	15.35	0.341	1.318	143.0	29.89	0.399	1.784	144.6	59.39	0.073	2.186	116.1

5	10.74	0.347	1.669	134.2	31.21	0.375	1.927	110.2	49.21	0.01	2.873	69.68
6	24.48	0.531	1.267	165	40.30	0.563	1.051	199.4	76.31	0.128	3.276	81.62
7	46.16	0.215	1.647	125.3	53.12	0.453	1.547	90.19	91.80	0.198	2.028	87.12
8	29.01	0.16	2.245	19.97	60.26	0.405	1.203	82.28	103.8	0.041	2.461	63.82

5. Conclusion

With the challenging of increasing dimensions of power load data and the time-series distortions, in this study, first, APAA with variable temporal resolution is adopted based on the curve fluctuation degree and curve shape characteristics. Then, the re-expressed low-dimension data will be clustered by k-Shape. Our method propose a distance measurement to weigh the similarity of the curves, which is scaling-, shift- and noise invariance. And also a way to extract meaningful centroids is proposed. The experiment testify that:

1) APAA with variable temporal resolution can reduce the data dimensions effectively without affecting the cluster results. It is not only able to improve the calculation speed of the algorithm, but also retain the main shape characteristic.
2) The shape extraction method mentioned in this study can better reflect the tendency of sharp increase or drop of the aggregated curves than the traditional arithmetic average method.
3) The algorithm performs much better than DTW on the calculation time and the cluster validity index.

Acknowledgments

The authors would like to appreciate the support by State Grid Corporation Scientific Research Project(SGSCJY00JHJS201700009) and National Natural Science Foundation of China(51437003).

References

[1] Zhao Wenqing, Gong Yaqiang.Load curve clustering based on Kernel K-means. *Electric Power Automation Equipment*, 2016, **36**(6) p203.

[2] Yi Wang, Qixin Chen, Chongqing Kang,et al. Load profiling and its application to demand response: a review. *Tsinghua Science and Technology*,2015,**20**(2): 117-129.

[3] Wang Chun, Gao Yuanhai. Dynamic Reconfiguration of Distribution Network Based on Optimal Fuzzy C-means Clustering and Improved Chemical Reaction Optimization. *Proceedings of the CSEE*,2014,**34**(10): 1682-1691.

[4] Chen Bocheng, Liang Bing, Zhou Yuebo, et al. An application of SOM neural network in customer classification. *Systems Engineering. Theory& Practice*,2004,**3**(8): 8-14(in Chinese).

[5] H. Ding, G. Trajcevski, P. Scheuermann, X. Wang, and E. Keogh. Querying and mining of time series data: Experimental comparison of representations and distance measures. *PVLDB*, 2008, **1**(2):1542–1552.

[6] Keogh E, Chu S, Hart D, et al. An Online Algorithm for Segmenting Time Series. *Proc of the IEEE International Conference on Data Mining*. San Jose,USA,2001: 289—296

[7] Keogh E, Chakrabarti K, Pazzani M, et al. Dimensionality Reduction for Fast Similarity Search in Large Time Series Databases. *Journal of Knowledge and Information Systems*,2001,**3**(3) : 263—286

[8] Ouyang Tinghui, Zha Xiaoming, Qin Liang, et al. Modeling on selection of the time window for ramp events Prediction. *Proceedings of the CSEE*,2015,**35**(13): 3204-3210.

[9] Wang X, Mueen A, Ding H, et al. Experimental comparison of representation methods and distance measures for time series data. *Journal of Data Mining and Knowledge Discovery*,2013,26: 275-309.

[10] R.L.Thorndike, Who belongs in the family? *Psychometrika,*1953, **18** (4) 267–276.

Distortion Characteristics and Experimental Research on Transformer Windings under Harmonic Condition

Zhongdong Yin[1] and Wensi Wei[1]

[1]State Key Laboratory of Alternate Electrical Power System with Renewable Energy Sources (North China Electric Power University), 102206 Changping District Beijing, China

1920115131@qq.com

Abstract. Firstly, the limitations of the existing transformer harmonic models both at home and abroad are analyzed. Based on this, the influence of the skin effect and the proximity effect on the windings under harmonics conditions is considered. Based on the electromagnetic field principle, the paper analyses the distortion characteristics of winding resistance parameters, introduces the AC power group coefficient and establishes of winding harmonic loss model. Compared with the traditional engineering calculation method, we can see that the theoretical model proposed in this paper can improve the accuracy and make the transformer winding loss calculation more accurate, Therefore, the model has certain guiding value for engineering calculation.

1. Introduction

The operation of power transformers is not only affected by the harmonics of the power grid, but also is one of the most important harmonic sources in the power system due to the nonlinearity of the iron core. Harmonic current flowing through the transformer will make the transformer temperature increases and the total loss increases, which not only will cause damage to the transformer insulation, but also lead to transformer load capacity decline, thus it affects its safety and stable operation. Due to the effective use of transformers for a long time, even very small losses, the annual energy consumption is also huge. The total transformer losses account for about 8% of the total generating capacity, and most of them are transformer harmonic losses. According to a Japanese company reports that when the transformer harmonic current reaches 10% of rated current, the transformer loss than the rated loss of 10%.

Research of China's harmonic loss model of the transformer is less, the main use of conventional IEEE transformer model[1-3]. These models do not deal with the resistance parameters of the transformer windings in the harmonic environment or do not handle well enough and are often calculated by multiplying the DC resistance instead of the AC resistance. This simplification has little effect on the circuit calculation with fundamental only, but under harmonic conditions the equivalent resistance of the transformer windings at different frequencies is much larger than the DC resistance due to skin and proximity effects, making the calculation model inaccurate. In addition, useing improved parameter method about the transformer harmonic loss calculation is easy to implement and more widely, the calculation is relatively simple, but can not well reflect the phase angle between the voltage and current Change, in practice, it can be used as a quick estimate method[4-8].

[1]Based on the Maxwell's equations and the Poynting's theory, this paper considers winding.

Project Supported by China Electric Power Research Institute, Laboratory Open Fund(YD83-17-003)

parameter distortion under harmonic current about skin effect and proximity effect, introduces the AC winding coefficient, and establishes the winding loss calculation model to make the harmonic loss more accurate calculate. In addition the paper sets up the experiment platform, carry out the transformer winding resistance measurement experiment and observe the variation of harmonic resistance with frequency, at last, compared with the calculation model and the DC resistance used to study the equivalent AC resistance model , We can see that the calculation model of harmonic loss in this paper is more accurate.

2. Analysis of Transformer Winding Harmonic Loss

Transformer windings through the exchange of current, if you need to accurately calculate the transformer losses according to the winding layer of wire as a whole analysis considering skin effect and proximity effects. Set the transformer winding radius of r, the axial direction of z, the current direction of flow ϕ, the establishment of coordinate system. as shown in figure 1. Such as i_k, i_ϕ, i_r. k, ϕ, r direction of the current, H_ and H_+ are the magnetic field strength which parallel to the transformer winding inner and outer surface. Assuming that the current flowing through the circular conductor inside the winding satisfies the relationship of the electric field intensity E (ϕ axis direction) and the magnetic field strength H of the winding layer in the cylindrical coordinates[9-11]

$$-\frac{\partial H_z}{\partial r} = \sigma E_\phi \tag{1}$$

$$\frac{1}{r}\frac{\partial}{\partial r}(rE_\phi) = -j\omega\mu H_z \tag{2}$$

Where: σ is the conductivity of copper conductor; μ is the permeability of copper conductor; ω is the angular frequency.

Since the magnetic field strength and electric field strength are all functions of the windings axis distance r at any point on the windings, the solution Bessel function equation is obtained:

$$H(\mathrm{r}) = \mathrm{c}_1 I_0(mr) + c_2 K_0(mr) \tag{3}$$

$$E(\mathrm{r}) = -\mathrm{m}/\gamma \times [\mathrm{c}_1 J_1(mr) - c_2 K_1(mr)] \tag{4}$$

Where J_0 and J_1 are Bessel functions of the first kind, K_0 and K_1 are Bessel functions of the second kind, and the coefficients c_1 and c_2 are determined by the boundary conditions.

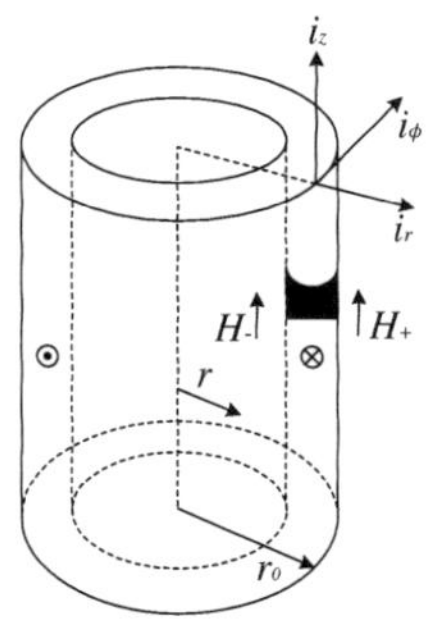

Figure 1. Hollow cylindrical conductor

Figure 1 is the coordinate system of hollow cylindrical conductor. For a layer of winding. Since the electric field strength E and the magnetic field strength H are orthogonal, the energy flux per unit area passing through a unit time is represented by the Poynting vector as E (r) × H (r). From the energy loss point of view, Layer winding unit length of energy:

The total dissipative power of the n_{th} winding is

$$P_n = P_{n_i} + P_{n_o} = \frac{N^2 I^2 m}{\sigma w}\left[(2n^2-2n+1)\coth(md_n) - \frac{n^2-n}{\sinh(md_n)}\left(\sqrt{1+{d_n}/{r_{n_i}}} + \sqrt{1-{d_n}/{r_{n_o}}}\right)\right] \quad (5)$$

Where W is the width of the winding; N is the number of turns of the coil contained in each winding; d_n is the thickness of the nth winding; and r_n is the distance between the nth winding and the winding axis. The use of Taylor expansion of equation (5) to simplify, available n-layer winding loss:

Assuming that the thickness of each layer of the winding is d, the loss of the n_{th}-layer winding can be obtained by using the Taylor expansion of equation (5).

$$P_n = \mathrm{Re}\left\{\frac{N^2 I^2 m}{\sigma w}\left[\coth(md) - 2(n^2-n)\tanh\left(\frac{md}{2}\right)\right]\right\} \quad (6)$$

Assuming that the direct current in the windings is I, the DC power loss per unit length of the n_{th} winding is:

$$F_{hr} = \frac{R_{ac}}{R_{dc}} \quad (7)$$

$$F_{hr} = \mathrm{Re}\left\{K_h(1+\mathrm{j})\left[\coth(K_h(1+\mathrm{j}) + \frac{2(n^2-1)}{3}\tanh\frac{K_h}{2}(1+\mathrm{j})\right]\right\} \quad (8)$$

Where: K_h is the ratio of the thickness d of the winding layer to the skin depth δ of the hth harmonic $K_h = \frac{d}{\delta_h}$, and N is the number of winding layers.

The formula (8) series expansion to simplify, get the first h harmonic harmonic AC resistivity:

$$F_{hr} = 1 + \frac{5n^2-1}{45}K_h^4 \quad (9)$$

The total harmonic loss caused by the transformer winding is:

$$P_{cu} = I_{dc}^2 R_{dc} + R_{dc}\sum_{h=1}^{\infty} F_{hr} I_h^2 \quad (10)$$

3. Experimental Research on Transformer Wingding Harmonic Resistance

In order to effectively carry out the transformer harmonic loss test research. This paper is based on the laboratory of power quality comprehensive experiment research platform. The platform grid simulator can simulate the output voltage harmonics. The tested transformer used in this experiment is SG-150/0.38 dry isolation transformer. Experiment platform is shown in Figure 2.

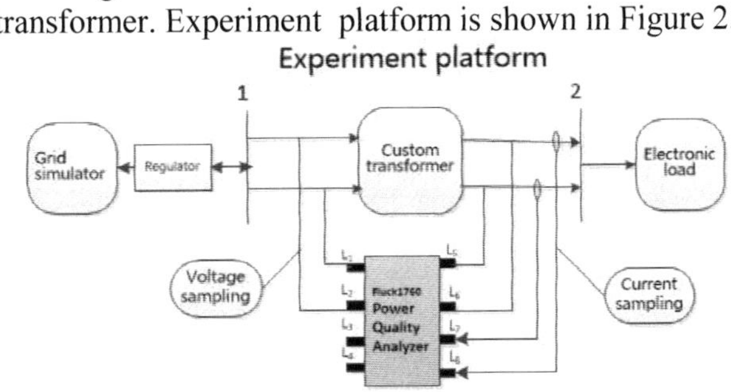

Figure 2. Experiment platform.

Carrying on short circuit test, the transformer side of the grid simulator, issued by the network simulator each harmonic voltage, the secondary short, respectively, measuring the number of different harmonics, different harmonic voltage transformer copper loss. In order to observe the change of the harmonic resistance of the transformer with the change of the frequency of the harmonic current, the harmonic voltages of 1, 3, 5, 7 ... 49 times are respectively applied to the transformer to measure the RMS current flowing through the transformer and the active current flowing through the primary side of the transformer Power, calculate the transformer active power loss and then calculate the harmonic resistance value. It is easy to carry out experiments and as far as possible to avoid errors. In this experiment, take the transformer through the ends of the active power in addition to the square of the current form of harmonic transformer winding resistance calculation. In this paper, the measurements are measured at room temperature 20 ℃, the experimental results are shown in Table 1.

Table 1. Wingding Harmonic Resistance Measuring Results

Frequency	Voltage RMS/V	Current RMS/A	power/W	Measuring resistance/Ω
1	7.380	4.892	32.984	1.379
3	10.35	4.521	30.56	1.489
5	12.266	4.410	28.756	1.479
7	17.3	4.741	40.404	1.794
9	15.67	3，86	33.44	2.3115
11	14.014	3.944	42.61	2.7393
13	9.150	4.305	62.021	3.3465
15	13.46	4.25	72.7	4.025
17	11.632	4.974	124.05	5.014
19	13.270	5.406	169.01	5.796
21	13.262	4.352	131.65	6.958
23	14.50	4.560	169.78	8.165
25	15.052	4.962	232.18	9.43
27	14.79	4.05	173.54	10.58
29	14.712	4.044	189.52	11.845
31	17.652	4.713	314.2	14.145
33	20.345	4.832	351.741	15.065
35	23.044	4.935	408.9074	16.79
37	18.780	4.166	323.334	18.63
39	18.5	4.0	314.824	19.6765
41	16.938	3.375	233.8211	20.5275
43	22.878	4.183	398.4178	22.77
45	23.502	3.994	391.112	24.518
47	24.051	3.904	410.3164	26.9215
49	24.444	3.886	443.184	29.348

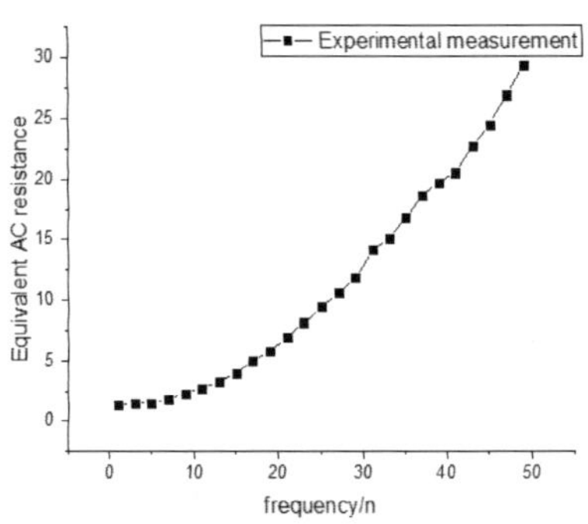

Figure 3. The relationship between harmonic equivalent resistance and frequency

Figure 3 shows the relationship between the equivalent harmonic resistance of the winding and the number of harmonics. It can be seen that the equivalent resistance of the winding increases obviously with the increase of the number of harmonics, showing a positive correlation, indicating that the resistance is affected by the skin Effects and proximity effects and gradually increase.

Table 2 is a partial treatment of the experimental measurements to calculate the AC group coefficient that is the ratio of the value of each harmonic resistance and DC resistance, and use the formula (9) to calculate the harmonic coefficient of each AC group, and compared.

Table 2. Comparison of Experiment and Theoretical Calculation of AC Power Factor

Frequency	Measuring resistance/Ω	ExperimentAC resistance coefficient	Skindepth /mm	TheoryAC resistance coefficient	$\sqrt{n}$
1	1.379	1.02	14.2	1.011	1
3	1.489	1.22	8.2	1.2	1.73205
5	1.479	1.286	6.35	1.276	2.2361
7	1.794	1.56	5.37	1.54	2.6457
9	2.3115	2.01	4.73	1.9	3
11	2.7393	2.382	4.28	2.34	3.3166
13	3.3465	2.91	3.94	2.865	3.6056
15	4.025	3.5	3.67	3.4774	3.873
17	5.014	4.36	3.444	4.2	4.123
19	5.796	5.04	3.26	4.98	4.359
21	6.958	6.05	3.1	5.8664	4.582576
23	8.165	7.1	2.96	6.8545	4.795832
25	9.43	8.2	2.84	7.91	5
27	10.58	9.2	2.733	9.056	5.196152
29	11.845	10.3	2.64	10.252	5.385165
31	14.145	12.3	2.55	11.63	5.567764
33	15.065	13.1	2.472	13.035	5.744563
35	16.79	14.6	2.4	14.546	5.91608
37	18.63	16.2	2.334	16.144	6.082763
39	19.6765	17.11	2.3	17.06	6.244998
41	20.5275	17.85	2.274	17.807	6.403124
43	22.77	19.80	2.22	19.503	6.557439
45	24.518	21.32	2.17	21.268	6.708204
47	26.9215	23.41	2.12	23.25	6.855655
49	29.348	25.52	2.07	25.478	7

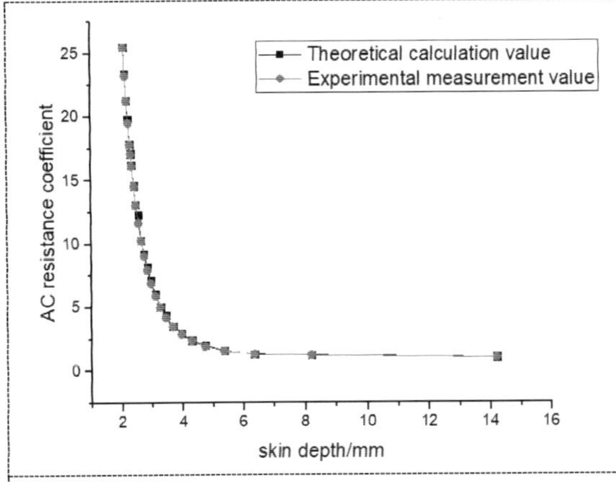

Figure 4. Comparison between the calculated value of the theoretical formula and the experimental measurement value.

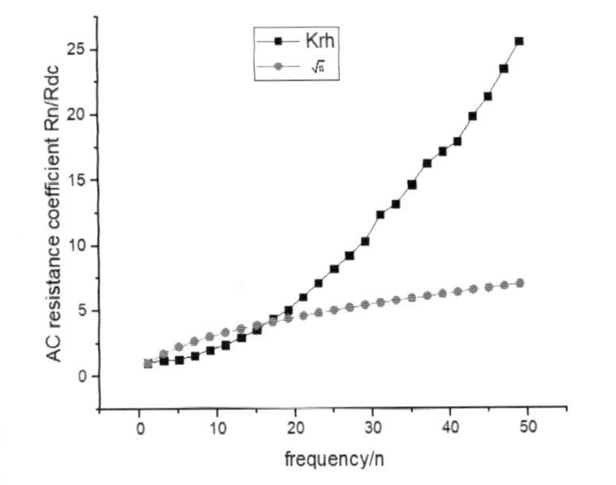

Figure 5. The calculated value of the theoretical formula is compared with the $\sqrt{n}$ times DC resistance value

Figure 4 compares the AC resistivity measured in the experiment with the AC resistivity in the theoretical formula proposed in this paper. As the number of harmonic currents flowing through the transformer windings increases, the harmonic resistance increases more significantly. The figure shows that although the theoretical formulas in this paper are underperforming at low frequencies, the overall trend is more in line with the facts. And at high frequencies, ie, the skin effects and proximity effects become more pronounced, the theoretical calculations proposed in this paper are more accurate.

Figure 5 is calculated by the theoretical formula of the AC power coefficient and the previous study of the DC resistance to the equivalent AC resistance comparison can be seen from the figure, the harmonic less than 20 times, the transformer winding resistance by The effect of skin effect and adjacent effect is not obvious, and the change of resistance value is insignificant, and the difference between the two methods is insignificant. However, when the harmonic frequency is higher than 20 times, as the harmonic frequency increases, the skin is affected by skin effect and The effect of the proximity effect is more pronounced, and the harmonic resistance increases significantly. It can be seen from the figure that the equivalent resistivity of thc theoretical calculation model is more accurate than the equivalent DC resistance is more reasonable.

4. Conclusion

Based on the Maxwell's equations and the Poynting's theorem, this paper considers the winding parameter distortion characteristics under harmonic currents and the skin effect and the proximity effect, introduces the AC winding coefficient and the consumption and builds the winding loss calculation model, Calculate. And set up the experimental platform, the harmonic current in the background of the transformer winding resistance measurement experiment, observation of harmonic resistance with frequency variation, compared with the calculation model and the model has been studied, we can see the harmonic loss calculation model more accurate This model has certain guiding value for engineering calculation.

Acknowledgments

Project Supported by China Electric Power Research Institute, Laboratory Open Fund(YD83-17-003)

References

[1] GAO Jie. Journal World. Comparison and Simulation of Transformer Harmonic Loss Calculation Methods . 14 (14): 133-134.(2016)

[2] Liu Shuming, Shi Hong, Feng Lei. Electric Power Automation Equipment. A model of transformer harmonic loss considering skin effect and proximity effect . 35 (3): 133-139.(2015)

[3] CAI Guowei, KONG Lingguo, PAN Chao, WANG Jiyuan, YANG Deyou. Power System Technology. Analysis of Harmonic Loss of Transformer Based on Frequency-varying Characteristics . 5 (11): 120-124.(2011)

[4] Li Qionglin, Zou Lei, Liu Huijin, Cui Xue, Zhang Zhenan, Fei Shangbei. Power System Technology, Calculation and Experimental Study of Harmonic Losses in Power Transformer. 37 (12): 3521-3527.(2013)

[5] Ding Yang, Xu Changbao, Li Huaiyu, Feng Cheng, Feng Donghan. Electric Energy and Energy Efficiency Technology, Computing Harmonic Loss of Transformer Considering Skin Effect. (23): 63-68. (2015)

[6] Zhao Jie. Electronic Technology and Software Engineering, Analysis of Harmonic Losses Based on Transformer and Influencing Factors. (23): 128. (2015)

[7] Jia Jingran, Duan Xiaobo, Lu Jinling, Zhou Wen. Hebei Electric Power Technology, A Review on the Calculation Methods of Transformer Harmonic Loss. 34 (04): 37-41. (2015)

[8] Xie Bingruo, Chen Qiefu, Kang Chonghao, et al. Proceedings of the CSEE, Modeling and impedance parameter design of multi-winding transformer based on combined field-circuit coupling method. 29 (9): 104-111.(2009)

[9] Makram E F, Thompson R L, Girgis A A. IEEE Trans on Power Systems, A new laboratory experiment for transformer modeling in the presence of harmonic distortion using a computer controlled harmonic generator. (1988)
[10] Wang X, Cheng Z, Lin L, et al. Compel International Journal of Computations & Mathematics in Electrical, Calculation and validation of iron loss in laminated core of power and distribution transformers. 33(1/2):137-146. (2014)
[11] Guo J, Li L. Electric Power Science & Engineering. Calculation and Measurement of Eddy Current and loss in Magnetic Shielding Model of Power Transformer. 2012.

IOP Conf. Series: Materials Science and Engineering **366** (2018) 012080 doi:10.1088/1757-899X/366/1/012080

DIGITAL PULSED SYSTEM OF SHADOW IMAGES RECORDING FOR RADIOGRAPHY OF OBJECTS OF LARGE OPTICAL DENSITY

K.I.Almazova, A.N.Belonogov, V.V.Borovkov, E.V.Gorelov, I.V.Morozov, V.D.Selemir and O.A.Shamro

FSUE "RFNC-VNIIEF", 37 Mira Ave., Sarov, Nizhniy Novgorod reg., Russia

E-mail: mailbox@ntc.vniief.ru

Abstract. The authors provide the development results of one channel of the digital multiframe recording system for the radiography of objects having large optical thicknesses. The recorder is created based on the modern CCD matrix and quick-acting unit that operates as an electron-optical shutter. Scintillators based on the monocrystal LYSO were used to transform the ionizing radiation into the visible spectrum. Spatio-time characteristics of the recording system were investigated in the laboratory; the system operability was also checked. It was shown that the recording system allows recording with time exposure $\geq$ 100 ns. The unit of the multiframe recording system was tested on the internal polygon of FSUE "RFNC-VNIIEF". Testing powering of the mobile cyclic accelerator (ironless pulsed betatron) was also performed. The X-ray patterns of the static test objects were obtained. The thickness of the X-rayed lead test object was ~90 mm at 4 m from the target at 11% contrast. The value of a capacitive storage of the pulsed powering system of the betatron electromagnet was C = 900 μF. The dimension of the tantalum target was 6x6 mm; the full width of the output gamma pulse at half maximum in a single frame mode was ~100 ns.

1. Introduction

Investigation of quick-acting processes such as forming and development of the cumulative jets, study of the state of substance in the conditions of the dynamic loading are very important and timely directions of the modern natural science. Characteristic velocities of these processes are from one up to several tens of km/s [1]. Pulsed radiography is one of the most informative techniques of such investigations.

Powerful sources of the ionizing radiation have been developed in FSUE "RFNC-VNIIEF" for many years. They allow performing the radiography of the dynamic objects with large optical thicknesses. The sources of such radiation are linear or cyclic electrons accelerators, particularly betatrons [2]. Such sources are large stationary complexes with complicated infrastructure. To reduce expenses and optimize the process of preparation and gas-dynamic experimentation, FSUE "RFNC-VNIIEF" began to apply the conception of using mobile radiographic complexes (MRC) [3]. They consist of mobile cyclic accelerators (MCA), an explosion-proof chamber (EPC) with a test object, system of X-rays collimation and shadow images recording system. Each of these sources allows getting from one up to three images of the test object with an interval from 0.5 up to 20 μs in an experiment.

Figure 1 presents one of the variants of the gas-dynamic experiment that allows obtaining up to three images in an experiment.

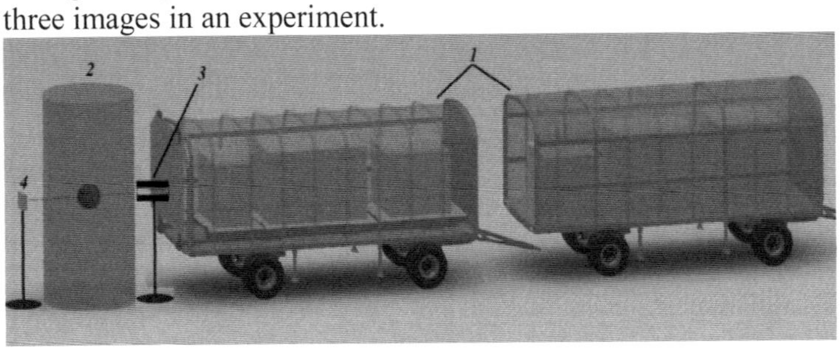

1 – mobile cyclic accelerator;
2 – EPC with a test object;
3 – X-rays collimation system;
4 – shadow images recording system

Figure 1. The experiment sketch with the usage of a single-beam three-frame mobile radiographic complex.

It is possible to increase the frames quantity and provide the small-angle tomography if to increase the MCA quantity [4].

The recording system is an important part of any radiographic facility. The employees of FSUE "RFNC-VNIIEF" develop multiframe electron-optical recording systems to equip protongraphic and radiographic complexes. The usage of modern electron-optical devices together with the effective scintillation converters allows creating multiframe recording of the dynamic objects images. Hereby, it increases the information volume during the experiment.

2. Shadow images recording system as a part of the radiographic complex

An X-ray sensitive film has been applied in the radiography as a recorder for more than sixty years. Currently, luminescent storage screens (Imaging Plate) are also widely used. They have got improved characteristics and better velocity of the data extraction. Such recorders have got a high resolution and allow getting multiframe shooting of the dynamic process in some tasks. However, a complex internal structure and dimension of the test object often cannot allow taking several frames using one and the same X-ray film or luminescent screen. The usage of modern electron-optical devices together with effective scintillation converters allows solving the tasks of the multiframe recording.

Nowadays the sphere of digital recorders is being developed very well and the choice of digital optical cameras is very wide. However, the experiment conditions at the radiographic complexes provide strict requirements for the used devices. As a rule, in the experiments soft images are created in optically dense objects. The radiography is also characterized by a limited number of data light quanta. Thus, the digital recording unit should have an electron-optical shutter of the required exposure, image intensifier and high spatial resolution. As a rule, a CCD chamber is joined to the imaging intensifier that operates as an electron-optical shutter and image intensifier.

According to the task requirements the authors created one channel of the multiframe recording system; its photo is presented in Figure 2.

1 – digital camera,
2 – remote control unit of a diaphragm and focus,
3 –fast lens.

Figure 2. A channel of the multiframe recording system of the radiographic complex.

The recording channel consists of a gating digital camera, remote control unit of a diaphragm and focus, and fast lens [5].

The digital camera consists of an imaging intensifier with a self-made generator of gate pulses and TV CCD chamber. The components are joined by means of a fiber optic washer. A microchannel plate (MCP) is used as an amplifier that can amplify the signal 10^3 times. The CCD chamber is used for image sensing. The matrix dimension of the CCD chamber is 16.6 × 18.6 mm, the pixel dimension is 16 × 16 μm. The digital camera is totally controlled by the personal computer by means of fiber optic interface and can operate both in the trigger mode and in the mode of internal synchronization.

In the created sample the exposure time of the registered image is varied in the diapason from 100 ns up to 5 μs by means of continuous adjustment. It is achieved by a noise-immune generator developed by the authors during this work. It should be noted that the provided normal generator of gate pulses had not got the claimed technical characteristics and had very weak magnetic pickup protection. At the generator development the inductive elements as well as digital and analog integrated circuits were excluded as much as possible from the electric circuit. The discrete semiconductor element circuit was used. The optimal design arrangement and electric circuit screening were performed. The generator provides the MCP gain control and a gate pulse formation. The generator control is performed from the remote panel by means of fiber optical lines with timing-tracking accuracy up to 1 ns. It also provides undisturbed operation of the recording system while measuring in the conditions of a higher level of the magnetic pickup. There is a photo of the gate pulse generator in Figure 3 (*a*) and a characteristic oscillogram of its output pulse (*b*).

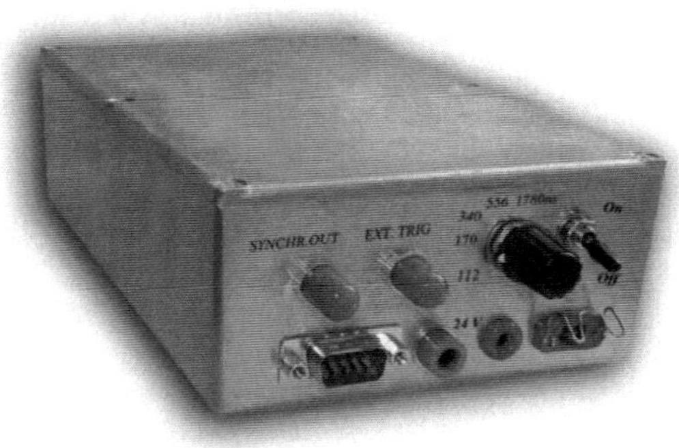

a) Photo of a gate pulse generator.

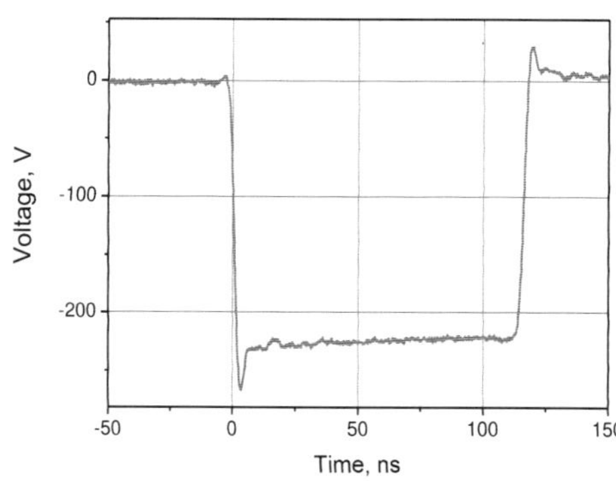

b) Oscillogram of an output pulse of the gate pulse generator.

Figure 3. Photo and oscillogram of an output pulse of the gate pulse generator.

The gate pulse generator with the obtained characteristics excels the present analogs in several parameters and can be effectively applied in the images recording systems based on the electron-optical devices. The generator usage provides new possibilities for data gathering at the dynamic objects radiography and imaging of fast processes in the conditions of electromagnetic disturbance and scattered X-ray radiation during the experiments at the physical facilities.

As component units are easily accessible and rather cheap, it is possible to use them in the multichannel digital systems of the images recording of the dynamic objects with minimal expenses.

The main characteristics of the recording channel are:

- quantum efficiency of imaging intensifier – at least 20%;
- spectral sensitivity from 350 up to 650 nm (sensitivity maximum is equal to the radiation spectrum of the used scintillator);
- spatial resolution of the photocathode – at least 25 lp/mm;
- number of pixels of the CCD matrix – > 10^6;
- dynamic diapason of the CCD chamber – 14 bit;

- amplifier diapason – $\leq 10^3$ times;
- exposure time – $\geq$100 ns;
- response accuracy – $\leq$ 5 ns;
- built-in high-speed series fiber-optic interface for synchronizing and system control, and data transmission into the computer.

The assembly and adjustment of the electron-optical recorder were performed in the laboratory. Spatial-time characteristics of the system were studied. It was defined that 25th field is resolved with at least 16% contrast with the use of an optical test pattern № 4 that has a limiting spatial frequency 25 lp/mm. The test pattern recording was fulfilled in the scale 1:1. There is a test pattern image (Fig. 4*a*), and its expanded 25th field (fig.4*b*).

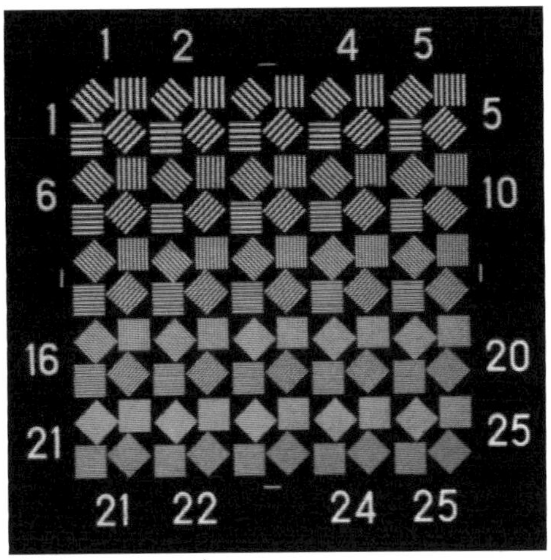

a) Test pattern image.

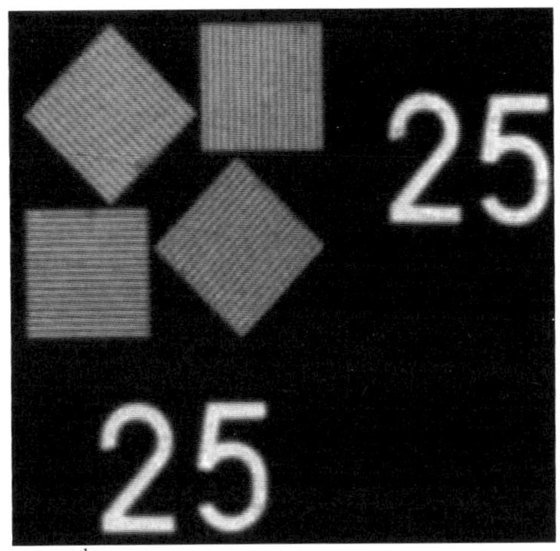

b) 25th test pattern field.

Figure 4. Study of the spatial characteristics of the recording channel.

The optical diagnostic system based on the pulsed laser was used to check time characteristics of the recording channel. As opposed to the technique based on the electric signal control on the photocathode imaging intensifier, this diagnostic system allows obtaining correct visual information. The pulse duration of the probe radiation is less than 10 ns. The typical dynamics of the recording channel operation is presented in Figure 5.

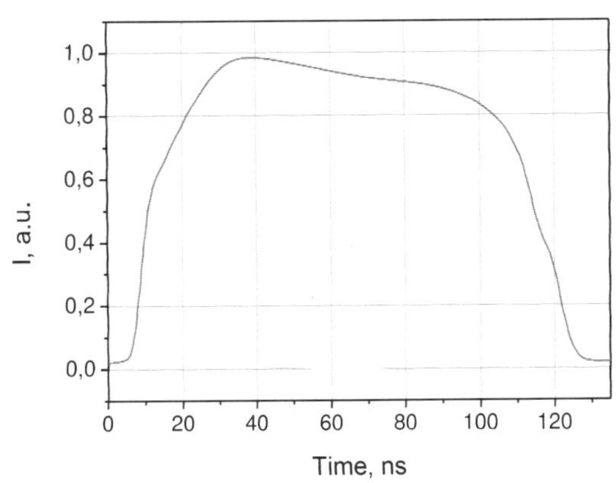

Figure 5. Dynamics of the recording channel operation.

The essential part of any recording system of the radiographic complex is the existence of an X-ray sensitive element, that is a converter of the X-ray radiation into the light one. In the presented recording system a scintillation crystal LYSO is chosen as a converter (lutetia – yttrium orthosilicate). This crystal has got high effective atom number (Z > 60) and density (ρ ~ 7 g/sm^3), that allows providing the required recording of high energy particles. The light output is in the frames from 27 000 photons/MeV up to 32 000 photons/MeV, the fall time constant is not more than 47 ns. At the same time, the crystal is not hygroscopic and is easily treated up to the thickness of 0.3 mm and even less. The distinguished properties make it rather attractive in the tasks of the pulsed radiography; with the use of this crystal it is possible to achieve rather high time resolution, required sensitivity and high spatial resolution of 2-D images.

Scintillation crystals of 78 mm diameter were used during the testing of the recording system. The scintillator photo is presented in Figure 6.

Density (g/sm^2)	7,1
Hygroscopicity	0
Absence of internal defects of the diameter more than (mm)	0,1
Radiation resistance (Grey)	10^6
Refractive index	1.81
Radiation spectrum maximum (nm)	420
Fall time (ns)	41
Specific light yield (photons/MeV)	32 000

Figure 6. Scintillation monocrystal LYSO Ø78 mm.

Preliminary, a recording channel was studied on the test bench in the laboratory aimed to compare the quality of images that are obtained at the usage of different lens and scintillation converters. The growth of noise immunity of the electron circuits in the conditions of strong magnetic pickup had to be also studied. The obtained results showed that it is possible to increase the coefficient of the optical system acquisition in the conditions of low light flux if to use faster lens; and hereby, the ratio signal/noise of the recording system will be also increased.

3. Testing of the recording system and its results.

The first full-scale test of the developed recording system was carried out during the testing powering of the mobile cyclic accelerator in September 2016. The value of the capacitive storage of the pulsed powering system of the electromagnet was C=300 μF.

The MCA BIM was moved to the internal polygon of FSUE "RFNC-VNIIEF" in June 2017. The testing powering was carried out. The value of the capacitive storage of the pulsed powering system of the electromagnet was C=900 μF. The dimension of the tantalum target was 6×6 mm. The full width of the output gamma pulse at half maximum in a single frame mode was ~120 ns. The recording was carried out with the use of the Imaging Plate system as well as with the use of the digital recording system. The photo of the recording channel and structure chart of the X-ray picture creation are presented in Fig.7.

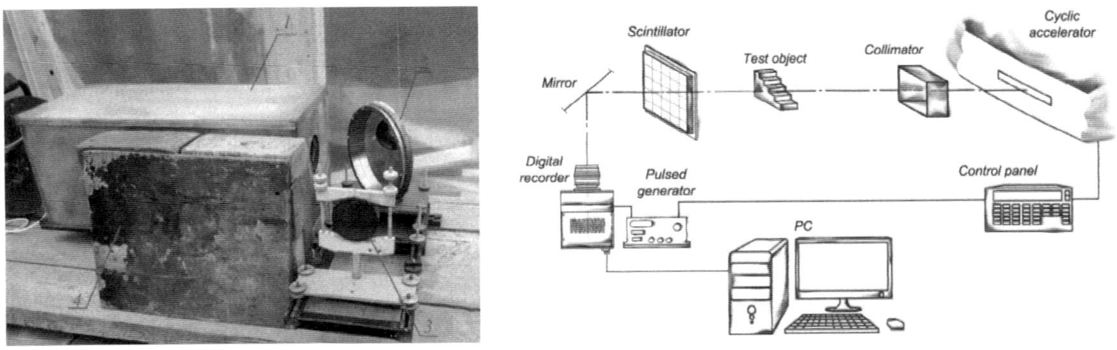

a) Photo of the recording channel *b*) Structure chart

Figure 7. Photo of the recording channel (*a*): 1 – recording channel in the screening can, 2 – sweep mirror, 3 – scintillator, 4 – ionizing radiation protection; structure chart of the test performance (*b*).

To estimate the parameters of the MCA BIM and recording system operation, the X-ray radiography of some test objects was carried out. The test objects were a lead step wedge and typesetting tantalum test pattern. The photos and sketches of these objects are presented in Figure 8. These objects allow obtaining a limiting spatial resolution and defining the X-ray ability, contrast sensitivity, and total system operability in general.

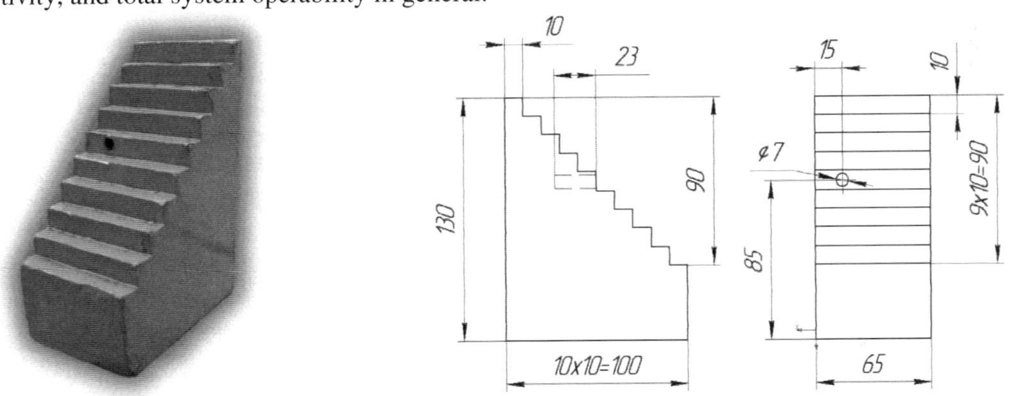

a) A lead wedge.

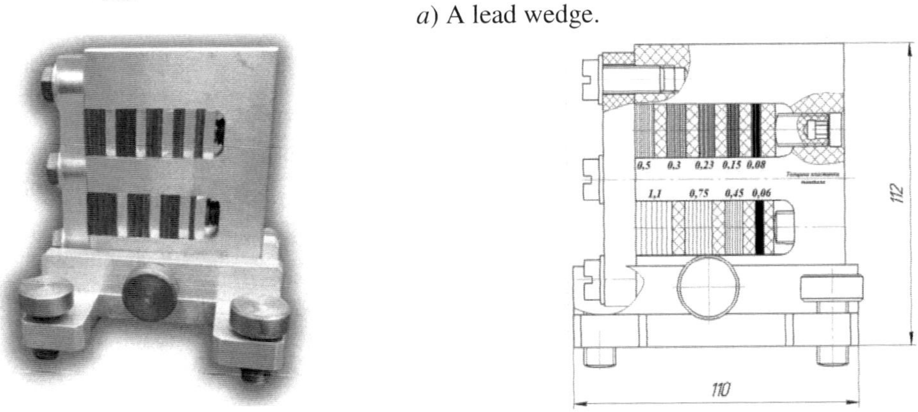

b) A typesetting test pattern.

Figure 8. Photos and sketches of the test-objects.

The recording of shadow images was carried out in the following geometry: the distance from the betatron target to the investigated test object was $L_1 \approx 400$ cm, from the object to the scintillator was

$L_2 \approx 50$ cm. Hereby, the geometrical factor was $K = 1.125$ and the optical system magnification was M = 0.189 (or 5.3).

The lead step wedge (Fig. 8*a*) was used to define a limiting the X-ray ability. The lead plate of 20 mm thickness was fixed before the test object to increase the optical thickness. The wedge is surrounded by the collimator made of the lead bricks. Figure 9 presents an X-ray picture and signal intensity profile along the vertical section of the wedge.

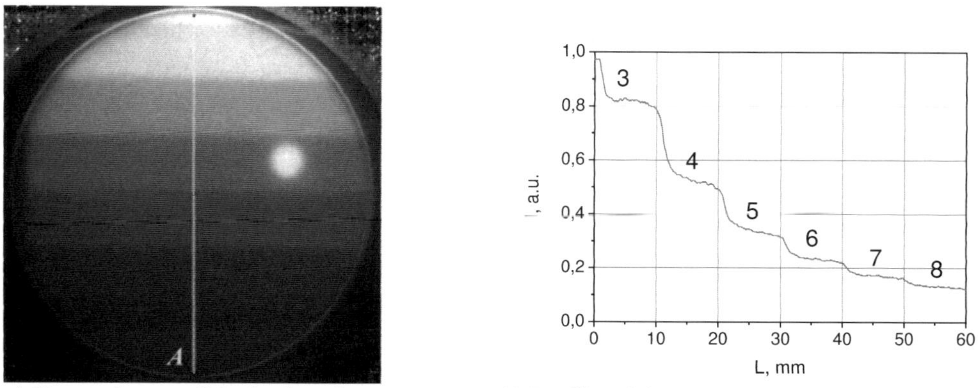

a) X-ray picture of the lead wedge. *b*) Profile of the signal intensity along A-A

Figure 9. Results of the data processing of the X-ray picture of the lead wedge.

While creating the intensity profile in the section A-A, the steps from the second one up to the eighth one were appearing. The seventh step relative to the eighth one is resolved with 11 % contrast. The thickness (along the beam) of the seventh step was 70 mm, 80 mm of the sixth step. Thus, with these parameters of the mobile complex at the usage of the proposed digital recorder ~ 90 mm of the lead was X-rayed with 11% contrast at 4 m from the target with a capacitive storage equal to C = 900 μF. It was done with the use of the fixed lead plate of 20 mm thickness.

A typesetting tantalum test pattern was used to perform the assessment of the radiographic complex resolution together with a recording channel. The photo of the test pattern is presented in Figure 8*b*. The X-ray picture of the test pattern and the densitogram of the field with the resolution of 0.5 lp/mm are provided in Fig. 10.

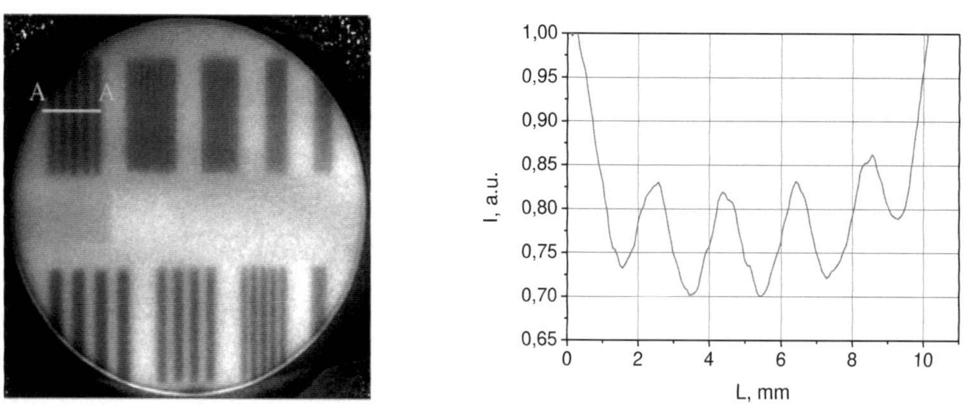

a) X-ray picture of the lead test pattern. b) Profile of the signal intensity along the section A-A.

Figure 10. Results of the data processing of the X-ray picture of the lead test pattern.

Analysis of the profile of the section A-A reveals that the test pattern area with the spatial frequency of 0.5 lp/mm is resolved with a ~8% contrast; and the area with less spatial frequency (0.9 lp/mm) is resolved with a contrast less than 3%.

According to the provided X-ray pictures and diagrams, it is evident that the boundaries of the thicknesses transition have got a blurring. The easiest explanation is not an exact positioning of a beam. One more reason can be the influence of the scattered radiation. It can be also mentioned that the fixed target in the betatron had not got optimal geometric dimensions (6×6 mm). Thus, it could lead to the additional blurring of the X-ray pictures.

In general, the experiments results are considered to be positive. The developed recording channel showed its operability and can be considered as a base of the multiframe recording system for small-angle X-ray radiography. The proposed recording channel of the multiframe recording system allows shooting of the dynamic objects in the multi-angle and multiframe mode. The experiments results showed a good correspondence with the previous images obtained on the radiographic film and can be compared with the results obtained in the Imaging Plate system.

4. Conclusion

The efficiency of the multi-angle radiography can be essentially increased using the multiframe recording system (at the operation of one or several betatrons in the mode of several pulses). In the frames of this task the authors developed the recording channel based on the digital camera and scintillation converter. The generator of gate pulses was created to control an electron-optical camera. This channel was previously tested on the test-bench and was used during the MCA BIM adjustment, and during the testing powering of the mobile X-ray radiographic complex on the internal polygon of FSUE "RFNC-VNIIEF" as well. The static test-objects (a lead wedge and typesetting tantalum test pattern) were used to estimate a limiting spatial resolution, transparent ability and contrast sensitivity.

At the usage of the digital recorder, ~ 90 mm of the lead was X-rayed with 11% contrast at 4 m from the tantalum target of the betatron. The value of a capacitive storage of the pulsed powering system of the betatron electromagnet was C = 900 μF. The dimension of the tantalum target was 6×6 mm; the full width of the output gamma pulse at half maximum in a single frame mode was ~100 ns.

In spite of the fact that the own resolution of the recording channel was at least 25 lp./mm, the resolution of the radiographic complex together with the recording channel was 0.5 lp./mm. in the performed experiment. Low index of the spatial resolution is explained by several reasons: non-optimized dimensions of the betatron target (6×6 mm) that lead to the additional blurring of the obtained shadow images, absence of the precise alignment platform for test objects.

In general, the results of the optical-dense objects investigation should be considered positive. The proposed digital recording channel showed its operability and can be used as a base at the creation of the multiframe recording system for small- angle X-ray radiography. At the same time the authors would like to note that they are always in the process of searching new fast acting electron optical devices and charge-coupled devices to improve the recording system and quality of experiment results.

References

[1] N.I. Egorov, G.V. Boriskov, A.I. Bykov, Yu.P. Kuropatkin, N.B. Lukynov, V.D. Mironenko, V.N. Pavlov, «Use of Pulsed Radiography for Investigation of Equation of State of Substances at Megabar Pressures», Contributions to Plasma Physics, 2011,vol.51,№4,p.p.333-338

[2] Mikhailov A.L., "Nonperturbative diagnostic techniques of fast processes", Sarov, 2015, pp.113-114 (in Russian).

[3] Pat. 2548585 C1 RU MPK G03B 42/02. "Mobile radiographic complex and radiation source of betatron type for radiographic complex" / Zenkov D.I., Kuropatkin Yu.P., Nizhegorodtsev V.I., Selemir V.D., Shamro O.A. // БИ. 2015, №11.

[4] S G Kozlov, Yu P Kuropatkin, V I Nizhegorodtsev, K V Savchenko, V D Selemir, E V Urlin and O A Shamro, « Mobile x-ray complex based on ironless pulsed betatrons. X-ray complex conception for small-angle tomography», IOP Conference Serials: Materials Scince and Engineeringe,199 (2017).

[5] Pat. for useful model RF 106819 K.I. Almazova, A.N. Belonogov, A.H., V.V. Borovkov, A.A. Volkov // БИ. 2011. №20.

IOP Conf. Series: Materials Science and Engineering **366** (2018) 012081 doi:10.1088/1757-899X/366/1/012081

Measuring Method of the Internal Temperature of UHV DC Voltage Measuring Device

XIE Tingting, YANG Zhongzhou, FENG Jianhua, ZHENG Wei and WANG Shan

XI'AN High Voltage Apparatus Research Institute, Xi 'an, Shaanxi, China 710077

yangzhongzhou@xihari.com

Abstract. For many years, it has been difficult to measure the internal temperatures of gas insulated equipments in high voltage apparatuses. This paper presents a method of measuring the internal temperature of UHV DC Voltage measuring device, which uses the temperature sensor of fluorescent decay time optical fiber transmission as the instrument. The method could accurately measure the internal temperature when the device is running. The paper mostly expounds the selecting temperature sensor and arrangement mode of sensor and detailed test method. The test method could determine the internal temperature of UHV DC voltage measuring device, as the same time could also provide references for measuring the internal temperatures of gas insulated equipments in high voltage apparatuses.

1. Introduction

UHV DC voltage measuring device is the key equipment in the UHV DC transmission system, which is mainly used for measuring the DC voltage of the bus bar and providing the voltage signal to the measuring instruments and secondary protection devices [1-3]. The main parameters of the UHV DC measuring device used in the project are nominal voltage of ±800kV, operating current of 2mA and rated power of 1600W. A prototype of ±800kV UHV DC voltage measuring device designed by a company is shown in figure 1. The prototype is mainly based on the principle of RC divider in parallel. SF_6 gas is used for insulation, the second output is digital, and the product height is about 10m.The DC voltage measuring accuracy meets the requirement of 0.2 %.

Transmission lines of UHV DC projects in China are long and span several provinces and cities. The internal temperature of the device during operation may vary greatly by region and seasonality [4-5]. The DC voltage measuring device designed based on the principle of resistance-capacitance partial pressure has a great relationship with the applied voltage and temperature. The influence of temperature changes on the measuring accuracy during the operation of the device cannot be ignored. Therefore, the relevant standard explicitly states that the measuring voltage divider resistance of the DC voltage measuring device should have sufficient thermal stability so as to ensure that the measuring accuracy of the device meets the design requirements within the specified temperature range [6]. The working current of ±800kV UHV DC voltage measuring device during normal operation is 2mA and the heating power is 1600W. The important part in the design and manufacture is to select the appropriate insulation medium and adopt reasonable internal structure so that the device can meet the requirements of insulation and effective cooling. And the device operates at different ambient temperatures, both to meet the requirements of measuring accuracy and long-term reliable operation [7-8]. To meet the above requirements, it is necessary to know the full temperature range of the device during operation. A method of measuring the internal temperature of UHV DC voltage measuring device is provided in the article, which can accurately obtain the internal temperature rise during operation of the device. If the internal temperature rise during operation of the device is measured, the full range of temperatures corresponding to the device operating in different regions and in different seasons can be calculated. In the manufacture of products by more appropriate selection of appropriate measuring devices to ensure that the measuring accuracy of the device meets the requirements within the full temperature, meanwhile reasonable control of manufacturing costs.

Figure 1. ±800kV UHV DC voltage measuring device.

2. A method of measuring the internal temperature of UHV DC voltage measuring device

2.1. The temperature sensor of fluorescent decay time optical fiber transmission

Inflatable sealed insulating structures are used in UHV DC voltage measuring devices and a built-in temperature sensor is used to measure the internal temperature of the device. The prerequisite for the implementation of this method is that the built-in temperature sensor should not affect the insulation of the device or affect the voltage division ratio of the device. Conventional platinum resistance and wireless temperature sensors cannot meet this requirement. The method in this paper is to select the temperature sensor of fluorescent decay time optical fiber transmission (hereinafter referred to as fluorescence temperature sensor) [9], which has the characteristics of good insulation, high voltage resistance, corrosion resistance and high temperature resistance, and can be used under the harsh conditions of strong electromagnetic. Fluorescent temperature sensor used to measure the temperature of the fluorescence probe is a fluorescent material temperature sensing material, and the material will emit fluorescence while excited by the excitation light. The length of the fluorescence lifetime is related to the temperature (the fluorescence decay time is called the fluorescence lifetime). The temperature measurement can be achieved by the fluorescence lifetime characteristics [10-13]. The biggest advantage of using the fluorescence lifetime to achieve temperature measurement is that the temperature of the target to be measured depends only on the time-constant characteristics of the fluorescent material regardless of other variables in the system, such as changes in the intensity of the laser light source, transmission loss, optical coupling efficiency, etc. This method has certain advantages over other temperature measurement methods [14].

Fluorescent temperature sensor temperature measurement principle is shown in figure 2. The microcontroller drives the optical signal processing system and emits a bunch of excitation light, which is transmitted to the fluorescence probe through the optical fiber. When the excitation light stops transmitting, the probe will emit fluorescence at this time, and the fluorescence will be returned to the optical signal processing system through the optical fiber. The processed system obtains and processes the data to calculate the temperature value.

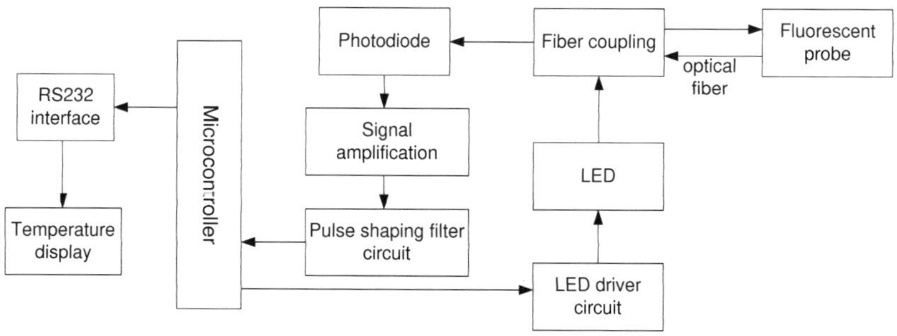

Figure 2. Temperature measurement principle of the optical fiber temperature sensor.

2.2. Implementation scheme

For descriptive convenience, a ±800kV UHV DC voltage measuring device is equivalent to a device consisting of a high voltage insulator and an internal first and second measuring voltage element. Insulation gas is filled between the high-voltage insulator and the internal measuring voltage element. Voltage measurement components are connected by the upper, middle and lower voltage flange connected in series. Fluorescent probes of the built-in fluorescent temperature sensors which are used to measure the temperature rise inside the device are attached to the three connecting voltage equalized flange surfaces. The top and bottom of the high-voltage insulator are fitted with the upper and lower flanges, and the upper flange is fitted with three (or more) optical connectors that are connected to their respective optical connectors by optical fibers. The optical fiber connector and the external optical signal processing unit are connected through their optical fibers, and the schematic diagram is shown in figure 3.

The height of ±800kV UHV DC voltage measuring device is about 10m. The following issues are focused on solving in the implementation scheme.

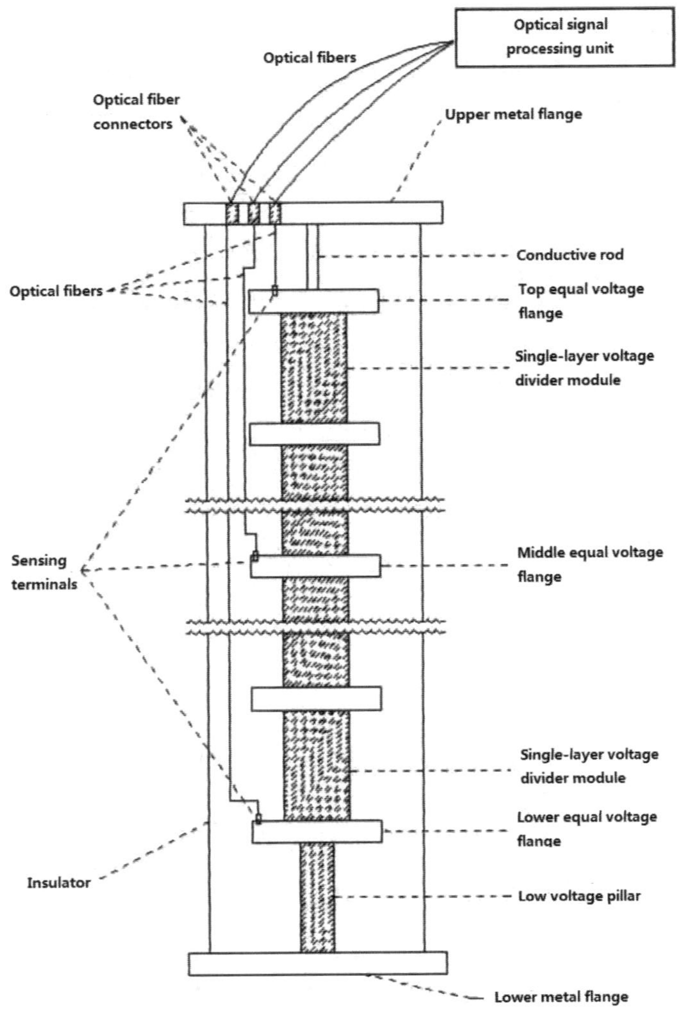

Figure 3. Implementation scheme of detecting the internal temperature rise of UHV DC voltage measuring device.

2.2.1. The laying of temperature fluorescence probes and optical fibers inside the device. The rated operating voltage of the UHV DC voltage measuring device is ±800 kV. The internal and external insulation of the device shall not be affected by the installation of the fluorescent temperature sensors. Prior to the implementation of the program, the selection of the optical fiber pressure-proof and the placement of the fluorescent probe must be done carefully to ensure that the buried probe and the optical fiber have no influence on the measurement and insulation characteristics of the device and can detect the temperature in different locations.

2.2.2. Tightness during the test. The internal components of the ±800kV UHV DC voltage measuring device operate in an SF6 gas-tight, hermetically sealed environment at a pressure of 0.35MPa. Therefore, one of the key issues that should be considered is how the temperature information of the internal temperature of the device detected by the fluorescent temperature sensor is transmitted to the outside through the gas inside the sealed device without affecting the gas-tightness of the device. The optical fiber connector selected during the test is shown in figure 4, and the sealing flange of the device with a plurality of optical fiber connectors designed according to the experiment is shown in figure 5. The device must be tested for air tightness before the test, requiring the optical connector and the sealing flange to meet the airtight requirements of the device during the internal temperature test.

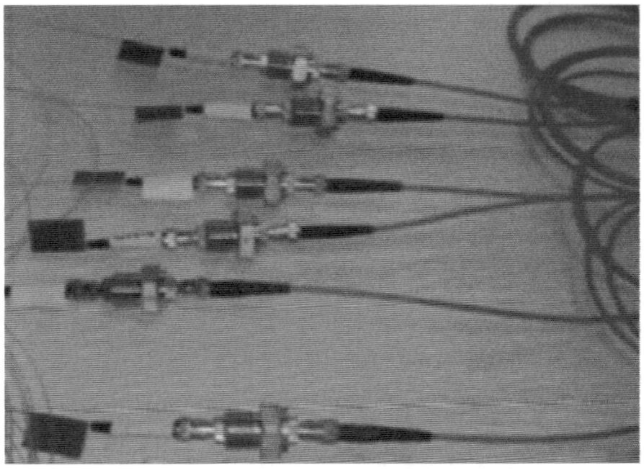

Figure 4. Optical fiber connector.

Figure 5. Optical fiber connector and sealing flange.

2.2.3. Installation and debugging. After the fluorescent probe and the optical fiber are installed inside the device, a sealing flange is arranged on the upper flange of the device and the internal optical fiber is connected with the internal optical interface of the optical fiber connector. The upper flange is sealed and assembled to ensure the device is airtight. The external optical interface of the optical fiber connector is coupled with the optical signal processing unit through the communication optical fiber. Site commissioning is carried out in the high voltage insulation hall of XI'AN High Voltage Apparatus Research Institute (XIHARI). The installation site is shown figure 6.

Figure 6. Debugging and installation.

3. Verification test

In order to verify this method, a test was conducted to measure the internal temperature of the device. The verification test was carried out in the high voltage insulation hall of XIHARI. The test system of ±1200kV DC voltage was used to apply the test voltage in the lab. Straight bus bars were used as high-voltage connections during the test to avoid corona generation during the test. In order to ensure the validity of the temperature measurement data, a fluorescence temperature sensor with a calibration certificate was used in the test. The rated voltage of +800kV was applied during the test. The application time was from the moment when the device prototype was applied until the device prototype reached thermal equilibrium. The criterion for the device prototype to reach thermal equilibrium conditions is that the temperature change within the device during the voltage application does not exceed ±0.5°C per hour.

After about 12 hours of +800 kV voltage test, the device reached the specified thermal equilibrium, and the embedding temperature data was obtained in table 1. Table 1 shows the internal temperature distribution of the unit and shows that the maximum internal temperature rise of the unit is about 29K.

Table 1. Temperature data of measuring points/°C.

Buried point No.1of upper metal flange	Buried point No.2of upper metal flange	Buried point No.3of middle metal flange	Buried point No.4of middle metal flange	Buried point No.5of lower metal flange	Buried point No.6of lower metal flange	Ambient tempera-ture
42.4	42.3	36.9	36.6	33.2	33.1	13.5

4. Conclusion

Through the above verification test, the internal temperature distribution of the device under rated voltage operation and the maximum temperature rise inside the device are obtained. From the experimental data, it can be deduced that the device operates in the operating temperature range of hot and cold regions in China, and provides the basis for selecting the measuring device suitable for the operating temperature range [15-19]. The proposed method of temperature measurement provides an intuitive view of the internal temperature of the unit during operation and also provides a means of measuring the internal temperature of gas-insulated products in high voltage applications.

References

[1] LIU Xing. Design and implementation of DC high voltage divider [J]. Metrology & Measurement Technology, 2008, 28(1): 17-19.
[2] LIU Bin, HUANG De-xiang. Thermal Design for High Precision DC Resistance Divider [J]. High Voltage Apparatus, 2011, 47(2): 31-33, 38.
[3] ZHANG Renyu, CHEN Changyu, WANG Changchang. High voltage testing technology [M]. Beijing: Tsinghua University Press, 2003.
[4] LIU Zhenya, SHU Yinbiao, ZHANG Wenliang. Study on voltage class series for HVDC

transmission system [J]. Proceedings of the CSEE, 2008, 28(10): 1-8.
[5] GAO Ximing, ZHANG Peng, HE Zhi. Protection behavior analysis of DC transmission line [J]. Automation of Electric Power Systems, 2005, 29(14): 96-99.
[6] LI Dengyun, LI Qian, LI He, et al. Calculating method for thermal equilibrium temperature of traditional DC TV [J]. High Voltage Engineering, 2010, 36(4): 994-999.
[7] YU Jianzu, GAO Hongxia, XIE Yongqi. Electronic equipment thermal design and analysis [M]. Beijing: Higher Education Press, 2002.
[8] ZHOU Yanming. Studies of optical fiber thermometer based on fluorescence excited by ultraviolet [D]. Changsha: Hunan University, 2004.
[9] LI Xin, ZHANG Liguo, WANG Yanling. Bio-medical temperature measurement system based on fiber fluorescence and wavelet analysis [J]. Chinese Journal of Sensors and Actuators, 2003, 16(1): 88-91.
[10] ZHANG Yang. Fluorescent fiber temperature [D]. Harbin:Harbin Engineering University, 2005.
[11] JIA Danping, JIA Ting, YUAN Weiqi, et al. Non-exponential component of fluorescence decay and its processing [J]. Chinese Journal of Scientific Instrument, 2008, 29(12): 2581-2586.
[12] JIA Danping, SHEN Yifeng, WU Jiang, et al. Study on fluorescent decay time thermometry by optical fiber [J]. Journal of Shenyang University of Technology, 1998, 20(2): 16-19.
[13] ZHANG Yan, LI Hongbin, ZHANG Xi, et al. Active optical fiber direct current sensor for HVDC system [J]. Chinese Journal of Scientific Instrument, 2008, 29(7): 1382-1386.
[14] ZHANG Yan, LI Hongbin. Reliability prediction of electronic direct current transformer in HDVC system [J]. High Voltage Engineering, 2007, 33(7): 119-123.
[15] LI Yan, ZHU Jing, ZHOU Xiaoqin, et al. Analysis of the current characteristic in HVDC converter station and the selection of current measure devices [J]. High Voltage Engineering, 2009, 35(5); 1183-1187.
[16] GAO Shuping, SUONAN Jiale, SONG Guobing, et al. A new current differential protection principle for HVDC transmission lines [J]. Automation of Electric Power Systems, 2010, 34(17): 45-49.
[17] YANG Guangliang, TAI Nengling, ZHENG Xiaodong, et al. Modular redundancy analysis of HVDC control and protection [J]. Automation of Electric Power Systems, 2010, 34(15): 85-90.
[18] CHEN Jun, WANG Zuowei. Principle analysis of DC current transformer based on HOLEC amplifier [J]. Central China Electric Power, 1996(4): 23-25.
[19] FEI Ye, WANG Xiaoqi, WANG Benjin, et al. Development on ± 1 000 kV UHV DC instrument transformer [J]. High Voltage Engineering, 2010(10): 2380-2387.

IOP Conf. Series: Materials Science and Engineering **366** (2018) 012082 doi:10.1088/1757-899X/366/1/012082

Business Value Evaluation of Electric Vehicle Charging Network-based Internet of Vehicles

H Yan[1], N Ding[1], W T Xiong[2,*], J Zhang[3], J Zhang[1] and T Y Li[3]

[1] State Grid Electric Vehicle Service Company, Ltd, Beijing 100053, China
[2] School of Electric and Electronic Engineering, North China Electric Power University, Beijing 102206, China
[3] Beijing Electric Vehicle Charging/Battery Swap Engineering and Technology Research Centre, China Electric Power Research Institute Co, Ltd, Beijing 100192, China

*13718899557@163.com

Abstract. Electric vehicle charging network develops rapidly, but now there is still lack of research about business value evaluation of electric vehicle charging network. This paper discusses the business value of electric vehicle charging network based on Internet of vehicles. By analyzing the business evaluation index, a three-dimensional value index system of the two business segments is set up from the aspects of economy, society and environmental benefits, which aimed at improving the business value of electric vehicle charging network.

1. Introduction

In recent years, electric vehicle users were explosive growing. Chinese government has introduced a series of policies to promote development of electric vehicle industry. As supporting facilities, charging network develops as well [1-2]. However, problems like single mode of the charging network, not interconnected of information and improper scheduling of management hinder the development of charging network market. To solve the problems above, the concept of Internet of vehicles emerges. Internet of vehicles, according to a variety of information sensing equipment, access technology and network services support technology, achieve to extract, use and provide integrated services in the information network platform for all vehicle [3]. In 2015, the State Grid Corporation started smart Internet of vehicles platform on line. Up to now, the number of charging pile is more than 16.7 million, and daily charge amount is more than 100 million kwh, which means it has become the world's most extensive coverage, most access equipment and highest level of skill Internet of vehicles. Internet of vehicles and its carrying intelligent traffic, as an intelligent information network service system, has brought historical opportunity of evolution and integration for the electric vehicle industry [4]. Since charging network based on Internet of vehicles is still emerging business [5], this paper starts from the current state of charging network, through the analysis of business evaluation indicators, builds integrated business value model combining three-dimensional value index system and two business plates, and has important guiding significance for development of electric vehicle industry [6].

2. Research status of electric vehicle charging network

2.1. Research status of electric vehicle based on Internet of vehicle

In the context of the development of Internet of vehicle and electric vehicle, in the area of business mode of Internet of vehicle and electric vehicle, some scholars starts from the relationship between Internet of vehicle and charging network, and comes up with the charging mode of electric vehicle based on Internet of vehicle [7]. Some scholars envisioned the electric vehicle - Internet of vehicle business mode, in which relevant parties can achieve optimal integration of resources and benefits [8-9]. In terms of specific applications, there are two kinds of user-based association methods to optimize the content sharing service of electric vehicles under the Internet of vehicle [10]. In summary, most of the research is from macro point of view, and rarely from business value evaluation.

2.2. Research state of business value

At present, there is a lack of research on business value of electric vehicle charging network at home and abroad. Among studies at abroad, there are methods of resource-centered and contingency-based, by using linear and non-linear means to analyze business value [11]. Among studies at home, a comprehensive evaluation method including value engineering, life cycle and index system is brought up [12], which can study the cost and economic benefits of business [13], market life [14] and comprehensive evaluation [15].

3. Business scope of electric vehicle charging network

Charging network business includes the basic interconnection of electric vehicle and power grid and Internet. This article divides electric vehicle charging network business into charging business and extra business. Learning from business model of Internet of vehicles, emerging business like route planning, intelligent charge and discharge are brought forward.

3.1. Charging business

With large-scale development of electric vehicle, charging demand gradually expands and charging business deepens. Electric vehicle charging network needs intelligent and interactive development to enhance service efficiency.

- Charging service: provide charging platform for electric vehicle
- Cost settlement: according to different regions and periods, provide charging price for charging facilities, calculate charging cost for electric vehicle, provide charging network data for operator to clear settlement

3.2. Extra business

Extra business is emerging business of electric vehicle charging network. As an important business plate, it includes charging and discharging, message pushing, statistical analysis, consumption drainage, information release, grid interaction, clean benefit analysis and etc.

- General: intelligent charging strategy, management of charge pile, route planning, safety monitoring, remote diagnosis, traffic data recording
- For individual: entertainment, electric vehicle rental
- For enterprise: vehicle information management, charging network platform operation
- For third party: user preferences mining, vehicle data analysis, traffic information statistics, highway tolls, electric vehicle cleaning efficiency analysis, charging load statistics and forecast, related business promotion

4. Integrated business value model of electric vehicle charging network

Since the value compositions of charging network business are complicated, in order to understand it better, this paper starts at the characteristics of the electric vehicle charging network business, sets economic, social and environmental benefits as indicators, builds two business evaluation index system respectively, and finally forms the integrated business value model of electric vehicle charging network through the comprehensive weight coefficient.

IOP Conf. Series: Materials Science and Engineering **366** (2018) 012082 doi:10.1088/1757-899X/366/1/012082

4.1. Establishment thoughts of evaluation index

In China, electric vehicle development considers not only economic benefits, but also social and environmental benefits. Among them, economic benefits are concerned to promote the upgrading of China's auto industry and cultivate new economic growth point. Social benefits are concerned to improve travel conditions and public satisfaction. Environmental benefits are mainly concerned to promote alternative energy, help transformation of energy consumption structure, reduce vehicle exhaust pollution emissions and ease the increasingly serious environmental problems.

Electric vehicle charging network business is the derivatives of electric vehicle industry. Since the two have inextricably connection, this article analyses electric vehicle charging network business value from economic, social and environmental value three aspects as well.

4.2. Establish of business value model

4.2.1. Establish basis of business value indicators. The business value of electric vehicle charging network is mainly reflected in its effectiveness. In general, benefits are beneficial effects of a particular activity, which also can be seemed as effectiveness and interests.

- Economic benefit analysis

  With the reason of charging business features, the performance of economic benefits is mainly affected by profitability and consumption of service. The indicators under profitability are used to evaluate business operation effects. The indicators under consumption consider cost and fixed assets turnover.

  Economic benefits of extra business are composed of cost and income. Cost concludes data mining analysis and so on. Income concludes electric vehicle rental, operating network platform and etc.
- Social benefit analysis

  The social benefits of electric vehicle charging network mainly consider effects in public and industry. In the part of public, except the shared user satisfaction, charging business concerns more about network construction, that is, the ratio of car and charging pile. Extra business is biased to measure service progress, and here it includes improve the reliability of the grid and digital process. And because measurement of industrial value is no different in the two, this paper selects advancement in technology and industrial agglomeration strength as its three level indicators.
- Environment benefit analysis

  Charging network business will have a certain impact on the environment, mainly in energy conservation and alternative. Since the electric vehicle has significant effects on environment. The larger number it is, the larger effects it applies on energy conservation. Electric vehicles replace traditional fuel with power. In this way, it promotes alternative energy and changes energy structure towards electrification.

4.2.2. Establishment of evaluation index system for charging business. According to benefit analysis, economic, social and environmental benefits are set as the first level indicators. Business operation results, consumption proportion and so on are the second level indicators. Business profit rate, cost profit rate and so on are the third indicators. All the factors comprise value indicators of electric vehicle charging business. As shown in Table 1:

Table 1. Value indicators of charging business

Serial number	First level indicators	Second level indicators	Third level indicators	Type	Unit
1	Economic benefit	Business operation results	Business profit margin	Numerical	%

IOP Conf. Series: Materials Science and Engineering **366** (2018) 012082 doi:10.1088/1757-899X/366/1/012082

2			Total rate of return on assets	Numerical	%
3		Consumption proportion	Cost profit margin	Numerical	%
4			Fixed asset turnover rate	Numerical	%
5		Public value	Degree of customer satisfaction	Numerical	High/medium/low
6	Social benefit		Ratio of car and charging pile	Numerical	%
7		Industry value	Advancement in technology	Logical	Yes/no
8			Advancement in industrial agglomeration strength	Logical	Yes/no
9			Unit mileage CO_2 emission reduction	Numerical	g/km
10	Environment benefit	Energy conservation	Unit mileage NO_x emission reduction	Numerical	g/km
11			Unit mileage SO_2 emission reduction	Numerical	g/km
12		Energy alternative	Electricity consumption proportion of charging network	Numerical	%

4.2.3. Establishment of evaluation index system for extra business. According to benefit analysis, economic, social and environmental benefits are set as the first level indicators. Cost, income and so on are the second level indicators. Data mining analysing cost and interactive mechanism research cost and so on are three indicators. All the factors comprise value indicators of electric vehicle extra business. As shown in Table 2:

Table 2. Value indicators of extra business

Serial number	First level indicators	Second level indicators	Third level indicators	Type	Unit
1			Data mining and analysing cost	Numerical	￥
2		Business cost	Interactive mechanism research cost	Numerical	￥
3			Clean benefit analysis cost	Numerical	￥
4	Economic benefit		Information service income	Numerical	￥
5			Financial business income	Numerical	￥
6		Business income	Operating network platform income	Numerical	￥
7			Electric vehicle rental income	Numerical	￥
8			Degree of customer satisfaction	Numerical	High/medium/low
9		Public value	Improve grid reliability	Logical	Yes/no
10	Social benefit		Improve the digital process	Logical	Yes/no
11		Industry value	Advancement in technology	Logical	Yes/no
12			Advancement in industrial agglomeration strength	Logical	Yes/no

13	Environment benefit	Energy conservation	Ratio of online service	Numerical	%
14		Energy alternative	Ratio of renewable energy generation	Numerical	%

4.3. Integrated business value model

Electric vehicle charging network business consists of charging business and extra business. Taking into account the difference between business positioning, the two plates should be distributed by weight when determining the integrated value of business plates. In addition, when deciding the weight of two services in integrated value, it is necessary to consider the degree of business development, and the most significant one among which is the user proportion of extra business. Extra business is the main part of integrated business value model.

This paper takes the comprehensive proportion to determine the value of business weight. According to company's internal judgments, quantify the impact of business positioning and get business positioning weight. According to the forecast data of network charging business, business development weight is determined by extra business proportion. Take the average of the two as comprehensive weight. According to above analysis, the establishment steps of integrated business value model of electric vehicle charging network are as follows:

4.3.1. Determine business position weight. Consult relevant departments of electric vehicle, and collect the position score of business. In this way, the original score of business position forms. Take the equalization ratio of business (charging business: extra business = M11: M12) as the business position weight.

4.3.2. Determine business development weight. According to the forecast data of business, take the proportion of extra business as extra business weight M22. On this basis, the charging business weight M21 = 1-M22. At last, ratio of the two is business development weight.

4.3.3. Determine business comprehensive weight. Take the average of business position weight and business development weight, recorded as M1: M2.

4.3.4. Establish integrated business value model. Record the value model of charging business as $f_1(x)$, the value model of extra business as $f_2(x)$. So the integrated business value model of electric vehicle charging network $f(x)$ is:

$$f(x) = M_1 f_1(x) + M_2 f_2(x)$$

5. Conclusion

This paper combines with the development state of electric vehicle charging network, and explores business value of electric vehicle charging network based on Internet of vehicles. First, through the comprehensive index analysis method, set economic, social and environmental benefits as three-dimensional value indicators. Then divide electric vehicle business into charging business and extra business, and build different business value index system respectively. Finally, according to business comprehensive weight consisting of business position weight and business development weight, combine two index systems, and form the integrated business value model of two business plates with three-dimensional value index. This paper provides a meaningful reference for the development of electric vehicles.

Acknowledgment

State Grid Corporation of China Science and Technology Project: Research and demonstration of key technologies of efficient-energy-service-oriented intelligent and interactive electric vehicle charging network, SGZJ0000KXJS1700329

References

[1] Ciwei G and Liang Z 2011 *J. Power Sys. Technol.* **35** 127 — 31
[2] Su W, Rahimi E H and Zeng W 2012 *J. IEEE Trans. Ind. Inf.* **8** 1-10
[3] Car Cloud Network 2014 *Internet of Vehicles: Decisive Battle of Fourth Screen* (Beijing: Electronic Industry Press) p 13-14
[4] Ruike Y, Xiumiao C and Fangsiyu Z 2012 *J. J. B. Inst. Technol.* **14** 39-44
[5] William S, Sjoerd B, Kees M and Bert van W 2012 *J. Environ. Innov. Societal Transit.* **2** 49–65
[6] Jian D, Dong X, Tao W, Bohua W, Fajing W and Ji Z 2012 *J. Cent. China Electr. Power.* **25** 96-102
[7] Mengdan C and Juanjuan L 2017 *J. Technol. Ind.* **17** 86-90
[8] Ruike Y, Fangsiyu Z, Fei F and Jianqiang B 2015 *J. Stud. Dialect. Nat.* **31** 76-80
[9] Jingjing J, Minjia G, Chunyan F, Xueying G, Changhua Y and Mingjian C 2016 *J. SX. Electr. Power.* **44** 46-50
[10] Chiti F, Fantacci R, Gu Y, et al. 2017 *J. Veh. Commun.* **8** 35-44
[11] Oh W and Pinsonneault A. 2007 *J. Manage. Int. Sys.* **31** 239-65
[12] Haichuan L. 2011 Study on Value Evaluation of Mobile Data Service *Beijing U. Post Telecommun.* D p 21-35
[13] Chifeng Z and Zhong C 2012 *J. Value Eng.* **31** 1-2
[14] Onat N C, Kucukvar M, Tatari O, et al. 2016 *J. Int. J. Life Cycle Ass.* **21** 1009-34
[15] Wei M. 2014 *D.* Research on the Evaluation Index System and Evaluation of Electric Vehicle Charging Station Project of Caofeidian *N. China Electr. Power U.* D p 6-7

IOP Conf. Series: Materials Science and Engineering **366** (2018) 012083 doi:10.1088/1757-899X/366/1/012083

Simulation Analysis Method of Internal Temperature Distribution in UHV DC Voltage Measuring Device

XIE Tingting, YANG Zhongzhou, FENG Jianhua, ZHANG Yuan and WANG Shan

XI'AN High Voltage Apparatus Research Institute, Xi 'an, Shaanxi, China 710077

yangzhongzhou@xihari.com

Abstract. The measuring accuracy of UHV DC voltage measuring device has a great relationship with the temperature. It is necessary to study the internal temperature distribution of the device to ensure that the operating device could meet the accuracy requirement in the temperature range. Firstly, the device dynamic thermal flow field is analyzed in this paper. The flow field model is established. Through analyzing the calculation model, the internal temperature distributions under different environment temperatures is obtained. And then, the internal temperature rise at the rated voltage is measured directly by experiment. And the validity of the model is verified by the experiment. This paper introduces the details of the method. By studying the internal temperature distribution of the device, the temperature range in which the device is running is gotten. Suitable resistive elements should be chosen during the design so as to meet the measuring accuracy of the device.

1. Introduction

UHV DC voltage measuring device is indispensable to the main equipment of converter station of DC transmission system, which undertakes the important functions of DC energy metering, power monitoring and signal protection for power system relay. In the design of UHV DC voltage measuring device, the parallel connection of resistance and capacitance is widely used [1].

In this paper, the UHV DC voltage measuring device rated voltage is ±800kV, the operating current is 2mA, DC voltage measuring accuracy is 0.2%, and the principle is the resistance and capacitance in parallel [2]. The structure of the device is an outdoor pillar type, which is composed of an external high-voltage insulator and an internal high-voltage resistance-capacitance device and is filled with SF_6 insulating gas. High and low voltage end shielding rings are arranged outside the insulator. High voltage double ring shielding and low voltage end single ring shielding are used to improve internal External electric field, the total height of the device is about 10 m, the structure diagram is shown in figure 1.

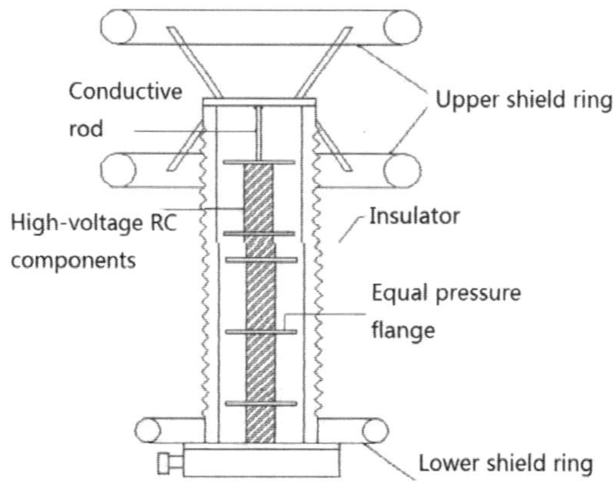

Figure 1. UHV DC voltage measuring device.

Relevant standards made it clear that the voltage divider resistor of measuring device should have sufficient thermal stability to ensure that the device within the specified temperature range of measurement accuracy to meet the requirements [3]. Measurement accuracy of the voltage divider depends largely on its voltage coefficient and temperature coefficient. It can be known from the design parameters that when the rated voltage of the device is operated, the power is 1600W, and the internal temperature of the device may have a certain temperature rise due to its own power heating [4]. By studying the internal temperature distribution of the device, the temperature range of the device operating at different ambient temperatures is obtained. In the design and manufacturing of the device, the suitable resistor and capacitor components can be selected more specifically. The temperature distribution inside the device is studied by establishing the flow field calculation model, analyzing and calculating the temperature distribution of the model under different ambient temperatures, and finally validating the calculation model of the flow field through experiments. Application of the manufacturing process model can be used for the selection of resistors, capacitors and other devices to provide reference to ensure that the device in the operating temperature range to meet the accuracy requirements to improve the reliability of product operation [5].

2. Establish a dynamic flow field model

2.1. Device thermal balance process

During the operation of the device, the internal resistance consumes power to generate heat. A part of the heat is first transferred by the internal gas to the inner walls of the upper and lower metal flanges of the device body in a manner of natural convection heat transfer, and then heat conduction from the inner wall of the upper and lower flanges to the outer wall, and ultimately convection heat transfer from the outer wall of the upper and lower flanges to the air outside the device; The other part of the heat is first transferred from the internal gas to the inner wall of the insulator of the device body, and then heat conduction from the inner wall of the insulator to the outer wall of the insulator, and finally from the outer wall of the insulator and the composite jacket to the air outside the device [6]. When the heat balance is reached, the internal temperature of the device would no longer changes, and all the heat generated internally is transmitted to the air outside the device.

2.2. Establishment and Analysis of Flow Field Model

SolidWorks and ANSYS DesignModeler are used to build the model [7]. Mesh is divided into two parts by using ICEM-CFD. FLUENT is used to calculate the pre-processing calculation. After calculating and solving, the data are processed. Finally, the flow velocity and temperature distribution of each part of the model are obtained. In the simulation calculation, CFD method is used to complete the flow heat transfer analysis, and FLUENT software is used to calculate.

To build the structural model used in the calculation, the original SolidWorks model was processed using ANSYS DesignModeler. The influence of umbrella skirt structure on overall flow and heat exchange is considered in the modeling, and the bracket to which the pressure equalizing ring is connected to the flange is simplified, and the features such as the smaller chamfer and the bolt hole on the flange at all levels are canceled. After the structural model is established, the model is imported into ICEM-CFD for meshing. The overall computational domain and the solid domain mesh are shown in figure 2 and figure 3, respectively.

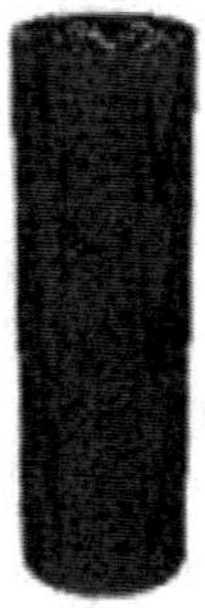

Figure 2. Whole structure mesh generation.

Figure 3. Solid region mesh generation.

According to Figure 1, the device consists of a high voltage RC unit (single-layer voltage divider module), brackets, flanges, insulators, umbrella skirts and equalizing rings. The material parameters of each component are shown in table 1, and the conservative 25 °C solid material property is used as the input of calculation for the analysis [8].

Table 1. Material parameters of components.

Parameters	Bushing/ Capacitor/ Bracket (Insulation Material)	Resistor	Flange (aluminum)	Umbrella skirt (silicone rubber)
Specific heat/ [J·(kg·°C)$^{-1}$]	800	1090	871	1700
Thermal conductivity/ [W·(m·°C)$^{-1}$]	0.175	20.00	202.40	0.270

The physical properties of the air and SF_6 used in the calculation of the fluid field are given in table 2. Due to gravity-driven natural convection in and out of the device body, natural convection is simulated using the actual temperature-density relationship of air and SF_6. For viscosity, specific heat capacity, thermal conductivity and other parameters, only the properties at room temperature are considered in the calculation [9].

Table 2. Physical properties of air and SF_6.

Physical properties	Air	SF_6
Viscosity/ [Pa·s^{-1}]	2.190×10^{-5}	1.211×10^{-5}
Specific heat/ [J·(kg·°C)$^{-1}$]	1009	634
Thermal conductivity/ [W·(m·°C)$^{-1}$]	0.0317	0.0106
Density (temperature/°C)/ [kg·m^{-3})]	1.25(10°C)/ 1.09(50°C) 0.95(100°C)/ 0.85(140°C)	24.56(-23°C)/ 23.44(27°C) 14.04(227°C)

The flowing process of the external flow field is analyzed by the heat transfer characteristics of the model device. The air is heated vertically by the outer wall of the device and flows at an increasing speed. When the air flows through the voltage equalizing ring, the flow channel is compressed, making the pressure ring near the air flow rate increased. Since the thermal conductivity of the insulator is inferior to that of the metal flanges at the upper and lower ends, and the air velocity is

IOP Conf. Series: Materials Science and Engineering **366** (2018) 012083 doi:10.1088/1757-899X/366/1/012083

relatively high, the temperature of the air outside the flow field is slightly higher than the ambient temperature. Outflow velocity distribution is shown in figure 4, the flow field temperature distribution is shown in figure 5.

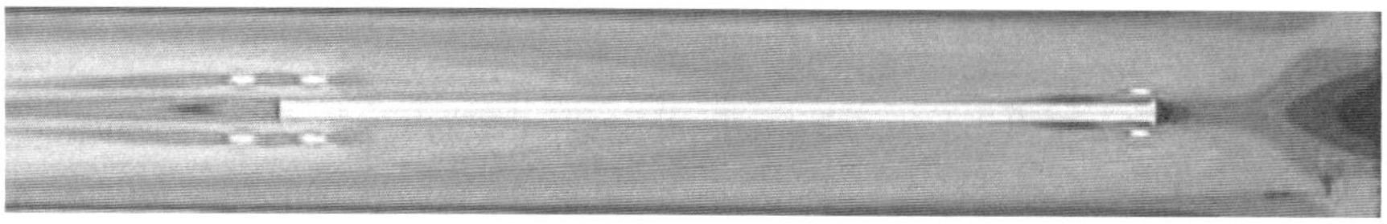

Figure 4. Velocity distribution of external flow field.

Figure 5. Temperature distribution of external flow field.

The flowing process of internal flow field: Due to the blocking effect of multi-layer voltage equalizing flanges, the SF_6 gas in the internal flow field presents a small cycle of disconnection so as to realize the heat exchange between the heat generating body and the inner wall of the insulator. The temperature of the resistor body in the single-layer voltage divider module is highest, followed by the flange and the capacitor, mainly because the resistor is the only heat-generating element. In addition, the temperature distribution in the vertical direction of the partial pressure modules is low in the lower part and high in the upper part. The main reason is that the hot SF_6 gas always flows upward from the lower part while the outside air temperature in the lower part of the device is lower than that in the upper part. The velocity distribution of the inner flow field is shown in figure 6, and the temperature distribution of the inner flow field is shown in figure 7.

Figure 6. Velocity distribution of internal flow field.

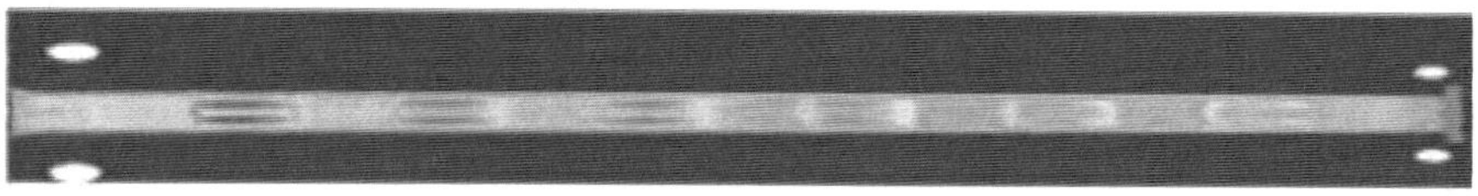

Figure 7. Temperature distribution of internal flow field.

3. Calculation results

Analysis shows that as the vertical height increases, the internal temperature of the device rises immediately, the temperature of the bottommost voltage divider module is the lowest, and the temperature of the topmost voltage divider module is the highest. In addition, as the ambient temperature increases, the internal temperature also increases. In this paper, the temperature distributions of the device are simulated and calculated at the ambient temperature of 20, 30, 40 and 50 °C respectively. The calculation results are shown in table 3 (The calculated values at the ambient temperature in cold region are not listed in table 3, and the highest temperature of the voltage divider module is numbered in the order of the topmost voltage divider module to the bottommost voltage divider module). It can be seen from table 3 that the temperature of the top single-layer voltage divider module is the highest and the lowest single-level voltage divider module is the lowest. From the vertical point of view, from the top of the single-level voltage divider module to the bottom of the single-level voltage divider module, the temperature distribution showed a decreasing trend. The temperature difference between the topmost single-layer voltage divider module and the bottommost single-layer voltage divider module is about 9 °C, and the temperature rise inside the device is about 28 K. It can also be seen from table 3 that if the device is operating in a hotter area in summer, it is necessary to check the thermal stability when the working temperature of the measuring resistor is up to about 80°C to meet the measurement accuracy requirements of the device operation.

Table 3. Calculated values of internal temperature distribution under environment temperature/ °C.

The number of voltage divider module	Ambient temperature			
	20	30	40	50
1	48.25	58.19	68.15	76.73
2	48.05	57.56	68.08	75.99
3	47.76	56.42	67.12	75.55
4	45.70	54.55	65.27	74.92
5	44.60	54.39	64.19	74.35
6	44.54	54.04	64.02	73.32
7	42.85	52.43	62.82	73.24
8	42.43	52.25	62.32	72.40
9	41.20	51.19	61.09	71.60
10	40.27	50.31	60.28	71.58
11	39.32	49.29	58.97	71.40
12	39.00	48.59	58.99	71.02
13	38.82	48.41	58.08	70.08

4. Verification test

In order to verify the correctness of the calculation results of the above model, further experimental verification was carried out. The verification test was carried out in the high-voltage insulation test hall of Xi'an High-voltage Electrical Apparatus Research Institute (XIHARI), and the verification test scheme is shown in figue.8 [10]. Test verification uses built-in temperature sensor, temperature sensor buried point should be under the premise of not affecting the conditions within the device can represent the key points of the internal temperature distribution device. Therefore, the test temperature sensors were buried in the voltage equalization flanges of the topmost voltage divider module, the bottommost voltage divider module and the middle voltage divider module. In addition, the device was charged with 0.35MPa insulation gas during operation. The optical connector was used to reliably transmit the internal temperature information to the outside during the test. The optical connector could also solve the problem of gas sealing during the test.

IOP Conf. Series: Materials Science and Engineering **366** (2018) 012083 doi:10.1088/1757-899X/366/1/012083

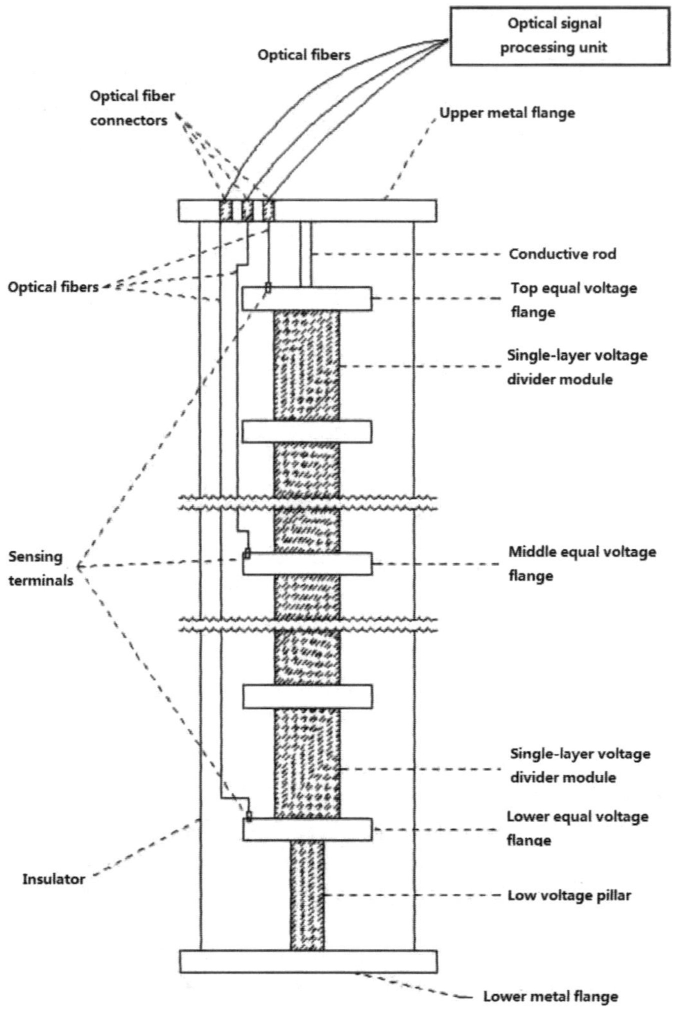

Figure 8. Implementation scheme for experiment.

The DC voltage generator used in the test can output ±1200kV DC voltage. The actual DC voltage applied was +800kV. After applying the voltage for about 12 hours at the ambient temperature, it is judged that the temperature rise within the device does not exceed ±0.5K per hour, and the heat balance is achieved within the device. The measured buried temperature data are shown in table 4, and the test site is shown in figure 9. The results in table 4 show that the maximum difference between the temperature of the topmost single-layer voltage divider module and the temperature of the bottommost single-level voltage divider module is 9.3 °C, and the maximum temperature rise inside the device is 28.9 K, which is consistent with the model calculation.

Table 4. Temperature data of measuring points/°C.

Buried point No.1of topmost voltage divider module	Buried point No.2of topmost voltage divider module	Buried point No.3of middle voltage divider module	Buried point No.4of middle voltage divider module	Buried point No.5of bottommost voltage divider module	Buried point No.6of bottommost voltage divider module	Ambient tempera-ture
42.4	42.3	36.9	36.6	33.2	33.1	13.5

Figure 9. Test site.

5. Conclusion

By studying the internal temperature distribution of the device, we can know the internal temperature distribution of the device and further verify by the experiment that the calculation results of the flow field model established in this paper are correct. Applying the calculation model, the internal temperature distribution of the device under different ambient temperatures can be analyzed and reference can also be provided for the design. At the same time, the design of the device can be more targeted selection of resistance, capacitance devices to ensure that the device in the operating temperature range to meet the accuracy requirements and to improve the reliability of the product operation.

References

[1] LIU Zhenya, SHU Yinbiao, ZHANG Wenliang, et al. Studyon voltage class series for HVDC transmission system [J]. Proceedings of the CSEE, 2008, 28(10): 1-8.
[2] FEI Ye, WANG Xiaoqi, WANG Benjin, et al. Developmenton ±1000kV UHV DC instrument transformer [J]. High Voltage Engineering, 2010, 36(10): 2380-2387.
[3] LIU Xing. Design and implementation of DC high voltage divider [J]. Metrology & Measurement Technology, 2008, 28(1): 17-19.
[4] YU Jianzu, GAO Hongxia, XIE Yongqi. Electronic equipment thermal design and analysis [M]. Beijing: Higher Education Press, 2002.
[5] LIU Bin, HUANG Dexiang. Thermal design for high precision DC resistance divider [J]. High Voltage Apparatus, 2011, 47(2): 31-33.
[6] LI Dengyun, LI Qian, LI He, et al. Calculating method for thermal equilibrium temperature of traditional DC TV [J]. High Voltage Engineering, 2010, 36(4): 994-999.
[7] HUANG Aixiang, ZHOU Tianxiao. Finite element theoryand methods [M]. Beijing: Science Press, 2009.
[8] MOU Qiuhong. Preparation and properties of thermal silicone rubber [J]. Materials Review, 2009, 23(6): 110.
[9] Shanxi Institute of Chemical Industry. Polyurethane Elastomer Manual [M]. Beijing: Chemical Industry Press, 2001.
[10] ZHANG Renyu, CHEN Changyu, WANG Changchang. High-voltage testing technology [M]. Beijing: Tsinghua University Press, 2003.

Application Research of Wireless Sensor Network in Building Structure Safety Monitoring

SHI Cheng[1] , WEI Rongtao[2] and ZHANG Yanling[3]

1 School of architecture, Guangzhou college, South China University of Technology, Guangzhou 510800, China
2 North China Electric Power University, Beijing 102206, China
3 North China Electric Power University, Beijing 102206, China

E mail: 1165700788@qq.com

Abstract. In order to ensure the safety of important buildings, the structural safety monitoring of buildings is indispensable. Compared with wired sensor network, wireless sensor network have the advantages of deployment flexibility, lower maintenance cost and higher monitoring accuracy. It has a better application prospect in the safety monitoring application of building structure. This paper studied the application of wireless sensor network in the safety monitoring of buildings by studying the application requirements in the safety monitoring of building structure in important buildings, proposed a routing algorithm to improve the network lifetime, and simulated the routing algorithm.The results show that the algorithm can effectively improve the network lifetime.

1. Introduction

The safety of buildings is related to people's life and property safety. During the service period of the building, due to the complex structure, corrosion, disasters and other reasons, it will cause potential safety hazard in the building structure. In order to ensure the safe service of the building, the safety monitoring of the structure of the building is very necessary.

The safety monitoring system of the building arranges the sensors in the key positions of the building to measure the real-time parameters of the building structure and the building environment. These parameters are transmitted to the data center through the communication technology, and the data center analyzes and stores the data. The data center observes changes in the structural system over time by analyzing dynamically collected dynamic response data and determines the health status of the building by analyzing the damage-sensitive eigenvalues extracted[1]. For long-term health monitoring of building structures, the effects of structural aging and poor service environment on the structure of the project are estimated mainly through periodic data updates and the ability of the building to continue its design capability. By monitoring the structural safety of the building, the problems existing in the building can be found in time to reduce the casualties and property losses[2].

In recent years, wireless sensor network technology has been developed rapidly, and its role in the bridge connecting the physical world and the computer world has been widely used. Wireless sensor networks not only have the advantages of low cabling cost, high monitoring accuracy and good fault tolerance, but also can be remotely monitored with outstanding advantages in fault diagnosis and safety monitoring[3].. At present, the main research directions of WSN are mainly in the field of military field,

field control of industrial automation and monitoring of the earth environment. There is not much research on the application of building safety monitoring, and there is a lack of practical wireless sensor network integrated system which accords with the characteristics of building safety monitoring[4].. With the continuous improvement of people's living standards, the modern intelligent building has made considerable progress. At the same time, the wireless sensor network system has gradually begun to infiltrate into the application of the modern intelligent building field because of its unique outstanding advantages. It has become a modern intelligent building Related areas of research focus.

2. Wireless Sensor Network Technology

The architecture of wireless sensor network has many kinds of forms, it can be divided into the following three kinds mainly: the centralized structure, the distributed structure and the network structure. There are no central nodes in a typical wireless sensor network. All the nodes self-organize into a distributed network to send the monitoring data to the sink node. The sink node analyzes the data and converges the data to the data center[5]..

The structure of the wireless sensor node consists of four functional modules: (1) sensor module: composed of sensor and analog-digital conversion unit, mainly responsible for information acquisition and data conversion; (2) processor module: composed of processor and memory, and Generally built-in a small embedded operating system, its main work is to control the overall operation of sensor nodes, running high-level network protocol, data calculation and processing; (3) communication module: the antenna and wireless transceiver controller, the module's main responsibility is Communicate with other sensor nodes; (4) Energy supply module[6].: This module usually consists of a micro-battery or a button battery with limited energy and is mainly responsible for providing energy to the sensor node, which determines the life of the sensor node.

The protocol stack of the wireless sensor network consists of the physical layer, data link layer, network layer, transport layer and application layer. In addition, the protocol stack also includes the energy management platform, the mobile management platform and the task management platform.

Routing protocol is one of the key technologies in wireless sensor network. It logically divides the network topology and is responsible for establishing the data transmission path from sensor node to sink node. In order to extend the network life cycle, the routing protocol must be able to reduce the energy consumption of each node and the energy consumption of the entire network, and achieve the purpose of balancing the energy consumption load of the entire network. The routing protocols of wireless sensor networks can be divided into planar routing protocols and clustering routing protocols according to the network topology[7]..

The routing protocol in wireless sensor network not only need to select the energy consumption of small message transmission path, but also choose the route that consumes the entire network energy balance from the perspective of the entire network. The sensor nodes have limited resources. The routing mechanism of wireless sensor network should be able to transmit information easily and efficiently.

3. Design of Wireless Sensor Network Routing Algorithm

LEACH is a classic routing protocol in wireless sensor network. In order to extend the lifetime of wireless sensor network, a new routing algorithm is proposed in this paper. The algorithm mainly includes five phases, which are clustering, node selection in a cluster Cluster head, node cluster, stable communication and communication radius adjustment. The specific process shown in Figure 1.

IOP Conf. Series: Materials Science and Engineering **366** (2018) 012084 doi:10.1088/1757-899X/366/1/012084

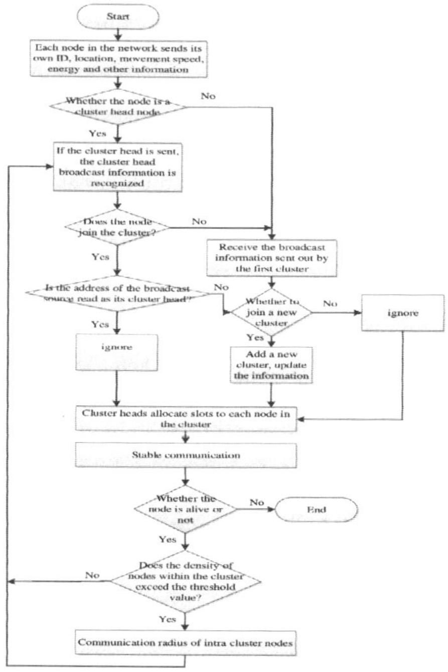

Figure 1. The wireless sensor network routing protocol flow chart based on node load balancing

3.1 Cluster head selection

In the same way as the LEACH protocol, in this algorithm, the nodes are self-organized into clusters, each cluster contains a cluster head node, which is responsible for converging and forwarding the data collected by nodes in the cluster to the Sink node. Due to the data fusion and forwarding, the cluster head consumes more energy than other nodes.

The node generates a random number between 0 and 1. For node n, if this random number is smaller than the threshold T (n), the node broadcasts itself as a cluster head to the entire network[8].

$$T(n) = \begin{cases} \dfrac{P}{1 - P \times (r \bmod \dfrac{1}{P})} & n \in G \\ 0 & otherwise \end{cases} \tag{1}$$

In the formula, P is the ratio of the preset cluster head node to all the nodes and the current round number, G is the set of candidate cluster head nodes, and also the set of nodes that have not elected cluster heads in the preceding consecutive P rounds, mod is modulo operation.

After the cluster head node is selected, the cluster head node will send the broadcast message. Once receiving the broadcast message sent by the first cluster head, the node joins the cluster. If the node can receive the broadcast message sent by other cluster heads , the node will be compared with the cluster head distance. If the new cluster head is closer to the node, the node will join the new cluster. Otherwise, the cluster head broadcast message will be ignored.

3.2 Energy level

The node neighborhood distance defined in this paper is for each node, which refers to the distance between the node in each node's neighbor node set and that node. The initial neighbor node set of each node is a set of nodes in the initially set transmission power range of the sensor nodes, that is the communication radius. During initialization, a test signal can be sent to the surrounding nodes through the node to record the number of neighboring nodes and the distance between each neighboring node and the node.

The radius of the power transmission range of each node at the initial time of the entire network is

IOP Conf. Series: Materials Science and Engineering **366** (2018) 012084 doi:10.1088/1757-899X/366/1/012084

R_0, which is the broadcast domain or communication radius of node i. The set of neighboring nodes of node i is the set of neighbor nodes of the current node i, and the distance between each neighboring node through the test signal node i is the node's neighbor distance d_i:

$$d_i = Dist(node\ i, node\ i_{neighbork}) \tag{2}$$

Where k represents the kth neighbor node.

Because the wireless sensor network is a self-organizing network, the remaining energy of each node in the network initialization is different. In order to save energy and increase the life cycle, load balancing is performed for each node, that is, the nodes with more residual energy can be more loaded, The remaining nodes with less energy load is relatively less, which can effectively extend the life cycle of the entire network to avoid excessive network load caused by local network collapse and consume more energy. In order to measure the residual energy difference between nodes, this paper introduces the node residual energy level REL, the introduction of the benefits of REL is equivalent to the normalization of energy to avoid tedious calculations for the sensor nodes to save storage space and Energy consumption, and use the number of loads corresponding to the discriminant energy level. Set an initial energy E_0 for the entire network and m nodes for each energy level, then $\mathrm{E_{average}} = \mathrm{E_0/m}$, for each level. The energy level m can be calculated based on the density of communication nodes in the entire network. Suppose the current remaining energy of each node i is $\mathrm{E_{i\_cur}}$, we can get the remaining energy level of a node i in the network at a certain moment as :

$$REL_{i\_cur} = [E_{i\_cur}/\mathrm{E_{average}}] \tag{3}$$

$[E_{i\_cur}/\mathrm{E_{average}}]$ represents the smallest integer greater than $\mathrm{E_{i\_cur}/E_{average}}$. From (3), we can get the remaining energy level of each node in the range of [1, m] positive integer. The selection of the energy level m has an important influence on the performance of the entire optimization protocol and can be set according to the average node density of the entire network. The requirements for communication quality m can be flexibly set according to different networks. If m is smaller, the remaining energy difference of each node can not be reflected. The number of loads set according to the remaining energy level will not change significantly, and the effect of load balancing will be very insignificant.

3.3 stable communication

After the network completes the clustering and the nodes are clustered, the network conducts a stable communication process. The sensor node collects the monitoring data and sends the data to the cluster head. After the cluster head finishes data fusion on the received data, the data Send to Sink node.

4.Communication radius adjustment strategy

Due to the randomness of the monitoring nodes, some nodes in some clusters have more nodes and fewer nodes in some clusters. In order to prolong the lifetime of cluster head node, the communication radius of cluster head will be dynamically adjusted when the density of nodes in a cluster is larger, and when the density of cluster heads is larger, the communication radius will be reduced appropriately , Reduce energy consumption, achieve load balancing, extend network life. The relationship between the communication radius before and after adjustment is as follows:

$$R_c = (1 - Q_i) * R \tag{4}$$

In (4), R_c is the adjusted communication radius, R is the communication radius before adjustment, and Q_i is the node density of nodes.

$$Q_i = \frac{N_{neighbor\_i}}{N_{wsn\_alive}} \tag{5}$$

The expression of Q_i is as shown in formula (5), $N_{neighbor\_i}$ is the number of neighbor nodes of the

IOP Conf. Series: Materials Science and Engineering **366** (2018) 012084 doi:10.1088/1757-899X/366/1/012084

node i, and N_{wsn_alive} is the number of nodes alive in WSN monitoring network.

5. Simulation analysis

In order to analyze and compare the performance of the routing algorithm, this paper carries on the data modeling of the LEACH algorithm and the proposed routing algorithm in the MATLAB environment. In the area of 1000m * 1000m, 400 sensor nodes are randomly deployed, as shown in FIG. 2.

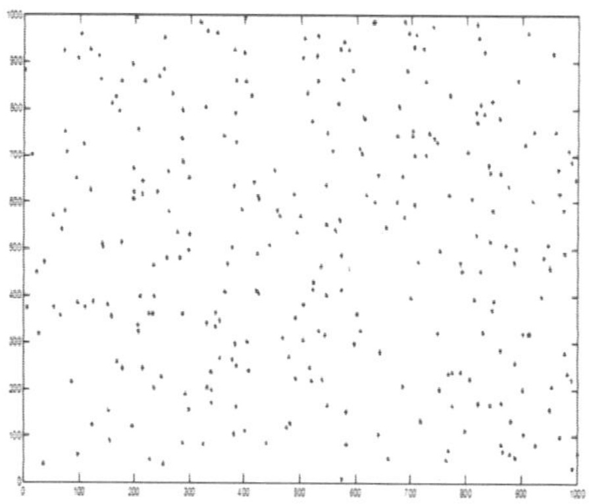

Figure 2. wireless sensor node distribution

Since the WSN is a self-organizing network, the initial energy of each node in the network initialization is different, and the sink node generally selected is a node with sufficient energy. Taking the above into account, the initial energy of the sink node is set as 1.00, In order to highlight differences in remaining energy levels, the initial energy of the remaining ordinary nodes is set to a random value between 0.50 and 1.00. Other simulation parameters can be set according to Table 1 below:

Table 1. simulation parameters

Parameter	Setting
Sensing area size	1000×1000
Number of sensing nodes	400
Sink node initial energy	1
Normal node initial energy	0.50~1.00
Broadcast energy consumption	0.0001
Send Message Energy Consumption	$\delta * Dist^3$

Send information energy consumption using the model of wireless communication distance and energy consumption E=δ*d3, and δ=0.0001*(1/(25^3)).The choice of energy level m has an impact on the performance of the protocol, so the value of m is crucial. In this paper, the value of m is chosen according to the average density of nodes in the entire sensing region. In this paper, m values 9, then REL has 9 levels .

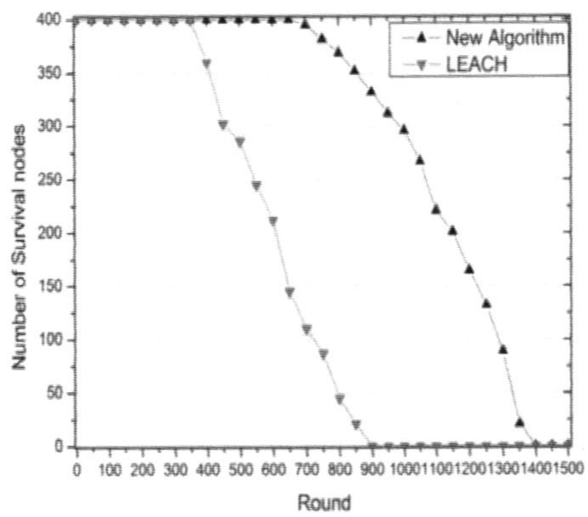

Figure 3. Comparison of the number of network survivors

Figure 3 shows the routing algorithm proposed in this paper and LEACH routing algorithm in the number of network nodes in the comparison. Network lifetime refers to the time when the first node in a network runs out of energy, expressed by the number of rounds. In order to fit an actual communication system, a node dies when the remaining energy is less than 5%. The number of surviving nodes in the network can also reflect the algorithm's lifespan of the entire network. It can be seen from the figure that LEACH algorithm will die at the first node in the 400th round, and the algorithm proposed in this paper will die at the 700th round. At the speed of node death, the proposed algorithm is also less effective than the LEACH algorithm. It can be seen from the figure that in the 900th round, all the nodes in the LEACH algorithm will die, and the algorithm proposed in this paper will die at 1400 rounds of all nodes. It proves that the proposed algorithm can indeed balance the load of the network and effectively extend the lifetime of the entire network.

6. Conclusion

Based on the analysis of the LEACH protocol, this paper proposes an improved routing algorithm for wireless sensor networks. In order to balance the load of each node, the energy steps are adopted in the algorithm. In order to reduce the node energy consumption, Used to adjust the communication radius of the method. The simulation results show that the algorithm can effectively improve the network lifetime.

References

[1] Yick J, Mukherjee B, Ghosal D. Wireless sensor network survey[J]. Computer Networks, 2008, 52(4): 2292-2330.
[2] Chouhan S, Bose R, Balarishnank M. Integrated energy analysis of error correcting codes and modulation for energy efficient wireless sensor nodes[J]. IEEE Transactions on Wireless Communica- tions, 2009, 8(10): 5348-5355.
[3] Kalis A, Kanatas A G, Efthymoglou G P. A co-operative beamforming solution for eliminating multi-hop communications in wireless sensor networks[J]. IEEE Journal on Selected Areas in Com- munications, 2010, 28(7): 1055-1062.
[4] Subramanian, A. P., Buddhikot, M. M., & Miller, S. (2006). Building structure condition monitoring system using wireless sensor network. Undergraduate Theses of Electrical Engineering Department, 55-63.
[5] Morello, R., C. D. Capua, and A. Meduri. "Remote monitoring of building structural integrity by a smart wireless sensor network." Instrumentation and Measurement Technology Conference IEEE, 2010:1150-1154.
[6] Dai, Zhicheng, S. Wang, and Z. Yan. "BSHM-WSN: A wireless sensor network for bridge structure health monitoring." Proceedings of International Conference on Modelling, Identification & Control IEEE, 2012:708-712.

[7] Gao, Huiyan. "Application of Zigbee Wireless Sensor Network in Safety Monitoring System." Computer Measurement & Control 16.6(2008):766-768.
[8] Gamwarige, S., and C. Kulasekere. "An energy efficient distributed clustering algorithm for ad hoc deployed wireless sensor networks in building monitoring applications." Electronic Journal of Structural Engineering 9(2009):11-27.

IOP Conf. Series: Materials Science and Engineering **366** (2018) 012085 doi:10.1088/1757-899X/366/1/012085

A review of the research on price-type demand response of industrial users

Jiancheng Yu[1], Guodong Li[2], Siyuan Li[3,4], Jing Chen[3] and Xiufan Ma[3]

[1] State Grid Tianjin Electric Power Company, Tianjin 300010, China
[2] State Grid Tianjin Electric Power Research Institute, Tianjin 300384, China.
[3] School of Electric and Electronic Engineering, North China Electric Power University, Beijing 102206, China

[4] lsy_94414@163.com

Abstract. With the continuous development and perfection of the power market, the interests of the power system are gradually diversified, and the role of demand-side resources in the competitive market is being reconsidered. This paper briefly introduces three kinds of price demand response projects in European and American countries, sums up and compared the research status of the demand response of industrial users in the Western countries and China, aiming at providing research directions for the research of Chinese demand response.

1. Introduction

Demand response means that when the cost of power supply market is high or the reliability of the power system is threatened, the electricity users in the direction of power supply will be given price guidance or economic incentive so as to make adjustments to their original power consumption modes so as to reduce the peak period of electricity load or make it the purpose of the migration period, thus ensuring the stable operation of the system, and curb short-term behaviour of rising electricity prices. From the perspective of the user's response motivation, the demand response items can be divided into “price-type demand response” and “incentive-type demand response”. Based on the research on demand response of industrial users in China and Western countries, this paper summaries the existing demand response projects and studies provided suggestions and ideas in order to solve the existing problems in the development of China's demand response.

2. Typical price-type demand response projects in Western countries

Since the concept of demand-side management emerged in the 1960s, Western power companies have been exploring the use of price-based demand responses in an attempt to find better and more user-friendly pricing models. Here are some typical price-based demand response projects in the exploration process.

2.1. TOU

Arizona Public Service Corporation of America offers two different pricing options for power users. 1. "time-benefit plan" is a one-part tariff suitable for users with shorter peak hours.2. “combined advantage plan” is a two-part tariff, which is composed of energy price and demand tariff, It is more suitable for users of peak-valley electricity difference.

2.2. CPP
The TEMPO project in France is a typical and successful critical peak pricing project. It sets a higher critical peak price on the basis of TOU. TEMPO project started in 1996, the year divided blue day, white day and red day 3 kinds of electricity prices, and each day has been divided into two kinds of price: peak load and nonpeak load.

2.3. RTP
In the past ten years, the two-part real-time pricing developed by Georgia Electric Power Company has been relatively mature. In this mechanism, the real-time electricity price is charged when the user load is higher than the baseline load set by the historical load data, the standard electricity price is executed when the user load is equal to or lower than the baseline load and the lower portion will be returned to consumers at the standard price.

3. The Research Status on Demand Response of Industrial Users in Western Countries
Among many power users, large industrial users are more sensitive to tariff signals and have more flexibility in load, able to respond quickly to price changes or incentives and cut back on significant electricity demands, making them ideal for demand response projects. At present, there are many studies on the demand response of industrial users in western countries.

3.1. The early basic research
The study of user response to price-based demand responses in western countries started earlier. A.K.David studied the optimal user response under spot electricity price, put forward the user load classification theory, and proposed the load classification optimization method under user response. At the same time, it analyzed the potential price response and response bottleneck of industrial users. Daryanian B subsequently established a mathematical model of the response of real-world electricity prices to energy-consuming industrial users and simulated the measures that users could take to respond to the demand response and the electricity costs they could save[1], but since the model was too general to be heavily referenced. J.G.Roos et al focused on the response behavior of industrial users under the single system real-time electricity price and its avoidable cost-effectiveness[2]. They optimize the load dispatching method to minimize the cost of industrial users and use the marginal price curve to decide whether to respond to real-time price and set up a corresponding model to calculate the resulting cost savings.

A.K.David holds that it is necessary to set a mathematic model for the user's electricity consumption characteristics and propose a user response model based on the theory of demand elasticity, user's rational consumption level and tariff signal[3]. It provides a clearer idea of the follow-up study.

3.2. Research on electricity behavior
It is an important way to analyze the power consumption behavior of industrial users to establish the mathematical model of the electrical behavior of various industrial equipment. In [4], the production power of manufacturing industry is modeled, and the mathematical model of industrial users as a whole to participate in demand-side bidding is constructed, but the management of the specific equipment is not analyzed systematically. Shah N states that the state task network (STN) can be used to describe the industrial production process[5], while Ding Y M establishes the power model of the production equipment by combining the parameters of the STN with the equipment of the production task[6].

Nolde.K takes the steel plant as the research object and uses a mixed integer linear programming to complete the production schedule on the basis of the given load curve[7]. Based on the above, Haït.A presents a continuous time model based on the description of the relative relations between the factory production tasks and the divided periods[8], which greatly improves the efficiency of the solution.

3.3. Research on Industrial user response behavior

In the study of industrial users, Bjørner T B uses the Panel unit root and the panel cointegration technology to do a deep research on the power demand of Danish industrial company[9]. The feasibility of cross-subsidy for power users in India is calculated by using the time-varying price elasticity of industrial power demand[10]. In [11], based on the principle of market equilibrium, the paper studies the response behavior of the user to electricity price under different marginal cost pricing rules.

In the aspect of formulating industrial electricity price plan, Geng J adopts the method of optimal load dispatching to minimize the electricity bill of industrial users[12]. According to whether the decision-making of continuous curve of hourly electricity price margins responds to the real-time electricity price, a corresponding model is established to calculate the cost- Side management and savings in electricity costs.

3.4. Research on Demand response assessment

In the research of demand response evaluation, many kinds of characteristic quantities are used to characterize the demand response behavior of industrial users from different perspectives. Literature [13] shows that it is very important to examine the adjustment of user load curve under demand response, the reliability and predictability of user's long-term and short-term behavior, and response speed when assessing the effectiveness of demand response in saving electricity generation investment. North American Electric Reliability Corporation summarizes the characteristics of demand response assessment should pay attention to, such as response power, response speed, implementation frequency and load recovery time[14]. These feature quantities are the focus of modeling demand response behavior.

It can be seen from the above studies that in the process of research on the field of industrial users' demand response, western scholars have established a systematic research system that includes user's electricity consumption behavior, demand price elasticity, electricity pricing scheme and response efficiency evaluation. The research ideas of this system are also applicable to China's exploration of the field of demand response. These studies provide a guiding ideology for the related research in China.

4. The Research Status on Demand Response of Industrial Users in China

In the field of domestic demand response research, Taiwan Province of China carried out the study and practice of time-of-use electricity price for industrial users earlier. J.N.Sheen and others analyzed the user's response to the selectable time-of-use electricity price from the nineties of last century for Taiwan's industrial power users[15]. Research on demand response started late in mainland China, and research on demand response of industrial users began to appear even after 2000[16].

4.1.Research on Electricity behavior

In the research on the inherent electricity consumption behavior of industrial users, L Kun takes into account the characteristics of high energy-consuming enterprises, such as the steel industry, that their production methods are special and the demand for electricity fluctuates with time. Aiming at the large random variation of industrial load, a deterministic model and a stochastic model for optimal scheduling of large consumers are established[17]. Y J Dang analyzes the impact of user industrial users on the production of the shift, process flow, product scale, energy consumption per unit of product output, characteristics of production machines and so on[18]. The research on the response of large industrial users after implementing the TOU price in Nanjing was conducted[19]. However, the analysis simply analyzed the data before and after the demand response and did not discuss how to control the possible influencing factors.

4.2.Research on Industrial user demand response model

Rational modeling of demand side's response behavior will instruct the formulation of time-of-use electricity price more reasonable. In different ideas of model building, literature [20] uses consumer psychology principles to establish a users' response model of time-of-use electricity price.A parameter identification model of users' response curve is established based on weighted least square method.For industrial users who have been implemented time-of-use electricity pricing, a simulation process of users' response behavior at the TOU price has been established.

J H Yuan adopts the method of multi-agent simulation and establishes the response rules of multi-users' decentralized decision based on fuzzy logic[21].Based on the previous ideas, J J Tan establishes an agent system including government, power supply department and user side, explore the mutual influence between them and establishes a model of multi-agent-based users' time-of-use price response[22]. The literature [23] presents the concept of price response lag and establishes a mathematical model based on this.

At present, many modeling methods are based on the analysis or simulation of the users' response behavior. Other modeling methods can also be used when a large amount of industrial user data is available. Such as using the expert system based on knowledge-based learning rules for large-users' time-of-use electricity price response modelling[24]. Using fuzzy comprehensive evaluation method which is based on users' data-based electricity attitude modelling[25]. Using statistical linear regression algorithm for real-time price response of air-conditioning load modelling[26] etc. Based on obtaining a large amount of data of user's electrical activity, the literature [27] conducts regression data mining using support vector machines to establish a model of response behavior of power users at time-of-use price. The model realizes the simulation of electric user's response behavior under time-sharing electricity price.

4.3.Research on Industrial User Elasticity Coefficient

In all kinds of response models, the user's demand elasticity coefficient is needed as the support to reflect the relationship between the valence signal and the user response behavior. In the earlier studies in China, the paper [25] first adopted the method of fitting the production cost function of various industries and using the coefficients of cost function to determine the price elasticity of time-sharing power demand, and made an assessment of the elasticity coefficient of electricity demand for the five most important industries of Taiwan Province. Such as textile, papermaking, petrochemical, nonmetal mineral and metal processing industry.

Z F Qin gives a complete establishment process of the electricity price elasticity matrix and analyzes the difference of the elastic matrix structure caused by the different electricity consumption characteristics of industrial users[28]. On this basis, Qin proposed a simplified method of elastic matrix.

Using the theory of factor derived demand in production economics, J Q Xin proves the characteristics of the short-term price elasticity of electricity demand to the peak-valley electricity price, and presents the zero-sum property of short-term time-share price elasticity of electricity demand[29] and proves the practical data of price elasticity which is applicable to the literature [25].

4.4.Research on pricing schemes for industrial users

In the literature [30], the price elasticity of demand is used to analyze the change of users' electricity consumption with electricity price. Taking the minimum difference between peaks and valleys as the goal, a constrained nonlinear programming single-layer model with a relatively simple peak-trough TOU is established considering the constraints of guaranteeing users' benefits and peak-to-valley electricity price ratio within a certain range.

In a more complicated study, literature[31] proposed a two-tier model of sales and purchasing decision-making for a sales company considering the user's demand response to the ever-increasing ability of users to respond to market prices and incentive information. In the upper model, corporate profit maximization as the goal, and the lower model to maximize user power utility as the goal. In the bivariate model of literature [32], the main power users of industrial power plants with self-provided power plants are the main body of the study, and the solution to such users' problems of using power supply and self-power generation is simulated to balance the problems of the power companies and power users Interests between.

The above documents respectively studied the demand response model and the demand-response time-demand price of industrial users from different perspectives. They also optimized and tested them according to the actual data, both of which have their own engineering application value. However, in general, compared with the developed theory and practice of industrial demand response in China, the theoretical level of China's research is still at the stage of conceptual interpretation. The institutional environment, technical means, implementation requirements and project evaluation for introducing demand response Tools and other core content has not been an in-depth study. In practice, most of them belong to incentive control for large industrial users. There are few demand response projects to achieve the goal through electricity price method, and the lack of analysis of the implemented electricity price projects. In these aspects, Chinese scholars need to conduct further research.

5. Conclusion and Prospect

The price-type demand side response is the direction and main form of the demand side response under the electricity market conditions. There are innumerable kinds of price theory designs in theory. There are many experiences in Europe and the United States in the practice and study of responding to the price-type demand side for learning in China. In contrast with the countries in Europe and the United States, the relevant theoretical research is constantly improving in various aspects under the continuous efforts of Chinese scholars. However, in addition to the time-of-use price mechanism, the application is still in its infancy. At present, the most important point in China is to set up a sound power market system as soon as possible and provide a platform for implementation of various new electricity price mechanism projects. Only when the actual interaction with users in the mechanism exploration gets the user's response can they be given more opportunities China's customer satisfaction effect better price mechanism.

Acknowledgments

2017 State Grid Corporation of China Science and Technology Project：Research and Application of Multiple Load Active Response and Predictive Control Technology for Stochastic Power Supply. Contract No: SGTJDK00DWJS1700034.

References

[1] Daryanian B, Bohn R E, Tabors R D. Optimal demand-side response to electricity spot prices for storage-type customers[J]. IEEE Transactions on Power Systems, 1989, 4:3(3):897-903.
[2] Roos J G, Lane I E. Industrial power demand response analysis for one-part real-time pricing[J]. IEEE Transactions on Power Systems, 1998, 13(1):159-164.
[3] David A K, Li Y Z. Consumer rationality assumptions in the real-time pricing of electricity[J]. IEE Proceedings C - Generation, Transmission and Distribution, 1992, 139(4):315-322.
[4] Mohagheghi S, Raji N. Managing Industrial Energy Intelligently: Demand Response Scheme[J]. Industry Applications Magazine IEEE, 2014, 20(2):53-62.
[5] Shah N, Pantelides C C, Sargent R W H. A general algorithm for short-term scheduling of batch operations—II. Computational issues[J]. Computers & Chemical Engineering, 1993, 17(2):229-244.
[6] Ding Y M, Hong S H, Li X H. A Demand Response Energy Management Scheme for Industrial Facilities in Smart Grid[J]. Industrial Informatics IEEE Transactions on, 2014, 10(4):2257-2269.
[7] Nolde K, Morari M. Electrical load tracking scheduling of a steel plant[J]. Computers & Chemical Engineering, 2010, 34(11):1899-1903.
[8] Haït A, Artigues C. On electrical load tracking scheduling for a steel plant[J]. Computers & Chemical Engineering, 2011, 35(12):3044-3047.

[9] Bjørner T B, Togeby M, Jensen H H. Industrial companies' demand for electricity: evidence from a micropanel[J]. Energy Economics, 2001, 23(5):595-617.
[10] Chattopadhyay P. Testing viability of cross subsidy using time-variant price elasticities of industrial demand for electricity: Indian experience[J]. Energy Policy, 2007, 35(1):487-496.
[11] Ding Z, Lee W J, Wang J. Stochastic Resource Planning Strategy to Improve the Efficiency of Microgrid Operation[J]. IEEE Transactions on Industry Applications, 2015, 51(3):1978-1986.
[12] Geng J, Wang X, Bai X, et al. Comparison of block bidding and hourly bidding based on case study[C]// International Conference on Power System Technology, 2002. Proceedings.Powercon. IEEE, 2002:1387-1391 vol.3.
[13] Woychik E C. Optimizing demand response[J]. Public Utilities Fortnightly, 2008(5): 52-56.
[14] North American Electric Reliability Corporation. Potential reliability impact of emerging flexible recourses, IVGTF task-1-5[R]. Princeton: NERC, 2011.
[15] Sheen J N, Chen C S, Yang J K. Time-of-use pricing for load management programs in Taiwan Power Company[J]. IEEE Transactions on Power Systems, 1994, 9(1):388-396.
[16] Y Li, Wang Zhihua, Lu Yi, etc. The implementation of peak-valley time-sharing Price and the response of large industrial users [J]. Power system Automation, 2001,25 (08): 45-48
[17] K Liu. Optimal scheduling model for high energy-consuming enterprises with uncertainties in their generation [A]. China Automation Institute control Theory Professional Committee, China Systems Engineering Society. Proceedings of the 32nd Session of China Control Conference (vol. b) [C]. China Automation Institute control Theory Professional Committee, China Systems Engineering Society: 2013,6.
[18] Dang Yujun. Strategic study on economic benefits between balanced power grid and self-owned power plant [D]. Power (Beijing), 2004.
[19] Liu Xiaolin, Wang Zhaojie, Gao Feng, Wujiang, Prof, Zhou. Power consumption response of high energy-consuming enterprises under time-sharing price [J]. Power system Automation, 2014,38 (08): 41-49.
[20] Liu Yan, Tan Zhong rich, begging to build a hoon. Optimization model of peak-Valley time-sharing pricing design [J]. Chinese Management Science, $number (5): 87-92
[21] J H Yuan, Li, Hu. Study on the response of large subscriber peak-valley electricity price based on multi-agent simulation [J]. Power grid Technology, 2005, 29 (23): 48-53
[22] J J Tan, Wang Beibei, Li. A multi-agent-based response model for user-timed pricing [J]. Power grid Technology, 2012,36 (2): 257-263.
[23] Yau Yu, Liu Junyong, Liu Youpo, etc. A method to classify the user's tariff response mode with the sensitivity of delay index [J]. Power grid Technology, 2010,34 (4): 30-36.
[24] He Yongxiu, Wang Bing, Sunwei, etc. The analysis and interaction mechanism design of residents ' intelligent power consumption based on fuzzy comprehensive evaluation [J]. Power grid Technology, 2012,36 (10): 247-252.
[25] Mathieu J L, Gadgil A J, Callaway D S, et al. Characterizing the Response of Commercial and Industrial Facilities to Dynamic Pricing Signals From the Utility[C]. ASME Energy Sustainability. 2010:1019-1028.
[26] Cortes C, Vapnik V. Support-vector networks[C]. Machine Learning. 1995:273-297.
[27] Liu Jidong, Han Coxan, Han Weiji, etc. Model and algorithm of user response behavior under time-sharing pricing [j]. Power grid Technology, 2013 (10): 2973-2978.
[28] Z F Qin, Yeshunmin, Yu Yu Xin, and so on. The electricity price elasticity matrix in the retail-end electricity market [J]. Power system Automation, 2004, 28 (5): 16-19.
[29] J Q Xin, Chenghaozhong. Characteristics of short-term price elasticity of power demand and its application [J]. Power system Automation, 2007, 31 (10): 32-35.
[30] Li Chunyan, approval, Ma Zhiyuan. An optimal model of time-sharing pricing with user demand response [J]. Journal of power System and automation, 2015, 27 (3): 11-16.

[31] Y Ning, M Zhou, Li Gengban. Two-tier model of purchase and sale decision of electric power company considering user demand response [J]. Power system automation, 2017 (14): 30-36.
[32] L Lin, the method of making time-sharing electricity price for large power users [D]. Power. 2016.

Research on the Economic Allocation Method of High-Penetrated Distributed Generations for Smoothing the Power Fluctuation of Active Distribution Network

FU Ming[1], CHEN Chun[1], ZHAO Jingtao[1], DU Jian[1], CAO Jing[1], YIN Hongyuan[2] and LIU Yujun[2]

[1] NARI Group Corporation, Nanjing 211000, China
[2] School of Electrical Engineering, Southeast University, Nanjing 210096, China

liuyujun0713@126.com

Abstract. With the increasing penetration of DG and electric vehicles in distribution network, ADN gradually takes the place of traditional distribution network, which also brings higher requirements to the configuration of distributed ESS; In this paper, we mainly discuss the problem of how to optimize the capacity of wind/photovoltaic/energy storage in ADN with high penetration of DG when the total power connected to the grid of DG is fixed, in order to realize the harmonization of the economic and stable operation in power grid. Firstly, the hybrid system is analysed, which demonstrated the feasibility and rationality of proposed scheme. Secondly, being in distributed network investors' shoes and considering the constraints of distributed network, DG and ESS, the wind/photovoltaic/energy storage capacity configuration bi-level optimization model is established with the objective function of maximum annual income and minimum fluctuation of DG output power. Then the genetic algorithm is used to solve this model. Finally, the experimental results of 33-bus system verify the effectiveness of the proposed method, and the influence of different penetration scenarios on the annual income of distributed network is also analysed to provide a new idea for the economic operation of ADN.

1. Introduction

With the access of distributed power supply, the distribution network is developing from the traditional form to the active distribution network. Active distribution network(ADN) means the distribution network with capacity of active control and operation, which has high penetration of distributed generation(DG). By the coordination of local energy Management, ADN ultimately realizes the effective integration of DG, energy storage systems(ESSs) and large-scale centralized power generation. However, wind power(WG), photovoltaic power(PV) have great volatility, which is bound to impact the stability and security of the power grid [1-3].

ESS is the key for ADN to realize flexible adjustment and network optimization operation [4-5]. Combined with WG and PV systems, ESSs can effectively improve the overall characteristics of output of active power. But the devices of EES are expensive and short-lived, and it is difficult to configurate widely [6]. Therefore, the rational distribution of the DGs with a high permeability in ADNs will affect the operation and management level of the ADNs and the economy of operation.

In recent years, many scholars have been carrying out many studies for the configuration of DGs. In the literature [7], the percentages of power complete compensation time and energy compensation are determined by combining the sliding average control and the slope limit control, and the average energy storage capacity is determined by these two indicators of power smoothing. In [8], a capacity optimization algorithm for the Fourier-Legendre series expansion of the state of charge(SOC) of the ESS is proposed. In the literature [9-10], the discrete Fourier transform is used to perform the spectral analysis on the output power to obtain the minimum storage capacity. Literature [11] studies the optimal allocation of energy storage in distribution network with high-PV permeability. However, the above literature does not take the economic problems of energy storage configuration into account. At present, some scholars have made useful attempts in terms of economic efficiency of energy storage planning: [12] establishes the hybrid optimization model based on the maximum benefit of the life cycle of ESSs, but it does not involve the configuration of DGs. Literature [13] considers the typical day as the analysis unit, and try to minimize the investment and operating costs of distribution energy storages. In [14], by considering the economy and reliability of distribution networks, the multi-objective optimal configuration model of EES with battery is established, and the improved Pareto algorithm is used to solve the Pareto frontier of the model. But most of the current economic researches consider the configuration of DGs as a known condition of the grid. So, they ignore the future ADN's requirements on the flexible control and switching of DGs.

This paper stands the point of view of the investment manager of distribution network, and defines the distributed generations and energy storage system as the main access form. The objective function is the maximum annual net profit after optimal allocation. By considering the safe operation of grid, DG outputs, SOC of storages and so on as the constraints, this paper establishes a two-tier optimization model to structure storage operation mode. Besides, the inner layer optimization model is based on the spectrum analysis results of DG power output. The frequency test method is applied to determine the required capacity of the ESSs and optimize the strategy of operation. To find the optimal economic ratio with the stable permeability of DG, the outer optimization model is used to calculate the mixing ratio of distributed generations and energy storage system by using genetic algorithm. Finally, an example based on IEEE33 node distribution network is given, and the change of annual income of system under different DG permeabilities is discussed in this paper.

2. Analysis of wind/photovoltaic/energy storage hybrid system

2.1. The analysis of output fluctuation in the distributed generation

With the increasing penetration of DGs, the impact of power fluctuations on the safe operation of the distribution network is growing. It is urgent for relevant aspects to make the standards.

The analysis of fluctuations is based on typical sample data from each DG, which are obtained from historical observations or typical daily sampling. The fluctuation degree of DG output is determined by observing DG maximum power fluctuation value $\Delta P_{\max}$ in a given time period [15]:

$$\Delta P_{\max} = \max(p_1, p_2, ..., p_n) - \min(p_1, p_2, ..., p_n) \tag{1}$$

Where, p_j (j=1,2, …, n) represents the output power value at time j for the given period. On this basis, this paper defines the rate of DG output power fluctuations during a given period of time is:

$$\xi = \frac{\Delta P_{\max}}{P_{avg}} \times 100\% \tag{2}$$

Where, P_{avg} is the average of DG output power.

2.2. Feasibility analysis of wind/photovoltaic/energy storage hybrid system's configuration

The devices of ESS generally have two kinds of configuration: the centralized and the distributed [16]. The centralized means that ESSs are installed intensively in the specific location, while the distributed means that each DG is configured with corresponding ESS which can only be used by its parallel DG.

This paper chooses the distributed configuration as the main research goal. The advantage of it is that the storage capacity is less than the centralized configuration, and its economy is higher and can be installed flexibly, which can effectively increase the use rate of DG power in the power network. As the battery technology matures, the future of electric vehicles can also manage charge and discharge reasonably to assume the distributed energy storage tasks [17]. If ESSs are applied by the distributed configuration, they can better solve the planning and operation problems of the different distributed power generation units, and can take the large-scale electric vehicle management into account. So, the distributed configuration has greater feasibility, and its prospect is more long-term.

3. Optimal allocation model of wind/photovoltaic/energy storage hybrid system

3.1. Objective function

3.1.1. The outer layer optimization. In this paper, the outer layer optimization mainly focuses on the distribution network's management. Taking profit and cost into account, the outer layer optimization analyses the economic benefits of wind/photovoltaic/energy storage system's capacity allocation after being used in the multi-node of the distribution network. The objective function is shown below. Since the initial investment is large, all costs are converted to one year from the long term.

$$\max\ F_{total} = f_{pur} + f_{save} + f_{loss} + f_{DR} - f_{cons} - f_{main} \tag{3}$$

Where, f_{pur} is the total saved cost of electricity after DGs connected; f_{save} is the saved cost of the grid expansion due to the ESS; f_{loss} is the reduced expense of net loss; f_{DR} is the arbitrage income by reserving less and using more energy; f_{cons} is the total cost of constructing the distributed power supplies and support equipment; f_{main} is the corresponding costs of operation and maintenance [18-20].

3.1.2. The inner layer optimization. In this paper, the main function of installing energy storages is to smooth DGs' output power. T is defined as the number of sampling data in the charge/discharge cycle, and the adjacent M data consist as a time window [21]. The main goal of the inner layer optimization is to find the minimum fluctuation variance of the output power of DG smoothed by the energy storage device in each time window. The objective function is as follows:

$$\min\ F_{flu} = \sum_{i=1}^{T-M+1}\sum_{t=i}^{i+M-1}(P_{DG}(t)+P_{ESS}(t)-P_{av}(i))^2 \tag{4}$$

Where, $P_{DG}(t)$ represents the value of DG's output power at time t of a time window. $P_{ESS}(t)$ represents the optimal output of the ESS at time t. $P_{av}(i)$ is the average of equivalent output over the time window. The equivalent output of the i-th time window is expressed as:

$$P_{av}(i) = \frac{1}{M}\sum_{t=i}^{i+M-1}(P_{DG}(t)+P_{ESS}(t)) \tag{5}$$

The maximum absolute value of the actual charge/discharge curves $P_{ESS}(t)$ in the ESS is the maximum charge/discharge power required by ESS, and that is the rated power value P_{ESSN} of the ESS. Based on the actual output power value of the ESS, the storage energy fluctuation $W(t)$ of the energy storage system can be obtained by accumulating the charge and discharge electricity only at each sampling time. Finally, taking SOC into account, coupled with the energy fluctuations of ESS in the period of corresponding time, the minimum capacity of ESS can be found, and that is rated capacity value of the energy storage system W_{ESSN}. The expression is as follows [9-11]:

$$\begin{cases} P_{ESSN} = \max\left\{\dfrac{P_{ESSd}(t)}{\eta_d}, |P_{ESSc}(t)|\eta_c\right\} \\ W(t) = \displaystyle\sum_{t=1}^{T}(P_{ESS}(t)\cdot\Delta t) \\ W_{ESSN} = \dfrac{W_d(x)+|W_c(x)|}{SOC_{\max}-SOC_{\min}} \end{cases} \tag{6}$$

Where, $P_{ESSd}(t)$ is the discharge power in the energy cycle. η_d is the discharge efficiency. $P_{ESSc}(t)$ is the charging power. η_c is the charging efficiency. Δt is the interval of adjacent data points. $W_d(t)$ is the maximum fluctuation value of the forward energy in the calculation period. $W_c(t)$ is the maximum fluctuation value of negative energy. SOC_{max} and SOC_{min} are the upper and lower bounds of SOC.

3.2. Constraints

The constraints are mainly considered from the grid side, DG and energy storage.

3.2.1. Grid side constraints.

(1) node power balance constraint

For the distribution network system, the active power and reactive power of each node of the distribution network including DG and energy storage system must meet the equilibrium constraint condition of the equation.

(2) node voltage constraints

DG and energy storage access to change the distribution network system within the trend, which will inevitably lead to voltage deviation. Node voltages, as an important measure of power quality indicators, need to be considered as a constraint.

3.2.2. DG constraints.

(1) DG output constraints

Subject to the natural environment, operating conditions of equipment and so on, the maximum DG output is certain. The constraint for DG output can be expressed as follows:

$$\begin{cases} 0 \le P_{\rm DG} \le P_{\rm DGmax} \\ Q_{\rm DG} \to 0 \end{cases} \tag{7}$$

Where, $P_{\rm DGmax}$ is the maximum output of DG after being connected to the grid. It is assumed that there is no reactive power exchange operation between DG and the grid, so $Q_{\rm DG}$ take 0.

(2) DGs output fluctuation constraint

Fluctuations of DGs output power should be guaranteed within a certain range, so as to ensure that DGs will not have a great impact on the safe operation of the grid, when they are connected to the grid.

$$\xi_i \le \xi_N \tag{8}$$

Where, ξ_i is the output fluctuation rate of the i-th DG, and ξ_N is the maximum output fluctuation rate allowed by the system.

3.2.3. Energy storage constraints.

(1) SOC equation constraint:

In the ESS, the state of charge represents the ratio of the remaining capacity of the energy storage device to the fully charged state capacity. The SOC of ESS at $t + 1$ is determined by the supply and demand at t and the charge / discharge state of ESS. The model is as follows:

$$SOC(t+1) = SOC(t) - (\alpha_{c,t}\frac{P_{ESS}(t)\eta_c}{W_m} + \alpha_{d,t}\frac{P_{ESS}(t)}{W_m\eta_d})\Delta t \tag{9}$$

Where, SOC(t) is the state of ESS at time t; $\alpha_{c,t}$, $\alpha_{d,t}$ is the variable of charging/discharging energy in the standardized display. At the time of charging, $\alpha_{c,t} = 1$, $\alpha_{d,t} = 0$. At the time of discharge, $\alpha_{c,t} = 0$, $\alpha_{d,t} = 1$. W_m is the rated capacity of ESS, and Δt is the time interval between charge and discharge.

(2) charge state amplitude constraint:

$$SOC_{\rm min} \le SOC_{\rm t} \le SOC_{\rm max} \tag{10}$$

Where, SOC_{min} and SOC_{max} are the lower and upper limits of the SOC of ESS respectively.

(3) continuous operation constraints:

$$\Delta E = \sum_{t=1}^{N} \frac{P_{\mathrm{ESS}}(t)}{3600} \Delta t = 0 \tag{11}$$

In its one cycle of operation (i.e., the time scale of the sample data), the charge and discharge energy of ESS should be substantially equal [22-23].

4. Solution of the model

4.1. Algorithm

4.1.1. The test frequency method. The sample output data of DG output power is subjected to do the discrete Fourier transform, and the width of the frequency band is gradually increased from high frequencies to low frequencies to determine the high-frequency part which should be filtered by the given power fluctuation constraint, and make the frequency band as small as possible. The filtered high-frequency power components are absorbed by the ESSs, and the remaining low-frequency power component is the smooth output of the DGs and the ESSs.

4.1.2. Genetic algorithm. Genetic algorithm has the characteristics of parallelism, adaptive optimization and so on. It is suitable for the request of using multiple distributed power supply.

In this paper, every individual generated by the genetic algorithm is encoded with a fixed number of binary numbers, and then the fitness function (objective function) is taken into the set scenario. According to genetic operation, increase the probability of individuals with high adaptability to enter the next generation. Through the cross, mutation and other operations to imitate the biological community to produce new individuals, the early precocious phenomenon, that entire population get into the local optimal, can be prevented. The next generation of the population is repeated according to the above steps until the termination condition is met.

4.2. Solution

In this paper, genetic algorithm is used to optimize the model of wind/photovoltaic/energy storage hybrid system built in Section 3. The inner layer optimization uses the test frequency method to reduce the fluctuation of the output power of the distributed power supply, and determines the storage capacity and its operation strategy according to the requirements of distributed power fluctuation. Then, the result of inner layer optimization is substituted into the outer objective function. According to the objective function, continue to screen and compare. Finally, the optimal configuration scheme is obtained under the condition of DG permeability determined.

5. Case study

5.1. Model and parameter settings

Select the IEEE33 node distribution network as the test system. The network has a total of 32 branches. The topology of the network structure shown in Figure 1. Set the Nom.19 and Nom.31 nodes of network accessed by WG, and Nom.20 and Nom.32 nodes accessed by PV.

The output data of a typical intraday WG and PV are selected as the research object, and the sampling period is 15min. Both the rated powers of WG and PV are 100 kW. The average output of the WG is 52.79kW and the maximum fluctuation rate of any 60min is 36.97%. The average output of PV in the typical daylight is 26.01kW and the maximum fluctuation rate of any 60min is 75.23%.

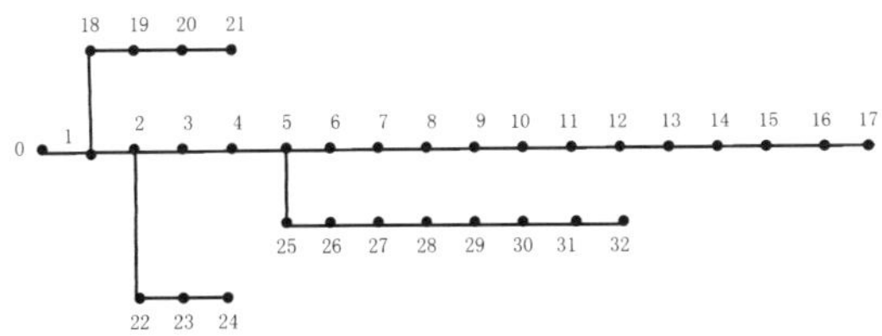

Figure 1. IEEE 33-bus system

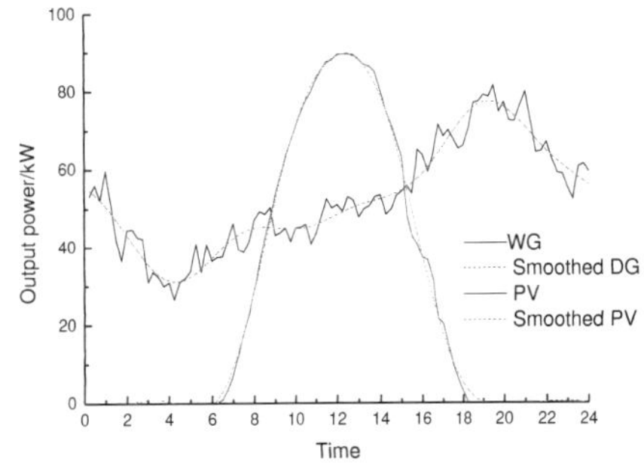

Figure 2. Curves of wind power, photovoltaic output and smooth result in typical day

In this case, the maximum fluctuation rate of the smoothed output of WG is 10% and the maximum fluctuation rate of PV output is 70% when calculating the capacity of ESS required to reduce the DG fluctuation. The capacity-type lithium-ion batteries are used as energy storage devices. Assume that both the charge and discharge efficiency are equal to 92.74%. The upper limit of SOC is 1 and the lower limit is 0.6. The characteristic curves of WG and PV output power in all the days are shown in Figure 2 respectively.

5.2. Results and analysis

5.2.1. Optimal configuration of wind-photovoltaic-storage mixed systems. Set the total capacity of the DGs connected to the grid to be 2MW, and that is about 40% of power generation. In the ADN with high permeability distributed power supply, the genetic algorithm evolves to 14 generations before convergence through the optimization method proposed in this paper. The optimal configuration scheme of the wind/photovoltaic/energy storage hybrid system is shown in Table 1.

Table 1. The optimal configuration scheme

Node	19	20	31	32
The rated power of DG /(kW)	292.68	351.22	600	756.10
Power of energy storage /(kW)	27.23	26.42	55.82	56.86
Capacity of energy storage /(kW·h)	46.52	37.92	95.32	81.63

In this scheme, the optimal matching coefficient of the WG systems and PV systems is about 0.81 through the flexible-switching technology of ADN, in which the wind power accounts for 44.63% of the total capacity of the grid-connected DGs, and the PV accounts for 55.37%. And the largest annual income of the distribution network is about 763,600 yuan.

After optimized, the fluctuation rate of WG power is reduced to 9.43% in the same time window, the fluctuation rate of PV power is reduced to 68.84% in the same time window, which satisfies the requirements of DG fluctuation in the inner layer optimization. After smoothed, the curves of the rated WG output and the rated photovoltaic output are shown in Figure 2, while the curve of energy storage device output is shown in Figure 3.

The fluctuation curve of SOC is shown in Figure 4. The WG-ESS represents the curve of the energy storage device used to smooth wind power output, and the PV-ESS represents the curve of the energy storage device used to smooth photovoltaic output. It is clear that they are normally operating within the energy storage constraints.

5.2.2. Study on changes of DG permeability. This section examines the relationship between the annual revenue of the distribution network and the change in the permeability of DGs on the grid.

When the permeability is too low, the scenery storage configuration of the power grid has little effect. Therefore, this paper mainly analyses the cases that DG's Permeability to the distribution network is from 30% to 50%, as shown in Figure 5.

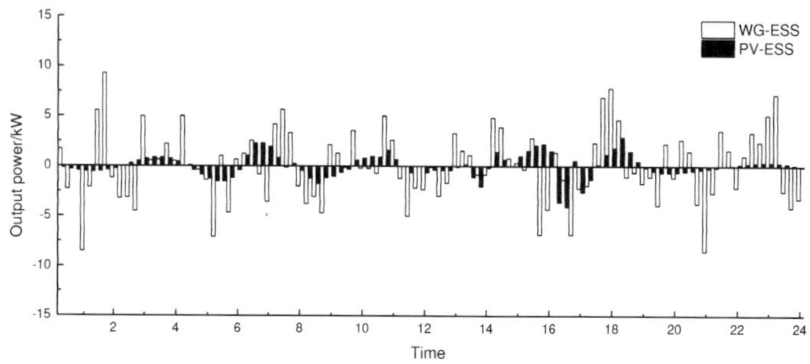

Figure 3. Curves of ESS output power.

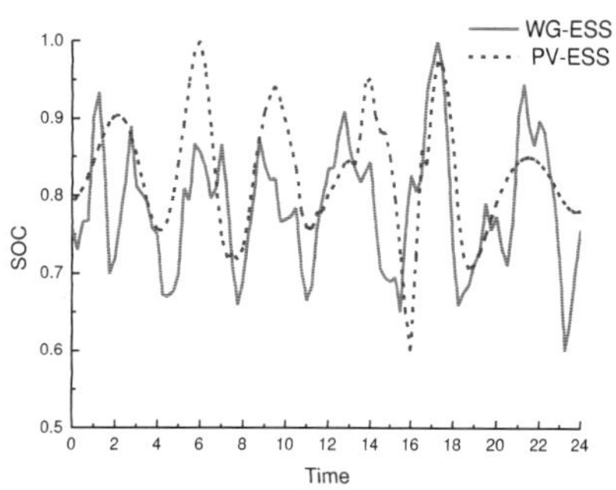

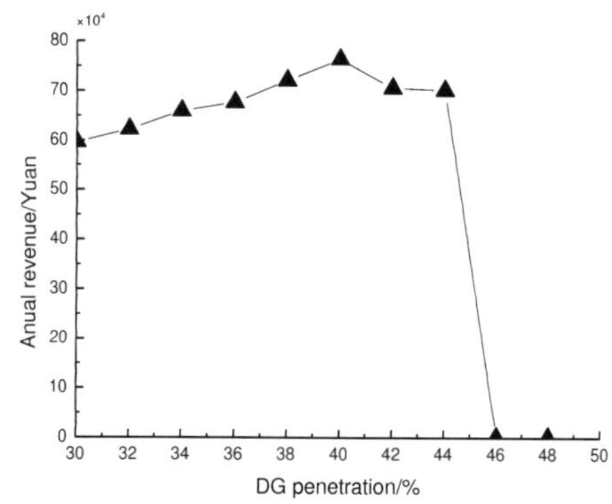

Figure 4. The change state of SOC.

Figure 5. The curve of annual income varies with penetration.

It can be found that the annual income of distribution network is gradually improved with the increase of DG's permeability. However, when the DG's permeability increased to 40%, the annual income began to decrease. Because with the DG access capacity becomes larger, the amplitude of node voltage and other constraints hinder the ratio of wind/photovoltaic/energy storage to obtain the maximum benefit. When the DG's permeability is greater than 44%, regardless of the ratio, the annual income is a minimum. Because the optimal configuration results do not meet the constraints, and the penalty is penalized by the algorithm penalty factor.

It can also be seen from the above analysis that the maximum permeability of the distributed power supply in this example is about 44%. When the permeability exceeds this value, the system node voltage amplitude will exceed the upper limit. At this point the grid will not be safe to run, and it is meaningless to discuss the economy.

6. Conclusion

The background of this paper is the ADN with high permeability of DGs. The object of this paper is to optimize the capacity of the wind/photovoltaic/energy storage hybrid system, considering the suppression of DGs' output volatility.

A double-layer optimization model is established to find the optimal economic configuration of the hybrid system's capacity. The optimization model of capacity is solved by genetic algorithm with good searching ability and convergence. The results of examples show that the model and method of

DGs and ESSs' configuration can get the reasonable scheme considering the economy and grid safety under the condition of DG's permeability determined. At the same time, the influence of different permeabilities of DG on the objective function is studied. At a later stage, the specific impacts and optimization strategy of the distributed energy storage network with the characteristics of electric vehicles will be further studied to match the development of flexible-switching technology of ADN.

Acknowledgments
Science and Technology Project of State Grid Corporation of China: Research and Application of Plug -and-Play Key Technology for Random Power.

References

[1] El-Khattam W, Salama M M A 2004 Distributed generation technologies, definitions and benefits *Electric Power Systems Research* **71** 119-28.
[2] Liu Y, Wu Z and TU Y 2008 A Survey on Distributed Generation and Its Networking Technology *Power System Technology* **32** 71-76.
[3] Dugan R C and McDermott T E 2002 Distributed generation *IEEE Industry Applications Magazine* **8** 19-25.
[4] You Y,Liu D and Yu W 2012 Technology and Its Trends of Active Distribution Network *Automa-tion of Electric Power Systems* **36** 10-6.
[5] Jim M 2008 Adaptive intelligent power systems: Active distribu-tion networks *Energy Policy* **36** 4346-51.
[6] Xiao H , Pei W and Yang Y 2015 Energy Storage Capacity Optimization for Microgrid Considering Battery Life and Economic Operation *High Voltage Engineering* **41** 3256-65.
[7] Han W, Liu J and Liu Y 2017 Storage Capacity Optimization in Active Distribution Net-work Based on Fourier-Legendre Series Expansion *Power System Technology* **41** 1123-33.
[8] Wang C, Yu Bo and Xiao Jun 2012 Sizing of Energy Storage Systems for Output Smoothing of Renewable Energy Systems *Proceedings of the CSEE* **32** 1-8.
[9] Sang B, Wang D and Yang B 2014 Optimal Allocation of Energy Storage System for Smoothing the Output Fluctuations of New Energy *Proceedings of the CSEE* **2014** 3700-06.
[10] Xu X, Martin Bishop and Donna G 2016 Application and Modeling of Battery Energy Storage in Power Systems *CSEE JOURNAL OF POWER AND ENERGY SYSTEMS* **2** 82-90.
[11] Tao Q, Sang B and Ye J 2016 Optimal Configu-ration Method of Distributed Energy Storage Systems in Distribution Network with High Penetration of Photovol-taic *High Voltage Engineering* **2016** 2158-65.
[12] Xiang Y, Wei Z and Sun G 2015 Life Cycle Cost Based Optimal Configuration of Battery Energy Storage System in Distribution Network *Power System Technology* **2015** 264-70.
[13] Tang W, Liang W and Cui R 2015 Optimal Allocation Method of Distributed Energy Storage System in Distribution Network *Electric Power Con-struction* **36** 38-45.
[14] Liu F, Li B and Zhang F 2015 Battery Energy Storage System Optimal Allocation Considering Economy and Re-liability of Distribution Network *Electric Power Con-struction* **2015** 76-83.
[15] You Y, Liu D and Zhong Q 2014 Multi-objective Optimal Placement of Energy Storage Systems in an Active Distribution Network *Automation of Electric Power Systems* 2014 46-52.
[16] Shen X, Shahidehpour M and Han Y 2016 Expansion Planning of Active Distribution Networks With Centralized and Dis-tributed Energy Storage Systems *IEEE Transactions on Sustainable Energy* **8** 126-34.
[17] Li Z, Zhao S and Liu Y 2016 Control Strategy and Application of Distributed Electric Vehicle Energy Storage *Power System Technology* **40** 442 -50.
[18] Yan Z, Wang C and Zheng Jian 2013 Value assessment model of battery energy storage system in dis-tribution network *Electric Power Automation Equipment* **2013** 57-61.
[19] Ted K A B, Alex Y and Annette J 2011 Optimal energy storage sizing and control for wind power applications *IEEE Transactions on Sustainable Energy* **2** 69-77.

[20] Chen H, Thang N and Yang W 2009 Progress in electrical energy storage system: A critical review *Progress in Natural Science* **19** 291-312.

[21] Yu D, Song S and Zhang B 2011 Impacts of Electric Vehicle on Power System as Well as the Associ-ated Dispatching and Control Problem *Electric Power Automation Equipment* **2011** 24-9.

[22] Nick M, Cherkaoui R and Paolone M 2017 Optimal Planning of Distributed Energy Storage Systems in Active Distribution Networks Embedding Grid Reconfiguration *IEEE Transactions on Power Systems* **99** 1.

[23] Wang Y, Tan K T and Peng X Y 2016 Coordinated Control of Distributed Energy-Storage Systems for Voltage Regulation in Distribution Networks *IEEE Transactions on Power Delivery* **31** 1132-41.

Research on application of lithium titanate battery in auxiliary system of rail vehicle

Qing Tian

Senior Engineer of CRRC QINGDAO SIFANG CO., LTD.

liqiwei@shenzhou-gaotie.com

Abstract. By real-time acquisition information of lithium titanate battery packs and single battery in the auxiliary battery system of rail vehicles, we have extracted the operating condition of the train auxiliary battery system, and emphatically established the battery float charging test under the influence of temperature, voltage and other factors. Then have analyzed the aging state information of the battery. Combined with the thermodynamic and dynamics characteristics of the power battery, established the law of change of the battery characteristic parameter with the battery float charge, and determined the recession mechanism of battery

1. Introduction

At present, power accumulator is widely used in auxiliary power system in rail transit vehicles, its main function is to act as an emergency backup power for the vehicle load (such as air conditioning, ventilation, lighting, lift bow, etc.) power supply [1, 2] in case of the vehicle power supply network fault, lack of catenary voltage or current transformer fault,. The lead acid and nickel cadmium battery are still mainly applied inrail vehicles auxiliary power system, however, the two types of battery exist problems such as larger mass volume, longer charging time, shorter service life and heavy metal pollution [3]. Power lithium-ion battery has advantages of light weight, small volume, long service life, high voltage, high efficiency of charge and discharge and no pollution, so it has been widely used in the field of new energy vehicles and power grid energy storage, but its application in the field of rail transit is in its infancy [4]. Among a variety of power type lithium-ion batteries, lithium titanate battery that applies lithium titanate materials as the cathode of batteries, has better low temperature properties, and the battery cathode surface does not form the SEI film, so it has a higher security and longer service life and make it become one of the most potential applications of battery in rail transit field [5]. In order to meet the requirements for lightweight, pollution-free and long life of the rail vehicle batteries, the DC 110V auxiliary power source of the rail vehicle is gradually using lithium titanate battery as the battery power supply system. On the basis of ensuring the emergency power supply function of the system, it also provides preparation for the large-scale promotion of future rail vehicles and the formulation of the standard of lithium battery system.

2. Analysis for operating conditions of rail vehicle auxiliary battery system

The main technical parameters of each set of batteries for rail vehicles are shown in Table 1. The auxiliary power supply system mainly supplies power in three phases in the actual vehicle operation applications,: first, the auxiliary power supply is required to provide the electric power to lift bow when the train starts; secondly, the auxiliary power supply is required to providepower for emergency load for the train in case of the vehicle power supply network fault, lack of catenary voltage or current transformer fault; Thirdly, when the train ends running and rests, the auxiliary power supply is required to provide the electric power to drop bow. After the lifting bow discharge and emergency discharge, the battery will be charged through the pantograph and into the float chargingstate after reaching the cut-off voltage.

Table 1. The main technical parameters of each set of batteries for rail vehicles

Battery type	Lithium titanatebattery	Max. charging current of battery pack	320A
Single voltage	2.3V	Undervoltage protection value of battery pack	67.5V (single 1.5V)
Battery group mode	45series (190Ah/110V)	Overvoltage protection value of battery pack	121.5V (single 2.7V)
Constant voltage control voltage of battery pack	112.5V	Battery operating temperature range	-40~55℃

Figure 1 is the operating conditions curve of two working days test forauxiliary battery system, the operating condition of Figure 1 (a) is relatively complex and changeable, wherein the symbol ① is the float charging condition, the battery voltage remains unchanged and the current is in a floating state; symbol ② is shelving conditions, the battery current is fixed value 0, and the voltage gradually diminishes due to polarization fading, the rest of the conditions is the battery discharging conditions; Figure 1 (b) is the second day's battery test conditions, the train is always in power state, so there is no shelving conditions, it is mainly float charging conditions. By analyzing the actual operating data collected, it is found that the auxiliary battery system operating conditions mainly include three conditions: shelving, float charging and emergency discharging, and the proportion of time spent on battery shelving and floating is relatively greater. Lithium titanate battery is used as an auxiliary battery system for rail vehicles, the train takes electricity from the catenary in normal operation, and the battery is generally in a float charging; After the train returning to the garage at night, the battery is disconnected from the load and will be in a shelved storage state. When the train needs lift and drop the bow or emergency situations, the battery should be discharged, after lifting bow, the battery will be fully charged and then into the float charging state. The operating conditions mainly include the above three types during the train actual operating. Due to discharge condition is less and the discharge depth is smaller, the conditions that affect the aging of the train battery are mainly two conditions of shelving and floating.

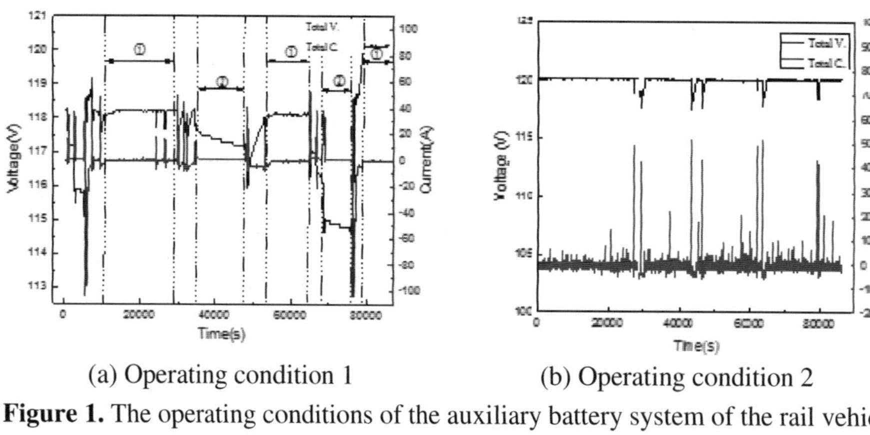

(a) Operating condition 1 (b) Operating condition 2

Figure 1. The operating conditions of the auxiliary battery system of the rail vehicle

At present, there are few research literatures on the aging life of lithium titanate batteries, and there are few literatureson the analysis of aging mechanism of batteries. This paper will focus on the test for lithium titanate battery floating condition, considering the influence of different temperature, voltage, such as stress; choose two different manufacturer of lithium titanate battery for test, comparative analysis of float charging condition battery aging characteristics and recession mechanism.

3. Lithium titanate battery float charging test

In the rail vehicle auxiliary battery system, the lithium titanate battery is in a float chargingstate for a long time when the train is in normal operation. Derived from use of the lead-acid battery system which has a large self-discharge, the definition of the float charging generally refers to that the charger is still recharging the battery with a small current after the battery is fully charged to make up the capacity loss of battery self-discharge. In the shelving process, battery still has a phenomenon of lessself-discharge, float charging method is used by lithium titanate battery system in the train driving process. As a long-life, high safety lithium-ion battery, lithium titanate battery is used as a train auxiliary battery system, its research literatures onfloat charging is less. Some scholars have researched that lithium titanate batteries float charging for 250 days at 35℃, 45℃, its capacity retention rate is 100% and 98%. Taking into account China's vast territory, greater temperature difference between North and South, so the designed test temperature is 50℃highest and -40℃ lowest, float charging to battery is carried outin extreme circumstances. The battery float charging test platform is shown in Figure 2, the setting temperature for left temperature box is 50℃, the right one is -40℃. The charging devices usedare from Ningbo BaTe Test Equipment and switching power supply.

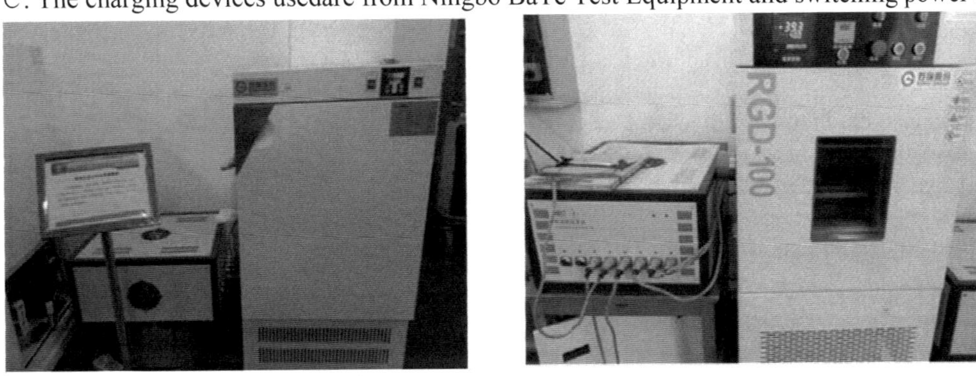

Figure 2. Battery float charging test platform

In this paper, two kinds of lithium titanate batteries were selected for testing, which were used to comparatively analyze the test results and aging mechanism of lithium titanate batteries for different anode materials. In addition, for the 20Ah lithium titanate batteries, different constant voltage points were selected for the float chargingtest to analyze influences of the battery aging under different constant voltage control voltage for float charging.

For 20Ah lithium titanate batteries, place the four batteries respectivelyfor float charging at different temperatures and constant voltages, analyze the influences of battery life in different circumstances for float charging, comparatively analyze the battery aging evolution by the battery capacity test and capacity increment curve before and after the float charging. The specific programs of battery float chargingtest are as follows: 1) The four batteries are chargedto the cut-off voltage of 2.7V in constant current and constant voltage (CCCV), constant current charging rate is 1C, the current in constant voltage phase drops to 2A, and then the voltage of 1C constant current was discharged to the cut-off voltage 1.5v, and the cycle was repeated three times, calibrate the battery initial capacity. At the same time, have a constant current charge and discharge capacity test for the battery at 0.05C rate to obtain the maximum battery capacity and capacity increment curve. 2) The # 1 and # 3 batteries were placed in a 50°C incubator and subjected to a float charging test at 2.7 V and 2.65 V respectively. After 60 days, the batteries were left at room temperature for 24 hours and subjected to 1C and 0.05 C capacity tests. 3) The # 2 and # 4 batteries were placed in a -40°C incubator and float charging at 2.7V and 2.65V respectively. After 60 days, the battery was allowed to stand at room temperature for 24 hours, and have respectively capacitytest at 1C and 0.05 C.

For 8.5Ah lithium titanate batteries, place the three batteriesforfloat charging test at different temperatures, analyze the influences of battery life in different ambient temperatures for battery charging and aging evolution. The specific programs of battery float charging test are as follows: 1) The three batteries are charged to the cut-off voltage2.8V in constant current and constant voltage (CCCV), constant current charging rate is 1C, the current in constant voltage phase drops to 0.85A, and then the voltage of 1C constant current was discharged to the cut-off voltage 1.5v, and the cycle was repeated three times, calibrate the battery initial capacity. At the same time, have a constant current charge and discharge capacity test for the battery at 0.05C rate to obtain the maximum battery capacity and capacity increment curve. 2) The three batteries were respectively placed in 50℃, -40℃ and room temperature environment forfloat chargingtest with 2.8V for 100 days, thenplace the batteryat room temperature for 24 hours, have a capacity test respectively by 1C and 0.05C .

4. Effect on float charging and aging of lithium titanate battery by different temperatures

4.1. Analysis for float charging test results and aging mechanism of 20Ah lithium titanate battery

Figure 3 shows the capacity test results before and afterthe battery experienced float charging, in which # 1 battery is floated at high temperature of 50°C., and # 2 battery is floated at a low temperature of -40°C., and the floating cycle is 60 days. Figure 3 (a) is the battery 1C rate capacity, the battery nominal capacity is 20Ah, but the actual test of the initial capacity of the battery were respectively 19.69Ah and 19.77Ah, after experiencing the float charging test, both batteries capacity were varying degrees of increase. Among them, the high temperature float battery capacity increases even more apparent. Figure 3 (b) shows the maximum usable capacity of the battery at 0.05C. Both batteries also show an increase in capacity. At high temperature, the capacity increases by 0.297 Ah, about 1.3%, and at low temperature the capacity increases by 0.21 Ah,about 0.9%.

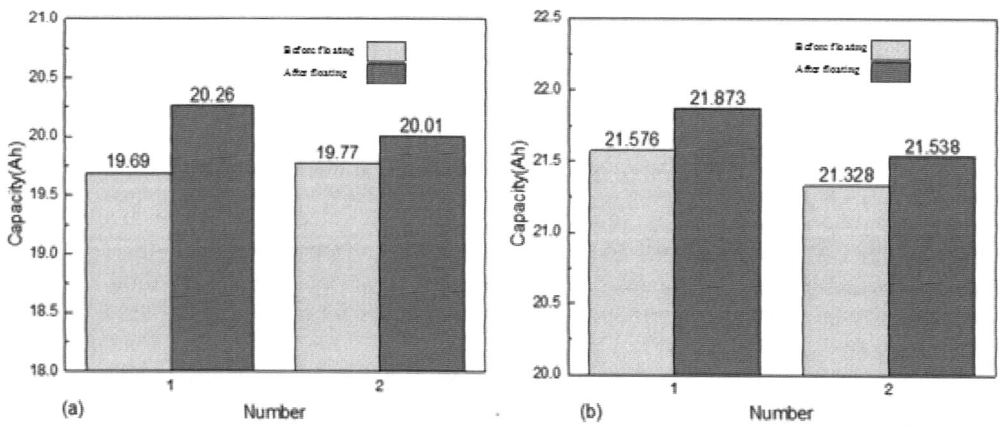

Figure 3. Test results of battery capacity before and after floating

Figure 4 shows the Incremental capacity (IC) curve and Differential voltage (DV) curves of # 1 battery before and after high-temperature floating. The left Figure shows the battery Incremental capacity curve. The anode material of the battery forms two IC peaks, the first IC peak corresponds to the platform of 2.2V voltage and corresponds to ternary material phase transition voltage. The second IC peak corresponds to the platform of 2.35V voltage and corresponds to the phase transformation platform of lithium cobalt oxide material. It is found in the Figure that the starting voltage ofthe first IC peak offsets to the right, but the entire peak did not offset, that indicatesthe offset was not caused by internal resistance increasing, but due to inconsistencies in the initial capacity of the battery, the initial SOC is too large when measuring after floating, resulting in a higher voltage starting point. The second IC peak appears to some degree of decline, which may bring capacity attenuation. The IC curve obviously rises near the solid solution zone, indicating that the float charging causes the increase of capacity near the solid solution zone. In the Differential voltage curve of the right Figure, it is obvious that the positions of several peaks are staggered, indicating a change of capacity before and after floating. The capacity starting point of first voltage platform is always 0, until the end of the first platform, that is, when maximum value of dV/dQ appears, the floatedcurve arrives in advance,indicating that the ternary material has capacity loss after floating , the capacity is 0.23Ah reduction. In the following voltage platform, the DV curve after floating appears that first offsetting to left and then to right, which shows the decrease of the battery second peak capacity and the increase of solid solution zone capacity, which corresponds to the analysis of IC curve. Among them, the lithium cobalt oxide material formed a capacity loss of 0.27Ah, while a 0.8Ah capacity increase occurred in the solid solution zone. In the case of battery floating at high temperatures, it caused reduction of capacity produced from lithium cobalt oxide material, and increment capacity from parts of the ternary and solid solution zone.

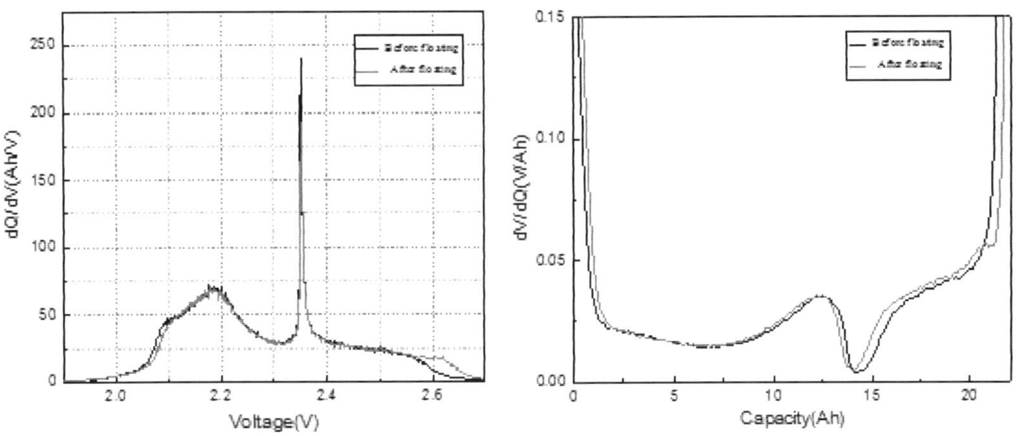

Figure 4. IC curve (left) and DV curve (right) of # 1 battery

The capacity of # 2 battery also increases after floating at low temperatures, the IC and DV curves of the battery as shown in Figure 5. Offset does not appear at the low end of the voltage in the IC curve of the battery, indicating that the initial SOC is consistent;offset does not appear in the overall curve, indicating the less influence on the internal resistance from low temperature floating. The coincidence of the first IC peak is higher, but different degrees of elevation of the second IC peak and the following solid solution zone appears, and the corresponding area increases. In the DV curve, the capacity of the ternary material increases by 0.07Ah, the capacity of the lithium cobalt oxide material changes less, reducing 0.01Ah, near the solid solution zone, the capacity of the battery increases by 0.16Ah.

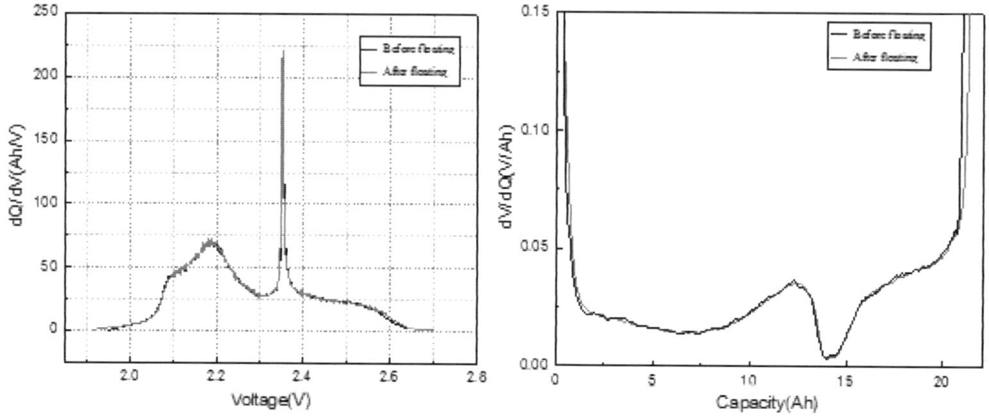

Figure 5. IC curve (left) and DV curve (right) of # 2 battery

After floating at different temperatures, the changes of capacity of the battery anode material are shown in Table 2. In terms of different materials, the lithium cobaltate shows a decreasing capacity after floating, the capacity increases near the solid solution zone, the ternary material is more complex, the capacity decreases under high temperature, but increases under low temperature, the increased value of capacity is less. On the other hand, the capacity of lithium cobalt oxide decreases under the high temperature environment, and the capacity of solid solution zone increases obviously. The temperature stress aggravates the progress and speed of the reaction and cause more anode material to have a side reaction in the negative terminal to form partial capacity increase in solid solution zone.

Table 2. Capacity changes of each part of the battery anode material

	Ternary material (Ah)	Lithium cobalt oxide (Ah)	Solid solution Zone (Ah)	Total capacity (Ah)
High temperature 2.7V floating	-0.23	-0.27	0.80	0.30
Low temperature 2.7V floating	0.07	-0.01	0.16	0.21

After floating, ternary materials and lithium cobalt oxide material of lithium titanate battery form a loss at high temperature, but the both materials do not obviouslychange at low temperature, due to in the reaction the currentis alwaysinjecting into high-end SOC near the solid solution zone, the lithium-ion that corresponds tointernal batteryanode moves out to the negative electrode, but the negative electrode is saturated due to lithium intercalation, so that the reaction is formed on the surface of the electrode, resulting in increasing the capacity of the solid solution zone. The temperature affects the activity of the battery material;reaction is more active at high temperatures, resulting in more capacity.

4.2. Analysis for float charging test results and aging mechanism of 8.5Ah lithium titanate battery

By studying the floating characteristics of 20Ah lithium titanate batteries, the corresponding capacity of lithium cobalt oxide materials appears to be reduced, and 8.5Ah lithium titanate batteries also contain lithium cobalt oxide material. In response to this phenomenon, carry out the full power state float charging test of 8.5Ah lithium titanate battery at different temperatures to compare the stability of the anode lithium cobaltate material and the change of the battery floating characteristics of different materials. According to the test plan, batteries labeled # 1, # 2 and # 3, will be tested respectively at the ambient temperature of high temperature 50°C, low temperature -40°C and room temperature for

100 days, due to the pre-data recording is not accurate, only selected later 60 days data to have quantitative analysis.

Figure 6 shows the capacity test results of the battery before and after floating, Figure 6 (a) is the battery capacity of 1C rate, it can be seen that capacitiesof three batteries are greater than the nominal capacity in the initial state, with the floatcharging, capacity of # 1 battery shows a declining trend, and capacity of # 2, # 3 increased. Figure 6 (b) is0.05C rate battery capacity, # 1 battery capacity decreased by 0.22Ah, # 2 battery capacity increased by 0.05Ah, # 3 battery capacity increased 0.2Ah, which is different with the 1C rate capacity increases, the greater battery resistanceleads to not sufficient reaction and the capacity is not fully released.

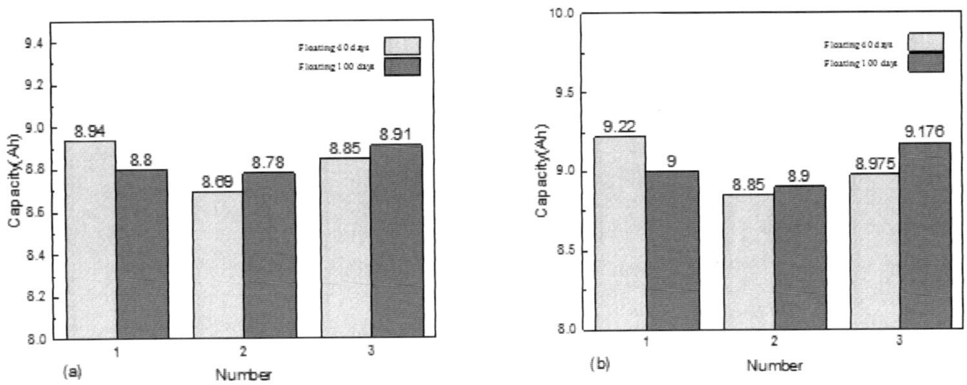

Figure 6. Test results of floating battery capacity: (a) 1C ratio and (b) 0.05C ratio

Figure 7 shows IC curve and DV curve of # 1 lithium titanate battery after floating for 40 days and 100 days at high temperature, three regions obviously appears. The IC curve in the left Figure shows that the height of the first IC peak decreases after floating and the second IC peak does not change significantly. After 100 days of floating, the capacity increases near the solid solution zone. The total capacity of the battery appears to decrease, which may be related to the lithium cobalt oxide capacity loss. In the right Figure, it can also be seen that change of the capacity in three stages. At the end of the peak 1, it can be seen that the reduction of the corresponding capacity is shown in the Figure with the capacity reduced by 0.32Ah, the corresponding to the capacity reduced from lithium cobalt oxide voltage platform on the IC curve. In the second stage, the capacity is reduced by 0.04Ah. In the third stage, the capacity is increased by about 0.14Ah, corresponding to an increase in the solid solution zone in the IC curve. From the perspective of the change of capacity of high-temperature floating, the floating has caused a significant drop in capacity of lithium cobaltate platform, while capacity increases in the other solid-solution platforms, but the volume decrease is more than that of incremental volume, and the overall performance is decreased. It can be seen from the above analysis that the lithium titanate battery will have partial capacity loss when floating at a high temperature. In contrast to the 20Ah lithium titanate battery, the loss of the lithium cobalt oxide material is even more obvious, near the solid solution zone will generate more capacity.

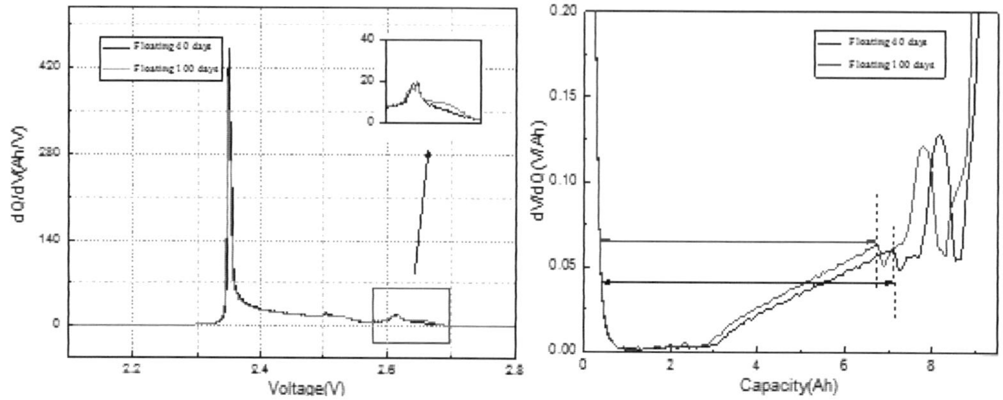

Figure 7. IC curve (left) and DV curve (right) of # 1 battery

Figure 8 shows the battery characteristics curve after floating at low temperature. In the IC curve, the first IC peak shows a slight decrease, corresponding to the reduction and offset of the DV curve. The corresponding capacity loss is 0.33Ah. The change of the second IC peak is not obvious; the change of DV curve is less, increasing 0.02Ah. Battery capacity near the solid solution zone increased with an increment of 0.37Ah, the final battery capacity change is less, indicating that the increased capacity of the solid solutionmake up for the capacity of corresponding loss of lithium cobalt oxide, compare to high temperature floating IC curve and find the capacity loss of lithium cobaltate decreases, which may be related to the activity at different temperatures. At the same time, the capacity of the solid solution zone increases less in the high temperature environment and different from the elevated position in the low temperature solid solution zone.

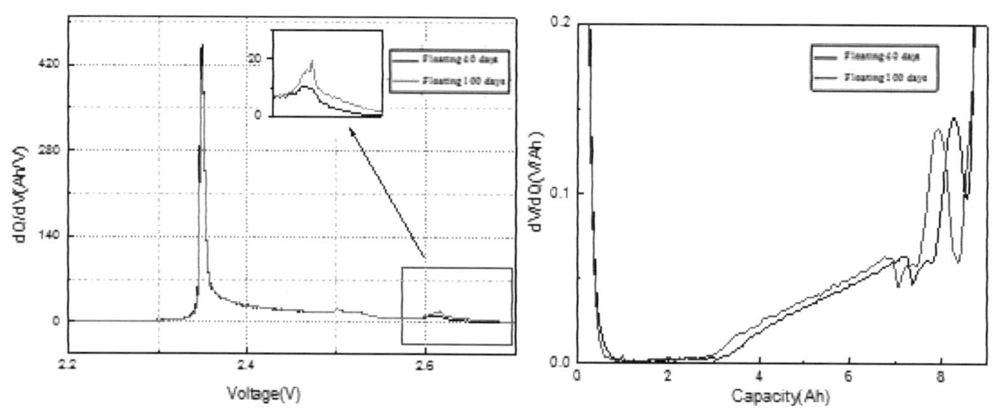

Figure 8. IC curve (left) and DV curve (right) of # 2 battery

Figure 9 shows the battery characteristics curve of # 3 battery after floating at room temperature. The height of the first IC significantly decreases, corresponding to the attenuation of capacity in the DV curve; however, the capacity decreases only by 0.09Ah. The corresponding capacity of the second IC peak increases by 0.05 Ah, and the battery capacity near the third peak significantly increase by 0.24Ah. Float charging at room temperature, the loss of lithium cobalt oxide material is less, and near the solid solution zonethe capacity also increases.

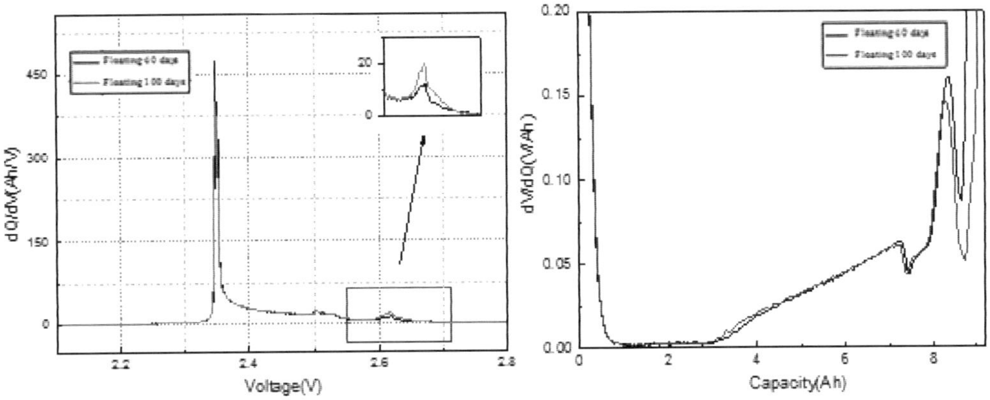

Figure 9. IC curve (left) and DV curve (right) of # 3 battery

In the floating test, the capacities of the lithium titanate batteries floated at various temperatures were significantly changed, the corresponding capacity changes of the three groups of battery materials are shown in Table 3, it is seen from the capacities of peak values, significant capacity attenuation occurred in the lithium cobalt oxide platform, the amount of attenuation is less at room temperature, which is the same as corresponds to phenomenon of 20Ah battery, indicating that the float process forms a significant capacity lossto lithium cobalt oxide material. Near the solid solution zone ②, the batteries capacity change is less at different temperatures, the role of float charging on the material is less, and the corresponding capacity change is not more. Near the solid solution zone ③, the floated capacities all increased at different temperatures.

Table 3. Capacity changes of each part of the battery anode material

	Lithium cobalt oxide (Ah)	Solid solution ②(Ah)	Solid solution ③ (Ah)	Total capacity (Ah)
High temperature 2.8V floating	-0.32	-0.04	0.14	-0.22
Low temperature 2.8V floating	-0.33	0.02	0.37	0.05
Room temperature 2.8V floating	-0.09	0.05	0.24	0.20

It may be foundbyfull charge statefloating at various temperatures that the batteries can form significant capacity loss on the lithium cobalt oxide materialand the results of the material loss are also verified by the 20Ah battery. Float charging has little effect on the battery material in the solid solution zone ②,but float chargingleads to the increase of capacity on the platform of solid solution zone③, indicating that temperature is an important factor affecting the float aging. Excessive temperature may cause the battery material side reaction intensified, while accelerating the reaction of high-potential platform, resulting in the battery to form more capacity at the end of charging. During use, the temperature balance in battery box of the rail vehicle should be kept as far as possible, not too high, so as to ensure the floating life of the battery pack.

IOP Conf. Series: Materials Science and Engineering **366** (2018) 012087 doi:10.1088/1757-899X/366/1/012087

5. Effect on float charging and aging of lithium titanate battery by different voltages

It is known from the above analysisthat full charge state floating of lithium titanate battery will cause changes between different materials of the battery, in which the corresponding capacity of lithium cobalt oxide material reduces, the capacity increasesnear the solid solution zone.Due to the voltage platform of the solid solution zone is near charging ends while continuingcharge at a low current underfull charge state may cause the material's reaction. Therefore, the floating test comparison shall be conducted for 20Ah lithium titanate batteries under different voltage points and at both high and low temperatures. The battery, as a backup power supply in the rail vehicle,its too low electricity cannot guarantee reliable power supply, so the floating voltage is set to 2.65V, floating time is 60 days, analyzeeffect of floating constant voltage point on the battery capacity decline.

Figure 10 shows the battery characteristics curve of the # 3 battery before and after floating at a high temperature 50°C. The float voltage is 2.65V. It can be seen that after the battery is floated at 2.65V, the loss of corresponding capacity of the ternary material does not appear, but at full charge state, the corresponding capacity of the ternary material decays, indicating that the partial reaction is related to the voltage level. Due to the fact that the capacity of the full charge state float at low temperature is not obvious, temperature isless effect on it. Battery floating at 2.65V voltage also caused the lithium cobalt oxide material capacity reduction. Comparison of the capacity between the anode materials in the composite Table 4 with different constant voltage points floating, we can see there is no differencebetween the loss of lithium cobalt oxide material and the full charge state of the floatcharging;it is causedby the high temperature conditions. Near the solid solution zone battery did not produce a significant increase in capacity, only increased by 0.04Ah, which is different with the full charge state floating. In full charge state, the battery internal reaction tends to saturation under the action of the float current;some materials of battery solid solution zone are involved in the reaction, resulting in an increase in capacity. However, when floating at 2.65V, the battery has not yet reached the full charge state, and there is still room for voltage increase, so that no capacity increase is formed near the solid solution zone.

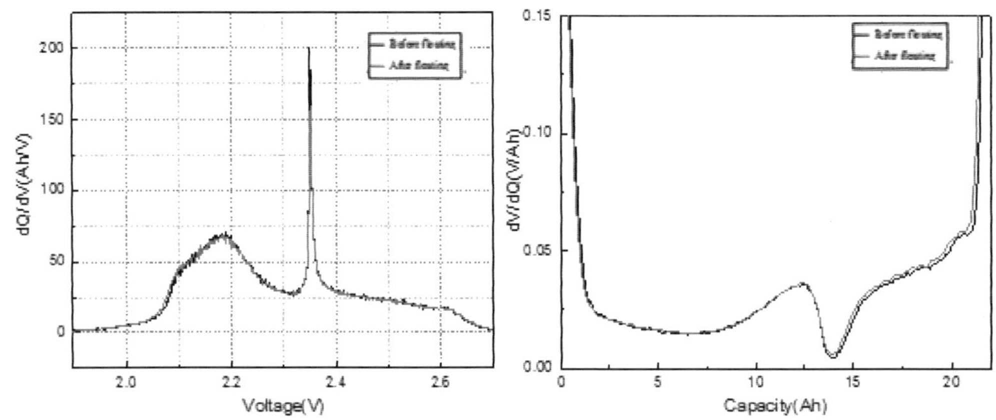

Figure 10. IC curve (left) and DV curve (right) of # 3 battery

Table 4. The corresponding capacity change of each part of materials after batteries floating at different voltage points

	Ternary material (Ah)	Lithium cobalt oxide (Ah)	Solid solution zone (Ah)	Total capacity (Ah)
High temperature 2.7V floating	-0.23	-0.27	0.80	0.30
High temperature 2.65V floating	0.03	-0.24	0.04	-0.17

Figure 11 shows the battery characteristics curve of the # 4 battery before and after floating at a low temperature, the float voltage is 2.65V, the test results are similar to those of full-charge state floating. Table 5 shows the capacity change of # 4 battery floated at different voltage points and at low temperature, it can be seen that the corresponding capacities of ternary materials all show an increase, indicating that the effect onternary materials is less from low temperature floating at different voltage points. The conclusion of high temperature floating test shows that the loss of ternary material is the result of float charging at high temperature and full charge state and is affected by the common stress of temperature and voltage. For the lithium cobalt oxide voltage platform, the corresponding capacity is alwaysloss when being floated under the high temperature environment. On the contrary, the capacity slightly increasesat low SOC state whenfloat charging at low temperature, which shows that high temperature environment is an important factor of lithium cobalt oxide material loss. Finally, in the solid solution zone, the full charge state floating in both high and low temperature environments increase in capacity,that is, the capacity increases slightly when floating at 2.65 V high temperature, while the corresponding capacity decreases when floating at 2.65 Vlow temperature, indicating that different voltage points areimportant factors that affect reaction in solid solution reaction zone, followed by the temperature.

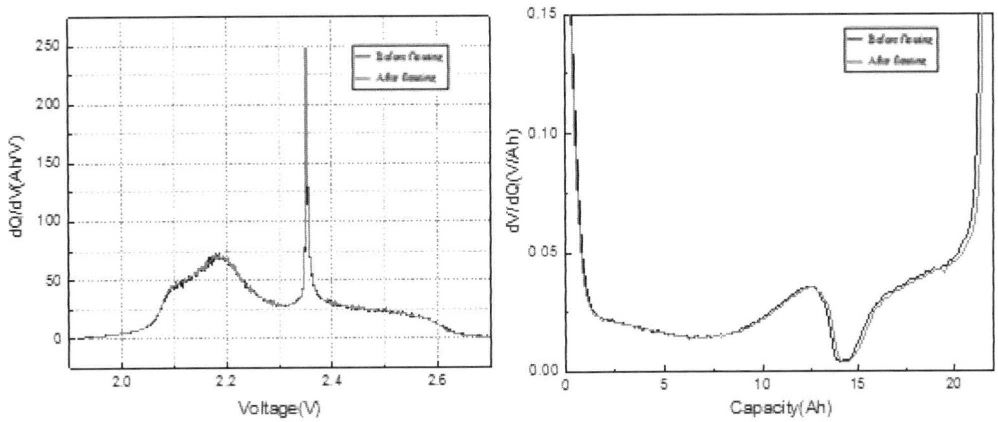

Figure 11. IC curve (left) and DV curve (right) of # 4 battery

Table 5. The corresponding capacity changes of each part of materials after battery floating at different voltage points

	Ternary material (Ah)	Lithium cobalt oxide (Ah)	Solid solution zone (Ah)	Total capacity (Ah)
Low temperature 2.7V floating	0.07	-0.01	0.16	0.21
Low temperature 2.65V floating	0.13	0.08	-0.05	0.16

Through the floating contrast of different voltage platform, we can draw the following conclusion: for 20Ah lithium titanate battery, different phase voltage platform will change; for the ternary material platform, at high temperature full charge state floating, there will be a small amount of attenuation of capacity, The rest of the environment is less impact on capacity; On the lithium cobalt oxide platform, high temperature is an important factor affecting the material, the impact of the constant voltage point of the float on the reaction is less; near the solid solution zone, the increase of capacity is synthetically affected by temperature and voltage, in which,the voltage point is the main affecting factor. When the float voltage is increased, the capacity of the solid solution zone will change obviously. The temperature is the secondary affecting factor; the increasing temperature will accelerate the reaction process, resulting in increased capacity. Therefore, in the actual use of rail vehicles, we shall appropriately reduce the constant voltage point and maintain the appropriate temperature to reduce the battery side reactions.

6. Effect on float charging and aging of lithium titanate battery by different temperatures

This paper has conducted decomposition and description for proportionof the actual vehicle data in the rail vehicle auxiliary battery system in different operating conditions, focusing on affect of the float charging conditions on the battery life. Two lithium titanate batteries made of different materials are carried out for floating test, have researched the recession mechanism of aging life of lithium titanate battery, coupled with the impact of ambient temperature and voltage points, analyzedthe battery recession under different factors, providedtheoretical and data support for formulating control strategy of the auxiliary battery system.

References

[1] Wang Zhanguo, Gong Minming. Lithium Titanate Battery Management System for EMU[J]. Chinese Railways,2011,(12):64-66.
[2] Su Jian, Li Hongbing. Lithium Titanate Battery in the Application of EMU [J]. Electric Drive for Locomotives,2011,(4):38-40.
[3] Yuan Fengbiao, Yang Jun. Lithium-ion Battery in the Application of the Railway Passenger Car [J]. Electric Drive for Locomotives,2011,(3):23-27.
[4] Gong Minming, Wang Zhanguo, Ma Zeyu. Design of Lithium-ion Battery System for Dual-powered Metro Metro [J]. Urban Rapid Rail Transit,2013,26(5):69-73.
[5] Wang liqiang, Wang Wei, Wang Zhanguo, Zhang Zhiru, Liu Sijia. Study on the Inconsistency of Lithium Titanate Battery in Rail Transit [J]. Journal of Power Sources,2017,41(2):195-197,218.

Effect of Atmospheric Conditions on Discharge Characteristics of Rod-plane Air Gap

Zhang Hao, Liu Hui, Duan Yubing, Wang Bin, Xu Ran, Lin Ying and Liu Rong

State Grid Shandong Electric Power Research Institute , Jinan 250002, China;

sdqzzh@163.com

Abstract. The rod-rod air gap electric fields are perfectly symmetrical in all kinds of extremely inhomogeneous electric field while the maximum asymmetry exists in the rod-plane air gap electric field. Other types of uneven electric field air gap breakdown characteristics fit in between. The rod-plane air gap electric field is an important reference for the designing of insulation strength. In this paper, the impact tests of rod-plate air gap are carried out in the field environment of different altitudes. Discharge characteristic curves of switching and lightning impulse at different environmental factors were given. The increases of the 50% discharge voltage of the rod- plane air gap switching impulse and rod- plane air gap lightning impact are larger. The influence of altitude on rod- plane air gap lightning impact is tiny. Through the analysis of test results, the influence mechanism of environmental factors on the discharge characteristics is also discussed.

1. Introduction

As the level of transmission voltage increases, the dielectric strength of the air gap becomes a major factor in the design of the outer insulation geometry[1].

In engineering practice, most of the electric fields are inhomogeneous electric fields. The rod-rod air gap electric fields are perfectly symmetrical in all kinds of extremely inhomogeneous electric field while the maximum asymmetry exists in the rod-plane air gap electric field. Other types of uneven electric field air gap breakdown characteristics fit in between. For a variety of very uneven electric field air gap that is encountered in the actual project, electrical strength can be estimated by the breakdown characteristics of the two typical air gap, rod-rod air gap and rod-plane air gap, according to the symmetry of their electrodes[2~4].

The air gap's withstand voltage value of external insulation of transmission project is closely related to the atmospheric conditions [5]. At present, the research on the discharge characteristics of long-air gap caused by environmental factors in domestic and foreign laboratories is mostly limited to small-sized low-pressure vessels, focusing only on humidity and pressure [6-9]. It not fully considers the actual temperature, humidity and strong radiation effects, and also ignores the factors such as scene of the electromagnetic environment in the real field. If condition allows, it would be a more effective way to carry out experiments in the high-altitude field environment.

In this paper, the impact discharge of rod-plane air gap is investigated in a place along the coast of China (altitude of 50m) and somewhere in the Qinghai-Tibet Plateau (altitude of 4300m). The effects of air gap distance, polarity effect, temperature and altitude and other factors on the impact discharge have been studied, and certain regular pattern about the Influence of different atmospheric environment on Impact discharge characteristics of rod-plane air gap can be drawn. This study provides a reference for other research about electrode used in high altitude area.

2. Test Equipment and Test Methods

2.1. Rod-plane Electrode Arrangement

The rod-plane air gap electrode is vertically arranged. The upper end of the rod electrode is suspended by a composite insulator which is 12m long while the lower end of the insulator is connected with a power supply wire and a 15m long galvanized steel pipe. The plane electrode is a stainless steel plate with a thickness of 3mm and is provided with a 20m*20m square grounding iron plate.

2.2. Basic Parameters of the Impact Generator

The impulse voltage is generated and measured by the impulse voltage generator and its measurement and control system. The basic parameters are shown as table 1.

Table 1 The Basic Parameters of The Impact Generator

	Altitude	Rated voltage	Rated energy	Output of the standard switching impulse			Output of the standard lightning impulse		
				Wave head time	Wave tail time	Effectiv -eness	Wave head time	Wave tail time	Effectiv-eness
a place along the coast of China	50m	7200 kV	480kJ	250μs	2500± 50μs	70%	1.2μs	50±1 μs	90%
somewhere in the Qinghai-Tibet Plateau	4300m	4200 kV	200kJ	240μs	2500± 60μs	75%	1.2μs	49±1 μs	90%

2.3. Environment conditions for test

When other environmental conditions are similar, the influence of other factors on gap discharge can be neglected, and only the effect of a single environmental variable on the discharge characteristics of the air gap is considered. Low-temperature (-5~5°C) and high-temperature (15~25°C) impact discharge tests conducted at a coastal area (50 m) and high-temperature (15~25°C) impact discharge tests conducted at a place in the Tibetan Plateau (4300 m).The absolute humidity of the environment is similar, hence, the impact of the two environmental parameters of temperature and altitude on the air gap discharge characteristics are analyzed by comparing the test data. The environment conditions of test are shown as table 2.

Table 2 The Environment Conditions of Test

Environment Conditions	Temperature (°C)	Average Temperature (°C)	Atmospheric Pressure (kPa)	Absolute Humidity (g/m3)	Average Absolute Humidity (g/m^3)	Altitude (m)
Low altitude and low temperature	-5~5	-1	101	2~5	4.1	50
Low altitude and high temperature	15~25	21	101	4~11	6.0	50
High altitude and high temperature	15~25	18	60.5	3~7	5.3	4300

2.4. Test Methods and Test Results

In this paper, the impact test is conducted by method recommended GB/T 16927.1-1997 high voltage test technology Part I: General test requirements [10] and GB311.1-1997 high voltage power transmission equipment with external insulation matching [11] .

In engineering practice, the breakdown voltage characteristic of the air gap is usually featured by the voltage at the 50% breakdown (U_{50}). The 50% impulse discharge voltage has the same changing rule as the nominal value and the per unit value. The impulse discharge voltage in this paper is expressed by the per unit value. The minimum value of the 50% impulse discharge voltage in each test is taken as the reference value, and the 50% impulse discharge voltage in the test is taken as the per unit value.

$$U_{50*} = \frac{U_{50}}{U_{50B}} \tag{1}$$

In the equation: U_{50} is the real value of the 50% impulse discharge voltage, kV; U_{50B} is the reference value of the 50% impulse discharge voltage, the minimum value in each test, kV.

The dispersion of the data of the test data can be showed by the coefficient of variation, while the greater the coefficient of variation, the greater the data dispersion gets. The coefficient of variation is calculated as follows:

$$z = \frac{\sigma}{U_{50}} \times 100\% \tag{2}$$

In the equation above: U_{50} is the 50% discharge voltage, kV; σ is the standard deviation.

3. Effect of Temperature on Discharge Characteristics of Rod-Plane Air Gap

The curve of lightning impulses and switching impulse (U_{50*}) of the rod-plane air gap at low temperatures (-5~5°C) and high temperatures (15~25°C) in a coastal area changes with gap distance d, as is shown in Figure 1. The comparison of the coefficient of variation of the two experimental data is shown in Table 3. Besides temperature, other environmental conditions for the impact test are basically the same.

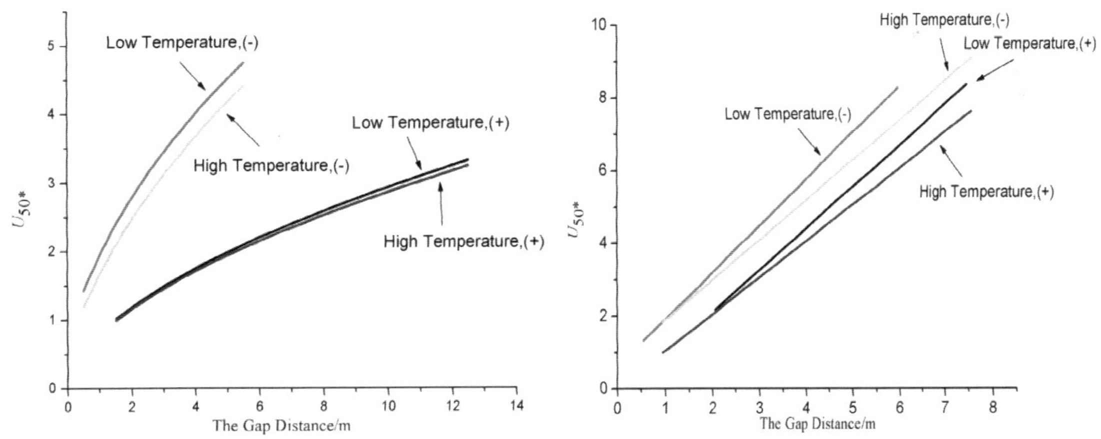

Figure 1. The relationship between switching impulseU_{50*} and the gap distance d as well as the relationship between lightning impact U_{50*} and the gap distance d in different temperature.

Table 3. Coefficient of Variation of Rod-plane Air Gap Impact U_{50}

Impact Type	Coefficient of Variation (z)
Switching Impulse(low temperature)	2.7%~4.8%
Switching Impulse(high temperature)	3.1%~4.6%
Lightning Impact(low temperature)	1.9%~3.0%
Lightning Impact(high temperature)	2.4%~2.8%

As can be drawn from Figure1 and Table3:

(1)The effect of temperature on the impact polarity of the rod-plane air gap is great: when the air gap distance is in the range of 1m~5m, the switching impulse at low temperature $U_{50(-)}$ increases by about 17% ~ 8% comparing to it at high temperature $U_{50(-)}$.The larger the gap distance, the smaller the increase of $U_{50(-)}$.The switching impulse $U_{50(+)}$ at low temperature increased about 1%~3% compared with the $U_{50(+)}$ at high temperature. The greater the gap distance, the greater the difference between them. When the air gap distance falls in the range of 2.5m~7m, the lightning impact at low temperature $U_{50(-)}$ increases by about 6%~12% comparing to it at high temperature $U_{50(-)}$.When the air gap distance falls between2.5m and 7m, the lightning impact at low temperature $U_{50(-)}$ increases by about10%~12% comparing to it at high temperature $U_{50(-)}$.

(2)When the temperature rises, the saturation tendency of the switching impulse voltage of the rod-plane air gap increases and the effect of temperature on the saturation trend is small.

(3)The temperature effect on the distribution of rod-plane air gap impact discharge U_{50} is more complicated. Within the selected test temperature, the dispersibility is no different.

4. Influence of Altitude on Rod-Plane Air Gap Discharge Characteristics

The curve showing the relationship between lightning impulses and switching impulse (U_{50*}) of the rod-plane air gap taken at a coastal area(50m) and a place in the Tibetan Plateau (4300 m)and gap distance d is shown in Figure 2.The comparison of the coefficient of variation of the two experimental data is shown in Table 4. Besides altitude, other environmental conditions for the impact test are basically the same.

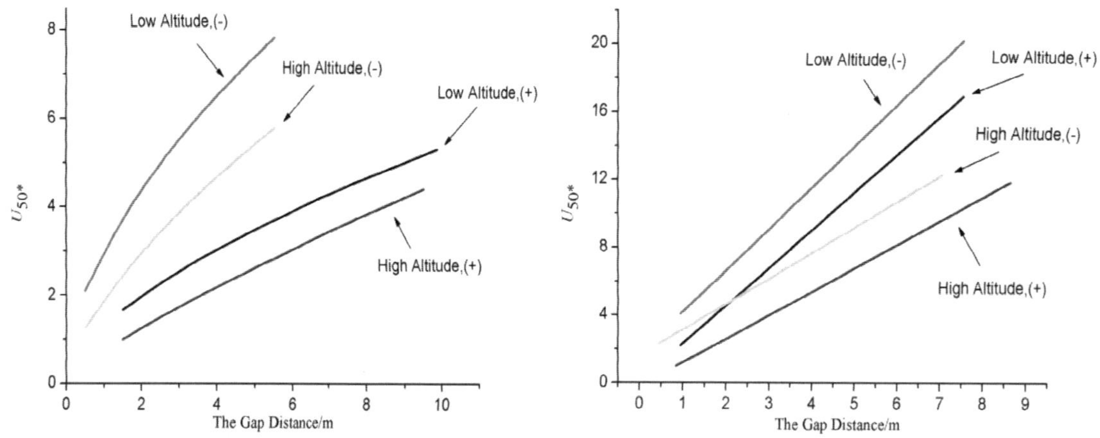

Figure 2. The relationship between switching impulseU_{50*} and the gap distance d as well as the relationship between lightning impact U_{50*} and in different altitude

Table 4. Coefficient of Variation of Rod-plane Air Gap Impact U_{50}

Impact Type	Coefficient of Variation (z)
Switching Impulse(50m)	2.9%~4.9%
Switching Impulse(4300m)	2.8%~4.8%
Lightning Impact(50m)	2.4%~2.8%
Lightning Impact(4300m)	2.6%~2.9%

As can be drawn from Figure 2 and Table 4:

(1)As altitude increases, the U_{50} of the rod- plane air gap decreases; the saturation trend of switching impulse U_{50} drops. Altitude has following effect on the impact discharge characteristics of the rod-plane air gap: U_{50} at an altitude of 4300 m is about 65%~85%of it at an altitude of 50 m for a positive-polarity switching impulse; for a negative-polarity switching impulse at an altitude of 4300 m U_{50} is about 60%~70% of it at the sea level of 50m. U_{50} at an altitude of 4300 m is about 60%of it at an altitude of 50 m for a positive-polarity lightning impact; for a negative-polarity lightning impact at an altitude of 4300 m U_{50} is about 70% of it at an altitude of 50m.

(2)As the altitude increases, the polar effect of the rod-plane air gap decreases.

(3)According to the test data dispersion analysis, it is found that the dispersion of the impact discharge U_{50} of the rod- plane air gap in the high altitude area is generally lower than that in the low altitude area.

5. Conclusion and Analysis

When the other environmental conditions are close to each other, temperature and altitude are chosen as two parameters for studying the influence of single environmental factor on the air-gap discharge voltage. The influence mechanism of environmental factors on the discharge characteristics is also discussed. The main conclusions of this thesis are as following:

(1) 50% of the switching impulse discharge voltage increases with decreasing temperature while the saturation tendency of the switching impulse U_{50} declines. The increases of the 50% discharge voltage of the rod- plane air gap switching impulse and rod- plane air gap lightning impact are larger.

(2) The increase of the altitude and the decrease of 50% impulse discharge voltage lead to the decrease of the polar effect of the rod-plane air gap. The saturation tendency of the switching impulse U_{50} decreases. The increase of U_{50} from the lightning impact becomes smaller. The influence of altitude on rod- plane air gap lightning impact is tiny.

(3) 50% impact discharge voltage dispersibility of the rod- plane air gap at the high altitude is slightly less than that in the low altitude area.

Pilot discharge is the main factor affecting the impact discharge characteristics of long air gap. As the temperature increases, the average kinetic energy of the gas molecules increases, the probability of collisional ionization increases, and the initial electrons are more likely to appear. When the altitude increases, the air density decreases, while the mean free path of free electrons in the air increases. As a result, the collision ionization is more likely to occur because of the increasing of the kinetic energy accumulated during free electrons acceleration. Temperature and altitude affect the impact discharge characteristics by affecting the pilot discharge process in long gap discharges.

References

[1] Sun Cai-xin, Jiang Xing-liang, Sima Wen-xia, etc.Ac Discharge Performance and Voltage Correction of Short Air Gaps in 4000m and Above Altitude DAistricts[J]. Proceedings of the CSEE, 2002, 22(10): 116-120.

[2] Li Erning, Liu Yangbing, Li Zhengying. Survey into Basic Problems on Gas Discharges[J]. High Voltage Engineering, 1997, 23(2): 95-96.

[3] Zhong Lianhong, Wang Fengming, Tang Hesheng. Experimenal Researeh of UHV Exernal in Sulaion under Siching Impulse[J]. High Voltage Engineering, 1995, 21(2): 7-10.
[4] The Calculation of Altitude Correction Coefficient and the idea of Insulation Coordination in Plateau Area[M]. Southwest Electric Survey and Design Institute, 1984: 89-98.
[5] Jiang Xing-liang, Wang Jun,Fan Jihe, etc Correction Method of Positive Switching Impulse Discharge Voltage for 0.5~1.0 m Air Gap Under Artificial and Natural Conditions [J]. Proceedings of the CSEE,2008,28(28):13-17.
[6] Chen Daohui. Switching Surge Breakdown Voltage of Long Air Gaps[J]. Transformer, 1997, 34(11): 87-93.
[7] Long Yuhua.Influence of Humidity on SI Disruptive Discharge Voltage in High Altitude Areas [J]. High Voltage Engineering 1994, 20(2): 58-60.
[8] Wang Xing-Sheng,Lu Chong-Hui The New Research of Air Density Correction Factor. Proceedings of The 3rd International Conference on Properties and Applications of Dielectric Materials July8-12, 1991 Tokyo, Japan.
[9] Impulse Flashover Characteristic of Long Air Gap and Atmospheric Correction IEEE PAS-97 No.2 1978.
[10] GB/T16927.1-1997 High-voltage test techniques. Part 1: General definitions and test requirement [S]. The People's Republic of China Standard. Put in practice in 1998.
[11] IEC, Insulation coordination Part 2 Application Guide, IEC600071-2-1996, 1996.